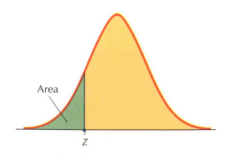

Area

Z

Table C Standard normal distribution

Z	0.00	0.01	0.02	0.03	0.04	0.05	0.06	0.07	0.08	0.09
−3.4	0.0003	0.0003	0.0003	0.0003	0.0003	0.0003	0.0003	0.0003	0.0003	0.0002
−3.3	0.0005	0.0005	0.0005	0.0004	0.0004	0.0004	0.0004	0.0004	0.0004	0.0003
−3.2	0.0007	0.0007	0.0006	0.0006	0.0006	0.0006	0.0006	0.0005	0.0005	0.0005
−3.1	0.0010	0.0009	0.0009	0.0009	0.0008	0.0008	0.0008	0.0008	0.0007	0.0007
−3.0	0.0013	0.0013	0.0013	0.0012	0.0012	0.0011	0.0011	0.0011	0.0010	0.0010
−2.9	0.0019	0.0018	0.0018	0.0017	0.0016	0.0016	0.0015	0.0015	0.0014	0.0014
−2.8	0.0026	0.0025	0.0024	0.0023	0.0023	0.0022	0.0021	0.0021	0.0020	0.0019
−2.7	0.0035	0.0034	0.0033	0.0032	0.0031	0.0030	0.0029	0.0028	0.0027	0.0026
−2.6	0.0047	0.0045	0.0044	0.0043	0.0041	0.0040	0.0039	0.0038	0.0037	0.0036
−2.5	0.0062	0.0060	0.0059	0.0057	0.0055	0.0054	0.0052	0.0051	0.0049	0.0048
−2.4	0.0082	0.0080	0.0078	0.0075	0.0073	0.0071	0.0069	0.0068	0.0066	0.0064
−2.3	0.0107	0.0104	0.0102	0.0099	0.0096	0.0094	0.0091	0.0089	0.0087	0.0084
−2.2	0.0139	0.0136	0.0132	0.0129	0.0125	0.0122	0.0119	0.0116	0.0113	0.0110
−2.1	0.0179	0.0174	0.0170	0.0166	0.0162	0.0158	0.0154	0.0150	0.0146	0.0143
−2.0	0.0228	0.0222	0.0217	0.0212	0.0207	0.0202	0.0197	0.0192	0.0188	0.0183
−1.9	0.0287	0.0281	0.0274	0.0268	0.0262	0.0256	0.0250	0.0244	0.0239	0.0233
−1.8	0.0359	0.0351	0.0344	0.0336	0.0329	0.0322	0.0314	0.0307	0.0301	0.0294
−1.7	0.0446	0.0436	0.0427	0.0418	0.0409	0.0401	0.0392	0.0384	0.0375	0.0367
−1.6	0.0548	0.0537	0.0526	0.0516	0.0505	0.0495	0.0485	0.0475	0.0465	0.0455
−1.5	0.0668	0.0655	0.0643	0.0630	0.0618	0.0606	0.0594	0.0582	0.0571	0.0559
−1.4	0.0808	0.0793	0.0778	0.0764	0.0749	0.0735	0.0721	0.0708	0.0694	0.0681
−1.3	0.0968	0.0951	0.0934	0.0918	0.0901	0.0885	0.0869	0.0853	0.0838	0.0823
−1.2	0.1151	0.1131	0.1112	0.1093	0.1075	0.1056	0.1038	0.1020	0.1003	0.0985
−1.1	0.1357	0.1335	0.1314	0.1292	0.1271	0.1251	0.1230	0.1210	0.1190	0.1170
−1.0	0.1587	0.1562	0.1539	0.1515	0.1492	0.1469	0.1446	0.1423	0.1401	0.1379
−0.9	0.1841	0.1814	0.1788	0.1762	0.1736	0.1711	0.1685	0.1660	0.1635	0.1611
−0.8	0.2119	0.2090	0.2061	0.2033	0.2005	0.1977	0.1949	0.1922	0.1894	0.1867
−0.7	0.2420	0.2389	0.2358	0.2327	0.2296	0.2266	0.2236	0.2206	0.2177	0.2148
−0.6	0.2743	0.2709	0.2676	0.2643	0.2611	0.2578	0.2546	0.2514	0.2483	0.2451
−0.5	0.3085	0.3050	0.3015	0.2981	0.2946	0.2912	0.2877	0.2843	0.2810	0.2776
−0.4	0.3446	0.3409	0.3372	0.3336	0.3300	0.3264	0.3228	0.3192	0.3156	0.3121
−0.3	0.3821	0.3783	0.3745	0.3707	0.3669	0.3632	0.3594	0.3557	0.3520	0.3483
−0.2	0.4207	0.4168	0.4129	0.4090	0.4052	0.4013	0.3974	0.3936	0.3897	0.3859
−0.1	0.4602	0.4562	0.4522	0.4483	0.4443	0.4404	0.4364	0.4325	0.4286	0.4247
−0.0	0.5000	0.4960	0.4920	0.4880	0.4840	0.4801	0.4761	0.4721	0.4681	0.4641

(Continued)

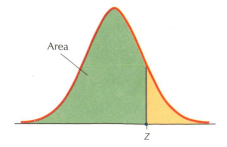

Area

Z

Table C Standard normal distribution (*continued*)

Z	0.00	0.01	0.02	0.03	0.04	0.05	0.06	0.07	0.08	0.09
0.0	0.5000	0.5040	0.5080	0.5120	0.5160	0.5199	0.5239	0.5279	0.5319	0.5359
0.1	0.5398	0.5438	0.5478	0.5517	0.5557	0.5596	0.5636	0.5675	0.5714	0.5753
0.2	0.5793	0.5832	0.5871	0.5910	0.5948	0.5987	0.6026	0.6064	0.6103	0.6141
0.3	0.6179	0.6217	0.6255	0.6293	0.6331	0.6368	0.6406	0.6443	0.6480	0.6517
0.4	0.6554	0.6591	0.6628	0.6664	0.6700	0.6736	0.6772	0.6808	0.6844	0.6879
0.5	0.6915	0.6950	0.6985	0.7019	0.7054	0.7088	0.7123	0.7157	0.7190	0.7224
0.6	0.7257	0.7291	0.7324	0.7357	0.7389	0.7422	0.7454	0.7486	0.7517	0.7549
0.7	0.7580	0.7611	0.7642	0.7673	0.7704	0.7734	0.7764	0.7794	0.7823	0.7852
0.8	0.7881	0.7910	0.7939	0.7967	0.7995	0.8023	0.8051	0.8078	0.8106	0.8133
0.9	0.8159	0.8186	0.8212	0.8238	0.8264	0.8289	0.8315	0.8340	0.8365	0.8389
1.0	0.8413	0.8438	0.8461	0.8485	0.8508	0.8531	0.8554	0.8577	0.8599	0.8621
1.1	0.8643	0.8665	0.8686	0.8708	0.8729	0.8749	0.8770	0.8790	0.8810	0.8830
1.2	0.8849	0.8869	0.8888	0.8907	0.8925	0.8944	0.8962	0.8980	0.8997	0.9015
1.3	0.9032	0.9049	0.9066	0.9082	0.9099	0.9115	0.9131	0.9147	0.9162	0.9177
1.4	0.9192	0.9207	0.9222	0.9236	0.9251	0.9265	0.9279	0.9292	0.9306	0.9319
1.5	0.9332	0.9345	0.9357	0.9370	0.9382	0.9394	0.9406	0.9418	0.9429	0.9441
1.6	0.9452	0.9463	0.9474	0.9484	0.9495	0.9505	0.9515	0.9525	0.9535	0.9545
1.7	0.9554	0.9564	0.9573	0.9582	0.9591	0.9599	0.9608	0.9616	0.9625	0.9633
1.8	0.9641	0.9649	0.9656	0.9664	0.9671	0.9678	0.9686	0.9693	0.9699	0.9706
1.9	0.9713	0.9719	0.9726	0.9732	0.9738	0.9744	0.9750	0.9756	0.9761	0.9767
2.0	0.9772	0.9778	0.9783	0.9788	0.9793	0.9798	0.9803	0.9808	0.9812	0.9817
2.1	0.9821	0.9826	0.9830	0.9834	0.9838	0.9842	0.9846	0.9850	0.9854	0.9857
2.2	0.9861	0.9864	0.9868	0.9871	0.9875	0.9878	0.9881	0.9884	0.9887	0.9890
2.3	0.9893	0.9896	0.9898	0.9901	0.9904	0.9906	0.9909	0.9911	0.9913	0.9916
2.4	0.9918	0.9920	0.9922	0.9925	0.9927	0.9929	0.9931	0.9932	0.9934	0.9936
2.5	0.9938	0.9940	0.9941	0.9943	0.9945	0.9946	0.9948	0.9949	0.9951	0.9952
2.6	0.9953	0.9955	0.9956	0.9957	0.9959	0.9960	0.9961	0.9962	0.9963	0.9964
2.7	0.9965	0.9966	0.9967	0.9968	0.9969	0.9970	0.9971	0.9972	0.9973	0.9974
2.8	0.9974	0.9975	0.9976	0.9977	0.9977	0.9978	0.9979	0.9979	0.9980	0.9981
2.9	0.9981	0.9982	0.9982	0.9983	0.9984	0.9984	0.9985	0.9985	0.9986	0.9986
3.0	0.9987	0.9987	0.9987	0.9988	0.9988	0.9989	0.9989	0.9989	0.9990	0.9990
3.1	0.9990	0.9991	0.9991	0.9991	0.9992	0.9992	0.9992	0.9992	0.9993	0.9993
3.2	0.9993	0.9993	0.9994	0.9994	0.9994	0.9994	0.9994	0.9995	0.9995	0.9995
3.3	0.9995	0.9995	0.9995	0.9996	0.9996	0.9996	0.9996	0.9996	0.9996	0.9997
3.4	0.9997	0.9997	0.9997	0.9997	0.9997	0.9997	0.9997	0.9997	0.9997	0.9998

The cover image for *Discovering the Fundamentals of Statistics,* 2nd Edition, shows a detail from the Shoshone beaded dress that belonged to Nahtoma, daughter of Chief Washakie of the Eastern Shoshone. This beaded dress plays an important role in the Chapter 9 Case Study, "The Golden Ratio." Examples of the golden ratio are found in art and architecture throughout the Western world—including the Parthenon, the *Mona Lisa,* and the great pyramids of Egypt. Some mathematicians have argued that the golden ratio is intrinsically pleasing to the human species. Support for this conjecture would be especially strong if evidence were found for the use of the golden ratio in non-Western artistic traditions. In this Case Study, we use hypothesis testing to determine whether the golden ratio is reflected in the non-Western beadwork of the Native American Shoshone tribe.

CASE STUDY

The Golden Ratio

What do Euclid's *Elements,* the Parthenon of ancient Greece, the *Mona Lisa,* and the beadwork of the Shoshone tribe of Native Americans have in common? An appreciation for the *golden ratio.* Suppose we have two quantities A and B, with $A > B > 0$. Then, A/B is called the golden ratio if

$$\frac{A+B}{A} = \frac{A}{B}$$

that is, if the ratio of the sum of the quantities to the larger quantity equals the ratio of the larger to the smaller.

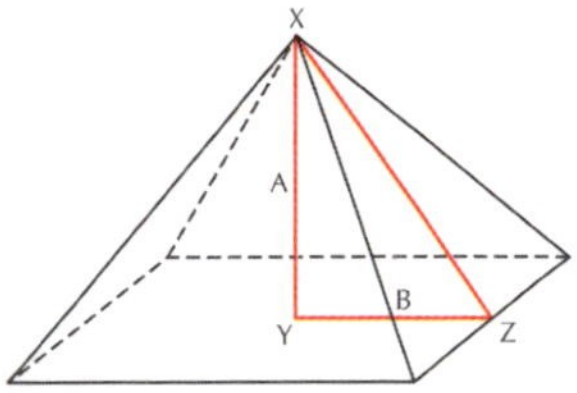

FIGURE 9.1

The golden ratio permeates ancient, medieval, Renaissance, and modern art and architecture. For example, the Egyptians constructed their great pyramids using the golden ratio. (Specifically, in Figure 9.1, if $A = \overline{XY}$ is the height from the top vertex to the base, and $B = \overline{YZ}$ is the distance from the center of the base to the edge, then $(A + B)/A = A/B$.) Some mathematicians have said that the golden ratio may be intrinsically pleasing to the human species. Support for this conjecture would be especially strong if evidence was found for the use of the golden ratio in non-Western artistic traditions. In the Case Study on page 445, we use hypothesis testing to determine whether the decorative beaded rectangles sewn by the Shoshone tribe of Native Americans follow the golden ratio. ■

Learn more as this Case Study unfolds in Chapter 9 (pages 405 and 445–447).

CASE STUDY   The Golden Ratio

Euclid's *Elements,* the Parthenon, the *Mona Lisa,* and the beadwork of the Shoshone tribe all have in common an appreciation for the *golden ratio.*

Suppose we have two quantities A and B, with $A > B > 0$. Then A/B is called the *golden ratio* if

$$\frac{A+B}{A} = \frac{A}{B}$$

that is, if the ratio of the sum of the quantities to the larger quantity equals the ratio of the larger to the smaller (see Figure 9.32).

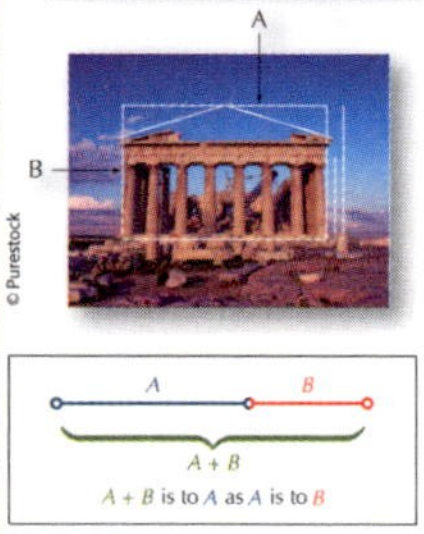

FIGURE 9.32 The golden ratio.

$A + B$ is to A as A is to B

Euclid wrote about the golden ratio in his *Elements,* calling it the "extreme and mean ratio." The ratio of the width A and height B of the Parthenon, one of the most famous temples in ancient Greece, equals the golden ratio (Figure 9.32). If you enclose the face of Leonardo da Vinci's *Mona Lisa* in a rectangle, the resulting ratio of the long side to the short side follows the golden ratio (Figure 9.33 on the next page). The golden ratio has a value of approximately 1.618.

Now we will test whether there is evidence for the use of the golden ratio in the artistic traditions of the Shoshone, a Native American tribe from the American West.

Discovering the Fundamentals of STATISTICS

Second Edition

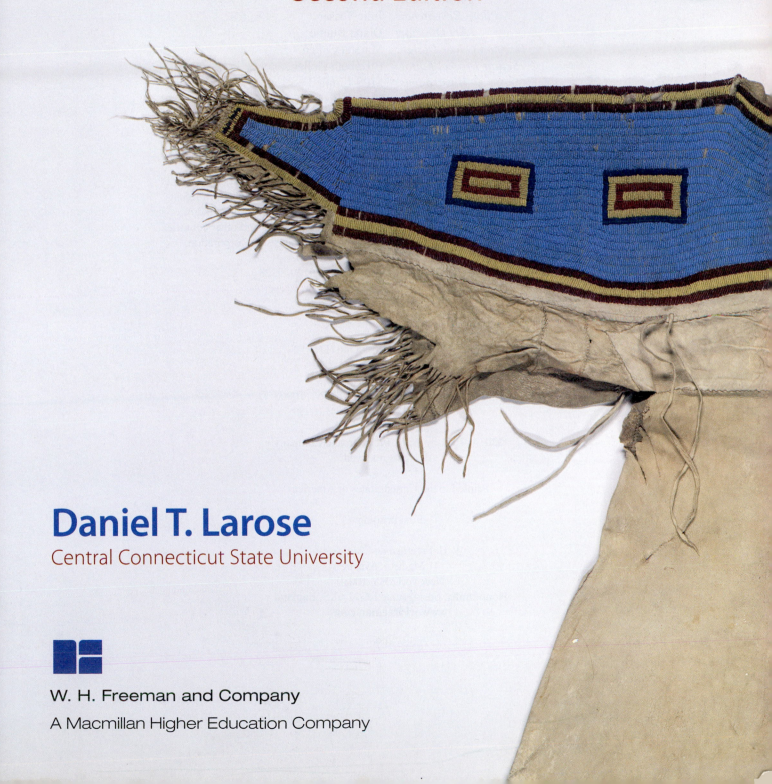

Daniel T. Larose

Central Connecticut State University

W. H. Freeman and Company

A Macmillan Higher Education Company

Publisher:	Ruth Baruth
Acquisitions Editor:	Karen Carson
Marketing Manager:	Steve Thomas
Marketing Assistant:	Alissa Nigro
Developmental Editor:	Andrew Sylvester
Senior Media Editor:	Roland Cheyney
Media Editor:	Laura Judge
Associate Editor:	Jorge Amaral
Associate Media Editor:	Courtney Elezovic
Editorial Assistant:	Liam Ferguson
Photo Editor:	Cecilia Varas
Photo Researcher:	Julie Tesser
Art Director:	Diana Blume
Text and Cover Design:	Marsha Cohen
Senior Project Editor:	Elizabeth Geller
Illustrations:	MPS Limited
Production Coordinator:	Paul W. Rohloff
Composition:	MPS Limited
Printing and Binding:	RR Donnelley

TI-83™ screen shots are used with permission of the publisher: ©1996, Texas Instruments Incorporated. TI-83™ Graphic Calculator is a registered trademark of Texas Instruments Incorporated. Minitab is a registered trademark of Minitab, Inc. Microsoft© and Windows© are registered trademarks of the Microsoft Corporation in the United States and other countries. Excel screen shots are reprinted with permission from the Microsoft Corporation.

Library of Congress Control Number: 2012949728

Paperback ISBN-13: 9781429289627
ISBN-10: 1429289627

Loose-Leaf ISBN-13: 9781464110832
ISBN-10: 1464110832

Instructor's Edition ISBN-13: 9781464110993
ISBN-10: 1464110999

Printed in the United States of America

First printing

W. H. Freeman and Company
41 Madison Avenue
New York, NY 10010
Houndmills, Basingstoke RG21 6XS, England
www.whfreeman.com

DETAILED TABLE OF CONTENTS

3

Describing Data Numerically   81

4

Correlation and Regression   149

7

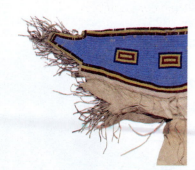

10

Two-Sample Inference   483

11

Further Inference Methods   529

Our 21st century world is flooded with data. Stock market returns and sports results snake across our TV screens in a nonstop stream. Grocery purchases are beep-beeped into data warehouses that enable the retailer to analayze the purchases and recommend individualized offers to their customers. Political candidates recite statistical facts and figures often massaged to support their positions on the issues. To develop a deeper sense of meaning and comprehension of data, students today need to turn to statistics: the art and science of collecting, analyzing, presenting, and interpreting data. *Discovering the Fundamentals of Statistics* will help you develop the quantitative and analytical tools needed to understand statistics in today's data-saturated world.

The Introductory Statistics Course

Discovering the Fundamentals of Statistics is intended for an algebra-based, undergraduate, one- or two-semester course in general introductory statistics for non-majors. The only prerequisite is basic algebra. *Discovering the Fundamentals of Statistics* will prepare you to work with data in fields such as psychology, business, nursing, education, and liberal arts, to name a few.

The GAISE guidelines, endorsed by the American Statistical Association, include the following recommendations:

1. Emphasize statistical literacy and develop statistical thinking
2. Use real data
3. Stress conceptual understanding rather than mere knowledge of procedures
4. Foster active learning in the classroom
5. Use technology for developing conceptual understanding and analyzing data
6. Use assessments to improve and evaluate student learning

Discovering the Fundamentals of Statistics adopts these guidelines verbatim as the course pedagogical objectives, with the following single adjustment: (3) Stress conceptual understanding *in addition to* knowledge of procedures. To these, the text adds two course pedagogical objectives:

7. Use case studies to show how newly acquired analytic tools may be applied to a familiar problem.
8. Encourage student motivation.

Approach of *Discovering the Fundamentals of Statistics,* Second Edition

Balanced analytical and computational coverage. The text integrates data interpretation and discovery-based methods with complete computational coverage of introductory statistics topics. Through unique and careful use of pedagogy, the text helps you develop your "statistical sense"—understanding the meaning behind the numbers. Equally, the text includes integrated and comprehensive computational coverage, including step-by-step solutions within examples. Select examples include screen shots and computer output from TI-83/84, Excel, Minitab, and CrunchIt!, with keystroke instructions located in the Step-by-Step Technology Guides at the ends of sections.

Communication of results. *Discovering the Fundamentals of Statistics,* Second Edition emphasizes how, in the real world and in your future careers, you will need to explain statistical results to others who have never taken a statistics course.

Emphasis on variability. The importance of variability in the introductory statistics curriculum cannot be overstated. Without a solid appreciation of how statistics may vary, there is little chance that you will be able to understand the crucial topic of sampling distributions.

Use of powerful, current examples with real data. The *Deepwater Horizon* oil spill, the use of cell-phone apps, and celebrity-followers on Twitter represent the variety of examples included in *Discovering the Fundamentals of Statistics*, Second Edition. Example and exercise topics reflect real-world problems and engage your interest in their solution. Real data (with sources cited) are frequently used to further demonstrate relevance of topics.

New to This Edition

- Additional topics have been added throughout the text. These additions include coverage of percentile ranks in Chapter 3, approximating probabilities for dependent events in Chapter 5, *t* inference for $\mu_1 - \mu_2$ using pooled variance, Z inference for $\mu_1 - \mu_2$, inference for two independent standard deviations in Chapter 10. For more information on content coverage, see "Key Chapter Changes" on page xiv.

- An increased number of examples and exercises offers extra support and provides a variety of relative examples to review and exercises to practice. Examples and exercises cover a wide range of applications and use updated, real data.

- **Now You Can Do Exercises feature**, found in the margin next to most examples, cues you to try related Practicing the Techniques exercises. These callouts are intended to prompt you toward practicing the techniques shown in the example. When working a particular exercise, you can also easily look back through the section to find the callout to a related example.

- **Bringing It All Together** exercises within each section offer a culmination of everything you have learned in a particular section, using a related set of Applying the Concepts exercises to tie together the main concepts and techniques learned.

- **Chapter 9, "Hypothesis Testing,"** has been rewritten to accommodate instructor preference with regard to teaching (a) the critical-value method only, (b) the p-value method only, or (c) both methods.

  (a) For those who like to cover the critical-value method but not the p-value method, simply cover Section 9.2 but not Section 9.3.

  (b) For those who like to cover the p-value method but not the critical-value method, cover only Objective 1 from Section 9.2, and then cover Section 9.3.

  (c) For those who like to cover both methods, simply cover both Section 9.2 and Section 9.3. For all hypothesis tests, coverage of the critical-value method has been moved ahead of the p-value method. This aligns our coverage with that of most of our competitors, making it easier for instructors who have previously taught using a different book, to use Discovering Statistics.

- In Chapters 9 and 10, the null hypothesis now always contains an equal sign. For example, the previous usage was:
$$H_0 : \mu \leq \mu_0 \text{ versus } H_a : \mu > \mu_0$$
The new notation is:
$$H_0 : \mu = \mu_0 \text{ versus } H_a : \mu > \mu_0$$

- The rejection rules is as follows, to be applied throughout the book:
  - Critical-value method (right-tailed test example): Changed from "Reject H_0 if test statistic > critical value" to "Reject H_0 if test statistic ≥ critical-value."
  - p-value method: Changed from "Reject H_0 if p-value < α" to "Reject H_0 if p-value ≤ α."

- **CrunchIt!® Statistical Software** is now included in the Step-by-Step Technology Guides at the end of select sections. This easily accessible and easy to use software offers all the basic statistical routines covered in introductory statistics courses.

- **Data sets,** available in a variety of software formats, are each named and marked with an icon in the text. You can locate the data sets on the CD in the back of the book or at **www.whfreeman.com/discofun2e.**

- The **Try This in Class** feature has been moved to the IRCD and is now integrated with the In-Class Activities for each chapter of the Instructor's Edition.

Key Chapter Changes

- **Chapter 2:** Crosstabulations and clustered bar graphs are now covered in Section 2.1, "Graphs and Tables for Categorical Data."

- **Chapter 3:** Section 3.1 now contains exercises covering the trimmed mean, the midrange, the harmonic mean, and the geometric mean. Section 3.2 now offers exercises on the coefficient of variation, the mean absolute deviation, and the coefficient of skewness. Quartiles and the interquartile range are now covered in Section 3.4 Measures of Position and Outliers. Chebyshev's Rule and the Empirical Rule have been moved to their more natural position as applications of the standard deviation in Section 3.2, "Measures of Variability."

- **Chapter 4** is newly titled "Correlation and Regression." Chapter 4 begins with a brand new case study, "Worldwide Patterns of Cell Phone Usage", where students use the methods learned in this chapter to examine whether residents of richer countries tend to use their cell phones to browse the Internet more often than residents of poor countries. Section 4.1 covers the closely related topics of scatterplots and the correlation coefficient. The regression equation has been changed from $\hat{y} = b_0 + b_1 x$ to $y = b_1 x + b_0$, so that instructors who also teach algebra may be comfortable moving from the $y = mx + b$ notation.

- **Chapter 6** is now titled "Probability Distributions." The chapter begins with a new case study, "Text Messaging," where students will learn that they must be careful what they assume. Section 6.2 offers new exercises on the geometric, hypergeometric, and multinomial distributions. Section 6.4 now covers the uniform probability distribution. NEW Section 6.6 covers the Normal Approximation to the Binomial Probability Distribution.

- **Chapter 7:** The *point estimate* topic has been moved to Section 8.1, where it appears more naturally just before confidence intervals. The awkward term *standard deviation of the sampling distribution of the sample mean* has been replaced with the more succinct *standard error of the mean*. Similarly, the *standard deviation of the sampling distribution of the sample proportion* is replaced with *standard error of the proportion*. Normal probability plots are now covered in Section 7.2, just in time for when they are needed. Overall, the coverage has been streamlined so that instructors may get to the Central Limit Theorem more quickly.

- **Chapter 8** opens with a NEW Case Study: Health Effects of the *Deepwater Horizon* Oil Spill. Section 8.1, "Z Interval for the Population Mean" now covers point estimates. The material on the Z confidence interval has been rewritten, making it simpler and increasing the pace.

- **Chapter 9:** The critical-value method is now covered before the p-value method. Starting in Section 9.4, in the Applying the Concepts exercises, the method to be used (critical value method or p-value method) is not specified. However, the Practicing the Techniques exercises continue to specify which method to be used. The null hypothesis and rejection rule formulas have been changed (see description above). There is a NEW Section 9.7 on probability of a type II error and the power of a hypothesis test.

- **Chapter 10:** The null hypothesis formula has been changed (see description above). Starting with this chapter, coverage of hypothesis testing is moved ahead of confidence intervals for the remainder of the book, in line with common practice. Section 10.2, "Inference for Two Independent Means," covers two new topics: (a) t inference for $\mu_1 - \mu_2$ using pooled variance and (b) Z inference for $\mu_1 - \mu_2$ when σ_1 and σ_2 are known.

Features of *Discovering the Fundamentals of Statistics,* Second Edition

The Second Edition retains many of the successful features from the First Edition.

Case Studies. A case study begins each chapter and is developed throughout the section examples, using the new set of tools that the section provides.

The Big Picture

Where we are coming from, and where we are headed . . .

- In Chapter 1 we learned the basic concepts of statistics, such as population, sample, and types of variables, along with methods of collecting data.

- Here, in Chapter 2, we learn about graphs and tables for summarizing qualitative data and quantitative data, and we examine how to prevent our graphics from being misleading.

- Later, in Chapter 3, we will learn how to describe a data set using numerical measures like statistics rather than graphs and tables.

The Big Picture. Brief, bulleted lists at the beginning of each chapter look at "where we are coming from, and where we are headed…". (Chapter 2, page 34)

Matched Objectives. Each section begins with a list of numbered objectives headed "By the end of this section, I will be able to…". The objective numbers are matched with the numbered topics within each section as well as the end-of-section summary. (Chapter 7, pages 332, 339)

7.2 CENTRAL LIMIT THEOREM FOR MEANS

OBJECTIVES By the end of this section, I will be able to . . .

1 Use normal probability plots to assess normality.

2 Describe the sampling distribution of $\bar{x}$ for skewed and symmetric populations as the sample size increases.

3 Apply the Central Limit Theorem for Means to solve probability questions about the sample mean.

SECTION 7.2 Summary

1. Normal probability plots are used to assess the normality of a data set.

2. A simulation study showed that the sampling distribution of $\bar{x}$ for a skewed population achieved approximate normality when n reached 30.

3. The Central Limit Theorem is one of the most important results in statistics and is stated as follows: given a population with mean μ and standard deviation σ, the sampling distribution of the sample mean $\bar{x}$ becomes approximately normal (μ, $\sigma/\sqrt{n}$) as the sample size gets larger, regardless of the shape of the population.

Developing Your Statistical Sense. This feature empowers students with some useful perspectives that real-world data analysts need to know. You will learn to think like real-world statistical analysts. This feature implements the GAISE guideline "develop statistical thinking." (Chapter. 9, page 411)

Developing Your Statistical Sense	**A Decision Is Not Proof**
	It is important to understand that the decision to reject or not reject H_0 does not prove anything. The decision represents whether or not there is sufficient evidence against the null hypothesis. This is our best judgment given the data available. You cannot claim to have *proven* anything about the value of a population parameter unless you elicit information from the entire population, which is usually not possible.

What Does This Mean? Feature boxes foster an intuitive approach and interpretation of results. Whenever a new formula or statistic is being introduced, the emphasis is on "What does this really mean?" Developing this understanding is just as important as getting the right answer, especially when the software can do the calculations. In the workplace, you may need to explain to your manager what the statistical results really mean. This feature helps to implement the GAISE guideline "stress conceptual understanding." (Chapter 8, page 358)

What Does This Confidence Interval Mean?	What does the 90% mean in the phrase *90% confidence interval*? If we take sample after sample for a very long time, then in the long run, the proportion of intervals that will contain the population mean μ will equal 90%.
	Interpreting Confidence Intervals
	You may use the following generic interpretation for the confidence intervals that you construct: "We are 90% (or 95% or 99% and so on) confident that the population mean _____ (for example, SAT Math score) lies between _____ (lower bound) and _____ (upper bound)."

 Give the Calculator a Rest

What If Scenarios offer you a chance to reflect on how changes in the initial conditions will percolate through the various aspects of a problem. The only requirement is to put your calculator down and think through the problem. You are asked to find the answers by using your knowledge of what the statistics *represent*.

Consider Example 3.6 once again. Now imagine: *what if* there was an incorrect data entry, such as a typo, and the number of Michael Jackson's music videos was greater than 31 by some unspecified amount.

Describe how and why this change would have affected the following, if at all:

a. The mean number of music videos

b. The median number of music videos

c. The mode number of music videos

What If Scenarios. The scenarios help you focus on statistical thinking rather than rote computation. Because of the availability of powerful statistical computer packages, statistical analysis is easy to do badly. The wrong analysis is worse than useless. It can cost companies lots of money, may convince lawmakers to pass legislation affecting millions of people, can incorrectly determine effects of pharmaceuticals or environmental pollution, and can have many other serious ramifications. The What If? scenarios are extensions of examples or exercises aimed at honing students' critical-thinking skills. In What If? exercises, the original problem set-up is altered in a specific but nonquantifiable way. You are then asked to think about how that change would percolate through the results, without recourse to calculations. The exercises as well as the scenarios are marked with the What If? icon. (Chapter 3, page 89)

Stepped Example Solutions. In selected examples, you are guided through the key steps needed to work through the calculations and find the solution. (Chapter 9, page 418)

Solution

We may apply the Z test because the sample is large ($n \geq 30$), and the population standard deviation σ is known.

STEP 1 **State the hypotheses.**

From Example 9.3, our hypotheses are

$$H_0 : \mu = 11 \quad \text{versus} \quad H_a : \mu > 11$$

where μ represents the population mean number of times people use their debit cards per month.

STEP 2 **Find Z_{crit} and state the rejection rule.**

We have a right-tailed test and level of significance $\alpha = 0.01$, which, from Table 9.4, tell us that $Z_{crit} = 2.33$. Because we have a right-tailed test, the rejection rule will be "Reject H_0 if $Z_{data} \geq Z_{crit}$," that is, "Reject H_0 if $Z_{data} \geq 2.33$" (see Figure 9.4).

STEP 3 **Find Z_{data}.**

From Example 9.3, we have $Z_{data} = 1$.

STEP 4 **State the conclusion and interpretation.**

Our rejection rule states that we will reject H_0 if $Z_{data} \geq 2.33$. Since $Z_{data} = 1$, which is *not* ≥ 2.33, the conclusion is to *not* reject H_0 (Figure 9.4). Even though the sample mean of 11.5 exceeds 11, it does not do so by a wide enough margin to overcome the

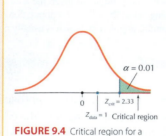

FIGURE 9.4 Critical region for a right-tailed test.

What Results Might We Expect?

Note from Figure 9.9 that the sample mean birth weight $\bar{x} = 3276$ grams is close to the hypothesized mean birth weight of $\mu_0 = 3200$ grams. This value of $\bar{x}$ is not extreme and thus does not seem to offer strong evidence that the hypothesized mean birth weight is wrong. Therefore, we might expect to *not reject* the hypothesis that $\mu_0 = 3200$ grams.

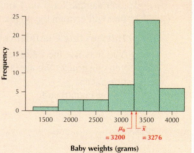

FIGURE 9.9 Sample mean, $\bar{x} = 3276$, is close to hypothesized mean, $\mu_0 = 3200$.

What Results Might We Expect? This feature, located in example solutions, challenges you to predict what the result of a particular problem will be. You are presented with a graphical view of the situation, and, before performing any calculations, you are asked to bring your intuition and common sense to bear on the problem and to state what results we might expect once we do the number crunching. (Chapter 9, page 426)

Definitions and Formulas. Easily located in highlighted boxes, key definitions and formulas are important for you to understand when working examples and exercises. Important vocabulary and formulas are also listed (with page references) at the end of each chapter. (Chapter 1, page 6)

The field of **statistics** is the *art* and *science* of

- collecting,
- analyzing,
- presenting, and
- interpreting data.

Exercises. *Discovering the Fundamentals of Statistics,* Second Edition, contains a rich and varied collection of section and chapter exercises.

- Clarifying the Concepts (conceptual)
- Practicing the Techniques (skill-based)
- Applying the Concepts (real-world applications)
- **NEW** Bringing It All Together

These exercises bring together everything you have learned in a particular section, using a related set of Applying the Concepts exercises to tie together the main concepts and techniques learned in the section.

- **NEW** **Now You Can Do Exercises** feature

Connects the Practicing the Techniques exercises to specific examples from the section. For example, in the margin at the end of Example 4.2 on page 152, you will find "Now You Can Do Exercises 13–18." This callout lets you know that you can use the example as a model when completing the exercise set.

| **EXAMPLE 4.2** | **CHARACTERIZE THE RELATIONSHIP BETWEEN TWO VARIABLES USING A SCATTERPLOT** |

Using Figure 4.1, characterize the relationship between lot square footage and lot price.

Solution

Now You Can Do Exercises 13–18.

The scatterplot in Figure 4.1 most resembles Figure 4.2a, where a positive relationship exists between the variables. Thus, smaller lot sizes tend to be associated with lower prices, and larger lot sizes tend to be associated with higher prices. Put another way, as the lot size increases, the lot price tends to increase as well.

Construct Your Own Data Sets

67. Construct your own data set with $n = 10$, where the mean, the median, and the mode are all the same. Yes, just make up your own list of numbers, as long as the mean, median, and mode are all the same. Draw a dotplot. Comment on the skewness of the distribution.

| **CHAPTER 3** | Quiz |

True or False

1. True or false: If two data sets have the same mean, median, and mode, then the two data sets are identical.

2. True or false: The variance is the square root of the standard deviation.

3. True or false: The Empirical Rule applies for any data set.

Fill in the Blank

4. An _____ is an extremely large or extremely small data value relative to the rest of the data set.

- **Construct Your Own Data Sets**

In these exercises, students are challenged to make up their own small set of numbers fulfilling some particular requirement, such as the mean being greater than the median. These exercises reinforce the statistical concepts beyond just rote calculation of the answers.

At the end of each chapter, **Review Exercises** and a **Chapter Quiz** help to test your overall understanding of each chapter's concepts and to practice for exams. The answers to odd-numbered exercises and all chapter quiz exercises are given in the back of the book.

Step-by-Step Technology Guide. This feature covers TI-83/84 calculators, Excel, Minitab, and CrunchIt!, providing stepped keystroke instructions for working through selected examples in the text. Screen shots of the results are often provided as well, either within the Step-by-Step Technology Guide or in the corresponding example. (Chapter 4, page 159)

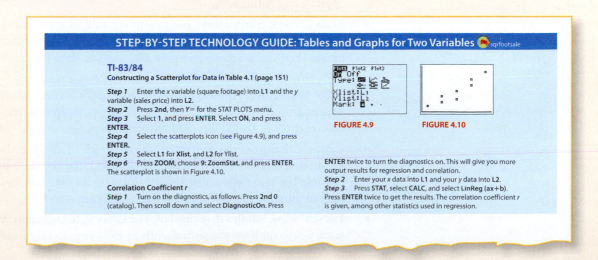

STEP-BY-STEP TECHNOLOGY GUIDE: Tables and Graphs for Two Variables sqrfootsale

TI-83/84
Constructing a Scatterplot for Data in Table 4.1 (page 151)

Step 1 Enter the *x* variable (square footage) into **L1** and the *y* variable (sales price) into **L2**.
Step 2 Press **2nd**, then **Y=** for the STAT PLOTS menu.
Step 3 Select **1**, and press **ENTER**. Select **ON**, and press **ENTER**.
Step 4 Select the scatterplots icon (see Figure 4.9), and press **ENTER**.
Step 5 Select **L1** for Xlist, and **L2** for Ylist.
Step 6 Press **ZOOM**, choose **9: ZoomStat**, and press **ENTER**. The scatterplot is shown in Figure 4.10.

Correlation Coefficient *r*
Step 1 Turn on the diagnostics, as follows. Press **2nd 0** (catalog). Then scroll down and select **DiagnosticOn**. Press

FIGURE 4.9   **FIGURE 4.10**

ENTER twice to turn the diagnostics on. This will give you more output results for regression and correlation.
Step 2 Enter your *x* data into **L1** and your *y* data into **L2**.
Step 3 Press **STAT**, select **CALC**, and select **LinReg (ax+b)**. Press **ENTER** twice to get the results. The correlation coefficient *r* is given, among other statistics used in regression.

Applets. Interactive statistical applets are located on the book's companion Web site: **www.whfreeman.com/discofun2e**. Applet icons in the text mark the related chapter material and exercises.

Caution notes. Signaled by the Caution icon, these warnings in the text help you avoid common errors and misconceptions.

Supplements

The following electronic and print supplements are available with *Discovering the Fundamentals of Statistics*, Second Edition:

STATSPORTAL **courses.bfwpub.com/discofun2e** (Access code required. Available packaged with *Discovering the Fundamentals of Statistics*, Second Edition, or for purchase online.) StatsPortal is the digital gateway to *Discovering the Fundamentals of Statistics*, Second Edition, designed to enrich the course and enhance your study skills through a collection of Web-based tools. StatsPortal integrates a rich suite of diagnostic, assessment, tutorial, and enrichment features, enabling you to master statistics at your own pace. StatsPortal is organized around the following learning components:

Interactive eBook offers a complete and customizable online version of the text, fully integrated with all the media resources available with *Discovering the Fundamentals of Statistics*, Second Edition. The eBook allows you to quickly search the text, highlight key areas, and add notes about what you are reading.

Resources organizes all the resources for *Discovering the Fundamentals of Statistics*, Second Edition, into one location for ease of use. These resources include the following:

- **NEW!** *LEARNINGCurve* is a formative assessment tool that tests your conceptual knowledge of the material in the text. As you progress through each Learning Curve activity, the system will customize the questions based on your performance so that you are tested more rigorously in those areas where you need the most work.

- **NEW! Stepped Tutorials** These new exercise tutorials (2-3 per chapter) feature algorithmically generated quizzing with step-by-step feedback and are easily assignable for homework.

- **Statistical Video Series** consisting of StatClips, StatClips Step-by-Step Examples, and Statistically Speaking "Snapshots." View animated lecture videos, whiteboard lessons, and documentary-style footage that illustrate key statistical concepts and help you visualize statistics in real world scenarios.

- **StatTutor Tutorials** offer over 150 audio-multimedia tutorials, including video, applets, and animations.

- **Stats@Work Simulations** put you in the role of a statistical consultant, helping you to better understand statistics interactively within the context of real-life scenarios. You are asked to interpret and analyze data presented in report form, as well as to interpret current events.

- **NEW! Statistical Applets** are interactive applications that allow you to work exercises from the text and practice key statistical procedures, such as correlation and regression, probability, and random sampling.

- **CrunchIt! Statistical Software** allows users to analyze data from any online location. Designed with the beginner in mind, the software is not only easily accessible but also easy to use. CrunchIt! offers all the basic statistical routines covered in introductory statistics courses and more.

- **EESEE Case Studies** developed by The Ohio State University Statistics Department, teach you to apply your statistical skills by exploring actual case studies using real data.

- **Student Solutions Manual** provides solutions to the odd-numbered exercises, with stepped out solutions to select problems.
- **WHFStat Macros for Excel**
- **Data sets** are available in ASCII, Excel, TI, Minitab, SPSS, and JMP formats.
- **Statistical Software Manuals** for TI-83/84, Excel, Minitab, SPSS, and JMP provide instruction, examples, and exercises using specific statistical software packages.
- **(Instructors Only) SolutionMaster** is a Web-based version of the instructor's solutions manual. This easy-to-use tool allows instructors to create homework assignments, quizzes, and tests from textbook exercises and generate a separate solution guide. Assignments and solutions can be downloaded in PDF format for convenient printing and posting. For more information or a demonstration, contact your local W. H. Freeman sales representative.

Assignment Center (for instructor use only) organizes assignments and grades through an easy-to-create assignment process providing access to questions from the Test Bank, Web Quizzes, and Exercises from *Discovering the Fundamentals of Statistics,* Second Edition.

Companion Web site: www.whfreeman.com/discostat2e is an open-access Web site includes statistical applets, data sets, and quizzes.

Printed Student Solutions Manual offers detailed solutions for key exercises from each section of *Discovering the Fundamentals of Statistics,* Second Edition. ISBN: 1464110808

EESEE (Electronic Encyclopedia of Statistical Examples and Exercises) Case Studies. Developed by The Ohio State University Statistics Department, these electronic case studies provide a wide variety of timely, real examples with real data. EESEE case studies are available via an access code-protected Web site. Access codes are included with new copies of *Discovering the Fundamentals of Statistics,* Second Edition, or subscriptions can be purchased online. Instructors can access EESEE through the companion Web site.

For Instructors Only

Instructor's Guide with Solutions The solutions manual offers teaching tips, chapter commentaries, lists of teaching resources, and solutions to all exercises from *Discovering the Fundamentals of Statistics,* Second Edition. Available electronically within the StatsPortal, the Online Study Center, and IRCD, as well as in print form.

Test Bank The Test Bank contains hundreds of multiple-choice questions to generate quizzes and tests. Available electronically on CD-ROM (for Windows and Mac), where questions can be downloaded, edited, and resequenced to suit each instructor's needs.

Enhanced Instructor's Resource CD-ROM Allows instructors to search and export (by key term or chapter) all the material from the student Web site, plus:

- All text images and tables
- Instructor's Guide with Solutions
- PowerPoint lecture slides
- Test bank files

ISBN: 1464110980

Course Management Systems W. H. Freeman and Company provides courses for Blackboard, WebCT (Campus Edition and Vista), and Angel course management systems. They are completely integrated courses that you can easily customize and adapt to meet your teaching goals and course objectives. Visit **http://www.macmillanhighered.com/Catalog/other/Coursepack** for more information.

i-clicker is a two-way radio-frequency classroom response solution developed by educators for educators. University of Illinois physicists Tim Stelzer, Gary Gladding, Mats Selen, and Benny Brown created the i-clicker system after using competing classroom response solutions and discovering they were neither classroom-appropriate nor student-friendly. Each step of i-clicker's development has been informed by teaching and learning. i-clicker is superior to other systems from both pedagogical and technical standpoints. To learn more about packaging i-clicker with this textbook, please contact your local sales rep or visit **www.iclicker.com**.

Acknowledgments

I would like to join W. H. Freeman and Company in thanking the reviewers who offered comments that assisted in the development and refinement of the second edition of *Discovering the Fundamentals of Statistics*:

Holly Ashton, *Pikes Peak Community College*
John Beyers, *University of Maryland University College*
Dean Burbank, *Gulf Coast State College*
Ferry Butar Butar, *Sam Houston State University*
Ann Cannon, *Cornell College*
Ayona Chatterjee, *University of West Georgia*
Zhao Chen, *Florida Gulf Coast University*
Geoffrey Dietz, *Gannon University*
Wanda Eanes, *Macon State College*
Elaine Fitt, *Bucks County Community College*
Elizabeth Flow-Delwiche, *Community College of Baltimore County*
Joe Gallegos, *Salt Lake Community College*
Dave Gilbert, *Santa Barbara City College*
Donna Gorton, *Butler Community College*
David Gurney, *Southeastern Louisiana University*
Steve Hundert, *College of Southern Maryland*
Andreas Lazari, *Valdosta State University*
Ananda Manage, *Sam Houston State University*
Christina Morian, *Lincoln University*

John Nardo, *Oglethorpe University*
Michael Nasab, *Long Beach City College*
Greg Perkins, *Hartnell College*
Rogelio Ruiz, *Riverside Community College*
Fary Sami, *Harford Community College*
Jason Samuels, *Borough of Manhattan Community College*
Mohammed Shayib, *Prairie View A&M University*
Kim Sheppard, *Cecil College*
Marcia Siderow, *California State University, Northridge*
Karen Smith, *University of West Georgia*
Tabrina Smith, *Lake Erie College*
Sherman Sowby, *Brigham Young University*
John Trimboli, *Macon State College*
Cameron Troxell, *Mt. San Antonio College*
Mahbobeh Vezvaei, *Kent State University*
Karin Vorwerk, *Westfield State University*
James Wan, *Long Beach City College*
Tanya Wojtulewicz, *Community College of Baltimore County*

I would also like to thank the many instructors from across the United States and Canada who offered comments on the first and second edition of the full version of *Discovering Statistics*, upon which *Fundamentals* is based:

ARKANSAS George Bratton, *University of Central Arkansas*
ARIZONA Cheryl Ossenfort, *Coconino Community College*
CALIFORNIA Christine Cole, *Moorpark College*; Carol Curtis, *Fresno City College*; Kevin Fox, *Shasta College*; Dave Gilbert, *Santa Barbara City College*; Kristin M. Hartford, *Long Beach City College*; Elizabeth Hamman, *Cypress College*; Sara Jones, *Santa Rosa Junior College*; Wendy Miao, *El Camino College*; Michael A. Nasab, *Long Beach City College*; Keith Oberlander, *Pasadena City College*; Greg Perkins, *Hartnell College*; Zika Perovic, *MiraCosta College*; Ladera Rosenburg, *Long Beach City College*; Rogelio Ruiz, *Riverside Community College*; Marcia Siderow, *California State University, Northridge*; Sherman Sowby, *California State University, Fresno*; Cameron Troxell, *Mt. San Antonio College*; James Wan, *Long Beach City College*; Michael Zeitzew, *El Camino College*

CANADA Susan Chen, *Camosun College*; Shaun Fallat, *University of Regina*; Dorothy Levay, *Brock University*
COLORADO Holly Ashton, *Pikes Peak Community College*; Dean Barchers, *Red Rocks Community College*; Nels Grevstad, *Metropolitan State College of Denver*; Jay Schaffer, *University of Northern Colorado*
DELAWARE Derald E. Wentzien, *Wesley College*
FLORIDA Abraham Biggs, *Broward Community College*; Lisa M. Borzewski, *St. Petersburg College*; Janette H. Campbell, *Palm Beach Community College*; Zhao Chen, *Florida Gulf Coast University*; Lani Kempner, *Broward Community College*; Nancy Liu, *Miami Dade College*; Panagiotis Nikolopoulos, *Nova Southeastern University*; William Radulovich, *Florida Community College at Jacksonville*; Traci M. Reed, *St. Johns River Community College*; Pali Sen, *University of North Florida*; Jerry Shawver, *Florida Community College at Jacksonville*; Deanna Voehl, *Indian River State College*

GEORGIA Donna Brouillette, *Georgia Perimeter College*; Ayona Chatterjee, *University of West Georgia*; Wanda Eanes, *Macon State College*; Todd Hendricks, *Georgia Perimeter College*; Shahryar Heydari, *Piedmont College*; Andreas Lazari, *Valdosta State University;* Barry J. Monk, *Macon State College*; John Nardo, *Ogelthorpe University*; Chandler Pike, *University of Georgia*; Kim Robinson, *Clayton State University*; Howard L. Sanders, *Georgia Perimeter College*; Karen H. Smith, *University of West Georgia*; Martha Tapia, *Berry College;* John Trimboli, *Macon State College*

HAWAII David Ching, University of Hawai'i at Manoa; Eric Matsuoka, *Leeward Community College*

ILLINOIS Virginia Coil, *College of Lake County*; James Cicarelli, *Roosevelt University*; Faye Dang, *Joliet Junior College*; Linda Hoffman, *McKendree University*; Glenn Jablonski, *Triton College*; Julius Nadas, *Wilbur Wright College*; Stephen G. Zuro, *Joliet Junior College*

INDIANA Ewa Misiolek, *Saint Mary's College*

IOWA Russell Campbell, *University of Northern Iowa*

KANSAS Donna Gorton, *Butler Community College*; Linda Herndon, *Benedictine College*; James Leininger, *MidAmerica Nazarene University*; Leesa Pohl, *Donnelly College*

KENTUCKY Brooke Buckley, *Northern Kentucky University*; Lloyd Jaisingh, *Morehead State University*; Christopher Schroeder, *Morehead State University*; Marlene Will, *Spalding University*

LOUISIANA Arun K. Agarwal, *Grambling State University*; David Busekist, *Southeastern Louisiana University*; Julien Doucet, *Louisiana State University at Alexandria*; Diane Fisher, *University of Louisiana at Lafayette*; David Gurney, *Southeastern Louisiana University*; Nabendu Pal, *University of Louisiana at Lafayette*; Victor S. Swaim, *Southeastern Louisiana University*

MARYLAND John Beyers, *University of Maryland University College*; Elizabeth Flow-Delwiche, *Community College of Baltimore County*; Cathy Hess, *Anne Arundel Community College*; Steven Hundert, *College of Southern Maryland*; Annette Noble, *University of Maryland Eastern Shore*; Steve Prehoda, *Frederick Community College*; Fary Sami, *Harford Community College*; Kim Sheppard, *Cecil College*; Tanya Wojtulewicz, *Community College of Baltimore County*

MASSACHUSETTS Mary Fowler, *Worcester State College*; LeRoy P. Hammerstrom, *Eastern Nazarene College*; Karin Vorwerk, *Westfield State University*; Bonnie Wicklund, *Mount Wachusett Community College*

MICHIGAN Jennifer Borrello, *Grand Rapids Community College*; Lorraine Gregory, *Lake Superior State University*; Linda Reist, *Macomb Community College*; Kathy Zhong, *University of Detroit Mercy*

MINNESOTA Ken Grace, *Anoka-Ramsey Community College*; Mezbahur Rahman, *Minnesota State University, Mankato*

MISSOURI Kathy Carroll, *Drury University*; Christina Morian, *Lincoln University of Missouri*

MONTANA Debra Wiens, *Rocky Mountain College*

NEBRASKA Polly Amstutz, *University of Nebraska at Kearney*; Kathy Woitaszewski, *Central Community College*

NEW JERSEY Robert Thurston, *Rowan University*; Cathleen Zucco-Teveloff, *Rowan University*

NEW YORK David Bernklau, *Long Island University*; Jadwiga Domino, *Medaille College*; Reva Fish, *University at Buffalo*; Maryann Justinger, *Erie Community College*; Michael Kent, *Borough of Manhattan Community College*; William Price, *North Country Community College*; Jason Samuels, *Borough of Manhattan Community College*; Sharon Testone, *Onondaga Community College*; Nicholas Zaino, *University of Rochester*

NORTH CAROLINA Emma B. Borynski, *Durham Technical Community College*; Ayesha Delpish, *Elon University*; Jackie MacLaughlin, *Central Piedmont Community College*; Jeanette Szwec, *Cape Fear Community College*; John Russell Taylor, *The University of North Carolina at Charlotte*; James Truesdell, *Chowan University*

OHIO G. Andy Chang, *Youngstown State University*; Don Davis, *Lakeland Community College*; Arjun Gupta, *Bowling Green State University*; William Huepenbecker, *BGSU Firelands*; Gaurab Mahapatra, *The University of Akron*; Tabrina Smith, *Lake Erie College*; Mahbobeh Vezvaei, *Kent State University*

OKLAHOMA Mickle Duggan, *East Central University*; John Nichols, *Oklahoma Baptist University*; William Warde, *Oklahoma State University*

OREGON Jong Sung Kim, *Portland State University*; Carrie Kyser, *Clackamas Community College*

PENNSYLVANIA Elaine Fitt, *Bucks County Community College*; Geoffrey Dietz, *Gannon University*; Linda M. Myers, *Harrisburg Area Community College*; Sandra Nypaver, *Mount Aloysius College*

SOUTH CAROLINA Diana J. Asmus, *Greenville Technical College*; Thomas Fitzkee, *Francis Marion University*; Erwin Walker, *Clemson University*

TENNESSEE Aniekan Ebiefung, *University of Tennessee at Chattanooga*; Frankie E. Harris, *Southwest Tennessee Community College*; Marc Loizeaux, *University of Tennessee at Chattanooga*; Mary Ella Poteat, *Northeast State Technical Community College*

TEXAS Ananda Bandulasiri, *Sam Houston State University*; Ferry Butar Butar, *Sam Houston State University*; Ola Disu, *Tarrant County College*; Emmett Elam, *Texas Tech University*; Maggie Foster, *Tarrant County College*; Grady Grizzle, *North Lake College*; Jada P. Hill, *Richland College*; Melinda Holt, *Sam Houston State University*; Jianguo Liu, *University of North Texas*; Amanda Manage, *Sam Houston State University*; David D. Marshall, *Texas Woman's University*; Melissa Reeves, *East Texas Baptist University*; Ricardo Rodriguez, *Eastfield College*; Mohammed Shayib, *Prairie View A&M University*; Daniela Stoevska-Kojouharov, *Tarrant County College*; Jo Tucker, *Tarrant County College*

UTAH Kari Arnoldsen, *Snow College*; Joe Gallegos, *Salt Lake Community College*; Sherman Sowby, *Brigham Young University*; Ruth Trygstad, *Salt Lake Community College*

VIRGINIA John Avioli, *Christopher Newport University*; Robert May, *Virginia Highlands Community College*; Mike Shirazi, *Germanna Community College*; Glenn Weber, *Christopher Newport University*; Ken Wissmann, *Shenandoah University*

WASHINGTON Margaret Balachowski, *Everett Community College*; Kelly Brooks, *Pierce College*; Abel Gage, *Skagit Valley College*; John Kellermeier, *Tacoma Community College*

WISCONSIN William K. Applebaugh, *University of Wisconsin Eau Claire*; David M. Reineke, *University of Wisconsin La Crosse*; Vicki Whitledge, *University of Wisconsin Eau Claire*

The Second Edition of *Discovering the Fundamentals of Statistics* owes much to the untiring efforts of the team of professionals at W. H. Freeman and Company. I would like to thank Elizabeth Widdicombe, Craig Bleyer, Andrew Sylvester, Karen Carson, Diana Blume, Elizabeth Geller, Paul Rohloff, Roland Cheyney, Laura Judge, Steve Thomas, Tony Palermino, Ann Cannon, Martha Solonche, and Christina Morian for contributing their talents to the creation of the book. Most especially, I would like to thank Ruth Baruth, Mathematics and Statistics Publisher, who recognized the need for a book like *Discovering the Fundamentals of Statistics* and helped make it a reality.

I also wish to thank Dr. Jeffrey McGowan and Dr. Chun Jin, Chair and Assistant Chair of the Department of Mathematical Sciences at Central Connecticut State University, Dr. Dipak K. Dey, Distinguished Professor and Associate Dean, College of Liberal Arts and Sciences at the University of Connecticut, and Dr. John Judge, Chair of the Department of Mathematics at Westfield State University. Thanks to my daughter and statistician-in-training Chantal Danielle (24) for carrying on the love of statistics to the next generation, and to my twin children Tristan Spring and Ravel Renaissance (13) for demonstrating that there is life beyond the computer screen. Above all, I extend my deepest gratitude to my darling wife of 27 years, Debra J. Larose, for her love, support, and understanding.

About the Author

Since his days of collecting baseball cards as a youngster and checking out the statistics of his favorite players, Dan Larose has loved statistics. He also loved language and writing, so when Dan went to college he majored in French, then philosophy, and finally, in linguistics and computer science. This background in the liberal arts honed his writing ability. However, his love of statistics never left him, so he went on to earn an M.S. (1993) and a Ph.D. in statistics (1996) from the University of Connecticut. Today, Dan is Professor of Statistics in the Department of Mathematical Sciences at Central Connecticut State University (CCSU).

At CCSU, Dan designed, developed, and now directs the world's first online Master of Science degree and Graduate Certificate program in data mining. He has published three books on data mining and one book on SAS programming. *Discovering Knowledge in Data: An Introduction to Data Mining* and *Data Mining Methods and Models* have been translated into French and Polish, while *Data Mining Methods and Models* and *Data Mining the Web* have been translated into Polish. He is the founder of DataMining-Consultant.com, and his consulting clients include *The Economist* magazine, Microsoft, *Forbes* magazine, the CIT Group, KPMG International, Computer Associates, Deloitte, Inc., Sonalysts, Inc., Booz Allen and Hamilton, and the Hospital for Special Care. His consulting work includes a $750,000 Phase II grant from the Air Force Office of Research, Storage Efficient Data Mining of High Speed Data Streams. He is the Series Editor for the Wiley series on Methods and Applications in Data Mining.

However, his favorite work is imparting a love of statistics to a new generation, and he trusts that *Discovering Statistics* and *Discovering the Fundamentals of Statistics* will help to do so.

Dan lives in Tolland, Connecticut, with his wife and children, including daughter Chantal, who is a PhD candidate in Statistics at the University of Connecticut.

Discovering
the Fundamentals of
STATISTICS

Second Edition

1 The Nature of Statistics

© Old Visuals/Alamy

CASE STUDY

Does Friday the 13th Change Human Behavior?

Superstitions affect most of us. Some people will never walk under a ladder, while others will alter their path to avoid a black cat. Do you think that people change their behavior on Friday the 13th? Perhaps, suspecting that it may be unlucky, some people might elect to stay home and watch television rather than venture outdoors or drive on the highway.

But how does one go about researching such a question? How would *you* do it? In this chapter, we will learn about a British study that considered this question. ■

The Big Picture

Where we are coming from, and where we are headed . . .

- The objective of *Discovering Statistics* is to help you understand how to analyze and interpret data, and thereby become a successful citizen in the Information Age.

- Chapter 1 introduces the basic ideas of the field of statistics and the methods for gathering data.

- In Chapter 2 we will learn to summarize the data we have gathered using graphs and tables.

1.1   DATA STORIES: THE PEOPLE BEHIND THE NUMBERS

OBJECTIVE   By the end of this section, I will be able to . . .

1 Realize that behind each data set lies a story about real people undergoing real-life experiences.

We begin *Discovering Statistics* by sharing some data stories. We hope that these stories will kindle a response in you, be it sympathy or curiosity or concern, for behind every data set lies a story about the lives of real people. Individual people are speaking to us from behind the numbers.

EXAMPLE 1.1   THE REASONS KATRINA SURVIVORS DID NOT EVACUATE

Hurricane Katrina was the costliest and one of the deadliest hurricanes in American history. Damages exceeded $50 billion and fatalities exceeded 1300, according to the National Oceanic and Atmospheric Administration. In September 2005, a survey was conducted of a group of hurricane survivors who had later been moved to shelters in the Greater Houston area. The respondents who did not evacuate were asked what was their most important reason for not evacuating. Figure 1.1 provides a bar graph of the responses, with Table 1.1 supplying more detailed information.

Table 1.1 Katrina survivors' most important reasons for not evacuating

Reason	Percent
I did not have a car or a way to leave	36
I thought the storm and its aftermath would not be as bad as they were	29
I just didn't want to leave	10
I had to care for someone who was physically unable to leave	7
All other reasons	18

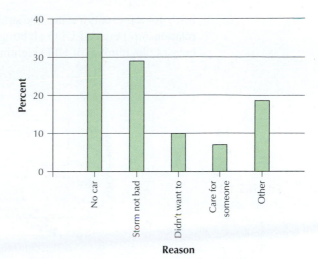

FIGURE 1.1
Bar graph of Katrina survivors' reasons for not evacuating.

EXAMPLE 1.2

WERE THERE GENDER DIFFERENCES IN THE EMOTIONS EXPERIENCED IMMEDIATELY AFTER SEPTEMBER 11, 2001?

On September 11, 2001, terrorists attacked New York City and Washington, DC. Do you think that men and women felt the same emotions about these attacks? In an NBC News Terrorism Poll conducted the day after the tragic events, the following question was asked: "Which one of the following emotions do you feel the most strongly in response to these terrorist attacks: sadness, fear, anger, disbelief, vulnerability?" Figure 1.2 is called a *clustered bar graph* and shows the results. The dominant emotion felt by the men was anger, while the women tended to feel either sadness, anger, or disbelief. Note how the bar graph makes these findings—that there were indeed systematic differences in the emotions felt by men and women regarding the events of September 11, 2001—crystal clear. We will learn how to construct bar graphs in Chapter 2, "Describing Data Using Graphs and Tables."

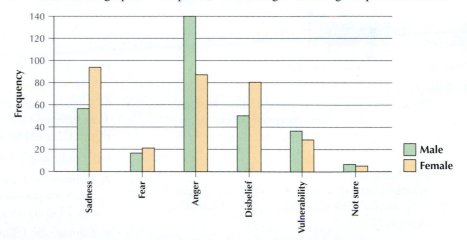

FIGURE 1.2 Clustered bar graph of strongest emotions felt regarding the September 11, 2001, attacks (by gender).

EXAMPLE 1.3

UFO SIGHTINGS

Have you or any of your friends sighted any unidentified flying objects (UFOs)? Americans in each of the 50 states have reported seeing UFOs. Figure 1.3 represents a scatterplot of the number of UFO sightings versus state population, for each of the 50 states.

Each dot represents a state. The straight line is a regression line which approximates the relationship between UFO sightings and state population. As the state population increases, the number of UFO sightings also tends to increase, which is not surprising.

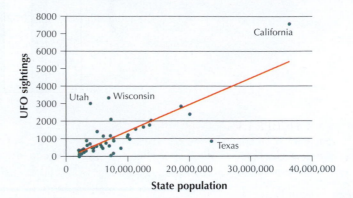

FIGURE 1.3

A scatterplot of the number of UFO sightings versus state population, showing that UFOs don't mess with Texas.

What may be surprising is that the UFOs seem to be attracted to certain states, and to avoid others. States considerably above the regression line have a larger than expected number of UFO sightings for their population size, while states below the line have a smaller than expected number of UFO sightings for their population size. So, there are more sightings than expected in California, Wisconsin, and Utah, given their population size, and fewer than expected in Texas. Why this might occur is open to discussion. Perhaps people in California are more likely to attribute unusual sightings to UFOs than most Americans; perhaps people in Texas are more pragmatic than most Americans. But if the sightings are valid (a big if!), it sure looks like the UFOs don't want to mess with Texas. We will learn how to construct scatterplots and how to quantify the relationship between two numerical variables in Chapter 4, "Correlation and Regression."

| SECTION 1.1 | **Exercises** |

Refer to Example 1.1 for Exercises 1–4.

1. Refer to Figure 1.1.
   a. What does the graph say was the most common reason why the Katrina survivors did not evacuate?
   b. What does Table 1.1 say was the most common reason?

2. Refer to Figure 1.1.
   a. Which is more descriptive, the table or the figure?
   b. Why do you think the text in Figure 1.1 has been shortened?

3. If you were writing a news story that sought to display the Katrina survivors in the most sympathetic light, which reasons from Table 1.1 might you emphasize?

4. If you were writing a news story that sought to display the Katrina survivors in a less favorable light, which reasons from Table 1.1 might you emphasize?

Refer to Example 1.2 for Exercises 5 and 6.

5. Do you think the emotions felt were different for men and women? If so, then what evidence from

Figure 1.2 would you offer in support of such a view?

6. Suppose you did not believe that the emotions felt were different for men and women. What evidence from Figure 1.2 could be offered in support of that position?

Refer to Example 1.3 for Exercises 7–10.

7. Estimate the following for the state of California.
   a. State population
   b. UFO sightings

8. Estimate the following for the state of Texas.
   c. State population
   d. UFO sightings

9. For a given population size, the expected number of UFO sightings falls on the regression line. For the state of California, what is the expected number of UFO sightings? (*Hint:* It's at the point on the line directly below the dot for California.)

10. For the state of Texas, what is the expected number of UFO sightings?

AN INTRODUCTION TO STATISTICS

OBJECTIVES By the end of this section, I will be able to . . .

1 Describe what the field of statistics is.

2 State the meaning of descriptive statistics.

3 Explain what is meant by inferential statistics.

1 WHAT IS STATISTICS?

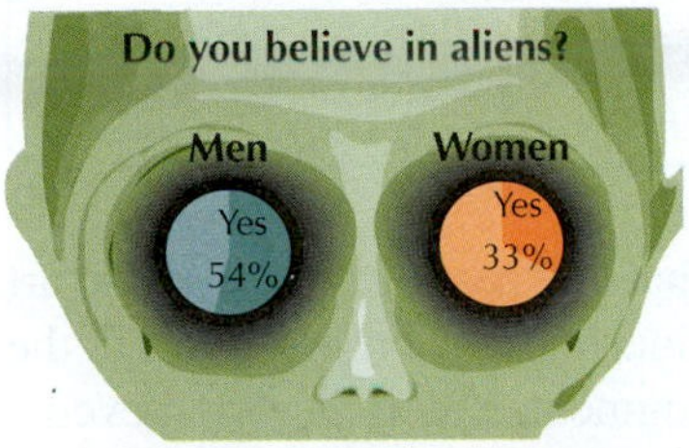

Do you believe in aliens? According to a recent survey, 54% of the men surveyed responded that they believed in aliens, and 33% of the women did (Figure 1.4). These numbers are examples of *statistics,* numbers that describe a group of people or things. Think about these numbers. Here are some questions we could ask about this survey:

- How did the pollsters arrive at these figures?

- Are the figures accurate? Could they be inaccurate?

- Why do pollsters never ask me my opinion about aliens?

- This survey found that more men than women believed in aliens. But is this difference meaningful or just a product of random chance?

These are some of the types of questions we shall be investigating throughout this book.

FIGURE 1.4 Graphs comparing percentages of men and women who believe in aliens. (© *USA Today*)

Examples of Statistics

Many people, including the author, first became interested in statistics as children collecting baseball cards. The back of each card contains the player's statistics season by season. Television networks routinely employ sports statisticians to collect and report statistics about sports figures. Table 1.2, for example, contains batting averages of the league-leading hitters from 2007 to 2011.

Table 1.2 Batting-average leaders, Major League Baseball, 2007–2011

Season	Player	Team	Batting average
2011	Miguel Cabrera	Detroit Tigers	.344
2010	Josh Hamilton	Texas Rangers	.359
2009	Joe Mauer	Minnesota Twins	.365
2008	Chipper Jones	Atlanta Braves	.364
2007	Magglio Ordonez	Detroit Tigers	.363

The informal meaning of the term *statistic* refers to a number that describes a person, a group, or a set of items. (On page 12, we provide a more precise definition of a statistic.) For example, Miguel Cabrera's batting average of .344 is a statistic, because it is a number that describes his batting performance for the entire 2011 season. Apart from sports, most people become familiar with statistics through exposure to media reports or advertising, such as

- "Polls indicate a majority of Democrats support stem cell research."
- "The median home sales price in Connecticut has climbed in recent months to $250,000."
- "Three out of four dentists surveyed recommend sugarless gum for their patients who chew gum."

You may have noticed that the section title, "What Is Statistics?" refers to statistics in the singular. Why? Because the *field of statistics* involves much more than just collecting and reporting numerical facts. The field of statistics may be defined as follows.

> The field of **statistics** is the *art* and *science* of
> - collecting,
> - analyzing,
> - presenting, and
> - interpreting data.

A statistician, then, is not simply a sports analyst but any person trained in the art and science of statistics. You may be surprised at the inclusion of the word *art* in the definition of statistics. But there is no question that judgment, experience, and even a little intuition are indispensable tools for any statistician's portfolio.

For today's college student, the field of statistics is especially relevant and useful.

- For example, a *business major* may be interested in whether she should consider diversifying her portfolio to tech stocks, based on their price/earnings ratio.

- A *psychology major* may be interested in determining whether there are differences in therapeutic outcomes between traditional counseling methods and a new cognitive approach.

- An *education major* may be interested in whether listening to a Mozart sonata before taking an exam can significantly improve your grade.

The field of statistics can help solve each of these puzzles.

CASE STUDY   ## Does Friday the 13th Change Human Behavior?

How would researchers go about studying whether superstitions change the way people behave? What kind of evidence would support the hypothesis that Friday the 13th causes a change in human behavior? T. J. Scanlon and his coresearchers thought that if there were fewer vehicles on the road on Friday the 13th than on the previous Friday, this would be evidence that some people were playing it safe on Friday the 13th and staying off the roads.[1] Note that the researchers didn't simply argue about the validity of the Friday the 13th superstition. Such discussions are interesting but largely subjective. What they deemed important is the effect of such a superstition on human behavior and how to measure such an effect as a change in behavior.

Phase 1 **Data collection.** The first phase of a statistical study, as in the definition of statistics, is to *collect* the data. The researchers obtained data kept by the British Department of Transport on the traffic flow through certain junctions of the M25 motorway in England.

Phase 2 **Data analysis.** Next comes the analysis of the data. The authors compared the number of vehicles passing through certain junctions on the M25 motorway on Friday the 13th and the previous Friday during 1990, 1991, and 1992.

Table 1.3 Traffic through M25 junctions, 1990–1992

Friday the 6th	Friday the 13th	Difference
139,246	138,548	698
134,012	132,908	1104
137,055	136,018	1037
133,732	131,843	1889
123,552	121,641	1911
121,139	118,723	2416
128,293	125,532	2761
124,631	120,249	4382
124,609	122,770	1839
117,584	117,263	321

Table 1.3 shows that, in every instance, the number of vehicles passing through these junctions on Friday the 13th was less than on the preceding Friday. Now, let's examine the data graphically. The clustered bar graph in Figure 1.5 illustrates the difference in the number of vehicles traveling on the M25 motorway on Friday the 6th (in green) and the subsequent Friday the 13th (in yellow) for 10 pairs (clusters) of dates. Note that, *in every instance,* the green bar is longer than its partner yellow bar. This indicates that the number of vehicles on the motorway *decreased* on Friday the 13th when compared with the previous Friday in every instance.

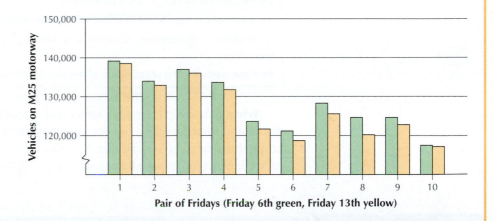

FIGURE 1.5
Clustered bar graph of motorway traffic.

Phase 3 **Data presentation.** The *presentation* of the results is important, and the researchers found a highly respectable journal, the *British Medical Journal,* in which to publish their findings. Other avenues for presentation are delivering a talk at a conference, writing up a report for one's supervisor, or presenting a class project.

(*continues*)

Phase 4 **Data interpretation.** Finally, the last facet in our definition of statistics is *interpretation*. It is crucial for those who are performing a statistical study to make their results understandable to nonstatisticians. It is not sufficient for the statistician alone to understand the results. Rather, the statistician must communicate the results clearly, whether in writing or orally. In this case, the researchers chose the decrease in number of vehicles as the criterion on which to base support for their hypothesis that people changed their behavior on Friday the 13th. Their finding of an observable decrease in traffic on Friday the 13th is consistent with their hypothesis. ∎

2 DESCRIPTIVE STATISTICS: THE BUILDING BLOCKS OF DATA ANALYSIS

Every data set holds within it a story waiting to be told, as we saw in the Friday the 13th Case Study. To provide us with the tools to uncover these stories we need to learn some simple concepts, *the building blocks of data analysis.*

> **Descriptive statistics** refers to methods for summarizing and organizing the information in a data set.

In **descriptive statistics** we use numbers (such as counts and percents), graphs, and tables to describe the data set, as a first step in data analysis. In Chapters 2 to 4, we will examine descriptive methods much more closely. But first we need to introduce a few terms. Suppose a data analyst for a health maintenance organization (HMO) is collecting data about the patients in a particular hospital, including the diagnosis, length of stay, gender, and total cost. The sources of the information (the patients) are called the **elements.** The patients' characteristics (for example, diagnosis, length of stay) are called the **variables.** Finally, the complete set of characteristics for a particular patient is called an **observation.**

> **Elements, Variables, and Observations**
>
> An **element** is a specific entity about which information is collected.
>
> A **variable** is a characteristic of an element, which can assume different values for different elements.
>
> An **observation** is the set of values of the variables for a given element.

When data are presented in tables and spreadsheets, it is typical practice to have the columns indicate the variables, and the rows to indicate the elements. So, for the hospital patients, the observation (specific values for the set of all the variables) for each element (patient) would appear as a row in the table.

EXAMPLE 1.4   **ELEMENTS, VARIABLES, AND OBSERVATIONS**

Information was collected on four students from two area colleges and is presented in Table 1.4.
a.   What are the elements?
b.   What are the variables?
c.   List the values that the variable *gender* takes.
d.   Provide the observation for Maria.

Table 1.4 Data set of four elements and seven variables

Student	Age	Gender	Ethnicity	No. of Children	Marital Status	GPA	College
Jamal	19	Male	African American	0	Single	4.00	Western CC
Maria	25	Female	Latina	2	Married	3.95	Northern State Univ.
Chang	20	Female	Asian	0	Single	3.90	Northern State Univ.
Michael	47	Male	European American	3	Divorced	3.75	Western CC

Solution

a. The elements are the students Jamal, Maria, Chang, and Michael.
b. The seven variables are age, gender, ethnicity, number of children, marital status, GPA, and college.
c. The variable *gender* takes values *female* and *male*.
d. Since the observation for Maria consists of the values for the variables in Maria's entire row, her observation is (see the following table)

Now You Can Do Exercises 11–18.

Student	Age	Gender	Ethnicity	No. of Children	Marital Status	GPA	College
Maria	25	Female	Latina	2	Married	3.95	Northern State Univ.

Notice that we have variables that can take on various types of values, some of which are numbers and some of which are categories. For example, Maria is 25 years old, has two children, and has a GPA of 3.95, each of which is numeric. On the other hand, Maria is Latina, married, and enrolled at Northern State University, characteristics that do not have numeric values but instead are categories. This leads us to define two types of variables: **qualitative** and **quantitative**.

> A **qualitative variable** is a variable that may be classified into categories. A **quantitative variable** is a variable that takes numeric values and upon which arithmetical operations such as addition or subtraction may be meaningfully performed.

Qualitative variables are also called *categorical variables*, because they can be grouped into categories. Maria's qualitative variables include her gender, ethnicity, marital status, and college. In contrast, Maria's grade point average is an example of a *quantitative variable*. Other quantitative variables include age and number of children.

EXAMPLE 1.5

QUALITATIVE OR QUANTITATIVE?

Some of the most widespread applications of statistical analysis occur in the business world. Managers examine patterns and trends in data, thereby hoping to increase profitability. Table 1.5 shows the five most active stocks on the New York Stock Exchange (NYSE) and NASDAQ, as reported by *USA Today* for September 10, 2010. (a) What are the elements and the variables of this data set? (b) Which variables are qualitative? Which are quantitative? (c) Provide the observation for Intel Corporation.

Xinhua/eyevine/Redux

NASDAQ (National Association of Securities Dealers Automated Quotations) is an American stock exchange that includes many technology companies.

Table 1.5 Most active stocks on NYSE and NASDAQ, September 10, 2010

Stock	Exchange	Last	Volume	Change
Citigroup, Inc.	NYSE	$ 3.91	256,441,698	0.00
Bank of America	NYSE	$13.55	85,884,565	+0.05
Intel Corporation	NASDAQ	$17.97	68,824,147	−0.17
Nokia Corporation	NYSE	$ 9.94	64,502,103	+0.18
Microsoft Corporation	NASDAQ	$23.85	58,293,790	−0.46

Solution

a. The *elements* are the five most active stocks traded on the NYSE and NASDAQ on this day in 2010. The *variables* are as follows:
   - Exchange: The exchange where the stock was traded.
   - Last: The most recent trading price for the stock.
   - Volume: How many shares of the stock were traded that day.
   - Change: The change in share price (in dollars) between the opening price and the closing price that day.

b. The exchange, since it can be categorized as either NYSE or NASDAQ, is qualitative. The other variables are quantitative.

c. The observation for Intel includes the exchange and the set of the day's stock data for that company. Intel is traded on the NASDAQ exchange. Its last share price was $17.97 per share, 68,824,147 shares of its stock were traded, and the price decreased by $0.17 per share.

Now You Can Do Exercises 19–22.

Stock	Exchange	Last	Volume	Change
Intel	NASDAQ	$17.97	68,824,147	−0.17

Quantitative variables can be classified as either **discrete** or **continuous.**

Hint: A quantitative variable that must be counted (not measured) is probably a discrete variable, while a quantitative variable that must be measured (not counted) is probably a continuous variable.

> A **discrete variable** can take either a finite or a countable number of values. Each value can be graphed as a separate point on a number line, with space between each point.
> A **continuous variable** can take infinitely many values, forming an interval on the number line, with no space between the points.

EXAMPLE 1.6

DISCRETE OR CONTINUOUS?

In Table 1.4, determine whether the following variables are discrete or continuous: (a) number of children and (b) GPA.

Solution

a. Since the number of children per student is finite, the variable *number of children* is discrete.

Now You Can Do Exercises 23–26.

b. Since GPA can take an infinite number of possible values, for example in the interval 0.0 to 4.0, the variable *GPA* is continuous.

Levels of Measurement

Data may be classified according to the following four *levels of measurement*.

- *Nominal data* consist of names, labels, or categories. There is no natural or obvious ordering of nominal data (such as high to low). Arithmetic cannot be carried out on nominal data.

- *Ordinal data* can be arranged in a particular order. However, no arithmetic can be performed on ordinal data.

- *Interval data* are similar to ordinal data, with the extra property that subtraction may be carried out on interval data. There is no *natural zero* for interval data.

- *Ratio data* are similar to interval data, with the extra property that division may be carried out on ratio data. There does exist a natural zero for ratio data.

| EXAMPLE 1.7 | LEVELS OF MEASUREMENT |

Identify which level of measurement is represented by the following data.
a. Years covered in European History 101: 1066–1492
b. Annual income of students in Statistics 101 class: $0–$15,000
c. Course grades in English 101: A, B, C, D, F
d. Student gender: male, female

Solution

a. The years 1066 to 1492 represent interval data. There is no natural zero (no "year zero"; the calendar goes from 1 B.C. to A.D. 1). Also, division (1492/1066) does not make sense in terms of years, so that the data are not ratio data. However, subtraction does make sense, in that the course covers $1492 - 1066 = 426$ years.

b. Student income represents ratio data. Here division does make sense. That is, someone who made $4000 last year made twice as much as someone who made $2000 last year. Also, some students probably had no income last year, so that $0, the natural zero, also makes sense.

c. Course grades represent ordinal data, since (a) they may be arranged in a particular order, and (b) arithmetic cannot be performed on them. The quantity $A - B$ makes no sense.

d. Student gender represents nominal data, since there is no natural or obvious way that the data may be ordered. Also, no arithmetic can be carried out on student gender.

Now You Can Do
Exercises 27–34.

3 INFERENTIAL STATISTICS: HOW DO WE GET THERE FROM HERE?

Descriptive methods of data analysis are widespread and quite informative. However, the modern field of statistics involves much more than simply summarizing a data set. For example, suppose a sociologist claims that one-third of American teenagers have been the targets of cyberbullying, that is, have received a threatening message or have had their emails or text messages forwarded without their consent, an embarrassing picture posted without permission, or rumors spread about them online. How should the sociologist go about collecting evidence to support her claim? One method would be to ask each and every person in the population of all American teenagers. In general, a **population** is the collection of *all* elements (persons, items, or data) of interest in a particular study.

However, to ask every teenager in America about his or her online experiences is a daunting task that is expensive, time-consuming, and, in the end, simply impossible. So, unfortunately, the population proportion of American teenagers who have been the targets of cyberbullying remains *unknown*. This proportion who have been targets of cyberbullying is one characteristic of the population of American teenagers. A characteristic of a population is called a **parameter.** The actual value of a population parameter is often unknown.

> **Population and Parameter**
>
> A **population** is the collection of *all* elements (persons, items, or data) of interest in a particular study. A **parameter** is a characteristic of a population.

A **sample** is a subset of the population from which information is collected. For example, from a sample of 100 teenagers at a local mall, 18 said they had been the targets of cyberbullying. That is, the sample proportion of students who had been targets is 18/100 = 18%. This proportion is a characteristic of the sample and is called a **statistic.** The advantage here is that, since the sample is relatively small, the characteristics of the sample can be determined.

> **Sample and Statistic**
>
> A **sample** is a subset of the population from which information is collected. A **statistic** is a characteristic of a sample.

Population   **Sample**

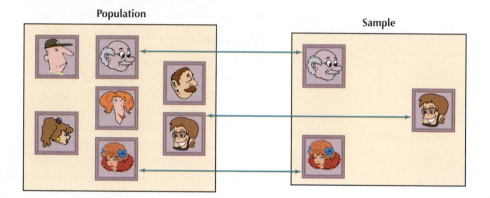

A sample is a subset of a population.

The U.S. Constitution requires that a census be conducted every 10 years. A **census** is the collection of data from every element in the population. As you can imagine, such a task is very difficult and very expensive. In fact, the Census Bureau estimates that the 2000 U.S. census "undercounted the actual U.S. population by over three million individuals."[2]

Because the population you are interested in may be too large to allow you to elicit information from every element, it is often best to gather data from a sample, a subset of that population. Also, time and money often constrain the researcher to choosing a sample rather than studying the entire population. Further, in some experiments, the resource is exhausted when testing is done, for example, in estimating the mean lifetime of light bulbs. Finally, it may be simply impossible to gather information from the entire population, such as when studying the quality of water in Lake Erie.

To estimate the proportion of all American teenagers who have been subjected to cyberbullying, we can use statistical inference. **Statistical inference** refers to learning about the characteristics of a population by studying those characteristics in a subset of the population (that is, in a *sample*). The Pew Internet and American Life Project conducted a survey of 886 teenagers and found that 284 of them (32%) said they had been the targets of cyberbullying.[3] These 886 teenagers represent a sample, and their characteristics can be known. For example, we know that 284 of the 886 teenagers in the sample said they have been subjected to cyberbullying. At this point, the sociologist can make the *inference* that the proportion of *all* American teenagers who have been subjected to cyberbullying is 32%, because this is the proportion in the sample. In doing this, the sociologist is performing *statistical inference.*

> **Statistical inference** consists of methods for estimating and drawing conclusions about population characteristics based on the information contained in a subset (sample) of that population.

"Now wait just a minute," you might object. "How can you say that the proportion of *all* American teenagers who have been subjected to cyberbullying is 32% just because your *sample* proportion is 32%?" Actually, you have a point. We *are* generalizing. We are taking what we know about a portion of the whole (a sample) and using it to draw a conclusion about the whole (the population). But even though the true proportion of American teenagers who have been the targets of cyberbullying is probably not exactly 32%, it is most likely not very far from 32%. The 32% is an *estimate,* an approximation based on sample data. In Chapter 8, we will learn how we can get the estimate as close as we wish to the actual value just by taking a large enough sample.

Finally, we need to point out one further attribute of parameters and statistics. The value of a parameter, even though it is unknown, is a fixed constant. For example, the average age of all persons in your home state (population) at noon today is unknown, but it still exists, and it is a single number. On the other hand, the value of a statistic depends on the sample. For example, a sample of 100 people in your hometown may produce an average age of 31. The average age of a sample of 100 people in a neighboring town may be 32. Later, we will learn that this is because a statistic is a *random variable*.

Of course, to deliver a valid estimate, the sample needs to be *representative* of the population. The sample should not differ systematically in any major characteristic from the population. We will learn more about this in Section 1.3, when we study sampling methods. Table 1.6 summarizes the attributes of a population and a sample.

Parameters are measures from a population, while statistics are measures from a sample. The characteristic associated with the population starts with the same letter, and the same is true for sample.

Table 1.6 Summary of attributes of population and sample

	Population	Sample
Thumbnail definition	All elements	Subset of population
Characteristic	Parameter	Statistic
Value	Usually unknown	Usually known
Status	Constant	Depends on sample

EXAMPLE 1.8

DESCRIPTIVE STATISTICS OR STATISTICAL INFERENCE?

State whether the following situations illustrate the use of descriptive statistics or statistical inference.

a. In Baltimore County, Maryland, the average amount spent per week on gasoline consumption in a sample of 500 commuters was $75. The county government infers that the average amount spent weekly by all Baltimore County commuters is $75.

b. A sample of 100 residents of Broward County, Florida, yielded 27 residents who work for the government at the local, state, or federal level. Thus, 27% of these 100 residents work for the government.

c. The average age of a sample of 200 residents of Garden City, New York, was 34 years old.

d. In a survey of 1000 citizens in the Seattle, Washington, metropolitan area, 570 said they would pay higher prices in order to reduce greenhouse emissions. City planners conclude that 57% of all Seattle citizens would do so.

Solution

a. **Statistical inference.** A sample was taken, and a sample statistic ($75 per week) was calculated. Then the county government used this statistic to make the *statistical inference* that this was the average amount spent by all Baltimore County commuters.

b. **Descriptive statistics.** Though a sample was taken, there was no attempt to make an inference from this sample of 100 workers to the entire population of Broward County, Florida. So, there is no statistical inference here.

c. **Descriptive statistics.** The average age of 34 years old is a descriptive statistic, since it describes the sample. But no inference is made regarding a larger population.

d. **Statistical inference.** The survey found that 57% of the sample of 1000 citizens would pay higher prices in order to reduce greenhouse emissions. This 57% is a statistic. Then the city planners used this statistic in order to perform statistical inference about the population of all Seattle citizens.

Now You Can Do Exercises 53–56.

A Statistical Literacy Quiz

Regardless of major, every student in America (indeed, every citizen) needs to become *statistically literate* in order to survive in today's wired society. Why not take this quiz to find out if you are statistically literate? Answer each question true or false.

1. A fair coin is tossed five times and comes up heads each time. That means that tails is "due" and the chances of tails on the next toss is increased.

2. One politician says that the mean income is rising, while another politician says that the median income is falling. One of them has to be lying.

3. Jim is tested for HIV and the test comes back positive. Thus, Jim is HIV-positive.

The correct answer to each question is *false*. Question 1 deals with something called "the Gambler's Fallacy," and we will cover this, along with the explanation for Question 3, in Chapter 5, "Probability." We will deal with Question 2, the quirks of means and medians, in Chapter 3, "Describing Data Numerically."

SECTION 1.2   **Summary**

1. The field of statistics is the art and science of collecting, analyzing, presenting, and interpreting data.

2. Descriptive statistics refers to methods for summarizing and organizing the information in a data set. Data sets include information collected on elements. Variables are characteristics of an element, and can take different values for different elements. Variables may be either quantitative or qualitative. A discrete variable is a quantitative variable that can take either a finite or a countable number of possible values. A continuous variable is a quantitative variable that can take an infinite number of possible values. A population is a collection of all elements of interest, while a sample is a subset of the population.

3. Inferential statistics consists of methods for estimating and drawing conclusions about population characteristics based on the information in the sample. The characteristics for a population are called parameters, while the characteristics for a sample are called statistics.

SECTION 1.2   **Exercises**

Clarifying the Concepts

1. Write a sentence describing in your own words the field of statistics.

2. True or false: Statistical inference refers to methods for summarizing and organizing the information in a data set.

3. What do we call the entities from which the data are collected?

4. Describe the difference between a qualitative and a quantitative variable.

5. What is another term for a qualitative variable?

6. True or false: The actual value of a population parameter is usually unknown.

7. What is the difference between a sample and a population?

8. Explain what a statistic is.

9. Describe one difference between a statistic and a parameter.

10. What is a census?

Practicing the Techniques

Refer to Table 1.7 for Exercises 11–14.

TABLE 1.7 **Information about four statistics students**

Student	Gender	Height	Class rank	Siblings	Math SAT
Michael	Male	67	Sophomore	2	510
Ashley	Female	67	Junior	1	520
Christopher	Male	70	Senior	0	490
Jessica	Female	66	Freshman	3	550

11. What are the elements?

12. List the variables.

13. List the values that the variable *class rank* takes.

14. Provide the observation for Jessica.

Refer to Table 1.8 for Exercises 15–18.

TABLE 1.8 **Information about five hospitals**

Hospital	Type	Number of floors	HMO ranking	Number of patients per nurse	Year opened
City	General	5	3rd	10.5	1999
Memorial	General	4	4th	12.7	1975
Children's	Specialized	3	1st	5.9	2005
Eldercare	Specialized	2	2nd	7.8	2009
County	General	6	5th	16.2	1967

15. What are the elements?

16. List the variables.

17. List the values that the variable *type* takes.

18. Provide the observation for Children's Hospital.

Refer to Table 1.7 for Exercises 19 and 20.

19. List the quantitative variables.

20. List the qualitative variables.

Refer to Table 1.8 for Exercises 21 and 22.

21. List the quantitative variables.

22. List the qualitative variables.

Refer to Table 1.7 for Exercises 23 and 24.

23. Which variables are discrete?

24. Which variables are continuous?

Refer to Table 1.8 for Exercises 25 and 26.

25. Which variables are discrete?

26. Which variables are continuous?

Refer to Table 1.7 for Exercises 27–30. Identify the variables that represent the following levels of measurement.

27. Nominal data

28. Ordinal data

29. Interval data

30. Ratio data

Refer to Table 1.8 for Exercises 31–34. Identify the variables that represent the following levels of measurement.

31. Nominal data

32. Ordinal data

33. Interval data

34. Ratio data

For Exercises 35–48:
   a. State whether the variable is qualitative or quantitative. If the variable is quantitative, state whether it is discrete or continuous.
   b. Identify the level of measurement represented by the data.

35. The year you were born

36. Whether you own a cell phone or not

37. The price of tea in China

38. The SAT Math score of the person sitting next to you (scores range from 200 to 800)

39. The winning score in next year's Super Bowl

40. The winning team in next year's Super Bowl

41. The rank of the winning Super Bowl team in its division

42. The number of friends on a student's Facebook page

43. Your favorite television show

44. How many contacts you have on your cell phone

45. Your favorite ice cream

46. Your credit card balance

47. How old your car is

48. What model your car is

For Exercises 49–52, identify the population and the sample.

49. A researcher is interested in the median home sales price in Tarrant County, Texas. He collects sales data on 100 home sales.

50. A psychologist is concerned about the health of veterans returning from war. She examines 20 veterans and assesses whether they show signs of post-traumatic stress disorder.

51. An educator asks a sample of students at Portland Community College whether they would be interested in taking a course online.

52. A financial adviser would like to assess the effect of mergers on price/earnings ratio. She collects data on 50 companies that recently underwent a merger.

For Exercises 53–56, state whether descriptive statistics or statistical inference was used, and explain why.

53. The average price in a sample of 15 homes sold in Jacksonville, Florida, for the week of April 21 was $253,200.

54. According to the Department of Transportation, 60% of all automobile passengers wear seat belts. This is based on a survey of 1000 automobile passengers, of whom 600 wore seat belts.

55. In a sample of 500 subjects, it was found that daily exercise lowered the average cholesterol level by 10%. A medical spokesperson then stated that daily exercise can lower everyone's cholesterol level by 10%.

56. The goals-against average for the Charlestown Chiefs hockey team in a sample of 20 games was 3.57 goals per game.

Applying the Concepts

For Exercises 57–62, do the following:
- **a.** List the elements and the variables.
- **b.** Identify the qualitative variables and the quantitative variables.
- **c.** For each variable, identify the level of measurement.
- **d.** For each quantitative variable, indicate whether it is discrete or continuous.
- **e.** Provide the observation for the indicated element.

57. Endangered Species. Refer to the following table, which lists four of the endangered animal species in the United States, as listed by **www.earthsendangered.com**. Do (**a**)–(**d**) and then provide the observation for the Florida panther.

Endangered species	Year listed as endangered	Estimated number remaining	Range
Pygmy rabbit	2001	20	Washington State
Florida panther	1973	50	Florida
Red wolf	1967	200	North Carolina
West Indian manatee	1967	2500	Florida

58. Top Five Employers in Santa Monica, CA. Refer to the following table. Do (**a**)–(**d**) and then provide the observation for the city of Santa Monica.

Company	Employees	Industry
City of Santa Monica	1892	Government
St. John's Health Center	1755	Health services
The Macerich Company	1605	Real estate
Fremont General Corp.	1600	Insurance
Entravision Corp.	1206	Media company

Source: Santa Monica Chamber of Commerce.

59. Genetically Engineered Crops. Genetically engineered (GE) crops are now planted on the majority of acreage in many states around the country. There are three varieties of GE corn: insect-resistant, herbicide-tolerant, and stacked genes. The following table contains the proportion of the corn grown in each of five states that is GE, along with the GE type most prevalent in each state, for 2007.[4] Do (**a**)–(**d**) and then provide the observation for the state of Texas.

State	Proportion of GE corn	Most prevalent type
Texas	79%	Herbicide-tolerant
Missouri	62%	Insect-resistant
Minnesota	86%	Herbicide-tolerant
Ohio	41%	Stacked genes
South Dakota	93%	Stacked genes

60. Crime Statistics for Stillwater, OK. Refer to the following table. Do (**a**)–(**d**) and provide the observation for motor thefts.

Crime type	2005 Total	Per 100,000 people	National per 100,000 people	Compared to national average
Robberies	10	24.4	195.4	Better
Assaults	83	202.4	340.1	Better
Burglaries	317	772.9	814.5	Better
Larceny/thefts	1147	2796.7	2734.7	Worse
Motor thefts	55	134.1	526.5	Better

61. Commodity Prices. The financial company Bloomberg (**www.bloomberg.com**) reported that, on November 24, 2011, the price and the change in price for the following commodities were oil ($107.60, +0.54%), gold ($1699.40, +0.04%), and coffee ($235.40, −0.63%). Do (**a**)–(**d**). What is the observation for gold?

62. Tornado Deaths. The Tornado Project (**www .tornadoproject.com**) reported the following list of the 10 years with the fewest tornado deaths. Do (**a**)–(**d**). What is the observation for 2004?

Year	Deaths	Year	Deaths
1910	12	1996	26
1986	15	1972	26
2004	16	1980	27
1981	24	1963	27
1962	25	1951	29

Light Bulb Lifetime. Use the following information for Exercises 63 and 64. An electrical company has developed a new form of light bulb that it claims lasts longer than current models. The company has 1 million bulbs in its inventory.

63. How do you think the company found evidence for its claim?

64. Suppose you take a representative sample of 100 of the new light bulbs and find the average lifetime to be 2000 hours.
   a. Is this a statistic or a parameter?
   b. Write a sentence that estimates the average lifetime of all the new light bulbs.

Bringing It All Together

Largest University Campuses. The National Center for Education Statistics reported that the university campuses with the largest enrollment in 2009 are as shown in the table. Use this information for Exercises 65 and 66.

Campus	Location	Enrollment	Rank
Arizona State	Tempe	55,552	1
Ohio State	Columbus	55,014	2
Central Florida	Orlando	53,537	3
Univ. of Minnesota	Taria Cities	51,659	4
Univ. of Texas	Austin	51,032	5

65. Do the following:
   a. List the elements.
   b. List the variables.
   c. Identify the qualitative variables.
   d. Identify the quantitative variables.
   e. For each variable, identify the level of measurement.

66. Answer the following:
   a. Do these five campuses represent a sample or a population?
   b. Could these five campuses be considered a representative sample of the enrollment for all university campuses in the United States? Explain.
   c. Provide the observation for Arizona State University.
   d. Write a sentence that describes Ohio State University using the information from its observation.

1.3   GATHERING DATA

OBJECTIVES   By the end of this section, I will be able to . . .

1   Explain what a random sample is, and why we need one.

2   Identify systematic sampling, stratified sampling, cluster sampling, and convenience sampling.

3   Explain selection bias and good questionnaire design.

4   Understand the difference between an observational study and an experiment.

1   RANDOM SAMPLING

We can use the information gathered from a sample to generalize about the population when it is impractical or impossible to take a census of the entire population. However, if we get a "bad" sample, the information gleaned from the sample will be misleading, with potentially catastrophic consequences. This section introduces a method of sampling that minimizes many potential biases, which could lead to incorrect generalizations about the population. This sampling method is called *random sampling*. Everyday examples of random sampling include

- randomly selecting lottery numbers from a basket which continuously churns the number-balls,

- randomly choosing one card from a deck of playing cards that has been well shuffled, and

- randomly pulling a name out of a hat, after the names have been well stirred.

Since random samples are not always practical or desirable, this section also discusses some of the many alternative sampling methods available, including stratified sampling and cluster sampling.

What Is a Random Sample, and Why Do We Need It?

Survey sampling, or polling, has now become so widespread that hardly a day goes by without the results of some new poll or survey making the headlines. Polls are a good example of statistical sampling at work. The pollsters canvass about 1000 or so respondents, analyze the sample results, and then report their statistical inference that, for example, "32% of Americans have used a cell phone to access the Internet."

Today many polls are conducted quite scientifically, and their results are usually very accurate. However, such was not always the case. In 1936, the *Literary Digest* had correctly predicted the past three presidential elections and went to work to predict the winner of the contest between Republican Alf Landon and Democrat Franklin Roosevelt. The magazine sent ballots to 10 million citizens. The results ran strongly in favor of Landon, leading the *Literary Digest* to predict Landon to win the election. About 25% of the ballots were returned, giving the newsweekly a sample size of 2.5 million. George Gallup, on the other hand, was working with a sample size that was much *smaller* than the *Literary Digest's*. However, Gallup predicted a victory for Roosevelt. Clearly, with more data, the *Literary Digest* should have been able to give a more accurate prediction, right? Not necessarily. Roosevelt won in a landslide, and the embarrassed *Literary Digest* later declared bankruptcy.

The problem stemmed from the way that the *Literary Digest* identified its sample. It used lists of people who owned cars and had telephones, which in the 1930s excluded millions of poor and underprivileged people, who overwhelmingly supported Roosevelt. Its sample of 2.5 million therefore was highly biased toward the richer folks, who were less likely to have any great fondness for Roosevelt and his New Deal policies. Gallup, on the other hand, chose his sample more scientifically, and even though his sample size was smaller, it was more representative of the population as a whole.

One inexpensive way of eliminating many types of bias is to make sure your sample is a **random sample.**

> A **random sample** (also known as a **simple random sample**) is a sample for which every element has an equal chance of being selected.

Note: When we take a sample, we usually discard any repeated elements because we already have their information.

How the Gallup Organization Obtains a Random Sample

The Gallup Organization (**www.gallup.com**) has been conducting polls since the 1930s. People often wonder how a random sample of 1000 adults can represent the sentiments of the more than 300 million American adults. How does Gallup obtain a random sample in the first place? Gallup's objective is to make sure that every American has an *equal probability of selection,* that is, an equal chance of being selected, for their poll.

In the early days, Gallup conducted interviews in person, going house to house. However, today it is much less expensive to conduct telephone interviews. How does Gallup help to ensure that its telephone sample is truly random? What about the Americans whose phone number is unlisted? The first step is to construct a table of all the telephone exchanges in America, along with an estimate of the proportion of Americans living in that exchange area and the broad characteristics of that population in terms of income, age, ethnicity, education, and so on. Gallup then uses *random digit dialing,* a computer program that generates random four-digit numbers, which are then appended to the telephone exchanges. Thus, each household phone number in America has an equal chance of being included in the sample, regardless of whether it is listed or unlisted. Finally, as of January 1, 2008, Gallup added a data base of cell phone numbers, in order to contact those who can more readily be reached via cell phone.

EXAMPLE 1.9

DO YOU PREFER WATCHING THE SUPER BOWL OR THE COMMERCIALS?

In February 2007, the Gallup Organization used random digit dialing in a poll of Americans who planned to watch the Super Bowl (Indianapolis Colts versus Chicago Bears). One question they asked was whether the subjects preferred to watch the game or the commercials. Does this represent a random sample?

Solution

Since random digit dialing ensures that each household phone number in America has an equal chance of being included in the sample, the sample is random.

A perhaps surprising 33% of respondents reported that they preferred watching the commercials, compared with 66% who preferred watching the game. Gender and age seemed to affect how one responded to this question. Twice the proportion of female viewers (44%) as male viewers (22%) preferred watching the commercials. Among females only, more than twice as many younger (aged 18 to 49) women preferred watching the commercials (56%) as older (aged 50 and over) women (26%).

Random samples may be generated using technology, using the *Simple Random Sample* applet, or using the random number table provided in Table A in the Appendix (page T-2). At the end of this section, we demonstrate how to generate random samples using the TI-83/84 graphing calculator, Excel, and Minitab.

EXAMPLE 1.10

GENERATING A RANDOM SAMPLE USING TECHNOLOGY

Recently, *Inc. Magazine* published a list of the top 25 cities for doing business, shown in Table 1.9. Use the TI-83/84, Excel, or Minitab to generate a random sample of 7 cities from this list.

Table 1.9 Top 25 cities for doing business, according to *Inc. Magazine*

1. Atlanta, GA	10. Suburban Maryland/DC	19. Austin, TX
2. Riverside, CA	11. Orlando, FL	20. Northern Virginia
3. Las Vegas, NV	12. Phoenix, AZ	21. Middlesex, NJ
4. San Antonio, TX	13. Washington, DC, metro area	22. Miami–Hialeah, FL
5. West Palm Beach, FL	14. Tampa–St. Petersburg, FL	23. Orange County, CA
6. Southern New Jersey	15. San Diego, CA	24. Oklahoma City, OK
7. Fort Lauderdale, FL	16. Nassau–Suffolk, NY	25. Albany, NY
8. Jacksonville, FL	17. Richmond–Petersburg, VA	
9. Newark, NJ	18. New Orleans, LA	

Solution

We used the instructions provided in the Step-by-Step Technology Guide at the end of this section (page 28) to create three random samples, listed on the next page. Note that each random sample is different, as yours will be.

Random sample 1 using the TI-83/84	Random sample 2 using Excel	Random sample 3 using Minitab
9. Newark, NJ	6. Southern New Jersey	3. Las Vegas, NV
25. Albany, NY	23. Orange County, CA	21. Middlesex, NJ
6. Southern New Jersey	11. Orlando, FL	18. New Orleans, LA
20. Northern Virginia	14. Tampa–St. Petersburg, FL	7. Fort Lauderdale, FL
24. Oklahoma City, OK	25. Albany, NY	2. Riverside, CA
10. Suburban Maryland/DC	7. Fort Lauderdale, FL	25. Albany, NY
1. Atlanta, GA	17. Richmond–Petersburg, VA	10. Suburban Maryland/DC

Now You Can Do Exercise 23.

2 MORE SAMPLING METHODS

In certain circumstances, simple random sampling can have shortcomings. A simple random sample may not provide sufficient information about subgroups within the population. For example, suppose you are interested in knowing the proportion of those of Latino descent in Walnut, California, who are registered Democrats. A random sample of size 100 of all the voters in Walnut may yield only 20 of Latino descent, which may be too small a sample to be useful for statistical inference. Therefore, the researcher needs other methods for obtaining samples, depending on the situation and the research question.

Systematic Sampling

Note: Most of the sampling methods mentioned here involve randomness. However, only the simple random sample is used throughout the text. Therefore, whenever you see the phrase *random sample*, it should be understood as *simple random sample.*

Perhaps the easiest method of sampling is systematic sampling, which is used when a random sample is unobtainable. In systematic sampling, each element of the population is numbered, and the sample is obtained by selecting every kth element, where k is some whole number. The first element selected corresponds to a random whole number between 1 and k. The ancient Romans understood well how to use systematic sampling. When a Roman legion mutinied or showed cowardice in battle, every 10th member was selected and summarily executed before his comrades. Literally, the legion was *decimated,* from the Latin *decem,* meaning "ten."

EXAMPLE 1.11

 20richest

SYSTEMATIC SAMPLING

Table 1.10 contains the top 20 richest people in the world for the year 2010, according to the annually published *Forbes 400* listing. Obtain a systematic sample from this list, using $k = 4$.

Table 1.10 Twenty richest people in the world

Rank	Name	Net Worth ($ billion)	Rank	Name	Net Worth ($ billion)
1	Bill Gates	54	11	Larry Page	15
2	Warren Buffett	45	12	Sergey Brin	15
3	Larry Ellison	27	13	Sheldon Adelson	14.7
4	Christy Walton	24	14	George Soros	14.2
5	Charles Koch	21.5	15	Michael Dell	14
6	David Koch	21.5	16	Steve Balmer	13.1
7	Jim Walton	20.1	17	Paul Allen	12.7
8	Alice Walton	20	18	Jeff Bezos	12.6
9	S. Robson Walton	19.7	19	Anne Cox Chambers	12.5
10	Michael Bloomberg	18	20	John Paulson	12.4

Source: Forbes magazine.

Solution

First we randomly select a whole number between 1 and $k = 4$. Suppose we select 2. Thus, our systematic sample will consist of every 4th person in Table 1.10, starting with the 2nd person. That is, our systematic sample will consist of the 2nd, 6th, 10th, 14th, and 18th persons, shown here:

Systematic sample: Warren Buffett, David Koch, Michael Bloomberg, George Soros, Jeff Bezos.

Now You Can Do Exercise 24.

Stratified Sampling

Often, researchers are interested in investigating characteristics of a certain subgroup of a population, such as those of Latino descent in Walnut, California. In cases like this, the researcher divides the population into subgroups, or *strata,* according to some characteristic, such as race or gender. Then a random sample is taken from each stratum. In this way, the researcher knows that a sample will be obtained from each stratum and that it will be large enough to provide reliable statistical inference for each stratum.

EXAMPLE 1.12

STRATIFIED SAMPLING

A researcher is interested in analyzing whether there are differences in scoring among the basketball teams in the three divisions of the Eastern Conference of the National Basketball Association (Table 1.11). Obtain a stratified sample of two teams from each division.

Table 1.11 Teams in the three divisions of the Eastern Conference of the National Basketball Association

Atlantic Division	Central Division	Southeast Division
Boston Celtics	Chicago Bulls	Atlanta Hawks
New Jersey Nets	Cleveland Cavaliers	Charlotte Bobcats
New York Knicks	Detroit Pistons	Miami Heat
Philadelphia 76ers	Indiana Pacers	Orlando Magic
Toronto Raptors	Milwaukee Bucks	Washington Wizards

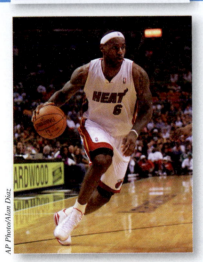

LeBron James, of the Miami Heat.

Solution

A random sample of size two was drawn from the teams in each of the three divisions. These six teams are then combined to form our stratified sample of basketball teams. Note that each random sample is different, as yours will be.

Atlantic Division	Central Division	Southeast Division
Boston	Chicago	Atlanta
New Jersey	Cleveland	Charlotte
New York	Detroit	Miami
Philadelphia	Indiana	Orlando
Toronto	Milwaukee	Washington

Stratified Sample of 6 Teams

Boston Celtics
Cleveland Cavaliers
Miami Heat
Milwaukee Bucks
New York Knicks
Orlando Magic

Now You Can Do Exercise 25.

Cluster Sampling

Cluster sampling is used when the population is widely scattered geographically or poses other logistical difficulties. For example, if we were interested in estimating the mean income of Manhattan residents, it would be time-consuming and expensive to visit 1000 different locations in Manhattan to elicit sample information. In cluster sampling, the population is divided into *clusters,* such as precincts or city blocks. Then several clusters are chosen at random, and all of the elements within the chosen clusters are selected for the sample. One disadvantage of cluster sampling is that the respondents from within a certain cluster will tend to be more similar to each other than the elements of a random sample would be. For example, if one of the clusters in the Manhattan income survey was a Fifth Avenue block, the mean income of residents there would be at the higher end of the income scale.

EXAMPLE 1.13

CLUSTER SAMPLING

Using Table 1.11, consider each division to be a cluster. Construct a cluster sample of the teams in the Eastern Conference by randomly selecting two of the three clusters (divisions).

Solution

Suppose that we randomly select our clusters to be the Atlantic Division and the Southeast Division. Our cluster sample then consists of *all* the teams in both of these divisions, as follows:

Atlantic Division	Central Division	Southeast Division
Boston	Chicago	Atlanta
New Jersey	Cleveland	Charlotte
New York	Detroit	Miami
Philadelphia	Indiana	Orlando
Toronto	Milwaukee	Washington

Cluster sample of 10 teams

Atlanta Hawks
Boston Celtics
Charlotte Bobcats
Miami Heat
New Jersey Nets
New York Knicks
Orlando Magic
Philadelphia 76ers
Toronto Raptors
Washington Wizards

Now You Can Do Exercise 26.

Developing Your Statistical Sense

Stratified Sampling versus Cluster Sampling

Stratified sampling and cluster sampling are sometimes confused. To obtain a stratified sample, we (a) divide the population into subgroups (strata, the divisions in Table 1.11), and (b) take a random sample from each subgroup, as shown by the shaded teams in Example 1.12. In cluster sampling, we (a) divide the population into subgroups (the divisions in Table 1.11, this time called clusters), (b) take a random sample of the clusters, as shown by the shaded divisions in Example 1.12, and (c) choose *all* the elements in the selected clusters for our cluster sample. In stratified sampling, we are randomly selecting elements from the subgroups; in cluster sampling, we are randomly selecting the clusters only, not the elements in the clusters.

Convenience Sampling

In convenience sampling, subjects are chosen based on what is convenient for the survey personnel. If you were to estimate the true proportion of females taking an introductory statistics course using only the people in your class, this would be considered a convenience sample. As we shall see in Example 1.14, convenience sampling usually does not result in a representative sample.

| EXAMPLE 1.14 | **CONVENIENCE SAMPLING USING ONLINE POLLS** |

An online newspaper reports that, in an online poll of its readership, 60% say that they get most of their news from online sources. Does this number accurately reflect the proportion of all Americans who get most of their news from online sources?

Solution

Caution: Surveys, like online polls, that use convenience sampling should be treated with a healthy dose of skepticism. They are not statistically sound.

No, the sample is not random. Only those Americans who are online already (and already using an online news source) can respond to this online poll. Therefore, the sample is not random, and it is biased. It overestimates the proportion of Americans who get their news from online sources. Further, there is no mechanism to guard against a single person responding repeatedly and getting his or her vote counted multiple times. Online polls are not scientific, and their results should not be considered a true reflection of the sentiments of all Americans.

| EXAMPLE 1.15 | **IDENTIFY THE SAMPLING METHOD** |

For each of the following, identify which type of sampling is represented.

a. Students in your class are divided into females and males. A random sample of size 5 is then drawn from each of the groups.
b. You are interested in estimating the average number of hours dormitory residents spend studying. In each dormitory, one floor is chosen at random and all the students on that floor are interviewed.
c. You are researching the proportion of college students who prefer country music to other forms of music. You obtain a listing of all the students at your college and contact every 20th student on the list.
d. Your campus statistical consulting center uses random digit dialing to locate potential subjects for a political survey.
e. A student is investigating the prevalence of flu on campus this semester, and asks 20 of his friends whether they have had the flu.

Solution

a. Stratified sampling: (a) the population was divided into subgroups (females and males), and (b) a random sample was drawn from each of the groups.
b. Cluster sampling: (a) the population was divided into clusters (dormitory floors), (b) a random sample of the clusters (floors) is taken, and (c) all students on that floor (cluster) were selected.
c. Systematic sampling, where every kth member of the population is taken, with $k = 20$.
d. An example of random sampling, as illustrated on pages 19–20.
e. Convenience sampling: the student is choosing a sample convenient for him.

Now You Can Do
Exercises 7–10.

3 SELECTION BIAS AND QUESTIONNAIRE DESIGN

Here we learn about some common pitfalls in the design and implementation of a survey, including selection bias and the wording of a questionnaire.

> The **target population** is the complete collection of all elements that we are interested in studying.
>
> The **potential population** is the collection of elements from the target population that had a chance of being sampled.
>
> **Selection bias** occurs when the population from which the actual sample is drawn is not representative of the target population, due to an inappropriate sampling method.

EXAMPLE 1.16 **SELECTION BIAS**

Suppose Ashley would like to estimate the proportion of American voters who would favor abandoning the present system of Social Security in favor of a system where retirement funds would be invested in the stock market. Ashley goes to the mall with her clipboard, and canvasses as many people as she can on Monday between 9 A.M. and 5 P.M. To each person, she asked the question "Do you favor or oppose abandoning the present Social Security system in favor of a system that invests retirement funds in the stock market?"

a. Identify Ashley's target population.
b. Identify Ashley's potential population.
c. Discuss any possible problems.

Solution

a. Ashley's **target population** is the population of all American voters.
b. The collection of all the American voters who visited the mall on Monday between 9 A.M. and 5 P.M. represent her **potential population**.
c. It appears that Ashley's survey may suffer from selection bias. The population of people who went to the mall on Monday between 9 A.M. and 5 P.M. is not representative of the target population of all American voters. Since many American voters work on Mondays between 9 A.M. and 5 P.M., they are not elements of the sampled population. Further, the proportion of retirees at the mall during that time was larger than in the target population of all American voters. These retirees tend to oppose strongly any tampering with the Social Security system and would probably tend to respond in the negative to the survey question.

Now You Can Do Exercises 11–14.

Five Factors for Good Questionnaire Design

You may have heard of the aphorism "Be careful what you ask for; you may get it." This warning is certainly relevant to the issue of questionnaire design. The wording of questions can greatly affect the responses. Here are several factors to consider when designing a questionnaire.

1. **Remember: simplicity and clarity.** Do not use four-syllable words when one-syllable words will do. Respondents will be shy about asking you to clarify the question. The result will be confused responses and muddled data.

2. **When reporting results, include the actual question asked.** Be careful about drawing generalizations. The conclusions you draw may not have been what your respondents had in mind when they answered the questions.

3. **Avoid leading questions.** The respondent is often eager to please and will try to tell you what he or she thinks you want to hear. For example, a researcher is interested in determining the proportion of Americans who favor preserving the welfare system. A leading question would be "A child growing up poor in America faces more than his fair share of crime and negligence. Do you support preserving the welfare safety net to help ensure that children are given a fair chance?"

4. **Avoid asking two questions in one.** Avoid questions like "Have you argued with your friends or family in the last month?" This is really two questions in one, and you will not know which question the respondents are answering.

5. **Avoid vague terminology.** Words mean different things to different people. Avoid using terminology like "often" or "sometimes." Instead, try to use specific terms such as "three times a week." If you use ambiguous terms, the data you collect will be ambiguous, and any conclusions you draw will probably not be valid.

EXAMPLE 1.17

QUESTIONNAIRE DESIGN

For each of the following questionnaire items, identify which of the five factors for good questionnaire design is violated, if any.

a.   Do you oppose the wasteful spending on foreign aid when so many problems confront us here at home?

b.   Do you often feel lonely?

c.   Do you espouse or disavow the conglomerative confluence of macroeconomic indicators?

d.   Have you watched television or downloaded music in the past 24 hours?

e.   Do you ever use a cell phone to access the Internet?

Solution

a.   This is a leading question, which is clearly trying to influence the respondent's answer.

b.   What is meant by "often"? Three times a week? Three times a day? This is vague terminology.

c.   This question would only be understood by those who have studied economics, and is neither simple nor clear.

d.   This is asking two questions in one. It is possible that respondents have done one or the other, or both.

e.   This question is fine. In fact, it is an actual survey question from the Pew Research Center.

Now You Can Do Exercises 15–18.

4   EXPERIMENTAL STUDIES AND OBSERVATIONAL STUDIES

Two major types of statistical studies are **experimental studies** and **observational studies.** We have seen that researchers can gather data by consulting existing sources, by distributing a questionnaire, or by taking a sample. However, you may not be able to obtain the information you require by using survey or sampling methods. In this case, you may prefer to conduct an experimental study.

Note: What is the difference between an element and a subject? *Subject* is a term usually reserved for statistical studies, while the term *element* can be used for any data set.

> **Experimental Studies**
>
> In an **experimental study**, researchers investigate how varying the predictor variable affects the response variable.
>
> A **predictor variable** (also called an **explanatory variable**) is a characteristic intended to explain differences in the response variable.
>
> A predictor variable that takes the form of a purposeful intervention is called a **treatment**.
>
> A **response variable** is an outcome, a characteristic of the subjects of the experiment presumably brought about by differences in the predictor variable or treatment.
>
> The **subjects** in a statistical study represent the elements from which the data are drawn.

EXAMPLE 1.18 NEWBORN BABIES AND A HEARTBEAT:
AN EXPERIMENTAL STUDY

Thinkstock

A psychologist wanted to test whether the sound of a human heartbeat would help newborn babies grow. A baby nursery at a hospital was set up so that the sound of a human heartbeat could be heard throughout the nursery. The heartbeat sound was played in the nursery for a large batch of newborn children, who were then weighed to determine their weight gain after four days in the nursery. Later, a second batch of children occupied the nursery, but no heartbeat sound was played. These children were also weighed after four days in the nursery. Babies were randomly placed into the two groups. Identify the following:

a. The subjects
b. The predictor variable
c. The treatment
d. The response variable

Solution

a. The babies were the subjects of this experimental study.
b. The **predictor variable** is whether or not the heartbeat sound was played in the nursery.
c. The treatment is the sound of the human heartbeat.
d. The **response variable** is the baby's weight gain, which is the outcome of the study.

The results were consistent with the psychologist's conjecture; the babies who listened to the heartbeat sound had a greater average weight gain than the babies for whom no heartbeat sound was played.

**Now You Can Do
Exercises 27–34.**

There are three main factors that should be considered when designing an experimental study: *control, randomization,* and *replication.*

Control. A control group is necessary to compare against the treatment group, if we wish the results of our experiment to be useful. The control group in the above example is the group of babies for whom the heartbeat sound was not played. Had the psychologist omitted this control group, there would have been nothing to compare his results against. In some experiments, especially in medicine, members of the control group receive a placebo, such as a sugar fill. Sometimes, the symptoms of the members of the control group improve simply by taking the placebo, a phenomenon known as the *placebo effect.*

Randomization. Many biases can be introduced into an experiment. For example, a well-meaning doctor may want to place underweight high-risk babies in the group with the heartbeat, in the hope that such babies will flourish. To eliminate biases like these, the placement of the subjects into the treatment and control groups should be done randomly.

Replication. One major theme of statistical investigation is that larger samples are usually better, because they allow more precise inference. In a statistical study, the treatment and the control groups each must contain a large enough number of subjects to allow detection of meaningful differences between the treatment and control. For example, if a researcher examined only three babies with the heartbeat sound and three babies without the heartbeat sound, this would not be a sufficient number of replications.

In Chapter 8, "Confidence Intervals," we will learn how large a sample size is sufficient for the needs of a particular study.

Observational Studies

There are circumstances where it is either impossible, impractical, or unethical for the researcher to place subjects into treatment and control groups. For example, suppose we are interested in whether women who work outside the home suffer less depression than women who remain at home with the children. The explanatory variable here is whether or not a woman works outside the home. However, it is not possible for the researcher to take women and randomly separate them into groups that either work outside the home or do not work outside the home.

Sometimes an experimental study is not possible for ethical reasons. Suppose you are interested in whether babies born to chemically dependent mothers display differences in cognitive skills from babies born to mothers who are not chemically dependent. It is clearly not ethical to randomly assign half of the mothers in the study to become chemically dependent during their pregnancy. Therefore, researchers need another type of statistical study: the observational study. In an observational study, the researcher observes whether the subjects' differences in the predictor variable are associated with differences in the response variable. No attempt is made to create differences in the predictor variable.

A sample survey is an example of an observational study. Data about a response variable may be obtained through the survey, along with information about possible predictor variables. No attempt is made to manipulate the variables. The researcher analyzes the information to determine whether differences in the predictor variable are associated with differences in the response variable.

EXAMPLE 1.19   ## IS ECSTASY TOXIC TO YOUR NEURONS?

According to the British medical journal *The Lancet,* experimental studies carried out on animals (nonhuman primates, squirrel monkeys, and rodents) have revealed that large doses of the drug Ecstasy (methylene-dioxy-methamphetamine, or MDMA) produce "large and possibly permanent damage" to neural axons in the brain. Explain why the researchers did not carry out their experiment on humans.

Solution

It is not ethical to randomly assign half of the human subjects to receive large doses of the drug Ecstasy, especially in view of its effect on animals. The difficulty of performing experimental studies on humans concerning the effects of controlled substances is addressed by the authors of the *Lancet* study:

> *Only a prospective[experimental] study . . . could definitively show that recreational MDMA use was neurotoxic in human beings. For ethical, political, and legal reasons such a study is unlikely to ever be done. Instead, we have to rely upon evidence from observational studies of recreational MDMA users.*[5]

**Now You Can Do
Exercise 42.**

The *Simple Random Sample* applet allows you to produce a random sample of up to 100 elements, in the form of a lotto.

STEP-BY-STEP TECHNOLOGY GUIDE: Generating a Random Sample

We illustrate using Example 1.10 (pages 19–20).

TI-83/84

Step 1   Enter a "seed," which can be any nonzero number.
Step 2   Press **STO** ⇒.
Step 3   Press **MATH**, highlight **PRB**, select **1: rand**, and press **ENTER** (see Figure 1.6, which uses 1776 for the seed). Your seed number is now in the calculator's memory.
Step 4   Press **MATH**, highlight **PRB**, and select **5: randInt(**.
Step 5   Enter **1, N**, *two times* **n**, where N = population size and n = sample size. We enter twice the sample size in case there are repeats. For Example 1.10, since **n** = 7, we enter **randInt(1, 25, 14)** and press **ENTER** (Figure 1.7).

Step 6   Store the random sample in list **L1** as follows: press **STO** ⇒, then **2ND**, then **L1** (Figure 1.7). Then press **ENTER**.
Step 7   View the random sample by pressing **STAT**, highlighting **EDIT**, and pressing **ENTER** (Figure 1.8). Note that there is a repeat (**6**). We therefore select the next number, **10**, to round out our sample. The random sample for Example 1.10 is therefore **9, 25, 6, 20, 24, 10, 1** (Figure 1.9).

FIGURE 1.6   **FIGURE 1.7**

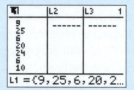

FIGURE 1.8

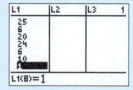

FIGURE 1.9

EXCEL

Step 1   Select cell **A1**. Click the **Insert Function** icon f_x.
Step 2   For "Search for a function," enter **randbetween**. Click **Go**, then **OK**.
Step 3   For **Bottom**, enter **1**. For **Top**, enter population size **N**. For Example 1.10, **N** = 25. Click **OK**.
Step 4   Cell **A1** now contains a random integer between **1** and **N**. Copy and paste cell **A1** into *twice* as many cells as needed for the sample size **n**, just in case there are repeats. For Example 1.10, copy and paste into cells **A2** to **A14**. The results are shown in Figure 1.10. Note that **8** is repeated, so that our random sample is **8, 2, 20, 16, 23, 7, 22**.

FIGURE 1.10 Excel random sample

MINITAB

Step 1   Click on **Calc > Random Data > Integer . . .**
Step 2   In the **Generate __ rows of data** section, enter *twice* your desired sample size, just in case there are repeats. For example, if your desired sample size is **7**, enter **14**.
Step 3   In the **Store in column __** section, enter whichever column is convenient for you, such as **C1**.
Step 4   For **Minimum value**, enter **1**. For **Maximum value**, enter your population size, **N**. Click **OK**.
Step 5   The random integers appear in column **C1**. Start from the top and go down the list, omitting any repeats, until you have your sample of size **n**. Our random sample (Figure 1.11) is therefore **3, 18, 2, 11, 21, 7, 25**.

Excel and Minitab base the seed on the current time, so that you need not set it yourself.

FIGURE 1.11 Minitab random sample

SECTION 1.3 **Summary**

1. A random sample is a sample for which every element has an equal chance of being included. A random sample can minimize many potential biases, which could lead to incorrect generalizations about the population.

2. Other sampling methods include stratified sampling, systematic sampling, cluster sampling, and convenience sampling.

3. When constructing a survey, avoid selection bias and follow the five factors for good questionnaire design.

4. There are two types of statistical studies: experimental studies and observational studies. In an experimental study, researchers investigate how varying the predictor variable affects the response variable. It is not always possible to conduct an experimental study, however, and sometimes an observational study is used instead.

SECTION 1.3 **Exercises**

Clarifying the Concepts

1. Explain why convenience sampling usually does not result in a representative sample.

2. What type of bias did the *Literary Digest* poll (page 18) exhibit? How did it affect the results?

3. How could the *Literary Digest* have decreased the bias in its poll?

4. Was the *Literary Digest* poll a random sample?

5. Describe what a random sample is.

6. Describe the difference between an observational study and an experimental study.

Practicing the Techniques

For Exercises 7–10, state which type of sampling is represented.

7. Students in your class are divided into freshmen, sophomores, juniors, and seniors. One of the groups is selected at random and all the students in that group are selected.

8. An instructor in a large lecture course of 300 students would like to get a student sample, and he selects every 10th name from the class roster.

9. You are researching the proportion of college students who prefer country music to other forms of music. You survey all the students in all the classes you are taking this semester.

10. An instructor in a large lecture course of 300 students (two lectures, one lab per week) would like to get a student sample. He takes a random sample of three of the 15 lab sections, and selects all of the students from those three sections.

Use the following information for Exercises 11 and 12. Brandon is trying to estimate the proportion of all college students who are physically fit. He obtains a sample of students working out at the gymnasium on Monday night.

11. Identify the target population and the potential population.

12. Does selection bias exist? Explain why or why not.

Use the following information for Exercises 13 and 14. Michelle would like to determine the proportion of small businesses that who employ at least one college student part-time. She obtains a sample of businesses near the state university.

13. Identify the target population and the potential population.

14. Does selection bias exist? Explain why or why not.

For Exercises 15–18, identify which of the five factors for good questionnaire design is violated, if any.

15. Do you sometimes feel anxiety about your health?

16. Do you support the valiant efforts of our mayor to dispel the lies spread by the corrupt opposition?

17. Do you espouse the diminution of the graduated income tax?

18. Do you support laws restricting invasion of privacy and locking up those responsible for doing so?

For Exercises 19–22, do the following: (**a**) State which type of study is involved, experimental or observational. (**b**) Identify the response variable and the predictor variable.

19. A sociologist would be interested in whether large families (at least four children) attend religious services more often than smaller families do.

20. A financial researcher would be interested in whether companies that give large bonuses to their chief executive officers (at least $1 million per year) have a higher stock price.

21. A manufacturer would be interested in whether a new computer processor will improve the performance of its electronics equipment.

22. A pharmaceutical company would like to see if its new drug will lower high blood pressure.

Applying the Concepts

Refer to Table 1.12 for Exercises 23–26.

TABLE 1.12   College football teams in four major conferences

Big Ten	Southeastern	Atlantic Coast	Pac 12
Illinois	Alabama	Boston College	Arizona
Indiana	Arkansas	Clemson	Arizona State
Iowa	Auburn	Duke	California
Michigan	Florida	Florida State	Colurado
Mich. State	Georgia	Georgia Tech	Oregon
Minnesota	Kentucky	Maryland	Oregon State
Nebraska	Louisiana State	Miami	Stanford
Northwestern	Mississippi	North Carolina	UCLA
Ohio State	Miss. State	NC State	USC
Penn State	South Carolina	Virginia	Utah
Purdue	Tennessee	Virginia Tech	Washington
Wisconsin	Vanderbilt	Wake Forest	Wash. State

23. Suppose that we ignore the different conferences, and think of all of these teams as belonging to one big Conference America. Obtain a simple random sample of size 5 teams from Conference America.

24. Suppose the conference chairperson for the Southeastern Conference would like to visit some campuses this year. Obtain a systematic sample of every third team from the Southeastern Conference.

25. Suppose the NCAA wants to form a committee to consider some rule changes for college football, and would like two teams randomly selected from each conference. Obtain a stratified sample of two teams from each conference.

26. Suppose the NCAA is considering a new playoff arrangement for the bowl games, and would like all the teams from two randomly selected conferences to be eligible for this playoff arrangement. Obtain a cluster sample of all the teams from two randomly selected conferences.

Use the following information for Exercises 27–30. Agricultural researchers are investigating whether a new form of pesticide will lead to lower levels of insect damage to crops than the traditional pesticide.

27. Identify the response variable.

28. Identify the predictor variable.

29. What is the treatment?

30. What is the control?

Use the following information for Exercises 31–36. Cholesterol researchers are investigating whether there is any difference between a new medication and a placebo (inactive pill) in lowering LDL cholesterol levels in the bloodstream.

31. Identify the response variable.

32. Identify the predictor variable.

33. What is the treatment?

34. What is the control?

35. Suppose there is a patient with very high LDL cholesterol levels, and so the doctor assigns this patient to the group of patients who receive the new medication rather than the placebo. Which of the experimental factors (control, randomization, replication) did the doctor violate?

36. Use the situation in the previous exercise to discuss why randomization is important.

37. Contradicting Ann Landers. "If you had to do it over again, would you have children?" This is the question that advice columnist Ann Landers once asked her readers. It turns out that nearly 70% of the 10,000 responses she received were "No." A professional poll by *Newsday* found that 91% of respondents would have children again. Explain the apparent contradiction between these two surveys using what you have learned in this section.

38. Living Below the Poverty Level. For the following survey, describe the target population and the potential population, and discuss the potential for selection bias. A sociologist is interested in the proportion of people living below the poverty level in Chicago. He takes a random sample of phone numbers from the Chicago phone directory and asks each respondent his or her annual household income.

39. Rap or Hip-Hop. Describe what is wrong, if anything, with the following survey question. "Do you enjoy listening to rap or hip-hop music?"

40. Financial Ruin. Describe what is wrong, if anything, with the following survey question: "Do you think that we should tax and spend our way into financial ruin?"

41. Mediterranean Diet. The American Heart Association reported the following results of an experimental study.[6] Patients who ate a Mediterranean diet had a significantly lower risk of having a second heart attack than did patients who ate a Western diet. Identify the response variable and the predictor variable in this experimental study.

42. Secondhand Smoking and Illness in Children. A Surgeon General's report found that "the evidence is sufficient to infer a causal relationship" between secondhand tobacco smoke exposure from parental smoking and respiratory illnesses in infants and children.[7]

   a. Given the health risks associated with tobacco use, discuss the ethics of forcing the parents of a treatment group to smoke tobacco.

   b. State whether this report was based on an experimental study or an observational study.

Bringing it all together

Evidence for an Alternative Therapy? Use the following information for Exercises 43–45. A company called QT, Inc. sells "ionized bracelets," called Q-Ray Bracelets, that it claims help to ease pain by balancing the body's flow of "electromagnetic energy." QT, Inc. claims that Q-Ray Bracelets can ease pain caused by cancer, restore well-being, and provide many other health benefits. The Mayo Clinic decided to conduct a statistical study to determine whether the extravagant claims for Q-Ray Bracelets were justified.[8] In the study, 305 subjects wore the Q-Ray "ionized" bracelet and 305 wore a placebo bracelet (identical to the ionized bracelet except for the ionization) for four weeks, at the end of which certain measures of pain were evaluated and compared between the treatments. The subjects, upon entry to the study, were randomly assigned to receive either the ionized bracelet or the placebo bracelet.

43. Identify the following aspects of this study.
  a. The control
  b. The randomization
  c. The replication

44. Identify the following aspects of this study.
  a. The predictor variable
  b. The treatment
  c. The response variable

45. Does this statistical study represent an experimental study or an observational study? Write a sentence explaining why.

Use the *Simple Random Sample* applet for Exercises 46–48.

46. Generate a random sample of 7 cities from Table 1.9 (page 19).

47. Generate another random sample of 7 cities from Table 1.9. Are all the cities in the two samples the same?

48. Before we generate a third sample of 7 cities, choose a city from Table 1.9.
  a. Will this city appear in the random sample?
  b. Is there any way of telling for certain in advance whether this city will appear in the random sample?
  c. Now go ahead and generate the third random sample of 7 cities. Is your city in the sample?

CHAPTER 1 | Vocabulary

Section 1.2
- **CENSUS** (p. 12)
- **CONTINUOUS VARIABLE** (p. 10)
- **DESCRIPTIVE STATISTICS** (p. 8)
- **DISCRETE VARIABLE** (p. 10)
- **ELEMENT** (p. 8)
- **OBSERVATION** (p. 8)
- **PARAMETER** (p. 12)
- **POPULATION** (p. 12)
- **QUALITATIVE VARIABLE** (p. 9)

- **QUANTITATIVE VARIABLE** (p. 9)
- **SAMPLE** (p. 12)
- **STATISTIC** (p. 12)
- **STATISTICAL INFERENCE** (p. 12)
- **STATISTICS** (p. 6)
- **VARIABLE** (p. 8)

Section 1.3
- **EXPERIMENTAL STUDY** (p. 25)
- **OBSERVATIONAL STUDY** (p. 27)

- **POTENTIAL POPULATION** (p. 24)
- **PREDICTOR VARIABLE (EXPLANATORY VARIABLE)** (p. 25)
- **RANDOM SAMPLE** (p. 18)
- **RESPONSE VARIABLE** (p. 25)
- **SELECTION BIAS** (p. 24)
- **SUBJECTS** (p. 25)
- **TARGET POPULATION** (p. 24)
- **TREATMENT** (p. 25)

CHAPTER 1 | Review Exercises

Section 1.2

Refer to Table 1.13 for Exercises 1–3. Table 1.13 contains information on some small sport utility vehicles (SUVs), as reported by *Consumer Reports* for model year 2010.

TABLE 1.13 2010 Small sport utility vehicles

Car	Cylinders	Passengers	Base price	Customer satisfaction
Subaru Forester	4	5	$20,295	Above average
Honda CR-V	4	5	$21,545	Above average
Nissan Rogue	4	5	$20,340	Average
Mitsubishi Outlander	6	7	$20,840	Average

1. Use Table 1.13 to find each of the following.
   a. List the elements.
   b. Identify the variables.
   c. Identify the qualitative variables.
   d. Identify the quantitative variables.
   e. For each variable, state the level of measurement.
2. Use Table 1.13 to answer the following.
   a. Which small SUV has the lowest base price? The highest?
   b. According to the data, what, if anything, is the difference between the Subaru Forester and the Honda CR-V?
3. Provide the observation for the Subaru Forester.
4. An electrical company has developed a new form of light bulb that it claims lasts longer than current models. The company has 1 million bulbs in its inventory. Consider the population average lifetime.
   a. What is the only way to find out the population average lifetime of the 1 million bulbs in the inventory?
   b. Suppose someone who worked for you wrote you a memo suggesting that it was crucial to know the exact value of the population average lifetime of all 1 million new light bulbs. How would you respond? What might you suggest instead?

Section 1.3
5. Refer to the *Literary Digest* poll discussed in Section 1.3.

   a. What was the target population?
   b. What was the potential population?
   c. What was the sample?
   d. Discuss whether the sample was similar to the target population in all important characteristics.
6. Suppose you are interested in finding out how the statistics grades for your class compare with those of the college as a whole.
   a. Would you use an experimental study or an observational study?
   b. Discuss how this study situation would preclude effective randomization.
7. A long-running television advertisement claimed that "3 out of 4 dentists surveyed recommend sugarless gum for their patients who chew gum."
   a. If in fact only 4 dentists were surveyed, which of the study factors were violated?
   b. Use this situation to discuss why replication is important.
8. Suppose we are interested in determining whether differences exist in the cognitive levels of children from single-parent families and those from two-parent families. Would we use an observational study or an experimental study? Clearly describe why.
9. Referring to the study in the previous exercise, suppose the children from single-parent families showed lower average cognitive skills than children from two-parent families. Does this mean that living in a one-parent family causes lower levels of cognitive skills? Why or why not?

CHAPTER 1   Quiz

True or False
1. True or false: Statistical inference consists of methods for estimating and drawing conclusions about sample characteristics based on the information contained in the population.
2. True or false: A parameter is a characteristic of a sample.

Fill in the Blank
3. Statistics is the art and science of _____, analyzing, presenting, and interpreting data.
4. An _____ is the set of values of all variables for a given element.
5. A statistic is a characteristic of a _____.

Short Answer
6. Is a sample survey examining the effects of secondhand smoke an example of an experimental study or an observational study?
7. State which type of statistical study is involved in the following. A large pharmaceutical company is interested in

whether a new drug will reduce Alzheimer's disease symptoms in elderly patients.
8. For the study in the previous exercise, identify the predictor variable and the response variable.

Calculations and Interpretations
9. Suppose we are interested in the proportion of left-handed statistics students, and we take a sample to estimate the percent of students in our class who are left-handed.
   a. What is the population?
   b. What is the sample?
   c. What is the variable? Is it quantitative or qualitative?
   d. Is the sample proportion likely to be exactly the same as the population proportion? Is it likely to be very far away from the population proportion? Explain.
10. Describe what is wrong, if anything, with the following survey question. "How often would you say that you attend the movie theater: often, occasionally, sometimes, seldom, or never?"

© Ancient Art & Architecture Collection Ltd./Alamy

2 Describing Data Using Graphs and Tables

CASE STUDY

The Caesar Cipher

Over two thousand years ago, Julius Caesar developed the Caesar Cipher, which was a means of encoding his messages so that enemy generals would not be able to understand the messages if they were intercepted. He did this by simply shifting each letter in the message a certain number of places. For example, if each letter is shifted one place to the right, then:

The message **MOVE THE ARMY NORTH INTO GAUL**	Would be encoded as **NPWF UIF BSNZ OPSUI JOUP HBVM**

Where does statistics come in? Well, what if you were an enemy general and you intercepted a message from Caesar to one of his generals? You would not know which shift was being used, so *how could you use statistics to decode the message?* The answer is to make use of your knowledge of modern English letter frequencies (for simplicity, we assume that Caesar was fluent in English, a language that wouldn't develop until hundreds of years later). This, along with the graphs and tables we will find in Section 2.1, will help us decode a secret message in the Case Study on page 42. ■

The Big Picture

Where we are coming from, and where we are headed . . .

- In Chapter 1 we learned the basic concepts of statistics, such as population, sample, and types of variables, along with methods of collecting data.

- Here, in Chapter 2, we learn about graphs and tables for summarizing qualitative data and quantitative data, and we examine how to prevent our graphics from being misleading.

- Later, in Chapter 3, we will learn how to describe a data set using numerical measures like statistics rather than graphs and tables.

2.1 GRAPHS AND TABLES FOR CATEGORICAL DATA

OBJECTIVES By the end of this section, I will be able to . . .

1 Construct and interpret a frequency distribution and a relative frequency distribution for qualitative data.

2 Construct and interpret bar graphs and Pareto charts.

3 Construct and interpret pie charts.

4 Construct crosstabulations to describe the relationship between two variables.

5 Construct a clustered bar graph to describe the relationship between two variables.

In Chapter 2, we apply the adage "A picture is worth a thousand words." The human mind can assess information presented in a graph or table better than it can through words and numbers alone. Psychologists sometimes call this innate ability *pattern recognition*. Statistical graphs and tables take advantage of this ability to quickly summarize data.

1 FREQUENCY DISTRIBUTIONS AND RELATIVE FREQUENCY DISTRIBUTIONS

Frequency Distributions

Recall from Chapter 1 that categorical (qualitative) data take values that are nonnumeric and are usually classified into categories. In this section we learn graphical and tabular methods for handling categorical data. Let us begin with an example.

Amazon.com tracks the best-selling merchandise on its Web site for many different categories. Table 2.1 shows the 20 best-selling video games of 2010, as reported by **Amazon.com**, along with the game console. We will analyze the variable *console*, which is a qualitative variable, not quantitative.

Table 2.1 Top 20 video games, September 2010, as reported by Amazon.com

Rank	Game	Console	Rank	Game	Console
1	Halo Reach	Xbox 360	11	New Super Mario Brothers	Wii
2	Final Fantasy XIII	PlayStation 3	12	Madden NFL 11	PlayStation 3
3	Alan Wake	Xbox 360	13	Sports Resort	Wii
4	Lego Rock Band	Xbox 360	14	Just Dance	Wii
5	Sid Meier's Civilization V	Windows	15	Fit	Wii
6	World of Warcraft: Cataclysm	Windows	16	Super Mario Galaxy 2	Wii
7	Call of Duty: Black Ops	Xbox 360	17	Starcraft II: Wings of Liberty	Windows
8	Final Fantasy XIV	Windows	18	Castlevania: Lords of Shadow	PlayStation 3
9	Bioshock 2	Xbox 360	19	Fable III	Xbox 360
10	Resonance of Fate	PlayStation 3	20	Medal of Honor	PlayStation 3

It is not immediately clear from this data set which game console is the most popular choice among the 20 games in the sample. That is why we need ways to summarize the values in a data set. One popular method used to summarize the values in a data set is the **frequency distribution** (or *frequency table*).

> The **frequency,** or **count,** of a category refers to the number of observations in each category. A **frequency distribution** for a qualitative variable is a listing of all the values (for example, categories) that the variable can take, together with the frequencies for each value.

EXAMPLE 2.1

WHICH IS THE MOST POPULAR GAME CONSOLE?

BLOOMimage/Punchstock Images

Create a frequency distribution for the variable *console* from Table 2.1.

Solution

For each game console, we compute the **frequency;** that is, we **count** how many games used that particular console. Table 2.2 shows the frequency distribution for the variable *console*. For example, five games used the PlayStation 3 game console. The frequency distribution summarizes the data set so that quick observations can be made, such as "The Xbox 360 was the game console used by the greatest number of games in the Amazon.com top 20."

Note: Check that the sum of the frequencies equals the sample size, *n*.

Table 2.2 Frequency distribution of console

Console	Tally	Frequency
Xbox 360	⦀⦀	6
PlayStation 3	⦀⦀	5
Wii	⦀⦀	5
Windows	⦀⦀	4

Now You Can Do
Exercises 11 and 15.

As the data set gets larger, the need for summarization gets more and more acute. (Imagine if the Amazon.com listing consisted of 1000 games rather than 20.) Take a moment to add up the frequencies in Table 2.2. What do they add up to? This number is the sample size: $n = 20$. Now, is this just a coincidence, or does this happen every time? Actually, this happens every time: the sum of the frequencies equals the sample size, n. One way to check if you made a mistake in forming your frequency distribution table is to add up the frequencies and see if the sum equals the sample size.

Relative Frequency Distributions

Next, suppose you didn't know the size of the sample in the survey. Suppose you were told only that 6 games ran on the Xbox 360. The logical question is "Is that a lot?" If our sample size was only 10 games, then 6 of those games using the Xbox 360 is certainly a lot. However, if our sample size was 1000 games, then only 6 of those games using the Xbox 360 is *not* a lot. So, the number's significance depends on what you compare the 6 games to—that is, "relative to what?" or "compared to what?" In statistics, we compare the frequency of a category with the total sample size to get the **relative frequency.**

> The **relative frequency** of a particular category of a qualitative variable is its frequency divided by the sample size. A **relative frequency distribution** for a qualitative variable is a listing of all values that the variable can take, together with the relative frequencies for each value.

EXAMPLE 2.2

RELATIVE FREQUENCY OF GAME CONSOLES

Create a **relative frequency distribution** for the variable *console* using Table 2.1.

Solution

The relative frequency of the Xbox 360 games is the frequency 6 divided by the sample size 20:

$$\text{Relative frequency of Xbox 360} = \frac{\text{Frequency}}{\text{Sample size}} = \frac{6}{20} = 0.30$$

The relative frequency of the games using the Xbox 360 is 0.30, or 30%. So, if someone told you that 30% of the games used the Xbox 360, without telling you the sample size, you would have a better idea of the relative popularity of that game console. To construct the relative frequency distribution in Table 2.3, divide each frequency in the frequency distribution in Table 2.2 by the sample size 20.

Note: The relative frequencies always add up to 1.00, which represents 100%.

Now You Can Do Exercises 12 and 16.

Table 2.3 Relative frequency distribution of console

Console	Relative frequency
Xbox 360	6/20 = 0.30
PlayStation 3	5/20 = 0.25
Wii	5/20 = 0.25
Windows	4/20 = 0.20

2 BAR GRAPHS AND PARETO CHARTS

Frequency distributions and relative frequency distributions are tabular, and thus useful for summarizing data sets. The graphical equivalent of a frequency distribution or a relative frequency distribution is called a **bar graph** (or **bar chart**).

> A **bar graph** is used to represent the frequencies or relative frequencies for categorical data. It is constructed as follows:
>
> 1.   On the horizontal axis, provide a label for each category.
>
> 2.   Draw rectangles (bars) of equal width for each category. The height of each rectangle represents the frequency or relative frequency for that category. Ensure that the bars are not touching each other.

EXAMPLE 2.3

CONSTRUCTING BAR GRAPHS

Construct a frequency bar graph and a relative frequency bar graph for the game console distributions in Tables 2.2 and 2.3.

Solution

The bar graphs are provided in Figures 2.1a and 2.1b. Across the horizontal axis are the four console categories. Next, draw rectangles, the heights of which represent either the frequency or the relative frequency for that category, represented on the vertical axis. For example, in Figure 2.1a, the first rectangle (Xbox 360) reaches a height of 6, while the second rectangle reaches only to 5. Note that the rectangles are of equal width, and none of them touch each other. Also notice that the two bar graphs are exactly alike except for the scale indicated on the vertical axis. This is because we divide each frequency by the same number, the sample size, to get the relative frequency.

**Now You Can Do
Exercises 13 and 17.**

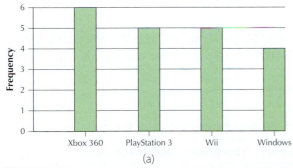

FIGURE 2.1 (a) Frequency bar graph; (b) relative frequency bar graph.

The bars in a bar graph may be presented horizontally, especially when the category names are long. Figure 2.2 contains a horizontal bar chart of the top five quarterbacks in the National Football League in 2009, in terms of passing yardage.

FIGURE 2.2
Horizontal bar chart of top
five passing quarterbacks,
2009. (*Source:* NFL.com)

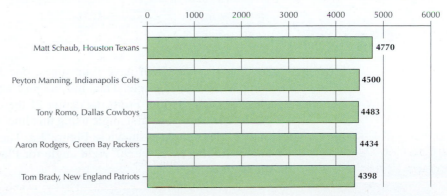

Both Figure 2.1a and Figure 2.1b are examples of **Pareto charts.**

> A **Pareto chart** is a bar graph in which the rectangles are presented in decreasing order from left to right.

Figures 2.5a and 2.5b (page 42) are examples of bar graphs that are not Pareto charts.

3 PIE CHARTS

Pie charts are a common graphical device for displaying the relative frequencies of a categorical variable.

> A **pie chart** is a circle divided into sections (that is, slices or wedges), with each section representing a particular category. The size of the section is proportional to the relative frequency of the category.

Pie charts are typically made using technology. However, one can construct a pie chart using a protractor and a compass. Since a circle contains 360 degrees, we need to multiply the relative frequency for each category by 360°. This will tell us how large a slice to make for each category, in terms of degrees.

EXAMPLE 2.4

CONSTRUCTING A PIE CHART

Construct a pie chart for the game console data from Example 2.2.

Solution

The relative frequencies from Example 2.2 are shown in Table 2.4. We multiply each relative frequency by 360° to get the number of degrees for that section (slice) of the pie chart.

Table 2.4 Finding the number of degrees for each slice of the pie chart

Variable: console	Relative frequency	Multiply by 360°	Degrees for that section
Xbox 360	6/20 = 0.30	0.30 × 360° =	108°
PlayStation 3	5/20 = 0.25	0.25 × 360° =	90°
Wii	5/20 = 0.25	0.25 × 360° =	90°
Windows	4/20 = 0.20	0.20 × 360° =	72°
Total	20/20 = 1.00		360°

Our pie chart will have four slices, one for each console category. Use the compass to draw a circle. Then use the protractor to construct the appropriate angles for each section. From the center of the circle, draw a line to the top of the circle. Measure your first angle using this line. For the Xbox 360, we need an angle of 108°. This angle is shown in Figure 2.3. Then, from there, measure your second angle—in this case, the 90° right angle for PlayStation 3. Continue until your circle is complete.

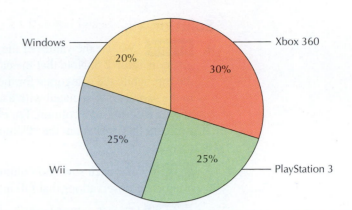

FIGURE 2.3
Pie chart of the video game console data.

Now You Can Do
Exercises 14 and 18.

4   CROSSTABULATIONS

So far, we have analyzed only one variable at a time. **Crosstabulation** is a tabular method for simultaneously summarizing the data for two categorical (qualitative) variables.

> **Steps for Constructing a Crosstabulation**
>
> **Step 1** Put the categories of one variable at the top of each column, and the categories of the other variable at the beginning of each row.
>
> **Step 2** For each row and column combination, enter the number of observations that fall in the two categories.
>
> **Step 3** The bottom of the table gives the column totals, and the right-hand column gives the row totals.

Crosstabulations are also known as **two-way tables** or **contingency tables.** We will introduce crosstabulations using an example.

EXAMPLE 2.5

CONSTRUCTING A CROSSTABULATION

  carsizegas

Table 2.5 contains information about the size (compact, midsize, or large) and the recommended gasoline (regular or premium) for a sample of ten 2011 automobiles
a.   Construct a crosstabulation of the variables *size* and *gasoline*.
b.   Identify any patterns.

Table 2.5 Size and recommended gasoline for ten 2011 automobiles

Car	Car size	Recommended gasoline
BMW 328i	Compact	Premium
Chevrolet Camaro	Compact	Regular
Honda Accord	Compact	Regular
Cadillac CTS	Midsize	Premium
Nissan Sentra	Midsize	Regular
Subaru Legacy AWD	Midsize	Premium
Toyota Camry	Midsize	Regular
Ford Taurus	Large	Regular
Hyundai Genesis	Large	Premium
Rolls-Royce	Large	Premium

Source: **www.fueleconomy.gov.**

Solution

a. **STEP 1** We use the values of the two variables to create the crosstabulation given in Table 2.6. Note that the categories for the variable *gasoline* are shown at the top, while the categories for the variable *size* are shown on the left. Each car in the sample is associated with a certain *cell* in the crosstabulation, in the appropriate row and column. For example, the Chevrolet Camaro is one of the two cars that appears in the "Compact" car size row and the "Regular" gasoline column.

STEP 2 For each row and column combination in the crosstabulation, enter the number of observations that fall in the two categories.

STEP 3 The "Total" column contains the sum of the counts of the cells in each row (category) of the *size* variable, and represents the frequency distribution for this variable. Similarly, the "Total" row along the bottom sums the counts of the cells in each column (category) of the *gasoline* variable, and represents the frequency distribution for this variable. In the lower right-hand corner we have the grand total, which should equal the sample size.

Table 2.6 Crosstabulation of car size and recommended gasoline

| | Recommended Gasoline | | |
Car size	Regular	Premium	Total
Compact	2	1	3
Midsize	2	2	4
Large	1	2	3
Total	5	5	10

b. We can use the crosstabulation to look for patterns in the data set. One possible pattern is the following: Compact cars tend to use regular gasoline while large cars tend to use premium gasoline. Of course, this sample size is too small to form any conclusions about such a relationship.

Now You Can Do Exercises 27 and 37.

5 CLUSTERED BAR GRAPHS

Clustered bar graphs are useful for comparing two categorical variables and are often used in conjunction with crosstabulations. Each set of bars in a clustered bar graph represents a single category of one variable across all the categories of the other categorical variable (see Figures 2.4a and 2.4b). This allows the analyst to make comparisons easily. One can construct clustered bar graphs using either frequencies or relative frequencies. To construct a clustered bar graph, identify which of the two categorical variables will define the cluster of bars. Then, for each category of the other variable, draw bars for each category of the clustering variable.

EXAMPLE 2.6

CONSTRUCTING CLUSTERED BAR GRAPHS

genderemotions

Recall Example 1.2, in Section 1.1 (page 3). (The original survey question read, "Which one of the following emotions do you feel the most strongly in response to these terrorist attacks: sadness, fear, anger, disbelief, vulnerability?")[1] The results are given in the crosstabulation in Table 2.7. Construct a clustered bar graph of the emotions felt, clustered by gender in order to illustrate any differences between males and females.

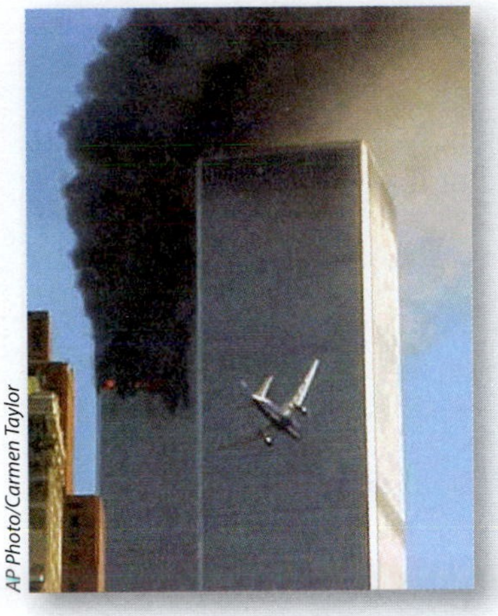

Table 2.7 Frequency of survey respondents expressing particular emotions, by gender

Gender	Emotion						Total
	Sadness	Fear	Anger	Disbelief	Vulnerability	Not sure	
Female	94	21	87	80	28	4	314
Male	56	16	141	50	36	5	304
Total	150	37	228	130	64	9	618

Solution

Gender is given as the clustering variable. Thus, for each category of the variable *emotion,* we will draw two bars, one representing males and the other representing females. For example, for the first emotion, sadness, we draw one rectangle going up to 56 on the vertical axis, and a separate rectangle going up to 94 on the vertical axis. These two rectangles should touch each other but should not touch any other rectangles. Continue to draw two rectangles for each emotion, one for each of the males' and females' frequencies. The resulting clustered bar graph is shown here as Figure 2.4a. We say that the emotions are *clustered* by gender.

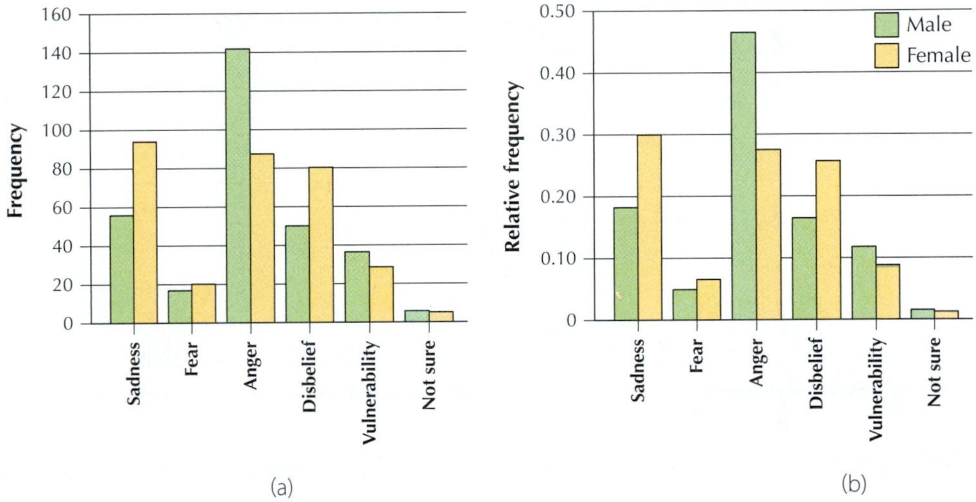

FIGURE 2.4 (a) Clustered bar graph using frequencies; (b) clustered bar graph using relative frequencies.

Note: We can use either *percentage* or *proportion* to describe relative frequency. For example, in Table 2.8, we can say either that the *percentage* of females who expressed sadness was 29.9% or that the *proportion* of females who expressed sadness was 0.299.

Now, what if females were underrepresented in this survey, so that there were only 100 females and 304 males? Then, direct comparison of the counts would be misleading. When the sample sizes are substantially different, one should use relative frequency clustered bar graphs. The relative frequencies for the frequencies in Table 2.7 are provided in Table 2.8, and the clustered bar graph is given in Figure 2.4b. Note that we divide the counts by the total for that gender, not by the total for the emotion.

Table 2.8 Relative frequencies of emotions, by gender

Gender	Emotion						Total
	Sadness	Fear	Anger	Disbelief	Vulnerability	Not sure	
Females	0.299	0.067	0.277	0.255	0.089	0.013	1.000
Males	0.184	0.053	0.464	0.164	0.118	0.016	1.000

Now You Can Do Exercises 28 and 38.

CASE STUDY | The Caesar Cipher

Recall the Caesar Cipher from the chapter introduction. Suppose we need to decipher the following secret message from Caesar to one of his generals:

LI ZH ZLQ, SLCCD IRU HYHUBRQH (HAWUD SHSSHURQL).

We will make a frequency distribution and bar graph of the letters in the message and then compare them with the bar graph of the letters in the English language given in Figure 2.5a.

We can observe in Figure 2.5a that the letter **E** far outstrips all other letters in the alphabet in frequency. Other high-frequency letters are **A, I, N, O, R, S,** and **T.** Compare this with the frequency distribution of the letters in the coded message, shown in Table 2.9. From this frequency distribution, we can see that **H** is the most frequently occurring letter in the coded message. Other frequently occurring letters are **L, Q, R, S, U,** and **W.** Since **E** is the most frequently occurring letter in English, perhaps this means that **E** is encoded as **H,** the most common letter in our message. The frequency bar graph of letters in the coded message, from Table 2.9, is shown in Figure 2.5b.

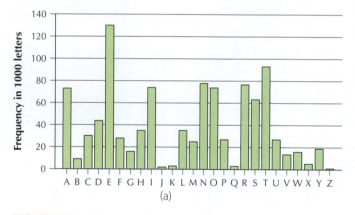

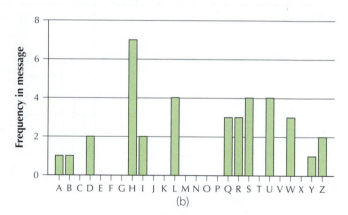

FIGURE 2.5 (a) Frequency bar graph of English letters; (b) frequency bar graph of letters in coded message.

codeletters

Table 2.9 Frequency distribution of letters in coded message

A	B	C	D	E	F	G	H	I	J	K	L	M
1	1	0	2	0	0	0	7	2	0	0	4	0
N	**O**	**P**	**Q**	**R**	**S**	**T**	**U**	**V**	**W**	**X**	**Y**	**Z**
0	0	0	3	3	4	0	4	0	3	0	1	2

Caesar used a simple shift of the letters for his code. If we substitute **H** for **E,** then the original letters have been shifted three places to the right (**E → F → G → H**). But this may just be an aberration. Is there further evidence for a "right shift of 3"? Let's see if this "right shift of 3" makes sense for the other high-frequency letters in the coded message. To undo a "right shift of 3," we would need to shift the letters in the coded message back three to the left to get the original letters. If the letter **L** is shifted back three places to the left, you get **I,** one of the high-frequency letters in English. Shift the letter **Q** three places, and you get **N,** another letter of high frequency in English. Shift the other letters of high frequency in our coded message, and you get **O, P, R,** and **T,** respectively, all high-frequency letters. There is a strong probability that we have found the correct decoding mechanism.

Let us now proceed to decode the message by shifting every letter in the coded message three places to the left (for example, **L → K → J → I**). It turns out that the decoded message reads

IF WE WIN, PIZZA FOR EVERYONE (EXTRA PEPPERONI).

Small wonder that Caesar went on to win an empire! We have gotten a taste of how the analysis of frequency distributions and bar graphs can be useful for solving problems. ■

STEP-BY-STEP TECHNOLOGY GUIDE: Frequency Distributions, Bar Graphs, and Pie Charts

We use the data set in Table 2.10 to demonstrate how to use technology to construct a frequency distribution, relative frequency distribution, bar graph, and pie chart. Table 2.10 lists the declared majors of 25 randomly selected students at a local business school. (MIS stands for management information systems.)

TABLE 2.10 Declared majors of business school students 🔴 studentmajor

Management	MIS	Management	MIS	Marketing
Marketing	Marketing	Management	Finance	Accounting
Accounting	Accounting	MIS	Management	MIS
Management	MIS	Management	Economics	Accounting
Finance	Management	Economics	Marketing	Finance

Excel

Frequency Distributions
Step 1 Enter the data in Column A, with the topmost cell indicating the variable name, *Major*.
Step 2 Select cells A1–A26, click **Insert > PivotTable**, and click **OK**.
Step 3 Under **Choose fields to add to report**, select *Major*.
Step 4 Click on *Major* and drag to the **Values** box at the lower right of the screen. The resulting frequency distribution is shown in Figure 2.6. In Excel, this takes the form of a *pivot table,* which is an interactive tabular format.

Clustered Bar Graphs
Step 1 Select the crosstabulation.
Step 2 Click **Insert > Column**.
Step 3 Click **Clustered column**.

Bar Graphs and Pie Charts
Note: Excel can make bar graphs or pie charts using frequency distributions but not from the raw data.
Step 1 Enter the frequency distribution as shown in Figure 2.7.
Step 2 Select cells A1 to B7. For a bar graph, click **Insert > Column**. For a pie chart, click **Insert > Pie**.
Step 3 The resulting frequency bar graph and pie chart are shown in Figures 2.8 and 2.9 on the next page.

3	Row Labels ▾	Count of Major
4	Accounting	4
5	Economics	2
6	Finance	3
7	Management	7
8	Marketing	4
9	MIS	5
10	**Grand Total**	**25**

FIGURE 2.6

	A	B
1	Major	Count
2	Accounting	4
3	Economics	2
4	Finance	3
5	Management	7
6	Marketing	4
7	MIS	5

FIGURE 2.7

(Continued)

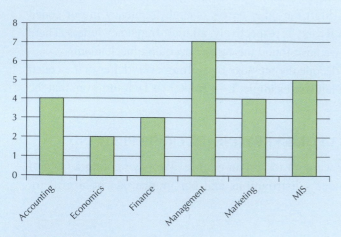

FIGURE 2.8 Excel frequency bar graph.

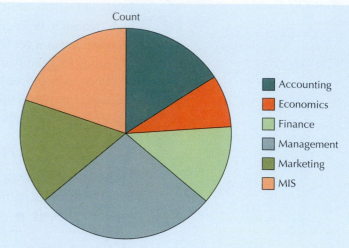

FIGURE 2.9 Excel pie chart.

CRUNCHIT!

Frequency Distributions
Step 1 Click **File** . . . then highlight **Load from Larose2e** . . . **Chapter 2** . . . and click on **Table 2.10**.
Step 2 Click **Statistics** and select **Frequency Table**. For **Sample** select **Major**. Then click **Calculate**.

Bar Graphs and Pie Charts
Step 1 Click **File** . . . then highlight **Load from Larose2e** . . . **Chapter 2** . . . and click on **Table 2.10**.
Step 2 Click **Graphics** and select **Bar Chart**. For a pie chart select **Pie Chart**.
Step 3 For **Sample** select **Major**. Then click **Calculate**. The resulting bar graph is shown here.

Crosstabulation
We use Table 2.5 from Example 2.5.

Step 1 Click **File** … then highlight **Load from Larose2e** . . . **Chapter 2** . . . and click on **Table 2.5**.
Step 2 Click **Statistics** . . . **Contingency Table** and select **Get frequencies**.

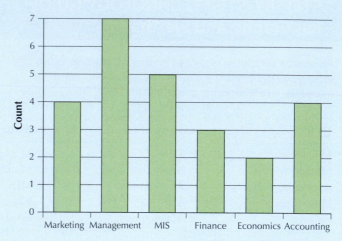

CrunchIt! bar graph.

Step 3 For **Row variable** select **Size**, and for **Column variable** select **gasoline**. Then click **Calculate**.

MINITAB

Frequency Distributions
Step 1 Name your variable *Major* and enter the data into the C1 column.
Step 2 Click **Stat** > **Tables** > **Tally**.
Step 3 Under **Display**, select **Counts** and **Percents**.
Step 4 Click inside the **Variables** box until you see your variable *major* listed. Select the variable **C1** *Major*, and click **Select**. Then click **OK**.

Bar Graphs
Step 1 Name your variable *Major* and enter the data into the C1 column.
Step 2 Click **Graph** > **Bar Chart**. For raw data select **Bars Represent: Counts of Unique Values,** select **Simple,** and click **OK**. (For summarized data such as a frequency distribution, select **Bars Represent: Values from a Table**, and select **Simple**. Then click **OK**.)

Step 3 In the **listing of variables** box, click on the *Major* variable to select it for analysis. Then click **OK**.

Pie Charts
Step 1 Name your variable *Major* and enter the data into the C1 column.
Step 2 Click **Graph** > **Pie Chart**. For raw data select **Chart Counts of unique values**. Then click in the **Variables** box to select the variable *Major*, and click **OK**. (For summarized data such as a frequency distribution, select **Chart Data from a Table**. Then select the category variable for **Categorical variable**, and select the variable with the frequencies or relative frequencies for the **Summary variable**. Then click **OK**.)

Crosstabulation of Career Data
Step 1 Enter the data from Table 2.6 (page 40) into two columns, named *size* and *gasoline*.
Step 2 Click **Stat** > **Tables** > **Cross-Tabulation and Chi-Square**.

Step 3 For **rows**, select *size*; for **columns**, select *gasoline*. Select **Counts** under **Display**. Then click **OK**.
Step 4 The resulting crosstabulation is shown here. The rows and columns are in alphabetical order.

```
Rows: Size
Columns: Gasoline

     Premium Regular  All
Compact    1       2    3
Large      2       1    3
Midsize    2       2    4
All        5       5   10
```

Clustered Bar Graphs
If you have the original data set:
Step 1 Click **Graph** > **Bar Chart**.
Step 2 Select **Bars Represent: Counts of Unique Values**, and select **Clustered**. Then click **OK**.
Step 3 Select your two categorical variables, and click **OK**.
If you have only the crosstabulation and not the original data:
Step 1 Click **Graph** > **Bar Chart**.
Step 2 Select **Bars Represent: Values from a Table**, and select **Clustered**. Then click **OK**.
Step 3 For **Graph Variables**, choose the variable that contains the frequencies or relative frequencies. For **Categorical Variables for Grouping**, choose your two categorical variables. Then click **OK**.

SECTION 2.1 **Summary**

In this section, we learned about tabular and graphical methods for summarizing qualitative (categorical) data.

1. Frequency distributions and relative frequency distributions list all the values that a qualitative variable can take, along with the frequencies (counts) or relative frequencies for each value.

2. A bar graph is the graphical equivalent of a frequency distribution or a relative frequency distribution. When the rectangles are presented in decreasing order from left to right, the result is a Pareto chart.

3. Pie charts are a common graphical device for displaying the relative frequencies of a categorical variable. A pie chart

is a circle divided into sections (that is, slices or wedges), with each section representing a particular category. The size of the section is proportional to the relative frequency of the category.

4. Crosstabulation summarizes the relationship between two categorical variables. A crosstabulation is a table that gives the counts for each row–column combination, with totals for the rows and columns.

5. Clustered bar graphs are useful for comparing two categorical variables, and are often used in conjunction with crosstabulations.

SECTION 2.1 **Exercises**

Clarifying the Concepts

1. Why do we use graphical and tabular methods to summarize data? What's wrong with simply reporting the raw data?

2. What's the difference between a frequency distribution and a relative frequency distribution?

3. True or false: For a given data set, a frequency bar graph and a relative frequency bar graph look alike except for the scale on the vertical axis.

4. True or false: A pie chart is used to represent quantitative data.

5. What should be the sum of the frequencies in a frequency distribution?

6. What should be the sum of the relative frequencies in a relative frequency distribution?

7. In a crosstabulation, the "Total" column represents what? How about the "Total" row?

8. What does the number in the lower right corner of the crosstabulation represent? What should this number be equal to?

9. When is it better to use a relative frequency (rather than a frequency) clustered bar graph?

10. Why can't we use crosstabulations for two numerical variables? Is there some way we could recode the variables in order to use crosstabulations?

Practicing The Techniques

The political party affiliations of a class of 20 statistics students are shown here. Use this information to construct the table or graph indicated in Exercises 11–14 (Dem = Democrat, Rep = Republican, Ind = Independent).

 politics

Dem	Rep	Ind	Rep	Ind	Dem	Rep	Dem	Ind	Ind
Dem	Rep	Dem	Rep	Ind	Rep	Ind	Dem	Dem	Rep

11. Frequency distribution

12. Relative frequency distribution

13. Bar graph

14. Pie chart

The blood types of a class of 25 nursing students are shown on the next page. The four categories are A, B, AB, and O. Use this information to construct the table or graph indicated in Exercises 15–18.

 bloodtypes

```
O   A   A   B   A   O   O   A   O
A   A   O   O   O   A   A   B   AB
O   A   A   O   O   B   A
```

15. Frequency distribution
16. Relative frequency distribution
17. Bar graph
18. Pie chart

The major and gender of a class of 12 statistics students are recorded here. Use this information to construct the table or graph in Exercises 19–28.

 gendermajor

Major	Gender	Major	Gender
Math	Female	Math	Female
Psychology	Female	Business	Female
Business	Male	Psychology	Male
Math	Male	Psychology	Male
Business	Male	Business	Female
Psychology	Female	Math	Female

19. Frequency distribution of *major*
20. Relative frequency distribution of *major*
21. Bar graph of *major*
22. Pie chart of *major*
23. Frequency distribution of *gender*
24. Relative frequency distribution of *gender*
25. Bar graph of *gender*
26. Pie chart of *gender*
27. Crosstabulation of *major* and *gender*
28. Clustered bar graph of *major*, clustered by *gender*

The class standing and handedness of a group of 14 students are shown here. Use this information to construct the table or graph in Exercises 29–38.

 classhands

Class	Handedness	Class	Handedness
Senior	Right	Junior	Right
Sophomore	Right	Freshman	Right
Senior	Right	Senior	Left
Sophomore	Right	Junior	Left
Sophomore	Right	Senior	Right
Freshman	Right	Junior	Right
Sophomore	Left	Freshman	Left

29. Frequency distribution of *class*
30. Relative frequency distribution of *class*
31. Bar graph of *class*

32. Pie chart of *class*
33. Frequency distribution of *handedness*
34. Relative frequency distribution of *handedness*
35. Bar graph of *handedness*
36. Pie chart of *handedness*
37. Crosstabulation of *class* and *handedness*
38. Clustered bar graph of *class*, clustered by *handedness*

Applying the Concepts

Cell Phone Ownership. Figure 2.10 shows the percentage of cell phone ownership, categorized by level of education. Use Figure 2.10 to answer Exercises 39 and 40.

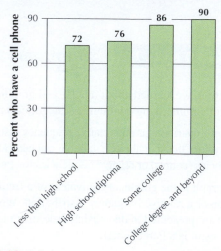

FIGURE 2.10 Cell phone ownership. (*Source*: Amanda Lenhart, *Cell Phones and American Adults*, Pew Internet and American Life Project, September 2, 2010.)

39. Can we use the information in Figure 2.10 to construct a pie chart? Explain why or why not.

40. Is Figure 2.10 a Pareto chart? Explain why or why not.

Cell Phones and the Internet. Figure 2.11 is a pie chart representing the percentage of Americans who access the Internet or email using their cell phones. Use Figure 2.11 to answer Exercises 41 and 42.

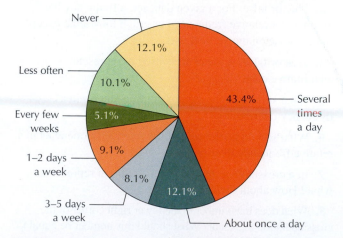

FIGURE 2.11 Percentage using cell phones for Internet or email. (*Source*: Amanda Lenhart, *Cell Phones and American Adults*, Pew Internet and American Life Project, September 2, 2010.)

41. According to this survey:
   a. What is the most common response? What percentage does this represent?
   b. What is the least common response? What percentage does this represent?

42. According to this survey:
   a. What percentage uses the cell phone to access the Internet or email about once a day?
   b. What percentage never uses the cell phone to access the Internet or email?

Sledding Injuries. Every year, about 20,000 children and teenagers visit the emergency room with injuries sustained from snow sledding. Use the horizontal bar graph in Figure 2.12 to answer Exercises 43 and 44.

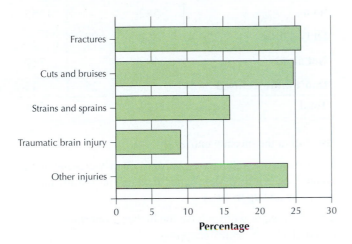

FIGURE 2.12 Most common injuries from sledding. (*Source:* Candace A. Howell, Nicolas G. Nelson, and Lara B. McKenzie, "Pediatric and adolescent sledding-related injuries treated in U.S. emergency departments 1997–2207," *Pediatrics*, 126 (2010): 517–514.)

43. According to this study:

   a. What is the most common category of injury? Estimate the percentage.
   b. Of the specific injuries shown, what is the least common category of injury? What is the percentage?
   c. Is it possible for there to be an injury type that has a lower percentage than traumatic brain injury? Explain.

44. According to this study:
   a. What is the percentage for cuts and bruises?
   b. What is the percentage for strains and sprains?

World Water Usage. See Table 2.11 for Exercises 45–48. For the indicated variable, construct the following:
🔴 worldwater
   a. Frequency distribution
   b. Relative frequency distribution
   c. Frequency bar graph

   d. Relative frequency bar graph
   e. Pareto chart, using relative frequencies
   f. Pie chart

TABLE 2.11   World water usage

Country	Continent	Climate	Main use
Iraq	Asia	Arid	Irrigation
United States	North America	Temperate	Industry
Pakistan	Asia	Arid	Irrigation
Canada	North America	Temperate	Industry
Madagascar	Africa	Tropical	Irrigation
North Korea	Asia	Temperate	Not reported
Chile	South America	Arid	Irrigation
Bulgaria	Europe	Temperate	Not reported
Afghanistan	Asia	Arid	Irrigation
Iran	Asia	Arid	Irrigation

45. The variable *continent*

46. The variable *climate*

47. The variable *main use*

48. Explain why it is not appropriate to construct a frequency distribution for *country*.

Use Table 2.11 for Exercises 49–54.

49. Construct a crosstabulation of the variables *continent* and *climate*.

50. Construct a crosstabulation of the variables *continent* and *main use*.

51. Construct a crosstabulation of the variables *climate* and *main use*.

52. Construct a clustered bar graph of *continent*, clustered by *climate*.

53. Construct a clustered bar graph of the variable *main use*, clustered by *continent*.

54. Construct a clustered bar graph of the variable *main use*, clustered by *climate*.

55. Vehicle Models. Table 2.12 on the next page shows the numbers of vehicle models, categorized by vehicle type, examined each year by the U.S. Department of Energy to determine vehicle gas mileage. Use Table 2.12 to construct the following:
🔴 cartypemodel
   a. Relative frequency distribution
   b. Frequency bar graph
   c. Relative frequency bar graph
   d. Pareto chart, using relative frequencies
   e. Pie chart of the relative frequencies

TABLE 2.12

Vehicle type	Number of models
SUVs	370
Compact cars	128
Midsize cars	120
Subcompact cars	110
Standard pickup trucks	106
Large cars	76
Station wagons	62
Small pickup trucks	59
Two-seaters	51
Minicompact cars	43
Vans	38
Minivans	19
Total	1182

Astrological Signs. Use the following information for Exercises 56–58. The General Social Survey collects data on social aspects of life in America. Here, 1464 respondents reported their astrological sign. A pie chart of the results is shown here.

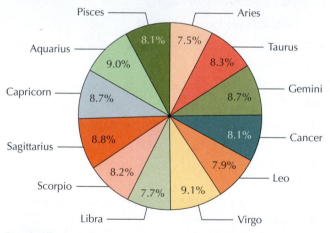

Pie chart of astrological signs.

56. Answer the following:
  a. What is the most common astrological sign?
  b. What is the least common astrological sign?

57. Use the percentages in the pie chart to do the following:
  a. Construct a relative frequency bar graph of the astrological signs.
  b. Construct a relative frequency bar graph, but this time have the *y* axis begin at 7% instead of zero. Describe the difference between the two bar graphs. When would this one be used as opposed to the earlier bar graph?

58. Construct a frequency distribution of the astrological signs. Which sign occurs the least? The most?

Bringing It All Together

Shopping Enjoyment and Gender. Use the information in the crosstabulation for Exercises 59–72. The Pew Internet and American Life Project surveyed 4514 American men and women and asked them, "How much, if at all, do you enjoy shopping?" The results shown in the crosstabulation are missing some entries.

Crosstabulation of shopping enjoyment by gender

| Response: "How much do you enjoy shopping?" | Gender | | Total |
	Male	Female	
A lot		950	1338
Some	582		1255
Only a little	662	497	
Not at all	497		717
Don't know/refused		25	45
Total	2149		4514

59. Fill in the missing entries.

60. Convert the table to a relative frequency crosstabulation. Make it so that the "Male" and "Female" proportions in each row add up to 1.0.

61. Did men or women have the higher proportion of respondents who enjoy shopping
  a. a lot?
  b. some?
  c. only a little?
  d. not at all?

62. Construct a frequency distribution of gender.

63. Construct a frequency distribution of response.

64. Construct a relative frequency distribution of gender.

65. Construct a relative frequency distribution of response.

66. Construct a bar graph of gender.

67. Construct a bar graph of response.

68. Construct a pie chart of gender.

69. Construct a pie chart of response.

70. Construct a clustered bar graph of gender, clustered by response.

71. Construct a clustered bar graph of response clustered by gender.

72. ❓ *What* if we doubled each cell count? How would that affect the following?
  a. Frequency distribution of gender
  b. Relative frequency distribution of gender
  c. Pie chart of gender

Educational Goals in Sports. Use your knowledge of technology to solve Exercises 73 and 74. Open the **Goals** data set. The subjects are students in grades four, five, and six from three school districts in Michigan. The students were asked which of the following was most important to them: good grades, sports, or popularity. Information about the students' age, gender, race, and grade was also gathered, as well as whether their school was in an urban, suburban, or rural setting.[2] 🌐 goals

73. Generate bar graphs for the following variables.
   a. *Gender*. Estimate the relative frequency of girls in the sample. Of boys.
   b. *Goals*. About what percentage of the students chose "grades" as most important? About what percentage chose "popular"? About what percentage chose "sports"?

74. Generate relative frequency distributions for the following variables.
   a. *Gender*. How close were your estimates in the previous exercise?
   b. *Goals*. How close were your estimates in the previous exercise?

Construct Your Own Data Sets

Environmental Club. Use the following information for Exercises 75–77. You are the president of the College Environmental Club, which has members among all four classes: freshmen, sophomores, juniors, and seniors. The total number of members in the club is 20.

75. Set the frequency of each class so that each class has an equal number of members.
   a. Construct a frequency distribution of the variable *class*.
   b. Construct a relative frequency distribution of the variable *class*.

76. Set the frequency of each class so that there are more sophomores than freshmen, more juniors than sophomores, and more seniors than juniors.
   a. Construct a Pareto chart of the variable *class*.
   b. Construct a pie chart of the variable *class*.

77. Set the frequency of each class so that there are more seniors than any other class while the other three classes have equal numbers.
   a. Construct a frequency bar graph of the variable *class*.
   b. Construct a relative frequency bar graph of the variable *class*.

2.2 GRAPHS AND TABLES FOR QUANTITATIVE DATA

OBJECTIVES   By the end of this section, I will be able to . . .

1   Construct and interpret a frequency distribution and a relative frequency distribution for discrete and continuous data.

2   Use histograms and frequency polygons to summarize quantitative data.

3   Construct and interpret stem-and-leaf displays and dotplots.

4   Recognize distribution shape, symmetry, and skewness.

1 FREQUENCY DISTRIBUTIONS AND RELATIVE FREQUENCY DISTRIBUTIONS

In Section 2.1, we introduced tables and graphs for summarizing qualitative data. However, most of the data sets that we will encounter in this book are quantitative rather than qualitative. Recall from Chapter 1 that quantitative data take on numerical values that arithmetic can be meaningfully performed on. We can apply frequency and relative frequency distributions to quantitative data, just as we did for the qualitative data in Section 2.1

EXAMPLE 2.7   **FREQUENCY DISTRIBUTION AND RELATIVE FREQUENCY DISTRIBUTION FOR DISCRETE DATA**

The National Center for Missing and Exploited Children (**www.missingkids.com**) keeps an online searchable data base of missing children nationwide. Table 2.13 contains a listing of the 50 children who have gone missing from California and who would have been between 1 and 9 years of age as of March 4, 2007. Suppose we are interested in analyzing the ages of these missing children. Use the data to construct a **frequency distribution** and a **relative frequency distribution** of the variable *age*.

Table 2.13 Missing children and their ages

Child	Age	Child	Age	Child	Age	Child	Age
Amir	5	Carlos	7	Octavio	8	Christian	8
Yamile	5	Ulisses	6	Keoni	6	Mario	8
Kevin	5	Alexander	7	Lance	5	Reya	5
Hilary	8	Adam	4	Mason	5	Elias	1
Zitlalit	7	Sultan	6	Joaquin	6	Maurice	4
Aleida	8	Abril	6	Adriana	6	Samantha	7
Alexia	2	Ramon	6	Christopher	3	Michael	9
Juan	9	Amari	4	Johan	6	Carlos	2
Kevin	2	Joliet	1	Kassandra	4	Lukas	4
Hazel	5	Christopher	4	Hiroki	6	Kayla	4
Melissa	1	Jonathan	8	Kimberly	5	Aiko	3
Kayleen	6	Emil	7	Diondre	4	Lorenzo	9
Mirynda	7	Benjamin	5				

Solution

We can construct the frequency distribution for the variable *age* and can construct the relative frequency distribution by dividing the frequency by the total number of observations, 50. See Table 2.14.

Table 2.14 Frequency distribution and relative frequency distribution of *age*

Age	Tally	Frequency	Relative frequency
1	III	3	0.06
2	III	3	0.06
3	II	2	0.04
4	HHI III	8	0.16
5	HHI IIII	9	0.18
6	HHI HHI	10	0.20
7	HHI I	6	0.12
8	HHI I	6	0.12
9	III	3	0.06
Total		50	1.00

Now You Can Do Exercises 9–12.

We can combine several ages together into "classes," in order to produce a more concise distribution. **Classes** represent a range of data values and are used to group the elements in a data set.

EXAMPLE 2.8	**FREQUENCY AND RELATIVE FREQUENCY DISTRIBUTIONS USING CLASSES**

Combine the age data from Example 2.7 into three classes, and construct frequency and relative frequency distributions.

Solution

Let us define the following classes for the age data: 1–3 years old, 4–6 years old, and 7–9 years old. For each class, we group together all the ages in the class. Table 2.15 provides the frequency distribution and relative frequency distribution for these three age classes.

Table 2.15 Distributions for the variable *age*, after combining into three classes

Class	Frequency	Relative frequency
1–3	8	0.16
4–6	27	0.54
7–9	15	0.30
Total	50	1.00

Now You Can Do Exercises 13–14.

Developing Your Statistical Sense	**Choosing Which Distribution to Use**

So which frequency distribution is the "right" one, Table 2.14 or Table 2.15? There is no absolute answer. It depends on the goals of the analysis, as well as other factors. For example, from Table 2.15, we can see that the majority (0.54 = 54%) of missing children are aged 4–6, an observation that was not immediately apparent from Table 2.14. So, combining data values into classes can lead to interesting overall findings. However, whenever data values are combined into classes, some information is lost. For example, it is not possible, using Table 2.15 alone, to determine that age 6 has the highest proportion of missing children.

We use the following to construct frequency distributions and histograms (for a discussion of histograms, see pages 54–55).

> The **lower class limit** of a class equals the smallest value within that class.
>
> The **upper class limit** of a class equals the largest value within that class.
>
> The **class width** equals the difference between the lower class limits of two successive classes.
>
> The **class boundary** of two successive classes is found by taking the sum of the upper class limit of a class and the lower class limit of the class to its right, and dividing this sum by two. The lower class boundary of the leftmost class equals its upper class boundary minus the class width. The upper class boundary of the rightmost class equals its lower class boundary plus the class width.

EXAMPLE 2.9

CLASS LIMITS, CLASS WIDTHS, AND CLASS BOUNDARIES

For the classes in Example 2.8, find the following:
a. The lower class limits and the upper class limits
b. The class width
c. The class boundaries

Solution

a. The following table shows the lower class limits and the upper class limits for the classes in Example 2.8.

Class	Lower class limit (smallest value)	Upper class limit (largest value)
1–3	1	3
4–6	4	6
7–9	7	9

b. Since our lower class limits are 1, 4, and 7, the class width of each class is 3 because the lower class limits differ by 3. For example, $4 - 1 = 3$.

c. To find the class boundary of the first and second class, we find the sum of the upper class limit of the first class and the lower class limit of the second class, and divide this sum by 2, giving us $(3 + 4)/2 = 3.5$. Similarly, the class boundary of the second class with the third class is $(6 + 7)/2 = 6.5$. The lower class boundary of the leftmost class equals its upper class boundary minus the class width, that is, $3.5 - 3 = 0.5$. The upper class boundary of the rightmost class equals its lower class boundary plus the class width, that is, $6.5 + 3 = 9.5$.

Next, we show how to construct frequency distributions for continuous data.

> To construct a frequency distribution for continuous data:
> 1. Choose the number of classes.
> 2. Determine the class width. It is best to use the same width for all classes.
> 3. Find the upper and lower class limits. Make sure the classes are nonoverlapping.
> 4. Calculate the class boundaries.
> 5. Find the frequencies of each class.

EXAMPLE 2.10

CONSTRUCTING A FREQUENCY DISTRIBUTION FOR CONTINUOUS DATA

Twenty management students, in preparation for graduation, took a course to prepare them for a management aptitude test. A simulated test provided the following scores:

77	89	84	83	80	80	83	82	85	92
87	88	87	86	99	93	79	83	81	78

mgmttest

Construct a frequency distribution of these management aptitude test scores.

Solution

STEP 1   Choose the number of classes.
It is generally recommended that between 5 and 20 classes be used, with the number of classes increasing with the sample size. A small data set such as this will do just fine with 5 classes. In general, choose the number of classes to be large enough to show the variability in the data set, but not so large that many classes are nearly empty.

STEP 2   Determine the class widths.
First, find the *range* of the data, that is, the difference between the largest and smallest data points. Then, divide this range by the number of classes you chose in Step 1. This gives an estimate of the class width. Here, our largest data value is 99 and our smallest is 77, giving us a *range* of $99 - 77 = 22$. In Step 1, we chose 5 classes, so that our estimated class width is $22/5 = 4.4$. We will use a convenient class width of 5. It is recommended that each class have the same width.

STEP 3   Find the upper and lower class limits.
Choose limits so that each data point belongs to only one class. For example, suppose we chose one class to be 75–80 and the next class to be 80–85. Then, to which class would a data value of 80 belong? The classes should not overlap. Therefore, we define the following classes:

| 75–79 | 80–84 | 85–89 | 90–94 | 95–99 |

Note that the lower class limit of the first class, 75, is slightly below that of the smallest value in the data set, 77. Also note that the class width equals $80 - 75 = 5$, as desired.

STEP 4   Calculate the class boundaries.
The class boundary for the first two classes is $(79 + 80)/2 = 79.5$. Similarly, we may calculate the other class boundaries to be 84.5, 89.5, and 94.5. The lower class boundary of the leftmost class is $79.5 - 5 = 74.5$. The upper class boundary of the rightmost class is $94.5 + 5 = 99.5$.

STEP 5   Find the frequencies for each class.
Using these five classes, we now proceed to construct the frequency and relative frequency distributions for the management aptitude test scores (see Table 2.16). We count the number of data values that fall into each class, and we divide each frequency by the sample size (20) to obtain the relative frequency. We see that the majority of the students ($0.40 + 0.30 = 0.70$) received scores between 80 and 89 and that only one received a score above 94.

Note: In this example, we have data values that are integers. If the data values, instead, had decimal values, then we would choose the class limits accordingly. For example, if the data values ranged from 75 to 100 but were of the form 75.6, we could choose the class limits of the first class to be 75.0–79.9, the second class to be 80.0–84.9, and so on.

Table 2.16 Distributions for the management aptitude test scores

Class	Tally	Frequency	Relative frequency
75–79	III	3	0.15
80–84	IIII III	8	0.40
85–89	IIII I	6	0.30
90–94	II	2	0.10
95–99	I	1	0.05
Total		20	1.00

Now You Can Do
Exercises 15 and 16.

2 HISTOGRAMS AND FREQUENCY POLYGONS

Histograms

There are many different methods of summarizing numeric data graphically. One example of a graphical summary for quantitative data is a **histogram.**

> A **histogram** is constructed using rectangles for each class of data. The heights of the rectangles represent the frequencies or relative frequencies of the class. The widths of the rectangles represent the class widths of the corresponding frequency distribution. The class boundaries are placed on the horizontal axis, so that the rectangles are touching each other.

EXAMPLE 2.11

CONSTRUCTING A HISTOGRAM

Construct a histogram of the frequency of the management aptitude test scores from Example 2.10.

Solution

STEP 1 **Find the class limits and draw the horizontal axis.**
Note that the class boundaries for these data were found in Example 2.10: 74.5, 79.5, 84.5, 89.5, 94.5, and 99.5. Draw the horizontal axis, with the numbers 74.5, 79.5, 84.5, 89.5, 94.5, and 99.5, equally spaced along it. The numbers indicate where the rectangles will touch each other.

STEP 2 **Determine the frequencies and draw the vertical axis.**
Use the frequencies given in Table 2.16. These will indicate the heights of the five rectangles along the vertical axis. Find the largest frequency, which is 8. It is a good idea to provide a little bit of extra vertical space above the tallest rectangle, so make 9 your highest label along the vertical axis. Then provide equally spaced labels along the vertical axis between 0 and 8.

STEP 3 **Draw the rectangles.**
Draw your first rectangle from 74.5 to 79.5, with height 3, the first frequency. Draw the remaining rectangles similarly. The resulting frequency histogram is shown in Figure 2.13a. The relative frequency histogram is shown in Figure 2.13b. Note that the two histograms have identical shapes and differ only in the labeling along the vertical axis.

**Now You Can Do
Exercises 17 and 18.**

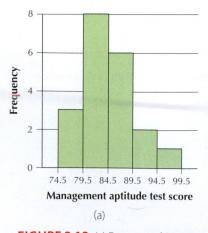

(a)

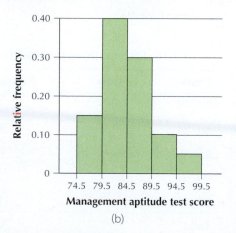

(b)

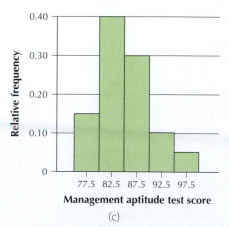

(c)

FIGURE 2.13 (a) Frequency histogram; (b) relative frequency histogram; (c) histogram using midpoints.

Note: Histograms are often presented using class midpoints rather than class boundaries. The class boundaries can be inferred by splitting the difference between the class midpoints. In Figure 2.13c, the upper class boundary for the leftmost class is halfway between 77.5 and 82.5, that is, 80. Otherwise, Figure 2.13c is equivalent to Figure 2.13b.

Note that the histogram, unlike the frequency distribution, provides us with a graphical impression of the data distribution. This characteristic will be crucial later on, when we evaluate the fitness of data sets to undergo certain data analysis methods. Also, notice that the rectangles are contiguous (touching), unlike the rectangles of the bar graphs in Section 2.1. Since the data are quantitative, the horizontal axis in a histogram should be considered as the number line. A **class midpoint** is the average of two consecutive lower class limits. For example, the class midpoint for the leftmost class in Figure 2.13c is $(75 + 80)/2 = 77.5$.

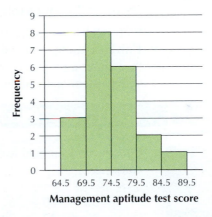

WHAT IF?

Shifting the Histogram to the Left

What if we subtracted ten points from each management aptitude test score; how would that affect the frequency histogram in Figure 2.13a? Assume that the number of classes and the class width would stay the same.

Solution

The new class limits and class boundaries would each be ten points lower than the corresponding class limits and class boundaries from Example 2.11. However, the frequencies for each corresponding class would be the same as those from Example 2.11. Thus, the rectangles would look the same, the only difference being that they are "shifted left" ten points along the number line. We discuss more about the shapes of histograms later in this section.

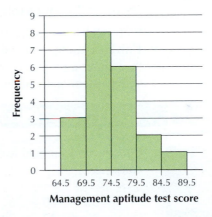

FIGURE 2.13A "Shape" of histogram is unchanged.

APPLET

The *One-Variable Statistical Calculator* applet can display histograms for a selection of data sets in this textbook, including the management aptitude test scores. The applet allows you to experiment with different class widths.

Frequency Polygons

Frequency polygons provide the same information as histograms, but in a slightly different format.

> A **frequency polygon** is constructed as follows. For each class, plot a point at the class midpoint, at a height equal to the frequency for that class. Then join each consecutive pair of points with a line segment.

EXAMPLE 2.12

CONSTRUCTING A FREQUENCY POLYGON

mgmttest

Construct a frequency polygon for the management aptitude test data in Example 2.10.

Solution

The midpoints for the classes were calculated for Figure 2.13c. Plot a point for each frequency above each midpoint, and join consecutive points. The result is the frequency polygon in Figure 2.14.

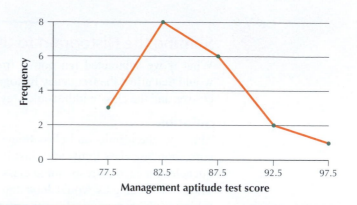

FIGURE 2.14
Frequency polygon.

**Now You Can Do
Exercises 19 and 26.**

---

3 STEM-AND-LEAF DISPLAYS AND DOTPLOTS

Stem-and-Leaf Displays

Stem-and-leaf displays were developed by Professor John Tukey of Princeton University in the late 1960s. This type of display generally contains more information than either a frequency distribution or a histogram. We will demonstrate how to construct a stem-and-leaf display in Example 2.13.

EXAMPLE 2.13

CONSTRUCTING A STEM-AND-LEAF DISPLAY

John Tukey, a statistician, who developed the stem-and-leaf display, is said to have coined the term "software."

Construct a stem-and-leaf display for the exam scores of 20 statistics students, given below:

| 57 | 60 | 61 | 65 | 69 | 73 | 74 | 75 | 75 | 75 |
| 76 | 77 | 78 | 81 | 82 | 82 | 85 | 91 | 95 | 98 |

Solution

First, find the leading digits of the numbers. Each number has one of the following as its leading digit: 5, 6, 7, 8, 9. Place these five numbers, called the **stems,** in a column:

stems

```
5
6
7
8
9
```

Each number represents the tens place of the exam scores. For example, 5 represents 5 tens. Now consider the ones place of each data value. For example, the first score, 57, has 5 in the tens place (the stem) and 7 in the ones place. Place this number, called the **leaf,** next to its stem:

```
        5 | 7
stem    6 |      leaf
        7 |
        8 |
        9 |
```

The second score, 60, has 6 in the tens place and 0 in the ones place, and the third score, 61, has 6 in the tens place and 1 in the ones place. Write the leaves 0 and 1 next to the stem 6:

```
5 | 7
6 | 01
7 |
8 |
9 |
```

Continue this process with the remaining data, placing each ones value next to its stem. Then, for each stem, order the leaves from left to right in increasing order. This produces the stem-and-leaf display:

```
5 | 7
6 | 0159
7 | 34555678
8 | 1225
9 | 158
```

Notice that the three 75s refer to three different students who happened to get the same grade on the exam. In general, the leaf units represent the smallest decimal place represented in the data values. Then the stem unit consists of the remainder of the number. For example, suppose we have a data value of 127. Then the 7 is the leaf unit, and the 12 is the stem. Or else, suppose our data value is 0.146. Then our leaf unit is the 6 and the stem is the 14. Note that the stem-and-leaf display contains all the information that a histogram turned on its side does. But it also contains more information than a histogram, because the stem-and-leaf display shows the original values.

Split stems may sometimes be used in a stem-and-leaf display to provide a clearer idea of the data distribution when too many data points fall on just a few stems. When using split stems, each stem appears twice, with the leaves 0 to 4 on the upper stem and the leaves 5 to 9 on the lower stem. The above stem-and-leaf display of statistics exam scores would appear as follows when using splits stems:

```
5 | 7
5 |
6 | 01
6 | 59
7 | 34
7 | 555678
8 | 122
8 | 5
9 | 1
9 | 58
```

Now You Can Do
Exercises 20 and 27.

The *One Variable Statistics and Graphs* applet can display stem-and-leaf displays for a selection of data sets in this textbook, including the statistics exam scores. The applet allows you to experiment with split stems if you like.

Dotplots

A simple but effective graphical display is a **dotplot.** In a dotplot, each data point is represented by a dot above the number line. When the sample size is large, each dot may represent more than one data point. Figure 2.15 is a dotplot of the 20 management aptitude test scores.

FIGURE 2.15

Dotplot of the managerial aptitude (MAT) test scores. The two dots above 87 indicate that two tests had the same score of 87. Which test score was the most common?

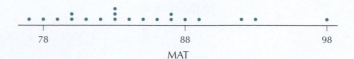

Dotplots are useful for comparing two variables. For example, suppose that an instructor taught two different sections of a management course and gave a simulated management aptitude exam in each section (MAT-1 and MAT-2). The instructor could then compare these two groups of scores directly, using a Minitab comparison dotplot, as in Figure 2.16. Although there is much overlap, Section 1 had the highest score, while Section 2 had the three lowest scores. Therefore, it looks as if Section 1 might have done better.

FIGURE 2.16

Comparison dotplot of MAT test scores for the two sections. Note that the two sections are graphed using the same number line, which makes comparison easier.

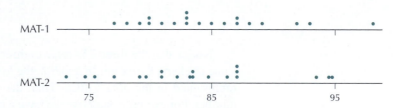

4 DISTRIBUTION SHAPE, SYMMETRY, AND SKEWNESS

Frequency distributions are tabular summaries of the set of values that a variable takes. We now generalize the concept of **distribution.**

> The **distribution** of a variable is a table, graph, or formula that identifies the variable values and frequencies for all elements in the data set.

For example, a frequency distribution is a distribution since it is a table that specifies each of the values that a variable can take, along with the frequencies. However, our definition of "distribution" also includes histograms, stem-and-leaf displays, dotplots, and other graphical summaries. (In Chapter 6, we will introduce distributions defined by formulas.) These graphical distributions invite us to consider the shape of a distribution. The *shape* of a distribution is the overall form of a graphical summary, approximated by a smooth curve.

The Bell-Shaped Curve

Figure 2.17 contains the relative frequency histogram of the heights of 1000 college women. Note that there are relatively fewer women in both the left-hand tail (shorter women) and the right-hand tail (taller women). Instead, as height increases from left to right, the relative frequency gradually increases until it reaches a peak near 65 inches tall and then gradually decreases. Thus, the distribution of heights is said to be *bell-shaped.*

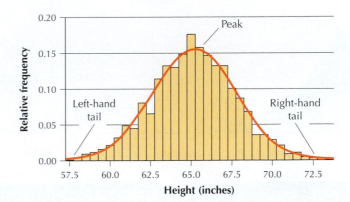

FIGURE 2.17
The bell-shaped curve superimposed on a histogram.

The rectangles represent the actual data. However, the smoothed curve represents an approximation of the overall form of the distribution, and thus the smoothed curve represents the shape of the distribution, which is bell-shaped. The formal name of this bell-shaped distribution is the *normal* distribution. In Chapter 6 we will learn much more about this important distribution, which occurs often in nature and the real world. For example, student heights (within a given gender) follow a bell-shaped distribution. In Chapter 7, we will learn how to assess whether or not a particular distribution is normal (bell-shaped). Starting in Chapter 8, many of the methods for statistical inference we will learn depend on this distribution.

Analyzing the Shape of a Distribution

We next learn some tools for analyzing the shape of a distribution. An image has *symmetry* (or is **symmetric**) if there is a line (axis of symmetry) that splits the image in half so that one side is the mirror image of the other. For example, the butterfly in Figure 2.18 has symmetry, since a line drawn down the middle of the butterfly would create two mirror images of each other. It is important to develop the talent for recognizing which distribution shapes are symmetric.

For example, the smoothed curve in Figure 2.17 is perfectly symmetric. However, the histogram rectangles reflecting the actual data are only nearly symmetric, since a vertical line drawn down the middle of the distribution would not result in two perfect mirror images. *Due to random variation, data from the real world rarely exhibit perfect symmetry.* With this in mind, the data analyst is usually content with the approximate symmetry exhibited by the data (the rectangles) in Figure 2.17.

However, not all distributions are symmetric. In Chapter 8 we will discuss a distribution called the chi-square distribution, which is not symmetric but is **skewed.** It often has a longer "tail" on the right than on the left (see Figure 2.19 on the next page). Since the right-hand tail is longer, we say that this distribution is *right-skewed.* Examples of right-skewed data are usually found when dealing with money. For example, if we graph the incomes of the families in your home state, the graph will probably be right-skewed. Most of us will lie somewhere in the middle or left with the bulk of the data, while the incomes of folks like Donald Trump and Bill Gates lie far out on the right of the graph, in the right-hand tail. Figure 2.20 on the next page shows a *left-skewed* distribution. Good examples of left-skewed data are retirement ages or death ages. Often, exam grade data can be left-skewed, as several students bump up against the 100% boundary on the right, most students are somewhere in the middle, and a few students stagger in with 40s and 50s in the left-hand tail.

FIGURE 2.18
This butterfly is symmetric.
© Burke/ Triolo/Jupiterimages

Note: Only quantitative data, not qualitative data, may be described as symmetric or skewed.

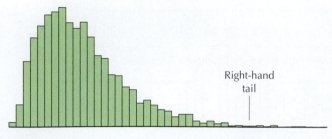

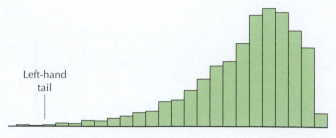

FIGURE 2.19 The chi-square distribution is right-skewed.

FIGURE 2.20 Some distributions are left-skewed.

EXAMPLE 2.14

CHOOSING THE APPROPRIATE GRAPHICAL SUMMARY

Statistically, literate citizens recognize that one may select different graphical summaries, depending on the intention of the presenter. Figures 2.21a, 2.21b, and 2.21c contain a dotplot, a histogram, and a stem-and-leaf display of the average size of households in the 50 states and the District of Columbia. Which graphical summary—the dotplot, the histogram, or the stem-and-leaf display—is most useful if our primary objective is

a. to assess symmetry and skewness?
b. to be able to construct it quickly using paper and pencil?
c. to retain complete knowledge of the original data set?
d. to give a presentation to people who have never had a stats course before?

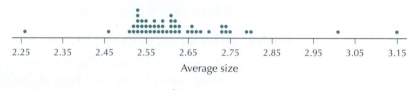

(a)

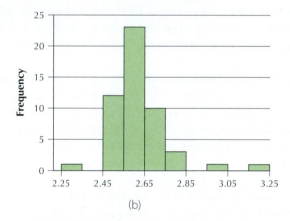

(b)

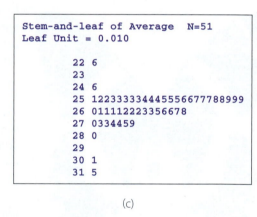

```
Stem-and-leaf of Average    N=51
Leaf Unit = 0.010

         22  6
         23
         24  6
         25  12233333444555667778889 99
         26  011112223356678
         27  0334459
         28  0
         29
         30  1
         31  5
```

(c)

FIGURE 2.21 (a) Dotplot; (b) histogram; (c) stem-and-leaf display. Which is most useful?

Solution

a. All three graphics are good at assessing symmetry and skewness.
b. The dotplot's great asset is its simplicity. It can be quickly drawn, with minimal preparation, in contrast to the other two summaries, which require some organization or calculation.
c. The stem-and-leaf display was invented in order to retain complete knowledge of the data set. Histograms are the least effective in this regard.
d. The histogram is widely used in the real world and is probably the best choice for a presentation in front of those who have not had a stats course before.

Now You Can Do
Exercise 36.

STEP-BY-STEP TECHNOLOGY GUIDE: Quantitative Data

Suppose we would like to produce a histogram of the management aptitude test scores from Example 2.10 (pages 52–53). 🔴 mgmttest

TI-83/84

Entering a Data Set
Step 1   Press **STAT**, then press **ENTER**. Highlight the **L1** list.
Step 2   Clear out any old data in **L1.** Press the **up arrow** key, then **CLEAR**, then **ENTER**.
Step 3   Enter the first data value **77** and press **ENTER**.
Step 4   Continue entering data until the entire data set is in **L1** (Figure 2.22).

Constructing a Histogram
Step 1   Press **2nd**, then Y=. In the STAT PLOTS menu, select **1**, and press **ENTER**.
Step 2   Select **ON**, and press **ENTER**. Select the histogram icon (Figure 2.23), and press **ENTER**.
Step 3   Press **ZOOM**, then select **9:ZOOMSTAT**.
Step 4   Press **TRACE**. Selecting each class in turn provides class limits and class frequency. The histogram is given in Figure 2.24.

FIGURE 2.22 All data entered.

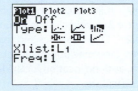

FIGURE 2.23 Selecting the histogram icon.

FIGURE 2.24 Histogram with leftmost class selected.

EXCEL

Constructing a Histogram
Make sure the Data Analysis package has been installed on your version of Excel.

Step 1   Click **Data > Data Analysis.**

Step 2   Select **Histogram** and click **OK.**
Step 3   For the *input range*, select the cells in which the data set resides. Then click **OK.**

MINITAB

Constructing a Histogram
Step 1   Enter the management aptitude test scores into column **C1**.
Step 2   Click **Graph** > **Histogram.**
Step 3   In the **Graph Variables** section, choose **Simple** and click **OK.** Select **C1 Scores**, and click **Select**. Then click **OK.**
Step 4   The histogram is shown in Figure 2.25. Note that by default Minitab uses midpoints rather than class limits to define the classes. Double-clicking anywhere on the midpoint values (78, 81, . . .) brings up a dialog box providing a wide range of options for changing the number of classes, class limits, etc.

Constructing a Stem-and-Leaf Display
Step 1   Enter the management aptitude test scores into column **C1**.
Step 2   Click **Graph** > **Stem-and-Leaf.**
Step 3   Click inside the space indicated **Variables**, select **C1 Scores**, and click **Select**. Then click **OK.**
Step 4   The output shown in Figure 2.26 tells us that the leaf unit is defined to be ones (1.0). Therefore, the stem unit is tens. (Ignore the leftmost column, which simply provides a cumulative count of the data points from the minimum and maximum.) The first row shows 7 7, indicating a single data point, 77. The second row shows 7 89, indicating two data points, 78 and 79.

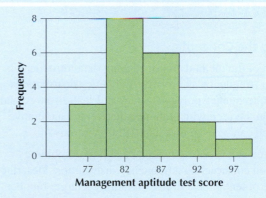

FIGURE 2.25 Minitab histogram.

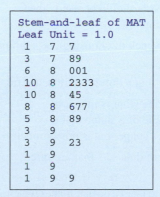

```
Stem-and-leaf of MAT
Leaf Unit = 1.0
  1    7   7
  3    7   89
  6    8   001
 10    8   2333
 10    8   45
  8    8   677
  5    8   89
  3    9
  3    9   23
  1    9
  1    9
  1    9   9
```

FIGURE 2.26 Minitab stem-and-leaf display.

Dotplots
Step 1 Enter the management aptitude test scores into column **C1**.

Step 2 Click **Graph . . . Dotplot**. Select **Simple**.
Step 3 In the **Graph Variables** section, select **C1** *Scores* and click **OK**.

CRUNCHIT!

Constructing a Histogram
Step 1 Click **File . . .** then highlight **Load from Larose2e . . . Chapter 2 . . .** and click on **Example 2.10**.
Step 2 Click **Graphics** and select **Histogram**. For **Sample** select **Scores**. (You may optionally select the number of bins, the bin width, and the location for the leftmost lower class limit.) Then click **Calculate**.

Constructing a Dotplot
Step 1 Click **File . . .** then highlight **Load from Larose2e . . . Chapter 2 . . .** and click on **Example 2.10**.
Step 2 Click **Graphics** and select **Dot Plot**. For **Sample** select **Scores**. Then click **Calculate**.

SECTION 2.2 Summary

In this section, we learned about using graphs and tables for summarizing quantitative (numerical) data.

1. Quantitative variables can be summarized using frequency and relative frequency distributions.

2. Histograms are a graphical display of a frequency or a relative frequency distribution with class intervals on the horizontal axis and the frequencies or relative frequencies on the vertical axis. A frequency polygon is constructed as follows: for each class, plot a point at the class midpoint, at a height equal to the frequency for that class; then join each consecutive pair of points with a line segment.

3. Stem-and-leaf displays contain more information than either a frequency distribution or a histogram, since they retain the original data values in the display. In a dotplot, each data point is represented by a dot above the number line.

4. An image or distribution has symmetry (or is symmetric) if there is a line (axis of symmetry) that splits the image in half so that one side is the mirror image of the other. Nonsymmetric distributions with a long right-hand tail are called right-skewed, while those with a long left-hand tail are called left-skewed.

SECTION 2.2 Exercises

Clarifying the Concepts

1. Which of the methods for displaying data introduced in this section (frequency and relative frequency distributions, histograms, frequency polygons, stem-and-leaf displays, and dotplots) can be used with both quantitative and qualitative data? Which can be used for quantitative data only?

2. Describe at least one potential benefit of combining classes when constructing a frequency distribution. Describe at least one potential benefit from retaining a larger number of classes.

3. In general, how many classes should be used when constructing a frequency distribution?

4. Describe at least one drawback of choosing class limits that overlap.

5. Describe at least one way that a dotplot may be useful.

6. In your own words, describe what is meant by "symmetry." Provide an example of a shape that is symmetric and an example of a shape that is not symmetric.

7. What are some examples of data sets that are often right-skewed? Left-skewed?

8. For a bar graph (not a Pareto chart), does it matter which order the bars are in? What does this mean for the relevance of symmetry and skewness for summaries of categorical data (such as we studied in Section 2.1)?

Practicing the Techniques

The following discrete data represent the number of game consoles owned by a random sample of college students. Use the data to construct the table or graph indicated in Exercises 9 and 10.

1	2	0	1	0	2	1	1	0	2	0	1
1	1	0	2	0	0	1	0	2	0	1	1

9. Frequency distribution

10. Relative frequency distribution

The following discrete data represent the ages of a random sample of college students. Use the data to construct the table or graph indicated in Exercises 11–14.

18	21	21	19	20	20	21	22	20
20	19	19	20	18	22	20	21	19

11. Frequency distribution

12. Relative frequency distribution

13. Define the following classes: 18–19, 20–21, and 22–23. Use these classes to construct a frequency distribution.

14. Using the classes in the previous exercise, construct a relative frequency distribution.

The following continuous data represent the pulse rates of a random sample of women. Use the data to construct the table or graph indicated in Exercises 15–21. 🔴 womenpulse

75 69 73 84 82 80 74 83 77 78 61 78 87 79 65 72 69 81 62 69

15. Frequency distribution

16. Relative frequency distribution

17. Frequency histogram

18. Relative frequency histogram

19. Frequency polygon

20. Stem-and-leaf display

21. Dotplot

The following continuous data represent the grades on a statistics quiz for a random sample of students. Use the data to construct the table or graph indicated in Exercises 22–28.
🔴 quizgrades

95 85 77 82 65 72 76 92 80 74 69 62 79 87 75 75 94 69 70 72

22. Frequency distribution

23. Relative frequency distribution

24. Frequency histogram

25. Relative frequency histogram

26. Frequency polygon

27. Stem-and-leaf display

28. Dotplot

Applying the Concepts

29. Die Roll. A fair die was thrown 100 times, and the values were recorded. The accompanying histogram shows the results.

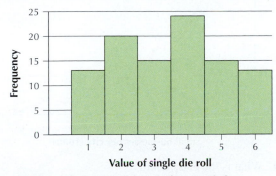

a. Which value occurred most frequently?

b. Which values occurred least frequently?

c. How often was a 3 observed?

d. What percentage of times was a 3 observed?

30. Police Citations. A random sample of 1000 police officers was taken, and the number of motor vehicle

citations each handed out in a particular week was recorded. The results are shown in the accompanying histogram.

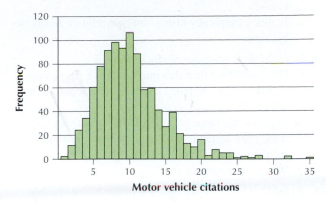

a. What was the greatest number of citations issued?

b. What was the fewest number of citations issued?

c. What was the most frequent number of citations issued? About how many police officers issued this many citations?

d. Describe the shape of the distribution.

31. Statistics Midterm Scores. A campus-wide statistics midterm worth 50 points resulted in the scores provided in the histogram below.

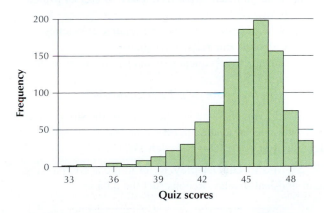

a. Which score occurred with the greatest frequency?

b. Which score occurred with the lowest frequency?

c. What is the highest score? Lowest score?

d. Describe the shape of the distribution.

32. Stock Prices. A portfolio contains stocks of 19 technology firms. The stock prices are shown in the accompanying histogram.

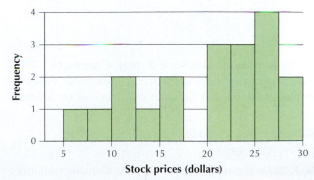

a. How many classes are there?
b. What is the class width? Is it the same for each class?
c. Is this a frequency histogram or a relative frequency histogram?

33. Refer to the histogram of stock prices.
a. How could we turn this into a relative frequency histogram? Would the classes or the rectangles be affected?
b. Suppose we were given a relative frequency histogram instead. How could we turn it into a frequency histogram?
c. What is the sample size?

34. Refer to the histogram of stock prices.
a. How many stocks were priced above $27.50?
b. What is the relative frequency of stocks priced above $27.50?
c. How many stocks had a price below $15?
d. What is the relative frequency of stocks with a price below $15?

35. Refer to the histogram of stock prices.
a. How many stocks are priced between $17.50 and $20?
b. What is the relative frequency of stocks priced below $5?
c. Which class has the largest relative frequency? Calculate this relative frequency.
d. What is the frequency of stocks priced between $10 and $15?
e. How many stocks had a price of $40?

36. Would you characterize the shape of the stock prices distribution as (a) tending to be symmetric, (b) tending to be right-skewed, (c) tending to be left-skewed?

37. Stem-and-Leaf Display. Refer to the accompanying stem-and-leaf display. Reconstruct the data set.

```
Stem-and-leaf of Data   N  = 20
Leaf Unit = 1.0

   2   3
   2   45
   2   67
   2   889
   3   011
   3   2223
   3   5
   3   67
   3   9
   4   0
```

38. Refer to the stem-and-leaf display. Construct a relative frequency distribution, using appropriate values for the class width and the lower class limit of the leftmost class.

39. Refer to the stem-and-leaf display. Construct a frequency histogram.

40. Refer to the stem-and-leaf display. Construct a dotplot.

41. Frequency Polygon. The following frequency polygon represents the quiz scores for a course in introductory statistics.

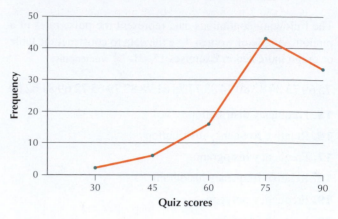

a. What is the class width?
b. What is the lower class limit of the class that has 45 as its midpoint?
c. What is the upper class limit of the class that has 45 as its midpoint?
d. Which class has the highest frequency?
e. Which class has the lowest frequency?

42. Refer to the frequency polygon of quiz scores.
a. About how many students scored higher than 82.5?
b. About how many students scored lower than 52.5?
c. Can we say how many students scored in the 90s? Why or why not?

43. Small Businesses. The U.S. Census Bureau tracks the number of small businesses per city. The accompanying frequency polygon represents the numbers of small businesses per city (in thousands) for 266 cities nationwide.

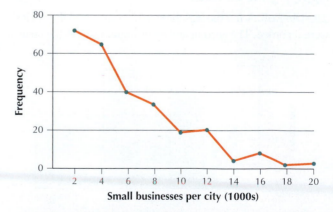

a. What is the class width?
b. What is the lower class limit of the leftmost class? (*Hint:* Don't forget about the units.)
c. Which class has the highest frequency?
d. Which class has the lowest frequency?

44. Refer to the frequency polygon of small businesses per city.

a. About how many cities have between 1000 and 3000 small businesses?

b. About how many cities have more than 19,000 small businesses?

c. About how many cities have between 9000 and 11,000 small businesses?

Miami Arrests. Answer Exercises 45–48 using the information in the following table. The table gives the monthly number of arrests made for the year 2005 by the Miami-Dade Police Department.

 miamiarrests

Jan.	751	May	919	Sept.	802
Feb.	650	June	800	Oct.	636
Mar.	909	July	834	Nov.	579
Apr.	881	Aug.	789	Dec.	777

45. Construct a relative frequency distribution of the monthly number of arrests. Use class width of 50 arrests, with the lower class limit of the leftmost class equal to 550.

46. Construct a frequency histogram and relative frequency histogram, using the same classes as in the previous exercise. Which class or classes have the highest frequency? Lowest?

47. Construct a dotplot.

48. Construct a frequency polygon.

Bringing It All Together

Statistics Exam Data. Use the following statistics exam data set from Example 2.13 for Exercises 49–52. 🔴 statsexam

57	60	61	65	69	73	74	75	75	75
76	77	78	81	82	82	85	91	95	98

49. Without using a computer, construct the following:
a. A frequency distribution
b. A relative frequency distribution
c. A relative frequency histogram

50. Without using a computer, construct the following:
a. A dotplot
b. A frequency polygon
c. Stem-and-leaf display

51. Compare and contrast the relative usefulness of each of four graphical presentation methods—dotplot, histogram, stem-and-leaf display, and frequency polygon—if our primary objective is
a. to assess symmetry and skewness.
b. to be able to construct it quickly using paper and pencil.
c. to retain complete knowledge of the data set.
d. to give a presentation to people who have never had a stats course before.

52. *What if* we subtract the same amount (say, 10) from each statistics exam score. Explain how this would affect the following. What would change? What would stay the same?
a. Relative frequency histogram
b. Dotplot
c. Stem-and-leaf display
d. Frequency polygon

Fats and Cholesterol. For Exercises 53–57, use your knowledge of Excel or Minitab. Open the **Nutrition** data set.
Nutrition

53. How many observations are there in the data set? How many variables?

54. The variable *fat* contains the fat content in grams for each food. Construct a histogram of *fat*. Comment on the symmetry or the skewness of the histogram.

55. Is there a particular type of food whose fat content is particularly large? Which type of food item is this (actually, a set of similar food items)?

56. The variable *cholesterol* contains the cholesterol content in milligrams for each food. Construct a histogram of *cholesterol*. Comment on the symmetry or the skewness of the histogram.

57. Which food item is highest in cholesterol?

Use the *One Variable Statistics and Graphs* applet for Exercises 58–60. Work with the **Earthquakes** data set, which shows the magnitude on the Richter scale of 57 earthquakes that occurred during the week of October 15–22, 2007.
Earthquakes

58. Click on the **Histogram** tab.
a. How many classes are there in the histogram?
b. What is the class width?

59. Click on the leftmost rectangle in the histogram.
a. What is the frequency for this class?
b. What are the lower and upper class limits?

60. Click on the number line and drag slowly all the way to the left.
a. What happens to the number of classes as you drag to the left?
b. What happens to the class widths as you drag to the left?

Construct Your Own Data Sets

61. Construct your own right-skewed data set of about 20 values. Just make up the data points, but be sure you know what the data represent (income, housing costs, etc.).
a. Construct a stem-and-leaf display of your data set.
b. Construct a dotplot of your data set.

62. Construct your own symmetric data set of about 20 values. Just make up the data points, but be sure you know what the data represent (for example, runs in a baseball game, number of right answers on a quiz).
a. Construct a stem-and-leaf display of your data set.
b. Construct a dotplot of your data set.

2.3 FURTHER GRAPHS AND TABLES FOR QUANTITATIVE DATA

OBJECTIVES   By the end of this section, I will be able to . . .

1   Build cumulative frequency distributions and cumulative relative frequency distributions.

2   Create frequency ogives and relative frequency ogives.

3   Construct and interpret time series graphs.

1   CUMULATIVE FREQUENCY DISTRIBUTIONS AND CUMULATIVE RELATIVE FREQUENCY DISTRIBUTIONS

Since quantitative data can be put in ascending order, we can keep track of the accumulated counts at or below a certain value using a **cumulative frequency distribution** or **cumulative relative frequency distribution.** For example, if we list the prices of homes for sale in a neighborhood, a cumulative frequency distribution tells us how many homes are priced at $300,000 or less.

For a discrete variable, a **cumulative frequency distribution** shows the total number of observations *less than or equal to* the category value. For a continuous variable, a **cumulative frequency distribution** shows the total number of observations *less than or equal to* the upper class limit.

A **cumulative relative frequency distribution** shows the proportion of observations less than or equal to the category value (for a discrete variable) or the proportion of observations less than or equal to the upper class limit (for a continuous variable).

EXAMPLE 2.15   CONSTRUCTING CUMULATIVE FREQUENCY AND CUMULATIVE RELATIVE FREQUENCY DISTRIBUTIONS

The first three columns in Table 2.17 contain the frequency distribution and relative frequency distribution for the total 2007 attendance for 25 Major League Baseball teams. Construct a cumulative frequency distribution and a cumulative relative frequency distribution for the attendance figures.

Solution

To find the cumulative frequency for a class, add the frequencies of the classes equal to or below the upper class limit of that class. For example, the cumulative frequency for the class 2.70–3.09 is the sum of the frequency for this class and the frequencies for the classes 1.90–2.29 and 2.30–2.69. The procedure for the cumulative relative frequencies is similar. The results are shown in the last two columns of Table 2.17, where we can see that more than two-thirds (0.68) of these teams had attendance of 3.09 million or less.

Now You Can Do Exercises 7 and 8.

Table 2.17 Cumulative frequency distribution and cumulative relative frequency distribution

Attendance (millions)	Frequency	Relative frequency	Cumulative frequency	Cumulative relative frequency
1.90–2.29	5	0.20	5	0.20
2.30–2.69	6	0.24	5 + 6 = 11	0.20 + 0.24 = 0.44
2.70–3.09	6	0.24	5 + 6 + 6 = 17	0.44 + 0.24 = 0.68
3.10–3.49	4	0.16	5 + 6 + 6 + 4 = 21	0.68 + 0.16 = 0.84
3.50–3.89	3	0.12	5 + 6 + 6 + 4 + 3 = 24	0.84 + 0.12 = 0.96
3.90–4.29	1	0.04	5 + 6 + 6 + 4 + 3 + 1 = 25	0.96 + 0.04 = 1.00
Total	25	1.00		

2 OGIVES

Just as histograms and frequency polygons are the graphical equivalent of frequency distributions, we have the following graphical equivalent of a cumulative frequency distribution.

> An **ogive** (pronounced "oh jive") is the graphical equivalent of a cumulative frequency distribution or a cumulative relative frequency distribution. Like a frequency polygon, an ogive consists of a set of plotted points connected by line segments. The x coordinates of these points are the upper class limits; the y coordinates are the cumulative frequencies or cumulative relative frequencies.

EXAMPLE 2.16

CONSTRUCTING AN OGIVE

bballattend

Construct a relative frequency **ogive** for the attendance data in Table 2.17.

Solution

For the x coordinates, we use the upper class limits for attendance, and for the y coordinates, we use the cumulative relative frequencies. The result is shown in Figure 2.27.

FIGURE 2.27
Ogive for baseball attendance.

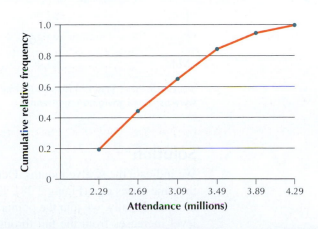

Now You Can Do
Exercises 9 and 10.

<table>
<tr><td>**What Does This Graph Mean?**</td><td>The ogive is a graphical representation of a cumulative relative frequency distribution. Thus, the first point (2.29, 0.2) indicates that 20% of the teams had total attendance at or below 2.29 million. The cumulative nature of the graph means that it can never decrease from left to right. The cumulative attendance increases until the rightmost point (4.29, 1.0) indicates that 100% (all) of the teams had total attendance at or below 4.29 million.</td></tr>
</table>

3 TIME SERIES GRAPHS

Data analysts are often interested in how the value of a variable changes over time. Data that are analyzed with respect to time are called *time series data*.

> A graph of time series data is called a **time series plot.** The horizontal axis of a time series plot represents time (for example, hours, days, months, years). The values of the time series data are plotted on the vertical axis, and line segments are drawn to connect the points.

EXAMPLE 2.17

CONSTRUCTING A TIME SERIES PLOT

Table 2.18 contains the amount of carbon dioxide in parts per million (ppm) found in the atmosphere above Mauna Loa in Hawaii, measured monthly from October 2006 to September 2007. Construct a **time series plot** of these data.

MaunaLoaBrief

Table 2.18 Atmospheric carbon dioxide at Mauna Loa, October 2006 to September 2007

Month	Carbon dioxide (ppm)	Month	Carbon dioxide (ppm)
Oct.	379.03	Apr.	386.37
Nov.	380.17	May	386.54
Dec.	381.85	June	385.98
Jan.	382.94	July	384.35
Feb.	383.86	Aug.	381.85
Mar.	384.49	Sept.	380.58

Source: Dr. Pieter Tans, Earth System Research Laboratory, National Oceanic and Atmospheric Administration, www.esrl.noaa.gov/gmd/ccgg/trends.

Solution

We indicate the twelve months, October through September, on the horizontal axis of the time series plot (Figure 2.28). Then, for each month, we plot the amount of carbon dioxide. Finally, we join the points using line segments. Note that the carbon dioxide level increases from the fall through the winter and peaks in the spring. It then decreases through the summer. In the Step-by-Step Technology Guide, we illustrate how to construct this time series graph using technology.

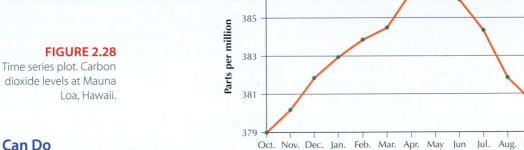

FIGURE 2.28

Time series plot. Carbon dioxide levels at Mauna Loa, Hawaii.

Now You Can Do Exercises 31 and 32.

| EXAMPLE 2.18 | **CONSTRUCTING A TIME SERIES PLOT USING TECHNOLOGY** |

 MaunaLoa

The data set **Mauna Loa** contains the carbon dioxide levels at Mauna Loa from September 1999 to September 2007. Use technology to construct a time series plot of the data.

Solution

We use the instructions provided in the Step-by-Step Technology Guide at the end of this section. The resulting time series plot is shown in Figure 2.29. (The year on the horizontal axis indicates September of each year. For example "1999" refers to September 1999.)

In Figure 2.29 we observe both a seasonal pattern and a long-term trend. Every autumn and winter, the carbon dioxide level increases, and every summer it decreases. In autumn and winter, leaves and other deciduous vegetation decays, releasing its store of carbon back into the atmosphere. In the spring and summer, the new year's leaves require carbon to grow and extract it from the atmosphere, thereby reducing the atmosphere's carbon dioxide level. Thus, the Earth "inhales" carbon each summer and "exhales" it each winter. However, the carbon dioxide level of each successive September does not quite reach the low level of the previous September. This leads to an overall increasing trend in the amount of carbon dioxide in the atmosphere as we move from 1999 to 2007.

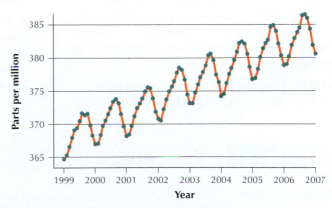

FIGURE 2.29 Watching the Earth breathe. Carbon dioxide levels at Mauna Loa, Hawaii.

STEP-BY-STEP TECHNOLOGY GUIDE: Time Series Plots

We illustrate how to construct a time series plot using Example 2.18 (page 69).

TI-83/84

Step 1 Enter your time index (integers 1, 2, . . .) into list **L1.**
Step 2 Enter the values of your time series variable into list **L2.**
Step 3 Press **2nd**, then **Y=**. In the STAT PLOTS menu, select **1**, and press **ENTER.**
Step 4 Select **ON**, and press **ENTER**. Select the time series icon (Figure 2.30), and press **ENTER.**
Step 5 Press **ZOOM** > **9:ZOOMSTAT** and press **ENTER**. The time series plot is shown in Figure 2.31.

FIGURE 2.30 Selecting the time series icon.

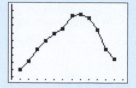

FIGURE 2.31 TI-83/84 time series plot.

EXCEL

Step 1 Enter the month data into column **A** (see Figure 2.32).
Step 2 Enter the values of your time series variable into column **B** (see Figure 2.32).
Step 3 Select cells A1–B12 and click **Insert** > **Line** (in the **Chart** section).
Step 4 Choose the type labeled "Line with markers."

	A	B
1	Oct	379.03
2	Nov	380.17
3	Dec	381.85
4	Jan	382.94
5	Feb	383.86
6	Mar	384.49
7	Apr	386.37
8	May	386.54
9	Jun	385.98
10	Jul	384.35
11	Aug	381.85
12	Sep	380.58

FIGURE 2.32

MINITAB

Step 1 Enter the values of your time series variable into column **C1.**
Step 2 Click **Graph** > **Time Series Plot . . .**
Step 3 Select **Simple** and click **OK.**
Step 4 For **Series,** double-click on **C1.**

Step 5 Click **Time/Scale**. Select **Calendar** > **Month.**
Step 6 For **Start value**, enter 10 (for October). For **Increment**, enter 1.
Step 7 Click **OK** and **OK.**

SECTION 2.3 Summary

1. A cumulative frequency distribution shows the total number of observations less than or equal to the category value (for a discrete variable) or the upper class limit (for a continuous variable). A cumulative relative frequency distribution shows the proportion of observations less than or equal to the category value (for a discrete variable) or the upper class limit (for a continuous variable).

2. An ogive is the graphical equivalent of a cumulative frequency distribution or a cumulative relative frequency distribution. The x coordinates of the points are the upper class limits; the y coordinates are the cumulative frequencies or cumulative relative frequencies.

3. Data that are analyzed with respect to time are called time series data. A graph of time series data is called a time series plot. The horizontal axis of a time series plot represents time (for example, hours, days, months, years). The values of the time series data are plotted on the vertical axis, and line segments are drawn to connect the points.

SECTION 2.3 Exercises

Clarifying the Concepts

1. Explain the difference between a frequency distribution and a cumulative frequency distribution.

2. Explain the difference between a cumulative frequency distribution and a cumulative relative frequency distribution.

3. What is the graphical equivalent of a cumulative frequency distribution?

4. Explain how to construct an ogive.

5. What do we call data that are analyzed with respect to time?

6. Explain how to construct a time series plot.

Practicing the Techniques

For Exercises 7–10, use the following relative frequency distribution of the age of students in a particular section of introductory statistics to construct the following graphical summaries of the variable *age*.

Age	Frequency	Relative frequency
17.0–18.9	4	0.2
19.0–20.9	10	0.5
21.0–22.9	6	0.3

7. Cumulative frequency distribution

8. Cumulative relative frequency distribution

9. Frequency ogive

10. Relative frequency ogive

For Exercises 11–14, use the following relative frequency distribution of the height of students in a particular section of introductory statistics to construct the following graphical summaries of the variable *height*.

Height (inches)	Frequency	Relative frequency
60.0–63.9	3	0.12
64.0–67.9	10	0.40
68.0–71.9	10	0.40
72.0–75.9	2	0.08

11. Cumulative frequency distribution

12. Cumulative relative frequency distribution

13. Frequency ogive

14. Relative frequency ogive

For Exercises 15–18, use the histogram from Exercise 29 in Section 2.2 to construct the indicated graph or table.

15. Cumulative frequency distribution

16. Cumulative relative frequency distribution

17. Frequency ogive

18. Relative frequency ogive

For Exercises 19–22, use the histogram from Exercise 32 in Section 2.2 to construct the indicated graph or table.

19. Cumulative frequency distribution

20. Cumulative relative frequency distribution

21. Frequency ogive

22. Relative frequency ogive

For Exercises 23–26, use the frequency distribution from Exercise 15 in Section 2.2 to construct the indicated graph or table.

23. Cumulative frequency distribution

24. Cumulative relative frequency distribution

25. Frequency ogive

26. Relative frequency ogive

For Exercises 27–30, use the frequency distributions from Exercise 22 in Section 2.2 to construct the indicated graph or table.

27. Cumulative frequency distribution

28. Cumulative relative frequency distribution

29. Frequency ogive

30. Relative frequency ogive

31. The following time series data represent the number of songs that Brandon downloaded per month last year, starting in January and ending in December. Construct the time series graph of the data.

🔴 songdownloads

2   5   7   10   8   10   18   20   15   12   10   8

32. The following time series data represent the number of friends that Kaitlyn had on her social networking page last year, starting in January and ending in December. Construct the time series graph of the data.

5   7   8   8   10   12   10   12   15   18   20   25

Applying the Concepts

33. Unemployment Rate. The frequency ogive below represents the unemployment rate (in percentages) for 367 cities nationwide.[3]

a. What is the class width?
b. What is the upper class limit of the leftmost class?
c. What is the class midpoint of the leftmost class?

34. Refer to the frequency ogive of unemployment rates.
   a. About how many cities have unemployment rates 3.99 and below?
   b. About how many cities have unemployment rates 5.59 and below?
   c. About how many cities have unemployment rates 5.6 and above?

Agricultural Exports. For Exercises 35–37, refer to Table 2.19. The table gives the value of agricultural exports (in billions of dollars) from the top 20 U.S. states in 2009.

 agriexports

TABLE 2.19 Agricultural exports (in billions of dollars)

State	Exports	State	Exports
California	12.5	Arkansas	2.6
Iowa	6.5	North Dakota	5.2
Texas	4.7	Ohio	2.7
Illinois	5.5	Florida	2.1
Nebraska	4.8	Wisconsin	2.2
Kansas	4.7	Missouri	2.7
Minnesota	4.3	Georgia	1.8
Washington	3.0	Pennsylvania	1.7
North Carolina	2.9	Michigan	1.6
Indiana	3.1	South Dakota	2.3

Source: U.S. Department of Agriculture.

35. Construct a cumulative frequency distribution of agricultural exports. Start at $0 and use class widths of $2 billion.
   a. How many states have exports of $4 billion or less?
   b. How many states have exports of $6 billion or less?
   c. How many states have exports of more than $6 billion?

36. Construct a cumulative relative frequency distribution of agricultural exports. Start at $0 and use class widths of $2 billion.
   a. What proportion of states have exports of $4 billion or less?
   b. What proportion of states have exports of $6 billion or less?
   c. What proportion of states have exports of more than $6 billion?

37. Use your cumulative relative frequency distribution to construct a relative frequency ogive of agricultural exports.

38. Interest Rates. The following data represent the prime lending rate of interest, as reported by the Federal Reserve, every six months from January 2003 to July 2010.

interestrates

Jan. 2003	4.25	Jan. 2007	8.25
July 2003	4.00	July 2007	8.25
Jan. 2004	4.00	Jan. 2008	6.98
July 2004	4.25	July 2008	5.00
Jan. 2005	5.25	Jan. 2009	3.25
July 2005	6.25	July 2009	3.25
Jan. 2006	7.25	Jan. 2010	3.25
July 2006	8.25	July 2010	3.25

   a. Construct a time series plot of the prime lending rate of interest.
   b. What trend do you see?

39. Rainfall in Fort Lauderdale. The following data represent the total monthly rainfall (in inches) in 2009 in Fort Lauderdale, Florida, as reported by the U.S. Historical Climatology Network.

flrainfall

Jan.	0.35	July	10.12
Feb.	0.35	Aug.	8.18
Mar.	7.09	Sept.	8.22
Apr.	0.73	Oct.	2.95
May	11.24	Nov.	4.63
June	7.58	Dec.	3.49

   a. Construct a time series plot of the data.
   b. Is it wetter in summer or winter in Fort Lauderdale?

40. In Exercise 39, *what if* we add 3 inches to each month's rainfall amount. Describe how this would affect the time series plot. What would change? What would stay the same?

2.4 GRAPHICAL MISREPRESENTATIONS OF DATA

OBJECTIVE By the end of this section, I will be able to . . .

1 Understand what can make a graph misleading, confusing, or deceptive.

In the Information Age, when our world is awash in data, it is important for citizens to understand how graphics may be made misleading, confusing, or deceptive. Such an understanding enhances our statistical literacy and makes us less prone to being deceived by misleading graphics.

Eight Common Methods for Making a Graph Misleading

1. Graphing/selecting an inappropriate statistic.
2. Omitting the zero on the relevant scale.
3. Manipulating the scale.
4. Using two dimensions (area) to emphasize a one-dimensional difference.
5. Careless combination of categories in a bar graph.
6. Inaccuracy in relative lengths of bars in a bar graph.
7. Biased distortion or embellishment.
8. Unclear labeling.

EXAMPLE 2.19

INAPPROPRIATE CHOICE OF STATISTIC

The United Nations Office on Drugs and Crime reports the statistics, given in Table 2.20, on the top five nations in the world ranked by numbers of cars stolen in 2000. The car thieves seem to be preying on cars in the United States, which has endured nearly as many cars stolen as the next four highest countries put together. (See also the bar graph in Figure 2.33.) However, the United States has a much greater population than these other countries. Is it possible that, *per capita* (per person), the car theft rate in the United States is not so bad?

Table 2.20 Total number of cars stolen

Country	Cars stolen
1. United States	1,147,300
2. United Kingdom	338,796
3. Japan	309,638
4. France	301,539
5. Italy	243,890

Table 2.21 Total number of cars stolen per capita

Country	Cars stolen per capita
1. Australia	0.00712
2. Denmark	0.00600
3. United Kingdom	0.00567
4. New Zealand	0.00563
5. Norway	0.00516

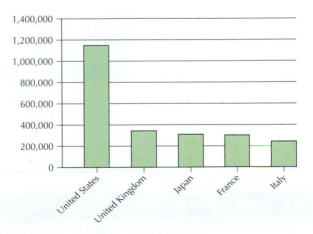

FIGURE 2.33 Bar graph of the top five nations for number of cars stolen in 2000.

Solution

In this case, the total number of cars stolen is an *inappropriate statistic* since the population of the United States is greater than the populations of the other countries. To find the per capita car theft rate, divide the number of cars stolen in a country by that

Now You Can Do
Exercises 3–5.

country's population. The resulting list in Table 2.21 of the top five countries for per capita car theft contains a few surprises. Note that the United States has disappeared from the list. It is found in ninth place, with 0.00409 car thefts per capita.

Developing Your Statistical Sense

Choose the Appropriate Statistic

The bottom line is that *we need to be careful how we use statistics*. Put in an extreme form, "Figures don't lie, but liars figure." One table of statistics tells us the car theft epidemic is striking the United States with special vehemence. The other table asserts the contrary. An American insurance company looking to increase car insurance rates could point to the first table to support its rate request. A citizens group opposing the request could cite the second table. Which table of statistics is true? They both are! We need to be careful how we phrase our research questions and how we choose the type of statistical evidence we use to investigate the research question.

EXAMPLE 2.20

OMITTING THE ZERO

MediaMatters.com reported that **CNN.com** used a misleading graph, reproduced here as Figure 2.34, to exaggerate the difference between the percentages of Democrats and Republicans who agreed with the Florida court's decision to remove the feeding tube from Terri Schiavo in 2005. Explain how Figure 2.34 is misleading.

Solution

Figure 2.34 is misleading because the vertical scale does not begin at zero. **MediaMatters.com** published an amended graphic, reproduced here as Figure 2.35, which includes the zero on the vertical axis and much reduces the apparent difference among the political parties.

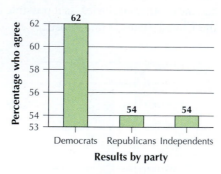

FIGURE 2.34 Omitting the zero is inappropriate.

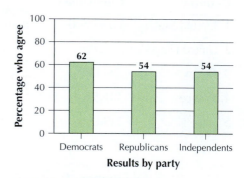

FIGURE 2.35 Appropriate graph.

EXAMPLE 2.21

MANIPULATING THE SCALE

Figure 2.36 shows a Minitab relative frequency bar graph of the majors chosen by 25 business school students. Explain how we could manipulate the scale to de-emphasize the differences.

Solution

If we wanted to de-emphasize the differences, we could extend the vertical scale up to its maximum, 1.0 = 100%, to produce the graph in Figure 2.37.

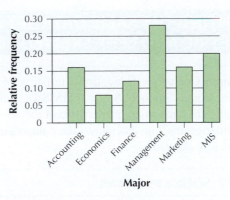

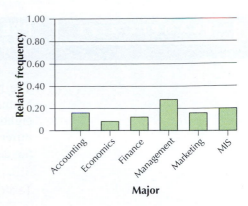

FIGURE 2.36 Well-constructed bar graph.

FIGURE 2.37 Inappropriate overextension of vertical scale.

EXAMPLE 2.22

USING TWO DIMENSIONS FOR A ONE-DIMENSIONAL DIFFERENCE AND UNCLEAR LABELING

Figure 2.38 compares the leaders in career points scored in the NBA All-Star Game among players active in 2007. Explain how this graphic may be misleading.

Solution

The height of the players is supposed to represent the total points, but this is not clearly labeled. Points should be indicated using a vertical axis, but there is no vertical axis at all. Further, note that Shaquille O'Neal dominates the graphic, because his body *area* is larger than the body areas of the other players. This is misleading. All four players should have the same body width, just as all bars in a bar graph have the same width.

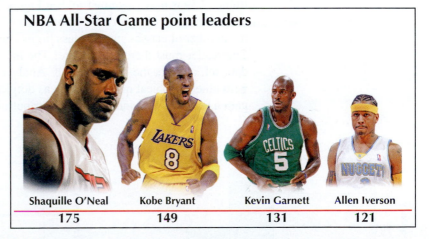

FIGURE 2.38 This graph uses two dimensions (height and width) to emphasize a one-dimensional (points) difference. (O'Neal: AP Photo/Alan Diaz; Bryant: AP Photo/Mark J. Terrill; Garnett: AP Photo/David Zalubowski; Iverson: AP Photo/David Zalubowski.)

When constructing a histogram, changing the number of classes or the width of the interval can sometimes lead to a completely different-looking distribution. Thus, we need to exercise care when someone shows us a histogram, since it presents, not the data themselves, but one of many ways of classifying the data.

EXAMPLE 2.23

PRESENTING THE SAME DATA SET AS BOTH SYMMETRIC AND LEFT-SKEWED

The National Center for Education Statistics sponsors the Trends in International Mathematics and Science Study (TIMSS). Science tests were administered to eighth-grade students in countries around the world (see Table 2.22). Construct two different histograms, one that shows the data as almost symmetric and one that shows the data as left-skewed.

Table 2.22 Science test scores

Country	Score	Country	Score	Country	Score
Singapore	578	New Zealand	520	Bulgaria	479
Taiwan	571	Lithuania	519	Jordan	475
South Korea	558	Slovak Republic	517	Moldova	472
Hong Kong	556	Belgium	516	Romania	470
Japan	552	Russian Federation	514	Iran	453
Hungary	543	Latvia	513	Macedonia	449
Netherlands	536	Scotland	512	Cyprus	441
United States	527	Malaysia	510	Indonesia	420
Australia	527	Norway	494	Chile	413
Sweden	524	Italy	491	Tunisia	404
Slovenia	520	Israel	488	Philippines	377

Solution

Figure 2.39 is nearly symmetric. But Figure 2.40 is clearly left-skewed. It is important to realize that *both figures are histograms of the very same data set.* Clever choices for the number of classes and the class limits can affect how a histogram presents the data. The reader must therefore beware! The histogram represents a summarization of the data set, and not the data set itself. Analysts may wish to supplement the histogram with other graphical methods, such as dotplots and stem-and-leaf displays, in order to gain a better understanding of the distribution of the data.

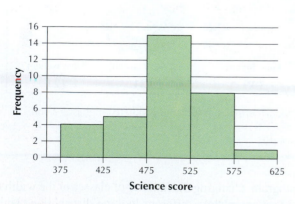

FIGURE 2.39 Nearly symmetric histogram of science test scores.

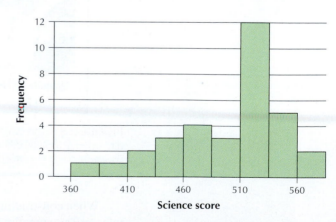

FIGURE 2.40 Left-skewed histogram of the same science test scores.

The *One-Variable Statistical Calculator* applet allows you to experiment with the class width and number of classes when constructing a histogram.

SECTION 2.4   **Summary**

1. Understanding how graphics are constructed will help you avoid being deceived by misleading graphics. Some common methods for making a graph misleading include manipulating the scale, omitting the zero on the relevant scale, and biased distortion or embellishment.

SECTION 2.4   **Exercises**

Clarifying the Concepts

1. Explain in your own words why it is important to be aware of the methods that can be used to make graphics misleading.

2. True or false: What we have learned in this chapter proves that all statistics are misleading.

Practicing the Techniques

Refer to Example 2.19 for the following exercises.

3. Which do you think is more effective at convincing the American public that a problem exists, Table 2.20 or Figure 2.33?

4. How would factoring in the *number of cars per country* affect the rankings, in your view?

5. If you were an insurance claims adjuster arguing for higher car insurance rates, would you prefer Table 2.20 or Table 2.21? Why?

Applying the Concepts

6. Eating Bread. Consider the accompanying graphic (similar to one found in *USA Today*) of the types of bread people eat.
   a. What type of graph is it supposed to represent, among the graphs that we have learned in this chapter?
   b. Consider how the *wheat* category dominates the graph. Which of the eight common methods for misrepresenting data is present here?
   c. Construct a graphic that is not misleading in this way.

7. Child-Rearing Costs. Consider the accompanying graphic (similar to one found in *USA Today*) of child-rearing costs by type of cost.
   a. Identify one problem with the graphic that makes it misleading.
   b. Construct a graphic that is not misleading in this way.

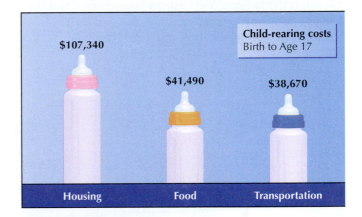

8. Going to the Game. Consider the accompanying graphic (similar to one found in *USA Today*) of the proportions of people who go to see professional sports events.
   a. Identify two problems with the graphic that make it misleading.
   b. Construct a graphic that is not misleading in these ways.

9. Living with AIDS. Consider the accompanying graphic.
 a. What point is the graphic trying to make?
 b. Which of the eight common problems is most obviously present here?
 c. Construct a graphic that is not misleading in this way.

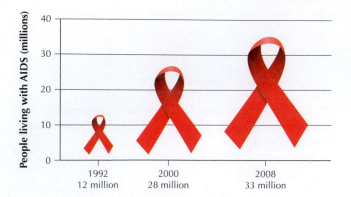

People living with AIDS (millions)

| 1992 | 2000 | 2008 |
| 12 million | 28 million | 33 million |

10. What's Your Sign? The General Social Survey collects data on social aspects of life in America. Consider the accompanying bar graph of the results of asking 1464 people what their astrological sign is.
 a. Which of the eight common problems is most obviously present here?
 b. Construct a graphic that is not misleading in this way.

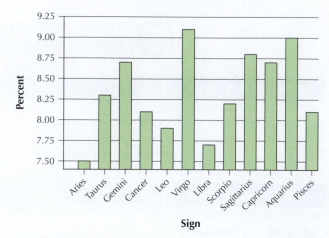

Sign

11. Video Game Consoles. Refer to the video game console data in Table 2.2 on page 35.
 a. Construct a bar graph that overemphasizes the difference among the game consoles.
 b. Which of the common methods for making graphics misleading are you using in **(a)**?
 c. Construct a bar graph that underemphasizes the difference among the game consoles.
 d. Which of the common methods for making graphics misleading are you using in **(c)**?

Use the *One-Variable Statistical Calculator* applet for Exercises 12–13. Work with the TIMSS scores from Example 2.23.

12. Click on the **Histogram** tab. Experiment with the class widths by clicking and dragging on the number line. Produce a histogram that is nearly symmetric, like Figure 2.39.

13. Produce a histogram that is somewhat left-skewed, like Figure 2.40.

CHAPTER 2   **Vocabulary**

Section 2.3

Section 2.4

CHAPTER 2	Review Exercises

Section 2.1

1. PARTS OF SPEECH. The accompanying bar graph summarizes the frequencies for the various parts of speech in a sample of English words. Should we be interested in determining whether this graph is symmetric or skewed? Clearly explain why or why not.

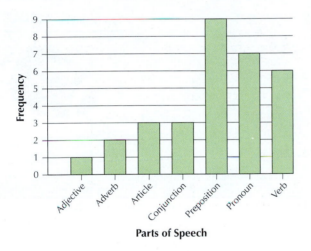

For Exercises 2–6, refer to the bar graph from Exercise 1 to construct the following for the variable *parts of speech.*
2. Relative frequency bar graph
3. Frequency distribution
4. Relative frequency distribution
5. Frequency pie chart
6. Relative frequency pie chart

HAPPINESS IN MARRIAGE. The General Social Survey tracks trends in American society through annual surveys. Use the following contingency table for Exercises 7–11.

Respondents' gender	Happiness of Marriage			Total
	Very happy	Pretty happy	Not too happy	
Male	242	115	9	366
Female	257	149	17	423
Total	499	264	26	789

7. What proportion of the males responded that they were very happy in their marriage?
8. What proportion of the females responded that they were very happy in their marriage?
9. What proportion of the males responded that they were not too happy in their marriage?
10. What proportion of the females responded that they were not too happy in their marriage?
11. Construct a clustered bar graph of the data.

Section 2.2

NEW YORK TOWNSPEOPLE. For towns in New York State, the accompanying histogram provides information on the percentage of the townspeople who are between 18 and 65 years old. Refer to the histogram for Exercises 12–14.

NewYork

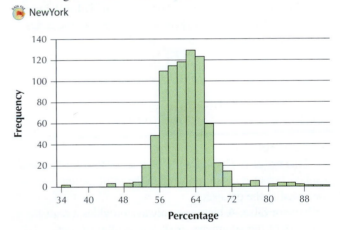

12. Would you characterize the distribution as left-skewed, right-skewed, or fairly symmetrical?
13. Provide an estimate of the "typical" percentage of townspeople who are between 18 and 65 years old. Is this typical value near the middle or near one of the "tails" of the distribution?
14. Would it be possible to construct a stem-and-leaf display, using the information from the histogram? Explain.

HOUSEHOLDS. Use the following information for Exercises 15–20. The data set **Household** contains eight variables' worth of information about the households in all 50 states plus the District of Columbia. The average size of the households is plotted in the accompanying dotplot, reproduced from Figure 2.21a on the next page.

 Household

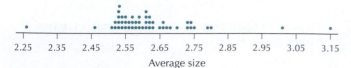

| 2.25 | 2.35 | 2.45 | 2.55 | 2.65 | 2.75 | 2.85 | 2.95 | 3.05 | 3.15 |

Average size

Dotplot of average household size.

Jan.	751	May	919	Sept.	802
Feb.	650	June	800	Oct.	636
Mar.	909	July	834	Nov.	579
Apr.	881	Aug.	789	Dec.	777

15. Construct a frequency distribution of the data.
16. Construct a relative frequency distribution of the data.
17. Construct a frequency histogram of the data.
18. Construct a relative frequency histogram of the data.
19. Construct a frequency polygon of the data.

(?) **20.** *What if* the data were faulty, and each data point should have had 0.5 added to it. How would that affect the shape of the distribution?

Section 2.3
21. Use the data from the stem-and-leaf display in Exercise 37 in Section 2.2 (page 64) to
  a. construct a cumulative frequency distribution.
  b. construct a cumulative relative frequency distribution.
22. STATISTICS EXAM DATA. Use the data from Exercises 49–52 in Section 2.2 (page 65) to
  a. construct a frequency ogive.
  b. construct a relative frequency ogive.
23. MIAMI ARRESTS. The Miami-Dade Police Department published the monthly number of arrests made for the year 2005, given in the following table. Construct a time series graph of the data.

  🔴 miamiarrests

Section 2.4
24. SPORTS CLOTHING. Consider the accompanying graphic of the types of sports clothing that children own.
  a. What type of graph does it represent, among the graphs that we have learned about in this chapter?
  b. Describe the difference between the representation of the NFL category versus the other categories.
  c. Which of the eight common methods for misrepresenting data is present here?
  d. Construct a graphic that is not misleading in this way.

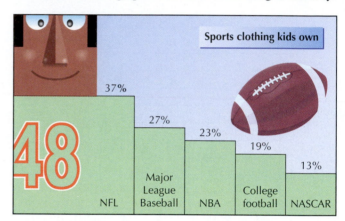

True or False
1. True or false: Histograms are superior to stem-and-leaf displays because histograms retain the information contained in the data set.
2. True or false: A histogram always provides a realistic summary of the symmetry or skewness of a data set.

Fill in the Blank
3. The frequencies in a frequency distribution must add up to the _____ _____ [two words].
4. A _____ _____ [two words] for a qualitative variable is a listing of all values that the variable can take, together with the frequencies for each value.

Short Answer
5. If there is a line that splits an image in half so that one side is the mirror image of the other, we say that the image is what?
6. If the right tail of a distribution is longer than the left tail, we say that the distribution is what?

Calculations and Interpretations
For Exercises 7–15, refer to the following table, which shows the life expectancy at birth in 2010, as reported by the World Health Organization.[5] 🔴 lifeexpect

Country	Life expectancy
Afghanistan	42
Canada	81
China	74
Ghana	62
India	64
Israel	81
Mexico	76
Russia	68
United Kingdom	80
United States	78

Construct the following:
7. Frequency distribution
8. Relative frequency distribution
9. Cumulative frequency distribution
10. Cumulative relative frequency distribution
11. Frequency bar graph
12. Relative frequency bar graph
13. Pie chart of the relative frequencies
14. Ogive of the frequencies
15. Relative frequency ogive of the frequencies

3 Describing Data Numerically

Mark Hooper/Getty Images

CASE STUDY

Can the Financial Experts Beat the Darts?

Have you ever wondered whether a bunch of monkeys throwing darts to choose stocks could select a portfolio that performed as well as the stocks carefully chosen by Wall Street experts? The *Wall Street Journal* (**www.wsj.com**) apparently believes that the comparison is worth a look. The *Journal* ran a contest between stocks chosen randomly by *Journal* staff members (rather than monkeys) throwing darts at the *Journal* stock pages (mounted on a board) and stocks chosen by a team of four professional financial experts.

At the end of six months, the *Journal* compared the percentage change in the price of the experts' stocks and the dartboard's stocks. So, who do you think did better? Did the six-figure-salary financial experts put the random dart selections to shame? We examine the results in the Case Study on pages 91 and 108. ■

> # The Big Picture
>
> ## *Where we are coming from, and where we are headed . . .*
>
> - Chapter 2 showed us graphical and tabular summaries of data.
>
> - Here, in Chapter 3, we "crunch the numbers," that is, develop numerical summaries of data. We examine measures of center, measures of variability, measures of relative position, and many other numerical summaries of data.
>
> - In Chapter 4, we will learn how to summarize the relationship between two quantitative variables.

3.1 MEASURES OF CENTER

OBJECTIVES By the end of this section, I will be able to . . .

1 Calculate the mean for a given data set.

2 Find the median, and describe why the median is sometimes preferable to the mean.

3 Find the mode of a data set.

4 Describe how skewness and symmetry affect these measures of center.

In Chapter 3 we learn how to summarize an entire data set with just a few numbers. For example, one numerical summary in baseball is a player's batting average (ratio of hits to at-bats). We know that Derek Jeter of the New York Yankees is a good hitter because his lifetime batting average is .314, which means that he gets hits 31.4% of the time. Most batters in Major League Baseball have a lower average. This simple number summarizes thousands of Jeter's at-bats over his long career. In Section 3.1, we will learn about three numerical measures that tell us where the center of the data lies: the mean, the median, and the mode.

1 THE MEAN

The mean is often called the arithmetic mean.

The most well known and widely used **measure of center** is the **mean.** In everyday usage, the word *average* is often used to denote the mean.

> To find the **mean** of the values in a data set, simply add up all the numbers and divide by how many numbers you have.

EXAMPLE 3.1 MOTOR VEHICLE THEFTS AT COMMUNITY COLLEGES

Recall from Chapter 1 that a random sample is a sample for which every element has an equal chance of being selected.

Table 3.1 contains a random sample of five community colleges from the thousands of U.S. community colleges, along with the number of motor vehicle thefts that took place in 2009 at each college. Find the mean number of motor vehicle thefts for these five colleges.

Table 3.1 Motor vehicle thefts at community colleges

Amarillo College	1
Columbus State Community College	2
Lone Star College System	1
Mesa Community College	4
Portland Community College	2

Source: U.S. Department of Education, The Campus Safety and Security Data Analysis Cutting Tool (**ope.ed.gov/security/**), 2010.

Solution

To find the mean, we add up the number of motor vehicle thefts for the five colleges, and divide by the number of colleges, 5:

$$\text{mean number of motor vehicle thefts} = \frac{1 + 2 + 1 + 4 + 2}{5} = 2$$

These five community colleges have a mean of 2 motor vehicle thefts.

Now You Can Do Exercises 13–16.

Notation

Statisticians like to use specialized notation. It is worth learning because it saves a lot of writing and because certain concepts can best be understood using this special notation.

- The **sample size,** which refers to how many observations you have in your sample data set, is denoted by n. Here, the five colleges from Table 3.1 represent a sample taken from the population (which in this case is all the community colleges in the United States). Thus, here, $n = 5$.

- We denote the ith data value by x_i, where i is simply an index or counter indicating which data point we are specifying. For example, in Table 3.1, $x_1 = 1$, $x_2 = 2$, $x_3 = 1$, and $x_4 = 4$. The last data value is $x_n = x_5 = 2$.

- The notation for "add them together" is Σ (capital sigma), the Greek letter for "S," because it stands for "Summation." To add up the number of cases for all five colleges, we could write out $1 + 2 + 1 + 4 + 2$, or we could simply represent this sum as Σx_i or, even more simply, as Σx.

- The **sample mean** is called $\bar{x}$ (pronounced "x-bar"). You should try to commit this to long-term memory, since $\bar{x}$ may be the most important symbol used in this book and will return again and again in nearly every chapter.

The **sample mean** can be written as $\bar{x} = \sum_{i=1}^{n} x_i/n = \Sigma x/n$. In plain English, this just means that, in order to find the mean $\bar{x}$, we

1. Add up all the data values, giving us Σx.
2. Divide by how many observations are in the data set, giving us $\Sigma x/n$.

So, for example, the sample mean number of motor vehicle thefts can be written as

$$\bar{x} = \frac{\Sigma x}{n} = \frac{1 + 2 + 1 + 4 + 2}{5} = \frac{10}{5} = 2$$

What Does This Number Mean?

The Mean as the Balance Point of the Data

Let's explore the vehicle theft data a bit further. Consider the dotplot of the number of motor vehicle thefts for each college, given in Figure 3.1. To find out where the mean of the number of motor vehicle thefts lies on this number line, imagine that the dots are little blocks on a ruler or a seesaw and that you must decide where to place the fulcrum so that the ruler balances perfectly. *The place where the data set balances perfectly is the location of the mean.* Placing the fulcrum too far to the right or left would create an imbalance. This data set balances precisely at the sample mean, $\bar{x} = 2$.

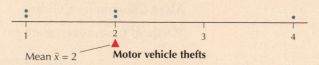

Mean $\bar{x} = 2$ **Motor vehicle thefts**

FIGURE 3.1

Developing Your Statistical Sense

Checking Your Results Against Experience and Common Sense

When you have found the balance point, you have found the mean. When you calculate the mean, or have a computer or calculator do it for you, don't just accept whatever value pops out. Make sure the result makes sense. Since the mean always indicates the place where the data values are in balance, the mean is often near the center of the data. If the value you have calculated lies nowhere near the center of the data, then you may want to check your calculations.

For example, suppose we were finding the mean of the vehicle theft data, and we accidentally entered 40 instead of 4 for the number of vehicle thefts for Mesa Community College. Then our value for the mean resulting from this *incorrect* calculation would be

$$\bar{x} = \frac{\sum x}{n} = \frac{1 + 2 + 1 + 40 + 2}{5} = \frac{46}{5} = 9.2$$

The mean number of thefts cannot equal 9.2 because all the values in the data set are less than 9.2. The mean can never be larger or smaller than all the values in the data set.

Don't automatically accept the result you get from a computer or calculator. Remember GIGO: Garbage In Garbage Out. If you enter the wrong data, the calculator or computer will not bail you out. Human error is one reason for the explosion of faulty statistical analyses in the newspapers and on the Internet. Now more than ever data analysts must use good judgment. When you calculate a mean, always have an idea of what you *expect* the sample mean to be, that is, at least a ballpark figure.

The Population Mean μ

The mean value of a population is usually unknown. For example, we cannot know the mean systolic blood pressure of *all* the residents in your hometown at noon today. Instead, data analysts use *estimation.* We could select a random sample of, say, 30 residents, find the mean systolic blood pressure $\bar{x}$ of this sample, and use this $\bar{x}$ as an estimate of the unknown population mean systolic blood pressure. We denote the **population mean** with μ (mu), which is the Greek letter for "m." The **population size** is denoted by N. When all the values of the population are known, the population mean is calculated as

Greek letters are sometimes used to represent the (usually unknown) population parameters (such as the population mean).

$$\mu = \frac{\sum x}{N}$$

EXAMPLE 3.2

CALCULATING THE POPULATION MEAN

Table 3.2 contains the number of victories per team in the Southeast Conference of NCAA football for the 2009 season. Since these teams represent *all* the teams in the conference, then they represent a population.

a. Calculate the population mean number of victories.
b. Suppose a random sample consists of the following four teams: Florida, Arkansas, Georgia, and Tennessee. Calculate the sample mean number of victories for that sample.

SEConference

Table 3.2 Victories for football teams in the Southeast Conference, 2009 season

Team	Victories	Team	Victories
Alabama	14	Georgia	8
Florida	13	Kentucky	7
Louisiana State	9	South Carolina	7
Mississippi	9	Tennessee	7
Arkansas	8	Mississippi State	5
Auburn	8	Vanderbilt	2

Solution

For calculating the mean, we will adopt the convention of rounding our final calculation to one more decimal place than that in the original data.

a. The population size is the number of teams in the conference, $N = 12$. The population mean number of victories is

$$\mu = \frac{\sum x}{N} = \frac{14 + 13 + 9 + 9 + 8 + 8 + 8 + 7 + 7 + 7 + 5 + 2}{12} = \frac{97}{12} \approx 8.0833 \approx 8.1$$

b. Our sample consists of the following teams: Florida (13 wins), Arkansas (8 wins), Georgia (8 wins), and Tennessee (7 wins), giving us the sample mean number of victories:

$$\bar{x} = \frac{\sum x}{n} = \frac{13 + 8 + 8 + 7}{4} = \frac{36}{4} = 9$$

Now You Can Do Exercises 17–20.

Of course, a different sample would have yielded a different value for $\bar{x}$.

The Mean Is Sensitive to Extreme Values

One drawback of using the mean to measure the center of the data is that the mean is sensitive to the presence of extreme values in the data set. We illustrate this phenomenon with the following example.

EXAMPLE 3.3

SENSITIVITY OF THE MEAN TO EXTREME VALUES

homesales

Table 3.3 contains a sample of six home sales prices for Broward County, Florida, for October 4, 2010. We would like to get an idea of the typical home sales price in Broward County. Find the mean sales price of the homes in this sample.

Solution

$$\bar{x} = \frac{\sum x}{n} = \frac{290{,}000 + 350{,}000 + 375{,}000 + 415{,}000 + 500{,}000 + 575{,}000}{6}$$

$$= \frac{2{,}505{,}000}{6} = \$417{,}500$$

Table 3.3 Home sales prices in Broward County, Florida

Location	Price
Pembroke Pines	$290,000
Weston	$350,000
Hallandale	$375,000
Miramar	$415,000
Davie	$500,000
Fort Lauderdale	$575,000

Source: **www.homes.com.**

Now, suppose that we append a seventh home to our sample, a home in Hillsboro Beach listed for $5,999,998, which is much more expensive than any of the other homes in the sample. Recalculating the mean, we get

$$\bar{x} = \frac{\sum x}{n}$$

$$= \frac{290,000 + 350,000 + 375,000 + 415,000 + 500,000 + 575,000 + 5,999,998}{7}$$

$$= \frac{8,504,998}{7} = \$1,215,000$$

Note that the mean sales price nearly tripled from $417,500 to $1,215,000 when we added this extreme value. Also, this new mean is much higher than every price in the original sample. Thus, it is highly unlikely that this new mean of about $1.2 million is representative of the *typical* sales price of homes in Broward County. This example shows how the mean is sensitive to the presence of extreme values. For situations like this, we prefer a measure of center that is not so sensitive to extreme values. Fortunately, the *median* is just such a measure.

2 THE MEDIAN

Recall that the median strip on a highway is the slice of land in the *middle* of the two lanes of the highway. In statistics, the **median** of a data set represents the middle of the data set when the data are put into ascending order. There are two cases, depending on whether the sample size is odd or even.

> **The Median**
>
> The **median** represents the middle of a data set when the data are put into ascending order. Half of the data values lie below the median, and half lie above.
> - If the sample size n is odd, then the median is the middle value and lies at the $\left(\frac{n+1}{2}\right)^{th}$ position when the data are put in ascending order.
> - If the sample size n is even, then the median is the mean of the two middle data values that lie on either side of the $\left(\frac{n+1}{2}\right)^{th}$ position.

The case when the sample size is even is clear if you hold up four fingers on one hand. Notice that there is no unique finger in the middle. Since there is no middle value when the sample size is even, we take the two data values in the middle and split the difference.

The Median Is Not Sensitive to Extreme Values

Unlike the mean, the median is not sensitive to extreme values. If someone purchases a very expensive house, the mean home sales price will jump, but the median home sales price will be less affected. Let's look at an example of how this would occur.

EXAMPLE 3.4

FINDING MEDIAN AND SHOWING IT IS NOT SENSITIVE TO EXTREME VALUES

Find the median home sales price for the following data:
a. Broward County data from Table 3.3
b. Broward County data from Table 3.3 with the Hillsboro Beach home costing $5,999,998

Solution

a. Fortunately, the data are already presented in ascending order in the table. Since $n = 6$ is even, the median is the mean of the two data values that lie on either side of the $\left(\frac{n+1}{2}\right)^{th} = \left(\frac{6+1}{2}\right)^{th} = 3.5$th position. That is, the median is the mean of the 3rd and 4th data values, \$375,000 and \$415,000. Splitting the difference between these two, we get

$$\text{median price} = \frac{\$375,000 + \$415,000}{2} = \$395,000$$

We note that in Table 3.3 there are exactly as many homes with prices lower than \$395,000 as there are homes with prices higher than \$395,000.

b. Now, what happens to the median when we add in the \$5,999,998 home from Hillsboro Beach? Since $n = 7$ is odd, the median is the unique $\left(\frac{n+1}{2}\right)^{th}$ $= \left(\frac{7+1}{2}\right)^{th} = 4$th observation, given by the home in Miramar for \$415,000. The extreme value increased the median only from \$395,000 to \$415,000. Recall that the mean nearly tripled to over \$1.2 million. Thus, the median home sales price is a better measure of center because it more accurately reflects the typical sales prices of homes in Broward County. Figure 3.2 shows how the mean (red triangles) changes significantly with the addition of the extreme value, while the median (green triangles) changes relatively little.

Because the median is not sensitive to extreme values, we say that it is a robust, or resistant, measure of center. The mean is neither robust nor resistant.

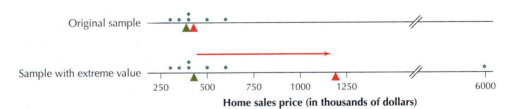

Now You Can Do Exercises 21–24.

FIGURE 3.2 The mean (red triangles) is sensitive to extreme values, but the median (green triangles) is not.

Note that the formula $\frac{n+1}{2}$ gives the *position*, not the *value*, of the median. For example, the median home sales price for Table 3.3 is *not* $\frac{n+1}{2} = \frac{6+1}{2} = 3.5$.

The *Mean and Median* applet allows you to insert your own data values and see how changes in these values affect both the mean and the median.

EXAMPLE 3.5

USING TECHNOLOGY TO FIND THE MEAN AND MEDIAN

homesales

Find the mean and median of the home sales prices in Table 3.3, using (a) the TI-83/84, (b) Excel, and (c) Minitab.

Solution

Using the instructions in the Step-by-Step Technology Guide on page 92, we get the following output.

a. The first TI-83/84 screen shows $\bar{x}$ = 417,500 and n = 6. The second screen shows the median Med 5 395,000.

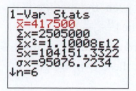

```
1-Var Stats
x̄=417500
Σx=2505000
Σx²=1.10008E12
Sx=104151.3322
σx=95076.7234
↓n=6
```

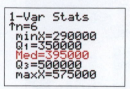

```
1-Var Stats
↑n=6
minX=290000
Q₁=350000
Med=395000
Q₃=500000
maxX=575000
```

b. The mean and median are shown in the Excel output.

```
Home Sales Price
Mean              417500
Standard Error    42519.6
Median            395000
Mode              #N/A
```

c. The mean and median are shown in the Minitab output.

```
Descriptive Statistics:
Home Price

Variable      Mean   Median
Home Price   417500  395000
```

3   THE MODE

Sometimes the mode does not indicate the center of a data set. See Exercise 25 on page 93 for an example.

A third measure of center is called the **mode.** French speakers will recognize that the term *mode* in French refers to *fashion*. The popularity of clothing, cosmetics, music, and even basketball shoes often depends on just which style is in fashion. In a data set, the value that is most "in fashion" is the value that occurs the most.

> The **mode** of a data set is the data value that occurs with the greatest frequency.

EXAMPLE 3.6

FINDING THE MEAN, MEDIAN, AND MODE: MUSIC VIDEOS

The Web site **MTV.com** contains music videos for many performers. Table 3.4 provides the number of music videos available for download for four performers, as of May 21, 2012.

Table 3.4 Music Videos for Four Performers

Performer	Music Videos
Michael Jackson	31
Taylor Swift	26
Usher	26
Katy Perry	15

AP Photo/Theron Kirkman

Taylor Swift

Find the (a) mean, (b) median, and (c) mode number of music videos.

Solution

a.   The sample mean number of followers is

$$\bar{x} = \frac{\sum x}{n} = \frac{31 + 26 + 26 + 15}{4} = 24.5$$

The mean number of music videos is 24.5.

b.   Since $n = 4$ is even, the median is the mean of the two middle data values:

$$\text{Median} = \frac{(26 + 26)}{2} = 26$$

The median number of music videos is 26.

c.   The mode is the data value that occurs with the greatest frequency. There are two performers with 26 music videos, Taylor Swift and Usher. No other data value occurs more than once. Therefore the mode is 26 music videos, as shown.

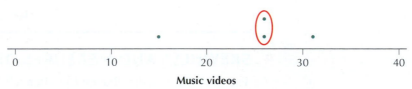

Music videos

Dotplot of music videos, showing 26 as the mode.

**Now You Can Do
Exercises 25–28.**

One of the strengths of the mode is that it can be used with categorical, or qualitative, data. Suppose you asked your friends to name their favorite flower. Six of them answered "rose," three answered "lily," and one answered "daffodil." Note that these data are categorical, not numerical. Since the most frequently occurring flower is "rose," the rose represents the mode of the variable *favorite flower*. Unfortunately, we cannot use arithmetic with categorical variables, and thus the mean or median for this variable cannot be found.

It may happen that no value occurs more than once, in which case we say there is *no mode*. Or else more than one data value could occur with the greatest frequency, in which case we would say there is more than one mode. Data sets with one mode are *unimodal*; data sets with more than one mode are *multimodal*.

WHAT IF?

Give the Calculator a Rest

What If Scenarios offer you a chance to reflect on how changes in the initial conditions will percolate through the various aspects of a problem. The only requirement is to put your calculator down and think through the problem. You are asked to find the answers by using your knowledge of what the statistics *represent*.

Consider Example 3.6 once again. Now imagine: *what if* there was an incorrect data entry, such as a typo, and the number of Michael Jackson's music videos was greater than 31 by some unspecified amount.

Describe how and why this change would have affected the following, if at all:

The Excel output on page 88 does not show a mode because no data value occurs more than once in Table 3.3.

a.   The mean number of music videos

b.   The median number of music videos

c.   The mode number of music videos

Solution

a. Consider Figure 3.3, a dotplot of the number of music videos, with the triangle indicating the mean or balance point, at 26. Recall that this represents the balance point of the data. As the number of Michael Jackson's music videos increases (arrow), the point at which the data balance (the mean) also moves somewhat to the right. Thus the mean number of followers will increase.

b. Recall from Example 3.6 that the median is the mean of the middle two data values. In other words, the median *ignores* most of the data values, including the largest value, which is the only one that has increased. Therefore, the median will remain unchanged.

c. The mode also remains unchanged, since the only data value that occurs more than once is the original mode, 26 music videos.

FIGURE 3.3 As the number of Michael Jackson's videos increases so does the mean.

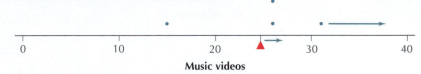

Music videos

4 SKEWNESS AND MEASURES OF CENTER

The skewness of a distribution can often tell us something about the relative values of the mean, median, and mode (see Figure 3.4).

FIGURE 3.4 How skewness affects the mean and median.

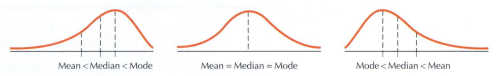

Mean < Median < Mode   Mean = Median = Mode   Mode < Median < Mean

How Skewness Affects the Mean and Median

- For a right-skewed distribution, the mean is larger than the median.
- For a left-skewed distribution, the median is larger than the mean.
- For a symmetric unimodal distribution, the mean, median, and mode are fairly close to one another.

EXAMPLE 3.7

MEAN, MEDIAN, AND SKEWNESS

The histogram of the average size of households in the 50 states and the District of Columbia from Example 2.14 (page 60) is reproduced here as Figure 3.5.

a. Based on the skewness of the distribution, state the relative values of the mean, median, and mode.

b. Use Minitab to verify your claim in (a).

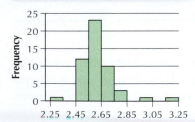

FIGURE 3.5 Household size is somewhat right-skewed.

Solution

a. The distribution of average household size is somewhat right-skewed. Thus, from Figure 3.4, we would expect the mean to be greater than the median, which is greater than the mode.

b. The Minitab descriptive statistics are shown here. Note that the mean is greater than the median, which is greater than the mode.

```
Descriptive Statistics: Size

Variable   Mean   Median   Mode
Size      2.619   2.590   2.530
```

Now You Can Do
Exercise 56.

| CASE STUDY | ## Can the Financial Experts Beat the Darts? |

Mark Hooper/Getty Images

Recall the contest held by the *Wall Street Journal* to compare the performance of stock portfolios chosen by financial experts and stocks chosen at random by throwing darts at the *Journal* stock pages. We will examine the results of 100 such contests in various ways, using the methods we have learned thus far, and will return to examine them further as we acquire more analysis tools. Let's start by reporting the raw result data. The percentage increase or decrease in stock prices was calculated for the portfolios chosen by the professional financial advisers and by the randomly thrown darts and was compared with the percentage net change in the Dow Jones Industrial Average (DJIA).

Exploratory Data Analysis

Note: In exploratory data analysis, we use graphical methods to compare numerical statistics.

Figure 3.6 shows comparative dotplots of the percentage net change in price for the professionally selected portfolio, the randomly selected darts portfolio, and the DJIA, over the course of the 100 contests. First, estimate the mean of each distribution by choosing the balance point of the data. This balance spot is the *mean*. For fun, write down your guess for the mean for the professionals so you can see how close you were when we provide the descriptive statistics later. Now compare this with where you would find the balance spot (mean) for the darts dotplot. Which numerical value is larger, the balance spot for the pros or the darts? Just think: you are comparing the mean portfolio performances for the professionals and the darts without using a formula or a calculator. This is *exploratory data analysis*. You are using *graphical* methods to compare *numerical* statistics.

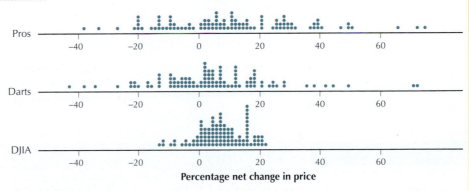

FIGURE 3.6
Dotplot of the percentage net price change for the professionally selected portfolio, the randomly selected darts portfolio, and the Dow Jones Industrial Average.

Hopefully, you discovered that the estimated mean for the pros is greater than the estimated mean for the darts. This is not particularly surprising, is it? Next, find the balance point for the DJIA dotplot. Compare the numerical value for the DJIA balance spot to the mean you found for the dotplot for the pros. Write down your estimate of the means for the DJIA and darts dotplots, so you can see how close you were later. Again, hopefully, you found that the estimated professionals' mean was higher than that of the DJIA. Now, a tougher comparison is to compare the estimated DJIA mean with that of the darts. Which of these two do you think is higher?

Remember: It is often helpful to have a "ballpark" estimate of the mean or other statistics, as a reality check of your calculations.

Finally, Minitab provides us with the mean percentage net price changes, as shown in Figure 3.7. Over the course of 100 contests, the mean price for the portfolios chosen by the professional financial advisers increased by 10.95%, by 6.793% for the DJIA, and by 4.52% for the random darts portfolio.

(continues)

FIGURE 3.7
Mean percentage net price change for the professionals, darts, and DJIA.

Variable	N	Mean
Pros	100	10.95
Darts	100	4.52
DJIA	100	6.79

This is evidence in support of the view that financial experts can consistently outperform the market. We return to this Case Study in Section 3.2 (page 108). ■

STEP-BY-STEP TECHNOLOGY GUIDE: Descriptive Statistics

TI-83/84

Step 1 Enter the data in **L1** using the instructions **(STAT > 1: Edit)** found in the Step-by-Step Technology Guide in Section 2.2.
Step 2 Press **STAT**. Use the **right arrow** button to move the cursor so that **CALC** is highlighted.

Step 3 Select **1-Var Stats,** and press **ENTER.**
Step 4 On the home screen, the command **1-Var Stats** is shown. Press **2nd,** then **L1** (above the **1** key) and press **ENTER.**

EXCEL

Step 1 Enter the data in column **A.**
Step 2 Select **Data > Data Analysis.**
Step 3 Select **Descriptive Statistics** and click **OK.**

Step 4 For the **Input Range,** click and drag to select the data in column A.
Step 5 Check **Summary Statistics** and click **OK.**

MINITAB

Step 1 Enter the data in column C1.
Step 2 Select **Stat > Basic Statistics > Display Descriptive Statistics . . .**
Step 3 The variable selection dialog box appears. Select the variable you want to summarize by double-clicking on it until it appears in the **Variables** box.

Step 4 Click **statistics.**
Step 5 Select the desired statistics and click **OK.**

CRUNCHIT!

We will use the data from Example 3.3 (page 85).

Step 1 Click **File . . .** then highlight **Load from Larose2e . . . Chapter 3 . . .** and click on **Example 3.3.**

Step 2 Click **Statistics** and select **Descriptive statistics.** For **Data** select **Prices.** Then click **Calculate.**

SECTION 3.1 Summary

1. Measures of center are introduced in Section 3.1. The sample mean ($\bar{x}$) represents the sum of the data values in the sample divided by the sample size (n). The population mean (μ) represents the sum of the data values in the population divided by the population size (N). The mean is sensitive to the presence of extreme values.

2. The median occupies the middle position when the data are put in ascending order and is not sensitive to extreme values.

3. The mode is the data value that occurs with the greatest frequency. Modes can be applied to categorical data as well as numerical data but are not always reliable as measures of center.

4. The skewness of a distribution can often tell us something about the relative values of the mean and the median.

SECTION 3.1 Exercises

Clarifying the Concepts

1. Explain what a measure of center is.

2. Which measure may be used as the balance point of the data set? Explain how this works.

3. Explain what we mean when we say that the mean is sensitive to the presence of extreme values. Explain whether the median is sensitive to extreme values.

4. What are the three measures of center that we learned about in this section?

For Exercises 5–12, either state what is being described or provide the notation.

5. The number of observations in your sample data set

6. The number of observations in your population data set

7. Notation for the *i*th data value in your data set

8. Notation denoting "add them together"

9. Notation for what we get when we add up all the data values in the sample, and divide by how many observations there are in the sample

10. Notation for what we get when we add up all the data values in the population, and divide by how many observations there are in the sample

11. The middle data value when the data are put in ascending order

12. The data value that occurs with the greatest frequency

Practicing the Techniques

For the sample data in Exercises 13–16:
　a. Find the sample size *n*.
　b. Calculate the sample mean $\bar{x}$.

13. 18, 15, 20, 20, 17

14. 3, 0, 5, −3, 0, −5

15. 75, 65, 90, 80, 85, 75, 100

16. 120, 155, 95, 155, 133

For the population data in Exercises 17–20:
　a. Find the population size *N*.
　b. Calculate the population mean μ.

17. 79, 92, 65, 75, 67, 59, 88

18. −50, −51, −45, 50, 45, 51

19. 1503, 1642, 1298, 1441, 2000

20. 9, 10, 9, 8, 6, 5, 8, 9, 6, 10, 8

For the data in Exercises 21–24, find the median.

21. 18, 15, 20, 20, 17

22. 3, 0, 5, −3, 0, −5

23. 75, 65, 90, 80, 85, 75, 100

24. 120, 155, 95, 155, 133

For the data in Exercises 25–28, find the mode.

25. 18, 15, 20, 20, 17

26. 3, 0, 5, −3, 0, −5

27. 75, 65, 90, 80, 85, 75, 100

28. 120, 155, 95, 155, 133

29. Five friends have just had dinner at the local pizza joint. The total bill came to $30.60. What is the mean cost of each person's meal?

30. Lindsay just bought four shirts at the boutique in the mall, costing a total of $84.28. What was the mean cost of the shirts?

31. The mean cost of a sample of five items is $20. The costs of four of the items are as follows: $25, $15, $15, $20. What is the cost of the fifth item?

32. The mean size of four downloaded music files is 3 Mb (megabytes). The sizes of three of the files are as follows: 5 Mb, 2 Mb, 3 Mb. What is the size of the fourth music file?

33. The median number of students in a sample of 7 statistics classes is 25. The ordered values are: 20, 22, 24, __, 27, 27, 28. What is the missing value?

34. The median number of academic credits taken in a sample of 6 students is 15. The ordered values are: 12, 12, 14, __, 17, 17. What is the missing value?

Applying the Concepts

Clickstream Analysis. Use the following information for Exercises 35–38. Clickstream analysis is the study of how humans behave on the Internet.[1] One measure is the number of new page requests (clicks) that the visitor makes. A sample of the visitors to a particular Web site had the following total numbers of clicks.

| 1 | 5 | 3 | 4 | 3 | 2 | 3 | 7 |

35. Find the sample size *n*.

36. Calculate the sample mean number of clicks $\bar{x}$.

37. Find the median.

38. Find the mode.

Fuel Economy. Table 3.5 contains the number of cylinders, the engine size (in liters), the fuel economy (miles per gallon, city driving), and the country of manufacture for six 2011 automobiles. Use this information for Exercises 39–42.
　cylinderengine

TABLE 3.5　Cylinders, engine size, and fuel economy for six cars

Vehicle	Cylinders	Engine size	City mpg	Country of manufacture
Cadillac CTS	6	3.0	18	USA
Ford Fusion Hybrid	4	2.5	41	USA
Ford Taurus	6	3.5	18	USA
Honda Civic	4	1.8	25	Japan
Rolls Royce	12	6.7	11	UK
Toyota Camry Hybrid	4	2.4	31	Japan

Source: **www.fueleconomy.gov.**

39. Find the following for the number of cylinders:
    a. Mean   **b.** Median   **c.** Mode

40. Find the following for the engine size:
    a. Mean   **b.** Median   **c.** Mode

41. Find the following for the fuel economy:
    a. Mean   **b.** Median   **c.** Mode

42. Find the mode for country of manufacture.

SAT Scores. Table 3.6 contains the SAT scores of students who took the SAT subject tests, Use this information for Exercises 43–46.

🔴 satsubject

TABLE 3.6 **SAT scores for students who took the SAT subject tests**

SAT Subject Test	SAT Mathematics	SAT Reading	SAT Writing
English Literature	585	599	596
Biology-E	614	594	595
U.S. History	615	617	609
Chemistry	674	615	620
Math Level 1	605	604	600
Math Level 2	655	598	605

43. Calculate the following for the SAT Mathematics test:
    a. Mean   **b.** Median

44. Find the following for the SAT Reading test:
    a. Mean   **b.** Median

45. Compute the following for the SAT Writing test:
    a. Mean   **b.** Median

46. Is there a mode score for any of the three tests? Explain.

47. Liberal Arts Majors. Here are the declared liberal arts majors for a sample of students at a local college:

English	History	Spanish	Art	Theater
Theater	Philosophy	English	Music	Math
Math	Math	History	English	Art
English	History	Spanish	Economics	Math
Music	English	Economics	Theater	Music

    a. What is the mode of this data set? Does this mean that most students at the college are majoring in this subject?
    b. Does the idea of the mean or median of this data set make any sense? Explain clearly why not.
    c. How would you respond to someone who claimed that economics was the most popular major?

For Exercises 48–52, refer to Table 3.7, which lists the top five paperback trade fiction books, for the week of October 9, 2010, as reported by the *New York Times*.

TABLE 3.7 **Top five best sellers in paperback trade fiction**

Rank	Title	Author	Price
1	The Girl with the Dragon Tattoo	Stieg Larsson	$14.95
2	The Girl Who Played with Fire	Stieg Larsson	$15.95
3	Little Bee	Chris Cleave	$14.00
4	Half Broke Horses	Jeanette Walls	$15.00
5	Cutting for Stone	Abraham Verghese	$15.95

48. Find the mean, median, and mode for the price of these five books on the best-seller list. Suppose a salesperson claimed that the price of a typical book on the best-seller list is less than $14. How would you use these statistics to respond to this claim?

49. Linear Transformations. Add $10 to the price of each book.
    a. Now find the mean of these new prices.
    b. How does this new mean relate to the original mean?
    c. Construct a rule to describe this situation in general.

50. Linear Transformations. Multiply the price of each book by 5.
    a. Now find the mean of these new prices.
    b. How does this new mean relate to the original mean?
    c. Construct a rule to describe this situation in general.

51. Find the mode for the following variables:
    a. Price
    b. Author

52. Explain whether it makes sense to find the mean or median of the variable *author*.

Car Model Years. Refer to Figure 3.8 for Exercises 53–55. The data represent the model year for a sample of cars in a used car lot.

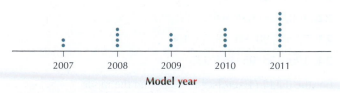

FIGURE 3.8 Dotplot of model year.

53. What are the mean, median, and mode of the model year?

54. Calculate a new statistic "age of the car in 2012" as follows: take the model year and subtract it from 2012.
    a. Find the mode of the car ages.
    b. Find the mean and median of the car ages.

55. What will be the mean, median, and mode of the ages of these cars in 2015?

56. Skewness and Symmetry. Consider the accompanying distributions. What can we say about the values of the mean, median, and mode in relation to one another?

A

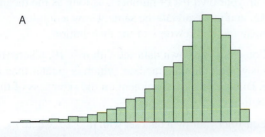

B

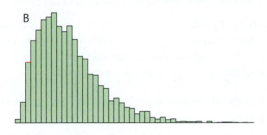

C

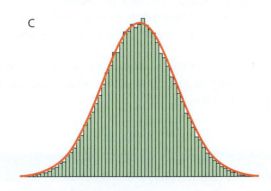

  a. The distribution in A
  b. The distribution in B
  c. The distribution in C

Bringing It All Together

Pulse Rates for Men and Women. To answer Exercises 57–60, refer to Figure 3.9, comparative dotplots of the pulse rates for males and females.[2]

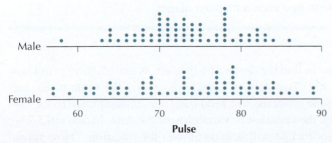

FIGURE 3.9 Comparative dotplots of pulse rates, by gender.

57. Examine Figure 3.9.
  a. Without doing any calculations, what is your impression of which gender, if any, has the higher overall pulse rate?
  b. Find the mean pulse rate for the males by estimating the location of the balance point.
  c. Find the mean pulse rate for the females by estimating the location of the balance point.
  d. Based on (**b**) and (**c**), which gender has the higher mean pulse rate? Does this agree with your earlier impression?

58. Find the following medians.
  a. The median pulse rate for the males
  b. The median pulse rate for the females
  c. Which gender has the higher median pulse rate? Does this agree with your findings for the mean earlier?

59. Find the following modes.
  a. The mode pulse rate for the males
  b. The mode pulse rate for the females
  c. Which gender has the higher mode pulse rate? Does this agree with your findings for the mean earlier?

60. *What if* the fastest pulse rate for the men was a typo and should have been an unspecified lower pulse rate. Describe how and why this change would have affected the following, if at all. Would they increase, decrease, or remain unchanged? Or is there insufficient information to tell what would happen? Explain your answers.
  a. The mean men's pulse rate
  b. The median men's pulse rate
  c. The mode men's pulse rate

61. Trimmed Mean. Because the mean is sensitive to extreme values, the *trimmed mean* was developed as another measure of center. To find the 10% trimmed mean for a data set, omit the largest 10% of the data values and the smallest 10% of the data values, and calculate the mean of the remaining values. Because the most extreme values are omitted, the trimmed mean is *less sensitive,* or *more robust* (*resistant*), than the mean as a measure of center. For the following random sample of women's pulse rates, calculate the following:
  a. The mean
  b. The 10% trimmed mean
  c. The 20% trimmed mean

75 69 73 84 82 80 74 83 77 78 61 78 87 79 65 72 69 81 62 69

62. Challenge Exercise. In general, would you expect the trimmed mean to be larger, smaller, or about the same as the mean, for data sets with the following shapes?
  a. Right-skewed data
  b. Left-skewed data
  c. Symmetric data

63. Midrange. Another measure of center is the *midrange*.

$$\text{midrange} = \frac{\text{largest data value} + \text{smallest data value}}{2}$$

Because the midrange is based on the maximum and minimum values in the data set, it is not a robust statistic, but is sensitive to extreme values. Calculate the midrange for the following data:

    a. The data from Table 3.7
    b. The data from Figure 3.8

64. Harmonic Mean. The *harmonic mean* is a measure of center most appropriately used when dealing with rates, such as miles per hour (mph). The harmonic mean is calculated as

$$\frac{n}{\sum \frac{1}{x}}$$

where *n* is the sample size, and the *x*'s represent rates, such as the speeds in mph. Emily walked five miles today, but her walking speed slowed as she walked farther. Her walking speed was 5 mph for the first mile, 4 mph for the second mile, 3 mph for the third mile, 2 mph for the fourth mile, and 1 mph for the fifth mile. Calculate her harmonic mean walking speed over the entire five miles.

65. Challenge Exercise. The (arithmetic) mean for Emily's five-mile walk in Exercise 64 is 3 mph. Explain clearly why the value you calculated for the harmonic mean in Exercise 64 makes more sense than this arithmetic mean of 3 mph. (*Hint:* Consider time.)

66. Geometric Mean. The *geometric mean* is a measure of center used to calculate growth rates. Suppose that we have *n* positive values; then the geometric mean is the *n*th root of the product of the *n* values. Jamal has been saving money in an account that has had 4% growth, 6% growth, and 10% growth over the last three years. Calculate the *average*

growth rate over these three years. (*Hint:* Find the geometric mean of 1.04, 1.06, and 1.10 and subtract 1.)

Construct Your Own Data Sets

67. Construct your own data set with $n = 10$, where the mean, the median, and the mode are all the same. Yes, just make up your own list of numbers, as long as the mean, median, and mode are all the same. Draw a dotplot. Comment on the skewness of the distribution.

68. Construct your own data set with $n = 10$, where the mean is greater than the median, which is greater than the mode. Draw a dotplot. Comment on the skewness of the distribution.

69. Construct your own data set with $n = 10$, where the mode is greater than the median, which is greater than the mean. Draw a dotplot. Comment on the skewness of the distribution.

70. Construct your own data set with $n = 3$. Let the mean and median be equal. Now, alter the three data values so that the mean of the altered data set has increased while the median of the altered data set has decreased.

 Use the *Mean and Median* applet for Exercises 71 and 72.

71. Insert three points on the line by clicking just below it, two near the left side and one near the middle.
    a. Click and drag the rightmost point to the right.
    b. Describe what happens to the mean when you do this.
    c. Describe what happens to the median when you do this.

72. Explain why each of the measures behaves the way it does in the previous exercise.

3.2   MEASURES OF VARIABILITY

OBJECTIVES   By the end of this section, I will be able to . . .

  1   Understand and calculate the range of a data set.

  2   Calculate the variance and the standard deviation for a population.

  3   Compute the variance and the standard deviation for a sample.

  4   Use the Empirical Rule to find approximate percentages for a bell-shaped distribution.

  5   Apply Chebyshev's Rule to find minimum percentages.

1   THE RANGE

In Section 3.1 we learned how to find the center of a data set. Is that all there is to know about a data set? Definitely not! Two data sets can have exactly the same mean, median, and mode and yet be quite different. We need measures that summarize the data set in a different way, namely, the variation or variability of the data. In Section 3.2 we will learn *measures of variability* that will help us answer the question: "How spread out is the data set?"

| EXAMPLE 3.8 | DIFFERENT DATA SETS WITH THE SAME MEASURES OF CENTER |

Table 3.8 contains the heights (in inches) of the players on two volleyball teams.

 volleyball

Table 3.8 Women's volleyball team heights (in inches)

Western Massachusetts University	Northern Connecticut University
60	66
70	67
70	70
70	70
75	72

a. Describe in words and graphs the variability of the heights of the two teams.
b. Verify that the means, medians, and modes for the two teams are equal.

Solution

a. There are some distinct differences between the teams. The Western Massachusetts (WMU) team has a player who is relatively short (60 inches; 5 feet tall) and a player who is very tall (75 inches; 6 feet, 3 inches tall). The Northern Connecticut (NCU) team has players whose heights are all within 6 inches of each other.

b. But despite the differences in (**a**), the mean, median, and mode of the heights for the two teams are precisely the same. As illustrated in Figure 3.10, the mean height (red triangle) for each team is 69 inches, the median height (green triangle) for each team is 70 inches, and the mode height (yellow triangle) for each team is 70 inches.

$$\bar{x}_{\text{WMU}} = \frac{60 + 70 + 70 + 70 + 75}{5} = \frac{345}{5} = 69$$

$$\bar{x}_{\text{NCU}} = \frac{66 + 67 + 70 + 70 + 72}{5} = \frac{345}{5} = 69$$

Clearly, these measures of location do not give us the whole picture. We need **measures of variability** (or **measures of spread** or **measures of dispersion**) that will describe how spread out the data values are. Figure 3.10 illustrates that the heights of the WMU team are *more spread out* than the heights of the NCU team.

FIGURE 3.10
Comparative dotplots of the heights of two volleyball teams.

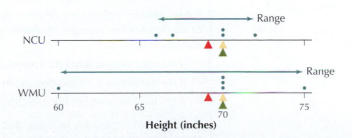

Just as there were several measures of the center of a data set, there are also a variety of ways to measure how spread out a data set is. The simplest measure of variability is the **range**.

> The **range** of a data set is the difference between the largest value and the smallest value in the data set:
>
> $$\text{range} = \text{largest value} - \text{smallest value} = \text{maximum} - \text{minimum}$$

A larger range is an indication of greater variability, or greater spread, in the data set.

EXAMPLE 3.9

RANGE OF THE VOLLEYBALL TEAMS' HEIGHTS

Calculate the range of player heights for each of the WMU and NCU teams.

Solution

What Results Might We Expect?

From Figure 3.10, it is intuitively clear that the heights of the WMU team are more spread out than the heights of the NCU team. Therefore, we would expect the range of the WMU team to be larger than the range of the NCU team, reflecting its greater variability.

$$\text{range}_{WMU} = \text{largest value} - \text{smallest value} = 75 - 60 = 15 \text{ inches}$$

$$\text{range}_{NCU} = \text{largest value} - \text{smallest value} = 72 - 66 = 6 \text{ inches}$$

Now You Can Do Exercises 11–22.

As we expected, the range for WMU is indeed larger than the range for NCU, reflecting WMU's greater variability in height.

The range is quite simple to calculate. However, it does have its drawbacks. For example, the range is quite sensitive to extreme values, since it is calculated from the difference of the two most extreme values in the data set. *It completely ignores all of the other data values in the data set.* We would prefer our measure of variability to quantify spread with respect to the center, as well as to actually use all of the available data values. Two such measures are the variance and the **standard deviation.**

2 POPULATION VARIANCE AND POPULATION STANDARD DEVIATION

Before we learn about the variance and the standard deviation, we need to get a firm understanding of what a **deviation** means, in the statistical sense.

> **Deviation**
>
> A **deviation** for a given data value x is the difference between the data value and the mean of the data set. For a sample, the deviation equals $x - \bar{x}$. For a population, the deviation equals $x - \mu$.
>
> - If the data value is larger than the mean, the deviation will be positive.
> - If the data value is smaller than the mean, the deviation will be negative.
> - If the data value equals the mean, the deviation will be zero.
>
> The deviation can roughly be thought of as the distance between a data value and the mean, except that the deviation can be negative while distance is always positive.

EXAMPLE 3.10 **CALCULATING DEVIATIONS**

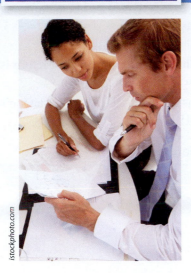

Ashley and Brandon, certified public accountants.

Ashley and Brandon are certified public accountants working for a large accounting firm, preparing tax returns for small business clients. Because tax returns are often filed close to the deadline, it is important that the returns be prepared in a timely fashion, with not a lot of variability in the length of time it takes to prepare a return. The Chief Accountant kept careful track of the amount of time (in hours, Table 3.9) for all the tax returns prepared by Ashley and Brandon in the last week of March.

a. Find the mean preparation time for each accountant.

b. Use comparative dotplots to compare the variability of Ashley and Brandon's tax preparation times.

c. Calculate the deviations for each of Ashley and Brandon.

Table 3.9 Preparation times (in hours) for Ashley and Brandon

Ashley	5	7	8	9	11
Brandon	3	5	7	11	14

Solution

Because the data represent *all* the tax returns for the indicated period, they may be considered a population.

a. For Ashley:

$$\mu = \frac{\sum x}{N} = \frac{5 + 7 + 8 + 9 + 11}{5} = 8 \text{ hours}$$

For Brandon:

$$\mu = \frac{\sum x}{N} = \frac{3 + 5 + 7 + 11 + 14}{5} = 8 \text{ hours}$$

So the two accountants spent the same mean amount of time in tax preparation.

b. Figure 3.11 contains comparative dotplots of Ashley and Brandon's tax preparation times. Note that Brandon's preparation times vary more than Ashley's. Compared to Ashley, we can say that Brandon's tax preparation times

- are *more spread out*,
- show *greater variability*,
- have *more variation*,
- are *more disperse*.

The Chief Accountant probably prefers a more consistent tax preparation time, with less variability.

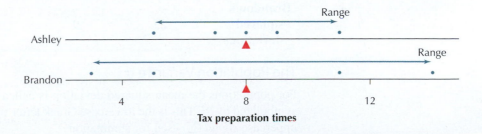

FIGURE 3.11 Brandon's tax preparation times are more spread out.

c. Here we find the deviations, $x - \mu$.

- Ashley's mean preparation time is $\mu = 8$ hours. Her first tax return took $x = 5$ hours, so the deviation for this first tax return is $x - \mu = 5 - 8 = -3$. Note that, when $x < \mu$, the deviation is negative.

- Ashley's last tax return took 11 hours, so the deviation for this last return is $x - \mu = 11 - 8 = 3$. Note that, when $x > \mu$, the deviation is positive.

- Continuing in this way, we find the deviations for all of Ashley's and Brandon's tax preparation time, as recorded in Table 3.10.

**Now You Can Do
Exercises 23–28.**

Table 3.10 Tax preparation times and their deviations

Ashley's times	5	7	8	9	11
Ashley's deviations	$5-8=-3$	$7-8=-1$	$8-8=0$	$9-8=1$	$11-8=3$
Brandon's times	3	5	7	11	14
Brandon's deviations	$3-8=-5$	$5-8=-3$	$7-8=-1$	$11-8=3$	$14-8=6$

These deviations are used for the most widespread measures of spread: the variance and the standard deviation. However, we cannot use the mean deviation, because the mean deviation always equals zero. For example,

- Ashley's mean deviation: $\dfrac{(-3) + (-1) + 0 + 1 + 3}{5} = 0$

- Brandon's mean deviation: $\dfrac{(-5) + (-3) + (-1) + 3 + 6}{5} = 0$

The mean deviation always equals zero for any data set because the positive and negative deviations cancel each other out. Thus, the mean deviation is not a useful measure of spread. To avoid this problem, we will work with the squared deviations.

Table 3.11 shows the squared deviations for Ashley and Brandon. Note that Brandon's squared deviations are on average larger than Ashley's, reflecting the greater spread in Brandon's preparation times. It is therefore logical to build our measure of spread using the *mean squared deviation*.

Table 3.11 Squared deviations of tax preparation times

Ashley's deviations	-3	-1	0	1	3
Ashley's squared deviations	9	1	0	1	9
Brandon's deviations	-5	-3	-1	3	6
Brandon's squared deviations	25	9	1	9	36

The Population Variance σ^2

For populations the mean squared deviation is called the **population variance** and is symbolized by σ^2. This is the lowercase Greek letter *sigma*, not to be confused with the uppercase *sigma* (Σ) used for summation.

> The **population variance** σ^2 is the mean of the squared deviations in the population and is given by the formula
>
> $$\sigma^2 = \frac{\sum(x - \mu)^2}{N}$$

Notice that the numerator in σ^2 is a sum of squares. Squared numbers can never be negative, so a sum of squares also can never be negative. The denominator, N, the population size, also can never be negative. Thus, σ^2 can never be negative. The only time $\sigma^2 = 0$ is when all the population data values are equal.

EXAMPLE 3.11

CALCULATING THE POPULATION VARIANCES FOR ASHLEY AND BRANDON

Calculate the population variances of the tax preparation times for Ashley and Brandon.

Solution

Using the squared deviations from Table 3.11, we have

$$\sigma^2 = \frac{\sum(x - \mu)^2}{N} = \frac{9 + 1 + 0 + 1 + 9}{5} = \frac{20}{5} = 4$$

for Ashley, and

$$\sigma^2 = \frac{\sum(x - \mu)^2}{N} = \frac{25 + 9 + 1 + 9 + 36}{5} = \frac{80}{5} = 16$$

Now You Can Do Exercises 29–34.

for Brandon. The population variance of the tax preparation times for Brandon is greater than that for Ashley, thus indicating that Brandon's tax preparation times are moral variable than Ashley's.

However, what is the *meaning* of the values we got for σ^2, 4 and 16, apart from their comparative value? The problem is that the units of these values represent *hours squared*, which is not a useful measure. Unfortunately, the intuitive meaning of the population variance is not self-evident.

The Population Standard Deviation σ

In practice, the *standard deviation* is easier to interpret than the variance. The standard deviation is simply the square root of the variance, and by taking the square root, we return the units of measure back to the original data unit (for example, "hours" rather than "hours squared"). The symbol for the **population standard deviation** is σ. Conveniently, $\sqrt{\sigma^2} = \sigma$.

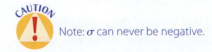
Note: σ can never be negative.

> The **population standard deviation** σ is the positive square root of the population variance and is found by
>
> $$\sigma = \sqrt{\frac{\sum(x - \mu)^2}{N}}$$

| EXAMPLE 3.12 | CALCULATING THE POPULATION STANDARD DEVIATIONS FOR ASHLEY AND BRANDON |

Calculate the population standard deviations of the tax preparation times for Ashley and Brandon.

Solution

Since Brandon's population variance of 16 is larger than Ashley's population variance of 4, Brandon's population standard deviation will also be larger, since we are simply taking the square root. We have

$$\sigma = \sqrt{\sigma^2} = \sqrt{4} = 2$$

for Ashley and

$$\sigma = \sqrt{\sigma^2} = \sqrt{16} = 4$$

for Brandon.

Now You Can Do Exercises 35–40.

The population standard deviation of Brandon's tax preparation times is 4 hours, which is larger than Ashley's 2 hours. As expected, the greater variability in Brandon's preparation times leads to a larger value for his population standard deviation σ.

| What Do These Numbers Mean? | **The Standard Deviation** |

So how do we interpret these values for σ? One quick thumbnail interpretation of the standard deviation is that it represents a "typical" deviation. That is, the value of σ *represents a distance from the mean that is representative for that data set.* For example, the typical distance from the mean for Ashley's and Brandon's tax preparation times is 2 hours and 4 hours, respectively.

| Developing Your Statistical Sense | **Communicating the Results** |

As you study statistics, keep in mind that during your career you will likely need to explain your results to others who have never taken a statistics course. Therefore, you should always keep in mind *how to interpret your results to nonspecialists.* Communication and interpretation of your results can be as important as the results themselves.

3 COMPUTE THE SAMPLE VARIANCE AND SAMPLE STANDARD DEVIATION

The Sample Variance s^2 and the Sample Standard Deviation s

Note: In this book, we will work with sample statistics unless the data set is identified as a population.

In the real world, we usually cannot determine the exact value of the population mean or the population standard deviation. Instead, we use the sample mean and **sample standard deviation** to estimate the population parameters. The **sample variance** also depends on the concept of the mean squared deviation. If the sample mean is $\bar{x}$, and the sample size is n, then we would expect the formula for the sample variance to resemble the formula for the population variance, namely

$$\frac{\sum(x - \bar{x})^2}{n}$$

However, this formula has been found to underestimate the population variance, so that we need to replace the n in the denominator with $n - 1$. We therefore have the following.

> The **sample variance s^2** is approximately the mean of the squared deviations in the sample and is found by
> $$s^2 = \frac{\sum(x - \bar{x})^2}{n - 1}$$

The sample standard deviation is perhaps the second most important statistic you will encounter in this book (after the sample mean $\bar{x}$). It is the most commonly used measure of spread. The sample standard deviation is simply the square root of the sample variance and takes as its symbol the letter s, which is the Roman letter for the Greek σ. Again, $s = \sqrt{s^2}$.

Neither s^2 nor s can ever be negative.

> The **sample standard deviation s** is the positive square root of the sample variance s^2:
> $$s = \sqrt{s^2} = \sqrt{\frac{\sum(x - \bar{x})^2}{n - 1}}$$
>
> The value of s may be interpreted as the typical distance between a data value and the sample mean, for a given data set.

EXAMPLE 3.13

CALCULATING THE SAMPLE VARIANCE AND THE SAMPLE STANDARD DEVIATION

Suppose we obtain a sample of size $n = 3$ from Ashley's population of tax preparation times, as follows: 5 hours, 8 hours, 11 hours, as shown.

Ashley's Population	5	7	8	9	11
	⬇		⬇		⬇
Ashley's Sample	5		8		11

a. Calculate the sample variance of the tax preparation times.
b. Compute the sample standard deviation of the tax preparation times.
c. Interpret the sample standard deviation.

Solution

a. We first find the sample mean $\bar{x} = \frac{\sum x}{n} = \frac{5 + 8 + 11}{3} = 8$. It so happens that the value for this sample mean equals the population mean $\mu = 8$, but this is only a coincidence.
Then the sample variance is

$$s^2 = \frac{\sum(x - \bar{x})^2}{n - 1} = \frac{(5 - 8)^2 + (8 - 8)^2 + (11 - 8)^2}{2} = \frac{9 + 0 + 9}{2} = 9$$

The sample variance is $s^2 = 9$ hours squared.

b. Then the sample standard deviation is

$$s = \sqrt{s^2} = \sqrt{9} = 3 \text{ hours.}$$

**Now You Can Do
Exercises 41–46.**

c. For this sample of Ashley's tax returns, the typical difference between a tax preparation time and the mean preparation time is 3 hours.

**Developing Your
Statistical Sense**

Less Variation Is Better

In most real-world applications, consistency is a great advantage. In statistical data analysis, less variation is often better even though variability is natural and cannot be eliminated. Throughout the text, you will find that smaller variability will lead to

- more precise estimates and
- higher confidence in conclusions.

In the exercises you will find alternative **computational formulas** for the variance and standard deviation.

EXAMPLE 3.14

USING TECHNOLOGY TO FIND THE SAMPLE VARIANCE AND SAMPLE STANDARD DEVIATION

gasmileage

Find the sample standard deviation and the sample variance of the city gas mileage for the 2011 cars shown in the following table. Use (a) the TI-83/84, (b) Excel, and (c) Minitab.

Vehicle	City mpg
Cadillac CTS	18
Ford Fusion Hybrid	41
Ford Taurus	18
Honda Civic	25
Rolls Royce Phantom	11
Toyota Camry Hybrid	31

Source: **www.fueleconomy.gov.**

Solution

Using the instructions in the Step-by-Step Technology Guide on page 92, we obtain the following output:

a. The TI-83/84 output is shown in Figure 3.12a. The sample standard deviation s is given as $Sx = 10.77032961$. The sample variance is $s^2 = (10.77032961)^2 = 116$.

b. The Excel output is provided in Figure 3.12b. The sample standard deviation and sample variance are highlighted.

c. The Minitab output is provided in Figure 3.12c. Note that Minitab rounds s to two decimal places.

CAUTION For the TI-83/84, do not confuse Sx, the TI's notation for the sample standard deviation, with σx, which the TI-83/84 uses to label the population standard deviation.

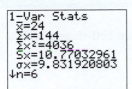

```
1-Var Stats
  x̄=24
  Σx=144
  Σx²=4036
  Sx=10.77032961
  σx=9.831920803
↓n=6
```

FIGURE 3.12a TI-83/84 output.

```
           City mpg
Mean                        24
Standard Error      4.396968653
Median                    21.5
Mode                        18
Standard Deviation 10.77032961
Sample Variance            116
```

FIGURE 3.12b Excel output.

Descriptive Statistics: City mpg				
Variable	Mean	StDev	Variance	Range
City mpg	24.00	10.77	116.00	30.00

FIGURE 3.12c Minitab output.

Next we turn to methods for applying the standard deviation.

4 THE EMPIRICAL RULE

If the data distribution is bell-shaped we may apply the Empirical Rule to find the approximate percentage of data that lies within k standard deviations of the mean, for $k = 1, 2,$ or 3.

> **The Empirical Rule**
>
> If the data distribution is bell-shaped:
> - About 68% of the data values will fall within 1 standard deviation of the mean.
>   - For a population, about 68% of the data will lie between $\mu - 1\sigma$ and $\mu + 1\sigma$.
>   - For a sample, about 68% of the data will lie between $\bar{x} - 1s$ and $\bar{x} + 1s$.
> - About 95% of the data values will fall within 2 standard deviations of the mean.
>   - For a population, about 95% of the data will lie between $\mu - 2\sigma$ and $\mu + 2\sigma$.
>   - For a sample, about 95% of the data will lie between $\bar{x} - 2s$ and $\bar{x} + 2s$.
> - About 99.7% of the data values will fall within 3 standard deviations of the mean.
>   - For a population, about 99.7% of the data will lie between $\mu - 3\sigma$ and $\mu + 3\sigma$.
>   - For a sample, about 99.7% of the data will lie between $\bar{x} - 3s$ and $\bar{x} + 3s$.
>
> Figure 3.13 illustrates these approximate percentages.
>
> About 99.7% of data lie within 3 standard deviations of mean
>
> About 95% within 2 standard deviations
>
> About 68% within 1 standard deviations
>
> 34%   34%
>
> 0.15%   13.5%   13.5%   0.15%
>
> 2.35%   2.35%
>
> $\mu-3\sigma$ $\mu-2\sigma$ $\mu-1\sigma$ μ $\mu+1\sigma$ $\mu+2\sigma$ $\mu+3\sigma$
>
> **FIGURE 3.13** Empirical Rule, with approximate percentages.

EXAMPLE 3.15

USING THE EMPIRICAL RULE TO FIND PERCENTAGES

Suppose we know that student grade point averages (GPAs) are bell-shaped with a mean of $\mu = 2.5$ and a standard deviation of $\sigma = 0.5$.

a. Find the percentage of GPAs between 2.0 and 3.0.
b. Compute the percentage of GPAs that are above 3.5.

Solution

a. We see that GPA = 2.0 represents 1 standard deviation below the mean, because

$$\mu - 1\sigma = 2.5 - 1(0.5) = 2.0$$

Similarly, GPA = 3.0 represents 1 standard deviation above the mean, since

$$\mu + 1\sigma = 2.5 + 1(0.5) = 3.0$$

Remember: That English word "about" is not optional; it is required. The Empirical Rule is an approximation of normal distribution probabilities that we will examine more closely in Chapter 6.

Thus, "GPAs between 2.0 and 3.0" represents between $\mu - 1\sigma$ and $\mu + 1\sigma$, that is, within 1 standard deviation of the mean. Since the data distribution is bell-shaped we may use the Empirical Rule. Therefore, about 68% of the GPAs lie between 2.0 and 3.0, as shown in Fig 3.14.

b. We note that GPA = 3.5 represents 2 standard deviations above the mean, because

$$\mu + 2\sigma = 2.5 + 2(0.5) = 3.5$$

We know from the Empirical Rule that about 95% of the GPAs lie within 2 standard deviations of the mean, so that about 95% of the GPAs lie between 1.5 and 2.5. The left-over area of about 5% in the two tails in Figure 3.14 is the percentage of GPAs above 3.5 or below 1.5. Because the bell-shaped curve is symmetric, the two tail areas are equal in area, which means that about 2.5% of the GPAs lie above 3.5 (Figure 3.14).

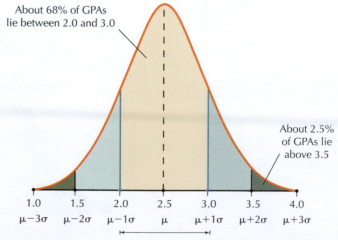

Now You Can Do Exercises 47–54.

FIGURE 3.14 Example of Empirical Rule applied to GAPs.

5 CHEBYSHEV'S RULE

P. L. Chebyshev (1821–94, Russia) derived a result, called **Chebyshev's Rule,** that can be applied to any data set whatsoever.

> **Chebyshev's Rule**
> The proportion of values from a data set that will fall within k standard deviations of the mean will be *at least*
>
> $$\left(1 - \frac{1}{k^2}\right)100\%,$$
>
> where $k > 1$. Chebyshev's Rule may be applied to either samples or populations. For example:
> - When $k = 2$, at least 3/4 (or 75%) of the data values will fall within 2 standard deviations of the mean.
> - When $k = 3$, at least 8/9 (or 88.89%) of the data values will fall within 3 standard deviations of the mean.

Because of the phrase "at least," we say that Chebyshev's Rule provides minimum percentages, rather than the approximate percentages provided by the Empirical Rule. The actual percentage may be much greater than the minimum percentage provided by Chebyshev's Rule.

EXAMPLE 3.16

USING CHEBYSHEV'S RULE TO FIND MINIMUM PERCENTAGES

An instructor giving an exam with an unknown data distribution knows that the mean is 70 and the standard deviation is 10. Find the minimum percentage of exam scores that is

a.   Between 50 and 90
b.   Between 55 and 85
c.   Between 60 and 80.

Solution

Since the data distribution is unknown, we cannot apply the Empirical Rule.

a.   Because 50 lies 2 standard deviations below the mean

$$\mu - 2\sigma = 70 - 2(10) = 50$$

and 90 lies 2 standard deviations above the mean

$$\mu + 2\sigma = 70 + 2(10) = 90$$

this question is really asking what is the minimum percentage within $k = 2$ standard deviations of the mean. From Chebyshev's Rule, the minimum percentage is

$$\left(1 - \frac{1}{k^2}\right)100\% = \left(1 - \frac{1}{2^2}\right)100\% = \left(\frac{3}{4}\right)100\% = 75\%$$

Thus, *at least* 75% of the exam scores will lie between 50 and 90.

b. The scores 55 and 85 lie $k = 1.5$ standard deviations below and above the mean, respectively. Therefore, at least

$$\left(1 - \frac{1}{1.5^2}\right)100\% = \left(1 - \frac{1}{2.25}\right)100\% = 55.6\%$$

of the exam scores will lie between 55 and 85.

Now You Can Do
Exercises 55–62.

c. The scores 60 and 80 lie $k = 1$ standard deviation below and above the mean, respectively. Unfortunately, Chebyshev's Rule is restricted to situations where $k > 1$. Thus, we cannot answer this question.

If a given data set is bell-shaped, either the Empirical Rule or Chebyshev's Rule may be applied to it.

CASE STUDY

Can the Financial Experts Beat the Darts?

Recall from Section 3.1 the *Wall Street Journal* competition between stocks chosen randomly by *Journal* staff members throwing darts and stocks chosen by a team of four financial experts. Note from Figure 3.15 that the DJIA exhibits less variability than the other two portfolios. This smaller variability is due to the fact that the DJIA is made up of 29 component stocks, whereas each portfolio is made up of only 4 stocks. Smaller sample sizes can be associated with increased variability, since an unusual result in one value has a relatively strong effect on the mean when it is not offset by a large sample.

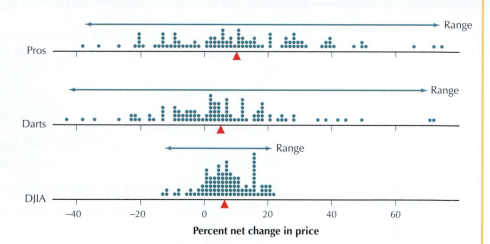

FIGURE 3.15
Comparative dotplots of the net change in prices.

Which of the portfolios, pros or darts, shows greater variability? It is difficult to determine just by examining Figure 3.15 which has the greater standard deviation. We therefore turn to the Minitab descriptive statistics in Figure 3.16. The range for the darts, 115.90, is greater than the range for the pros, 112.80. But the standard deviation for the darts (19.39) is less than that of the pros (22.25).

FIGURE 3.16
Descriptive statistics for the portfolios.

Descriptive Statistics: Pros, Darts, DJIA				
Variable	Mean	StDev	Variance	Range
Pros	10.95	22.25	494.91	112.80
Darts	4.52	19.39	375.91	115.90
DJIA	6.793	8.031	64.505	35.600

Measures of spread may disagree about which data set is more variable. However, since the range takes into account only the two most extreme data values, the standard deviation is the preferred measure of spread, since it uses all the data values. Our conclusion, therefore, is that the returns for the professionals exhibit the greater variability.

Why did the pros have more variability than the darts? After all, in finance, high variability is not necessarily advantageous because it is associated with greater *risk*. The professionals evidently chose higher-risk stocks with greater potential for high returns—but also greater potential for losing money. ■

SECTION 3.2   Summary

1. The simplest measure of variability, or measure of spread, is the range. The range is simply the difference between the maximum and minimum values in a data set, but the range has drawbacks because it relies on the two most extreme data values.

2. The variance and standard deviation are measures of spread that utilize all available data values. The population variance can be thought of as the mean squared deviation. The standard deviation is the square root of the variance. We interpret the value of the standard deviation as the typical deviation, that is, the typical distance between a data value and the mean.

3. The variance and standard deviation may also be calculated for a sample. We interpret the value of the

standard deviation as the typical deviation, that is, the typical distance between a data value and the mean.

4. For bell-shaped distributions, the Empirical Rule may be applied. The Empirical Rule states that, for bell-shaped distributions, about 68%, 95%, and 99.7% of the data values will fall within 1, 2, and 3 standard deviations of the mean, respectively.

5. Chebyshev's Rule allows us to find the minimum percentage of data values that lie within a certain interval. Chebyshev's Rule states that the proportion of values from a data set that will fall within k standard deviations of the mean will be at least $[1 - 1/(k)^2]100\%$ where $k > 1$.

SECTION 3.2   Exercises

Unless a data set is identified as a population, you can assume that it is a sample.

Clarifying the Concepts

1. Explain what a deviation is.

2. What is the interpretation of the value of the standard deviation?

3. State one benefit and one drawback of using the range as a measure of spread.

4. True or false: If two data sets have the same mean, median, and mode, then they are identical.

5. What is one benefit of using the standard deviation instead of the range as a measure of spread? What is one drawback?

6. Which measure of spread represents the mean squared deviation for the population?

7. True or false: Chebyshev's Rule provides exact percentages.

8. When can the sample standard deviation s be negative?

9. When does the sample standard deviation s equal zero?

10. When may the Empirical Rule be used?

Practicing the Techniques

Find the range of the data in Exercises 11–22.

11. 5, 25, 0, 10

12. 40, 40, 60, 80, 80

13. 10, 10, 10, 10, 10

14. –5, –7, –4, –8, –6

15. 1.0, 3.0, 4.0, 2.0

16. 40, 60, 60, 60, 80

17. 3.14159, 3.14159, 3.14159, 3.14159

18. 3, 0, 5, –3, –5

19. 15, 20, 10, 15, 10, 20, 15, 10, 20, 10

20. 79, 92, 65, 75, 67, 59, 88, 100, 85, 60

21. –15, –20, –10, –15, –10, –20, –15, –10, –20, –10

22. 69, 82, 55, 65, 57, 49, 78, 90, 75, 50

For the population data in Exercises 23–28, do the following:

    a. Find the population mean μ.
    b. Calculate the deviations $x - \mu$.

23. 5, 25, 0, 10

24. 40, 40, 60, 80, 80

25. 10, 10, 10, 10, 10

26. –5, –7, –4, –8, –6

27. 1.0, 3.0, 4.0, 2.0

28. 40, 60, 60, 60, 80

For the population data in Exercises 29–34, do the following:

    a. Using the deviations you computed in Exercises 23–28, find the squared deviations.
    b. Find the mean of the squared deviations. This is the population variance.

29. Data from Exercise 23

30. Data from Exercise 24

31. Data from Exercise 25

32. Data from Exercise 26

33. Data from Exercise 27

34. Data from Exercise 28

For the population data in Exercises 35–40, use your work from Exercises 29–34 to help calculate the population standard deviation.

35. Data from Exercise 23

36. Data from Exercise 24

37. Data from Exercise 25

38. Data from Exercise 26

39. Data from Exercise 27

40. Data from Exercise 28

For the sample data in Exercises 41–46, do the following:

    a. Calculate the sample variance.
    b. Compute the sample standard deviation.
    c. Interpret the sample standard deviation.

41. 3.14159, 3.14159, 3.14159, 3.14159

42. 3, 0, 5, –3, –5

43. 15, 20, 10, 15, 10, 20, 15, 10, 20, 10

44. 79, 92, 65, 75, 67, 59, 88, 100, 85, 60

45. –15, –20, –10, –15, –10, –20, –15, –10, –20, –10

46. 69, 82, 55, 65, 57, 49, 78, 90, 75, 50

For Exercises 47–50, use the following information. A data distribution is bell-shaped, and has a mean of 100 and a standard deviation of 10. Use the Empirical Rule to approximate the percentage of data.

47. Between 90 and 110

48. Between 80 and 120

49. Between 70 and 130

50. Greater than 110

For Exercises 51–54, use the following information. A data distribution is bell-shaped, and has a mean of 500 and a standard deviation of 100. Use the Empirical Rule to approximate the percentage of data.

51. Between 300 and 700

52. Greater than 700

53. Less than 300

54. Between 300 and 500

For Exercises 55–58, use the following information. A data set has an unknown distribution, with a mean of 10 and a standard deviation of 2. Use Chebyshev's Rule to estimate the minimum possible percentage of data.

55. Between 6 and 14

56. Between 4 and 16

57. Between 2 and 18

58. Between 3 and 17

For Exercises 59–62, use the following information. A data set has an unknown distribution, with a mean of 50 and a standard deviation of 5. If possible, use Chebyshev's Rule to estimate the minimum possible percentage of data.

59. Between 40 and 60

60. Between 35 and 65

61. Between 37.5 and 62.5

62. Between 45 and 55

63. Match the histograms in **(a)–(d)** to the statistics in **(i)–(iv)**.

    i. Mean = 75, standard deviation = 20
    ii. Mean = 75, standard deviation = 10
    iii. Mean = 50, standard deviation = 20
    iv. Mean = 50, standard deviation = 10

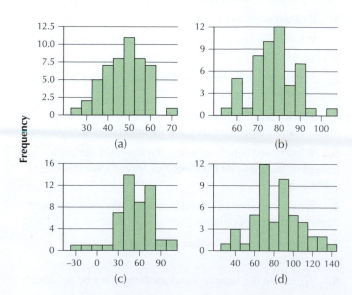

64. Match the histograms in (**a**)–(**d**) (see the next page) to the statistics in (**i**)–(**iv**).

    i. Mean = 1, standard deviation = 1
   ii. Mean = 1, standard deviation = 0.1
  iii. Mean = 0, standard deviation = 1
  iv. Mean = 0, standard deviation = 0.1

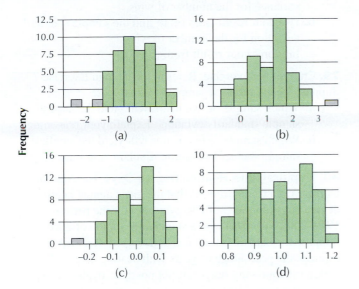

(a)   (b)

(c)   (d)

Applying the Concepts

For the following exercises, make sure to state your answers in the proper units, such as "years" or "years squared."

Fuel Economy. Refer to Table 3.5 from the Section 3.1 exercises on page 94 to answer Exercises 65–68. The data represent a sample.

65. Find the following measures of spread for the number of cylinders:
    a. Range
    b. Variance
    c. Standard deviation

66. Find the following measures of spread for the engine size:
    a. Range
    b. Variance
    c. Standard deviation

67. Find the following measures of spread for the fuel economy:
    a. Range
    b. Variance
    c. Standard deviation

68. Is "cylinders squared" easy for nonstatisticians to understand? Which measure do you find to be more easily understood and interpreted for these data, the variance or the standard deviation?

SAT scores. Refer to Table 3.6 on page 94 of the Section 3.1 exercises to answer Exercises 69–72. The data represent a sample.

69. Find the following measures of spread for the SAT Mathematics scores:

    a. Range
    b. Variance
    c. Standard deviation

70. Find the following measures of spread for the SAT Reading scores:
    a. Range
    b. Variance
    c. Standard deviation

71. Find the following measures of spread for the SAT Writing scores:
    a. Range
    b. Variance
    c. Standard deviation

72. Is "SAT scores: squared" easy for nonstatisticians to understand? Which measure do you find to be more easily understood and interpreted for these data, the variance or the standard deviation?

Zooplankton and Phytoplankton. Refer to the table below for Exercises 73 and 74. *Meta-analysis* refers to the statistical analysis of a set of similar research studies. In a meta-analysis, each data value represents an effect size calculated from the results of a particular study. The table contains effect sizes calculated in a meta-analysis for zooplankton and phytoplankton.[3]

 plankton

Zooplankton		Phytoplankton	
−2.37	−3.00	10.61	3.04
−0.64	−0.68	2.97	0.65
−2.05	−1.39	1.58	2.55
−1.54	−0.64	2.55	1.05
−6.60	−3.88	5.67	2.11
0.26		1.57	

73. Calculate the ranges for the zooplankton and the phytoplankton.
    a. Which has the greater range?
    b. Which plankton group has the greater variability according to the range?

74. Calculate the standard deviations for the zooplankton and the phytoplankton.
    a. Which has the greater standard deviation?
    b. Which plankton group has the greater variability according to the standard deviation? Does this concur with your answer from the previous exercise?
    c. Without calculating the variances, say which group has the greater variance. How do you know this?

Ant Size. Use the following information for Exercises 75 and 76. A study compared the size of ants from different colonies. The masses (in milligrams) of samples of ants from two different colonies are shown in the accompanying table.[4]

 antcolony

Colony A		Colony B	
109	134	148	115
120	94	110	101
94	113	110	158
61	111	97	67
72	106	136	114

75. Calculate the range for each ant colony.
  a. Which has the greater range?
  b. Which colony has the greater variability according to the range?

76. Calculate the variance for each colony.
  a. Which has the greater variance?
  b. Which colony has the greater variability according to the variance? Does this concur with your answer from the previous exercise?
  c. Without calculating the standard deviations, say which colony has the greater standard deviation. How do you know this?

77. Computational Formula for the Population Variance and Standard Deviation: Wins in Baseball. The following table provides the number of wins for all the teams in the American League East Division for the 2011 season, which we can consider to be a population.

Team	Wins
New York Yankees	97
Tampa Bay Rays	91
Boston Red Sox	90
Toronto Blue Jays	81
Baltimore Orioles	69

An alternative computational formula for the population variance is as follows.

$$\sigma^2 = \frac{\sum x^2 - (\sum x)^2/N}{N}$$

  a. Use the computational formula to find the population variance for the number of wins.
  b. Use your result from (a) to find the population standard deviation for the number of wins.

(*Note:* $\sum x^2$ means that you square each data value and then add up the squared data values, and $(\sum x)^2$ means that you add up all the data values and then square the sum.)

78. Computational Formula for the Sample Variance and Standard Deviation. Refer to the previous exercise. Suppose a random sample of size $n = 3$ from these teams yields the New York Yankees, the Tampa Bay Rays, and the Baltimore Orioles.

An alternative computational formula for the sample variance is as follows.

$$s^2 = \frac{\sum x^2 - (\sum x)^2/n}{n - 1}$$

  a. Use the computational formula to find the sample variance for the number of wins.
  b. Use your result from (a) to find the sample standard deviation for the number of wins.
  c. Interpret your result from (b).

79. Challenger Exercise. Refer to the table in Exercise 77. Suppose we are taking a sample of size $n = 2$.
  a. Which sample of two teams will yield the largest sample standard deviation. Explain your reasoning.
  b. Which sample of two teams will yield the smallest sample standard deviation. Explain your reasoning.

80. Empirical Rule: Heating Degree-Days. The National Climate Data Center reports that the mean annual heating degree-days (an index of energy usage) for the period 1949–2006 was 4500 with a standard deviation of 200. Suppose the data distribution is bell-shaped. If possible, estimate the percentage of years with heating degree-days within the following ranges. If not possible, explain why.
  a. Between 4100 and 4900 heating degree-days
  b. Between 3900 and 5100 heating degree-days
  c. Between 4300 and 4700 heating degree-days

81. Empirical Rule: Solar Power Production. The U.S. Department of Energy reports that the mean annual production of solar power in the United States for the years 1989–2006 was 66 trillion Btu (British thermal units) with a standard deviation of 4 trillion Btu. Suppose the data distribution is bell-shaped. If possible, estimate the percentage of years with solar power production within the following ranges. If not possible, explain why.
  a. Between 62 trillion and 70 trillion Btu
  b. Between 60 trillion and 72 trillion Btu
  c. Above 72 trillion Btu

82. Chebyshev's Rule. Refer to Exercise 80. Suppose that we did not know that the distribution of heating degree-days is bell-shaped. If possible, find minimums for the three percentages in Exercise 80.

83. Chebyshev's Rule. Refer to Exercise 81. Suppose that we did not know that the distribution of solar power production is bell-shaped. If possible, find minimums for the three percentages in Exercise 81.

SAT Scores. Refer to Table 3.6 (page 94) for Exercises 84–87.

84. Construct dotplots of the SAT Mathematics, the SAT Reading, and the SAT Writing tests. Which test data would you say has the greatest spread (variability)? Why?

85. Find the range and variance for the SAT Mathematics, the SAT Reading, and the SAT Writing tests. Do your findings agree with your judgment from the previous exercise?

86. *Without performing any calculations,* use your results from the previous exercise to state which test has (a) the largest standard deviation, and (b) the smallest standard deviation.

⚠ 87. Now suppose we omit the Reading test from the data.

    a. *Without recalculating them,* describe how this would affect the values of the measures of spread you found for the SAT Mathematics test and the SAT Writing test.

    b. Now recalculate the three measures of spread for the SAT Mathematics and the SAT Writing test. Was your judgment in (**a**) supported?

Women's Volleyball Team Heights. Refer to Table 3.8 (page 97) for Exercises 88–90.

88. Suppose a new player joins the NCU team. She is 7 feet tall (84 inches) and replaces the 72-inch-tall player.

    a. Would you expect the standard deviation to go up or down, and why?

    b. Now find the standard deviation for the team including the new player. Was your intuition correct?

89. Linear Transformations. Add 4 inches to the height of each player on the WMU team.

    a. Recalculate the range and standard deviation.

    b. Formulate a rule for the behavior of these measures of variability when a constant (like 4) is added to each member of the data set.

90. Linear Transformations Starting with the original data, double the height of each player on the NCU team.

    a. Recalculate the range and standard deviation.

    b. Formulate a rule for the range and standard deviation when the data values are doubled.

Coefficient of Variation. The *coefficient of variation* enables analysts to compare the variability of two data sets that are measured on different scales. The coefficient of variation (CV) itself does not have a unit of measure. Larger values of CV indicate greater variability or spread. The coefficient of variation is given as

$$CV = \frac{\text{standard deviation}}{\text{mean}} \cdot 100\%$$

Use this measure of variability for Exercises 91 and 92.

91. Coefficient of Variation for Cylinders, Engine Size, and City MPG. Refer to Table 3.5 on page 94.

    a. Calculate the coefficient of variation for the following variables: *cylinders, engine size,* and *city mpg.*

    b. According to the coefficient of variation, which variable has the greatest spread? The least variability?

92. Coefficient of Variation for the SAT Scores. Refer to Table 3.6 on page 94.

    a. Calculate the coefficient of variation for the SAT Mathematics, the SAT Reading, and the SAT Writing tests.

    b. According to the coefficient of variation, which test data has the greatest spread?

Mean Absolute Deviation. Recall that the variance and standard deviation use squared deviations because the mean deviation for any data set is zero. Another way to avoid negative deviations offsetting positive ones is to use the absolute value of the deviations. The *mean absolute deviation (MAD)* is a measure of spread that looks at the average of the absolute values of the deviations:

$$\text{MAD} = \frac{\sum |x_i - \bar{x}|}{n}$$

Use this measure of variability for Exercises 93 and 94.

93. Mean Absolute Deviation for the Fuel Economy Data. Refer to Table 3.5 on page 94.

    a. Find the mean absolute deviation for *cylinders, engine size,* and *city mpg.*

    b. According to the mean absolute deviation, which variable has the greatest variability? The least variability?

94. Mean Absolute Deviation for the SAT Scores. Refer to Table 3.6 on page 94.

    a. Calculate the mean absolute deviation for the SAT Mathematics, the SAT Reading, and the SAT Writing tests.

    b. According to the mean absolute deviation, which test data has the greatest spread?

Coefficient of Skewness. The coefficient of skewness quantifies the skewness of a distribution. It is defined as

$$\text{skewness} = \frac{3(\text{mean} - \text{median})}{\text{standard deviation}}$$

Most skewness values lie between -3 and 3. Negative values of skewness are associated with left-skewed distributions, while positive values are associated with right-skewed distributions. Values close to zero indicate distributions that are near by symmetric. Use this information for Exercises 95–97.

95. Coefficient of Skewness. For the following distributions, compute the coefficient of skewness and comment on the skewness of the distribution.

    a. Mean = 0, Median = 0, Standard deviation = 1

    b. Mean = 1, Median = 0, Standard deviation = 1

    c. Mean = 0, Median = 1, Standard deviation = 1

    d. Mean = 75, Median = 80, Standard deviation = 10

    e. Mean = 100, Median = 100, Standard deviation = 15

    f. Mean = 3.2, Median = 3.0, Standard deviation = 1.0

96. What is the coefficient of skewness for any distribution where the mean equals the median, regardless of the value of the standard deviation?

97. Coefficient of Skewness for the Case Study Data. The median price change for the Professional analysts is 9.60, the median for the Darts is 3.25, and the median for the DJIA is 7.00. Use this information, along with the information in Figure 3.16 on page 108 to answer the following.

  a. Calculate the coefficient of skewness for each of the Pros, the Darts, and the DJIA.

  b. Comment on the skewness of each distribution.

Bringing It All Together

98. Fuel Economy Data. You calculated the range, variance, and standard deviation for this data in Exercises 65–67. You calculated the coefficient of variation in Exercise 91 and the mean absolute deviation in Exercise 93. Use this information to do the following.

  a. Construct a table of the five measures of dispersion (range, sample variance, sample standard deviation, coefficient of variation, and mean absolute deviation) for the number of cylinders, the engine size, and the city mpg.

  b. Which measures of dispersion suggest that the *city mpg* is the most dispersed variable? *Engine size?* Number of *cylinders?*

99. SAT Scores Data. You calculated the range and variance for this data in Exercise 85. You calculated the coefficient of variation in Exercise 92 and the mean absolute deviation in Exercise 94. Use this information to do the following.

  a. Using the variance, calculate the standard deviation for the SAT Mathematics, the SAT Reading, and the SAT Writing tests.

  b. Construct a table of the five measures of spread (range, sample variance, sample standard deviation, coefficient of variation, and mean absolute deviation) for the SAT Mathematics, the SAT Reading, and the SAT Writing tests.

  c. Do the measures of spread agree on which distribution has the greatest variability?

  d. Bringing together all your statistics about measures of spread, what is your conclusion about the variability in the SAT Mathematics test, compared to the other two tests?

Construct Your Own Data Sets

100. Construct two data sets, A and B, that you make up on your own, so that the range of A is greater than the range of B. Verify this.

101. Construct two data sets, A and B, that you make up on your own, so that the standard deviation of A is greater than the range of B. Verify this.

102. Construct two data sets, A and B, that you make up on your own, so that the mean of A is greater than the mean of B, but the standard deviation of B is greater than that of A. Verify this.

103. Construct two data sets, A and B, that you make up on your own, so that the mean of A is greater than the mean of B, and the standard deviation of A is greater than that of B. Verify this.

104. Construct two data sets, A and B, that you make up on your own, so that the range of A is greater than the range of B, but the standard deviation of B is greater than that of A. Verify this. (*Hint:* Remember the sensitivity of the standard deviation to extreme values.)

3.3 WORKING WITH GROUPED DATA

OBJECTIVES By the end of this section, I will be able to . . .

  1 Calculate the weighted mean.

  2 Estimate the mean for grouped data.

  3 Estimate the variance and standard deviation for grouped data.

1 THE WEIGHTED MEAN

Note: Before tackling this section, you may wish to review Section 2.2, "Graphs and Tables for Quantitative Data" (page 49).

Sometimes not all the data values in a data set are of equal importance. Certain data values may be assigned greater imporantance or weight than others when calculating the mean. For example, have you ever figured out what your final grade for a course was based on the percentages listed in the syllabus? What you actually found was the **weighted mean** of your grades.

> **Weighted Mean**
>
> To find the **weighted mean:**
>
>   1. Multiply each weight w by its corresponding data value x.
>   2. Add up the products, to get $\sum(w \cdot x)$.
>   3. Divide the result by the sum of the weights $\sum w$.
>
> $$\bar{x} = \frac{\sum(w \cdot x)}{\sum w}$$

EXAMPLE 3.17

WEIGHTED MEAN OF COURSE GRADES

The syllabus for the Introduction to Management course at a local college specifies that the midterm exam is worth 30%, the term paper is worth 20%, and the final exam is worth 50% of your course grade. Now, say you did not get serious about the course until Halloween, so that you got a 40 on the midterm. You then began working harder, and got a 70 on the term paper. Finally, you remembered that you had to pay for the course again if you did not pass and had to retake it, and so you worked really hard for the last month of the course and got a 90 on the final exam. Calculate your course average, that is, the weighted mean of your grades.

Solution

Note: The weights w do not have to be percentages that add up to 1.

The data values are 40, 70, and 90. The weights are 0.30, 0.20, and 0.50. Your course weighted mean is then calculated as follows:

$$\bar{x} = \frac{\sum(w \cdot x)}{\sum w} = \frac{(0.30)(40) + (0.20)(70) + (0.50)(90)}{0.30 + 0.20 + 0.50} = \frac{71}{1.0} = 71$$

Now You Can Do
Exercises 4–8.

Because the final exam had the most weight, you were able to raise your course weighted mean to 71, and pass the course.

---

2 ESTIMATING THE MEAN FOR GROUPED DATA

Thus far in Chapter 3, we have computed measures of center and spread from a raw data set. However, data are often reported using grouped frequency distributions. Without the original data, we cannot calculate the exact values of the measures of center and spread. The remainder of this section examines methods for approximating the mean, variance, and standard deviation of *grouped data*—that is, population data summarized using frequency distributions.

For each class in the frequency distribution, we estimate the class mean using the class *midpoint*. The class midpoint, denoted x, is defined as the mean of two adjoining lower class limits.

EXAMPLE 3.18

FINDING THE CLASS MIDPOINTS

There were 1150 children adopted in the state of Georgia in 2006, according to the Administration for Children and Families.[5] The frequency distribution of the ages of the children at adoption is shown in Table 3.12. Find the class midpoints.

Table 3.12 Frequency distribution of children adopted in Georgia, by age

Class: age	Frequency f	Midpoint x
$0 \leqslant$ age < 1	12	0.5
$1 \leqslant$ age < 6	611	3.5
$6 \leqslant$ age < 11	320	8.5
$11 \leqslant$ age < 16	161	13.5
$16 \leqslant$ age < 18	46	17.0

Solution

The midpoint for the first class (ages 0–1) is the mean of the lower class limits for this class (0) and the adjoining class (1). That is, the midpoint is $(0 + 1)/2 = 0.5$. Similarly, the midpoint for the second class (ages 1–6) is $(1 + 6)/2 = 3.5$. The remainder of the class midpoints are shown in Table 3.12.

**Now You Can Do
Exercises 9 and 10.**

The product of the class frequency f and class midpoint x is used as an estimate of the sum of the data values within that class. Summing these products across all classes and dividing by the size of the data set thus provides us with an **estimated mean for data grouped into a frequency distribution.**

> **Estimated Mean for Data Grouped into a Frequency Distribution**
>
> Given a frequency distribution, the **estimated mean** for the variable is given by
>
> $$\bar{x} = \frac{\sum(f \cdot x)}{\sum f}$$
>
> where x and f represent the midpoint and frequency of the ith class, respectively.

EXAMPLE 3.19

CALCULATING THE ESTIMATED MEAN FOR GROUPED DATA

Calculate the estimated mean age of the adopted children in Table 3.12.

Solution

The midpoints x and frequencies f are provided in Table 3.12. We calculate the sum of the products as follows:

$$\sum(f \cdot x) = (0.5)(12) + (3.5)(611) + (8.5)(320) + (13.5)(161) + (17)(46)$$

$$= 6 + 2138.5 + 2720 + 2173.5 + 782 = 7820$$

Next we calculate the sum of the frequencies;

$$\sum f_i = 12 + 611 + 320 + 161 + 46 = 1150$$

The estimated mean is therefore

$$\bar{x} = \frac{\sum(f \cdot x)}{\sum f} = \frac{7820}{1150} = 6.8$$

**Now You Can Do
Exercises 11, 12, and 15b.**

The estimated mean age of the children adopted in Georgia in 2006 is 6.8 years.

3 ESTIMATING THE VARIANCE AND STANDARD DEVIATION FOR GROUPED DATA

We also use class midpoints and class frequencies to calculate the **estimated variance for data grouped into a frequency distribution** and the **estimated standard deviation for data grouped into a frequency distribution.**

> **Estimated Variance and Standard Deviation for Data Grouped into a Frequency Distribution**
>
> The **estimated variance** for data grouped into a frequency distribution is given by
>
> $$s^2 = \frac{\sum (x - \bar{x})^2 \cdot f}{\sum f}$$
>
> and the **estimated standard deviation** is given by
>
> $$s = \sqrt{s^2} = \sqrt{\frac{\sum (x - \bar{x})^2 \cdot f}{\sum f}}$$
>
> where x represents the class midpoints, f represents the class frequencies, and $\bar{x}$ is the estimated mean.

You should carry as many decimal places as you can for the value of $\bar{x}$ when calculating, s^2, and for s^2 when calculating s.

EXAMPLE 3.20

CALCULATING THE ESTIMATED VARIANCE AND STANDARD DEVIATION FOR GROUPED DATA

Calculate the estimated variance and standard deviation of the ages of the adopted children in Table 3.12.

Solution

Table 3.13 contains the calculations required for finding $\sum (x - \bar{x})^2 \cdot f = 20{,}068$. The variance is therefore estimated as

$$s^2 = \frac{\sum (x - \bar{x})^2 \cdot f}{\sum f} = \frac{20{,}068}{1150} = 17.45043478$$

and the standard deviation is estimated as

$$s = \sqrt{s^2} = \sqrt{17.45043478} \approx 4.177371755 \approx 4.2$$

Table 3.13 Calculating $\sum (x - \bar{x})^2 \cdot f$

Class: age	Midpoint x	Frequency f	$\bar{x}$	$x - \bar{x}$	$(x - \bar{x})^2 \cdot f$
0–1	0.5	12	6.8	−6.3	476.28
1–6	3.5	611	6.8	−3.3	6653.79
6–11	8.5	320	6.8	1.7	924.8
11–16	13.5	161	6.8	6.7	7227.29
16–18	17.0	46	6.8	10.2	4785.84

$$\sum (x - \bar{x})^2 \cdot f = 20{,}068$$

**Now You Can Do
Exercises 13 and 14.**

In other words, the age of the adopted children typically differs from the mean age of 6.8 years by about 4.2 years.

EXAMPLE 3.21 | **USING TECHNOLOGY TO FIND THE ESTIMATED MEAN, VARIANCE, AND STANDARD DEVIATION FOR GROUPED DATA**

georgiaadopt

Use the TI-83/84 calculator to find the estimated mean, variance, and standard deviation for the frequency distribution in Table 3.13.

Solution

Following the instructions in the Step-by-Step Technology Guide, we get the estimated mean $\bar{x} = 6.8$, the estimated standard deviation s (shown in the output as σ_x) = 4.177371755, and the estimated variance to be $(4.177371755)^2 = 17.45043478$.

```
1-Var Stats
x̄=6.8
Σx=7820
Σx²=73244
Sx=4.179189189
σx=4.177371755
↓n=1150
```

STEP-BY-STEP TECHNOLOGY GUIDE: Estimating the Mean, Variance, and Standard Deviation for Grouped Data

TI-83/84

Step 1   Press **STAT** and select **1:Edit.** Enter the class midpoints in **L1** and the frequencies or relative frequencies in **L2.**
Step 2   Press **STAT**, select the **CALC** menu, and choose **1: 1-Var Stats.**

Step 3   Press **2nd 1 Comma 2nd 2,** so that the following appears on the home screen: **1-Var Stats L1, L2.**
Step 4   Press **ENTER.**

SECTION 3.3 | **Summary**

1. The weighted mean is the sum of the products of the data points with their respective weights, divided by the sum of the weights.

2. Since we do not have access to the original raw data, it is not possible to find exact values for the mean, variance, and standard deviation of data that have been grouped into a

frequency distribution. The estimated mean $\bar{x}$ in this case is the sum of the products of the class frequencies f and class midpoints x, divided by the sum of the frequencies Σf.

3. Class midpoints and class frequencies are also used to find the estimated variance s^2 and estimated standard deviation s of grouped data.

SECTION 3.3 | **Exercises**

Clarifying the Concepts

1. Explain why the formula for the mean of grouped data will provide an estimate only and not the exact value of the mean if the data were not grouped.

2. Describe how the weighted mean is calculated.

3. Suppose we calculate the weighted mean of the following data 2, 7, 4. Let each of the weight equal 1. What measure of center from Section 3.1 does this weighted simplify to when all the weights equal 1?

Practicing the Techniques

For Exercises 4–8, the data values and weights are provided. Find the weighted mean.

4. $x_1 = 50, x_2 = 60; x_3 = 70; w_1 = 0.25, w_2 = 0.50,$ $w_3 = 0.25.$

5. $x_1 = 50, x_2 = 80, x_3 = 70; w_1 = 0.25, w_2 = 0.40,$ $w_3 = 0.35.$

6. $x_1 = 100, x_2 = 120, x_3 = 150; w_1 = 10, w_2 = 20, w_3 = 5.$

7. $x_1 = 3.0$, $x_2 = 2.5$, $x_3 = 3.5$, $x_4 = 4.0$, $x_5 = 3.0$; $w_1 = w_2 = w_3 = w_4 = 3$, $w_5 = 4$.

8. $x_1 = 70$, $x_2 = 80$, $x_3 = 85$, $x_4 = 95$; $w_1 = 0.20$, $w_2 = 0.30$, $w_3 = 0.25$, $w_4 = 0.25$.

For Exercises 9 and 10, the class limits are provided. Find the class midpoints.

9.

0–1.99	6–7.99
2–3.99	8–9.99
4–5.99	

10.

0–4.99	20–29.99
5–9.99	30–49.99
10–14.99	50–99.99
15–19.99	100–199.99

For Exercises 11 and 12, find the estimated mean for the frequency distribution.

11.

Midpoint x	Frequency f
5	10
10	20
15	20
20	10
25	10

12.

Midpoint x	Frequency f
−10	3
−5	2
0	5
5	12
10	8
15	10

For Exercises 13 and 14, find the estimated variance and standard deviation

13. For the frequency distribution in Exercise 11.

14. For the frequency distribution in Exercise 12.

Applying the Concepts

15. Dupage County Age Groups. The Census Bureau reports the following frequency distribution of population by age group for Dupage County, Illinois, residents less than 65 years old.

 dupageage

Age	Residents
0–4.99	63,422
5–17.99	240,629
18–64.99	540,949

a. Find the class midpoints.
b. Find the estimated mean age of residents of Dupage County.
c. Find the estimated variance and standard deviation of ages.

16. Broward County House Values. Table 3.14 gives the frequency distribution of the dollar value of the owner-occupied housing units in Broward County, Florida.

 browardhouse

TABLE 3.14 Broward County house values

Dollar value	Housing units
0–49,999	5,430
50,000–99,999	90,605
100,000–149,999	90,620
150,000–199,999	54,295
200,000–299,999	34,835
300,000–499,999	15,770
500,000–999,999	5,595

a. Find the class midpoints.
b. Find the estimated mean dollar value for housing units in Broward County.
c. Find the estimated variance and standard deviation of the dollar value.

17. Lightning Deaths. Table 3.15 gives the frequency distribution of the number of deaths due to lightning nationwide over a 67-year period. Find the estimated mean and standard deviation of the number of lightning deaths per year.

 lightningdeath

TABLE 3.15 Lightning deaths

Deaths	Years
20–59.99	13
60–99.99	21
100–139.99	10
140–179.99	6
180–259.99	10
260–459.99	7

Source: National Oceanic and Atmospheric Administration.

18. Calculating a Course Grade. An introductory statistics syllabus has the following grading system. The weekly quizzes are worth a total of 25% toward the final course grade. The midterm exam is worth 32%; the final exam is worth 33%; and attendance/participation is worth 10% toward the final course grade. Anthony's weekly quiz average is 70. He got an 80 on the midterm and a 90 on the final exam. He got 100 for attendance/participation. Calculate Anthony's final course grade.

19. Wages for Computer Managers. The U.S. Bureau of Labor Statistics (BLS) publishes wage information for various occupations. For the occupation "computer and information systems management," Table 3.16 gives the wages reported by the BLS for the top-paying states. Find the weighted mean wage across all five states, using the employment figures as weights.

 compwage

20. Salaries of Scientists and Engineers. The National Science Foundation compiles statistics on the annual salaries of full-time employed doctoral scientists and engineers in universities and four-year colleges. The mean annual salary for the fields of science, engineering, and health are $67,000, $82,200, and $70,000, respectively. Suppose we have a sample of 10 professors, 5 of whom are in science, 2 in engineering, and 3 in health, and each of whom is making the mean salary for his or her field. Find the weighted mean salary of these 10 professors.

21. Challenge Exercise. Assign the weights w to show that the formula for the sample mean from Section 3.1 $\bar{x} = \sum x_i / n$ is a special case of the formula for the weighted mean $\bar{x} = \sum (w \cdot x) / \sum w$.

Table for Exercise 19

TABLE 3.16   Wages for computer managers

State	Employment	Hourly mean wage
New Jersey	12,380	$60.32
New York	18,580	$60.25
Virginia	9,540	$59.39
California	35,550	$57.98
Massachusetts	10,130	$55.95

3.4   MEASURES OF RELATIVE POSITION AND OUTLIERS

OBJECTIVES By the end of this section, I will be able to . . .

1 Calculate z-scores, and explain why we use them.

2 Detect outliers using the z-score method.

3 Find percentiles and percentile ranks for both small and large data sets.

4 Compute quartiles and the interquartile range.

In this section we learn about *measures of relative position,* which tell us the position that a particular data value has relative to the rest of the data set. For example, a prestigious nursing school may grant admission to only the top 10% of applicants. How high a score would you need to enter? This is one type of question we will answer in this section.

1   Z-SCORES

Our first measure of relative position is the z-score. Recall that the standard deviation is a common measure of the variability, or spread, of a data set. The value of the standard deviation is interpreted as a typical deviation from the mean. Many students take the Scholastic Aptitude Test (SAT) when preparing to apply for college admission. The SAT is designed so that the distribution of scores is bell-shaped with a mean of 500 and a standard deviation of 100. Note in Figure 3.17 that we can measure the distance from a particular SAT score to the mean in terms of standard deviations. For example, an SAT score of 600 lies 1 standard deviation above the mean, while an SAT score of 300 lies 2 standard deviations below the mean.

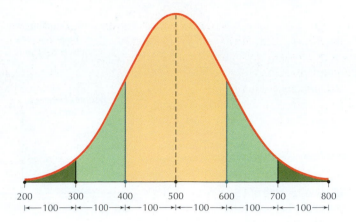

FIGURE 3.17 The distribution of SAT scores.

The term **z-score** indicates how many standard deviations a particular data value is from the mean. If the z-score is positive, then the data value is above the mean. If the z-score is negative, then the data value is below the mean.

---

z-Score

The **z-score** for a particular data value from a *sample* is

$$z\text{-score} = \frac{\text{data value} - \text{mean}}{\text{standard deviation}} = \frac{x - \bar{x}}{s}$$

where $\bar{x}$ is the sample mean, and s is the sample standard deviation.

The z-score for a particular data value from a *population* is

$$z\text{-score} = \frac{\text{data value} - \text{mean}}{\text{standard deviation}} = \frac{x - \mu}{\sigma}$$

where μ is the population mean, and σ is the population standard deviation.

Recall that the standard deviation is a common measure of the variability, or spread, of a data set, and its value is interpreted as a typical deviation from the mean.

In this section, we will use the sample z-score unless otherwise indicated.

EXAMPLE 3.22

MEANING OF A Z-SCORE

Suppose the mean score on the Math SAT is $\mu = 500$, with a standard deviation of $\sigma = 100$ points. Suppose Jasmine's Math SAT score is 650. How many standard deviations is Jasmine's score from the mean? Note that here we have population values.

Solution

Here $\mu = 500$, $\sigma = 100$, and Jasmine's score is $x = 650$. Her z-score is

$$z\text{-score} = \frac{\text{data value} - \text{mean}}{\text{standard deviation}} = \frac{x - \mu}{\sigma} = \frac{650 - 500}{100} = 1.5$$

Jasmine's z-score of 1.5 indicates that her Math SAT is 1.5 standard deviations from the mean of 500. Z-scores can be positive or negative. Jasmine's z-score is positive (1.5), which means that her Math SAT score falls above the mean. Bright lady! Consider Figure 3.18, which shows the distribution of SAT scores, with a mean of 500 and a standard deviation of 100. The arrows represent "units" of 1 standard deviation each, that is, each arrow is 100 SAT points long. Counting the arrows as you go above or below the mean is thus the same as counting the number of standard deviations above or below the mean. Jasmine's SAT score lies between 600 and 700, an area with z-scores ranging from 1 to 2.

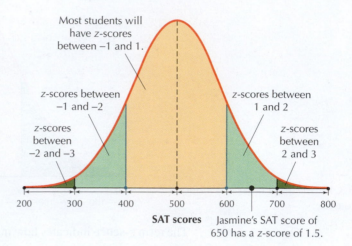

FIGURE 3.18
Jasmine's z-score of 1.5 places her 1.5 standard deviations above the mean.

In Example 3.22, since the standard deviation equals 100, the z-score represents units of 100. That is, a z-score of 1 represents 1 standard deviation above the mean, which is 100 points above the mean. Thus, the *scale* of the z-scores for the SAT scores in Figure 3.18 is in units of 100, since the standard deviation equals 100. However, if the standard deviation was, say, $\sigma = 50$, then the scale would be different.

EXAMPLE 3.23

CALCULATING Z-SCORES GIVEN DATA VALUES

Note: This use of "μ" for "micro" in the measure "micrograms per deciliter" is not related to our use of μ as the population mean.

A study of workers who were exposed to lead at their jobs found that their mean blood lead level was 31.4 μg/dl (micrograms per deciliter) with a standard deviation of 14.2 μg/dl.[6]

a. If we calculate z-scores, what is the scale?
b. Calculate the z-scores for the following workers:
   i. Ryan, with a blood lead level of 78.26 μg/dl
   ii. Megan, with a blood lead level of 1.58 μg/dl
   iii. Kyle, with a blood lead level of 55.54 μg/dl
c. For each worker, interpret the value of the z-score.

Solution

a. If we calculate z-scores for the workers' lead levels, the scale of the z-scores will be 14.2 μg/dl, since that is the value of the standard deviation.
b. Here are the workers' lead levels.
   i. Ryan:

$$z\text{-score} = \frac{x - \bar{x}}{s} = \frac{78.26 - 31.4}{14.2} = \frac{46.86}{14.2} = 3.3$$

   ii. Megan:

$$z\text{-score} = \frac{x - \bar{x}}{s} = \frac{1.58 - 31.4}{14.2} = \frac{-29.82}{14.2} = -2.1$$

   iii. Kyle:

$$z\text{-score} = \frac{x - \bar{x}}{s} = \frac{55.54 - 31.4}{14.2} = \frac{24.14}{14.2} = 1.7$$

c. Ryan's lead level lies 3.3 standard deviations above the mean; Megan's lead level lies 2.1 standard deviations below the mean; and Kyle's lead level lies 1.7 standard deviations above the mean.

Now You Can Do
Exercises 12–15.

Alternatively, we may be given a z-score, and asked to find its associated data value x. To do so, use the following formulas.

Note: We arrive at these formulas simply by taking the z-score formula and using algebra to solve for x.

> Given a z-score, to find its associated data value x:
>
> For a sample: $\qquad\qquad\qquad\qquad\qquad$ $x = z\text{-score} \cdot s + \bar{x}$
>
> For a population: $\qquad\qquad\qquad\qquad$ $x = z\text{-score} \cdot \sigma + \mu$
>
> where μ is the population mean, $\bar{x}$ is the sample mean, σ is the population standard deviation, and s is the sample standard deviation.

EXAMPLE 3.24

FINDING DATA VALUES GIVEN Z-SCORES

Continuing with the blood lead level data from Example 3.23, find the blood lead levels associated with the following z-scores:

a. –1 $\qquad$ **b.** 0 $\qquad$ **c.** 3

Solution

We have $\bar{x} = 31.4\ \mu\text{g/dl}$ and $s = 14.2\ \mu\text{g/dl}$.

a. For a z-score of -1, we have

$$x = z\text{-score} \cdot s + \bar{x} = (-1) \cdot 14.2 + 31.4 = 17.2\ \mu\text{g/dl}.$$

A blood lead level of $17.2\ \mu\text{g/dl}$ lies 1 standard deviation below the mean.

b. For a z-score of 0, we have

$$x = z\text{-score} \cdot s + \bar{x} = (0) \cdot 14.2 + 31.4 = 14.2\ \mu\text{g/dl}.$$

A blood lead level of $14.2\ \mu\text{g/dl}$ lies exactly on the mean.

c. For a z-score of 3, we have

$$x = z\text{-score} \cdot s + \bar{x} = (3) \cdot 14.2 + 31.4 = 74\ \mu\text{g/dl}.$$

Now You Can Do
Exercises 16 and 17.

A blood lead level of $74\ \mu\text{g/dl}$ lies 3 standard deviations above the mean.

EXAMPLE 3.25

USING THE Z-SCORE TO COMPARE DATA FROM DIFFERENT DATA SETS

Andrew is bragging to his friend Brittany that he did better than she did on the last statistics test. Andrew got a 90 while Brittany got an 80. Andrew's class mean was 80 with a standard deviation of 10. Brittany's class mean was 60 with a standard deviation of 10. The professors in both classes grade "on a curve" using z-scores. Who did better relative to his or her class?

Solution

Brittany can use z-scores to show that she did better *relative to her class*. Figure 3.19 shows comparative dotplots of the scores in the two classes. The red dots represent Brittany's and Andrew's scores. Brittany found her z-score by subtracting her class mean from her score of 80 and then dividing by the standard deviation $s = 10$:

$$z\text{-score}_{\text{Brittany}} = \frac{x - \bar{x}}{s} = \frac{80 - 60}{10} = 2$$

z-Scores enable the data analyst to compare data values from two different distributions.

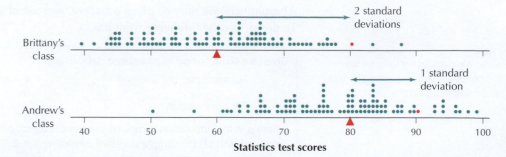

FIGURE 3.19 Brittany actually did better relative to her class.

Brittany's *z*-score is 2. What does that mean? It means that *Brittany scored 2 standard deviations above the mean* of 60. Brittany then found the *z*-score for Andrew:

$$z\text{-score}_{\text{Andrew}} = \frac{x - \bar{x}}{s} = \frac{90 - 80}{10} = 1$$

Andrew's *z*-score was 1, which means that Andrew scored 1 standard deviation above the mean. From Figure 3.19 we can observe that Andrew's exam score of 90 lies closer to the mean exam score of 80 for his class. That is, the arrow is shorter for Andrew than for Brittany. Finally, note that 10 of the 100 students who took the exam in his class did better than he did, whereas only 2 did better than Brittany in her class. So, relative to her class, Brittany did better than Andrew, even though Andrew got a higher score. The *z*-scores allowed her to compare their grades, even though they were in different classes.

Now You Can Do
Exercises 18 and 19.

2 DETECTING OUTLIERS USING THE *Z*-SCORE METHOD

An **outlier** is a data value that is very much greater than or less than the mean. It may represent a data entry error, or it may be genuine data. One way of identifying an outlier is to determine whether it is farther than 3 standard deviations from the mean, that is, its *z*-score is less than −3 or greater than 3. Figure 3.20 illustrates the following guidelines for identifying outliers using *z*-scores.

Note: If an outlier is detected, it does not automatically follow that it should be discarded. Outliers often indicate the presence of something interesting going on in the data that would call for further investigation. On the other hand, it could simply be a typo. The analyst should check with the data source.

Guidelines for Identifying Outliers

1. A data value whose *z*-score lies in the following range is not considered to be unusual:

$$-2 < z\text{-score} < 2$$

2. A data value whose *z*-score lies in either of the following ranges may be considered moderately unusual:

$$-3 < z\text{-score} \leq -2 \quad \text{or} \quad 2 \leq z\text{-score} < 3$$

3. A data value whose *z*-score lies in either of the following ranges may be considered an outlier:

$$z\text{-score} \leq -3 \quad \text{or} \quad z\text{-score} \geq 3$$

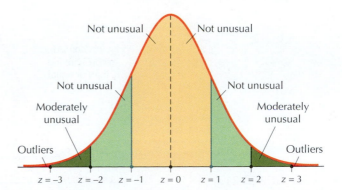

FIGURE 3.20
z-Scores help to identify outliers.

| EXAMPLE 3.26 | **DETECTING OUTLIERS USING THE Z-SCORE METHOD** |

For the three workers in Example 3.23 on page 122, determine whether each of their blood lead levels represent an outlier.

Solution

Ryan's z-score is 3.3, which is greater than 3. Thus, Ryan's lead level of 78.26 μg/dl represents an outlier. Megan's z-score is -2.1, which lies between -3 and -2. Hence, Megan's lead level of 1.58 μg/dl may be considered moderately unusual but is not an outlier. Kyle's z-score is 1.7, which lies between -2 and 2. Thus, Kyle's lead level of 55.54 is not considered unusual.

In Section 3.5 we will learn about the IQR method of detecting outliers.

**Now You Can Do
Exercises 20–23.**

---

3   PERCENTILES AND PERCENTILE RANKS

Some analysts prefer to define the pth percentile to be a data value at which at least p percent of the values in the data set are less than or equal to this value, *and* at least $(1 - p)$ percent of the values are greater than or equal to this value.

The next measure of relative position we consider is the **percentile**, which shows the location of a data value relative to the other values in the data set.

> **Percentile**
>
> Let p be any integer between 0 and 100. The pth **percentile** of a data set is the data value at which p percent of the values in the data set are less than or equal to this value.

| EXAMPLE 3.27 | **MEANING OF A PERCENTILE** |

After taking the SAT, students receive test results that include not only their score, but also the percentile that this score represents. Jasmine's Math SAT score was 650, which represents the 90th percentile. What does "90th percentile" mean?

Solution

To say that 650 is the 90th percentile means that 90% of all scores on the Math SAT fell at or below Jasmine's score of 650. We call the percentile a *measure of relative position* since it indicates the position of Jasmine's Math SAT score relative to all other Math SAT scores. Clearly, Jasmine is good at math. Figure 3.21 indicates the position of Jasmine's score relative to the rest of the test takers.

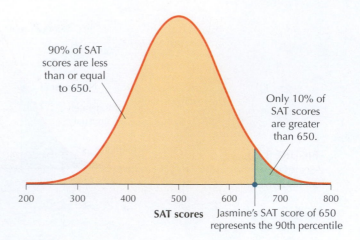

90% of SAT scores are less than or equal to 650.

Only 10% of SAT scores are greater than 650.

FIGURE 3.21
The 90th percentile is the score with 90% of the data values at or below its value.

200   300   400   500   600   700   800

SAT scores   Jasmine's SAT score of 650 represents the 90th percentile

For large data sets, calculation of the percentiles is best left to computers. However, for small data sets, we can use the following step-by-step method to calculate the related position of any percentile.

STEP 1 Sort the data into ascending order (from smallest to largest).

STEP 2 Calculate

$$i = \left(\frac{p}{100}\right)n$$

where p is the particular percentile you wish to calculate, and n is the sample size.

CAUTION These steps do not give the value of the pth percentile itself, but rather the *position* of the pth percentile in the data set when the data set is in ascending order.

STEP 3 **a.** If i is an integer (a whole number with no decimal part), the pth percentile is the mean of the data values in positions i and $i + 1$.
b. If i is not an integer, round up to the next integer and use the value in this position.

E X A M P L E   3 . 2 8

FINDING PERCENTILES

dancescore

Yolanda would like to go to a prestigious graduate school of the arts. She knows that this school accepts only those students who score at the 75th percentile or higher in a grueling dance audition. The following data represent the dance audition scores of Yolanda's group. Yolanda scored 85. Find the 75th percentile of the data set. Will Yolanda be accepted at the prestigious graduate school of the arts?

78   56   89   44   65   94   81   62   75   85   30   68

Solution

STEP 1 Sort the data into ascending order:

30   44   56   62   65   68   75   78   81   85   89   94

STEP 2 The particular percentile we wish to calculate is the 75th percentile, so $p = 75$. There are 12 scores in our data set, so $n = 12$. Calculate

$$i = \left(\frac{p}{100}\right)n = \left(\frac{75}{100}\right)12 = 9$$

So, $i = 9$.

STEP 3 Here, since i is an integer, the 75th percentile is the mean of the data values in positions 9 and 10.

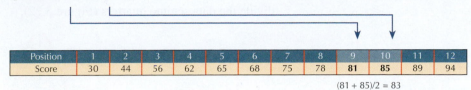

$(81 + 85)/2 = 83$

Position	1	2	3	4	5	6	7	8	9	10	11	12
Score	30	44	56	62	65	68	75	78	**81**	**85**	89	94

Now You Can Do Exercises 24–29.

Counting from left to right, the data value in the ninth position is 81, and the data value in the tenth position is 85. The mean of these two values is 83. Thus, the 75th percentile is 83. Yolanda's dance score of 85 is therefore above the 75th percentile. She will be accepted to the prestigious graduate school.

Remember: A percentile is a data value, while a percentile rank is a percentage.

> The **percentile rank** of a data value x equals the percentage of values in the data set that are less than or equal to x. In other words:
>
> $$\text{percentile rank of data value } x = \frac{\text{number of values in data set} \leq x}{\text{total number of values in data set}} \cdot 100$$

EXAMPLE 3.29

missingchild

FINDING PERCENTILE RANKS

In Example 2.7 (page 49), we were introduced to a data set of 50 missing and exploited children in California. Table 3.17 shows the ages of those children, sorted into ascending order. Find the percentile ranks for the following ages:

a. 2 years old and
b. 5 years old

Table 3.17 Ages of 50 missing and exploited children in California

1	1	1	2	2	2	3	3	4	4	4	4	4	4	4	4	5	5	5	5	5	5	5	5	5
6	6	6	6	6	6	6	6	6	6	7	7	7	7	7	7	8	8	8	8	8	8	9	9	9

Solution

a. Here $x = 2$. There are 3 two-year old children and 3 one-year old children, so the percentile rank of two-year old children is

$$\text{percentile rank of } (x = 2) = \frac{\text{number of values in data set} \leq 2}{\text{total number of values in data set}} \cdot 100$$

$$= \frac{6}{50} \cdot 100 = 12\%$$

b. Here $x = 5$. There are 9 five-year old children, and 16 children less than five years old.

$$\text{percentile rank of } (x = 5) = \frac{\text{number of values in data set} \leq 5}{\text{total number of values in data set}} \cdot 100$$

Now You Can Do Exercises 30–35.

$$= \frac{25}{50} \cdot 100 = 50\%$$

4 QUARTILES AND THE INTERQUARTILE RANGE

Just as the median divides the data set into halves, the **quartiles** are the percentiles that divide the data set into quarters (Figure 3.22).

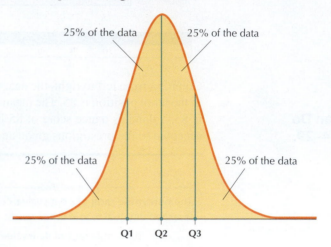

25% of the data 25% of the data

25% of the data 25% of the data

Q1 Q2 Q3

FIGURE 3.22
The quartiles Q1, Q2, and Q3 divide the data set into four quarters.

> **The Quartiles**
>
> The **quartiles** of a data set divide the data set into four parts, each containing 25% of the data.
> - The *first quartile* (Q1) is the 25th percentile.
> - The *second quartile* (Q2) is the 50th percentile, that is, the median.
> - The *third quartile* (Q3) is the 75th percentile.
>
> For small data sets, the division may be into four parts of only approximately equal size.

EXAMPLE 3.30

FINDING THE QUARTILES FOR A SMALL DATA SET

Note: It may be helpful to note that the phrase *third quartile* is akin to the phrase *three quarters,* which is 75%, representing the 75th percentile. Also, the phrase *first quartile* is akin to the phrase *one quarter,* which is 25%, representing the 25th percentile.

In Example 3.28 (pages 126–127), we examined the dance scores of 12 students auditioning for admission into a prestigious graduate school of the arts. Recall that we found the 75th percentile of the dance audition scores to be 83. By definition, the 75th percentile is the third quartile Q3. Therefore, this score of 83 is also the third quartile (Q3) of the audition scores. Now we will find the first quartile and the median (second quartile).

Solution

To find the quartiles, we use the steps for finding percentiles (page 126). First, arrange the data set in ascending order, as follows:

30   44   56   62   65   68   75   78   81   85   89   94

Here, $n = 12$. To find Q1, plug $p = 25$ into the equation $i = \left(\dfrac{p}{100}\right)n$, where $n = 12$. We get $i = \left(\dfrac{p}{100}\right)n = \left(\dfrac{25}{100}\right)12 = 3$. Since 3 is an integer, we know that the 25th percentile is the mean of the dance scores in the 3rd and 4th positions. The score of 56 is in the 3rd position, while 62 is in the 4th position. Since $(56 + 62)/2 = 59$, we get the 25th percentile of the dance scores to be 59 (Figure 3.23).

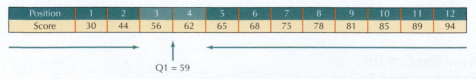

Position	1	2	3	4	5	6	7	8	9	10	11	12
Score	30	44	56	62	65	68	75	78	81	85	89	94

Q1 = 59

FIGURE 3.23 The 25th percentile splits the difference between 56 and 62.

To find the median (the second quartile, Q2), plug $p = 50$ into your steps for finding the percentiles: $i = \left(\dfrac{p}{100}\right)n = \left(\dfrac{50}{100}\right)12 = 6$. Since 6 is an integer, we know that the 50th percentile is the mean of the dance scores in the 6th and 7th positions, that is, 68 and 75. Since $(68 + 75)/2 = 71.5$, the 50th percentile of the dance scores is 71.5 (Figure 3.24). This agrees with the method we learned for finding the median, on page 86.

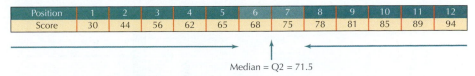

Position	1	2	3	4	5	6	7	8	9	10	11	12
Score	30	44	56	62	65	68	75	78	81	85	89	94

Median = Q2 = 71.5

FIGURE 3.24 The 50th percentile splits the difference between 68 and 75.

In Example 3.28, we determined that the 75th percentile was 83. Therefore, the quartiles for the dance score data set are Q1 = 59, median = Q2 = 71.5, and Q3 = 83. Note that these quartiles divide the data set into four equal sections, of three observations each (Figure 3.25).

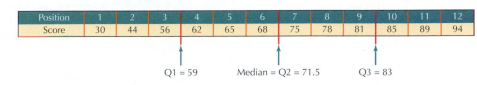

Position	1	2	3	4	5	6	7	8	9	10	11	12
Score	30	44	56	62	65	68	75	78	81	85	89	94

Q1 = 59   Median = Q2 = 71.5   Q3 = 83

FIGURE 3.25 The quartiles for the dance audition data.

Of course, for small data sets, the division into quarters is not always exact. For example, what if one dancer had sprained her ankle that morning and could not make the audition? Then there would have been only 11 dance scores, which cannot be divided equally into four quarters. In this case, therefore, the quartiles would divide the data set up into four sections of approximately equal size. However, for large data sets, which the data analyst most often encounters, this becomes less of an issue.

EXAMPLE 3.31

FINDING QUARTILES OF A LARGE DATA SET: CHOLESTEROL LEVELS IN FOOD

Nutrition

The U.S. Department of Agriculture recommends a diet low in cholesterol, to reduce the risk of heart disease. The data set **Nutrition** contains information on the cholesterol content (in milligrams) of 961 different foods. Find the mean, standard deviation, and quartiles.

Solution

The Minitab descriptive statistics for the cholesterol data are shown in Figure 3.26. Note that the mean cholesterol content is 32.55 mg and that the standard deviation is about 120 mg. Recall that a standard deviation that is much larger than the mean may be associated with strongly skewed distributions. Compare the value for the mean with the values for the quartiles.

- Q1, the first quartile, or 25th percentile, is 0 mg of cholesterol.

- The median, or Q2, the second quartile (50th percentile), is also 0 mg of cholesterol.

- Q3, the third quartile, or 75th percentile, is 20 mg of cholesterol.

Variable	N	Mean	StDev	Min	Q1	Median	Q3	Max
Cholesterol	961	32.55	119.96	0	0	0	20	2053

FIGURE 3.26 Descriptive statistics for the cholesterol data.

The quartiles may be found on the TI-83/84 by using the instructions for descriptive statistics shown on page 92.

Now You Can Do
Exercises 36–38.

Note: Minitab uses a different way to calculate the quartiles than the way we have learned, which results in different values than our hand-calculation methods. However, for large data sets, the difference is minimal.

Figure 3.27 shows that the data distribution is extremely right-skewed. There are only a few foods with over 1000 mg cholesterol, and another handful with over 500 (see data on disk). Therefore, it appears that we have outliers in this data set. What is the effect of these outliers on the mean and standard deviation? Does the mean represent a truly typical cholesterol content level for the data set, or is its value unduly increased by the outliers? Let's find out.

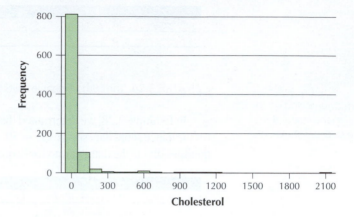

FIGURE 3.27
Cholesterol content (mg) of 961 foods.

Developing Your Statistical Sense

The Mean Is Not Always Representative

Note that the median is 0 mg of cholesterol, meaning that at least half of the food items tested by the USDA in this data set had no cholesterol at all. We are intrigued by this result and ask Minitab to provide us with a frequency distribution for the cholesterol content, along with the cumulative percentages ("CumPct"). Figure 3.27 provides a portion of this frequency distribution, with the following results:

- 61.91% of the food items have no cholesterol at all, which explains why Q1 and the median are both zero.

- The 75th percentile, Q3, is verified to be 20 mg cholesterol.

- The 81st percentile of the data set is 32 mg cholesterol.

61.91% of food items had zero cholesterol. Thus, Q1 = 0 and median = 0.

Chol	Count	CumPct		Chol	Count	CumPct	
0	595	61.91		20	5	75.23	**75th percentile (Q3) = 20 mg cholesterol**
1	12	63.16		21	5	75.75	
2	8	64.00		22	6	76.38	
3	7	64.72		23	3	76.69	
4	11	65.87		24	4	77.11	
5	10	66.91		25	3	77.42	
6	6	67.53		26	4	77.84	
7	5	68.05		27	9	78.77	
8	7	68.78		28	3	79.08	
9	4	69.20		29	4	79.50	
10	8	70.03		30	3	79.81	
11	3	70.34		31	6	80.44	
12	4	70.76		32	5	80.96	**81st percentile is 32 mg. The mean is 32.55 mg.**
13	4	71.18		33	3	81.27	
14	5	71.70		34	4	81.69	
15	7	72.42		35	2	81.89	
16	5	72.94		36	1	82.00	
17	6	73.57		37	4	82.41	
18	7	74.30		38	1	82.52	
19	4	74.71		40	1	82.62	
⋮	⋮	⋮		⋮	⋮	⋮	

FIGURE 3.28 Partial frequency distribution of cholesterol content.

Think about these results for a moment. We found that the 81st percentile is 32 mg cholesterol. In other words, 81% of the food items have a cholesterol content of 32 mg or less. And yet, this 32 mg is still *less than the mean* cholesterol content, reported by Minitab to be 32.55 mg. In other words, the mean of this data set is larger than 81% of the data values in the data set.

It seems clear, then, that *the mean 32.55 mg cannot be considered as typical or representative* of the data set. Its value has been exaggerated by the presence of the outliers, to such an extent that it is now larger than 81% of the data. We need another, more *robust* measure of center, one that is *resistant* to the undue influence of outliers, such as the median. Here, the value of the median is 0 mg cholesterol. An argument may certainly be made that this is indeed typical and representative of the data set, since 61.91% of the food items have no cholesterol content at all.

Recall from Section 3.2 that the variance and standard deviation are measures of spread that are sensitive to the presence of extreme values. A more robust (less sensitive) measure of variability is the **interquartile range, or IQR.**

Interquartile Range

The **interquartile range (IQR)** is a robust measure of variability. It is calculated as

$$IQR = Q3 - Q1$$

The interquartile range is interpreted to be the spread of the middle 50% of the data.

The Latin word *inter* means "between," so the *inter*quartile range is the *difference between* the quartiles Q3 and Q1. The IQR represents how spread out the "middle half" of the data set is. A larger IQR implies a greater degree of variability, or spread, in the data set. Since the IQR ignores both the highest 25% and the lowest 25% of the data set, it is completely unaffected by outliers and is thus quite robust.

EXAMPLE 3.32

FINDING THE INTERQUARTILE RANGE

In Example 3.30, we found that, for the dance audition score data, Q1 = 59 and Q3 = 83. Find the IQR for the dance score data and explain what it means.

Solution
Since Q1 = 59 and Q3 = 83, the IQR is IQR = Q3 − Q1 = 83 − 59 = 24. We would say that the middle 50%, or middle half, of the dance audition scores ranged over 24 points (see Figure 3.29).

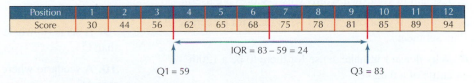

Position	1	2	3	4	5	6	7	8	9	10	11	12
Score	30	44	56	62	65	68	75	78	81	85	89	94

IQR = 83 − 59 = 24

Q1 = 59   Q3 = 83

FIGURE 3.29 The interquartile range for the dance audition data.

Now You Can Do Exercise 39.

What would happen if we introduced an outlier into this data set? For example, what if we changed the lowest score from 30 to 3? The IQR would remain completely unaffected, as it would even if we changed the 44 to a 4. However, if we changed the 56, then the IQR would be affected, since Q1 would then change.

STEP-BY-STEP TECHNOLOGY GUIDE: Percentiles and Quartiles

TI-83/84
The quartiles are provided using the instructions for descriptive statistics shown on page 92.

EXCEL
Step 1   Enter the data into column **A.**
Step 2   Select **Data . . . Data Analysis.**

Step 3   Select **Rank and Percentile** and click **OK.**
Step 4   Click in the **Input Range** cell. Then highlight the data in column **A.** Click **OK.**

CRUNCHIT!
We will use the data from Example 3.29 (page 127).

Step 1   Click **File . . .** then highlight **Load from Larose2e . . . Chapter 3 . . .** and click on **Example 3.29.**
Step 2   Click **Statistics** and select **Descriptive statistics.** For **Data,** select **Scores.**

Step 3   In the **Percentiles (comma-separated)** cell, enter the percentiles that you would like to find. For example, to find the 5th and 95th percentiles, enter **5, 95.**
Step 4   Click **Calculate.**

| SECTION 3.4 | Summary |

1. In this section, we learned about measures of relative position, which tell us the position that a particular data value holds relative to the rest of the data set. The *z*-score indicates how many standard deviations a particular data value is from the mean. The *z*-score equals the data value minus the mean, divided by the standard deviation. We may also calculate a data value, given its *z*-score.

2. An outlier is a value that is very much greater than or less than the mean. An outlier can be identified when its *z*-score is less than −3 or greater than 3.

3. The *p*th percentile of a data set is the value at which *p* percent of the values in the data set are less than or equal to this value. The percentile rank of a data value equals the percentage of values in the data set that are less than or equal to that value.

4. Quartiles divide the data set into approximately equal quarters. The interquartile range (IQR) is a measure of spread found by subtracting the first quartile from the third quartile.

| SECTION 3.4 | Exercises |

Clarifying the Concepts

1. What does it mean for a *z*-score to be positive? Negative? Zero?

2. Explain in your own words why *z*-scores are useful.

3. Explain in your own words what the 95th percentile of a data set means.

4. Why doesn't it make sense for there to be a 120th percentile of a data set?

5. Is it possible for the 1st percentile of a data set to equal the 99th percentile? Explain when this would happen.

6. Explain the difference between a percentile and a percentile rank.

7. True or false: The IQR is sensitive to the presence of outliers.

For Exercises 8–11, consider whether the scenarios are possible. If it is possible, then clearly describe what the data set would look like. If it is not possible, why not?

8. A scenario where the first and second quartiles of a data set are equal

9. A scenario where the mean of a data set is larger than Q3

10. A scenario where the median of a data set is smaller than Q1

11. A scenario where the IQR is negative

Use the following information for Exercises 12–17. Suppose the mean blood sugar level is 100 mg/dl (milligrams per deciliter), with a standard deviation of 10 mg/dl.

12. Alyssa has a blood sugar level of 90 mg/dl. How many standard deviations is Alyssa's blood sugar level below the mean?

13. Benjamin has a blood sugar level of 135 mg/dl. How many standard deviations is Benjamin's blood sugar level above the mean?

14. Chelsea has a blood sugar level of 125 mg/dl.
   a. If we calculate Chelsea's z-score, what is the scale?
   b. Calculate Chelsea's z-score.
   c. Interpret her z-score.

15. David has a blood sugar level of 85 mg/dl.
   a. Calculate David's z-score.
   b. Interpret his z-score.

16. Find the blood sugar level associated with a z-score of 1.

17. Find the blood sugar level associated with a z-score of −2.

18. Elizabeth's statistics class had a mean quiz score of 70 with a standard deviation of 15. Fiona's statistics class had a mean quiz score of 75 with a standard deviation of 5. Both Elizabeth and Fiona got an 85 on the quiz. Who did better relative to her class?

19. Juan's business class had a mean quiz score of 60 with a standard deviation of 15. Luis's business class had a mean quiz score of 70 with a standard deviation of 5. Both Juan and Luis got a 75 on the quiz. Who did better relative to his class?

For Exercises 20–23, determine whether the person's blood sugar level represents an outlier, using the z-score method.

20. Alyssa from Exercise 12

21. Benjamin from Exercise 13

22. Chelsea from Exercise 14

23. David from Exercise 15

Use the following set of stock prices (in dollars) for Exercises 24–39.

   10   7   20   12   5   15   9   18   4   12   8   14

For Exercises 24–29, find the stock price representing the indicated percentiles.

24. 50th   **25.** 75th   **26.** 25th

27. 10th   **28.** 5th   **29.** 95th

For Exercises 30–35, calculate the percentile rank for the indicated stock price.

30. $12   **31.** $20   **32.** $7

33. $4   **34.** $18   **35.** $5

36. Find Q1, the first quartile.

37. Calculate Q2, the second quartile.

38. Compute Q3, the third quartile.

39. Calculate the IQR.

Applying the Concepts

Breakfast Calories. Refer to Table 3.18 for Exercises 40–47.

🍎 breakfastcal

TABLE 3.18   Calories in 12 breakfast cereals

Cereal	Calories
Apple Jacks	110
Basic 4	130
Bran Chex	90
Bran Flakes	90
Cap'n Crunch	120
Cheerios	110
Cinammon Toast Crunch	120
Cocoa Puffs	110
Corn Chex	110
Corn Flakes	100
Corn Pops	110
Count Chocula	110

40. Find the z-scores for the calories for the following cereals.
   a. Corn Flakes   **b.** Basic 4
   c. Bran Flakes   **d.** Cap'n Crunch

41. Find the number of calories associated with the following z-scores:

   a. 0   **b.** 1   **c.** 21   **d.** 0.5

42. Determine whether any of the cereals is an outlier.

43. Find the following percentiles:

   a. 25th   **b.** 50th   **c.** 75th   **d.** 95th

44. Find the percentile rank for each of the following:
   a. 90 calories   **b.** 120 calories
   c. 110 calories   **d.** 100 calories

45. Find the following:

   a. Q1   **b.** Q2   **c.** Q3   **d.** IQR

46. Explain what the IQR value from Exercise 45(**d**) means.

47. Suppose that a weight-control organization recommended eating breakfast cereals with the lowest 10% of calories.
   a. How many calories does this cutoff represent?
   b. Which cereals are recommended?

Dietary Supplements. Refer to Table 3.19 for Exercises 48–55. The table gives the number of American adults who have used the indicated "nonvitamin, nonmineral, natural products." 🌸 dietarysupp

TABLE 3.19 Use of dietary supplements

Product	Usage (in millions)	Product	Usage (in millions)
Echinacea	14.7	Ginger	3.8
Ginseng	8.8	Soy	3.5
Ginkgo biloba	7.7	Chamomile	3.1
Garlic	7.1	Bee pollen	2.8
Glucosamine	5.2	Kava kava	2.4
St. John's wort	4.4	Valerian	2.1
Peppermint	4.3	Saw palmetto	2.0
Fish oil	4.2		

Source: Centers for Disease Control and Prevention, Vital and Health Statistics, 2004.

48. Find the *z*-scores for usage for the following products:
   a. Echinacea   **b.** Saw palmetto
   c. Valerian   **d.** Ginseng

49. Find the usage associated with each of the following *z*-scores.
   a. 0   **b.** 3   **c.** –3   **d.** 1

50. Identify any outliers in the data set.

51. Find the following percentiles:
   a. 10th   **b.** 90th   **c.** 5th   **d.** 95th

52. Find the percentile rank for each of the following usages:
   a. 14.7 million   **b.** 2.0 million
   c. 8.8 million   **d.** 2.1 million

53. Find the following:
   a. Q1   **b.** Q2   **c.** Q3   **d.** IQR

54. Interpret the IQR value from Exercise 53(**d**) so that a nonspecialist could understand it.

55. Suppose an advertising agency is interested in the top 15% of supplements.
   a. What usage does this represent?
   b. Which supplements would be of interest?

56. Expenditure per Pupil. The 5th percentile expenditure per pupil nationwide in 2005 was $6381, the 50th percentile was $8998, and the 95th percentile was 17,188.[7]
   a. Determine whether the distribution of expenditures is symmetric, left-skewed, or right-skewed.
   b. Would we expect the mean expenditure per pupil to be less than, equal to, or greater than $8998? Explain.
   c. Draw a distribution curve that matches this information.

Bringing It All Together

Twitter Followers. Refer to the following table for Exercises 57–64. 🔴 twitterceleb

Celebrity	Twitter followers (millions)
Lady Gaga	6.6
Britney Spears	6.1
Ashton Kutcher	5.9
Justin Bieber	5.6
Ellen DeGeneres	5.3
Kim Kardashian	5.0

57. Find the *z*-scores for the number of Twitter followers for the following celebrities.
   a. Kim Kardashian   **b.** Lady Gaga   **c.** Justin Bieber

58. Find the number of followers indicated by the following *z*-scores.
   a. −2   **b.** 1   **c.** 3

59. Determine whether the number of followers for any of the celebrities represents an outlier.

60. If the number of followers for Lady Gaga and Kim Kardashian do not represent outliers, explain why we need not check whether the numbers of followers for the other celebrities are outliers.

61. Find the indicated percentiles.
   a. 50th   **b.** 75th   **c.** 25th

62. Calculate the percentile rank for the following.
   a. 5.0 million followers   **b.** 5.3 million followers
   c. 6.6 million followers

63. Find the following for the number of followers.
   a. Q1   **b.** Q2   **c.** Q3   **d.** IQR

64. Interpret the IQR value from Exercise 63(**d**) so that a nonspecialist could understand it.

3.5 FIVE-NUMBER SUMMARY AND BOXPLOTS

OBJECTIVES By the end of this section, I will be able to . . .

   1 Calculate the five-number summary of a data set.

   2 Construct and interpret a boxplot for a given data set.

   3 Detect outliers using the IQR method.

1   THE FIVE-NUMBER SUMMARY

Because the mean and the standard deviation are sensitive to the presence of outliers, data analysts sometimes prefer a less sensitive set of statistics to summarize a data set. The **five-number summary** is an alternative method of summarizing a data set. It includes the median and the qualitiles, which are less sensitive to the preserved of out liers than are the mean and standrd deviation. On the other hand, it also includes the minimum and maximum data values, which are very sensitive to outliers. The five-number summary consists of five measures we have already seen.

> The **five-number summary** consists of the following set of statistics:
>
> 1.   Minimum; the smallest value in the data set
>
> 2.   First quartile, Q1
>
> 3.   Median, Q2
>
> 4.   Third quartile, Q3
>
> 5.   Maximum; the largest value in the data set

EXAMPLE 3.33

THE FIVE-NUMBER SUMMARY FOR A SMALL DATA SET: THE DANCE AUDITION SCORES

dancescore

Find the five-number summary for the dance audition data from Example 3.30 on page 128.

Solution

Examining Figure 3.30, we can without difficulty find the five-number summary for the dance audition data.

Position	1	2	3	4	5	6	7	8	9	10	11	12
Score	30	44	56	62	65	68	75	78	81	85	89	94

$Q1 = 59$ $Q2 = 71.5$ $Q3 = 83$

FIGURE 3.30 The quartiles for the dance audition data.

1.   Minimum $= 30$
2.   First quartile, Q1 $= 59$
3.   Median $=$ Q2 $= 71.5$
4.   Third quartile, Q3 $= 83$
5.   Maximum $= 94$

Now You Can Do Exercises 9, 15, and 21.

More succinctly, the five-number summary is often reported as Min $= 30$, Q1 $= 59$, Med $= 71.5$, Q3 $= 83$, Max $= 94$.

EXAMPLE 3.34

THE FIVE-NUMBER SUMMARY FOR A LARGE DATA SET: CHOLESTEROL LEVELS IN FOOD

dancescore

Find the five-number summary for the cholesterol data from Example 3.31 on page 129.

Solution

Minitab's reporting of the descriptive statistics makes it particularly straightforward to report the five-number summary, as here in Figure 3.31 (repeated from page 129) for the cholesterol data.

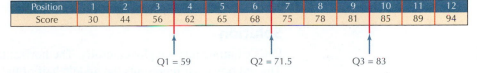

Variable	N	Mean	StDev	Min	Q1	Median	Q3	Max
Cholesterol	961	32.55	119.96	0	0	0	20	2053

FIGURE 3.31 Descriptive statistics for the cholesterol data.

The five-number summary for the cholesterol data set is
1. Smallest value in the data set = Min = 0
2. First quartile, Q1 = 0
3. Median = 0
4. Third quartile, Q3 = 20
5. Largest value in the data set = Max = 2053

Or, simply, Min = 0, Q1 = 0, Med = 0, Q3 = 20, Max = 2053.

The five-number summary is associated with a certain type of graphical summary of data, called a *boxplot*, which we examine next.

2 THE BOXPLOT

The **boxplot** (sometimes called a box-and-whisker plot) is a convenient graphical display of the five-number summary of a data set. The boxplot allows the data analyst to evaluate the symmetry or skewness of a data set.

EXAMPLE 3.35

THE CHARACTERISTICS OF A BOXPLOT

Interpret the boxplot for the audition scores in Figure 3.32.

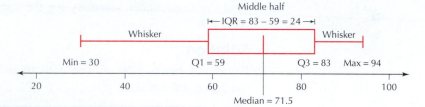

FIGURE 3.32 Boxplot of the dance score data.

Solution

Let's examine this boxplot carefully. The horizontal axis represents the dance scores. The red box itself represents the middle half of the data set. The right-hand side of the box, called the *upper hinge,* is located at Q3, which is 83. The left-hand side of the box, called the *lower hinge,* is located at Q1, which is 59. The solid vertical line inside the box is located at the median, which is 71.5. The horizontal lines emanating from the left and right of the box are called the *whiskers.* If there are no outliers, the whiskers extend as far as the maximum and minimum values of the data set, which are represented by the vertical lines at Max = 94 and Min = 30.

Constructing a Boxplot by Hand

1. Determine the lower and upper fences:
   a. Lower fence = Q1 − 1.5(IQR)
   b. Upper fence = Q3 + 1.5(IQR), where IQR = Q3 − Q1
2. Draw a horizontal number line that encompasses the range of your data, including the fences. Above the number line, draw vertical lines at Q1, the median, and Q3. Connect the lines for Q1 and Q3 to each other so as to form a box.
3. Temporarily indicate the fences as brackets ([and]) above the number line.
4. Draw a horizontal line from Q1 to the smallest data value greater than the lower fence. This is the lower whisker. Draw a horizontal line from Q3 to the largest data value smaller than the upper fence. This is the upper whisker.
5. Indicate any data values smaller than the lower fence or larger than the upper fence using an asterisk (*). These data values are outliers. Remove the temporary brackets.

EXAMPLE 3.36   **CONSTRUCTING A BOXPLOT BY HAND**

On page 141, we demonstrate how to create a boxplot using technology. Construct a boxplot for the dance score data.

Solution

From Example 3.33, the five-number summary for the dance score data is Min = 30, Q1 = 59, Med = 71.5, Q3 = 83, Max = 94. The interquartile range for the dance score data is IQR = Q3 − Q1 = 83 − 59 = 24.

STEP 1   Determine the lower and upper fences:
a. Lower fence = Q1 − 1.5(IQR) = 59 − 1.5(24) = 59 − 36 = 23
b. Upper fence = Q3 + 1.5(IQR) = 83 + 1.5(24) = 83 + 36 = 119

STEP 2   Draw a horizontal number line that encompasses the range of your data, including the fences. Above the number line, draw vertical lines at Q1 = 59, median = 71.5, and Q3 = 83. Connect the lines for Q1 and Q3 to each other so as to form a box, as shown in Figure 3.33A.

FIGURE 3.33A Constructing a boxplot by hand: Steps 1 and 2.

STEP 3   Temporarily indicate the fences (lower fence = 23 and upper fence = 119) as brackets above the number line. (See Figure 3.33B.)

FIGURE 3.33B Constructing a boxplot by hand: Step 3.

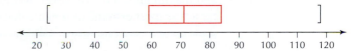

STEP 4   Draw a horizontal line from Q1 = 59 to the smallest data value greater than the lower fence. The lowest data value is Min = 30. This is greater than the lower fence = 23. So draw the line from 59 to 30. Draw a horizontal line from Q3 = 83 to the largest data value smaller than the upper fence. The largest data value is Max = 94, which is smaller than the upper fence. So draw the line from 83 to 94. (See Figure 3.33C.)

FIGURE 3.33C Constructing a boxplot by hand: Step 4.

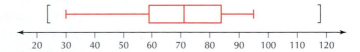

STEP 5   There are no data values lower than the lower fence or greater than the upper fence. Thus, there are no outliers in this data set. Therefore, simply remove the temporary brackets, and the boxplot is complete, as shown in Figure 3.33D.

FIGURE 3.33D The completed boxplot.

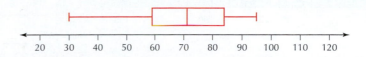

Now You Can Do
Exercises 12, 18, and 24.

The next examples show how to recognize when boxplots indicate that a data set is right-skewed, left-skewed, or symmetric.

EXAMPLE 3.37

BOXPLOT FOR RIGHT-SKEWED DATA

The number of strikeouts per player in the 2007 American League season is a right-skewed distribution, as shown in histogram of the data in Figure 3.34. The five-number summary is Min = 0, Q1 = 9, Med = 21, Q3 = 47, and Max = 111. How is this skewness reflected in a boxplot (Figure 3.35)? Well, in right-skewed data, the median is closer to Q1 than to Q3, and the lowest non-outlier is closer to Q1 than the highest non-outlier is to Q3. This means that the median is closer to the lower hinge than the upper hinge, and the upper whisker is much longer than the lower whisker. This combination of characteristics indicates a right-skewed data set.

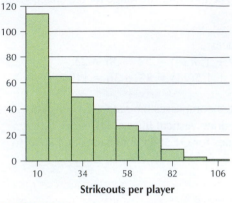

FIGURE 3.34 Strikeouts are right-skewed.

FIGURE 3.35 TI-83/84 boxplot of strikeouts: right-skewed.

The two little boxes at the right represent outliers. (The TI-83/84 uses little boxes rather than asterisks.) These players are David Ortiz of the Boston Red Sox, who led the league that year with 111 strikeouts, and Jack Cust of the Oakland Athletics, with 105 strikeouts. When there are no outliers, the whiskers extend as far as the minimum and maximum values. However, when there are outliers, the whiskers extend only as far as the most extreme data value that is not an outlier.

EXAMPLE 3.38

BOXPLOT FOR LEFT-SKEWED DATA

Figure 3.36 is a histogram of 650 exam scores. Clearly, the data are left-skewed, with many students getting scores in the 90s, and fewer getting grades in the 70s or 80s. Now, with right-skewed data, remember that the median was closer to Q1 than to Q3. What do you think will happen for left-skewed data?

Solution

The five-number summary is Min = 70, Q1 = 86, Med = 94, Q3 = 98, and Max = 100. So, this time, with left-skewed data, the median is closer to Q3 than to Q1. Bet you guessed it!

In the boxplot (Figure 3.37), notice that the median (94) is closer to the upper hinge (Q3, 98) than to the lower hinge

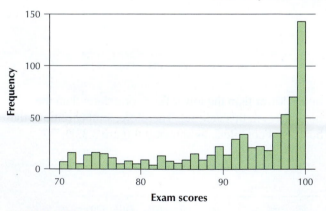

FIGURE 3.36 Histogram of exam scores.

FIGURE 3.37 TI-83/84 boxplot of the exam scores.

(Q1, 86), and the lower whisker is much longer than the upper whisker. This combination of characteristics indicates a left-skewed data set.

Symmetric Data and Boxplots

So, can you now predict how a boxplot of *symmetric* data will look? The median will be about the same distance from Q1 (lower hinge) and Q3 (upper hinge). And the upper and lower whiskers will be about the same length. An example of a boxplot of symmetric data is shown in Figure 3.38

FIGURE 3.38 Boxplot of symmetric data.

3 DETECTING OUTLIERS USING THE IQR METHOD

When using the mean and standard deviation as your summary measures, in most cases outliers occur more than 3 standard deviations from the mean. However, due to the sensitivity of these measures to the outliers themselves, we often use a more robust method of detecting outliers. Earlier we mentioned that, when constructing a boxplot, data values lower than the lower fence and higher than the upper fence are considered outliers. We can use this method to detect outliers without constructing a boxplot.

IQR Method to Detect Outliers

A data value is an outlier if
    a. it is located 1.5(IQR) or more below Q1, or
    b. it is located 1.5(IQR) or more above Q3.

EXAMPLE 3.39

IQR METHOD FOR DETECTING OUTLIERS

Determine if there are any outliers in the dance score data.

Solution

Recall for the dance score data set that IQR = 24, Q1 = 59, and Q3 = 83. So we have 1.5(IQR) = 1.5(24) = 36. The first step is to find the two quantities Q1 − 1.5(IQR) and Q3 + 1.5(IQR):

$$Q1 - 1.5(IQR) = Q1 - 36 = 59 - 36 = 23$$

$$Q3 + 1.5(IQR) = Q3 + 36 = 83 + 36 = 119$$

**Now You Can Do
Exercises 25–28.**

Thus, for this data set, a data value would be an outlier if it were 23 or less or 119 or more. Since there are no data values that are 23 or less or 119 or more in the data set, no outliers are identified by the IQR method.

IQR Method for Outlier Detection

What if the minimum dance score of 30 is changed to 23. Based on Example 3.39, this new value should be detected as an outlier. Note that changing the minimum value does not affect the calculation of Q1, Q3, the IQR, or the thresholds for outlier detection.

Figure 3.39 shows that the box, hinges, and whiskers are all located at precisely the same spots as in the boxplot of the original dance score data. However, the software has calculated, using the robust detection method, that the new data value of 23 is an outlier and indicates it as such with a blue dot. Comparing this boxplot to the earlier one (see Figure 3.32), we notice that the lower whisker is shorter. In Figure 3.39, the whisker terminates at the dance score of 44 instead of 30.

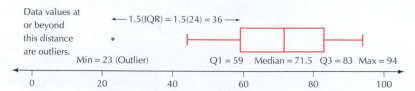

FIGURE 3.39 Boxplot of dance score data showing presence of outlier, after change.

The next example shows how comparison boxplots may be used to compare two data sets side-by-side.

EXAMPLE 3.40

COMPARISON BOXPLOTS: COMPARING BODY TEMPERATURES FOR WOMEN AND MEN

Determine whether the body temperatures of women or men exhibit greater variability.

Solution

Consider the comparison boxplots in Figure 3.40. The box for females (on top) lies slightly to the right of that for the males, meaning that the first quartile, the median, and the third quartile are each higher for the women than the men. Therefore, the middle 50% of the body temperatures is higher for women than men.

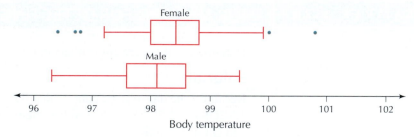

FIGURE 3.40 Comparison of boxplots of female and male body temperatures.

We will formally test whether there is a difference in the true mean body temperature between women and men in Chapter 10.

This figure seems to offer some evidence that the mean body temperature for women may be higher than that for men. The location of the box is an indication of the center of the data. But where would we look for a difference in the variability of body temperatures between women and men? From Figure 3.41, for the females we have

$$IQR = Q3 - Q1 = 98.8 - 98.0 = 0.8$$

For the males we have

$$IQR = Q3 - Q1 = 98.6 - 97.6 = 1.0$$

So the IQR for males is greater.

Let's determine which data set has greater variability based on the three different measures of spread that we have learned: the range, the standard deviation, and the IQR.

Gender	–	Mean	Median	StDev	Min	Max	Q1	Q3
female	65	98.394	98.4	0.743	96.4	100.8	98.0	98.8
male	65	98.105	98.1	0.699	96.3	99.5	97.6	98.6

FIGURE 3.41 Descriptive statistics for body temperature, by gender.

Now You Can Do Exercises 10, 11, 16, 17 and 22, and 23.

Range for women = 100.8 − 96.4 = 4.4   Range for men = 99.5 − 96.3 = 3.2
Standard deviation for women = 0.743   Standard deviation for men = 0.699
IQR for women = 0.8   IQR for men = 1.0

Developing Your Statistical Sense

When Measures of Spread Disagree

Two measures of spread that are sensitive to the presence of extreme values—range and standard deviation—find that the female body temperatures are more variable. The measure of spread that is resistant to the effects of extreme values—IQR—finds that the male body temperatures are more variable. How do we resolve this apparent inconsistency? What appears to be happening is that, for the middle 50% of each data set, the men are more variable, but as we move toward the tails, the women are more spread out.

Note that there are outliers for the women but not for the men. In part, this may be because the IQR for the women is smaller, and thus the distance 1.5(IQR) is smaller as well. For example, the woman whose body temperature is 100 degrees is identified as an outlier because 100 is the same as the outlier threshold Q3 + 1.5(IQR) = 98.8 + 1.5(0.8) = 100. *The same temperature in a man would not be classified as an outlier, even though the male temperatures are lower overall* (and Q3, specifically, is lower). This is because the temperature of 100 is not higher than Q3 + 1.5(IQR) = 98.6 + 1.5(1.0) = 100.1, the male outlier threshold. Thus, the measures of spread that are sensitive to outliers indicate that women have greater variability, while the measure of spread that is not sensitive to outliers indicates that men have greater variability.

STEP-BY-STEP TECHNOLOGY GUIDE: Boxplots

We will make boxplots for the data in Example 3.30 (page 128).

TI-83/84

Step 1   Enter the data in list **L1**.
Step 2   Press **2nd Y =**, and choose **1: Plot 1**.
Step 3   Turn plots On. Highlight the boxplot icon, as shown in Figure 3.41.
Step 4   Press **ZOOM**, and choose **9: ZoomStat**.
A boxplot similar to Figure 3.32 in Example 3.35 is then produced.

FIGURE 3.42

MINITAB

Step 1 Enter the data in column **C1**, and name your data Scores.

Step 2 Click **Graph** > **Boxplot**.

Step 3 Select **Simple** and click **OK**.

Step 4 Select the variable Scores, and click **OK**, as shown in Figure 3.43. A boxplot similar to Figure 3.32 in Example 3.35 is then produced.

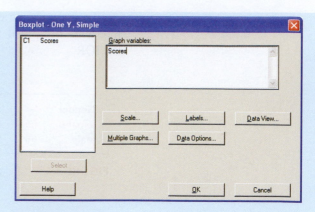

FIGURE 3.43

CRUNCHIT!

Step 1 Click **File . . .** then highlight **Load from Larose2e . . . Chapter 3 . . .** and click on **Example 3.33**.

Step 2 Click **Graphics** and select **Box plot**. For **Data** select **Scores**. Click **Calculate**.

SECTION 3.5 Summary

1. The five-number summary is an alternative to the usual mean-and-standard-deviation method of summarizing a data set. It consists of simply reporting the minimum, first quartile, median, third quartile, and maximum of the data set.

2. A boxplot is a graphical representation of the five-number summary and is useful for investigating skewness and the presence of outliers.

3. The IQR method of detecting outliers is to consider a data value an outlier if it is located 1.5(IQR) or more below Q1, or it is located 1.5(IQR) or more above Q3.

SECTION 3.5 Exercises

Clarifying the Concepts

1. True or false: The five-number summary consists of the minimum, Q1, Mean, Q3, Maximum.

2. Explain what we mean when we say that the five-number summary is associated with the boxplot.

3. Explain how we can use a boxplot to recognize the following:
   a. Symmetric distribution
   b. Right-skewed distribution
   c. Left-skewed distribution

4. When is it possible for outliers to be found inside the box of a boxplot?

5. Explain the IQR method for detecting outliers.

6. Why do we need the IQR method for detecting outliers when we already have the z-score method?

Practicing the Techniques

Use the following set of 10 student heights (in inches) to answer Exercises 7–12.

64 64 65 66 68 68 70 70 71 78

7. Find the quartiles.

8. Calculate the interquartile range.

9. Compute the five-number summary.

10. Use the IQR method to determine whether 71 inches is an outlier.

11. Use the IQR method to determine whether 78 inches is an outlier.

12. Construct a boxplot for student height.

Use the following data, the commuting times (in minutes) for 12 community college students, to answer Exercises 13–18.

10 15 10 20 15 15 25 50 15 20 25 15

13. Find the quartiles.

14. Calculate the interquartile range.

15. Compute the five-number summary.

16. Use the IQR method to determine whether 10 minutes is an outlier.

17. Use the IQR method to determine whether 50 minutes is an outlier.

18. Construct a boxplot for commuting time.
Here are the final-exam scores for 20 psychology students. Use this data set to answer Exercises 19–24.

75 81 82 70 60 59 94 77 68 98
86 68 85 72 70 91 78 86 41 67

19. Find the quartiles.

20. Calculate the interquartile range.

21. Compute the five-number summary.

22. Use the IQR method to determine whether a score of 41 is an outlier.

23. Use the IQR method to determine whether a score of 98 is an outlier.

24. Construct a boxplot for final-exam score.

For Exercises 25 and 26, do the following:
   a. Identify the shape of the distribution.
   b. Use the boxplot to find the five-number summary.

25.

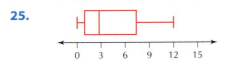

26.

Use the comparison boxplots shown to answer Exercises 27–30.

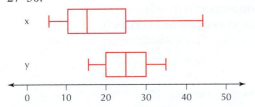

27. For the variable *x*:
   a. Identify the shape of the distribution.
   b. Use the boxplot to find the five-number summary.

28. For the variable *y*:
   a. Identify the shape of the distribution.
   b. Use the boxplot to find the five-number summary.

29. Which variable has greater variability, according to the IQR?

30. Which variable has greater variability, according to the range?

Applying the Concepts

Most Active Stocks. Use Table 3.20 for Exercises 31–38. These companies represent the 10 most actively traded stocks on the New York Stock Exchange for March 9, 2012. Variables include the stock price and the net change in stock price, with both variables in dollars.

 nysestock

TABLE 3.20   The most active stocks on the NYSE

Company	Price	Change
Bank of America Corp	8.15	+0.09
Sprint Nextel Corp	2.80	+0.20
Citigroup Inc	34.73	+0.73
Ford Motor Co	12.61	+0.15
JPMorgan Chase and Co	41.23	+0.79
General Electric Co	19.08	+0.05
Freeport Copper & Gold Inc	39.39	−0.11
Microsoft Corp	31.99	−0.02
Pfizer Inc	21.70	+0.25
Oracle Corp	30.25	+0.18

31. Find the five-number summary for *price*.

32. Find the interquartile range for *price*. Interpret what this value actually means, so that a nonspecialist could understand it.

33. Use the IQR method to investigate the presence of outliers in *price*.

34. Construct a boxplot for *price*.

35. Find the five-number summary for *change*.

36. Find the interquartile range for *change*. Interpret what this value actually means, so that a nonspecialist could understand it.

37. Use the IQR method to investigate the presence of outliers in *change*.

38. Construct a boxplot for *change*.

Dietary Supplements. Refer to Table 3.19 (page 134) for Exercises 42–47.

 dietarysupp

39. Find the five-number summary for *usage*.

40. Find the interquartile range for *usage*. Interpret what this value actually means, so that a nonspecialist could understand it.

41. Use the IQR method to investigate the presence of outliers in *usage*.

42. Construct a boxplot for *usage*.

43. Calculate the mean and standard deviation of *usage*.

44. Find the z-score for echinacea, and use it to determine whether the product is an outlier. Compare the result with that from the IQR method.

Bringing It All Together

Zooplankton and Phytoplankton. For Exercises 45–53, refer to the zooplankton and phytoplankton meta-analysis effect size data from the Section 3.2 exercises (page 111).

🌎 plankton

45. Compute the five-number summary for each of the zooplankton and phytoplankton data.

46. Construct comparison boxplots for the zooplankton and phytoplankton data.

47. Describe the shapes of the distribution for the zooplankton and phytoplankton data.

48. Based on your descriptions in the previous exercise, would you expect the mean to be larger or smaller or about the same as the median for the zooplankton data? The phytoplankton data?

49. Calculate the mean for the zooplankton data and the phytoplankton data. Do they concur with your expectations from the previous exercise?

50. Describe the difference between the effect sizes between the zooplankton and phytoplankton, in terms of the location of the box. Which type of plankton seems to have the greater overall effect sizes? Does this agree with what a comparison of the means from the previous exercise is telling you?

51. Describe the difference between the effect sizes between the zooplankton and phytoplankton, in terms of the IQR measure of spread. Which type of plankton has greater variability?

52. Identify any outliers for the zooplankton data and the phytoplankton data, using the IQR method.

53. Challenge Exercise. Identify any outliers for the phytoplankton data using the z-score method. Compare the outliers identified using the IQR method and the z-score method. Clearly explain why the two methods disagree.

Nutrition. Use the data set Nutrition for Exercises 54–57.

🌎 Nutrition

54. Open the data set Nutrition.
  a. How many observations are in the data set?
  b. How many variables?

55. Use a statistical computing package (like Minitab) to explore the variable *iron*.
  a. Find the mean and standard deviation for the amount of iron in the food.
  b. Find the five-number summary, the range, and the interquartile range.

56. Which food item has the maximum amount of iron? Does this surprise you?

57. Use the computer to generate a boxplot. Also, comment on the symmetry or the skewness of the boxplot.

CHAPTER 3	**Formulas and Vocabulary**

Section 3.1

- **MEAN** (p. 82)
- **MEASURE OF CENTER** (p. 82)
- **MEDIAN** (p. 86)
- **MODE** (p. 88)
- **POPULATION MEAN** (p. 84). $\mu = \sum x / N$.
- **POPULATION SIZE** (p. 84). Denoted by N.
- **SAMPLE MEAN** (p. 83). $\bar{x} = \sum x / n$.
- **SAMPLE SIZE** (p. 83). Denoted by n.

Section 3.2

- **CHEBYSHEV'S RULE** (p. 107). The proportion of values from a data set that will fall within k standard deviations of the mean will be *at least* $\left(1 - \frac{1}{k^2}\right) 100\%$, where $k > 1$.
- **DEVIATION** (p. 98). $x - \bar{x}$.

- **EMPIRICAL RULE** (p. 105). If the data distribution is bell-shaped:

  About 68% of the data values will fall within 1 standard deviation of the mean.

  About 95% of the data values will fall within 2 standard deviations of the mean.

About 99.7% of the data values will fall within 3 standard deviations of the mean.

- **MEASURE OF VARIABILITY (MEASURE OF SPREAD, MEASURE OF DISPERSION)** (p. 97)
- **POPULATION STANDARD DEVIATION** (p. 101).

$$\sigma = \sqrt{\frac{\sum(x - \mu)^2}{N}}$$

- **POPULATION VARIANCE** (p. 101).

$$\sigma^2 = \frac{\sum(x - \mu)^2}{N}$$

- **RANGE** (p. 98)
- **SAMPLE STANDARD DEVIATION** (p. 103).

$$s = \sqrt{\frac{\sum(x - \bar{x})^2}{n - 1}}$$

- **SAMPLE VARIANCE** (p. 103).

$$s^2 = \frac{\sum(x - \bar{x})^2}{n - 1}$$

- **STANDARD DEVIATION** (p. 101)

Section 3.3
- **ESTIMATED MEAN FOR DATA GROUPED INTO A FREQUENCY DISTRIBUTION** (p. 116).

$$x = \frac{\sum (f \cdot x)}{\sum f}$$

- **ESTIMATED STANDARD DEVIATION FOR DATA GROUPED INTO A FREQUENCY DISTRIBUTION** (p. 117).

$$s = \sqrt{s^2} = \sqrt{\frac{\sum (x - \bar{x})^2 \cdot f}{\sum f}}$$

- **ESTIMATED VARIANCE FOR DATA GROUPED INTO A FREQUENCY DISTRIBUTION** (p. 117).

$$s^2 = \frac{\sum (x - \bar{x})^2 \cdot f}{\sum f}$$

- **WEIGHTED MEAN** (p. 115).

$$x = \frac{\sum (w \cdot x)}{\sum w}$$

Section 3.4
- **FINDING A DATA VALUE *X* GIVEN ITS *Z*-SCORE** (p. 123)

$$\text{Sample}: x = z\text{-score} \cdot s + \bar{x}$$
$$\text{Population}: x = z\text{-score} \cdot \sigma + \mu$$

- **INTERQUARTILE RANGE (IQR)** (p. 131).

$$IQR = Q3 - Q1$$

- **OUTLIER** (p. 124)
- **PERCENTILE** (p. 125)
- **PERCENTILE RANK** (p. 127)
- **QUARTILES** (p. 128)
- **z-SCORE** (p. 121)
    a. Sample:

$$z\text{-score} = \frac{\text{data value} - \text{mean}}{\text{standard deviation}} = \frac{x - \bar{x}}{s}$$

    b. Population:

$$z\text{-score} = \frac{\text{data value} - \text{mean}}{\text{standard deviation}} = \frac{x - \mu}{\sigma}$$

Section 3.5
- **BOXPLOT** (p. 136)
- **FIVE-NUMBER SUMMARY** (p. 135)
- **IQR METHOD OF DETECTING OUTLIERS** (p. 139)

CHAPTER 3   Review Exercises

Section 3.1
PHYTOPLANKTON. Refer to the phytoplankton data from Exercise 51 in Section 3.2 (page 111) for Exercises 1–3.
1. Find the mean.
2. Find the median.
3. Find the mode, if any.

CALORIES IN CEREAL. For Exercises 4–6, refer to the calories in breakfast cereals gives in Table 3.18 (page 133).
4. Which is the largest, the mean, median, or mode? How do you know?
5. If we eliminated the cereals with 90 or less calories from the sample, which measure would not be affected at all? Why?
6. If we added 10 calories to each cereal, how would that affect the mean, median, and mode? Would it affect each of the measures equally?

Section 3.2
COMMON SYLLABLES IN ENGLISH. Refer to the table shown here of some common syllables in English for Exercises 7–10.
🔴 syllables

Syllable	Frequency
an	462
bi	621
sit	104
ed	907
its	293
est	186
wil	470
tiv	136
en	675
biz	114

7. Find the mean and the range of the syllable frequencies.
8. Would you say that a typical distance from the mean for the frequencies is about 900, about 500, about 300, or about 100?

9. What is your best guesstimate of the value of a typical distance from the mean for the syllable frequencies?

10. Find the sample variance and the sample standard deviation of syllable frequencies.

   a. How far is each from your estimate of the typical deviation earlier?

   b. Interpret the meaning of this value for the standard deviation so that someone who has never studied statistics would understand it.

Section 3.3

11. CALCULATING A GRADE POINT AVERAGE. At a certain college in Texas, student grade point averages are calculated as follows. For each credit hour, an A is worth 4.0 quality points, an A− is worth 3.7 quality points, a B+ is worth 3.3 quality points, a B is worth 3.0, a B− is worth 2.7, a C+ is worth 2.3, and so on. To find the grade point average, the number of credits for each course is multiplied by the quality points earned for that course; the results are added together; and the sum is divided by the number of credits. This semester, Angelita's grades are as follows. She got an A in her four-credit honors biology course, an A− in her three-credit calculus course, a B+ in her three-credit English course, a B− in her three-credit anthropology course, and a C+ in her two-credit physical education course. Calculate Angelita's grade point average for this semester.

12. AIDS CASES BY AGE. The National Center for Health Statistics reported the number of cases of acquired immunodeficiency syndrome (AIDS) by age of patient in 2004.[8] Find the estimated mean and standard deviation of the age of AIDS patients. 🔴 aidsbyage

Class: age	Frequency f_i
0–12.99	48
13–14.99	60
15–24.99	2,114
25–34.99	9,361
35–44.99	16,778
45–54.99	10,178
55–64.99	3,075
65–74.99	901

Section 3.4

RAGWEED POLLEN. Use the table of ragweed pollen index in New York localities for Exercises 13–25. Do you suffer from ragweed pollen? You are not alone. The American Academy of Allergy maintains the ragweed pollen index, which details the severity of the pollen problem for hundreds of communities across the nation. The following table contains the ragweed pollen index on a particular day for 10 localities in New York State.

🔴 ragweed

Locality	Ragweed pollen index
Albany	48
Binghamton	31
Buffalo	59
Elmira	43
Manhattan	25
Rochester	60
Syracuse	25
Tupper Lake	8
Utica	26
Yonkers	38

Find the following percentiles of total ragweed pollen index.

13. 10th percentile

14. 50th percentile

15. 90th percentile

For Exercises 16–18, find the z-scores for the following localities for the ragweed pollen index.

16. Albany

17. Rochester

18. Tupper Lake

19. Identify any outliers or moderately unusual observations in the ragweed pollen index.

For Exercises 20–22, find the percentile rank for the given ragweed pollen index.

20. 25

21. 59

22. 48

23. Find the first, second, and third quartiles of the ragweed pollen index.

24. Find the interquartile range. Interpret what this value actually means, so that a nonspecialist could understand it.

25. Detect any outliers using the IQR method.

Section 3.5

26. Let's draw a boxplot of the ragweed pollen index.

   a. What is the five-number summary?

   b. By hand, draw a boxplot.

   c. Is the data set left-skewed, right-skewed, or symmetric?

   d. What should the symmetry or skewness mean in terms of the relative values of the mean and median?

   e. Find the mean and standard deviation. Is your prediction in **(d)** supported?

27. Detect any outliers using the IQR method. Compare with Exercise 25. Do the two methods concur or disagree?

28. Suppose the ragweed pollen index in Rochester were 600 instead of 60. How would this outlier affect the quartiles and the IQR? What property of these measures is this behavior an example of?

CHAPTER 3 **Quiz**

True or False

1. True or false: If two data sets have the same mean, median, and mode, then the two data sets are identical.

2. True or false: The variance is the square root of the standard deviation.

3. True or false: The Empirical Rule applies for any data set.

Fill in the Blank

4. An _____ is an extremely large or extremely small data value relative to the rest of the data set.

5. The mean can be viewed as the _____ point of the data.

6. The measure of center that is sensitive to the presence of extreme values is the _____.

Short Answer

7. What do we call summary descriptive measures that are not sensitive to the presence of outliers?

8. Which of the mean, median, and mode may be used for categorical data?

9. For any data set, what is the average of the deviations?

10. What do we use to estimate the mean for each class in a frequency distribution?

Calculations and Interpretations

AIRLINE PASSENGERS. Refer to the following table for Exercises 11 and 12.

 portlandair

Passengers arriving at Portland International Airport, January–April 2007, by airline

Airline	Passengers
Alaska Airlines	98,008
Delta Air Lines	31,054
Horizon Air	117,964
Southwest Airlines	106,178
United Airlines	84,059

11. Calculate the following:
 a. Sample mean
 b. Sample median

12. Calculate the following:
 a. Range
 b. Sample standard deviation

13. DEATHS DUE TO HEAT. The following frequency distribution contains the numbers of deaths due to heat, by age group, as reported by the National Weather Service for 2006. Find the estimated mean and standard deviation of age.

🔴 heatdeath

Age	Deaths due to heat
0–39.99	22
40–49.99	31
50–59.99	51
60–69.99	47
70–79.99	44
80–89.99	44

14. A sample of 30 Americans yielded a sample mean consumption of carbonated beverages this year of 60 gallons with a sample standard deviation of 40 gallons. Find the z-scores for the following amounts of carbonated beverage consumption.
 a. 120 gallons
 b. 20 gallons
 c. 100 gallons
 d. 0 gallons
 e. 60 gallons

15. Refer to the information in Exercise 14. Assume the distribution is bell-shaped. (*Note:* Use your knowledge about the Empirical Rule to give a range for the proportions in parts **(b)** and **(d)**).
 a. Find the 50th percentile.
 b. Estimate the proportion of Americans who drink between 20 and 100 gallons per year.
 c. Discuss whether we could find the estimate in **(b)** without assuming that the distribution is bell-shaped.
 d. Estimate the proportion of Americans who drink more than 100 gallons per year.

Use the following SAT 1 Math score for Exercises 16–20.

510,   515,   523,   514,   521,   501,   502,   499

🔴 satmath

16. Find the following quartiles for SAT 1 Math score:
 a. Q1
 b. Q2
 c. Q3

17. Find the interquartile range of SAT 1 Math score.

18. Find the five-number summary for SAT 1 Math score.

19. Use robust methods to investigate the presence of outliers.

20. Construct a boxplot for SAT 1 Math score.

4 Correlation and Regression

Clockwise from top: © Visions of America, LLC/Alamy; © Dinodia Photos/Alamy; © Dinodia Photos/Alamy; © RubberBall/Alamy; © RubberBall/Alamy; © VStock/Alamy.

CASE STUDY

Worldwide Patterns of Cell Phone Usage

Cell phones can be used to send text messages, browse the Internet, take photos, record video, or even to make phone calls. But what are the patterns of cell phone usage worldwide? For example, would you expect that residents of richer countries tend to use their cell phones to browse the Internet more often than do residents of poorer countries? The Pew Global Attitudes Project conducted a study[1] of cell phone usage in countries around the world. In the Chapter 4 Case Study, we explore the relationship between some quantitative variables measured in this study, such as the percentage of cell phone owners who use their cell phones to browse the Internet, with a measure of their countries wealth, the per capita gross domestic product. ■

The Big Picture

Where we are coming from, and where we are headed . . .

- Chapter 3 showed us methods for summarizing data using descriptive statistics, but only one variable at a time.

- In Chapter 4, we learn how to analyze the relationship between two quantitative variables using scatterplots, correlation, and regression.

- In Chapter 5, we will learn about probability, which we will need in order to perform statistical inference.

4.1 SCATTERPLOTS AND CORRELATION

OBJECTIVES By the end of this section, I will be able to . . .

1. Construct and interpret scatterplots for two quantitative variables.

2. Calculate and interpret the correlation coefficient.

3. Determine whether a linear correlation exists between two variables.

So far, most of our work has looked at ways to describe only one quantitative variable at a time. But there may exist a relationship between two quantitative variables, say, *height* and *weight,* that we would like to graph or quantify. We may also want to use the value of one variable, say, *height,* to predict the value of the other variable, *weight.* In Section 4.1 we explore scatterplots, which are graphs of the relationship between two quantitative variables, and we learn about correlation, which quantifies this relationship.

1 SCATTERPLOTS

Whenever you are examining the relationship between two quantitative variables, your best bet is to start with a scatterplot. A **scatterplot** is used to summarize the relationship between two quantitative variables that have been measured on the same element. An example of a scatterplot is given in Figure 4.1.

> A **scatterplot** is a graph of points (x, y), each of which represents one observation from the data set. One of the variables is measured along the horizontal axis and is called the *x variable.* The other variable is measured along the vertical axis and is called the *y variable.*

Often, the value of the *x* variable can be used to predict or estimate the value of the *y* variable. For this reason, the *x* variable is referred to as the *predictor* variable, and the *y* variable is called the *response* variable.

EXAMPLE 4.1

sqrfootsale

CONSTRUCTING A SCATTERPLOT

Suppose you are interested in moving to Glen Ellyn, Illinois, and would like to purchase a lot upon which to build a new house. Table 4.1 contains a random sample of eight lots for sale in Glen Ellyn, with their square footage and prices. Identify the predictor variable and the response variable, and construct a scatterplot.

Note: The square footage is expressed in 100s of square feet, so that "90" represents 90 × 100 = 9000 square feet. Similarly, the sales price is expressed in $1000s, so that "200" = 200 × 1000 = $200,000.

Table 4.1 Lot square footage and sales price

Lot	x = square footage (100s of sq. ft.)	y = sales price ($1000s)
Harding St.	75	155
Newton Ave.	125	210
Stacy Ct.	125	290
Eastern Ave.	175	360
Second St.	175	250
Sunnybrook Rd.	225	450
Ahlstrand Rd.	225	530
Eastern Ave.	275	635

Note: The predictor variable and response variable are sometimes referred to as the independent variable and dependent variable, respectively. This textbook avoids this terminology, since it may be confused with the definition of independent and dependent events and variables in probability (Chapter 5) and categorical data analysis (Chapter 11).

Solution

It is reasonable to expect that the price of a new lot depends in part on how large the lot is. Thus, we define our predictor variable x to be $x = $ *square footage* and our response variable y to be $y = $ *sales price*. Next we construct the scatterplot using the data from Table 4.1. Draw the horizontal axis so that it can contain all the values of the predictor (x) variable, and similarly for the vertical axis. Then, at each data point (x, y), draw a dot. For example, for the Harding Street lot, move along the x axis to 75, then go up until you reach a spot level with $y = 155$, at which point you draw a dot. Proceed similarly for all eight properties. The result should look similar to the scatterplot in Figure 4.1.

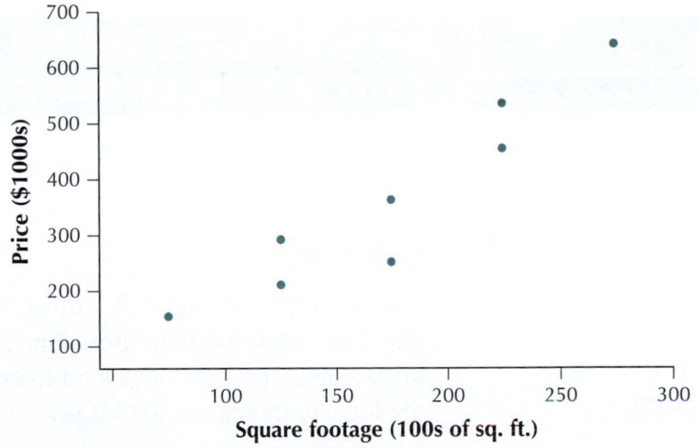

FIGURE 4.1
Scatterplot of sales price versus square footage.

From this scatterplot, we can see that there is a tendency for larger lots to have higher prices. This is not the case for each observation. For example, the Second Street property is larger than the Stacy Court property but has a lower price. Nevertheless, the overall tendency remains.

Now You Can Do
Exercises 9–12.

Scatterplot Terminology

Note the terminology in the caption to Figure 4.1. When describing a scatterplot, always indicate the *y* variable first and use the term *versus* (*vs.*) or *against* the *x* variable. This terminology reinforces the notion that the *y* variable depends on the *x* variable.

The relationship between two quantitative variables can take many different forms. Four of the most common relationships are shown in Figures 4.2a–4.2d.

- **Positive linear relationship** between *x* and *y* (Figure 4.2a): Smaller values of the *x* variable are associated with smaller values of the *y* variable; larger values of *x* are associated with larger values of *y*. In other words, as *x* increases, *y* also tends to increase.

- **Negative linear relationship** between *x* and *y* (Figure 4.2b): Smaller values of the *x* variable are associated with larger values of the *y* variable; larger values of *x* are associated with smaller values of *y*. In other words, as *x* increases, *y* tends to decrease.

Note: the phrase, "as *x* increases in value . . ." When interpreting scatterplots, we always move from left to right.

- **No apparent relationship** (Figure 4.2c): The values of the *x* variable are not associated with any particular range of values of the *y* variable. In other words, as *x* increases, *y* tends to remain unchanged.

- **Nonlinear relationship** (Figure 4.2d): The *x* variable and the *y* variable are related, but not in a way that can be approximated using a straight line.

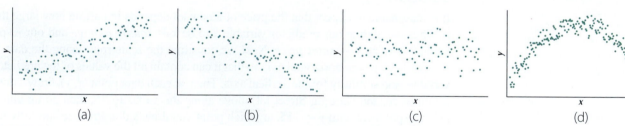

(a)   (b)   (c)   (d)

FIGURE 4.2 Scatterplots of (a) a positive relationship; (b) a negative relationship; (c) no apparent relationship; (d) a nonlinear relationship.

EXAMPLE 4.2

CHARACTERIZE THE RELATIONSHIP BETWEEN TWO VARIABLES USING A SCATTERPLOT

Using Figure 4.1, characterize the relationship between lot square footage and lot price.

Solution

The scatterplot in Figure 4.1 most resembles Figure 4.2a, where a positive relationship exists between the variables. Thus, smaller lot sizes tend to be associated with lower prices, and larger lot sizes tend to be associated with higher prices. Put another way, as the lot size increases, the lot price tends to increase as well.

**Now You Can Do
Exercises 13–18.**

2 CORRELATION COEFFICIENT

Scatterplots provide a visual description of the relationship between two quantitative variables. The *correlation coefficient* is a numerical measure for quantifying the linear relationship between two quantitative variables. Table 4.2 contains the low and high temperatures in degrees Fahrenheit (°F) for 10 American cities on a particular winter day. The variables are *x = low temperature* and *y = high temperature*. Applying what we have just learned, we construct a scatterplot of the data set, which is presented in Figure 4.3.

Table 4.2 Low and high temperatures, in degrees Fahrenheit, of 10 American cities

City	x = low temp.	y = high temp.	City	x = low temp.	y = high temp.
Minneapolis	10	29	Washington, DC	40	50
Boston	20	37	Las Vegas	40	58
Chicago	20	43	Memphis	50	64
Philadelphia	30	41	Dallas	50	70
Cincinnati	30	49	Miami	60	74

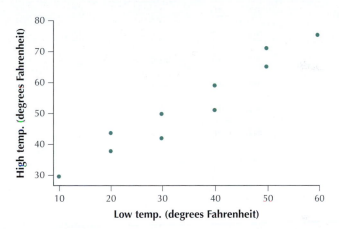

FIGURE 4.3
Scatterplot of high versus low temperatures for 10 American cities.

Figure 4.3 shows us that there is a positive relationship between the high temperature and the low temperature of a city. That is, colder low temperatures are associated with colder high temperatures. Warmer low temperatures are associated with warmer high temperatures. In this section we seek to *quantify this relationship* between two numerical variables, using the **correlation coefficient *r***. The correlation coefficient *r* (sometimes known as the *Pearson product moment correlation coefficient*) measures the strength and direction of the linear relationship between two variables. By *linear*, we mean *straight line*. The correlation coefficient does not measure the strength of a curved relationship between two variables.

> The **correlation coefficient *r*** measures the strength and direction of the linear relationship between two variables. The correlation coefficient *r* is
>
> $$r = \frac{\sum(x - \bar{x})(y - \bar{y})}{(n - 1)s_x s_y}$$
>
> where s_x is the sample standard deviation of the *x* data values, and s_y is the sample standard deviation of the *y* data values.

EXAMPLE 4.3

CALCULATING THE CORRELATION COEFFICIENT r

highlowtemp

Find the value of the correlation coefficient r for the temperature data in Table 4.2.

Solution

We will outline the steps used in calculating the value of r using the temperature data.

STEP 1 Calculate the respective sample means, $\bar{x}$ and $\bar{y}$.

$$\bar{x} = \frac{\sum x}{n} = \frac{10 + 20 + 20 + 30 + 30 + 40 + 40 + 50 + 50 + 60}{10} = \frac{350}{10} = 35.0$$

$$\bar{y} = \frac{\sum y}{n} = \frac{29 + 37 + 43 + 41 + 49 + 50 + 58 + 64 + 70 + 74}{10} = \frac{515}{10} = 51.5$$

STEP 2 Construct a table, as shown here in Table 4.3.

Table 4.3 Calculation table for the correlation coefficient r

City	x	y	$(x - \bar{x}) =$ $(x - 35)$	$(x - \bar{x})^2 =$ $(x - 35)^2$	$(y - \bar{y}) =$ $(y - 51.5)$	$(y - \bar{y})^2 =$ $(y - 51.5)^2$	$(x - \bar{x})(y - \bar{y}) =$ $(x - 35)(y - 51.5)$
Minneapolis	10	29	−25	625	−22.5	506.25	562.5
Boston	20	37	−15	225	−14.5	210.25	217.5
Chicago	20	43	−15	225	−8.5	72.25	127.5
Philadelphia	30	41	−5	25	−10.5	110.25	52.5
Cincinnati	30	49	−5	25	−2.5	6.25	12.5
Washington, DC	40	50	5	25	−1.5	2.25	−7.5
Las Vegas	40	58	5	25	6.5	42.25	32.5
Memphis	50	64	15	225	12.5	156.25	187.5
Dallas	50	70	15	225	18.5	342.25	277.5
Miami	60	74	25	625	22.5	506.25	562.5
				$\sum(x - \bar{x})^2$ $= 2250$		$\sum(y - \bar{y})^2$ $= 1954.5$	$\sum(x - \bar{x})(y - \bar{y})$ $= 2025$

Note on Rounding: Whenever you calculate a quantity that will be needed for later calculations, do not round. Round only when you arrive at the final answer. Here, since the quantities s_x and s_y are used to calculate the correlation coefficient r, neither of them is rounded until the end of the calculation.

STEP 3 Calculate the respective sample standard deviations s_x and s_y. Using the sums calculated from Table 4.3, we have

$$s_x = \sqrt{\frac{\sum(x - \bar{x})^2}{n - 1}} = \sqrt{\frac{2250}{10 - 1}} \approx 15.8113883 \qquad \text{and}$$

$$s_y = \sqrt{\frac{\sum(y - \bar{y})^2}{n - 1}} = \sqrt{\frac{1954.5}{10 - 1}} \approx 14.73657581$$

STEP 4 Put these values all together in the formula for the correlation coefficient r:

$$r = \frac{\sum(x - \bar{x})(y - \bar{y})}{(n - 1)s_x s_y} = \frac{2025}{(9)(15.8113883)(14.73657581)} \approx 0.9656415205 \approx 0.9656$$

Now You Can Do Exercises 19–22.

The correlation coefficient r for the high and low temperatures is 0.9656.

What Does This Formula Mean?

The Correlation Coefficient *r*

Let's analyze the definition formula for the correlation coefficient r. When would r be positive, and when would it be negative? We see that the formula

$$r = \frac{\sum(x - \bar{x})(y - \bar{y})}{(n - 1)s_x s_y}$$

consists of a ratio. Note that the denominator can never be negative, since it is the product of three non-negative values (standard deviations can never be negative). Therefore, the numerator determines whether r will be positive or negative. We know that $x - \bar{x}$ is positive whenever the data value x is greater than $\bar{x}$, and negative when x is less than $\bar{x}$. Similarly for $y - \bar{y}$. The numerator of r is the sum of the products $(x - \bar{x}) \cdot (y - \bar{y})$. There are four cases (or regions, illustrated in Figure 4.4) that describe when the product $(x - \bar{x})(y - \bar{y})$ will be positive or negative. Note that Figure 4.4 is centered at the point $(\bar{x}, \bar{y})$.

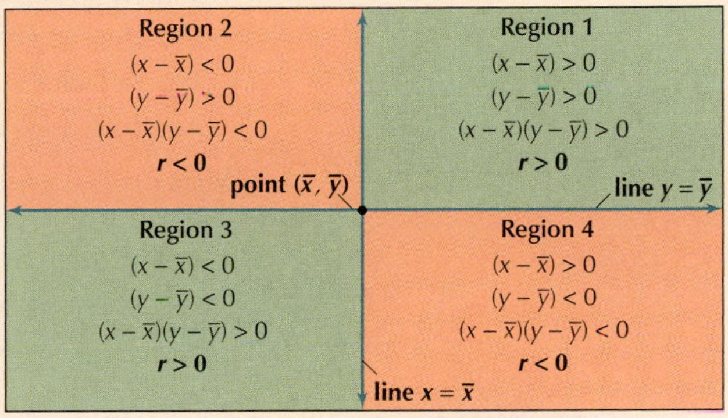

FIGURE 4.4
The four regions for determining whether *r* will tend to be positive or negative.

Region 2	Region 1
$(x - \bar{x}) < 0$	$(x - \bar{x}) > 0$
$(y - \bar{y}) > 0$	$(y - \bar{y}) > 0$
$(x - \bar{x})(y - \bar{y}) < 0$	$(x - \bar{x})(y - \bar{y}) > 0$
$r < 0$	$r > 0$
Region 3	**Region 4**
$(x - \bar{x}) < 0$	$(x - \bar{x}) > 0$
$(y - \bar{y}) < 0$	$(y - \bar{y}) < 0$
$(x - \bar{x})(y - \bar{y}) > 0$	$(x - \bar{x})(y - \bar{y}) < 0$
$r > 0$	$r < 0$

point $(\bar{x}, \bar{y})$ · line $y = \bar{y}$ · line $x = \bar{x}$

Data values that fall in Regions 1 and 3 will tend to make the value of r positive, while data values that fall in Regions 2 and 4 will tend to make the value of r negative. The summation in the numerator of r acts as a blender, combining the contributions of all the various data values falling in all the various regions.

- If most of the data values fall in Regions 1 and 3, then r will tend to be positive.
- If most of the data values fall in Regions 2 and 4, then r will tend to be negative.
- If the four regions share the data values more or less equally, then r will be near zero.

Let's explore how our high and low temperature data fit into the above framework. The mean low temperature is $\bar{x} = 35°F$, while the mean high temperature is $\bar{y} = 51.5°F$. We find the point $(\bar{x}, \bar{y}) = (35, 51.5)$ in our scatterplot of the high and low temperatures, draw the lines $x = \bar{x} = 35$ and $y = \bar{y} = 51.5$, and mark out our four regions, as shown in Figure 4.5. Note that nine of the ten data points fall in Regions 1 and 3. Therefore, we expect the value of r for this data set to be positive, which is indeed the case, since we observed $r = 0.9656$ in Example 4.3.

Next we outline the properties of the correlation coefficient r.

1. The correlation coefficient r always takes on values between -1 and 1, inclusive. That is, $-1 \leq r \leq 1$.
2. When $r = +1$, a perfect positive relationship exists between x and y. See Figure 4.6a.

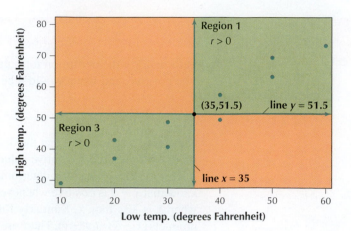

FIGURE 4.5
Nearly all of the temperature data points lie in Regions 1 and 3, making r positive.

3. Values of r near $+1$ indicate a positive relationship between x and y (Figures 4.6b and 4.6c):
   - The closer r gets to $+1$, the stronger the evidence for a positive relationship.
   - The variables are said to be **positively correlated.**
   - As x increases, y tends to increase.

4. When $r = -1$, a perfect negative relationship exists between x and y. See Figure 4.6d.

5. Values of r near -1 indicate a negative relationship between x and y (Figures 4.6e and 4.6f):
   - The closer r gets to -1, the stronger the evidence for a negative relationship.
   - The variables are said to be **negatively correlated.**
   - As x increases, y tends to decrease.

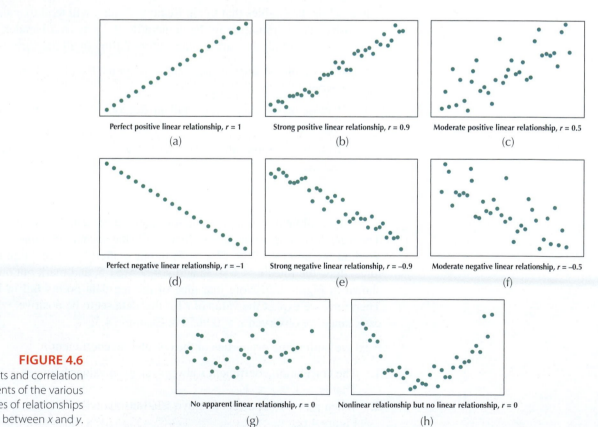

FIGURE 4.6
Scatterplots and correlation coefficients of the various types of relationships between x and y.

6. Values of r near 0 indicate there is no linear relationship between x and y (Figure 4.6g):
- The closer r gets to 0, the weaker the evidence for a linear relationship.
- The variables are **not linearly correlated.**
- A *nonlinear* relationship may exist between x and y. See Figure 4.6h.

Developing Your Statistical Sense

Correlation Is Not Causation

If we conclude that two variables are correlated, it does not necessarily follow that one variable *causes* the other to occur. For example, in the late 1940s, prior to the development of a vaccine for the disease polio, analysts noticed a strong correlation between the amount of ice cream consumed nationwide and higher levels of the onset of polio. Some doctors went on to recommend eliminating ice cream as a way to fight polio. But did ice cream really *cause* polio? No. Ice cream consumption and polio outbreaks both peaked in the hot summer months, and so were *correlated* seasonally. Ice cream did not cause polio. After the development of the polio vaccine by Jonas Salk in the 1950s, the disease disappeared from most countries in the world.

EXAMPLE 4.4   INTERPRETING THE CORRELATION COEFFICIENT

Interpret the correlation coefficient found in Example 4.3.

Here we have made a judgment that 0.9656 is close to 1. Later in this section, we will learn a more precise method for making such decisions.

Solution

In Example 4.3, we found the correlation coefficient for the relationship between high and low temperatures to be $r = 0.9656$. This value of r is very close to the maximum value $r = 1$. We would therefore say that high and low temperatures for these ten American cities are positively correlated. As low temperature increases, high temperatures also tend to increase.

Now You Can Do Exercises 23–26.

The following computational formula may be used as an equivalent of the definition formula for the correlation coefficient r.

> **Equivalent Computational Formula for Calculating the Correlation Coefficient r**
> $$r = \frac{\sum xy - \left(\sum x \sum y\right)/n}{(n-1)\, s_x\, s_y}$$

EXAMPLE 4.5   USING THE COMPUTATIONAL FORMULA TO CALCULATE r

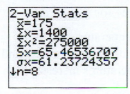

FIGURE 4.7 Statistics for x = square footage.

Note: that this numerator of r equals 72,000. We shall use this fact for Example 4.9 in Section 4.2.

Use the computational formula and the TI-83/84 to calculate the correlation coefficient r for the relationship between square footage and sales price of the eight home lots for sale in Glen Ellyn from Example 4.1 (page 151).

From Figures 4.7 and 4.8 we have $n = 8$, $n - 1 = 7$, $\sum x = 1400$, $\sum y = 2880$, $s_x = 65.46536707$, $s_y = 166.5404284$, and $\sum xy = 576,000$.

Substituting into the computational formula, we have

$$r = \frac{\sum xy - \left(\sum x \sum y\right)/n}{(n-1)\, s_x\, s_y} = \frac{576{,}000 - (1400)(2880)/8}{(7)(65.46536707)(166.5404284)} \approx 0.9434$$

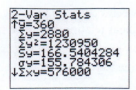

FIGURE 4.8 Statistics for
y = sales price.

The value of r is close to 1, so it appears that square footage and sales price are positively correlated. But we need the next topic, the *comparison test*, to determine this conclusively.

3 TEST FOR LINEAR CORRELATION

We have seen that values of the correlation coefficient r that are close to $+1$ indicate a positive linear relationship between x and y. However, what do we mean by "close to $+1$"? There is a simple comparison test that will tell us whether or not a positive correlation exists between the variables. In general, the comparison test will help us determine whether the correlation coefficient is strong enough to conclude that the variables are correlated.

Comparison Test for Linear Correlation

1. Find the absolute value of the correlation coefficient r, denoted as $|r|$. For example, $|0.5| = 0.5$ and $|-0.4| = 0.4$.

2. Turn to the Table of Critical Values for the Correlation Coefficient (Table G in the Appendix), and select the row corresponding to the sample size n.

3. Compare the absolute value of your correlation coefficient $|r|$ from Step 1 to the critical value from the table from Step 2,

   a. If $|r|$ is greater than the critical value, then you can conclude that x and y are **linearly correlated**.

      i. If $r > 0$, then x and y are **positively correlated**.

      ii. If $r < 0$, then x and y are **negatively correlated**.

   b. If $|r|$ is not greater than the critical value, then x and y are **not linearly correlated**.

EXAMPLE 4.6

DETERMINING WHETHER x AND y ARE CORRELATED, AND INTERPRETING THE RESULTS

For the data from the following examples, determine whether x and y are correlated, and interpret the results.

a. The temperature data from Example 4.3

b. The square footage and sales price data from Example 4.5

Solution

a. From Example 4.3, we have $r = 0.9656$ and $n = 10$.

> **STEP 1** $|r| = |0.9656| = 0.9656$.

> **STEP 2** From Table G in the Appendix, the critical value for $n = 10$ is 0.632.

> **STEP 3** 0.9656 is > 0.632, so we conclude that low temperature and high temperature are correlated. Since $r > 0$, we can state that low temperature and high temperature are positively correlated. As low temperatures increase, high temperatures tend to increase.

b. From Example 4.5, we have $r = 0.9434$ and $n = 8$.

> **STEP 1** $|r| = |0.9434| = 0.9434$.

> **STEP 2** From Table G in the Appendix, the critical value for $n = 8$ is 0.707.

Now You Can Do
Exercises 27–30.

STEP 3 0.9434 is > 0.707, so we conclude that square footage and sales price are correlated. Since $r > 0$, we can state that square footage and sales price are positively correlated. As square footage increases, sales price tends to increase.

 The *Correlation and Regression* applet allows you to insert your own data values and see how the value of the correlation coefficient changes.

STEP-BY-STEP TECHNOLOGY GUIDE: Tables and Graphs for Two Variables sqrfootsale

TI-83/84

Constructing a Scatterplot for Data in Table 4.1 (page 151)

Step 1 Enter the *x* variable (square footage) into **L1** and the *y* variable (sales price) into **L2**.
Step 2 Press **2nd**, then **Y=** for the STAT PLOTS menu.
Step 3 Select **1**, and press **ENTER**. Select **ON**, and press **ENTER**.
Step 4 Select the scatterplots icon (see Figure 4.9), and press **ENTER**.
Step 5 Select **L1** for **Xlist**, and **L2** for Ylist.
Step 6 Press **ZOOM**, choose **9: ZoomStat**, and press **ENTER**. The scatterplot is shown in Figure 4.10.

Correlation Coefficient *r*
Step 1 Turn on the diagnostics, as follows. Press **2nd 0** (catalog). Then scroll down and select **DiagnosticOn**. Press

FIGURE 4.9   **FIGURE 4.10**

ENTER twice to turn the diagnostics on. This will give you more output results for regression and correlation.
Step 2 Enter your *x* data into **L1** and your *y* data into **L2**.
Step 3 Press **STAT**, select **CALC**, and select **LinReg (ax+b)**. Press **ENTER** twice to get the results. The correlation coefficient *r* is given, among other statistics used in regression.

EXCEL

Scatterplots
Step 1 Enter your *x* variable and your *y* variable in two neighboring columns, with the *x* variable on the left. Make sure the first entry in each column is the variable name. Select the two columns.
Step 2 Click **Insert > Scatter** (in **Chart** section). See Figure 4.11.

Correlation Coefficient *r*
Step 1 Make sure the Data Analysis add-in is activated. Click on the **Data** tab, then the **Data Analysis** package, then select **Correlation** and click **OK**.
Step 2 Click on the box next to **Input Range**, then highlight the data, and click **OK**.

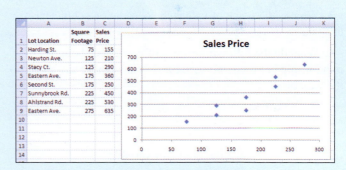

FIGURE 4.11 Excel Scatterplot.

MINITAB

Scatterplots
Step 1 Enter the data into two columns.
Step 2 Click **Graph > Scatterplot**.
Step 3 Click on the cell under **Y**, and double-click on your *y* variable; then click on the cell under **X**, and double-click on your *x* variable. Then click **OK**.

Correlation Coefficient *r*
Step 1 Enter your *x* data into column **C1** and your *y* data into column **C2**.
Step 2 Click on **Stat**, highlight **Basic Statistics**, and select **Correlation**.
Step 3 Choose the variables you wish to analyze and click **OK**.

CRUNCHIT!

We will use the data from Example 4.1 (page 151).

Scatterplots
Step 1 Click **File** ... then highlight **Load from Larose2e** ...
Chapter 4 ... and click on **Example 4.1**.
Step 2 Click **Graphics** and select **Scatterplot**. For **X** select the predictor (x) variable **Square feet**. For **Y** select the response (y) variable **Price**. Then click **Calculate**.

Correlation Coefficient r
Step 1 Click **File** ... then highlight **Load from Larose2e** ...
Chapter 4 ... and click on **Example 4.1**.
Step 2 Click **Statistics** and select **Correlation**.
Step 3 Click the boxes next to **Square feet** and **Price**. Then click **Calculate**.

SECTION 4.1 Summary

1. For two quantitative variables, scatterplots summarize the relationship by plotting all the (x, y) points.

2. The correlation coefficient r is a measure of the strength of linear association between two numeric variables. Values of r close to 1 indicate that the variables are positively correlated. Values of r close to -1 indicate that the variables

are negatively correlated. Values of r close to 0 indicate that the variables are not linearly correlated.

3. A comparison test may be used to determine whether the value of the correlation coefficient r is strong enough to conclude that x and y are correlated.

SECTION 4.1 Exercises

Clarifying the Concepts

1. When investigating the relationship between two quantitative variables, what graph should you use first?

2. In your own words, explain what the correlation coefficient measures. What is the symbol that we use for the correlation coefficient?

3. What is the range of values the correlation coefficient can take?

4. What do the following values of r indicate about the relationship between two variables? What can we say about the variables?
   a. A value of r close to 1
   b. A value of r close to -1
   c. A value of r close to 0

5. Why do we call x the predictor variable?

6. Suppose two quantitative variables have a positive relationship. What can we say about the values of the y variable as the x variable increases?

7. Suppose two quantitative variables have a negative relationship. What can we say about the values of the y variable as the x variable increases?

8. Suppose that the correlation coefficient r equals 0. Does this mean that x and y have no relationship? Explain.

Practicing the Techniques

For Exercises 9–12, construct a scatterplot of the relationship between x and y.

9.

x	y
1	2
2	2
3	3
4	4
5	4

10.

x	y
10	10
20	9
30	8
40	8
50	7

11.

x	y
-3	-5
-1	-15
3	-20
4	-25
5	-30

12.

x	y
0	11
20	11
40	16
60	21
80	26

For Exercises 13–18, do the following:
   a. Characterize the relationship between x and y.
   b. State what happens to the values of the y variable as the x-values increase.

13.

14.

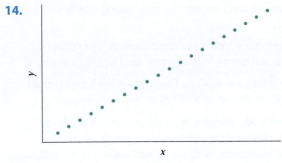

15.

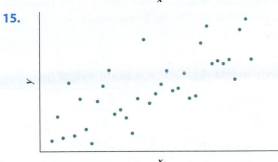

16.

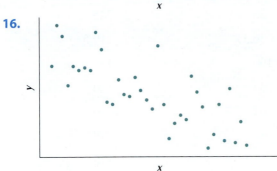

17.

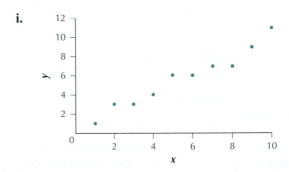

18.

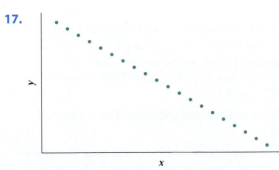

For Exercises 19–22, calculate the correlation coefficient r for the indicated data.

19. The data from Exercise 9

20. The data from Exercise 10

21. The data from Exercise 11

22. The data from Exercise 12

For Exercises 23–26, interpret the value of the correlation coefficient r in the indicated exercise.

23. From Exercise 19

24. From Exercise 20

25. From Exercise 21

26. From Exercise 22

For Exercises 27–30, use the comparison test to determine whether x and y are correlated, using the data and the value of r from the indicated exercises.

27. The data from Exercise 9 and the value of r from Exercise 19

28. The data from Exercise 10 and the value of r from Exercise 20

29. The data from Exercise 11 and the value of r from Exercise 21

30. The data from Exercise 12 and the value of r from Exercise 22

For Exercises 31–34, identify which of the scatterplots in **i–iv** represents the data set with the following correlation coefficients:

i.

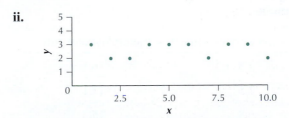

ii.

iii.

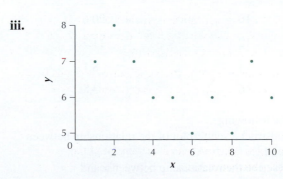

iv.

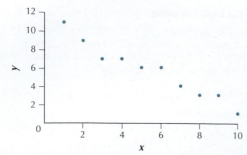

31. Near 1

32. Near zero

33. Near −0.5

34. Near −1

In Exercises 35–38, the values for x and y in each scatterplot are integer-valued. For each scatterplot, (a) reconstruct the original data set, and (b) calculate the correlation coefficient for the data.

35. The data in scatterplot i

36. The data in scatterplot ii

37. The data in scatterplot iii

38. The data in scatterplot iv

For Exercises 39–42, determine whether the correlation coefficient r is strong enough to conclude that x and y are correlated for the indicated data.

39. The data in scatterplot i

40. The data in scatterplot ii

41. The data in scatterplot iii

42. The data in scatterplot iv

Applying the Concepts

Education and Unemployment. Does it pay to stay in school? Use the table of U.S. Census Bureau data for Exercises 43–46.

 edunemploy

x = years of education	y = unemployment rate
5	16.8
7.5	17.1
8	15.3
10	20.6
12	11.7
14	8.1
16	3.8

43. Do the following.
   a. Construct a scatterplot of the relationship between x and y.
   b. Describe the relationship between x and y.

44. Calculate and interpret the correlation coefficient for x and y.

45. Determine whether we can conclude that years of education and unemployment rate are linearly correlated.

46. Based on your work in Exercises 43–45, in general, does it pay to stay in school? State your evidence.

Brain and Body Weight. A study compared the body weight (in kilograms) and brain weight (in grams) for a sample of mammals, with the results shown in the following table. Use the following data for Exercises 47–49.

 brainbody

x = body weight (kg)	y = brain weight (g)
52.16	440
60	81
27.66	115
85	325
36.33	119.5
100	157
35	56
62	1320
83	98.2
55.5	175

47. Construct a scatterplot of the data. Describe the apparent relationship, if any, between the variables. Based on the scatterplot alone, would you say that x and y are positively correlated, negatively correlated, or not correlated?

48. Calculate and interpret the value of the correlation coefficient r.

49. Does this agree with your judgment from Exercise 49? Determine whether we can conclude that body weight and brain weight are correlated.

Bringing It All Together

Country and Hip-Hop CDs. Use the information in the table for Exercises 50–54. The table contains the number of country music CDs and the number of hip-hop CDs owned by six randomly selected students.

 countryhip

		Student				
	1	2	3	4	5	6
Hip-hop CDs owned (y)	10	12	1	3	6	1
Country CDs owned (x)	1	3	11	8	5	27

50. Investigate the relationship.
   a. Construct a scatterplot of the variables. Make sure the *y* variable goes on the *y* axis.
   b. What type of relationship do these variables have: positive, negative, or no apparent linear relationship?
   c. Will the correlation coefficient be positive, negative, or near zero?

51. Calculate and interpret the correlation coefficient.
   a. Compute the value of the correlation coefficient.
   b. Does this value for *r* concur with your judgment in part (**a**) of the previous exercise?
   c. Interpret the meaning of this value of the correlation coefficient.

52. Determine whether we can conclude that *x* and *y* are correlated.

53. Transformation. Add 5 to each value for *y*.
   a. Redraw the scatterplot. Comment on the similarity or difference from the scatterplot in Exercise 50(**a**).
   b. Recalculate the correlation coefficient.
   c. Compare your answers from Exercises 53(b) and 51(a).
   d. Compose a rule that states the behavior of the correlation coefficient *r* when a constant is added to each *y* data value.

54. Transformation. Suppose that, starting with the original data in the table, we added a certain unknown constant amount to each value for *x*.
   a. *Without redrawing the scatterplot,* describe how this change would affect the scatterplot you drew in Exercise 50(**a**).
   b. *Without recalculating the correlation coefficient,* state what you think the effect of this change would be on the correlation coefficient. Why do you think that?
   c. Compose a rule that states the behavior of the correlation coefficient *r* when a constant is added to each *x* data value.

SAT Scores. Refer to the following table for Exercises 55–57.

 statesat

Mean SAT scores for the five states with the best participation rate

State	SAT Reading	SAT Math
New York	497	510
Connecticut	515	515
Massachusetts	518	523
New Jersey	501	514
New Hampshire	522	521

55. Construct a scatterplot of the data, with *x* = SAT Reading and *y* = SAT Math. Describe the apparent relationship, if any, between the variables. Based on the scatterplot, would you say that *x* and *y* are positively correlated, negatively correlated, or not correlated?

56. Calculate the value of the correlation coefficient, using the following steps.
   a. Calculate the respective sample means $\bar{x}$ and $\bar{y}$.
   b. Construct a table like Table 4.3, as follows.
      i. For each observation, calculate the deviations $(x - \bar{x})$ and $(y - \bar{y})$.
      ii. For each observation, calculate $(x - \bar{x})^2$, $(y - \bar{y})^2$, and $(x - \bar{x})(y - \bar{y})$.
      iii. Calculate the following sums: $\Sigma(x - \bar{x})^2$, $\Sigma(y - \bar{y})^2$, and $\Sigma(x - \bar{x})(y - \bar{y})$.
   c. Calculate the respective sample standard deviations s_x and s_y.
   d. Put these all together in the formula for the correlation coefficient *r*.
   e. Using technology, confirm the value you calculated in (**d**).

57. Interpret the meaning of the correlation coefficient you found in Exercise 56, using at least two sentences. Does this agree with your judgment from Exercise 55?

Construct Your Own Data Sets

58. Describe two variables from real life that would have a value of *r* close to 1. Explain why they are positively correlated.

59. Create a sample of five observations from each of your variables in the previous exercise, and put them into a table similar to Table 4.2 (page 153). Next, construct a scatterplot of the variables. Finally, draw a single straight line through the data points in the plot in a manner that you think best approximates the relationship between the variables.

 Use the *Correlation and Regression* applet for Exercises 60–62.

60. Create a set of *n* = 10 points such that the correlation coefficient *r* takes approximately the following values. Note that you can drag points up or down to adjust your value of *r*.
   a. *r* = 0.90
   b. *r* = −0.90
   c. *r* = 0.00

61. Describe the relationship between the variables for each of the sets of points in the previous exercise.

62. Select "Show mean X and mean Y lines." Create a set of *n* = 4 points such that the correlation coefficient *r* takes approximately the following values. Note that you can drag points up or down to adjust your value of *r*.
   a. *r* = 0.70
   b. *r* = −0.70
   c. *r* = 0.00

OBJECTIVES By the end of this section, I will be able to . . .

1 Calculate the value and understand the meaning of the slope and the y intercept of the regression line.

2 Predict values of y for given values of x, and calculate the prediction error for a given prediction.

1 THE REGRESSION LINE

In Section 4.1 we learned about the *correlation coefficient*. Here, in Section 4.2, we will learn how to approximate the linear relationship between two numerical variables using the regression line and the regression equation. For convenience, we repeat Table 4.2 here.

Consider again Figure 4.3 (page 153), the scatterplot of the high and low temperatures for ten American cities, from Table 4.2. The data points generally seem to follow a roughly linear path. We may in fact draw a straight line from the lower left to the upper right to approximate this relatively linear path. Such a straight line, called a **regression line,** is shown in Figure 4.12.

Table 4.2 Temperature data

City	x = low temp.	y = high temp.
Minneapolis	10	29
Boston	20	37
Chicago	20	43
Philadelphia	30	41
Cincinnati	30	49
Wash., DC	40	50
Las Vegas	40	58
Memphis	50	64
Dallas	50	70
Miami	60	74

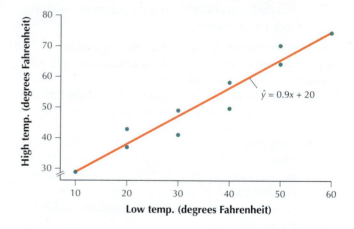

FIGURE 4.12
Scatterplot of high versus low temperatures, with regression line.

  highlowtemp

As you may recall from high school algebra, the equation of a straight line may be written as $y = mx + b$. We will write the **equation of the regression line** similarly as $\hat{y} = b_1x + b_0$.

The "hat" over the y (pronounced "y-hat") indicates that this is an estimate of y and not necessarily an actual value of y.

Equation of the Regression Line

The **equation of the regression line** that approximates the relationship between x and y is

$$\hat{y} = b_1x + b_0$$

where the *regression coefficients* are the **slope**, b_1, and the **y intercept**, b_0. Do not let $\hat{y}$ and $\bar{y}$ be confused. $\hat{y}$ is the predicted value of y from the regression equation. $\bar{y}$ represents the mean of the y values in the data set. The equations of these coefficients are

$$b_1 = r \cdot \frac{S_y}{S_x} \qquad\qquad b_0 = \bar{y} - (b_1 \cdot \bar{x})$$

Where S_x and S_y represent the sample standard deviation for the x and y data, respectively.

There are an infinite number of different straight lines that could approximate the relationship between high and low temperatures. Why did we choose this one? Because this is the *least-squares* regression line, which is the most widely used linear approximation for bivariate relationships. We will learn more about least squares in Section 4.3.

EXAMPLE 4.7

The scatterplot is shown in Figure 4.12 on page 164.

CALCULATING THE REGRESSION COEFFICIENTS b_0 AND b_1

Find the value of the regression coefficients b_0 and b_1 for the temperature data in Table 4.2.

Solution

We will outline the steps used in calculating the value of b_1 using the temperature data.

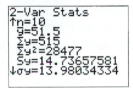

STEP 1 Calculate the respective sample means $\bar{x}$ and $\bar{y}$. We have already done this in Example 4.3: $\bar{x} = 35$ and $\bar{y} = 51.5$.

STEP 2 Calculate the respective sample standard deviations s_x and s_y. We have already done this in Example 4.3: $s_x \approx 15.8113883$ and $s_y \approx 14.73657581$.

STEP 3 Find the correlation coefficient r. This was computed in Example 4.3: $r \approx 0.9656415205$.

Regression summary statistics.

STEP 4 Combine the statistics from Steps 2 and 3 to calculate b_1:

$$b_1 = r \cdot \frac{s_y}{s_x} = 0.9656415205 \cdot \frac{14.73657581}{15.8113883} = 0.9$$

STEP 5 Use the statistics from Steps 1–4 to calculate b_0:

$$b_0 = \bar{y} - (b_1 \cdot \bar{x}) = 51.5 - (0.9)(35) = 20$$

Thus, the equation of the regression line for the temperature data is

$$\hat{y} = 0.9x + 20$$

Now You Can Do Exercises 13–20.

Since y and x represent high and low temperatures, respectively, this equation is read as follows: "The estimated high temperature for an American city is 0.9 times the low temperature for that city plus 20 degrees Fahrenheit."

What Do These Numbers Mean?

Interpreting the Slope and the y Intercept

- In statistics, we interpret the slope of the regression line as the *estimated change in y per unit increase in x*. In our temperature example, the units are degrees Fahrenheit, so we interpret our value $b_1 = 0.9$ as follows:

  "For each increase of 1°F in low temperature, the estimated high temperature increases by 0.9°F."

- The y intercept is interpreted as the *estimated value of y when x equals zero*. Here, we interpret our value $b_0 = 20$ as follows:

  "When the low temperature is 0°F, the estimated high temperature is 20°F."

Now You Can Do Exercises 21–28.

Recall from Section 4.1 that the correlation coefficient for the temperature data is $r = 0.9656$. Is it a coincidence that both the slope and the correlation coefficient are positive? Not at all.

This relationship holds because $b_1 = r \cdot \dfrac{S_y}{S_x}$ and neither S_y nor S_x can be negative.

Relationship Between Slope and Correlation Coefficient

The slope b_1 of the regression line and the correlation coefficient r always have the same sign.

- b_1 is positive if and only if r is positive.
- b_1 is negative if and only if r is negative.

Hence, when we found in Section 4.1 that the correlation coefficient between high and low temperatures was positive, we could have immediately concluded that the slope of the regression line was also positive.

EXAMPLE 4.8   CORRELATION AND REGRESSION USING TECHNOLOGY

Other ways to describe regression include

- "Perform a regression of the y variable versus the x variable."
- "Regress the y variable on the x variable."

Note that the first variable is always the y variable and the second variable is always the x variable. For example, in Example 4.8 we could write, "Perform a regression of high temperature against low temperature."

Use technology to find the correlation coefficient r and the regression coefficients b_1 and b_0 for the temperature data in Example 4.3.

Solution

The instructions for using technology for correlation and regression are provided in the Step-by-Step Technology Guide at the end of this section (page 173). The TI-83/84 scatterplot is shown in Figure 4.13, and the TI-83/84 results are shown in Figure 4.14. (Note that the TI-83/84 indicates the slope b_1 as a, and the y intercept b_0 as b.) Figures 4.15a and 4.15b show the Excel results, with the y intercept ("Intercept") and the slope ("Low") highlighted. Figure 4.16 shows the Minitab results, with the y intercept ("Constant") and the slope ("Low") highlighted.

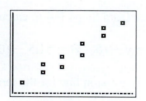

FIGURE 4.13 TI-83/84 scatterplot.

Slope b_1
y intercept b_0
(Coefficient of Determination, Section 4.3)
Correlation coefficient r

```
LinReg
y=ax+b
a=.9
b=20
r²=.9324635457
r=.9656415203
```

FIGURE 4.14 TI-83/84 correlation and regression results.

	Low	High
Low	1	
High	**0.965642**	1

FIGURE 4.15a Excel correlation results.

	Coefficients	Standard error	t stat	P-value
Intercept	20	3.260879227	6.133315161	0.000279056
Low	0.9	0.085634884	10.50973575	5.84917E-06

FIGURE 4.15b Excel regression results.

Correlations: Low, High
```
Pearson correlation of Low and High = 0.966
P-Value = 0.000
```

Regression Analysis: High versus Low
```
The regression equation is High = 20.0 + 0.900 Low
```

Predictor	Coef	SE Coef	T	P
Constant	20.000	3.261	6.13	0.000
Low	0.90000	0.08563	10.51	0.000

FIGURE 4.16 Minitab correlation and regression results.

The following computational formula is equivalent to the definition formula for the slope b_1.

Equivalent Computational Formula for Calculating the Slope b_1

$$b_1 = \frac{\sum xy - \left(\sum x \sum y\right)/n}{\sum x^2 - \left(\sum x\right)^2/n}$$

EXAMPLE 4.9

USING THE COMPUTATIONAL FORMULA TO CALCULATE THE SLOPE b_1

Use the computational formula to calculate the slope b_1 for the relationship between square footage and sales price of the eight home lots for sale in Glen Ellyn from Example 4.1 in Section 4.1. Then find the y intercept b_0 and the regression equation.

Solution

To save time, we could have remembered that the numerator for the computational formula for r from Example 4.5 is 72,000.

From Example 4.5 (pages 157–158), we have $n = 8$ and the following summations: $\Sigma x = 1400$, $\Sigma y = 2880$, $\Sigma xy = 576,000$, and $\Sigma x^2 = 275,000$. Substituting into the computational formula, we have

$$b_1 = \frac{576,000 - (1400)(2880)/8}{275,000 - 1400^2/8} = \frac{72,000}{30,000} = 2.4$$

To find b_0, we first calculate

$$\bar{y} = \frac{\Sigma y}{n} = \frac{2880}{8} = 360 \text{ and } \bar{x} = \frac{\Sigma x}{n} = \frac{1400}{8} = 175$$

Then

$$b_0 = \bar{y} - (b_1 \cdot \bar{x}) = 360 - (2.4)(175) = -60$$

This gives us the following regression equation:

$$\hat{y} = b_1 x + b_0 = 2.4x - 60$$

The Sensitivity of the Regression Line to Extreme Values

What if the sales price of the largest lot for sale (27,500 square feet) was not $635,000 but $120,000. What would happen to the slope and the y intercept of the regression line?

Solution

The correlation coefficient and the regression line are both sensitive to extreme values. As shown in Figure 4.17, the change to a much lower price for the largest lot acts as a weight pulling down on the upper (right-hand) end of the regression line. The slope decreases from $b_1 = 2.4$ to $b_1 = 0.683$.

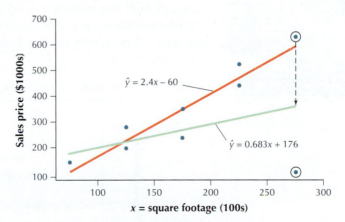

FIGURE 4.17
Regression line is sensitive to extreme values.

Consequently, the y intercept increases from $b_0 = -60$ to $b_0 = 176$, giving us the new regression equation:

$$\hat{y} = 0.683x + 176$$

Also, the correlation coefficient falls from $r = 0.9434$ to $r = 0.3130$.

2 PREDICTIONS AND PREDICTION ERROR

We can use the regression equation to make estimates or predictions. For any particular value of *x*, the predicted value for *y* lies on the regression line.

EXAMPLE 4.10

USING THE REGRESSION EQUATION TO MAKE A PREDICTION

Suppose we are moving to a city that has a low temperature of 50°F on this particular winter's day. Use the regression equation in Example 4.7 to find the predicted high temperature for this city.

Solution

To generate an estimate of the high temperature, we plug in the value of 50°F for the *x* variable *low*:

$$\hat{y} = 0.9(low) + 20 = 0.9(50) + 20 = 65$$

Now You Can Do Exercises 29–36.

We would say, "The estimated high temperature for an American city with a low temperature of 50°F is 65°F."

Developing Your Statistical Sense

Actual Data versus Predicted (Estimated) Data

We have two cities in our data table (Table 4.2) whose low temperature is 50°F: Dallas, Texas, and Memphis, Tennessee. For simplicity, we will illustrate using Dallas only. The actual high temperature for Dallas is 70°F, but our predicted high temperature is $\hat{y} = 65°F$. The actual high temperature in Dallas is an established fact: real, observed data. On the other hand, our prediction $\hat{y}$ is just an estimate based on a formula, the regression equation.

Prediction Error

The actual data point for Dallas is shown circled in the scatterplot in Figure 4.18. The predicted high temperature $\hat{y} = 65°F$ is the *y*-value of the point on the regression line where it intersects $x = 50$. Notice that the point with the predicted high tempera-ture value $\hat{y}$ lies directly on the regression line vertically below the Dallas data point. This is true for all values of $\hat{y}$:

All values of $\hat{y}$ (the predicted values of *y*) lie on the regression line.

iStockphoto

Dallas, Texas

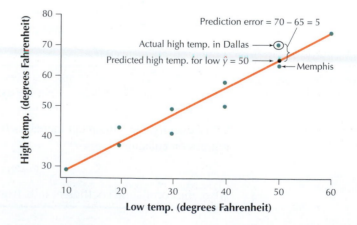

FIGURE 4.18
Prediction error for Dallas high temperature

Our prediction's position in the graph is at $(x, \hat{y}) = (50, 65)$, compared to $(x, y) = (50, 70)$ for Dallas. Our prediction for Dallas was too low by

$$y - \hat{y} = 70 - 65 = 5°F$$

The difference $y - \hat{y}$ is the vertical difference from the Dallas data point to the regression line. This difference is called the *prediction error.*

The **prediction error** or **residual** $(y - \hat{y})$ measures how far the predicted value $\hat{y}$ is from the actual value of y observed in the data set. The prediction error may be positive or negative.

- Positive prediction error: The data value lies *above* the regression line, so the observed value of y is greater than predicted for the given value of x.

- Negative prediction error: The data value lies *below* the regression line, so the observed value of y is lower than predicted for the given value of x.

- Prediction error equal to zero: The data value lies *directly* on the regression line, so the observed value of y is exactly equal to what is predicted for the given value of x.

Of course, we need not restrict our predictions to values of x (low temperature) that are in our data set (though see the warning on extrapolation below). For example, the estimated high temperature for a city in which $low = 25°F$ is

$$\hat{y}\,y = 0.9(low) + 20 = 0.9(25) + 20 = 42.5°F$$

Note that we cannot calculate the prediction error for this estimate, since we do not have a city with a low temperature of 25°F to compare it to.

| EXAMPLE 4.11 | **CALCULATING AND INTERPRETING PREDICTION ERRORS (RESIDUALS)** |

Use the regression equation from Example 4.10 to calculate and interpret the prediction error (residual) for the following cities.
a. Cincinnati: Low = 30, high = 49
b. Philadelphia: Low = 30, high = 41

Solution

a. The actual high temperature in Cincinnati that day was $y = 49$. Using the regression equation, the predicted high temperature is $\hat{y} = 0.9(30) + 20 = 47$. So the prediction error is $y - \hat{y} = 49 - 47 = 2°F$. The data point lies above the regression line, so that its actual high temperature of 49°F is greater than predicted given its low temperature of 30°F.
b. Philadelphia: Actual high $= y = 41$. Predicted high $= \hat{y} = 0.9(30) + 20 = 47$. So the residual is $y - \hat{y} = 41 - 47 = -6°F$. Philadelphia's data point lies below the regression line, so that its actual high temperature of 41°F is lower than predicted given its low temperature of 30°F.

Now You Can Do Exercises 37–42.

Extrapolation

The y intercept b_0 is the estimated value for y when x equals zero. However, in many regression problems, a value of zero for the x variable would not make sense. For example, a lot for sale of $x = 0$ square feet does not make sense, so the y intercept would not be meaningful. On the other hand, a value of zero for the low temperature does make sense. Therefore, we would be tempted to predict $\hat{y} = 0.9(0) + 20 = 20°F$ as the high temperature for a city with a low of zero degrees. However, *low* = 0°F is not within the range of the data set. Making predictions based on *x*-values that are beyond the range of the *x*-values in our data set is called *extrapolation*. It may be misleading and should be avoided.

> **Extrapolation** consists of using the regression equation to make estimates or predictions based on *x*-values that are outside the range of the *x*-values in the data set.

Extrapolation should be avoided, if possible, because the relationship between the variables may no longer be linear outside the range of *x*. A regression line based solely on the available data (white background) and ignoring the hidden data (gray background) is shown in Figure 4.19. Since the regression line is based on incomplete data, in this case, predicting *y* at the point *x* = *a* resulted in a large difference between the predicted value $\hat{y}$ and the actual value *y*, called the *prediction error,* or *residual.*

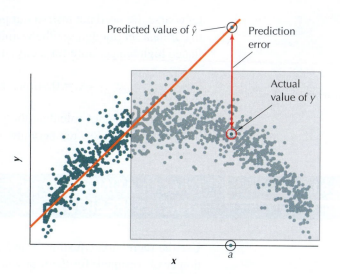

FIGURE 4.19
Dangers of extrapolation.

EXAMPLE 4.12

IDENTIFYING WHEN EXTRAPOLATION OCCURS

Using the regression equation from Example 4.10, estimate the high temperature for the following low temperatures. If the estimate represents extrapolation, indicate so.
a. 60°F
b. 70°F

Solution
From Table 4.2, the smallest value of *x* is 10°F and the largest is 60°F, so estimates for any value of *x* between 10°F and 60°F, inclusive, would not represent extrapolation.

a. $\hat{y} = 0.9(60) + 20 = 74°F$. Since *x* = 60°F lies between 10°F and 60°F, inclusive, this estimate does not represent extrapolation.

**Now You Can Do
Exercises 43–50.**

b. $\hat{y} = 0.9(70) + 20 = 83°F$. Since *x* = 70°F does not lie between 10°F and 60°F, this estimate represents extrapolation.

CASE STUDY   Worldwide Patterns of Cell Phone Usage

In this case study, we bring together many of the correlation and regression ideas we have covered thus far in this chapter.

Would you expect that residents of richer countries tend to use their cell phones to browse the Internet more often than do residents of poorer countries? The Pew Global Attitudes Project conducted a study[2] of cell phone usage in countries around the world. Table 4.3 shows x = the per capita gross domestic product (GDP, a measure of the wealth of the country), and y = the percentage of cell phone owners who use their cell phones to browse the Internet for a random sample of 10 countries. We can use this data to answer questions **(a)**–**(h)** below.

Table 4.3 Percentage who use cell phone to browse the Internet and per capita gross domestic product for 10 countries

Nation	X = Per Capita GDP($)	Y = Percentage who use cell phone to browse Internet
USA	48,147	43
Britain	35,974	38
France	35,048	28
Russia	16,687	27
Poland	20,136	30
Israel	31,004	47
China	8,394	37
Japan	34,362	47
India	3,703	10
Mexico	15,121	18

a. Construct and interpret a scatterplot of the data in Table 4.3.

b. Based on your interpretation in **(a)**, would the value for the correlation coefficient r be positive or negative?

c. Calculate the correlation coefficient r.

d. Use the comparison test to determine whether x and y are correlated, and interpret the results.

e. Find the slope and y intercept of the regression line. Write the regression equation in a sentence.

f. Interpret the values of the slope and the y intercept. Determine whether the interpretation of the y intercept represents extrapolation in this case.

g. Calculate the estimated percentage using their cell phones to browse the Internet for a nation with a per capita GDP of $48,147.

(continues)

h. Identify the country with a per capita GDP of $48,147. Calculate and interpret the prediction error for this country.

Solution

a. Figure 4.20 shows a scatter plot of this data.

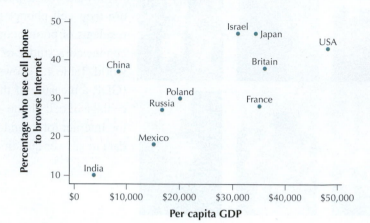

FIGURE 4.20
Scatterplot
of data in Table 4.3

Based on the scatterplot in Figure 4.20, we can state that there is a positive relationship between the x variable and the y variable. That is, as the per capita gross domestic product increases, the percentage of people who use their cell phone to browse the Internet also increases.

b. Since the relationship is positive, the correlation coefficient r must therefore be positive, $0 < r < 1$.

c. Excel provides the correlation coefficient in Figure 4.21, $r = 0.6958$. As expected, the correlation coefficient is positive.

	Percentage using Internet	Per capita GDP
Percentage using Internet	1	
Per capita GDP	**0.6958**	1

FIGURE 4.21
Excel correlation results

d. For a sample size of $n = 10$, the critical value from Table G in the Appendix is 0.632. Since $|r| = |0.6958| = 0.6958 > 0.632$, we can conclude that the x variable and the y variable are positively correlated. An increase in gross domestic product is associated with an increase in the percentage who use their cell phone to browse the Internet.

e. Turning to regression, the y intercept and the slope of the regression line are shown in the Excel output in Figure 4.22, giving us the regression line

$$\hat{y} = 0.0006 \text{ (per capita GDP)} + 17.4976$$

The estimated percentage using the Internet equals 0.0006 times the per capita GDP plus 17.4976.

	Coefficients	
Intercept	**17.4976**	**y intercept**
Per capita GDP	**0.0006**	**slope**

FIGURE 4.22
Excel regression results

f. We interpret the slope as follows: an increase of $1 in the per capita GDP is associated with an estimated increase in the percentage using the Internet of 0.0006. We may also say that an estimated increase of $10,000 is associated with an increase in the percentage using the Internet of $(10,000)(0.0006) = 6$, that is, 6 percentage points. The *y* intercept is interpreted as follows: when the per capita GDP equals $0, the estimated percentage using the Internet is 17.4976. But since no value of *x* in our data set is as low as $0, this would represent extrapolation.

g. For a country with a per capita GDP of $48,147, the estimated percentage who use their cell phones to browse the Internet is

$$\hat{y} = 0.0006\,(48,147) + 17.4976 = 46.3858$$

h. The nation with a per capita GDP of $48,147 is the United States, so we may proceed to calculate the prediction error for the estimated percentage in (**g**). The actual percentage of Americans who use their cell phones to browse the Internet is 43, as shown in Table 4.3. Thus the prediction error is

$$(y - \hat{y}) = (43 - 46.3858) = -3.3858$$

In other words, the percentage of people in the United States who use their cell phones to browse the Internet is lower than predicted by 3.3858 percentage points, given the American per capita GDP of $48,147 in the United States. ∎

The *Correlation and Regression* applet allows you to insert your own data values and see how the regression line changes.

STEP-BY-STEP TECHNOLOGY GUIDE: Regression

We illustrate using Example 4.3, the temperature data (page 154).

TI-83/84

Step 1   Turn diagnostics on as follows. Press **2nd 0**. Scroll down and select **DiagnosticOn** (Figure 4.23). Press **ENTER** twice to turn diagnostics on.
Step 2   **Enter** the **X (Low Temp)** data in **L1**, and the **Y (High Temp)** data in **L2**.
Step 3   Press **STAT** and highlight **CALC**.
Step 4   Select **LinReg(ax + b)**.
Step 5   On the home screen, **LinReg(ax1b)** appears. Press **ENTER**.

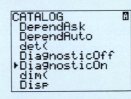

FIGURE 4.23

EXCEL

Step 1   Enter the *x* variable in column **A** and the *y* variable in column **B**.
Step 2   Click on **Data > Data Analysis > Regression** and click **OK**.

Step 3   For **Input Y Range**, select cells **B1–B10**. For **Input X Range**, select cells **A1–A10**. Click **OK**.

MINITAB

Regression
Step 1 Enter the x variable in **C1** and the y variable in **C2**.
Step 2 Click on **Stat > Regression > Regression**.

Step 3 Select the y variable for the **Response Variable** and the x variable for the **Predictor Variable**. Click **OK**.

CRUNCHIT!

Step 1 Click **File** ... then highlight **Load from Larose2e** ... **Chapter 4** ... and click on **Example 4.3**.
Step 2 Click **Statistics** ... **Regression** ... **Simple linear regression**.

Step 3 For **Response variable y** select **High temp**. For **Predictor variable x** select **Low Temp**.
Step 4 For **Display** make sure **Numerical results** is selected. Then click **Calculate**.

SECTION 4.2   Summary

1. Section 4.2 introduces regression, where the linear relationship between two numerical variables is approximated using a straight line, called the regression line. The equation of the regression line is written as $\hat{y} = b_1 x + b_0$, where the regression coefficients are the y intercept, b_0, and the slope, b_1.

2. The regression equation can be used to make predictions about values of y for particular values of x.

SECTION 4.2   Exercises

Clarifying the Concepts

1. What is the objective of regression analysis?

2. What is the regression equation?

3. Describe how we use the regression equation to make predictions.

4. Explain the difference between y and $\hat{y}$.

5. Describe what is meant by extrapolation.

6. What is the relationship between the slope of the regression line and the correlation coefficient?

Practicing the Techniques

Exercises 7–12 refer to scatterplots in the Section 4.1 exercises. For each indicated scatterplot, state whether the slope b_1 of the regression line would be positive, negative, or near zero.

7. Exercise 13
8. Exercise 14
9. Exercise 15
10. Exercise 16
11. Exercise 17
12. Exercise 18

For Exercises 13–20, do the following:
   a. Calculate the slope b_1 of the regression line.
   b. Calculate the y intercept b_0 of the regression line.
   c. Write the regression equation.

13.

x	1	2	3	4
y	2	5	9	12

14.

x	0	2	4	6
y	5	6	5	4

15.

x	−5	−4	−3	−2	−1
y	0	8	8	16	16

16.

x	−3	−1	1	3	5
y	−5	−15	−20	−25	−30

17.

x	5	10	15	20	25	30
y	2	3	3	3	2	3

18.

x	6	7	8	9	11	13
y	4	4	4	4	4	4

19.

x	0	10	20	30	40	50	60	70
y	5	10	15	20	25	30	35	40

20.

x	−30	−23	−15	−12	−1	5	14	29
y	93	78	66	52	44	37	20	10

For Exercises 21–28, do the following for the indicated data:
   a. Interpret the value for the slope b_1 of the regression line.
   b. Interpret the value for the y intercept b_0 of the regression line.

21. Data from Exercise 13

22. Data from Exercise 14

23. Data from Exercise 15

24. Data from Exercise 16

25. Data from Exercise 17

26. Data from Exercise 18

27. Data from Exercise 19

28. Data from Exercise 20

For Exercises 29–36, predict the value of y for the given value of x for the indicated data.

29. Data from Exercise 13; $x = 3$

30. Data from Exercise 14; $x = 0$

31. Data from Exercise 15; $x = -2$

32. Data from Exercise 16; $x = 8$

33. Data from Exercise 17; $x = 10$

34. Data from Exercise 18; $x = 4$

35. Data from Exercise 19; $x = 0$

36. Data from Exercise 20; $x = 40$

For Exercises 37–42, do the following:
  a. Calculate the prediction error.
  b. Interpret the prediction error.

37. Prediction from Exercise 29

38. Prediction from Exercise 30

39. Prediction from Exercise 31

40. Prediction from Exercise 33

41. Prediction from Exercise 34

42. Prediction from Exercise 35

For Exercises 43–50, for the prediction from the indicated exercise, state whether or not the prediction represents extrapolation.

43. Prediction from Exercise 29

44. Prediction from Exercise 30

45. Prediction from Exercise 31

46. Prediction from Exercise 32

47. Prediction from Exercise 33

48. Prediction from Exercise 34

49. Prediction from Exercise 35

50. Prediction from Exercise 36

Applying the Concepts

For Exercises 51–54, do the following
  a. Calculate the slope b_1 and the y intercept b_0 of the regression line.
  b. State the regression equation in words that a non-specialist would understand, as shown at the end of Example 4.7.

  c. Interpret the value for the slope b_1 of the regression line, in terms of the variables from the particular exercise.
  d. Interpret the value for the y intercept b_0 of the regression line, in terms of the variables from the particular exercise.

51. Education and Unemployment. The U.S. Census Bureau published the following data on years of education and unemployment rate. Use your calculations from Exercise 44 in Section 4.1.

 edunemploy

x = years of education	y = unemployment rate
5	16.8
7.5	17.1
8	15.3
10	20.6
12	11.7
14	8.1
16	3.8

52. NASCAR Wins. Refer to the following table of NASCAR wins in super speedway races and short track races.

 nascar

Driver	x = short track wins	y = super speedway wins
Darrell Waltrip	47	18
Dale Earnhardt	27	29
Jeff Gordon	15	15
Cale Yarborough	29	15
Richard Petty	23	19

53. SAT Scores. Refer to the following table of SAT Reading scores and SAT Math scores.

satesat

State	x = mean SAT Reading score	y = mean SAT Math score
New York	497	510
Connecticut	515	515
Massachusetts	518	523
New Jersey	501	514
New Hampshire	522	521

54. Brain and Body Weight. Refer to the following table of brain and body weight for a sample of mammals. Use your calculations from Exercise 48 in Section 4.1.

 brainbody

x = body weight (kg)	y = brain weight (g)
52.16	440
60	81
27.66	115
85	325
36.33	119.5
100	157
35	56
62	1320
83	98.2
55.5	175

55. Education and Unemployment. Refer to your work from Exercise 51. For parts (a)–(c), if appropriate, use your regression equation to estimate the unemployment for individuals with the following years of education. If it is not appropriate, clearly state why not.

    a. 10 years   **b.** 15 years   **c.** 20 years

    d. Calculate the prediction error for your prediction in part (a). Does this data point lie above or below the regression line, and what does that mean?

56. NASCAR Wins. Refer to your work from Exercise 52. For parts (a)–(c), if appropriate, use your regression equation to estimate the number of super speedway wins for drivers with the following numbers of short track wins. If it is not appropriate, clearly state why not.

    a. 30 short track wins

    b. 47 short track wins

    c. 50 short track wins

    d. Calculate the prediction error for your prediction in part (b). Does this data point lie above or below the regression line, and what does that mean?

57. SAT Scores. Refer to your work from Exercise 53.

    a. Estimate the mean SAT Math score for a state with a mean SAT Reading score of 501.

    b. Is the interpretation of the y intercept from Exercise 53 useful? Explain.

    c. Is it OK, or is it misleading to use the regression equation to predict the mean SAT Math score for a state with a mean SAT Reading score of 400? Explain.

    d. What is the distinction between your result from part (a) and the mean SAT Math score for New Jersey?

    e. Calculate and interpret the prediction error for your prediction in part (a).

58. Brain and Body Weight. Refer to your work from Exercise 54.

    a. Estimate the brain weight for a mammal with a body weight of 100 kilograms.

    b. Is the interpretation of the y intercept from Exercise 54 useful? Explain.

    c. Is it OK, or is it misleading to use the regression equation to predict the brain weight for a mammal with body weight of 10 kg? Explain.

    d. Explain the distinction between your result from part (a) and the actual brain weight of 157 grams for the mammal from the data table.

    e. Calculate and interpret the prediction error for your prediction in part (a).

59. Consider again the temperature data in Example 4.7. What if there was a typo, and all the low temperatures in the data set needed to be adjusted downward by the same amount. Explain how this change would affect the following, and why. Increase, decrease, or no change?

    a. $\bar{x}$

    b. $\bar{y}$

    c. y intercept b_0

    d. Slope b_1

    e. Correlation coefficient r

DC Households. Use the following information for Exercises 60–62. The data set **Households,** located on your CD and companion Web site, contains information on the number and type of households in the fifty states and the District of Columbia. For each state, there are seven variables. Two of these variables are the percentage of households headed by women ($y = $ HHLD_WOMEN) and the total number of households in the state ($x = $ TOT_HHLD). Minitab provides the following regression equation:

Households

> **Regression Analysis**
> The regression equation is
> HHLD_Women = 10.5 + 2.82E-07 TOT_HHLD

Note: Minitab shows its regression equations as $y = b_0 + b_1x$ rather than $\hat{y} = b_1 x + b_0$. Also, the notation 2.82E-07 refers to the scientific notation method of writing numbers. Often, software and calculators will present you with this type of notation, so you need to know how to read it. The number 2.82E-07 represents 2.82 times 10^{-7}, or 0.000000282.

60. In this exercise, we explore the regression coefficients and the regression equation.

    a. Find and interpret the meaning of the value for the y intercept. Does it make sense?

    b. Would the estimate in (a) be considered extrapolation? Why or why not?

    c. Find and interpret the meaning of the slope coefficient as the total number of households in the state increases.

    d. Write the regression equation. Now state in words what the regression equation means.

    e. Is the correlation coefficient positive or negative? How do you know?

61. Estimate the increase or decrease in the percentage of households headed by women, using a sentence, for the following situations.
  a. Suppose State A has 1 million more households than State B.
  b. Suppose State C has 5 million fewer households than State D.

62. The number of households per state ranges from about 170,000 to about 10 million.
  a. Estimate the percentage of households headed by women for a state with 7 million households, if appropriate.
  b. Estimate the percentage of households headed by women for a state with 100,000 households, if appropriate.

Bringing It All Together

Fuel Economy. Refer to the following table of fuel economy data for a sample of 10 vehicles for Exercises 63–67. The predictor variable is x = engine size, expressed in liters; the response variable is y = combined (city/highway) gas mileage, expressed in miles per gallon (mpg).

 enginempg

Vehicle	x = engine size (liters)	y = combined mpg
Mini Cooper	1.6	31
Ford Focus	2.0	28
Toyota Camry	2.5	26
Subaru Forester	2.5	23
Honda Accord	2.4	26
Toyota Highlander	2.7	22
Chevrolet Equinox	3.0	19
Ford Taurus	3.5	20
Dodgo Nitro	4.0	17
Cadillac Limousine	4.6	14

63. Exploring the Data.
  a. Look at the data table. As the engine size values increase, what seems to be happening to the combined mpg?
  b. Construct a scatterplot of the data.
  c. Interpret the scatterplot. Is your insight from part (a) supported?

64. What Results Do You Expect? Based on your scatterplot in Exercise 64, answer the following.
  a. Will the correlation coefficient be positive or negative?
  b. Do you expect that the correlation will be closer to −0.9 or −0.5? Why?

  c. Would you predict that our comparison test will allow us to conclude that engine size is correlated with combined mpg?
  d. Do you think that the slope b_1 will be positive or negative? Why?

65. Correlation. Do the following.
  a. Calculate the correlation coefficient r. Does this concur with your predictions from Exercises 64(**a**) and 64(**b**)?
  b. Test whether we may conclude that engine size is correlated with combined mpg. Does this agree with your prediction from Exercise 64(**c**)?
  c. Interpret the correlation between engine size and combined mpg.

66. Regression. Answer the following.
  a. Calculate the slope b_1 of the regression equation. Does the sign of b_1 agree with your prediction from Exercise 64(**d**)?
  b. Calculate the y intercept b_0.
  c. Interpret the values you calculated in parts (**a**) and (**b**) so that a nonstatistician would understand them.

67. Making Predictions. Answer the following.
  a. Predict the combined mpg for a vehicle with an engine size of 3 liters.
  b. Is your prediction error positive or negative? Hence, does the data value lie above or below the regression line? What does this mean?

Construct Your Own Data Sets

68. Describe two variables from real life whose regression line would have a positive slope b_1.
  a. Explain why the y variable depends on the x variable.
  b. Explain why the slope is positive.

69. Create a sample of five observations from each of your variables from Exercise 68, and put them into a table similar to Table 4.1 in Section 4.1.
  a. Construct a scatterplot of the variables.
  b. Draw a single straight line through the data points in the plot in a manner that you think best approximates the relationship between the variables.
  c. Using your regression line from (**b**), estimate the slope b_1 and the y intercept b_0.
  d. Write your results from (**c**) in the form of a regression equation.

Use the *Correlation and Regression* applet for Exercises 70 and 71.

70. Create a set of n = 10 points such that the slope of the regression line has the following characteristic. (Note that you can drag points up or down to adjust your regression line.)
  a. The slope is positive.
  b. The slope is negative.
  c. The slope is neither positive nor negative.

71. Describe the relationship between the variables for each of the sets of points in the previous exercise.

4.3   FURTHER TOPICS IN REGRESSION ANALYSIS

OBJECTIVES   By the end of this section, I will be able to . . .

1 Calculate the sum of squares error (SSE), and use the standard error of the estimate s as a measure of a typical prediction error.

2 Describe how total variability, prediction error, and improvement are measured by the total sum of squares (SST), the sum of squares error (SSE), and the sum of squares regression (SSR).

3 Explain the meaning of the coefficient of determination r^2 as a measure of the usefulness of the regression.

In Section 4.2 we were introduced to regression analysis, which uses an equation to approximate the linear relationship between two quantitative variables. Here in Section 4.3, we learn some further topics that will enable us to better apply the tools of regression analysis for a deeper understanding of our data.

1 SUM OF SQUARES ERROR (SSE) AND STANDARD ERROR OF THE ESTIMATE s

Getty Images/Fuse

Table 4.4 shows the results for ten student subjects who were given a set of short-term memory tasks to perform within a certain amount of time. These tasks included memorizing nonsense words and random patterns. Later, the students were asked to repeat the words and patterns, and the students were scored according to the number of words and patterns memorized and the quality of the memory. Partially remembered words and patterns were given partial credit, so the score was a continuous variable. Figure 4.24 displays the scatterplot of y = score versus x = time, together with the regression line $\hat{y} = 7 + 2x$ (that is, $\hat{y} = 2x + 7$), as calculated by Minitab.

Table 4.4 Results of short-term memory test

Student	Time to memorize (in minutes) (x)	Short-term memory score (y)
1	1	9
2	1	10
3	2	11
4	3	12
5	3	13
6	4	14
7	5	19
8	6	17
9	7	21
10	8	24

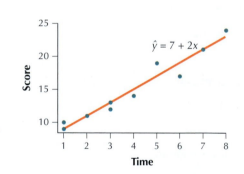

FIGURE 4.24
Scatterplot with regression line.

The regression equation is
Score = 7.00 + 2.00 Time

Minitab regression results (excerpt).

shortmemory

In Section 4.2, we learned that the difference $y - \hat{y}$ represented the prediction error or residual between the actual data value y and the predicted value $\hat{y}$. For example, for a student who is given $x = 5$ minutes to study, the predicted score is $\hat{y} = 2(time) + 7 = 17$.

For Student 7, who was given 5 minutes to study and got a score of 19, the prediction error is $y - \hat{y} = 19 - 17 = 2$.

We can calculate the prediction errors for every student who was tested. If we wish to use the regression to make useful predictions, we would like to keep all our prediction errors small. To measure the prediction errors, we calculate the sum of squared prediction errors, or more simply, the **sum of squares error (SSE):**

> **Sum of Squares Error (SSE)**
>
> $$\text{SSE} = \sum(y - \hat{y})^2 = \sum(\text{residual})^2 = \sum(\text{prediction error})^2$$

Since we want our prediction errors to be small, it follows that we want SSE to be as small as possible.

> **Least-Squares Criterion**
>
> The least-squares criterion states that the regression line will be the *line for which the SSE is minimized.* That is, out of all possible straight lines, the least-squares criterion chooses the line with the smallest SSE to be the regression line.

EXAMPLE 4.13

CALCULATING SSE, THE SUM OF SQUARES ERROR

a. Construct a scatterplot of the memory score data, indicating each residual.
b. Calculate the sum of squares error (SSE) for the memory score data.

Solution

a. The brackets (}) in the scatterplot in Figure 4.25 indicate the residual for each student's score. The quantities represented by these brackets are the residuals $y - \hat{y}$.
b. Table 4.5 shows the $\hat{y}$-values and residuals for the data in Table 4.4. The sum of squares error is then found by squaring each residual and taking the sum. Thus

$$\text{SSE} = \sum(y - \hat{y})^2 = 12$$

Now You Can Do Exercises 11–16.

Since we know that $\hat{y} = 2x + 7$ is the regression line, according to the least-squares criterion, no other possible straight line would result in a smaller SSE.

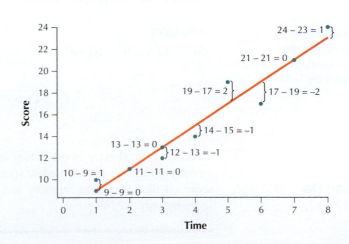

FIGURE 4.25
Scatterplot showing the prediction errors or residuals $y - \hat{y}$.

Table 4.5 Calculation of the SSE for the short-term memory test example

Student	Time (x)	Actual score (y)	Predicted score (ŷ = 2x + 7)	Residual (y − ŷ)	(Residual)² (y − ŷ)²
1	1	9	9	0	0
2	1	10	9	1	1
3	2	11	11	0	0
4	3	12	13	−1	1
5	3	13	13	0	0
6	4	14	15	−1	1
7	5	19	17	2	4
8	6	17	19	−2	4
9	7	21	21	0	0
10	8	24	23	1	1

$$\text{SSE} = \sum(y - \hat{y})^2 = 12$$

A useful interpretive statistic is s, the **standard error of the estimate.** The formula for s follows.

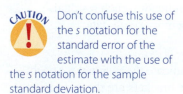

CAUTION Don't confuse this use of the s notation for the standard error of the estimate with the use of the s notation for the sample standard deviation.

> **Standard Error of the Estimate s**
>
> $$s = \sqrt{\frac{\text{SSE}}{n - 2}}$$

The standard error of the estimate gives a measure of the typical residual. That is, s is a measure of the size of the *typical prediction error,* the typical difference between the predicted value of y and the actual observed value of y. If the typical prediction error is large, then the regression line may not be useful.

EXAMPLE 4.14

CALCULATING AND INTERPRETING s, THE STANDARD ERROR OF THE ESTIMATE

Calculate and interpret the standard error of the estimate s for the memory score data.

Note: Here we are rounding $s = 1.2247$ for reporting purposes. However, when we use s for calculating other quantities later, we will not round until the last calculation.

Solution

SSE = 12 and $n = 10$, so

$$s = \sqrt{\frac{\text{SSE}}{n - 2}} = \sqrt{\frac{12}{8}} \approx 1.2247$$

Thus, the typical error in prediction is 1.2247 points. In other words, if we know the amount of time (x) a given student spent memorizing, then our estimate of the student's score on the short-term memory test will typically differ from the student's actual score by only 1.2247 points.

Now You Can Do Exercises 17–22.

2 SST, SSR, AND SSE

The coefficient of determination r^2 depends on the values of two new statistics, SST and SSR, which we learn next. The least-squares criterion guarantees that the value of SSE = 12 that we found in Example 4.13 is the smallest possible value for SSE, given the data in Table 4.4. However, this guarantee in itself does not tell us that the regression is useful. For the regression to be useful, the prediction error (and therefore SSE) must be small. But, we cannot yet tell whether the value of SSE = 12 is indeed small, since we have nothing to compare it against.

Suppose for a moment that we want to estimate short-term memory scores but have no knowledge of the amount of time (x) for memorizing. Then the best estimate for y is simply $\bar{y} = 15$, the mean of the sample of short-term memory test scores. The graph of $\bar{y} = 15$ is the horizontal line in Figure 4.26.

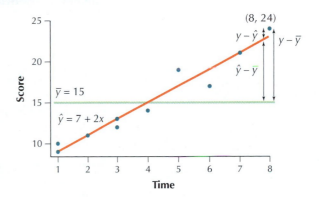

FIGURE 4.26
Comparing $(y - \hat{y})$ and $(y - \bar{y})$.

In general, the data points are closer to the regression line than they are to the horizontal line $\bar{y} = 15$, indicating that the errors in prediction are smaller when using the regression equation. Consider Student 10, who had a short-term memory score of $y = 24$ after memorizing for $x = 8$ minutes. Using $\bar{y} = 15$ as the estimate, the error for Student 10 is

$$(y - \bar{y}) = 24 - 15 = 9$$

This error is shown in Figure 4.26 as the vertical distance $(y - \bar{y})$.

Suppose we found this value $(y - \bar{y})$ for every student in the data set and summed the squared $(y - \bar{y})$, just as we did for the $(y - \hat{y})$ when finding SSE. The resulting statistic is called the **total sum of squares (SST)** and is a measure of the total variability in the values of the y variable:

$$SST = \sum (y - \bar{y})^2$$

Developing Your Statistical Sense	**Relationship Between SST and the Variance of the y's**

Note that SST ignores the presence of the x information; it is simply a measure of the variability in y. Recall (see page 103) that the *variance* of a sample of y-values is given by $s^2 = \sum (y - \bar{y})^2 / (n - 1)$. Thus

$$SST = (n - 1) s^2$$

Hence, SST is proportional to the variance of the y's and, as such, is a measure of the variability in the y data.

EXAMPLE 4.15 | **CALCULATING SST, THE TOTAL SUM OF SQUARES, IN TWO WAYS**

Calculate SST, the total sum of squares, for the memory score data in two ways:
a. By using Table 4.6.
b. By using the fact that the sample variance of the score data (the y values) equals $25\frac{1}{3}$.

Solution

a. Table 4.6 shows the values for $(y - \bar{y}) = (y - 15)$ for the data in Table 4.4. Thus, $\text{SST} = \Sigma(y - \bar{y})^2 = 228$.

Table 4.6 Calculation of SST

Student	Score (y)	($y - \bar{y}$)	($y - \bar{y}$)^2
1	9	−6	36
2	10	−5	25
3	11	−4	16
4	12	−3	9
5	13	−2	4
6	14	−1	1
7	19	4	16
8	17	2	4
9	21	6	36
10	24	9	81

$$\text{SST} = \Sigma(y - \bar{y})^2 = 228$$

Now You Can Do
Exercises 23a,b–28a,b.

b. When we are given the variance of y, we may calculate SST as follows:

$$\text{SST} = (n - 1)s^2 = (10 - 1)(25\tfrac{1}{3}) = 228$$

Consider Figure 4.26 once again. For Student 10, note that the error in prediction when ignoring the x data is $(y - \bar{y}) = 9$, while the error in prediction when using the regression equation is $(y - \hat{y}) = 1$. (Recall that $\hat{y} = 2(8) + 7 = 23$, since Student 10's time is $x = 8$.) The amount of improvement (that is, the amount by which the prediction error is diminished) is the difference between $\hat{y}$ and $\bar{y}$:

$$(\hat{y} - \bar{y}) = 23 - 15 = 8$$

Once again, we can find $(\hat{y} - \bar{y})$ for each observation in the data set, square them, and sum the squared results to obtain $\Sigma(\hat{y} - \bar{y})^2$. The resulting statistic is **SSR, the sum of squares regression.**

$$\text{SSR} = \Sigma(\hat{y} - \bar{y})^2$$

SSR measures the amount of *improvement* in the accuracy of our estimates when using the regression equation compared with relying only on the y-values and ignoring the x information. Note in Figure 4.26 that the distance $(y - \bar{y})$ is the same as the sum of the distances $(\hat{y} - \bar{y})$ and $(y - \hat{y})$. It can be shown, using algebra, that the following also holds true.

Note: None of these sums of squares can ever be negative.

> **Relationship Among SST, SSR, and SSE**
> $$SST = SSR + SSE$$

If any two of these sums of squares are known, the third can be calculated as well, as shown in the following example.

EXAMPLE 4.16

USING SST AND SSE TO FIND SSR

Use SST and SSE to find the value of SSR for the data from Example 4.15.

Solution

From Example 4.13, we have SSE = 12, and from Example 4.15, we have SST = 228. That leaves us with just one unknown in the equation SST = SSR + SSE, so we can solve for the unknown SSR:

Now You Can Do
Exercises 23c–28c.

$$SSR = SST - SSE = 228 - 12 = 216$$

---

3 COEFFICIENT OF DETERMINATION r^2

SSR represents the amount of variability in the response variable that is accounted for by the regression equation, that is, by the linear relationship between y and x. SSE represents the amount of variability in the y that is left unexplained after accounting for the relationship between x and y (including random error). Since we know that SST represents the sum of SSR and SSE, it makes sense to consider the *ratio* of SSR and SST, called the **coefficient of determination r^2**.

> The **coefficient of determination r^2** = SSR/SST measures the goodness of fit of the regression equation to the data. We interpret r^2 as the proportion of the variability in y that is accounted for by the linear relationship between y and x. The values that r^2 can take are $0 \leq r^2 \leq 1$. Note that $\pm \sqrt{r^2} = r$, the correlation coefficient.

EXAMPLE 4.17

CALCULATING AND INTERPRETING THE COEFFICIENT OF DETERMINATION r^2

Calculate and interpret the value of the coefficient of determination r^2 for the memory score data.

Solution

From Example 4.15 we have SST = 228, and from Example 4.16 we have SSR = 216. Hence,

$$r^2 = \frac{SSR}{SST} = \frac{216}{228} \approx 0.9474$$

Now You Can Do
Exercises 29a,b–34a,b.

Thus, 94.74% of the variability in the memory test score (y) is accounted for by the linear relationship between score (y) and the time given for study (x).

What does the value of $r^2 \approx 0.9474$ mean? Consider that the memory test scores have a certain amount of variability: some scores are higher than others. In addition to the amount of time (x) given for memorizing, there may be several other factors that might account for variability in the scores, such as the memorizing ability of the students, how much sleep the students had, and so on. However, $r^2 \approx 0.9474$ indicates that 94.74% of this variability in memory scores (y) is explained by the single factor "amount of time given for study" (x). All other factors (including factors like amount of sleep) account for only 100% − 94.74% = 5.26% of the variability in the memory test scores.

Suppose that the regression equation was a perfect fit to the data, so that every observation lay exactly on the regression line. Since there would be no errors in prediction, SSE would equal 0, which would imply that

$$SST = SSR + 0 = SSR$$

Since in this case SST = SSR, then

$$r^2 = \frac{SSR}{SST} = \frac{SST}{SST} = 1$$

Conversely, if SSR = 0, then *no improvement at all* is gained by using the regression equation. That is, the regression equation accounts for no variability at all, and $r^2 = 0/SST = 0$.

The closer the value of r^2 is to 1, the better the fit of the regression equation to the data set. A value near 1 indicates that the regression equation fits the data extremely well. A value near 0 indicates that the regression equation fits the data extremely poorly.

Here are the alternate computational formulas for finding SST and SSR.

Computational Formulas for SST and SSR

$$SST = \sum y^2 - \left(\sum y\right)^2/n \qquad SSR = \frac{\left[\sum xy - \left(\sum x\right)\left(\sum y\right)/n\right]^2}{\sum x^2 2 \left(\sum x\right)^2/n}$$

EXAMPLE 4.18

CALCULATING SSR AND SST USING THE COMPUTATIONAL FORMULAS

Use the computational formulas to find SSR and SST for the memory score data. Assume we have the following summary statistics: $\sum x = 40$, $\sum y = 150$, $\sum xy = 708$, $\sum x^2 = 214$, $\sum y^2 = 2478$.

Solution

$$SST = \sum y^2 - \left(\sum y\right)^2/n = 2478 - (150)^2/10 = 228$$

$$SSR = \frac{\left[\sum xy - \left(\sum x\right)\left(\sum y\right)/n\right]^2}{\sum x^2 - \left(\sum x\right)^2/n} = \frac{[708 - (40)(150)/10]^2}{214 - (40)^2/10} = [108]^2/54 = 216$$

Then SSE = SST − SSR = 228 − 216 = 12. This value SSE = 12 agrees with the value we calculated earlier using Table 4.5.

Recall from Section 4.1 that the correlation coefficient r is given by

$$r = \frac{\sum(x - \bar{x})(y - \bar{y})}{(n - 1)\, s_x s_y}$$

where s_x and s_y represent the sample standard deviation of the x data and the y data, respectively. We can express the correlation coefficient r as

$$r = \pm\sqrt{r^2}$$

where r^2 is the coefficient of determination. The correlation coefficient r takes the same sign as the slope b_1. If the slope b_1 of the regression equation is positive, then $r = \sqrt{r^2}$; if the slope b_1 of the regression equation is negative, then $r = -\sqrt{r^2}$.

EXAMPLE 4.19

CALCULATE AND EVALUATE THE CORRELATION COEFFICIENT USING r^2

a. Use r^2 to calculate the value of the correlation coefficient r for the memory score data.

b. Perform the comparison test to determine whether x and y are correlated.

Solution

a. The slope $b_1 = 2$, which is positive, tells us that the sign of the correlation coefficient r is positive. Hence

$$r = \sqrt{r^2} = \sqrt{0.9474} \approx 0.9733$$

b. From Table G in the Appendix, the critical value for the correlation coefficient for $n = 10$ is 0.632. We have $r \approx 0.9733$, which is greater than 0.632.

**Now You Can Do
Exercises 29c–34c.**

Thus, student scores on the short-term memory test are strongly positively correlated with the amount of time allowed for memorization.

SECTION 4.3   **Summary**

1. The sum of squared prediction errors is referred to as the sum of squares error, $\text{SSE} = \sum(y - \hat{y})^2$. The standard error of the estimate, $s = \sqrt{\dfrac{\text{SSE}}{n - 2}}$, is an indicator of the precision of the estimates derived from the regression equation, since it provides a measure of the typical residual or prediction error.

2. The total variability in the y variable is measured by the total sum of squares, $\text{SST} = \sum(y - \bar{y})^2$, and may be divided into the sum of squares regression, $\text{SSR} = \sum(\hat{y} - \bar{y})^2$, and the sum of

squares error, $\text{SSE} = \sum(y - \hat{y})^2$. SSR measures the amount of improvement in the accuracy of estimates when using the regression equation compared with ignoring the x information.

3. The coefficient of determination, $r^2 = \text{SSR}/\text{SST}$, measures the goodness of fit of the regression equation as an approximation of the relationship between x and y. Finally, the correlation coefficient r may be expressed as $r = \pm\sqrt{r^2}$, taking the positive or negative sign of the slope b_1.

SECTION 4.3   **Exercises**

Clarifying the Concepts

1. What does s measure? Would we want s to be large or small? Why?

2. How does the least-squares criterion choose the "best" line to approximate the relationship between x and y?

3. What does SSE measure? Would we want SSE to be large or small? Why?

4. What does SSR measure? Would we want SSR to be large or small? Why?

5. What does SST measure? What statistic is it proportional to?

6. What does it mean when r^2 is close to 1? How about when it is close to 0?

7. Do the values of x affect SST at all?

8. Suppose we performed a regression analysis that resulted in $r^2 = 0.64$. Without further information, would it be possible to calculate the correlation coefficient r? Explain.

9. Suppose we performed a regression analysis on a data set that resulted in $r^2 = 0.64$. Interpret this statistic in terms of the amount of variance in y explained by the linear relationship between x and y.

10. True or false: When the prediction errors are too small, the sum of squared error SSE can be negative.

Practicing the Techniques

For Exercises 11–16, use the regression equations you calculated in Exercises 13–18 in Section 4.2. Do the following.

a. Construct a table like Table 4.5, and calculate the following quantity for each observation.
   i. $\hat{y}$, the estimated value of y
   ii. $y - \hat{y}$, the prediction error or residual
   iii. $(y - \hat{y})^2$, the squared residual
b. Calculate SSE, the sum of squares error.

11.

x	1	2	3	4
y	2	5	9	12

12.

x	0	2	4	6
y	5	6	5	4

13.

x	−5	−4	−3	−2	−1
y	0	8	8	16	16

14.

x	−3	−1	1	3	5
y	−5	−15	−20	−25	−30

15.

x	5	10	15	20	25	30
y	2	3	3	3	2	3

16.

x	6	7	8	9	11	13
y	4	4	4	4	4	4

For Exercises 17–22, calculate the standard error of the estimate, s, for the indicated data.

17. Data from Exercise 11

18. Data from Exercise 12

19. Data from Exercise 13

20. Data from Exercise 14

21. Data from Exercise 15

22. Data from Exercise 16

For Exercises 23–28, follow these steps.
   a. Compute the sample variance of the y data, $s^2 = \Sigma(y - \bar{y})^2/(n - 1)$.
   b. Use s^2 to calculate the total sum of squares, SST $= (n - 1)s^2$.
   c. Then use the relationship between the three sums of squares to find SSR, based on part **(b)** and your work in Exercises 11**(b)**–16**(b)**.

23. Data in Exercise 11

24. Data in Exercise 12

25. Data in Exercise 13

26. Data in Exercise 14

27. Data in Exercise 15

28. Data in Exercise 16

For Exercises 29–34, do the following.
   a. Using the results from Exercises 23**(b, c)**–28**(b, c)**, calculate the coefficient of determination, r^2.
   b. Interpret r^2 in terms of the proportion of variance in y accounted for by the linear regression between x and y.
   c. Use your work from part **(a)** and from Exercises 13**(a)**–18**(a)** of Section 4.2 to calculate the correlation coefficient, r.

29. Data in Exercise 11

30. Data in Exercise 12

31. Data in Exercise 13

32. Data in Exercise 14

33. Data in Exercise 15

34. Data in Exercise 16

Applying the Concepts

For Exercises 35–38, follow these steps. You have already calculated the regression equation in Exercises 51–54 in Section 4.2.
   a. Compute the residual for each data value. Form a table similar to Table 4.5 of the residuals and squared residuals. Sum the squared residuals to get SSE.
   b. Calculate and interpret s, the standard error of the estimate.

35. Education and Unemployment. Refer to the education and unemployment data from Exercise 51 in Section 4.2.

36. NASCAR Wins. Refer to the NASCAR data from Exercise 52 in Section 4.2.

37. SAT Scores. Refer to the SAT data from Exercise 53 in Section 4.2.

38. Brain and Body Weight. Refer to the brain and body weight data from Exercise 54 in Section 4.2.

For Exercises 39–42, follow these steps. Use your calculations from Exercises 35–38.
   a. Calculate the sample variance of the y data, s^2. Then use s^2 to calculate SST.

b. Use SST and SSE to find SSR.
c. Calculate and interpret the coefficient of determination r^2.
d. Use r^2 and b_1 to find the correlation coefficient r.

39. Education and Unemployment. Refer to your calculations in Exercise 35 above and Exercise 44 in Section 4.1.

40. NASCAR Wins. Refer to your calculations in Exercise 36 above.

41. SAT Scores. Refer to your calculations in Exercise 37 above.

42. Brain and Body Weight. Refer to your calculations in Exercise 38 above and Exercise 48 in Section 4.1.

For Exercises 43–44 the regression equation is provided. Follow these steps.
a. Compute the residual for each data value. Form a table similar to Table 4.5 of the residuals and squared residuals. Sum the squared residuals to get SSE.
b. Calculate and interpret s, the standard error of the estimate.
c. Calculate the sample variance of the y data, s^2. Then use s^2 to calculate SST.
d. Use SST and SSE to find SSR.
e. Calculate and interpret the coefficient of determination, r^2.
f. Use r^2 and b_1 to find the correlation coefficient r.

43. World Temperatures. Listed in the table are the low (x) and high (y) temperatures for a particular day measured in degrees Fahrenheit, for a random sample of cities worldwide. The regression equation is $\hat{y} = 1.05x + 11.9$.

worldtemp

City	Low (x)	High (y)
Kolkata, India	57	77
London, England	36	45
Montreal, Quebec	7	21
Rome, Italy	39	55
San Juan, Puerto Rico	70	83
Shanghai, China	34	45

44. Teenage Birth Rate. The National Center for Health Statistics publishes data on state birth rates. The table contains the overall birth rate and the teenage birth rate for ten randomly chosen states. The overall birth rate is defined by the NCHS as "live births per 1000 women," and the teenage birth rate is defined as "live births per 1000 women aged 15–19." The regression equation is $\hat{y} = 5.39x - 34.3$.

 teenbirth

State	x = overall birth rate	y = teenage birth rate
California	15.2	39.5
Florida	12.5	42.4
Georgia	15.7	53.4
New York	13.0	26.9
Ohio	13.0	38.5
Pennsylvania	11.7	30.5
Texas	17.0	62.6
Virginia	13.9	35.2

Education and Unemployment. Refer to your work in Exercise 35 for Exercises 45 and 46.

45. Answer the following.
a. Which data value has the largest residual? Describe what is unusual about this observation.
b. Suppose a public figure stated that 50% of the variability in the unemployment rate was due to competition from abroad. How would you use the regression results to respond to this claim?
c. Suppose a politician claimed that using the years of education alone could allow us to predict the unemployment rate to within 1%. How would you use the regression results to respond to this claim?
d. Suppose a newspaper claimed that each additional year of education brought down the unemployment rate by "more than 1%." How would you use the regression results to either support or refute this claim?

46. *What if* the unemployment rate for those with 5 years of education was not 16.8% but a much higher percentage. Describe how this would affect the slope and y intercept of the regression line. Explain your reasoning. (*Hint:* Consult the *What If* Scenario in Section 4.2, page 167.)

Bringing It All Together

Fuel Economy. For Exercises 47–54, refer to the table of fuel economy data from Exercises 63–67 in Section 4.2. The predictor variable is x = engine size, expressed in liters; the response variable is y = combined (city/highway) gas mileage, expressed in miles per gallon (mpg).

47. Calculating and interpreting the residuals and SSE and s.
a. Compute the residual for each data value. Form a table similar to Table 4.5 of the residuals and squared residuals. Sum the squared residuals to get SSE.

b. What is it that SSE is measuring? At this point, do we know whether SSE is large or small? Why or why not?

c. Which vehicle has the largest absolute residual? Clearly explain why this vehicle is unusual.

48. Calculating and Interpreting s.

a. Calculate the value of s, the standard error of the estimate.

b. Interpret the value of s so that a nonstatistician could understand it.

49. Computing and Interpreting SST, SSR, and r^2.

a. Calculate the sample variance of the y data, s^2. Then use s^2 to calculate SST.

b. Use SSE and SST to find SSR. Explain clearly what it is that SSR is measuring.

c. Calculate and interpret the coefficient of determination, r^2.

50. Correlation. Do the following.

a. Use r^2 and b_1 to find the correlation coefficient r.

b. Use the comparison test to determine whether we may conclude that engine size is correlated with combined mpg. Interpret the correlation between engine size and combined mpg.

51. *What if* we added one new vehicle to the data set, and its value was exactly $(\bar{x}, \bar{y})$. How would this affect the slope and the y intercept?

52. Refer to the previous exercise. *What if* we added an unknown amount to the engine size of the new vehicle. Describe how this change would affect the slope and the y intercept.

53. Challenge Exercise. Suppose we increased the combined mpg for the Cadillac Limousine so that the slope of the regression line would be exactly zero. What would the combined mpg for the Cadillac Limousine have to be to accomplish this?

54. Challenge Exercise. Refer to the previous exercise. Describe how this change to the fuel economy of the Cadillac Limousine would affect each of the following, and why: SSE, SSR, SST, s, r^2, r.

For Exercises 55–57, use technology and follow steps **(a)–(e)**.

a. Construct the scatterplot.

b. Compute and interpret the regression equation.

c. Calculate and interpret the coefficient of determination, r^2.

d. Compute and interpret s, the standard error of the estimate.

e. Find r, using r^2.

55. Open the **darts** data set, which we used for the Chapter 3 Case Study. Let x = the Dow Jones Industrial Average, and let y = the pros' performance.
🔴 darts

56. Open the **Nutrition** data set. Let x = the amount of fat per gram, and let y = the number of calories per gram.
🔴 Nutrition

57. Open the **pulse and temp** data set. Let x = heart rate, and let y = body temperature.
🔴 pulseandtemp

Construct Your Own Data Sets

Suppose we have a tiny data set with the following (x, y) pairs.

x	y
1	?
2	?
3	?

For Exercises 58–62, create a set of y-values that would fulfill each specification.

58. The slope of the line is positive.

59. The slope of the line is negative.

60. The slope of the line is 0.

61. The slope of the line is equal to 2.

62. The slope of the line is equal to -3.

Use the *Correlation and Regression* applet for Exercises 63–65.

63. In these applet exercises, use the "thermometer" above the graph (where it says "Sum of squares =") to help find the least-squares regression line interactively.

a. Select 5 points so that the correlation coefficient is about 0.8. Then select "Draw line."

b. Make your best guess about where the least-squares regression line should be, and draw the line there.

64. The blue section of the thermometer is a measure of the sum of squares error, the total squared vertical distance from the data points to the actual regression line. Recall that the least-squares regression line minimizes this distance. The green section of the thermometer tells you how much "extra" squared error you get from using the line you constructed in Exercise 63(a).

a. Adjust the line you drew in Exercise 63(a) by clicking and dragging on the points until the green section of the thermometer has disappeared.

b. What does the disappearance of the green part tell you about the adjusted line you constructed?

c. Will the line now coincide with the least-squares regression line?

65. Verify that your adjusted line from Exercise 64 coincides with the least-squares regression line by selecting "Show least-squares line."

CHAPTER 4 — Formulas and Vocabulary

Section 4.1
- **COMPARISON TEST FOR LINEAR CORRELATION** (p. 158)
- **CORRELATION COEFFICIENT r** (p. 153).

Definition formula:

$$r = \frac{\sum(x - \bar{x})(y - \bar{y})}{(n - 1)s_x s_y}$$

Computational formula:

$$r = \frac{\sum xy - \left(\sum x \sum y\right)/n}{\sqrt{\left[\sum x^2 - \left(\sum x\right)^2/n\right]\left[\sum y^2 - \left(\sum y\right)^2/n\right]}}$$

- **POSITIVE AND NEGATIVE CORRELATION** (p. 156)
- **SCATTERPLOT** (p. 150)

Section 4.2
- **EXTRAPOLATION** (p. 170)
- **PREDICTION ERROR, OR RESIDUAL** (p. 169).

$$(y - \hat{y})$$

- **REGRESSION EQUATION (REGRESSION LINE)** (p. 164).

$$\hat{y} = b_1 x + b_0$$

- **SLOPE OF THE REGRESSION LINE** (p. 164).

Definition formula:

$$b_1 = \frac{\sum(x - \bar{x})(y - \bar{y})}{\sum(x - \bar{x})^2}$$

Computational formula (p. 166):

$$b_1 = \frac{\sum xy - \left(\sum x \sum y\right)/n}{\sum x^2 - \left(\sum x\right)^2/n}$$

- **y INTERCEPT OF THE REGRESSION LINE** (p. 167).

$$b_0 = \bar{y} - (b_1 \cdot \bar{x})$$

Section 4.3
- **COEFFICIENT OF DETERMINATION r^2** (p. 183).

$$r^2 = SSR/SST$$

- **LEAST-SQUARES CRITERION** (p. 179)
- **SSE, SUM OF SQUARES ERROR** (p. 179).

$$SSE = \sum(y - \hat{y})^2$$

- **STANDARD ERROR OF THE ESTIMATE s** (p. 180).

$$s = \sqrt{\frac{SSE}{n - 2}}$$

- **SSR, SUM OF SQUARES REGRESSION** (p. 182).

Definition formula:

$$SSR = \sum(\hat{y} - \bar{y})^2$$

Computational formula (p. 184):

$$SSR = \frac{\left[\sum xy - \left(\sum x\right)\left(\sum y\right)/n\right]^2}{\sum x^2 - \left(\sum x\right)^2/n}$$

- **SST, TOTAL SUM OF SQUARES** (p. 181).

Definition formula:

$$SST = \sum(y - \bar{y})^2$$

Computational formula (p. 184):

$$SST = \sum y^2 - \left(\sum y\right)^2/n$$

CHAPTER 4 — Review Exercises

Section 4.1
MIDTERM EXAMS AND OVERALL GRADE. Use the data in the following table to answer Exercises 1–5. Can you predict how you will do in a course based on the result of the midterm exam only? The midterm exam score and the overall grade were recorded for a random sample of 12 students in an elementary statistics course. The results are shown in the following table.

 midexam

Midterm exams and overall grades

Student	Midterm exam score (x)	Overall grade (y)
1	50	65
2	90	80
3	70	75
4	80	75
5	60	45
6	90	95
7	90	85
8	80	80
9	70	65
10	70	70
11	60	65
12	50	55

1. Construct a scatterplot of overall grade versus midterm exam score.

2. Refer to your scatterplot from Exercise 1.
   a. Characterize the relationship as positive, negative, or not apparent.
   b. Write a sentence that describes the behavior of the overall grade as the midterm exam score increases.

3. Calculate the value of the correlation coefficient r between midterm exam scores and overall grades.

4. Use the comparison test to determine whether we may conclude x and y are correlated.

5. Interpret the value for r.

Section 4.2
For Exercises 6–12, refer to the table of midterm exams (x) and overall grades (y).

6. Calculate the regression coefficients b_0 and b_1, and write the regression equation.

7. State the regression equation in words, as shown at the end of Example 4.7 (page 165).

8. Interpret the value of the slope b_1.

9. Interpret the value of the y-intercept b_0.

10. Use the regression equation to predict the overall grades for the following midterm exam scores.

    a. $x = 50$ **b.** $x = 100$

11. Calculate and interpret the prediction error for each prediction in Exercise 10.

12. For each prediction in Exercise 10, state whether the prediction represents extrapolation.

Section 4.3
Refer to the midterm exam and overall grade data, to answer Exercises 13–17.

13. Calculate SSE.

14. Calculate s, the standard error of the estimate. What does this number mean?

15. Calculate SST. Then use SSE and SST to find SSR.

16. Calculate r^2, the coefficient of determination. Comment on how useful midterm exam scores are for predicting overall grades.

17. Use r^2 to calculate the correlation coefficient. Comment on the relationship between midterm exam scores and overall grades.

CHAPTER 4   **Quiz**

True or False

1. True or false: Scatterplots are constructed with the y variable on the horizontal axis and the x variable on the vertical axis.

2. True or false: The y intercept measures the strength of the linear relationship between two numerical variables.

Fill in the Blank

3. The "hat" over the y in $\hat{y}$ indicates that it is an _____ of y.

4. We interpret the slope of the regression line as the estimated change in y per _____ increase in x.

Short Answer

5. Making predictions based on x-values that are beyond the range of the x-values in our data set is called what?

6. Values of r close to -1 indicate what type of relationship between the two variables?

Calculations and Interpretations

VIOLENT CRIME. Use the following information for Exercises 7–14. The Federal Bureau of Investigation publishes crime statistics, including those in the following table, which shows the percentage of violent crime committed per month nationwide for the years 2002 and 2004.[3]

🔴 violentcrime

Month	2002	2004
January	7.9	7.8
February	6.8	7.0
March	7.9	8.3
April	8.1	8.2
May	8.7	9.0
June	8.8	8.6
July	9.3	9.2
August	9.3	9.0
September	9.2	8.5
October	8.6	8.6
November	7.7	7.8
December	7.7	7.9

7. Construct a scatterplot of 2004 monthly crime versus 2002 monthly crime.

8. Based on your scatterplot, would you characterize the linear relationship, if any, as positive or negative?

9. Compute the regression equation.

10. Calculate the three sums of squares: SSR, SST, and SSE.

11. Calculate s, the standard error of the estimate. What does this number mean?

12. Calculate r^2. Comment on how useful the 2002 percentages are in predicting the 2004 percentages.

13. Use r^2 to calculate and interpret the correlation coefficient.

14. Find the prediction error for the following percentages:
   a. 7.9   **b.** 9.3   **c.** 8.1

5 Probability

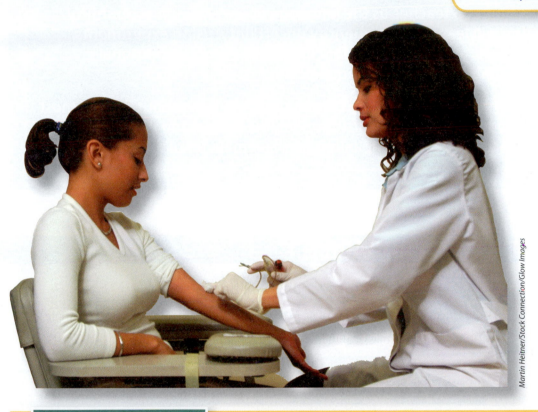

Martin Heitner/Stock Connection/Glow Images

CASE STUDY

The ELISA Test for the Presence of HIV

If someone suspects that he or she is at increased risk of HIV infection, then he or she might be interested in going for an HIV ELISA test. The ELISA test is used to screen blood for the presence of HIV. Sometimes called an HIV enzyme immunoassay (EIA), an HIV ELISA is the most basic test for finding out if an individual is carrying a particular pathogen, such as HIV.

Like most diagnostic procedures, the ELISA test is not foolproof. In this chapter's Case Study we study the types of errors the ELISA test can make and what this means for those who carry the HIV virus and for those who do not. For example, did you know that if your ELISA test comes back positive, then the chances are eight out of ten that you do *not* carry the virus? ∎

The Big Picture

Where we are coming from, and where we are headed . . .

- Chapters 1–4 dealt with descriptive statistics that summarize data. In later chapters, we will learn inferential statistics, which generalize from a sample to a population. But generalizing involves *uncertainty*.

- Chapter 5 teaches us the language of uncertainty: **probability**. We will learn how to quantify uncertainty, using experiments, events, outcomes, rules for combining events, conditional probability, and counting methods.

- In Chapter 6, "Probability Distributions," we learn about the two most important probability distributions, the normal and the binomial, which will be our companions for the remainder of the text.

5.1   INTRODUCING PROBABILITY

OBJECTIVES   By the end of this section, I will be able to . . .

1   Understand the meaning of an experiment, an outcome, an event, and a sample space.

2   Describe the classical method of assigning probability.

3   Explain the Law of Large Numbers and the relative frequency method of assigning probability.

Imagine you are striding down the midway of your local town fair, when a particular game of chance catches your eye. The object of this game is to roll a 6 on a single roll of a single fair die. If you do so, you win $5. It costs $1 to play the game. What is the likelihood of winning?

To show how to solve this problem, we must first introduce the building blocks of probability.

1 BUILDING BLOCKS OF PROBABILITY

Our daily lives are filled with *uncertainty*, seemingly governed by *chance*. We try to cope with uncertainty by estimating the *chances* that a particular event will occur. We are daily called on to make intelligent decisions about probabilities. Consider the following scenarios, and think about how the italicized words all refer to uncertainty.

- What is the *chance* that there will be a speed trap on this stretch of I-95 on a particular day?

- What is the *likelihood* that this lottery ticket will make me rich?

- What is the *probability* that this throw of the dice will come up a seven?

Sometimes, the amount of uncertainty in our daily lives is so great that there appears to be no order to the world whatsoever. However, if you look closely, there are *patterns in randomness*. In this chapter, we learn to become better decision makers by becoming acquainted with the tools of probability in order to quantify many of the uncertainties of everyday life.

Developing Your Statistical Sense

A Different Perspective

As you read this chapter, notice that the perspective differs from that in previous chapters. Earlier, we were looking at a data set and trying to describe it graphically and numerically. Now, instead of trying to describe a data set, we are faced with an experimental situation, and our task is to calculate probabilities associated with various outcomes in the experiment.

> The **probability** of an outcome represents the chance or likely hood that the outcome will occur.

Let us acquaint ourselves with the building blocks of probability, starting with the concept of an experiment. In probability, an **experiment** is any activity for which the outcome is uncertain. Consider the stock market, for example. Suppose you own 100 shares of Consolidated Widgets and are interested in what the share price will be at the end of trading tomorrow. Will the share price increase or decrease? The actual result is uncertain, so this is an example of an experiment. Each of the possible results of the experiment is called an **outcome.** Another example of an experiment is when you toss a coin. In the coin-toss experiment, the result may be heads or it may be tails. The collection of all possible outcomes is called the **sample space.** The sample space for the coin-toss experiment is {heads, tails} or {H, T}. Following are some common experiments, together with their sample spaces.

We use braces, { }, to enclose a set of outcomes.

Experiment	Sample space
Roll a single six-sided die	{1, 2, 3, 4, 5, 6}
Toss two coins	{HH, HT, TH, TT}
Play a video game	{win, lose}

We use the building blocks of probability to investigate the likelihood of an outcome or **event.**

> **Building Blocks of Probability**
>
> An **experiment** is any activity for which the outcome is uncertain.
>
> An **outcome** is the result of a single performance of an experiment.
>
> The collection of all possible outcomes is called the **sample space.** We denote the sample space S.
>
> An **event** is a collection of outcomes from the sample space. To find the probability of an event, add up the probabilities of all the outcomes in the event.

When we talk about the probability of some outcome, we are referring to a number that indicates how likely the particular outcome is. The notation $P(A)$ stands for "the probability that outcome *A* occurred." Say we define outcome *W* to be "you win the video game." Then "the probability that you win the video game" can be denoted as $P(W)$. Probabilities abide by the following rules.

> **Rules of Probability**
>
> 1. The probability $P(E)$ for any event *E* is always between 0 and 1, inclusive. That is, $0 \le P(E) \le 1$.
> 2. **Law of Total Probability:** For any experiment, the sum of all the outcome probabilities in the sample space must equal 1.

CAUTION If the probability that you calculated is negative or greater than 1, then you should try again.

Now You Can Do Exercises 11–16.

From the definition, the probability of an event is a proportion, so the probability cannot be negative because proportions cannot be negative and it cannot be greater than 1 (100%) because an event cannot occur more than 100% of the time. A **probability model** is a table or listing of all the possible outcomes of an experiment, together with the probability of each outcome. A probability model must follow the Rules of Probability.

Throughout the remainder of this book, you will often be asked to calculate the probability of various events. Following are the meanings of some probabilities.

Probability value	Meaning
Near 0	Outcome or event is very unlikely.
Equal to 0	Outcome or event cannot occur.
Near 1	Outcome or event is nearly certain to occur.
Equal to 1	Outcome or event is certain to occur. It's "a sure thing."
Low	Outcome or event is unusual.
High	Outcome or event is not unusual.

The threshold of an unusual event depends on the specific experiment; the 0.05 is not set in stone.

Higher probability values are associated with higher likelihood of occurrence. An outcome with probability 0.5 will happen about half of the time. An outcome with probability 0.95 is very likely. We say that an outcome or event is *unusual* if its probability is below a certain threshold, say, 0.05. When we perform an experiment, it is a "sure thing" that one of the outcomes in the sample space will occur. For example, when you toss a coin, you know that it will be either heads or tails. Put into probability terms, the sum of the probabilities of all the individual outcomes must equal 1, the **Law of Total Probability.**

The following table shows some typical events for the experiments in the table on page 195.

Experiment	Sample space	Typical events
Roll a single die	{1, 2, 3, 4, 5, 6}	*E*: roll an even number = {2, 4, 6} *L*: roll a 4 or larger = {4, 5, 6}
Toss two coins	{HH, HT, TH, TT}	*H*: exactly one head = {HT, TH} *T*: at most one tail = {HH, HT, TH}
Play a video game	{win, lose}	*W*: win = {win} *L*: lose = {lose}

2   CLASSICAL METHOD OF ASSIGNING PROBABILITY

Many people have a certain degree of intuition when it comes to assigning probabilities. For example, when asked what the chances are of rolling a 6 on a single toss of a fair die, many people would quite correctly answer 1/6. However, intuition can often let us down. For example, when asked what the chances are of observing two heads when you toss a fair coin twice, many people would incorrectly respond 1/3 ("Well, it's either both heads or both tails or one of each." The correct answer is in fact 1/4.) In this section, we learn how to quantify our methods of assigning probabilities so that we don't have to depend on intuition alone.

There are three methods for assigning probabilities:

Reunion des Musées Nationaux/Art Resource, NY

Did you know? People have been tossing dice for a long time. Archaeologists have dug up dice from Roman ruins looking just the same as ours. These three dice were uncovered from the ruins of Pompeii buried by the eruption of Mount Vesuvius in the first century A.D.

- Classical method

- Relative frequency method

- Subjective method

We first take a close look at the classical method. Later in this section, we will examine the relative frequency method and the subjective method.

Many experiments are structured so that each experimental outcome is equally likely. *Equally likely outcomes* are outcomes that have the same probability of occurring. For example, if you toss a fair coin, the probability of observing either of the outcomes heads or tails is the same. The **classical method of assigning probabilities** is used when an experiment has equally likely outcomes.

Classical Method of Assigning Probabilities

Let $N(E)$ and $N(S)$ denote the number of outcomes in event E and the sample space S, respectively. If the experiment has equally likely outcomes, then the probability of event E is

$$P(E) = \frac{\text{number of outcomes in } E}{\text{number of outcomes in sample space}} = \frac{N(E)}{N(S)}$$

EXAMPLE 5.1

PROBABILITY OF DRAWING AN ACE

Find the probability of drawing an ace when drawing a single card at random from a deck of cards.

Solution

The sample space for the experiment where a subject chooses a single card at random from a deck of cards is given in Figure 5.1. If the card is chosen truly at random, then it is reasonable to assume that each card has the same chance of being drawn. Since each card is equally likely to be drawn, we can use the classical method to assign probabilities.

There are 52 outcomes in this sample space, so $N(S) = 52$. Let E be the event that an ace is drawn. Event E consists of the four aces $\{A\heartsuit, A\diamondsuit, A\clubsuit, A\spadesuit\}$, so $N(E) = 4$. Therefore, the probability of drawing an ace is

$$P(E) = \frac{N(E)}{N(S)} = \frac{4}{52} = \frac{1}{13}$$

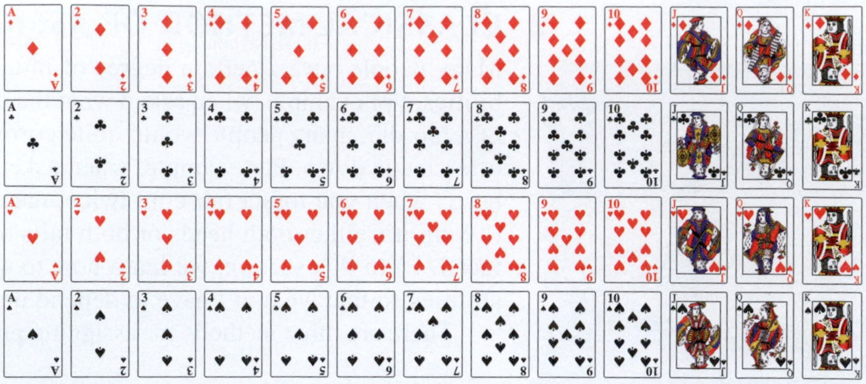

**Now You Can Do
Exercises 17–20.**

FIGURE 5.1
Sample space for drawing a
card at random from a deck
of cards.

EXAMPLE 5.2

FAIR DIE TOSS OUTCOMES ARE EQUALLY LIKELY

Recall the town fair example (at the top of page 194). In the game, you win if you roll a 6 on a single roll of a single fair die. Find the probability of winning the game.

Solution

The sample space for a single die toss consists of six outcomes, $\{1, 2, 3, 4, 5, 6\}$. When the six outcomes are equally likely, we say that the die is *fair*. If the outcomes are not equally likely, then the die is *loaded* or defective. If we assume the die is fair, then, since the sum of the probabilities of the $n = 6$ outcomes must equal 1, the probability of any particular outcome must equal 1/6, using the classical method. We write

$$\text{probability of winning} = P(W) = 1/6$$

**Now You Can Do
Exercises 21–26.**

Tree Diagrams

A **tree diagram** is a graphical display that allows us to list all the outcomes in the sample space of a multistage experiment. The next example shows how to construct a tree diagram.

EXAMPLE 5.3

LIST ALL OUTCOMES IN A SAMPLE SPACE USING A TREE DIAGRAM

Suppose our experiment is to toss a fair coin twice.
a. Construct a tree diagram.
b. Use the tree diagram to list all the outcomes in the sample space.

Solution

a. Think of this experiment as a two-stage process:

- Stage 1: Toss the coin the first time.
- Stage 2: Toss the coin the second time.

Figure 5.2 shows the tree diagram for the experiment of tossing a fair coin twice. Note the branches for Stage 1: the first time the coin is tossed, it can come up heads or tails. At Stage 2, the tree diagram again has branches for either heads or tails.

b. The sample space for the experiment of tossing a coin twice is $\{HH, HT, TH, TT\}$. There are $N(S) = 4$ outcomes in the sample space.

FIGURE 5.2

Tree diagram for the experiment of tossing a fair coin twice.

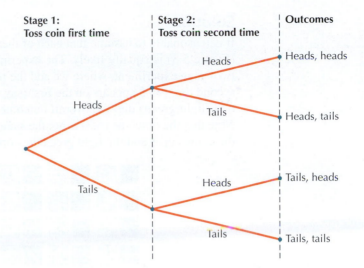

Stage 1: Toss coin first time	Stage 2: Toss coin second time	Outcomes
Heads	Heads	Heads, heads
	Tails	Heads, tails
Tails	Heads	Tails, heads
	Tails	Tails, tails

Now You Can Do Exercises 29 and 30.

Note that there are two possible outcomes at Stage 1 of this two-stage experiment and two possible outcomes when flipping the coin at Stage 2. To determine how many outcomes there are in the entire experiment, the *counting rule* is simply to multiply the number of possible outcomes at each stage. In this two-stage experiment, 2 times 2 equals 4 possible outcomes, which is the number of outcomes we see in the sample space.

EXAMPLE 5.4

FINDING PROBABILITIES FOR THE EXPERIMENT OF TOSSING A COIN TWICE

Find the probability of obtaining one heads and one tails when a fair coin is tossed twice.

Solution

It is reasonable to assume that the $N(S) = 4$ outcomes in the sample space {HH, HT, TH, TT} are equally likely. The coin doesn't remember what occurred at Stage 1, so the probabilities at Stage 2 are precisely the same as at Stage 1. Also, recall from the Law of Total Probability that the sum of the probabilities of all the outcomes in the sample space must equal 1. Thus, each of the four outcomes must have probability 1/4. Let E be the event that one heads and one tails is obtained. Then $E = \{HT, TH\}$, so $N(E) = 2$. Thus,

Now You Can Do Exercises 31–33.

$$P(E) = \frac{\text{number of outcomes in } E}{\text{number of outcomes in sample space}} = \frac{N(E)}{N(S)} = \frac{2}{4} = \frac{1}{2}$$

EXAMPLE 5.5

FINDING PROBABILITIES FOR THE EXPERIMENT OF TOSSING TWO FAIR DICE

Punchstock/CutandDeal

Imagine that you are playing Monopoly with your dormitory roommate, and the loser has to do the laundry for both of you for the rest of the semester. You have a hotel on Boardwalk, and if your roommate lands on it, you will surely win. Right now your roommate's piece is on Short Line: if he or she rolls a 4, you will win and get your laundry done free for the remainder of the semester. Put into statistical terms, the experiment is to toss two fair dice and observe the sum of the two dice. Find the probability of rolling a sum of 4 when tossing two fair dice.

Solution

It is reasonable to assume that each of these $N(S) = 36$ outcomes in the sample space (Figure 5.3) is equally likely. The experiment of tossing two dice can be viewed as a two-stage experiment, where we add the result from the first die to the result from the second die. If a 5 appears on the first (say, dark green) die, and a 3 appears on the second (light green) die, the overall outcome is (5,3), with the resulting sum equal to 8. Note that the outcome (5,3) is not the same as the outcome (3,5), where the dark green die comes up 3 and the light green die comes up 5.

FIGURE 5.3
Sample space for tossing two fair dice.

Let E denote the event that your roommate rolls a sum equal to 4. Then the outcomes that belong in this event are E: {(3,1) (2,2) (1,3)}, so $N(E) = 3$. Since the outcomes are equally likely, we can use the classical method for finding probabilities of events.

$$P(E) = \frac{\text{number of outcomes in } E}{\text{number of outcomes in sample space}} = \frac{N(E)}{N(S)} = \frac{3}{36} = \frac{1}{12}$$

Now You Can Do Exercises 49–53.

The probability that your roommate will land on Boardwalk on this throw of the dice is 1/12.

EXAMPLE 5.6

INAPPROPRIATE USE OF THE CLASSICAL METHOD

A recent study[1] showed that 59% of teenagers owned a computer (either a desktop or a laptop). Suppose we choose one teenager at random. Define the following events:

  C: The randomly chosen teenager owns a computer.
  D: The randomly chosen teenager does not own a computer.

Determine whether the classical method can be used to assign probability to events C and D.

Solution

The proper method for solving this problem is the relative frequency method, which we discuss next.

Because more than half of teenagers own a computer, if we choose a teenager at random, we are more likely to select a teenager who owns a computer than to select one who does not. Therefore, the events C and D are not equally likely. It would be inappropriate to use the classical method of assigning probabilities for this experiment because the classical method can be used only when all the outcomes of an experiment are equally likely.

---

3 RELATIVE FREQUENCY METHOD

In Example 5.2, we need the classical method to find that the probability of rolling a 6 with a fair die is 1/6. What does this probability mean? Remember that the definition of *probability* included the phrase "long-term proportion." The next example demonstrates what we mean by "long-term."

EXAMPLE 5.7

SIMULATING THE LONG-TERM PROPORTION OF 6S IN A FAIR DIE ROLL

Suppose we would like to investigate the proportion of 6s we observe if we roll a fair die 100 times. We can use technology, such as the TI-83/84 used here, to help us simulate rolling a fair die a large number of times. A **simulation** uses methods such as rolling dice or computer generation of random numbers to generate results from an experiment. The actual die rolls from our simulation are shown here, in order, with the 6s in boldface.

1 **4 4 6** 2 4 3 2 1 3 4 3 3 4 3 3 **6** 3 5 5 1 5 3 5 5 2 1 3 1 1 1 5 5 **6 3 6** 2 1 **6** 5 5 4 4 **6** 5 4 1 1 4 **6**
4 2 2 2 **6** 3 2 5 5 **6** 1 1 3 1 **6** 5 4 **6 6** 5 5 5 2 5 5 3 4 2 4 **6** 4 5 5 1 **6** 3 1 1 1 3 5 4 2 3 3 3 **6** 2 5 3

Thus, the first die roll was a 1, so the proportion of 6s was 0/1. The second and third die rolls were 4s, so the proportion of 6s after 3 rolls was 0/3. On the fourth roll a 6 appeared, so the proportion of 6s after the fourth roll was 1/4. Figure 5.4 provides a graph of the proportion of 6s in this simulation as the number of die rolls increased. Note that as the number of die rolls increases, the proportion of 6s tends to get closer to the horizontal line, $0.1667 \approx 1/6$.

The simulation was rerun, this time with 1000 die rolls. The resulting graph of the proportion of 6s is provided in Figure 5.5. Note that as the number of die rolls increases, the proportion of 6s approaches the line $0.1667 \approx 1/6$, and the fit is tighter with 1000 die rolls than with 100. This is what we mean by "long-term proportion."

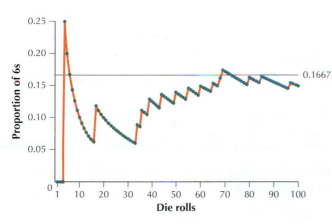

FIGURE 5.4 Proportion of 6s, 100 die rolls.

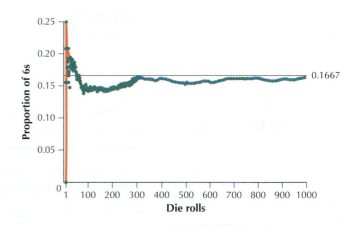

FIGURE 5.5 Proportion of 6s, 1000 die rolls.

This example leads directly to the following law.

> **Law of Large Numbers**
>
> As the number of times that an experiment is repeated increases, the relative frequency (proportion) of a particular outcome tends to approach the *probability* of the outcome.
>
> • For quantitative data, as the number of times that an experiment is repeated increases, the mean of the outcomes tends to approach the population mean.
>
> • For categorical (qualitative) data, as the number of times that an experiment is repeated increases, the proportion of times a particular outcome occurs tends to approach the population proportion.

The *Law of Large Numbers for Proportions* applet allows you to simulate coin tossing and observe the proportion of heads as the number of tosses increases.

Relative Frequency Method

If we can't use the classical method for assigning probabilities, then the **Law of Large Numbers** gives us a hint about how we can estimate the probability of an event. It often happens that previous information is available about the relative frequency of an event. Relative frequency information can be used to estimate the probability of the event.

Note: Tree diagrams can be used for the relative frequency method as well as the classical method of assigning probability.

> **Relative Frequency Method of Assigning Probabilities**
>
> The probability of event E is approximately equal to the relative frequency of event E. That is,
>
> $$P(E) \approx \text{relative frequency of } E = \frac{\text{frequency of } E}{\text{number of trials of experiment}}$$
>
> The relative frequency method is also known as the **empirical method.**

EXAMPLE 5.8 RELATIVE FREQUENCY METHOD: TEEN BLOGGERS

A recent study found that 35% of all online teen girls are bloggers, compared to 20% of online teen boys. Suppose that the 35% came from a random sample of 100 teen girls who use the Internet, 35 of whom are bloggers. If we choose one teen girl at random, find the probability that she is a blogger.

Solution

Define the event.

B: The online girl is a blogger.

We use the relative frequency method to find the probability of event *B*:

Now You Can Do Exercises 57–60.

$$P(B) \approx \text{relative frequency of } B = \frac{\text{frequency of } B}{\text{number of trials in experiment}} = \frac{35}{100} = 0.35$$

We can also use the relative frequency method to build a probability model with data that have been summarized in a table.

EXAMPLE 5.9 PROBABILITY MODELS BASED ON FREQUENCY TABLES

Table 5.1 Employment types

Employment type	Count
Private company	597
Federal government	141
Self-employed	97
Private nonprofit	92
Local government	59
State government	12
Other	2

Table 5.1 contains the employment type for a sample of 1000 employed citizens of Fairfax County, Virginia.[2] Use the data to construct the probability model by generating the relative frequencies and using the relative frequencies to estimate the probabilities for each employment type.

Solution

We calculate the relative frequencies of each employment group by dividing the count (frequency) for each group by the sample size 1000. For example, the relative frequency for "Private Company" is $\frac{597}{1000} = 0.597$. The relative frequency is then used to estimate the probability of selecting citizens who work at private companies in Fairfax County, Virginia. Filling in the remaining calculations produces the *probability model* in Table 5.2. Note that the table follows the Rules of Probability in that (a) each outcome has probability between 0 and 1 and (b) the sum of the probabilities of all the outcomes equals 1.0.

fairfaxemploy

Table 5.2 Probability model

Employment type	Probability
Private company	0.597
Federal government	0.141
Self-employed	0.097
Private nonprofit	0.092
Local government	0.059
State government	0.012
Other	0.002

Now You Can Do
Exercises 67 and 68.

EXAMPLE 5.10

RANDOM DRAWS USING A PROBABILITY MODEL

Suppose we consider the probabilities in Table 5.2 as population values. Use technology to simulate random draws using the probability model in Table 5.2.

Solution
Using the Step-by-Step Technology Guide on the next page, we drew samples of sizes 10, 100, 1000, and 10,000 from the probability model in Table 5.2. The results are shown in Table 5.3.

Table 5.3 Relative frequencies from random draws of different sizes

Employment type	Rel freq $n = 10$	Rel freq $n = 100$	Rel freq $n = 1000$	Rel freq $n = 10,000$
Private company	0.60	0.62	0.566	0.596
Federal government	0.20	0.15	0.15	0.143
Self-employed	0.10	0.11	0.109	0.991
Private nonprofit	0.10	0.07	0.106	0.914
Local government	0.00	0.04	0.055	0.056
State government	0.00	0.01	0.012	0.012
Other	0.00	0.00	0.002	0.002

Note that each relative frequency tends to approach its respective probability as the sample sizes grow larger.

Subjective Method

There are cases where the outcomes are not equally likely (so the classical method does not apply) and there has been no previous research (so the relative frequency approach does not apply). For example, what is the probability that the Dow Jones Industrial Average will decrease today? In cases like this, there is no absolutely correct probability. Reasonable people can disagree reasonably over these probabilities. The idea is to consider all available information, tempered by our experience and intuition, and then assign a probability value that expresses our estimate of the likelihood that the outcome will occur. For example, we might say, "The Chairman of the Federal Reserve warned against inflation in a major speech yesterday, so we expect that the probability that the Dow Jones Industrial Average will go down today is about 90%." Finally, it should be noted that the subjective method should be used when the event is not (even theoretically) repeatable.

**Now You Can Do
Exercise 72.**

> **Subjective probability** refers to the assignment of a probability value to an outcome based on personal judgment.

STEP-BY-STEP TECHNOLOGY GUIDE: Probability Simulations Using Technology

TI-83/84

Simulating 100 Die Rolls
Step 1 Set the random number seed as follows. (The random number seed is a number that the calculator uses to generate random numbers.) Enter any number on the home screen. Press **STO→**, then **MATH**, highlight **PRB**, select **1: rand**, and press **ENTER**. On the home screen press **ENTER**.
Step 2 Press **MATH**, highlight **PRB**, select **5: randInt(**, and press **ENTER**.
Step 3 Enter **1**, comma, **6**, comma, **100**, close parenthesis (Figure 5.6).
Step 4 Store the data in list **L1** as follows. Press **STO→**, then **2nd**, then **1**, then press **ENTER**.
Step 5 To examine the die rolls, press **STAT**, select **1: EDIT**, and press **ENTER** (Figure 5.7).

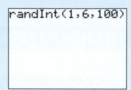

FIGURE 5.6   **FIGURE 5.7**

Simulating Coin Flips
You can simulate coin flips instead of die rolls by coding "heads" as 1 and "tails" as 0. Use the instructions for simulating 100 die rolls with the following changes: Enter **0**, comma, **1**, comma, **100**, close parenthesis, so that the home screen shows **randInt(0, 1, 100)**.

EXCEL

Simulating 100 Die Rolls
Step 1 Select cell **A1**. Click the **Insert Function** icon f_x.
Step 2 For **Search for a Function**, type **randbetween** and click **OK**.
Step 3 For **Bottom**, enter **1**. For **Top**, enter **6** (Figure 5.8). Click **OK**. Cell **A1** now contains a simulated random die roll.
Step 4 Select cell **A1**, copy it, and paste the contents into cells **A2** through **A100**.

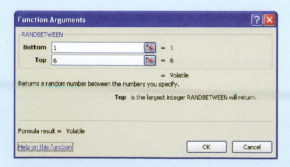

FIGURE 5.8 Random die rolls in Excel.

Simulating the Sum of Two Dice
Step 1 Generate 100 die rolls in column A and another 100 die rolls in column B.
Step 2 Select cell **C1**. Enter **= (A1+B1)**, and press **ENTER**.

Step 3 Select cell **C1**, copy it, and paste the contents into cells **C2** through **C100**. Column C then represents 100 randomly generated sums of two dice.

Simulating Random Draws from a Probability Table
We illustrate using Example 5.10 (page 203). Excel and Minitab both require that the categories in the probability model be coded as numeric. We therefore code "Private company" as 1, "Federal government" as 2, and so on.
Step 1 Type the model categories (for example, "Employment type") in column A, their numeric codes in column B, and the respective probabilities in column C.
Step 2 Click **Data > Data Analysis > Random Number Generation**, then **OK**.
Step 3 For **Number of Variables**, enter **1**.
Step 4 For **Number of Random Numbers**, enter the desired sample size.
Step 5 For **Distribution**, select **Discrete**.
Step 6 For **Value & Prob. Input Range**, click and drag to select the coded categories and their probabilities, for example, **B1:C7**.

Repeat Steps 1–6 for increasing sample sizes.

Simulating Coin Flips Using Technology
You can simulate coin flips instead of die rolls by coding "heads" as 1 and "tails" as 0. Use the die roll instructions with the following changes: For **Bottom**, enter **0**. For **Top**, enter **1**.

MINITAB

Simulating 100 Die Rolls

Step 1   Click on **Calc > Random Data > Integer**.

Step 2   For **Generate ___ rows of data**, enter **100**.

Step 3   For **Store in column(s)**, select **C1**.

Step 4   For **Minimum value**, enter **1**. For **Maximum value**, enter **6**.

Step 5   Click **OK**.

Simulating the Sum of Two Dice

Step 1   Generate 100 die rolls in **C1** and another 100 die rolls in **C2**.

Step 2   Click **Calc > Calculator**. For **Store result in variable**, enter **C3**. For **Expression**, enter **C1 + C2**. Click **OK**. Column **C3** then represents 100 randomly generated sums of two dice.

Simulating Random Draws from a Probability Table

Step 1   Type the model categories in **C1**, their numeric codes in **C2**, and the respective probabilities in **C3** (Figure 5.9).

Step 2   Click on **Calc > Random Data > Discrete**.

Step 3   For **Generate ___ rows of data**, enter the desired sample size.

Step 4   For **Store in column(s)**, select the next available column, such as **C4**.

Step 5   For **Values in**, enter the column with the numerically coded categories, such as **C2**.

	C1-T	C2	C3	C4	C5	C6	C7
	Type	Type-N	Prob	10	100	1000	10000
1	Private Company	1	0.597	1	3	1	1
2	Federal Government	2	0.141	1	5	1	4
3	Self-Employed	3	0.097	1	1	1	2
4	Private Nonprofit	4	0.092	1	1	4	4
5	Local Government	5	0.059	1	1	1	1
6	State Government	6	0.012	3	1	2	3
7	Other	7	0.002	4	1	1	2
8				2	1	1	1
9				2	1	3	1
10				1	4	1	6
11					1	1	5
12					1	3	1
13					4	1	3

FIGURE 5.9 *Random draws in Minitab.*

Step 6   For **Probabilities in**, enter the column with the probabilities, such as **C3**.

Step 7   Click **OK**.

Repeat Steps 1–7 for increasing sample sizes, as shown in Figure 5.9.

Simulating Coin Flips

You can simulate coin flips instead of die rolls by coding "heads" as 1 and "tails" as 0. Use the die roll instructions with the following changes: For **Minimum value**, enter **0**. For **Maximum value**, enter **1**.

SECTION 5.1	**Summary**

1. Section 5.1 introduces the building blocks of probability, including the concepts of probability, outcome, experiment, and sample space. Probabilities always take values between 0 and 1, where 0 means that the outcome cannot occur and 1 means that the outcome is certain.

2. The classical method of assigning probability is used if all outcomes are equally likely. The classical method states that the probability of an event *A* equals the number of outcomes in *A* divided by the number of outcomes in the sample space.

3. The Law of Large Numbers states that, as an experiment is repeated many times, the relative frequency (proportion) of a particular outcome tends to approach the probability of the outcome. The relative frequency method of assigning probability uses prior knowledge about the relative frequency of an outcome. The subjective method of assigning probability is used when the other methods are not applicable.

SECTION 5.1	**Exercises**

Clarifying the Concepts

1. Describe in your own words how chance and uncertainty affect you in your life. List some synonyms that we use in everyday life for the word *probability*.

2. Why do you think we use numerical values for probability rather than only qualitative terms such as "likely" or "impossible"?

3. Give three examples from your own life of experiments, as the term is used in this chapter.

   a. For each experiment, what are some of the outcomes?

   b. Write out the sample space of one of these experiments.

   c. Describe how the Law of Total Probability applies to the sample.

4. List the three methods for assigning probabilities.

5. What assumption do we need to make to use the classical method?

6. When can we use the relative frequency method?

7. If we can't use either the classical method or the relative frequency method, explain how we go about using the subjective method.

8. The experiment is to toss 10 fair coins 25 times each. Which methods can we use to assign probabilities?

9. How would you find the probability that a randomly chosen student at your college likes hip-hop music? What method would you use?

10. Describe the meaning of the following probabilities.
   a. Near 0   **b.** 0
   c. Near 1   **d.** 1

Practicing the Techniques

Determine whether each table in Exercises 11–16 is a probability model. If not, clearly explain why it is not a probability model.

11. Customers at a clothing store at the mall

Gender	Probability
Females	1.5
Males	0.2

12. Singers in the church choir

Voice	Probability
Soprano	0.25
Alto	0.25
Tenor	−0.25
Bass	0.50

13. Voters at a town meeting

Party	Probability
Democrat	0.3
Republican	0.25
Independent	0.25
Green	0.1
Libertarian	0.1
Other	0.1

14. Majors of students taking introductory statistics

Major	Probability
Business	0.75
Nursing	0.25
Social sciences	0.20
Science	0.20
Math	0.10

15. Students taking undergraduate introductory statistics

Class	Probability
Freshmen	0.15
Sophomores	0.25
Juniors	0.40
Seniors	0.20

16. Reasons why Hurricane Katrina survivors did not evacuate

Reason	Probability
I did not have a car or a way to leave	0.36
I thought the storm and its aftermath would not be as bad as they were	0.29
I just didn't want to leave	0.10
I had to care for someone who was physically unable to leave	0.07
All other reasons	0.18

For Exercises 17–20, the experiment is to draw a card at random from a shuffled deck of 52 cards. Find the following probabilities.

17. Drawing a king

18. Drawing a heart

19. Drawing the king of hearts

20. Drawing a black card

For Exercises 21–26, the experiment is to roll a fair die once. Find the following probabilities.

21. Observing a 3

22. Observing an even number

23. Observing a number greater than 3

24. Observing a number less than 3

25. Observing a 3 or a 5

26. Observing a 3 and a 5

For Exercises 27 and 28, refer to Exercises 21–26.

27. For each of Exercises 21–26, was the probability you found for an event or an outcome?

28. Explain in your own words why the probability of observing a 3 cannot be more than the probability of observing a 3 or a 5.

For Exercises 29–34, consider the experiment of tossing a fair die two times, with the outcomes being observing either an even number or an odd number.

29. Construct a tree diagram for the experiment.

30. Construct the sample space for the experiment.

31. Find the probability of observing zero even numbers.

32. Find the probability of observing one even number and one odd number.

33. Find the probability of observing two even numbers.

34. Use your results from Exercises 31–33 to construct the probability model for the number of even numbers observed.

For Exercises 35–38, let the experiment be tossing a fair die two times, with the outcomes being observing either a number less than 4 or a number greater than or equal to 4.

35. Construct a tree diagram for the experiment.

36. Construct the sample space for the experiment.

37. What is the probability of observing both outcomes being less than 4?

38. What is the probability of observing both outcomes being 4 or greater?

For Exercises 39–48, consider the experiment of tossing a fair coin three times, and observing either heads or tails.

39. Construct a tree diagram for the experiment.

40. Construct the sample space for the experiment.

41. How does the tree diagram help to construct the sample space?

42. How do we find each outcome using the tree diagram?

43. Find the probability of zero heads.

44. What is the probability of exactly one head.

45. Calculate the probability of exactly two heads.

46. Find the probability of exactly three heads.

47. Use your results from Exercises 43–46 to construct a probability model for the number of heads observed.

48. For Exercises 43–46, which method of assigning probability are you using?

For Exercises 49–56, consider the experiment of tossing two fair dice, and observing the sum of the two dice. (*Hint:* Use the sample space in Figure 5.3 on page 200.)

49. What is the probability that the sum of the dice equals 5?

50. Find the probability that the dark green die equals 5.

51. Calculate the probability that the sum of the dice equals 12.

52. Find the probability that the light green die equals 6.

53. What is the probability that the sum of the dice equals 1?

54. Construct the probability model for the sum of the dice.

55. Use the probability model to find which event has the greatest probability.

56. Which events have the lowest probability?

For Exercises 57–62, suppose that, in a sample of 100 students who drink hot caffeinated beverages, 40 preferred regular coffee, 25 preferred latte, 20 preferred cappuccino, and 15 preferred tea. Find the probability that a randomly selected student prefers the following.

57. Regular coffee

58. Latte

59. Cappuccino

60. Tea

61. For Exercises 57–60, which method of assigning probability are you using?

62. Construct the probability model for hot caffeinated beverages.

For Exercises 63–66, suppose that, in a sample of 200 college students, 100 live on campus, 60 live with family off campus, and 40 live in an apartment off campus. Find the probability that a randomly selected student lives in the following places.

63. On campus

64. With family off campus

65. In an apartment off campus

66. Construct the probability model for where these students live.

67. Use the following frequency table to estimate the probabilities for each color and construct the probability model. A sample of 100 students were asked to name their favorite color.

Favorite color	Frequency
Red	30
Blue	25
Green	20
Black	10
Violet	10
Yellow	5

68. Use the following frequency table to estimate the probabilities for each season and construct the probability model. A sample of 200 students were asked to name their favorite season.

Favorite season	Frequency
Summer	80
Spring	60
Autumn	40
Winter	20

Applying the Concepts

69. Picnic Lunch. Picnickers at the Fourth of July Fair have the following preferences for grilled lunch: cheeseburger 50%, hot dog 25%, veggieburger 25%. Consider the experiment of two picnickers chosen at random choosing their preferred lunch.

    a. Construct the tree diagram for the experiment.

    b. What is the sample space?

70. Video Games. The following percentages of students at a local high school express preference for the following game consoles: PlayStation 3, 40%; Xbox 360, 35%; Wii, 25%. Consider the experiment of choosing three students at random.

   a. Construct the tree diagram for the experiment.
   b. What is the sample space?

71. Rainy Days. Students at the local middle school have been keeping track of the number of days it has rained. In the past 100 days, it rained on 33 days.

   a. What is the probability that it rains on a randomly chosen day?
   b. What is the probability that it doesn't rain on a randomly chosen day?
   c. Which method of assigning probability did you use?

72. Basketball. Your college's basketball team is playing a game next week.

   a. What is the probability that the team will win the game?
   b. Which method did you use?

73. Brisbane Babies. The table shows the births of babies at a Brisbane, Australia, hospital on a particular day.

Girl	Girl	Boy	Boy	Boy	Girl	Girl	Boy	Boy
Boy	Boy	Boy	Girl	Girl	Boy	Girl	Girl	Boy
Boy	Boy	Boy	Girl	Girl	Girl	Girl	Boy	Boy
Boy	Girl	Boy	Girl	Boy	Boy	Boy	Boy	Boy
Girl	Boy	Boy	Boy	Boy	Girl	Girl	Girl	

   a. Construct a relative frequency distribution of the numbers of girls and boys born.
   b. Use the relative frequencies to construct a probability model.
   c. Confirm that your probability model follows the Rules of Probability.

74. Draw an Ace. If you draw the ace of spades from a deck of cards, you win $100.

   a. What is the probability of winning this game?
   b. What would be a fair price for playing this game? (*Hint:* A fair price might be determined by balancing out the winnings and the price in the long run.)

75. A Bazaar Game. Lenny has gone to the church bazaar with his family. In one of the games at the bazaar, if Lenny rolls two dice and gets a sum of at least 9, he wins $5; otherwise, he wins nothing.

   a. Find the probability of winning $5.
   b. Find the probability of winning nothing.
   c. What would you suggest would be a fair (break-even) price for playing this game?

76. Fairfax County Income. The following table contains a probability model for the distribution of income in Fairfax County, Virginia.

   🌎 fairfaxincome
   a. Use technology to draw random samples of sizes 10, 100, 1000, and 10,000 from this probability model.

   b. What can you conclude about the relative frequencies as the sample size increases?

Annual income	Probability
Under $25,000	0.083
$25,000 to $49,999	0.166
$50,000 to $74,999	0.169
$75,000 to $99,999	0.160
$100,000 to $149,999	0.200
$150,000 or more	0.222

Bringing It All Together

Use the following information for Exercises 77–82. Consider the experiment where a fair die is rolled twice. Define the following events for each roll: low = {1, 2}, medium = {3, 4}, high = {5, 6}, odd = {1, 3, 5}, even = {2, 4, 6}.

77. Construct a tree diagram for this experiment. Make sure you use the outcomes and not the events.

78. Use the tree diagram to construct the sample space. Which sample space discussed in Section 5.1 is the sample space for this experiment similar to? Explain why this is so.

79. The sample space is the collection of all possible outcomes of an experiment. Explain why the sample space is not defined as the collection of all possible events.

80. Find the probability of observing a 1, followed by another 1. What method of assigning probability are you using? Why?

81. Find the probability of observing two high die rolls. What method of assigning probability are you using? Why?

82. Find the following probabilities.
   a. Two high die results
   b. Exactly one medium die result
   c. No low die results
   d. At least one high die result
   e. At most one medium die result

🅰 Use the *Law of Large Numbers for Proportions* applet for Exercises 83 and 84.

83. Set the probability of heads to 0.5 and the number of tosses to 40. Click **Toss.**
   a. Record the proportion of heads observed.
   b. Without pressing **Reset,** continue to click **Toss** until the total number of tosses is 120. Again record the proportion of heads.
   c. Without pressing **Reset,** continue to click **Toss** until the total number of tosses is 240. Again record the proportion of heads.
   d. Without pressing **Reset,** continue to click **Toss** until the total number of tosses is 480. Again record the proportion of heads.

84. The proportions you recorded in Exercise 83 are relative frequencies of heads. What can you conclude about the relative frequencies as the sample size increases?

5.2

COMBINING EVENTS

OBJECTIVES By the end of this section, I will be able to . . .

1 Understand how to combine events using complement, union, and intersection.

2 Apply the Addition Rule to events in general and to mutually exclusive events in particular.

1 COMPLEMENT, UNION, AND INTERSECTION

In Example 5.5, if your roommate rolled a 4, then your roommate was to do your laundry for the rest of the semester. Your roommate is keenly interested in *not* rolling a 4. If A is an event, then the collection of outcomes not in event A is called the **complement of A,** denoted A^C. The term *complement* comes from the word "to complete," meaning that any event and its complement together make up the complete sample space.

EXAMPLE 5.11

FINDING THE PROBABILITY OF THE COMPLEMENT OF AN EVENT

If A is the event "observing a sum of 4 when the two fair dice are rolled," then your roommate is interested in the probability of A^C, the event that a 4 is not rolled. Find the probability that your roommate does not roll a 4.

Solution

Which outcomes belong to A^C? By the definition, A^C is all the outcomes in the sample space that do not belong in A. There are the following outcomes in A: $\{(3,1), (2,2), (1,3)\}$.

Figure 5.10 shows all the outcomes except the outcomes from A in the two-dice sample space. There are 33 outcomes in A^C and 36 outcomes in the sample space. The classical probability method then gives the probability of not rolling a 4 to be

$$P(A^C) = \frac{N(A^C)}{N(S)} = \frac{33}{36} = \frac{11}{12}$$

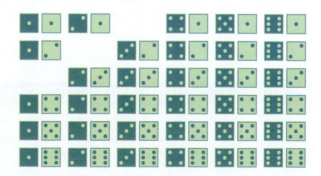

FIGURE 5.10
Outcomes in A^C.

Now You Can Do
Exercises 9–11.

The probability is high that, on this roll at least, your roommate will not land on Boardwalk.

For event A in Example 5.11, note that

$$P(A) + P(A^C) = \frac{1}{12} + \frac{11}{12} = 1$$

Is this a coincidence, or does the sum of the probabilities of an event and its complement always add to 1? Recall the Law of Total Probability (Section 5.1), which states that the sum of all the outcome probabilities in the sample space must be equal to 1. Since any event A and its complement A^C together make up the entire sample space, then it always happens that $P(A) + P(A^C) = 1$.

> **Probabilities for Complements**
>
> For any event A and its complement A^C, $P(A) - P(A^C) = 1$. Applying a touch of algebra gives the following:
> - $P(A) = 1 - P(A^C)$
> - $P(A^C) = 1 - P(A)$

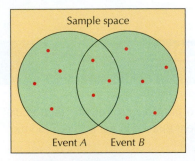

FIGURE 5.11 Union of events A and B.

Sometimes we need to find the probability of a combination of events. For example, consider the casino game of craps where you roll two dice. One way of winning is by rolling the sum 7 or 11. We can find the probability of the following two events: the sum is 7 or the sum is 11. First, we need some tools for finding the probability of a combination of events.

> **Union and Intersection of Events**
>
> The **union** of two events A and B is the event representing all the outcomes that belong to A or B or both. The union of A and B is denoted as $A \cup B$ and is associated with "or."
>
>   The **intersection** of two events A and B is the event representing all the outcomes that belong to both A and B. The intersection of A and B is denoted as $A \cap B$ and is associated with "and."

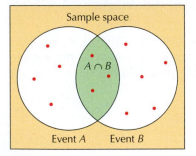

FIGURE 5.12 Intersection of events A and B.

If you are asked to find the probability of "A or B," you should find the probability of $A \cup B$. Figure 5.11 shows the union of two events, with the red dots indicating the outcomes. Note from Figure 5.11 that the union of the events A and B refers to all outcomes in A or B or both. Figure 5.12 shows that the intersection of the two events is the part where A and B overlap. Both union and intersection are commutative. That is, $A \cup B = B \cup A$ and $A \cap B = B \cap A$.

EXAMPLE 5.12 | **UNION AND INTERSECTION**

Let our experiment be to draw a single card at random from a deck of cards. Define the following events:

  A: The card drawn is an ace.
  H: The card drawn is a heart.

a. Find $A \cup H$.
b. Find $A \cap H$.

Solution

a. The union of A and H is the event containing all the outcomes that are either aces or hearts or both (the ace of hearts). That is, the event $A \cup H$ consists of the set of outcomes (the cards) shown in Figure 5.13.

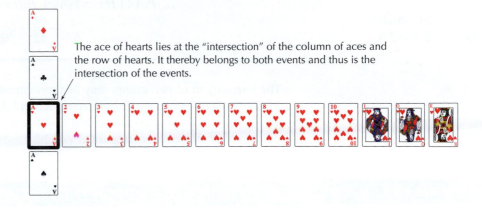

The ace of hearts lies at the "intersection" of the column of aces and the row of hearts. It thereby belongs to both events and thus is the intersection of the events.

FIGURE 5.13 The ace of hearts is the intersection of the events "ace" and "hearts."

Now You Can Do
Exercises 13–18.

b. The intersection of A and H is the event containing the outcomes that are common to both A and H. There is only one such outcome: the ace of hearts (see Figure 5.13).

2 ADDITION RULE

We are often interested in finding the probability that either one event *or* another event may occur. The formula for finding these kinds of probabilities is called the **Addition Rule.**

> **Addition Rule**
>
> $$P(A \text{ or } B) = P(A \cup B) = P(A) + P(B) - P(A \cap B)$$

What Does the Addition Rule Mean?

We can use Figure 5.13 to understand the Addition Rule. We are trying to find the probability of all the outcomes in A or B or both. The first part of the formula says to add the probabilities of the outcomes in A to those of the outcomes in B. But what about the overlap between A and B, outcomes that belong to both events? To avoid counting the outcomes in the overlap (intersection) twice, we have to subtract the probability of the intersection, $P(A \cap B)$.

EXAMPLE 5.13

ADDITION RULE APPLIED TO A DECK OF CARDS

Suppose you pay \$1 to play the following game. You choose one card at random from a deck of 52 cards, and you will win \$3 if the card is either an ace *or* a heart. Find the probability of winning this game.

Solution

Using the same events defined in Example 5.12, we find $P(A \text{ or } H) = P(A \cup H)$. By the Addition Rule, we know that

$$P(A \cup H) = P(A) + P(H) - P(A \cap H)$$

There are 4 aces in a deck of 52 cards, so by the classical method (equally likely outcomes), $P(A) = 4/52$. There are 13 hearts in a deck of 52 cards, so $P(H) = 13/52$. From Example 5.12, we know that $A \cap H$ represents the ace of hearts. Since each card is equally likely to be drawn, then $P(\text{ace of hearts}) = P(A \cap H) = 1/52$. Thus,

Now You Can Do Exercises 19–24.

$$P(A \cup H) = P(A) + P(H) - P(A \cap H)$$
$$= \frac{4}{52} + \frac{13}{52} - \frac{1}{52} = \frac{16}{52} = \frac{4}{13}$$

The intersection of two events may be represented by the intersection of a row and a column in a two-way table. Recall from Section 2.1 (pages 38–40) that a *two-way table* (also known as a *crosstabulation* or a *contingency table*) is a tabular summary of the relationship between two categorical variables.

EXAMPLE 5.14

ADDITION RULE APPLIED TO A TWO-WAY TABLE

A study of online dating behavior found that users of a particular online dating service self-reported their physical appearance according to the counts given in Table 5.4.[3]

Students may wish to refresh their knowledge of crosstabulation form (pages 38–40).

Table 5.4 Gender and self-reported physical appearance

	Physical Appearance				
Gender	**Very attractive**	**Attractive**	**Average**	**Prefer not to answer**	**Total**
Female	3113	16,181	6,093	3478	28,865
Male	1415	12,454	7,274	2809	23,952
Total	4528	28,635	13,367	6287	52,817

Using this information, find the probability that a randomly chosen online dater has the following characteristics.
a. Is female
b. Self-reported as attractive
c. Is a female who self-reported as attractive
d. Is a female *or* self-reported as attractive

Solution

a. There are a total of $N(S) = 52,817$ online daters in the entire data set. Of these, 28,865 are female, denoted as event F. Therefore,

$$P(F) = P(\text{Female}) = \frac{N(\text{Female})}{N(S)} = \frac{N(F)}{N(S)} = \frac{28,865}{52,817} \approx 0.5465$$

b. There are 28,635 people who self-reported their physical appearance as attractive, denoted as event A. Therefore,

$$P(A) = P(\text{Self-reported attractive}) = \frac{N(\text{Self-reported attractive})}{N(S)} = \frac{N(A)}{N(S)}$$
$$= \frac{28,635}{52,817} \approx 0.5422$$

c. The online daters who are both female and self-reported as attractive are shown in the highlighted cell in Table 5.4. This cell is located at the *intersection* of the row of females and the column of people who self-reported as attractive. Therefore, this cell reports the frequency of people who belong to both events. Thus,

$$P(F \text{ and } A) = P(F \cap A) = P(\text{Female and self-reported attractive}) = \frac{N(F \cap A)}{N(S)}$$

$$= \frac{16{,}181}{52{,}817} \approx 0.3064$$

Now You Can Do
Exercises 25–32.

d.   Here we seek $P(F \text{ or } A) = P(F \cup A)$. By the Addition Rule,

$$P(F \cup A) = P(F) + P(A) - P(F \cap A) = 0.5465 + 0.5422 - 0.3064 = 0.7823$$

Mutually Exclusive Events

When drawing a card at random from a deck of 52 cards, the events "a heart is drawn" and "a diamond is drawn" have no outcomes in common. That is, no card is both a heart and a diamond. We say that these two events are **mutually exclusive.**

> Two events are said to be **mutually exclusive,** or **disjoint,** if they have no outcomes in common.

Note that any event and its complement are always mutually exclusive. Other examples of mutually exclusive events are given in Table 5.5.

Table 5.5 Examples of mutually exclusive events

Experiment	Mutually exclusive events
Toss fair coin	Observe heads; observe tails
Draw a single card from a deck of 52 cards	Card is red; card is a spade
Select a student at random	Student is female; student is male
Choose a digit at random	Digit is even; digit is odd

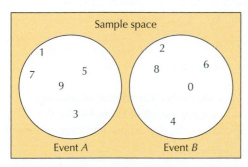

FIGURE 5.14 Even and odd digits are mutually exclusive.

Figure 5.14 shows how mutually exclusive events are represented graphically. It shows the events

$$A = \{1, 3, 5, 7, 9\} \qquad \text{and} \qquad B = \{0, 2, 4, 6, 8\}$$

Note that there is no overlap between the two events. When two events are mutually exclusive, they share no outcomes, and therefore the intersection of mutually exclusive events is empty. Since the intersection $(A \cap B)$ is empty, then for mutually exclusive events, $P(A \cap B) = 0$. Therefore, we can formulate a special case of the **Addition Rule for Mutually Exclusive Events** A and B:

$$P(A \cup B) = P(A) + P(B) - P(A \cap B) = P(A) + P(B) - 0 = P(A) + P(B)$$

> **Addition Rule for Mutually Exclusive Events**
> If A and B are mutually exclusive events, $P(A \cup B) = P(A) + P(B)$.

| EXAMPLE 5.15 | ADDITION RULE FOR MUTUALLY EXCLUSIVE EVENTS |

Using Table 5.4 from Example 5.14, find the probability that a randomly chosen online dater self-reported as either attractive or very attractive.

Solution

From Table 5.4, there are 28,635 online daters who self-reported as attractive and 4528 who self-reported as very attractive, yielding the following probabilities:

$$P(A) = P(\text{Self-reported attractive}) = \frac{N(A)}{N(S)} = \frac{28,635}{52,817} \approx 0.5422$$

$$P(V) = P(\text{Self-reported very attractive}) = \frac{N(V)}{N(S)} = \frac{4528}{52,817} \approx 0.08573$$

Since no online daters self-reported as both attractive and very attractive, the two groups are mutually exclusive. Thus, by the Addition Rule for Mutually Exclusive Events,

**Now You Can Do
Exercises 33 and 34.**

$$P(A \cup V) = P(A) + P(V) = 0.5422 + 0.08573 = 0.62793$$

| SECTION 5.2 | Summary |

1. Combinations of events may be formed using the concepts of complement, union, and intersection.

2. The Addition Rule provides the probability of event A or event B to be the sum of their two probabilities minus the probability of their intersection. Mutually exclusive events have no outcomes in common.

| SECTION 5.2 | Exercises |

Clarifying the Concepts

1. Describe in your own words what it means for two events to be mutually exclusive.

2. Describe the intersection of two mutually exclusive events.

3. Describe the union of two mutually exclusive events.

4. Is it true that the union of two events always contains at least as many outcomes as the intersection of two events? Use Figures 5.11 and 5.12 to help you visualize this problem.

5. If we choose a student at random from your college or university, is it more likely that we choose a male or a male football player? Why?

6. What is your personal estimate of the probability that it will rain on any given day? How about the probability that it won't rain? Why do these numbers have to add up to 1 (or 100%)?

Practicing the Techniques

For Exercises 7–12, consider the experiment of rolling a fair die once. Find the indicated probabilities.

7. Observing a number that is not 4

8. Observing some other number than 5

9. The complement of the event E, where E: {2, 4, 6}

10. L^C, where L: {1, 2}

11. E^C, where E: {2, 4, 6}

12. Not rolling an odd number

For Exercises 13–18, consider the experiment of drawing a single card at random from a deck of cards. Define the following events. Find the indicated unions and intersections.

　K: The card is a king.

　R: The card is a red suit.

　H: The card is a heart.

13. $K \cup R$　　　**15.** $R \cup H$　　　**17.** $K \cap H$

14. $K \cup H$　　　**16.** $K \cap R$　　　**18.** $R \cap H$

For Exercises 19–24, consider the experiment of drawing a single card at random from a deck of cards. Define the following events. Find the indicated probabilities.

K: The card is a king.
R: The card is a red suit.
H: The card is a heart.

19. $P(K \cup R)$ **21.** $P(R \cup H)$ **23.** $P(K \cap H)$

20. $P(K \cup H)$ **22.** $P(K \cap R)$ **24.** $P(R \cap H)$

For Exercises 25–32, refer to Table 5.4 in Example 5.14 on page 212. Find the probability that a randomly chosen dater has the following characteristics.

25. Is male

26. Self-reported as average

27. Is a male who self-reported as average

28. Is a male or self-reported as average

29. Self-reported as very attractive

30. Is a female who self-reported as very attractive

31. Is a male who self-reported as very attractive

32. Self-reported as prefer not to answer

For Exercises 33–44, consider the experiment of rolling a single die once. Define the following events: *L*: {1, 2, 3}, *H*: {4, 5, 6}, *E*: {2, 4, 6}, *O*: {1, 3, 5}. Find the following probabilities.

33. $P(L \cup H)$ **37.** $P(H \cup E)$ **41.** $P(L \cap E)$

34. $P(E$ or $O)$ **38.** $P(H$ or $O)$ **42.** $P(L$ and $O)$

35. $P(L$ or $E)$ **39.** $P(L$ and $H)$ **43.** $P(H$ and $E)$

36. $P(L \cup O)$ **40.** $P(E \cap O)$ **44.** $P(H \cap O)$

For Exercises 45–50, consider the experiment of rolling a fair die twice. Find the indicated probabilities.

45. Exactly one of the dice is a 4

46. Neither die is a 4

47. Sum of the two dice equals 3

48. Sum of the two dice equals 3 and one of the dice is a 4

49. Sum of the two dice equals 3 or one of the dice is a 4

50. Sum of the two dice equals 3 or neither of the dice is a 4

For Exercises 51–56, consider the experiment of drawing a card at random from a shuffled deck of 52 cards. Find the indicated probabilities.

51. Drawing a king and a black card

52. Drawing a king or a black card

53. Drawing a card that is neither a king nor a black card

54. Drawing a heart or a spade

55. Drawing a heart and a spade

56. Drawing a card that is not the king of hearts

For Exercises 57–62, consider the experiment of drawing a card at random from a shuffled deck of 52 cards. Find the indicated probabilities.

57. Drawing a face card (king, queen, or jack)

58. Drawing a card that is not red

59. Drawing a card that is not a face card

60. Drawing a face card that is not a diamond

61. Drawing a face card or a diamond

62. Drawing a face card and a diamond

For Exercises 63–66, consider the experiment of tossing a fair coin three times. Find the indicated probabilities. (*Hint:* Use a tree diagram similar to the one in Figure 5.2 in Section 5.1 [page 199] but adding one more stage.)

63. Observing 3 heads

64. Not observing 3 heads

65. Observing 2 tails

66. Not observing 2 tails

For Exercises 67–72, imagine that your sister is going to have triplets. Assume that the probability of a baby boy or a baby girl is equally likely. (In fact, it is not quite.)

67. Construct the sample space.

68. Find the probability of 1 girl and 2 boys.

69. Find the probability of 1 boy and 2 girls.

70. Find the probability of 2 of one gender and 1 of the other gender.

71. Find the probability of 1 girl or 1 boy.

72. Find the probability of getting 3 girls.

Applying the Concepts

73. Game of Craps. You win the casino game of craps if you roll a 7 or 11. Find the probability of rolling a sum of 7 or 11 when two fair dice are rolled.

Trout Fishing. Use the following information for Exercises 74 and 75. Of the 20 fish Brent has caught at his favorite fishing spot this season, 5 have been trout and 7 have been bass.

74. Find the following probabilities.
  a. Catching a trout
  b. Catching a bass

75. Find the following probabilities.
  a. Catching a trout or a bass
  b. Catching a fish that is not a trout
  c. Catching a fish that is neither a trout nor a bass

76. Traffic Lights. Let *A* be the event that you encounter a green light at your next traffic light.
  a. What outcomes make up A^C?
  b. What is the probability of *A*? Which method did you use?
  c. What is the probability of A^C?

77. High School Students. In a local high school of 500 students, there are 200 females, 100 sophomores, and 50 female sophomores.
   a. If we choose a student at random, what is the probability that we choose a female or a sophomore?
   b. Find the probability that a randomly chosen student is a male or a sophomore.
   c. Find the probability that a randomly chosen student is a female or is not a sophomore.

78. Halloween Candy. In a sample of 100 children, 70 like chocolate bars, 60 like peanut butter cups, and 50 like both.
   a. If we choose one child at random, find the probability that the child likes either chocolate bars or peanut butter cups.
   b. In (**a**), suppose you forgot to subtract the probability of the intersection. How would you know that your answer is wrong?

79. Pick a Card. If we draw a single card at random from a deck of 52 playing cards, find the probability that the card is
   a. a heart or a diamond.
   b. a red card or a jack.
   c. a club or a face card (king, queen, jack).
   d. a heart and a diamond.
   e. not a spade.

80. Online Dating Data. Refer to Table 5.4 (page 212). Find the probability that a randomly selected online dater has the following characteristics.
   a. Prefers not to describe physical appearance
   b. Is male and prefers not to describe physical appearance
   c. Is male or prefers not to describe physical appearance

81. Social Networking Apps. The Nielsen Apps Playbook was a survey taken in 2010 of 3692 males and females on the use of social networking apps (mobile software applications). The results are shown in the following table.

	Has downloaded a social networking app in the last 30 days	
	Yes	No
Male	884	841
Female	1220	746

Source: Kristen Purcell, Roger Entner, and Nichole Henderson, *The Rise of Apps Culture,* Internet and American Life Project, Pew Research Center, September 15, 2010.

Find the probability that a randomly chosen person has the following characteristics.
   a. Is female
   b. Has downloaded a social networking app in the last 30 days
   c. Is a female who has downloaded a social networking app in the last 30 days
   d. Is a female *or* has downloaded a social networking app in the last 30 days

Causes of Death. Refer to Table 5.6 for Exercises 82–84.

TABLE 5.6   Causes of death

Cause of death	Deaths
Heart disease	654,092
Cancer	550,270
All other causes	1,194,003
Total	2,398,365

Source: Centers for Disease Control and Prevention.

82. Find the following probabilities.
   a. The cause of death was heart disease.
   b. The cause of death was not heart disease.

83. Find the following probabilities.
   a. The cause of death was heart disease and cancer.
   b. The cause of death was heart disease or cancer.

84. Are the causes of death mutually exclusive?

Bringing It All Together

Don't Mess with Texas. Don't Mess with Texas (**dontmesswithtexas.org**) is a Texas statewide antilittering organization. Its 2005 report, *Visible Litter Study 2005*, identified paper, plastic, metal, and glass as the top four categories of litter by composition. The report also identified tobacco, household/personal, food, and beverages as the top four categories of litter by use. Assume that a sample of 12 items of litter had the following characteristics. Use Table 5.7 for Exercises 85–87.

TABLE 5.7   Litter composition and use

Litter item	Composition	Use
1	Paper	Tobacco
2	Plastic	Household/personal
3	Glass	Beverages
4	Paper	Tobacco
5	Metal	Household/personal
6	Plastic	Food
7	Glass	Beverages
8	Paper	Household/personal
9	Metal	Household/personal
10	Plastic	Beverages
11	Paper	Tobacco
12	Plastic	Food

85. A litter item is chosen at random.
  a. Find the probability that the composition of the item is paper.
  b. Find the probability that the composition of the item is not paper. Calculate this probability in two different ways.

86. A litter item is chosen at random.
  a. Find the probability that the use of the item is tobacco.

  b. Find the probability that the use of the item is not tobacco. Calculate this probability in two different ways.

87. A litter item is chosen at random.
  a. Find the probability that the composition of the item is paper *and* its use is tobacco.
  b. Find the probability that the composition of the item is paper *or* its use is tobacco.

5.3 CONDITIONAL PROBABILITY

OBJECTIVES By the end of this section, I will be able to . . .

1 Calculate conditional probabilities.

2 Explain independent and dependent events.

3 Solve problems using the Multiplication Rule and recognize the difference between sampling with replacement and sampling without replacement.

4 Approximate probabilities for dependent events.

1 INTRODUCTION TO CONDITIONAL PROBABILITY

As we progress through this book, you will notice a recurring theme: *the more information available, the better.* Very often, when we are investigating the probability of a certain event *A,* we learn that another event *B* has occurred. If events *A* and *B* are related, then the occurrence of event *B* often influences the probability that event *A* will occur.

EXAMPLE 5.16   **HAVING MORE INFORMATION OFTEN AFFECTS THE PROBABILITY OF AN EVENT**

In Section 5.1, we found that the probability of rolling a sum of 4 when tossing two dice is 3/36 ≈ 0.0833. But what if we were told that at least one of the dice shows a 1. How does this extra information affect the probability of rolling a 4?

Solution

Figure 5.15 shows the 11 outcomes from the two-dice sample space in which at least one die shows a 1. The extra information reduces the number of possible outcomes in the sample space from 36 to 11. We see that two of these outcomes have a sum equal to 4. Thus, the probability of observing a sum of 4, *given that* at least one of the dice shows a 1, is 2/11 ≈ 0.1818.

FIGURE 5.15
Using the extra knowledge changes the probability.

The extra information about a related event changed the probability of the event of interest. This type of probability is an example of what is called **conditional probability.**

For two related events *A* and *B*, the probability of *B* given *A* is called a **conditional probability** and denoted $P(B|A)$.

Thus, if we let *A* represent the event that one of the dice shows a 1, and let *B* represent the event that the sum of the two dice equals 4, then

$$P(B) = \frac{3}{36} \approx 0.0833 \qquad \text{but} \qquad P(B|A) = \frac{2}{11} \approx 0.1818$$

Figure 5.16 can help us visualize how conditional probability works. The idea is that, *once event A has occurred*, the only chance for event *B* to occur is in the overlap, the intersection $A \cap B$. Therefore, the conditional probability that *B* will occur, given that event *A* has already taken place, is found by taking the ratio $P(A \cap B)/P(A)$.

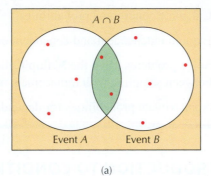
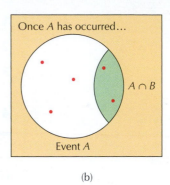

(a)   (b)

FIGURE 5.16
How conditional probability works.

Calculating Conditional Probability
The conditional probability that *B* will occur, given that event *A* has already taken place, equals

$$P(B|A) = \frac{P(A \cap B)}{P(A)} = \frac{N(A \cap B)}{N(A)}$$

EXAMPLE 5.17   CALCULATING CONDITIONAL PROBABILITY

Table 5.8 is adapted from a study on direct mail marketing. It contains the numbers of customers who either responded or did not respond to a direct mail marketing campaign, along with whether they had a credit card on file with the company. The two events are

    R: Responded to direct mail marketing campaign
    C: Has a credit card on file

Table 5.8 Credit card status and marketing response

Response	Credit Card on File?		Total
	No	**Yes**	
Did not respond	161	79	240
Did respond	17	31	48
Total	178	110	288

Source: Daniel Larose, *Data Mining Methods and Models* (Wiley Interscience, 2006).

a.   Find the probability that a randomly chosen customer responded to the marketing campaign.

b.   Find the conditional probability that a randomly selected customer responded, given that the customer has a credit card on file.

Solution

a. $P(R) = \dfrac{N(R)}{N(S)}$. There are $N(R) = 48$ customers who did respond, and there are $N(S) = 288$ customers in this experiment. Thus,

$$P(R) = \frac{N(R)}{N(S)} = \frac{48}{288} \approx 0.1667$$

b.   We will use $P(R|C) = N(R \cap C)/N(C)$ because in this example it is easier to work directly with the numbers of outcomes rather than the probabilities. Now, $R \cap C$ represents customers who did respond and had a credit card on file. From Table 5.8, there are $N(R \cap C) = 31$ such customers. Also, there are $N(C) = 110$ customers total who had a credit card on file. Therefore,

$$P(R|C) = \frac{N(R \cap C)}{N(C)} = \frac{31}{110} \approx 0.2818$$

Now You Can Do Exercises 9–24.

That is, the probability that a randomly chosen customer responded to the direct mail marketing campaign, given that the customer had a credit card on file, is 0.2818.

What Do These Numbers Mean?

Conditional Probability

Conditional probabilities can often be interpreted as percentages of some *subset* of a population. For example, the conditional probability that a customer responded, given that the customer has a credit card on file, may be interpreted as the percentage of customers with credit cards who responded.

Students sometimes confuse the meanings of $P(B \cap A)$ and $P(B|A)$. For $P(B|A)$, we *assume that the event A has occurred* and now need to find the probability of event B, given event A. On the other hand, for $P(B \cap A)$, we do not assume that event A has occurred and instead need to determine the probability that both events occurred.

2   INDEPENDENT EVENTS

Since having a credit card on file increased the probability of a customer responding from 0.1667 to 0.2818, we can therefore say that the probability of responding *depends* in part on whether the customer has a credit card on file. In other words, the events R and C are *dependent events*.

On the other hand, if the probability of responding had been unaffected by whether the customer had a credit card on file, then we would have said that R and C were **independent events.** That is, R and C would have been independent events had $P(R|C)$ equaled $P(R)$. In general, if the occurrence of an event does not affect the probability of a second event, then the two events are independent.

Events *A* and *B* are **independent** if

$$P(A \mid B) = P(A) \qquad \text{or if} \qquad P(B \mid A) = P(B)$$

Otherwise the events are said to be **dependent.**

Alternatively, in Step 1 you can find $P(A)$ and in Step 2 you can find $P(A \mid B)$. Then compare these two quantities for Step 3.

Strategy for Determining Whether Two Events Are Independent

1. Find $P(B)$.
2. Find $P(B \mid A)$.
3. Compare the two probabilities. If they are equal, then *A* and *B* are independent events. Otherwise, *A* and *B* are dependent events.

EXAMPLE 5.18

DETERMINING WHETHER TWO EVENTS ARE INDEPENDENT

Suppose our experiment is to toss two fair dice, so that our sample space is given in Figure 5.3 on page 200. Define the following events.

  X: Roll a sum equal to 7.
  Y: Roll a sum equal to 6.
  Z: Dark green die equals 1.

a. Determine whether events *X* and *Z* are independent.
b. Determine whether events *Y* and *Z* are independent.

Solution

For both **(a)** and **(b)** we use the strategy for determining whether two events are independent.

a. **STEP 1** Find P(Rolling a sum equal to 7) = $P(X)$. There are 36 outcomes in the sample space, 6 of which have a sum equal to 7. Thus, $P(X) = 6/36 = 1/6$.

 STEP 2 We need to find $P(X \mid Z)$, which is the probability that the sum equals 7, given that the dark green die equals 1. Figure 5.17 shows how the sample space is reduced when we know that *Z* has occurred (dark green die equals 1). There are 6 outcomes where the dark green die equals 1, of which 1 has a sum equal to 7. Thus, $P(X \mid Z) = 1/6$.

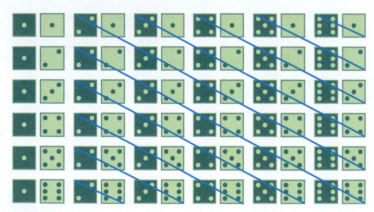

FIGURE 5.17 When we know the dark green die equals 1, the sample space is reduced.

 STEP 3 Since $P(X) = P(X \mid Z)$, we conclude that *X* and *Z* are independent events.

b. **STEP 1** Find P(Rolling a sum equal to 6) = $P(Y)$. There are 36 outcomes in the sample space, 5 of which have a sum equal to 6. Thus, $P(Y) = 5/36$.

 STEP 2 There are 6 outcomes where the dark green die equals 1, of which 1 has a sum equal to 6. Thus, $P(Y \mid Z) = 1/6$.

Now You Can Do
Exercises 29–34.

 STEP 3 From Step 1, $P(Y) = 5/36$. From Step 2, $P(Y \mid Z) = 1/6$. Since $P(Y) \neq P(Y \mid Z)$, we conclude that *X* and *Z* are dependent events.

Developing Your Statistical Sense

Don't Confuse Independent Events and Mutually Exclusive Events

It is important to stress the difference between independent events and mutually exclusive events. Mutually exclusive events have no outcomes in common. For two events to be independent means that the occurrence of one does not affect the probability of the other. The concepts are different.

EXAMPLE 5.19

GAMBLER'S FALLACY

Suppose we have tossed a fair coin ten times and have observed heads come up every time. Find the probability of tails on the next toss.

Solution

Since we have observed an unusual number of heads, we might think that the probability of tails on the next toss is increased. However, the short answer is "Not so." Successive tosses of a fair coin are independent because the coin has no memory of its previous tosses. Thus, what happened on the first ten tosses has no effect on the next toss. Probability theory tells us that, in the long run, the proportion of heads and tails will eventually even out if the coin is fair. Therefore, the probability of tails on the next toss is 0.5. This is an example of the Gambler's Fallacy.

3 MULTIPLICATION RULE

Just as the Addition Rule is used to find probabilities of unions of events, the **Multiplication Rule** is used to find probabilities of intersections of events. Recall the formula for the conditional probability of event B given event A:

$$P(B|A) = \frac{P(A \cap B)}{P(A)} \qquad \text{where } P(A) \neq 0$$

We solve for $P(A \cap B)$ by multiplying each side by $P(A)$:

$$P(A \cap B) = P(A) \, P(B|A)$$

Similarly, consider the conditional probability of event A given event B:

$$P(A|B) = \frac{P(A \cap B)}{P(B)} \qquad \text{where } P(B) \neq 0$$

Solving for $P(A \cap B)$ gives us a second equation for $P(A \cap B)$:

$$P(A \cap B) = P(B) \, P(A|B)$$

The two equations for $P(A \cap B)$ lead directly to the Multiplication Rule.

Multiplication Rule

$$P(A \cap B) = P(A) \, P(B|A) \qquad \text{or equivalently} \qquad P(A \cap B) = P(B) \, P(A|B)$$

Alamy

EXAMPLE 5.20

MULTIPLICATION RULE

According to the Pew Internet and American Life Project,[4] 35% of American adults have cell phones with apps, but only 68% of those who have apps on their cell phones actually use the apps. Define the following events:

    A: American adult has a cell phone with apps.
    U: American adult uses the apps on his or her cell phone.

a. Find $P(A)$.
b. Find $P(U|A)$, the probability that an American adult uses the apps, given that he or she has a cell phone with apps.
c. Use the multiplication rule to calculate $P(A \text{ and } U)$, the probability that an American adult has a cell phone with apps *and* uses the apps on his or her cell phone.

Solution

a. According to the study, 35% of American adults have a cell phone with apps. So $P(A) = 0.35$.
b. The research says that 68% of those who have apps actually use them, so $P(U|A) = 0.68$.
c. Using the Multiplication Rule, we have

$$P(A \text{ and } U) = P(A \cap U) = P(A)P(U|A) = 0.35(0.68) = 0.238$$

**Now You Can Do
Exercises 35–38.**

The probability that an American adult has a cell phone with apps *and* uses them is 0.238.

When events *A* and *B* are independent, $P(A|B) = P(A)$ or $P(B|A) = P(B)$. Using these identities, we can formulate a special case of the Multiplication Rule. Using $P(A|B) = P(A)$, we can write the Multiplication Rule as

$$P(A \cap B) = P(B)\,P(A|B) = P(B)\,P(A) = P(A)\,P(B)$$

Equivalently, the Multiplication Rule also states that $P(A \cap B) = P(A)\,P(B|A)$, but if *A* and *B* are independent, $P(B|A) = P(B)$, so, again, $P(A \cap B) = P(A)\,P(B)$.

> **Multiplication Rule for Two Independent Events**
> If *A* and *B* are any two independent events, $P(A \cap B) = P(A)\,P(B)$.

EXAMPLE 5.21

MULTIPLICATION RULE FOR TWO INDEPENDENT EVENTS

Suppose the experiment is to toss a fair die twice. If you roll a 6 on both tosses, you will win $18.
a. What is the probability that you will win this game?
b. What is a "fair price" to play this game? A fair price might be determined by balancing out the winnings and the price in the long run.

Solution

a. Define the following events:
    A: Roll a 6 on the first toss.
    B: Roll a 6 on the second toss.

From Example 5.2, $P(A) = P(B) = 1/6$. It is reasonable to assume that successive die rolls are independent, since the die has no memory of its previous tosses. Then, from the Multiplication Rule for Two Independent Events:

$$P(\text{Winning}) = P(A \text{ and } B) = P(A \cap B) = P(A)\,P(B) = \left(\frac{1}{6}\right)\left(\frac{1}{6}\right) = \left(\frac{1}{36}\right)$$

b.   In the long run, you would win $18 on average once every 36 games, so your long-run average winnings would be $\left(\frac{1}{36}\right)$ ($18) = 0.50. Thus, a fair price to play this game would be 50 cents.

We investigate this idea of a fair price, known as your *expected winnings*, much more in Chapter 6.

Now You Can Do Exercises 39–44.

Sampling With and Without Replacement

The relationship between two events can be determined by the way the samples are chosen. Two methods of choosing samples are *sampling with replacement* and *sampling without replacement*.

> In **sampling with replacement,** the randomly selected unit is returned to the population after being selected. When sampling with replacement, it is possible for the same unit to be sampled more than once.
>
> In **sampling without replacement,** the randomly selected unit is not returned to the population after being selected. When sampling without replacement, it is not possible for the same unit to be sampled more than once.

EXAMPLE 5.22   **SAMPLING WITH REPLACEMENT**

We draw a card at random from a shuffled deck, observe the card, and return it to the deck. The deck is then reshuffled, and we draw another card at random. What is the probability that both cards we select will be aces?

Solution

Define the following events:

A: Observe an ace on the first draw.
B: Observe an ace on the second draw.

We want to find $P(A \cap B)$, the probability of observing an ace on the first draw *and* an ace on the second draw. From the Multiplication Rule, $P(A \cap B) = P(A)\,P(B|A)$. To find $P(A)$, recall that there are 4 aces in the deck of 52 cards. It is reasonable to assume that all cards are equally likely to be selected, so using the classical method, $P(A) = 4/52$. Similarly, $P(B) = 4/52$.

Next we need to find $P(B|A)$, the probability of observing an ace on the second draw, given that we observe an ace on the first draw. Since *the deck of 52 cards has not changed* (except for shuffling), there are still 52 cards, 4 of which are aces. Therefore, $P(B|A) = 4/52$. Thus, the probability that both cards we select will be aces is $P(A \cap B) = P(A)P(B|A) = (4/52)(4/52) \approx 0.0059$.

Note that $P(B|A) = P(B) = 4/52$. Thus, by the alternative method for determining independence, *A* and *B* are independent events when sampling with replacement.

Now You Can Do Exercises 45 and 46.

We can generalize this result as follows.

> When sampling with replacement, successive draws can be considered **independent.**

Punchstock/Charles Sturge

SAMPLING WITHOUT REPLACEMENT

Suppose we alter the experiment in Example 5.22 as follows: We draw a card at random from a shuffled deck, hold onto the card (do not replace it) while the deck is reshuffled, and then select another card at random. What is the probability that both cards we select will be aces?

Solution

Define events A and B as in Example 5.22. Again we use the Multiplication Rule to find $P(A \cap B)$. The difference in this experiment comes when finding $P(B|A)$, the probability of observing an ace on the second draw given an ace on the first draw. Once we select the first ace, we do not replace it in the deck. Therefore, when the deck is reshuffled, it has only 51 cards left, only 3 of which are aces. The classical method then gives the probability of observing an ace on the second draw:

$$P(B|A) = \frac{\text{number of aces in the deck}}{\text{number of cards in the deck}} = \frac{3}{51}$$

Thus, the probability that both cards we select will be aces is

$$P(A \cap B) = P(A)P(B|A) = \frac{4}{52} \cdot \frac{3}{51} = \frac{12}{2652} \approx 0.0045$$

**Now You Can Do
Exercises 47 and 48.**

This probability is somewhat less than the probability that both cards will be aces when sampling with replacement. Note that here we found that $P(B|A)$ was not equal to $P(B)$. Thus, by the alternative method for determining independence, A and B are not independent events; they are dependent events.

We can generalize this result as follows.

> When sampling without replacement, successive draws should be considered **dependent.**

4 APPROXIMATING PROBABILITIES FOR DEPENDENT EVENTS

In some instances we can estimate the probability of a dependent event as if it were independent. The next example shows this can be done.

APPROXIMATING PROBABILITIES FOR DEPENDENT EVENTS

Imagine that we are fraud investigators sifting through a database (population) of 1 million financial transactions. Suppose we know that 100 of these transactions are fraudulent. We select two transactions without replacement, so that the successive draws are dependent.
a. Find the probability that both transactions are fraudulent.
b. Approximate this probability, on the assumption that the successive draws are independent, and compute the approximation error.
c. Draw a conclusion about the relationship between sample size and the size of the error using this approximation.

Solution

a.   Define the following events:

   A: First transaction is fraudulent.
   B: Second transaction is fraudulent.

   Then, $P(A) = \dfrac{100}{1,000,000}$ and $P(B|A) = \dfrac{99}{999,999}$. Thus, similar to Example 5.23, we have

$$P(\text{both fraudulent}) = \frac{100}{1,000,000} \cdot \frac{99}{999,999} = 0.0000000099$$

b.   Suppose that we estimate this probability using the assumption that the successive draws were independent. Then

$$P(\text{both fraudulent}) = \frac{100}{1,000,000} \cdot \frac{100}{1,000,000} = 0.00000001$$

   The approximation error is the difference between these two probabilities, which is very small:

$$0.00000001 - 0.0000000099 = 0.0000000001$$

   This means that our approximation is rather good.

c.   Our sample size is small (2) compared to our population (1,000,000). Thus, the error in using the independence assumption to approximate dependent successive draws is very small.

The question is: How small is a small sample? We shall use the following *1% Guideline*.

> **The 1% Guideline**
>
> Suppose successive draws, such as those for a random sample, are being made from a population. If the sample size is no larger than 1% of the size of the population, then the probability of dependent successive draws from the population may be approximated using the assumption that the draws are independent.

EXAMPLE 5.25

APPLYING THE 1% GUIDELINE

Metropolitan Washington, D.C., has the highest proportion of female top-level executives in the United States: 27%.[5] Suppose there are 1000 top-level executives in the area, and we take a random sample of size 2. Approximate the probability that both top-level executives are female, using the 1% Guideline.

Solution

Define the following events:

   A: First top-level executive is female.
   B: Second top-level executive is female.

The 1% Guideline is also helpful when we do not know the size of the population, but may presume that the population is very large compared to the sample size.

The sample of size 2 represents $\dfrac{2}{1000} = 0.002 = 0.2\%$ of the population. Thus, the 1% Guideline applies, and we may treat the successive draws as independent. Thus, we can use the Multiplication Rule for Independent Events to solve this problem.

$$P(A \cap B) = P(A)P(B) = (0.27)(0.27) = 0.0729$$

Now You Can Do
Exercises 49–52.

Note that the Multiplication Rule for Independent Events provides us with an alternative method for determining whether two events are indeed independent.

> **Alternative Method for Determining Independence**
> If $P(A) P(B) = P(A \cap B)$, then events A and B are **independent.**
> If $P(A) P(B) \neq P(A \cap B)$, then events A and B are **dependent.**

EXAMPLE 5.26

DETERMINING INDEPENDENCE USING THE ALTERNATIVE METHOD

We return to the direct mail marketing data from Example 5.17, reproduced here in Table 5.9. Use the alternative method for determining independence to determine whether the following two events are independent.

R: Responded to direct mail marketing campaign.
C: Has a credit card on file.

Table 5.9 Credit card status and marketing response

Response	Credit Card on File? No	Yes	Total
Did not respond	161	79	240
Did respond	17	31	48
Total	178	110	288

Source: Daniel Larose, *Data Mining Methods and Models* (Wiley Interscience, 2006).

Solution

Using Table 5.9, we may find the following probabilities:

$$P(R) = \frac{48}{288} \quad P(C) = \frac{110}{288} \quad P(R \cap C) = \frac{31}{288} \approx 0.1076$$

$$P(R)P(C) = \frac{48}{288} \cdot \frac{110}{288} \approx 0.0637$$

Now You Can Do Exercises 53–56.

Since $0.0637 \neq 0.1076$, we have $P(R)P(C) \neq P(R \cap C)$, and therefore, R and C are dependent.

EXAMPLE 5.27

CONDITIONAL PROBABILITY FOR MUTUALLY EXCLUSIVE EVENTS

Suppose two events A and B are mutually exclusive, with $P(A) > 0$ and $P(B) > 0$.
a. Find $P(B|A)$.
b. Are events A and B independent or dependent?

Solution

a. Since A and B are mutually exclusive, $P(A \cap B) = 0$. Then

$$P(B|A) = \frac{P(A \cap B)}{P(A)} = 0$$

That is, *if event A has occurred, then event B cannot occur.* This is a natural consequence of events A and B being mutually exclusive.

What Results Might We Expect?

Two events are independent if the occurrence of one does not affect the probability that the other will occur. However, as we saw in (a), if event A occurs, then the probability that event B will occur is 0. Thus, we would expect events A and B to be dependent.

In other words, if two events are mutually exclusive, then they are dependent.

Now You Can Do
Exercises 57–60.

b. We are given that $P(A) > 0$ and $P(B) > 0$. Hence the product $P(A)\,P(B)$ is also greater than 0. However, from (a), $P(A \cap B) = 0$. Thus, $P(A)\,P(B) \neq P(A \cap B)$, and from the alternative method for determining independence, we conclude that events A and B are dependent.

We can extend the Multiplication Rule to cover n independent events.

> **Multiplication Rule for n Independent Events**
> If $A, B, C, \ldots$ are independent events, then $P(A \cap B \cap C \cap \ldots) = P(A)\,P(B)\,P(C)\ldots$

EXAMPLE 5.28

MULTIPLICATION RULE FOR n INDEPENDENT EVENTS

According to the National Health Interview Survey, 24% of Americans aged 18–44 smoke tobacco.
a. In a random sample of $n = 3$ Americans aged 18–44, find the probability that all 3 smoke.
b. In a random sample of $n = 10$ Americans aged 18–44, find the probability that all 10 smoke.

Solution

The US Census Bureau estimates that there are over 100 million Americans aged 18–44. Thus, by the 1% Guidelines it is reasonable to assume that the successive draws are independent. Let S_i denote the event that the ith American aged 18–44 smokes.
a. $P(S_1) = P(S_2) = P(S_3) = 0.24$. Then, using the Multiplication Rule for n Independent Events,

$$P(S_1 \cap S_2 \cap S_3) = P(S_1) \cdot P(S_2) \cdot P(S_3) = (0.24)(0.24)(0.24) = (0.24)^3 = 0.013824$$

Now You Can Do
Exercises 61–64.

b. $P(S_1) = P(S_2) = \ldots = P(S_{10}) = 0.24$. Then, using the Multiplication Rule for Independent Events,

$$P(S_1 \cap S_2 \cap \ldots \cap S_{10}) = P(S_1) \cdot P(S_2) \cdot \ldots \cdot P(S_{10}) = (0.24)^{10} \approx 0.0000006$$

EXAMPLE 5.29

SOLVING AN "AT LEAST" PROBLEM

Using information in Example 5.28, find the probability that, in a random sample of three Americans aged 18–44, at least one of them smokes.

Solution

The phrase "at least" means that one or more of the three Americans smoke. Using the complement, the probability for this event may be written

$$P(\text{At least one of the three Americans smokes})$$

$$= P(\text{One or more of the three Americans smoke})$$

$$= 1 - P(\text{None of the three Americans smokes})$$

The probability of not smoking for the first American is

$$P(N_1) = 1 - P(S_1) = 1 - 0.24 = 0.76$$

and similarly for each American in the sample. Thus,

$$P(\text{None of the three Americans smokes}) = P(N_1) \cdot P(N_2) \cdot P(N_3) = (0.76)^3 = 0.438976$$

Hence, the probability that at least one of the three Americans smokes is

**Now You Can Do
Exercises 65–68.**

$$1 - P(\text{None of the three Americans smokes}) = 1 - 0.438976 = 0.561024$$

WHAT IF?

Give the Calculator a Rest

Suppose that the percentage of Americans aged 18–44 who smoke tobacco this year has decreased to less than 24%, though we are not sure how much less. Determine whether the following quantities will increase or decrease from the values calculated in Examples 5.28 and 5.29.

a. In a random sample of $n = 3$ Americans aged 18–44, the probability that all 3 smoke

b. In a random sample of $n = 3$ Americans aged 18–44, the probability that none of them smokes

Solution

a. Let $P(S_1^*) < 0.24$ represent the revised probability that an American aged 18–44 smokes. Then $P(S_1^* \cap S_2^* \cap S_3^*) = P(S_1^*) \cdot P(S_2^*) \cdot P(S_3^*) < P(S_1) \cdot P(S_2) \cdot P(S_3) = P(S_1 \cap S_2 \cap S_3)$
Thus, the probability that all three will smoke will decrease.

b. If $P(S_1^*) < 0.24$, then $P(N_1^*) = 1 - P(S_1^*) > 1 - 0.24 = P(N_1)$; that is, the probability that an American aged 18–44 doesn't smoke has increased. Thus,

$$P(\text{None of the three Americans smokes}) = P(N_1^*) \cdot P(N_2^*) \cdot P(N_3^*) > (0.76)^3 = 0.438976.$$

Therefore, the probability that none of the three Americans aged 18–44 smokes will increase.

CASE STUDY | **The ELISA Test for the Presence of HIV**

The ELISA test is used to screen blood for the presence of HIV. Like most diagnostic procedures, the test is not foolproof.

• When a blood sample contains HIV, the ELISA test will give a positive result 99.6% of the time. That is, the *false-negative rate,* the percentage of tests returning a negative result when the HIV virus is actually present, is $1 - 0.996 = 0.004$.

• When the blood does not contain HIV, the ELISA test will give a negative result 98% of the time. That is, the *false-positive rate,* the percentage of tests returning a positive result when the HIV virus is not actually present, is $1 - 0.98 = 0.02$.

A positive result means that the test says that the person has the HIV infection. A negative result means that the test says that the person does not have the virus. The *prevalence rate* for HIV in the general population is 0.5%. That is, 5 of 1000 persons in the general population have HIV.

Suppose we have samples of blood from 100,000 randomly chosen people.

Problem 1. How many people in the sample of 100,000 have HIV? How many do not?

Solution

The prevalence rate of 0.5% means that 0.005 (100,000) = 500 people in the sample have HIV. The remainder—99,500—do not.

Problem 2. A positive result is given 99.6% of the time for blood containing HIV. For the 500 people with HIV, how many positive results will the ELISA test return? How many of the 500 people with HIV will receive a negative result?

Solution

The ELISA test will return a positive result for 0.996 (500) = 498 of the 500 people. Thus, two people who actually have HIV will receive a test result indicating that they do not have the virus.

Problem 3. A negative result is given 98% of the time for blood without HIV. For the 99,500 people without HIV, how many negative results will the ELISA test return? Positive results?

Solution

The ELISA test will return a negative result for 0.98 (99,500) = 97,510 of the 99,500 people without HIV. The remaining 2%, or 1990 people, will receive positive ELISA test results, even though they do not have the virus.

We can use the counts we found to fill in the following table.

	In Reality		
ELISA test results	**Person has HIV**	**Person does not have HIV**	**Total**
Positive	498	1,990	2,488
Negative	2	97,510	97,512
Total	500	99,500	100,000

We will use the information in the ELISA test contingency table to solve Problems 4 and 5. If a person is chosen at random from the sample of 100,000, define the following events:

A: Person has HIV.

A^C: Person does not have HIV.

Pos: ELISA test returned positive results.

Neg: ELISA test returned negative results.

Problem 4. What is the probability that a randomly chosen person actually does have HIV, given that the ELISA results are negative? In other words, find $P(A\,|\,\text{Neg})$.

Solution

$$P(A\,|\,\text{Neg}) = \frac{N(A \cap \text{Neg})}{N(\text{Neg})} = \frac{2}{97,512} \approx 0.0000205$$

Problem 5. What is the probability that a randomly chosen person actually does not have HIV, given that the ELISA test results are positive? In other words, find $P(A^C\,|\,\text{Pos})$.

Solution

$$P(A^C\,|\,\text{Pos}) = \frac{N(A^C \cap \text{Pos})}{N(\text{Pos})} = \frac{1990}{2488} \approx 0.7998 \approx 0.80$$

Developing Your Statistical Sense

Which Error Is More Dangerous?

In Problems 4 and 5, we examined the probabilities of the two ways that the ELISA test can be wrong. Which error do you think is more dangerous? $P(A\,|\,\text{Neg})$ represents the probability that HIV is present, even though the ELISA test says otherwise. $P(A^C\,|\,\text{Pos})$ represents the probability that HIV is not present, even though the ELISA test says it is present. The designers of the ELISA test worked hard to reduce the false-negative rate $P(A\,|\,\text{Neg})$ to as low a level as possible. They rightly considered that it is the more dangerous type of error because of the epidemic nature of the illness. A person who receives a false-negative ELISA result could spread the infection further. Therefore, the designers tried to keep this probability as low as they could.

There is a price to be paid, however, which is the high false-positive rate, $P(A^C\,|\,\text{Pos})$, a very high 80%. Thus, if a random person receives a positive ELISA test result, the probability that the person does *not* have HIV is 80%. When the ELISA test comes back positive, a second batch of tests that have a more reasonable false-positive rate is usually administered.

SECTION 5.3 **Summary**

1. Section 5.3 discusses conditional probability $P(B\,|\,A)$, the probability of an event B given that an event A has occurred.

2. We can compare $P(B\,|\,A)$ to $P(B)$ to determine whether the events A and B are independent. Events are independent if the occurrence of one event does not affect the probability that the other event will occur.

3. The Multiplication Rule for Independent Events is the product of the individual probabilities. Sampling with replacement is associated with independence, while sampling without replacement means that the events are not independent.

4. We can use the 1% Guideline for approximating probabilities of dependent events.

SECTION 5.3 **Exercises**

Clarifying the Concepts

1. Suppose you are the coach of a football team, and your star quarterback is injured.
   a. Does the injury affect the chances that your team will win the big game this weekend?
   b. How would you describe this situation in the terminology presented in this section?

2. Write a sentence or two about a situation in your life similar to Exercise 1, where the probability of some event was affected by whether or not some other event occurred.

3. Explain clearly the difference between $P(A \cap B)$ and $P(A\,|\,B)$.

4. Give an example from your own experience of two events that are independent. Describe how they are independent.

5. Picture yourself explaining to your friends about the Gambler's Fallacy. How would you explain the Gambler's Fallacy in your own words?

6. Explain why two events A and B cannot have the following characteristics: $P(A) = 0.25$, $P(B) = 0.25$, and $P(A \cap B) = 0.30$. (*Hint:* Figure 5.16b might help.)

7. Explain why each of the following events is either dependent or independent.
   a. Drawing a ball from a box, replacing it, and then drawing a second ball
   b. Drawing a ball from a box, not replacing it, and then drawing a second ball

8. Explain why the following events are either dependent or independent, and provide support for your assertion.
   a. Tossing a coin and drawing a card from a deck of playing cards
   b. Drawing a card from a deck, not replacing it, and drawing another card

Practicing the Techniques

A sample of 200 students was asked to state whether they prefer the color pink or blue. Define the following events:

P: Prefers pink, B: Prefers blue, F: Female, M: Male. Use the results compiled in Table 5.9 to find the probabilities indicated in Exercises 9–24.

TABLE 5.9 Color preference and gender

Gender	Pink	Blue
Female	40	60
Male	10	90

9. $P(P)$

10. $P(B)$

11. $P(F)$

12. $P(M)$

13. $P(P$ and $F)$

14. $P(P$ and $M)$

15. $P(B$ and $F)$

16. $P(B$ and $M)$

17. $P(P \mid F)$

18. $P(P \mid M)$

19. $P(B \mid F)$

20. $P(B \mid M)$

21. $P(F \mid P)$

22. $P(M \mid P)$

23. $P(F \mid B)$

24. $P(M \mid B)$

For Exercises 25–28, refer to Table 5.8 on page 218 to find the probability that a randomly selected customer had the following characteristics.

25. Did not have a credit card on file

26. Did not respond to the direct mail marketing

27. Did not respond, given that he or she did not have a credit card on file

28. Did not have a credit card on file, given that he or she did not respond to the direct mail marketing

For Exercises 29–34, let the experiment be to toss two fair dice. Use the sample space in Figure 5.3 on page 200. Define the following events.

  X: Roll a sum equal to 7.
  Y: Roll a sum equal to 6.
  Z: Roll doubles, where the dark green die equals the light green die.
  W: Light green die equals 6.

Use the strategy for determining whether two events are independent (page 220) to determine whether the following pairs of events are independent.

29. X and Z

30. Y and Z

31. X and W

32. Y and W

33. X and Y

34. Z and W

For Exercises 35–38, use the Multiplication Rule to find the indicated probability.

35. Thirty percent of students at a particular college take statistics. Ninety percent of students taking statistics at the college pass the course. What is the probability that a student will take statistics and pass the course?

36. Fifty percent of students at a particular college are commuters. Of those, 10% bike to school. Find the probability that a student is a commuter and bikes to school.

37. Twenty-five percent of the nursing students at a particular college are male. Of these, 50% are taking a biology course this semester. Calculate the probability that a nursing student is a male and is taking a biology course this semester.

38. Thirty percent of the statistics students at a particular college have taken advantage of the college tutoring program. After doing so, 80% of them received a higher score on the next exam. Find the probability that a statistics student has taken advantage of the college tutoring program and has received a higher score on the next exam.

For Exercises 39–44, let A, B, C, and D be independent events such that $P(A) = 0.5$, $P(B) = 0.4$, $P(C) = 0.2$, and $P(D) = 0.1$. Use the Multiplication Rule for Two Independent Events to find the following probabilities.

39. $P(A$ and $B)$

40. $P(A \cap C)$

41. $P(A \cap D)$

42. $P(B$ and $C)$

43. $P(B$ and $D)$

44. $P(C \cap D)$

For Exercises 45 and 46, suppose we sample two cards at random and with replacement from a deck of cards. Define the following events. $R1$: Red card observed on the first draw, $R2$: Red card observed on the second draw, $H1$: Heart observed on the first draw, $H2$: Heart observed on the second draw.

45. Find $P(R1$ and $R2)$.

46. Find $P(H1 \cap H2)$.

For Exercises 47 and 48, suppose we sample two cards at random and without replacement from a deck of cards. Define the same events as for Exercises 45 and 46.

47. Find $P(R1$ and $R2)$.

48. Find $P(H1 \cap H2)$.

Use the following information for Exercises 49–52. Suppose 25% of the 2000 students at a local college use Gmail as their primary email account.

49. If we take a sample of 2 students, verify that the 1% Guideline applies.

50. If we take a sample of 2 students, use the 1% Guideline to approximate the probability that both students use Gmail as their primary email account.

51. If we take a sample of 3 students, approximate the probability that all 3 students use Gmail as their primary email account.

52. If we take a sample of 4 students, approximate the probability that all 4 students use Gmail as their primary email account.

For Exercises 53–56, use Table 5.9 and the alternative method for determining independence (page 226) to determine whether the following pairs of events are independent.

53. *P* and *F*

54. *P* and *M*

55. *B* and *F*

56. *B* and *M*

57. Suppose *P*(*X* and *Y*) = 0, for events *X* and *Y*. State whether X and Y are independent.

58. Define the following events. *A* = salary $50,000 or more, *B* = salary less than $50,000. Are *A* and *B* independent? Why?

59. The intersection between events *W* and *Z* is empty. Then is it true or not true that *P*(*W* and *Z*) = *P*(*W*) · *P*(*Z*)? Explain.

60. Define event *W*: team wins. Are *W* and *W*^C independent? Why?

For Exercises 61–64, use the Multiplication Rule for *n* Independent Events to find the probabilities. Define *L*: observe either a 1 or a 2 on a toss of a fair die.

61. *L* occurs on three successive tosses.

62. *L* occurs on four successive tosses.

63. *L* occurs on five successive tosses.

64. *L* occurs on ten successive tosses.

For Exercises 65–68, define *H*: observe a number greater than 2 on a toss of a fair die. Find the following probabilities.

65. That *H* occurs at least once in three tosses

66. That *H* occurs at least once in four tosses

67. That *H* occurs at least once in five tosses

68. That *H* occurs at least once in ten tosses

69. *Calculate the probability of* observing tails on each of five successive tosses of a fair coin.

70. *Compute the probability of* observing tails on each of ten successive tosses of a fair coin.

For Exercises 71–74, let *A* and *B* be two independent events, with *P*(*A*) = 0.6 and *P*(*B*) = 0.4. Find the indicated probabilities.

71. $P(A \cap B)$

72. $P(A|B)$

73. $P(B|A)$

74. $P(A \cup B)$

For Exercises 75–78, let *A* and *B* be two independent events, with *P*(*A*) = 0.5 and *P*(*B*) = 0.2. Find the indicated probabilities.

75. $P(A \cap B)$

76. $P(A|B)$

77. $P(B|A)$

78. $P(A \cup B)$

79. Suppose that *A* and *B* are two events with *P*(*A*) = 0.3 and $P(A \cap B) = 0.05$. Find $P(B|A)$.

80. Suppose that *A* and *B* are two events, with *P*(*A*) = 0.9 and $P(B|A) = 0.6$. Find $P(A \cap B)$.

For Exercises 81–86, let *A* and *B* be independent events such that *P*(*A*) = 0.4 and *P*(*B*) = 0.5. Find the indicated probabilities.

81. *P*(*A* and *B*)

82. $P(A|B)$

83. $P(B|A)$

84. *P*(*A* or *B*)

85. *P*(*A* and *B*)^C

86. *P*(*A* or *B*)^C

For Exercises 87–90, let *C* and *D* be events such that *P*(*C*) = 0.7, *P*(*D*) = 0.3, and *P*(*C* and *D*) = 0.21.

87. Find $P(C|D)$.

88. Find $P(D|C)$.

89. Are events *C* and *D* independent? How can you tell?

90. Are events *C* and *D* mutually exclusive? How can you tell?

For Exercises 91 and 92, let *E* and *F* be events such that *P*(*E*) = 0.5 and *P*(*F*) = 0.6.

91. What further information do we need to know to determine whether events *E* and *F* are independent?

92. What further information do we need to know to determine whether events *E* and *F* are mutually exclusive?

For Exercises 93–96, a single fair die is rolled twice in succession. Find the indicated probabilities.

93. Observe a 1 on the second roll

94. Observe an even number on the second roll

95. Observe an even number on the second roll, given that you observe an even number on the first roll

96. Based on the probabilities in Exercises 93–95, what can you say about the dependence or independence of successive rolls of a single fair die?

Applying the Concepts

97. Teen Birth Rate. The Federal Interagency Forum on Child and Family Statistics (**www.childstats.gov**) reported that the teenage birth rate in 2010 was 0.04.
   a. Find the probability that two randomly selected births are to teenagers.
   b. Find the probability that five randomly selected births are to teenagers.
   c. Find the probability that at least one of four randomly selected births is to a teenager.

98. Balls in a Box. A box contains four blue balls and three red balls. If we select two balls at random, what is the probability that both balls will be blue if
   a. we sample with replacement.
   b. we sample without replacement.

99. Acceptance Sampling. You are in charge of purchasing for a large computer retailer. Your wholesaler delivers computers to you in batches of 100. You either accept or reject an entire batch based on a random sample of two computers: if both computers you sample are defective, then you reject the entire batch. Suppose that (unknown to you, of course) there are 10 defective computers in the batch of 100 computers.
   a. Should you conduct your sampling with or without replacement? Why?
   b. What is the probability that the first computer you select is defective?
   c. What is the probability that the second computer you select is defective, given that the first was defective, if you sample without replacement?
   d. What is the probability that you will accept the batch?
   e. What is the probability that you will reject the batch?
   f. Usually you accept each batch of computers from this wholesaler. Do you think that is a wise move, considering that 10% of their product is defective? How could you make your test stricter so that there is a smaller chance of accepting a batch with 10% defectives?

100. Treasury Bonds. One of the most important tasks for economists is to make forecasts for the performance (up or down) of investments such as 30-year Treasury bonds. *The Journal of Investing* (Vol. 6, No. 2, page 8, 1997) reports that, in a sample of 30 six-month surveys, the consensus estimate of performance for the 30-year Treasury bond has been wrong 20 out of the 30 times!
   a. Find the probability that two randomly selected consensus estimates were correct.
   b. Find the probability that three randomly selected consensus estimates were wrong.
   c. If we choose two consensus estimates and if we sample with replacement, find the probability that the second consensus estimate was right, given that the first consensus estimate was right. Are the successive draws independent? Why or why not?
   d. Repeat **(c)**, this time sampling without replacement. Are the successive draws independent? Why or why not?

101. Adjustable Rate Mortgages. Half of the 20 mortgages provided by a certain mortgage lending company last week are adjustable rate mortgages (ARMs). Suppose we sample three mortgages without replacement. Find the following probabilities.
   a. The first mortgage is an ARM.
   b. The second mortgage is an ARM, given that the first mortgage is an ARM.
   c. The third mortgage is an ARM, given that the first two mortgages are ARMs.

9/11 and Pearl Harbor. What were the feelings of Americans in the days immediately following the events of September 11, 2001? In an NBC News Terrorism Poll, the following question was asked: "Would you say that Tuesday's attacks are more serious than, equal to, or not as serious as the Japanese attack on Pearl Harbor?" This poll was conducted on September 12, 2001. Use the following crosstabulation of the poll results for Exercises 102–105.

	Sex		
	Male	**Female**	**Total**
More serious than Pearl Harbor	200	212	412
Equal to Pearl Harbor	70	84	154
Not as serious as Pearl Harbor	23	6	29
Not sure	11	12	23
Total	304	314	618

102. Find the probabilities that a randomly chosen person has the following characteristics.
   a. Is female, $P(F)$
   b. Is male, $P(M)$
   c. Believes September 11 is more serious than Pearl Harbor, $P(\text{More})$

103. Find the probability that a randomly chosen person has the following characteristics.
   a. Is female and believes September 11 is more serious than Pearl Harbor, $P(F \cap \text{More})$
   b. Is male and believes September 11 is more serious than Pearl Harbor, $P(M \cap \text{More})$

104. Find the following conditional probabilities for a randomly chosen person.
   a. Given that the person is female, believes September 11 is more serious than Pearl Harbor, $P(\text{More} \mid F)$
   b. Given that the person is male, believes September 11 is more serious than Pearl Harbor, $P(\text{More} \mid M)$

105. Are gender and the belief whether September 11 was more or less serious than Pearl Harbor independent? Why or why not?

Bringing It All Together

Gender and Pet Preference. Use Table 5.10 for Exercises 106–110. Do you think your gender affects what type of pet you own?

TABLE 5.10 Pet preference

Gender	Cats	Dogs	Other pets	Total
Female	100	50	30	180
Male	50	50	20	120
Total	150	100	50	300

106. Find the probabilities that a randomly chosen person has the following characteristics.
   a. Is female, $P(F)$
   b. Is male, $P(M)$
   c. Owns a cat, $P(C)$
   d. Owns some other kind of pet, $P(O)$

107. Find the probability that a randomly chosen person has the following characteristics.
   a. Is female and owns a cat, $P(F \cap C)$
   b. Is female and owns some other kind of pet, $P(F \cap O)$
   c. Is male and owns a cat, $P(M \cap C)$
   d. Is male and owns some other kind of pet, $P(M \cap O)$

108. Find the following conditional probabilities for a randomly chosen person.

   a. Owns a cat, given that the person is female, $P(C|F)$
   b. Owns a cat, given that the person is male, $P(C|M)$
   c. Owns some other kind of pet, given that the person is female, $P(O|F)$
   d. Owns some other kind of pet, given that the person is male, $P(O|M)$

109. Are gender and pet preference independent? Why or why not?

110. If you were a cat-food manufacturer, would you advertise more in men's magazines or women's magazines? Why? Cite your evidence.

5.4 COUNTING METHODS

OBJECTIVES By the end of this section, I will be able to . . .

1 Apply the Multiplication Rule for Counting to solve certain counting problems.

2 Use permutations and combinations to solve certain counting problems.

3 Compute probabilities using combinations.

Counting methods allow us to solve a range of problems, including how to compute certain probabilities.

1 MULTIPLICATION RULE FOR COUNTING

Let us begin with an example illustrating a general rule of counting.

EXAMPLE 5.30   **DESIGN YOUR OWN T-SHIRT**

Bruce Laurance/The Image Bank/Getty Images

A store at the local mall allows customers to design their own T-shirts. The store offers the following options to its customers:
- **Sleeve type:** Long-sleeve or short-sleeve
- **Color:** White, black, or red
- **Image:** Stock picture or uploaded photo

List the possible T-shirt options.

Solution

Figure 5.18 is a tree diagram that shows all the different T-shirts that can be designed. There are two choices for type of sleeve. For each sleeve type, there are three choices for color. For each color, there are two choices of image: stock picture or uploaded photo. All together, customers have a choice from among

$$2 \cdot 3 \cdot 2 = 12$$

different T-shirt options.

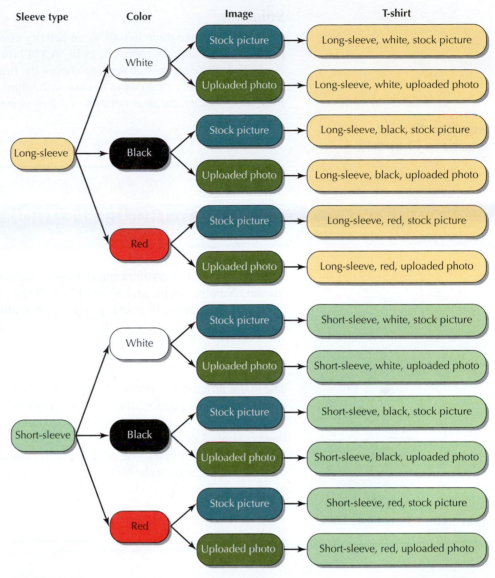

FIGURE 5.18 Tree diagram for the different T-shirt options.

Now You Can Do Exercises 7–10.

We can generalize from Example 5.30 the result as the **Multiplication Rule for Counting.**

> **Multiplication Rule for Counting**
>
> Suppose an activity consists of a series of events in which there are a possible outcomes for the first event, b possible outcomes for the second event, c possible outcomes for the third event, and so on. Then the total number of different possible outcomes for the series of events is
>
> $$a \cdot b \cdot c \cdot \dots$$

EXAMPLE 5.31   **COUNTING WITH REPETITION: FAMOUS INITIALS**

Some Americans in history are uniquely identified by their initials. For example, "JFK" stands for John Fitzgerald Kennedy, and "FDR" stands for Franklin Delano Roosevelt. How many different possible sets of initials are there for people with a first, middle, and last name?

Solution

Let us consider the three initials as an activity consisting of three events. Note that a particular letter may be repeated, as in "AAM" for A. A. Milne, author of *Winnie the Pooh*. Then there are $a = 26$ ways to choose the first initial, $b = 26$ ways to choose the second initial, and $c = 26$ ways to choose the third initial. Thus, by the Multiplication Rule for Counting, the total number of different sets of initials is

**Now You Can Do
Exercises 11 and 12.**

$$26 \cdot 26 \cdot 26 = 17{,}576$$

EXAMPLE 5.32   **COUNTING WITHOUT REPETITION: INTRAMURAL SINGLES TENNIS**

A local college has an intramural singles tennis league with five players, Ryan, Megan, Nicole, Justin, and Kyle. The college presents a trophy to the top three players in the league. How many different possible sets of three trophy winners are there?

Solution

The major difference between Examples 5.31 and this example is that in Example 5.31 there can be no repetition. Ryan cannot finish in first place *and* second place. So we proceed as follows. Five possible players could finish in first place, so $a = 5$. Now there are only four players left, one of whom will finish in second place, so $b = 4$. That leaves only three players, one of whom will finish in third place, giving $c = 3$. Thus, by the Multiplication Rule for Counting, the number of different possible sets of trophy winners is

Note: To summarize the key difference between Examples 5.31 and 5.32: if repetitions are allowed, then $a = b = c$. If repetitions are not allowed, then the numbers being multiplied decrease by one from left to right.

**Now You Can Do
Exercises 13 and 14.**

$$5 \cdot 4 \cdot 3 = 60$$

EXAMPLE 5.33   **TRAVELING SALESMAN PROBLEM**

A Southeast regional salesman has eight destinations that he must travel to this month: Atlanta, Raleigh, Charleston, Nashville, Jacksonville, Richmond, Mobile, and Jackson. How many different possible routes could he take?

Solution

The salesman has $a = 8$ different choices for where to go first. Once the first destination has been chosen, there are only $b = 7$ choices for where to go second. And once the first two destinations have been chosen, there are only $c = 6$ choices for where to go third, and so on. Thus, by the Multiplication Law for Counting, the number of different possible routes for the salesman is

**Now You Can Do
Exercises 15 and 16.**

$$a \cdot b \cdot c \cdot d \cdot e \cdot f \cdot g \cdot h = 8 \cdot 7 \cdot 6 \cdot 5 \cdot 4 \cdot 3 \cdot 2 \cdot 1 = 40{,}320$$

The calculation in Example 5.33 leads us to introduce the **factorial symbol,** which is used for the counting rules we will learn in the remainder of this section.

> For any integer $n \geq 0$, the **factorial symbol n!** is defined as follows:
> - $0! = 1$
> - $1! = 1$
> - $n! = n(n-1)(n-2)\dots 3 \cdot 2 \cdot 1$

For example:

$$
\begin{aligned}
2! &= && 2 \cdot 1 = 2 \\
3! &= && 3 \cdot 2 \cdot 1 = 6 \\
4! &= && 4 \cdot 3 \cdot 2 \cdot 1 = 24 \\
5! &= && 5 \cdot 4 \cdot 3 \cdot 2 \cdot 1 = 120 \\
6! &= && 6 \cdot 5 \cdot 4 \cdot 3 \cdot 2 \cdot 1 = 720 \\
7! &= && 7 \cdot 6 \cdot 5 \cdot 4 \cdot 3 \cdot 2 \cdot 1 = 5040 \\
8! &= 8 \cdot 7 \cdot 6 \cdot 5 \cdot 4 \cdot 3 \cdot 2 \cdot 1 &&= 40{,}320, \text{ as in Example 5.33}
\end{aligned}
$$

Now You Can Do
Exercises 17–22.

2   PERMUTATIONS AND COMBINATIONS

EXAMPLE 5.34

TRAVELING TO SOME BUT NOT ALL OF THE CITIES

Example 5.33 calculated the number of possible routes for traveling to $n = 8$ cities. However, suppose we are interested in traveling to *some but not all* of the cities? For example, suppose that the salesman is traveling to three of the eight cities. Find the number of possible routes.

Solution

There are eight choices for the first city, seven choices for the second city, and six choices for the third city. Since the salesman is traveling to three cities only, the number of possible routes is thus

$$8 \cdot 7 \cdot 6 = 336$$

This result may be rewritten using factorial notation, as follows:

Now You Can Do
Exercises 23 and 24.

$$8 \cdot 7 \cdot 6 = \frac{8 \cdot 7 \cdot 6 \cdot (5 \cdot 4 \cdot 3 \cdot 2 \cdot 1)}{(5 \cdot 4 \cdot 3 \cdot 2 \cdot 1)} = \frac{8!}{5!} = \frac{8!}{(8-3)!}$$

Example 5.34 leads us to the following definition.

> **Permutations**
> A **permutation** is an arrangement of items, such that
> - r items are chosen at a time from n distinct items.
> - repetition of items is not allowed.
> - the order of the items is important.
>
> The number of permutations of n items chosen r at a time is denoted as $_nP_r$ and given by the formula
>
> $$_nP_r = \frac{n!}{(n-r)!}$$

In Example 5.34, we are looking for the number of permutations of 8 cities taken 3 at a time. We have $n = 8$, $r = 3$:

$$_nP_r = {}_8P_3 = \frac{n!}{(n-r)!} = \frac{8!}{(8-3)!} = \frac{8!}{5!} = 8 \cdot 7 \cdot 6 = 336$$

EXAMPLE 5.35

CALCULATING NUMBERS OF PERMUTATIONS

Find the following numbers of permutations.
a. $_5P_2$ **b.** $_6P_2$ **c.** $_6P_6$

Solution

a. $_5P_2 = \dfrac{5!}{(5-2)!} = \dfrac{5 \cdot 4 \cdot 3!}{3!} = 20$

b. $_6P_2 = \dfrac{6!}{(6-2)!} = \dfrac{6 \cdot 5 \cdot 4!}{4!} = 30$

**Now You Can Do
Exercises 25–32.**

c. $_6P_6 = \dfrac{6!}{(6-6)!} = \dfrac{6 \cdot 5 \cdot 4 \cdot 3 \cdot 2 \cdot 1}{0!} = 720$

EXAMPLE 5.36

COUNTING PERMUTATIONS: SECRET SANTAS

"Secret Santa" refers to a method whereby each member of a group anonymously buys a holiday gift for another member of the group. Each person is secretly assigned to buy a gift for another randomly chosen person in the group. Suppose Jessica, Laverne, Samantha, and Luisa share a dorm suite and would like to do Secret Santa this holiday season.
a. Verify that in this instance one woman purchasing a gift for another woman represents a permutation.
b. Calculate how many possible different permutations of gift buying there are for the four women.

Solution

a. • There are $n = 4$ women, and $r = 2$ people are associated with each gift, the giver and the receiver.
    • Each person can buy only one gift, so repetition is not allowed.
    • Finally, there is a difference between Jessica buying for Laverne and Laverne buying for Jessica. Thus, order is important, and thus, buying a gift represents a permutation.
b. The number of permutations is calculated as follows:

$$_nP_r = {}_4P_2 = \frac{4!}{(4-2)!} = \frac{4 \cdot 3 \cdot 2!}{2!} = 12$$

In a permutation, order is important. For example, in Example 5.36, there was a difference between Jessica buying a gift for Laverne and Laverne buying one for Jessica. However, what if we consider shaking hands instead? Then Jessica shaking hands with Laverne is considered the same as Laverne shaking hands with Jessica. Hence, sometimes order is not important. What is important here is the **combination** of Jessica and Laverne.

> **Combinations**
>
> A **combination** is an arrangement of items in which
> - r items are chosen from n distinct items.
> - repetition of items is not allowed.
> - the order of the items is not important.
>
> The number of combinations of r items chosen from n different items is denoted as
>
> $$_nC_r$$

EXAMPLE 5.37   **HOW MANY COMBINATIONS IN THE INTRAMURAL TENNIS LEAGUE?**

We return to the intramural singles tennis league at the local college. There are five players: Ryan, Megan, Nicole, Justin, and Kyle. Each player must play each other once.

a. Confirm that a match between two players represents a combination.

b. How many matches will be held?

Solution

a. Let {Ryan, Megan} denote a tennis match between Ryan and Megan. *Note:*
- There are $r = 2$ players chosen from $n = 5$ players.
- Each player plays each other player once, so repetition is not allowed.
- There is no difference between {Ryan, Megan} and {Megan, Ryan}, so order is not important.

Thus, a tennis match between two players represents a combination.

b. The list of all matches is as follows.

{Ryan, Megan}	{Megan, Nicole}	{Nicole, Justin}
{Ryan, Nicole}	{Megan, Justin}	{Nicole, Kyle}
{Ryan, Justin}	{Megan, Kyle}	{Justin, Kyle}
{Ryan, Kyle}		

Thus there are $_5C_2 = 10$ possible matches of $r = 2$ players chosen from $n = 5$ players.

We saw in Example 5.35 that $_5P_2 = 20$ and in Example 5.37 that $_5C_2 = 10$. Permutations and combinations differ only in that ordering is ignored for combinations. To calculate the number of combinations $_nC_r$, we simply do not count; however, many rearrangements there are of the same items. For example, in Example 5.37, there are $r! = 2! = 2$ rearrangements of the same players, such as {Ryan, Megan} and {Megan, Ryan}. Thus,

$$_5C_2 = \frac{_5P_2}{2!} = \frac{20}{2} = 10$$

In general, the *number of combinations* can be computed as the number of permutations divided by the factorial of the number of items chosen.

Note: Following are some special combinations you may find useful. For any integer n:

$$_nC_n = 1$$
$$_nC_0 = 1$$
$$_nC_1 = n$$
$$_nC_{n-1} = n$$

> **Formula for the Number of Combinations**
>
> The number of combinations of r items chosen from n different items is given by
> $$_nC_r = \frac{n!}{r!(n-r)!}$$

For instance, in Example 5.37, the formula for the number of combinations is

$$_5C_2 = \frac{5!}{2!(5-2)!} = \frac{5!}{2!\,3!} = \frac{5 \cdot 4 \cdot 3!}{2 \cdot 1 \cdot 3!} = \frac{20}{2} = 10$$

Thus the relation: $_5C_2 = {_5P_2}/2!$ is verified.

EXAMPLE 5.38

CALCULATING NUMBERS OF COMBINATIONS

Find the following numbers of combinations.
a. $_6C_2$ b. $_6C_3$ c. $_6C_4$

Solution

a. $_6C_2 = \dfrac{6!}{2!(6-2)!} = \dfrac{6 \cdot 5 \cdot 4!}{2 \cdot 1 \cdot 4!} = \dfrac{30}{2} = 15$

b. $_6C_3 = \dfrac{6!}{3!(6-3)!} = \dfrac{6 \cdot 5 \cdot 4 \cdot 3!}{3 \cdot 2 \cdot 1 \cdot 3!} = \dfrac{120}{6} = 20$

c. $_6C_4 = \dfrac{6!}{4!(6-4)!} = \dfrac{6!}{(6-4)!\,4!} = \dfrac{6 \cdot 5 \cdot 4!}{2 \cdot 1 \cdot 4!} = \dfrac{30}{2} = 15$

Now You Can Do Exercises 33–40.

Note that in (c) we used the commutative property of multiplication ($a \cdot b = b \cdot a$) and found that $_6C_4 = {_6C_2} = 15$. In general, $_nC_r = {_nC_{n-r}}$ for this reason.

EXAMPLE 5.39

CALCULATING THE NUMBER OF PERMUTATIONS AND COMBINATIONS USING TECHNOLOGY

Use the TI-83/84 and Excel to calculate the following.
a. $_9P_6$ b. $_{10}C_7$

Solution

We use the instructions provided in the Step-by-Step Technology Guide at the end of this section (page 244).

a. From Figures 5.19 and 5.20, we find that $_9P_6 = 60{,}480$.

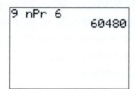

FIGURE 5.19 TI-83/84 permutation results.   **FIGURE 5.20** Excel permutation results.

b. From Figures 5.21 and 5.22, we find that $_{10}C_7 = 120$.

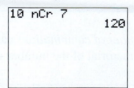

FIGURE 5.21 TI-83/84 combination results.   **FIGURE 5.22** Excel combination results.

Sometimes we wish to find the number of permutations of items where some of the items are not distinct.

EXAMPLE 5.40

PERMUTATIONS WITH NONDISTINCT ITEMS

How many distinct strings of letters can we make by using all the letters in the word STATISTICS?

Solution

Each string will be 10 letters long and include 3 S's, 3 T's, 2 I's, 1 A, and 1 C. The 10 positions shown here need to be filled.

$$\underline{\quad}\ \underline{\quad}\ \underline{\quad}\ \underline{\quad}\ \underline{\quad}\ \underline{\quad}\ \underline{\quad}\ \underline{\quad}\ \underline{\quad}\ \underline{\quad}$$
$$\textbf{1}\quad \textbf{2}\quad \textbf{3}\quad \textbf{4}\quad \textbf{5}\quad \textbf{6}\quad \textbf{7}\quad \textbf{8}\quad \textbf{9}\quad \textbf{10}$$

The string-forming process is as follows:

STEP 1 Choose the positions for the three S's.

STEP 2 Choose the positions for the three T's.

STEP 3 Choose the positions for the two I's.

STEP 4 Choose the position for the one A.

STEP 5 Choose the position for the one C.

There are $_{10}C_3$ ways to place the three S's in Step 1. Once Step 1 is done, there are seven slots left, leaving $_7C_3$ positions for the three T's. Once Step 2 is done, there are four slots left, so there are $_4C_2$ ways to place the two I's. Once Step 3 is done, there are only 2 slots left, so there are only $_2C_1$ ways to position the A. Finally, there is only $_1C_1$ way to place the C.

Putting Steps 1–5 together, we calculate the number of distinct letter strings as

$$_{10}C_3 \cdot {_7}C_3 \cdot {_4}C_2 \cdot {_2}C_1 \cdot {_1}C_1 = \frac{10!}{3!\,7!} \cdot \frac{7!}{3!\,4!} \cdot \frac{4!}{2!\,2!} \cdot \frac{2!}{1!\,1!} \cdot \frac{1!}{1!\,0!}$$

$$= \frac{10!}{3!\,3!\,2!\,1!\,1!} = \frac{3{,}628{,}800}{72}$$

$$= 50{,}400$$

There are 50,400 distinct strings of letters that can be made using the letters in the word STATISTICS.

This example can be generalized in the following result.

Permutations of Nondistinct Items

The number of permutations of n items of which n_1 are of the first kind, n_2 are of the second kind, . . . , and n_k are of the kth kind is calculated as

$$\frac{n!}{n_1! \cdot n_2! \cdot \ldots \cdot n_k!}$$

where $n = n_1 + n_2 + \cdots + n_k$.

EXAMPLE 5.41

NUMBER OF PERMUTATIONS OF NONDISTINCT ITEMS

Brandon brings a healthy snack to school each day, consisting of 5 carrot sticks, 4 celery sticks, and 2 cherry tomatoes. If Brandon eats one item at a time, in how many different ways can he eat his snack?

Solution

We are seeking the number of permutations of $n = 11$ items, of which $n_1 = 5$ are carrot sticks, $n_2 = 4$ are celery sticks, and $n_3 = 2$ are cherry tomatoes. Using the formula for the number of permutations of nondistinct items,

$$\frac{n!}{n_1! \cdot n_2! \cdot n_3!} = \frac{11!}{5! \cdot 4! \cdot 2!} = \frac{39,916,800}{120 \cdot 24 \cdot 2} = 6930$$

**Now You Can Do
Exercises 41 and 42.**

There are 6930 distinct ways in which Brandon can eat his snack.

Acceptance sampling refers to the process of (1) selecting a random sample from a batch of items, (2) evaluating the sample for defectives, and (3) either accepting or rejecting the entire batch based on the evaluation of the sample.

EXAMPLE 5.42   **ACCEPTANCE SAMPLING USES COMBINATIONS**

Suppose we have a batch of 20 cell phones, of which, unknown to us, 3 are defective and 17 are nondefective. We will take a random sample of size 2 and evaluate both items once.

a. Are the arrangements in acceptance sampling permutations or combinations?
b. Find the number of ways that both sampled cell phones are defective.

Solution

a. Both permutations and combinations require the following:
   - r items are chosen from n distinct items. Here we are selecting $r = 2$ phones from a batch of $n = 20$.
   - Repetition of the items is not allowed. Each item is evaluated only once.

The difference between permutations and combinations is that, for permutations order is important while for combinations order is not important. In acceptance sampling, the order of the items is not important. Thus, acceptance sampling uses combinations.

b. The number of ways of choosing 2 of the 3 defectives is

$$_3C_2 = \frac{3!}{2!(3-2)!} = \frac{3 \cdot 2!}{2! \cdot 1!} = 3$$

Selecting 2 defectives means that we are choosing 0 of the 17 nondefectives. The number of ways this can happen is

$$_{17}C_0 = \frac{17!}{0!(17-0)!} = \frac{17!}{1 \cdot 17!} = 1$$

By the Multiplication Rule for Counting, the number of ways that both sampled cell phones are defective is

$$_3C_2 \cdot {_{17}C_0} = 3 \cdot 1 = 3$$

---

3 COMPUTING PROBABILITIES USING COMBINATIONS

The counting methods we have learned in this section may be used to compute probabilities. We assume that each possible outcome in a random sample is *equally likely*, and thus we use the classical method for assigning the probability of an event E:

$$P(E) = \frac{\text{number of outcomes in } E}{\text{number of outcomes in sample space}} = \frac{N(E)}{N(S)}$$

EXAMPLE 5.43

PROBABILITY USING COMBINATIONS: ACCEPTANCE SAMPLING

Continuing with Example 5.42, if both cell phones in the sample of size 2 are defective, we will reject the batch and cancel our contract with the supplier.
a. What is the number of ways that both cell phones will be defective?
b. What is the number of outcomes in this sample space?
c. What is the probability that both cell phones will be defective?

Solution

a. From Example 5.42, the number of ways that both cell phones will be defective is

$$_3C_2 \cdot {_{17}C_0} = 3 \cdot 1 = 3$$

b. The number of outcomes in the sample space is given by the number of ways of selecting 2 cell phones out of a batch of 20, that is,

$$N(S) = {_{20}C_2} = \frac{20!}{2!(20-2)!} = \frac{20 \cdot 19 \cdot 18!}{2! \cdot 18!} = \frac{380}{2} = 190$$

c. Therefore, the probability that both cell phones will be defective is given by

$$P(\text{Both defective}) = \frac{\text{number of ways both defective}}{\text{number of outcomes in sample space}} = \frac{3}{190} \approx 0.01579$$

EXAMPLE 5.44

FLORIDA LOTTO

You can win the jackpot in the Florida Lotto by correctly choosing all 6 winning numbers out of the numbers 1–53.
a. What is the number of ways of winning the jackpot by choosing all 6 winning numbers?
b. What is the number of outcomes in this sample space?
c. If you buy a single ticket for $1, what is your probability of winning the jackpot?
d. If you mortgage your house and buy 500,000 tickets, what is your probability of winning the jackpot (assuming that all the tickets are different)?

Solution

a. The number of ways of winning the jackpot by correctly choosing all 6 of the winning numbers and none of the losing numbers is

$$N(\text{Jackpot}) = {_6C_6} \cdot {_{47}C_0} = 1 \cdot 1 = 1$$

b. The size of the sample space is

$$N(S) = {_{53}C_6} = \frac{53!}{6!(53-6)!} = \frac{53 \cdot 52 \cdot 51 \cdot 50 \cdot 49 \cdot 48 \cdot 47!}{6! \cdot 47!}$$

$$= \frac{16,529,385,600}{720} = 22,957,480$$

c. Therefore, if you buy a single ticket for $1, your probability of winning the jackpot is given by

$$P(\text{Jackpot}) = \frac{1}{22,957,480} \approx 0.00000004356$$

d. If you buy 500,000 tickets and they are all unique, then your probability of winning becomes

$$P(\text{Jackpot}) = \frac{500,000}{22,957,480} \approx 0.02178$$

This is because the unique tickets are mutually exclusive, and the Addition Rule for Mutually Exclusive Events allows us to add the probabilities of the 500,000 tickets. After mortgaging your $500,000 house and buying lottery tickets with the proceeds, there is a better than 97% probability that you will *not* win the lottery.

STEP-BY-STEP TECHNOLOGY GUIDE: Factorials, Permutations, and Combinations

TI-83/84

Factorials n!

Step 1   On the home screen, enter the value of n.

Step 2   Press **MATH**, highlight **PRB**, and select **4: !** (Figure 5.23).

Step 3   Press **ENTER**.

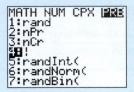

FIGURE 5.23

Permutations $_nP_r$ and Combinations $_nC_r$

Step 1   On the home screen, enter the value of n.

Step 2   **a.** For permutations, press **MATH**, highlight **PRB**, and select **2:nPr**.

b. For combinations, press **MATH**, highlight **PRB**, and select **3:nCr**.

Step 3   On the home screen, enter the value of r.

Step 4   Press **ENTER** (see Figure 5.19 and Figure 5.21 in Example 5.39 [page 240]).

EXCEL

Factorials n!

Calculate **9!**

Step 1   Select an empty cell, and type = **FACT(9)**.

Step 2   Press **ENTER**.

Permutations $_nP_r$

We illustrate Example 5.39a (page 240): $_9P_6$.

Step 1   Select an empty cell and type = **PERMUT(9,6)**.

Step 2   Press **ENTER**. See Figure 5.20 in Example 5.39 for the result.

Combinations $_nC_r$

We illustrate Example 5.39b (page 240): $_{10}C_7$.

Step 1   Select an empty cell and type =**COMBIN(10,7)**.

Step 2   Press **ENTER**. See Figure 5.22 in Example 5.39 for the result.

SECTION 5.4   Summary

1. The Multiplication Rule for Counting provides the total number of different possible outcomes for a series of events.

2. A permutation $_nP_r$ is an arrangement in which
- r items are chosen from n distinct items.
- repetition of items is not allowed.
- the order of the items is important.

In a permutation, order is important. In a combination, order does not matter. A combination $_nC_r$ is an arrangement in which
- r items are chosen from n distinct items.
- repetition of items is not allowed.
- the order of the items is not important.

3. Combinations may be used to calculate certain probabilities. For such problems, use the following steps.

Step 1   Confirm that the desired probability involves a combination.

Step 2   Find $N(E)$, the number of outcomes in event E.

Step 3   Find $N(S)$, the number of outcomes in the sample space.

Step 4   Assuming that each possible combination is equally likely, find the probability of event E as follows:

$$P(E) = \frac{N(E)}{N(S)}$$

Exercises

Clarifying the Concepts

1. What type of diagram is helpful in itemizing the possible outcomes of a series of events?

2. Explain in words how 5! is calculated.

3. What is the difference between a permutation and a combination?

4. Does $_8P_9$ make sense? Explain why or why not.

5. Describe in your own words what is meant by acceptance sampling.

6. The counting methods that we have learned in this section may be used to compute probabilities.
   a. For assigning probability, which method is used: classical, relative frequency, or subjective?
   b. Referring to part **(a)**, what assumption must be made to apply the method?

Practicing the Techniques

7. A pizza store offers the following options to its customers. Use a tree diagram to list all the possible options that a customer may choose from.
   - Cheese: no cheese, regular cheese, double cheese
   - Pepperoni: no pepperoni, regular pepperoni, double pepperoni

8. An ice cream shop offers the following options to its customers. Use a tree diagram to list all the possible options that a customer may choose from.
   - Ice cream: vanilla, chocolate, mint chocolate chip
   - Toppings: hot fudge, butterscotch, sprinkles

9. A particular baseball pitcher has to choose from the following options on each pitch. Use a tree diagram to list all the possible options.
   - Type of pitch: fastball, curve, slider
   - Horizontal position: inside corner, over the plate, outside corner
   - Vertical position: high, low

10. A women's clothing store tracks its sales transactions according to the following options. Use a tree diagram to list all the possible options.
   - Payment method: credit card, debit card, check, cash
   - Size category: Juniors, Misses, Women's
   - Type of clothing: top, pants

11. Our 41st president, George Herbert Walker Bush, had four names, with initials GHWB. How many different possible sets of initials are there for people with four names?

12. NCAA ice hockey games can have the following outcomes: win (W), lose (L), or tie (T). In a tournament of five games, how many different possible sets of outcomes are there? (*Hint:* LLTWW is one possible set.)

13. A college dining service conducted a survey in which it asked students to select their first and second favorite flavors of ice cream from a list of five flavors: vanilla, chocolate, mint chocolate chip, strawberry, maple walnut. How many different possible sets of two favorites are there?

14. A town library is considering loaning video games, and surveyed its membership to ask their four favorite PlayStation 3 games from among the following six games: Gran Turismo, Call of Duty 4, Metal Gear Solid 4, Little Big Planet, Grand Theft Auto IV, and Final Fantasy XIII. How many different possible sets of four favorites are there?

15. A woman is considering four sororities to rush this year. How many possible orderings are there?

16. Students working for the college newspaper have six drop locations around campus at which they must drop off newspapers. How many different possible routes are there for the students to do so?

For Exercises 17–22, find the value of each factorial.

17. 6!   **19.** 0!   **21.** 1!

18. 9!   **20.** 11!   **22.** 15!

23. A woman is considering four sororities to rush this year, but only has time to rush two. How many possible orderings are there?

24. Students working for the college newspaper have six drop locations around campus at which they must drop off newspapers, but they only have enough time to get to four locations. How many different possible routes are there for the students to do so?

For Exercises 25–32, find the value of each permutation $_nP_r$.

25. $_7P_3$ **28.** $_8P_3$ **31.** $_{100}P_{100}$

26. $_7P_4$ **29.** $_{100}P_1$ **32.** $_{100}P_{99}$

27. $_8P_5$ **30.** $_{100}P_0$

For Exercises 33–40, find the value of each combination $_nC_r$. Then answer Exercises 43 and 44.

33. $_7C_3$ **36.** $_{11}C_9$ **39.** $_{100}C_0$

34. $_7C_4$ **37.** $_{11}C_{10}$ **40.** $_{100}C_1$

35. $_{11}C_8$ **38.** $_{11}C_{11}$

41. How many distinct strings of letters can we make by using all the letters in the word PIZZA?

42. How many distinct strings of letters can we make by using all the letters in the word PEPPERONI?

43. Explain why the answers to Exercises 33 and 34 are equal. Use the commutative property of multiplication (for example, $2 \cdot 7 = 7 \cdot 2$) in your answer.

44. Use the idea behind your answer to Exercise 43 to find a combination that is equal to $_{11}C_8$. Verify your answer.

45. List all the permutations of the following people taken three at a time: Amy, Bob, Chris, Danielle. What is $_4P_3$?

46. List all the combinations of the following people taken three at a time: Amy, Bob, Chris, Danielle. What is $_4C_3$?

47. Explain in your own words why $_4P_3$ is larger than $_4C_3$.

48. What quantity do we divide $_4P_3$ by to get $_4C_3$? Express this quantity as a factorial. (*Hint:* For example, if the quantity were 120, we would express it as 5!)

49. In general, what do we divide $_nP_r$ by to get $_nC_r$?

Applying the Concepts

50. Fast Food. A fast-food restaurant has three types of sandwiches: chicken sandwich, fish sandwich, and beef burger. The restaurant has two types of side dishes: French fries and salad.

   a. Draw a tree diagram to find all the different meals a customer can order at this restaurant.

   b. How many different meals can a customer order at this restaurant?

51. What to Eat? A sit-down restaurant has two types of appetizers: garden salad and buffalo wings. It has three entrees: spaghetti, steak, and chicken. And it offers three kinds of desserts: ice cream, cake, and pie.

   a. Draw a tree diagram to find all the different meals a customer can order at this restaurant.

   b. How many different meals can a customer order at this restaurant?

52. Greek Alphabet. The ancient Greek alphabet had 24 letters. How many different possible initials are there for people with a first and last name?

53. Facebook Friends. A student has 10 friends on her Facebook page. How many ways can she arrange her 10 friends top to bottom?

54. Document Delivery. A document delivery person must deliver documents to five different destinations within a particular city. How many different routes are possible?

55. Traveler Fellow. A corporate sales executive must travel to the following countries this quarter: China, Russia, Germany, Brazil, India, and Nigeria. How many different routes are possible?

56. Sales Traveler. A corporate sales executive has the choice of traveling to four of the following six countries this quarter: China, Russia, Germany, Brazil, India, and Nigeria. How many different routes are possible?

57. Playing Catch. Five children are playing catch with a ball. How many different ways can one child throw a ball to another child once?

58. Chimp Grooming. Six chimpanzees are grooming each other at the city zoo. In how many different ways can one chimp groom another?

59. Shake Hands. In an ice-breaker exercise, each of 25 students is asked to shake hands with each of the other students. How many handshakes will there be in all?

60. Statistics Competition. Three students from the Honors Statistics class of 15 students will be chosen to represent the school at the state statistics competition. How many different possible groupings of 3 students are there?

61. How many random samples of size 1 can be chosen from a population of size 20?

62. How many random samples of size 20 can be chosen from a population of size 20?

63. How many random samples of size 10 can be chosen from a population of size 20?

64. How many distinct strings of letters can be made using all the letters in the word MATHEMATICS?

65. How many distinct strings of letters can be made using all the letters in the word BUSINESS?

66. Acceptance Sampling. A shipment of 25 personal digital assistants (PDAs) contains 3 that are defective. A quality control specialist inspects 2 of the 25 PDAs. If both are defective, then the shipment is rejected.

   a. Explain whether a permutation or a combination is being used.

   b. Find the number of ways that both PDAs will be defective.

   c. Find the probability of rejecting the shipment.

CHAPTER 5   **Formulas and Vocabulary**

Section 5.1

- **CLASSICAL METHOD OF ASSIGNING PROBABILITIES** (p. 197).

$$P(E) = \frac{\text{number of outcomes in } E}{\text{number of outcomes in sample space}} = \frac{N(E)}{N(S)}$$

- **EVENT** (p. 195)
- **EXPERIMENT** (p. 195)
- **LAW OF LARGE NUMBERS** (p. 201)
- **LAW OF TOTAL PROBABILITY** (p. 186)

- **OUTCOME** (p. 195)
- **PROBABILITY** (p. 195)
- **PROBABILITY MODEL** (p. 196)
- **RELATIVE FREQUENCY METHOD OF ASSIGNING PROBABILITIES** (Also known as the **EMPIRICAL METHOD**) (p. 202).

$$P(E) \approx \frac{\text{frequency of } E}{\text{number of trials of experiment}}$$

- **SAMPLE SPACE** (p. 195)

- **SIMULATION** (p. 201)
- **SUBJECTIVE PROBABILITY** (p. 204)
- **TREE DIAGRAM** (p. 198)

Section 5.2
- **ADDITION RULE** (p. 211).
$P(A \text{ or } B) = P(A \cup B) = P(A) + P(B) - P(A \cap B)$
- **ADDITION RULE FOR MUTUALLY EXCLUSIVE EVENTS** (p. 213). If A and B are mutually exclusive, then $P(A \cup B) = P(A) + P(B)$.
- **COMPLEMENT OF AN EVENT A** (p. 209). Denoted as A^C.
- **INTERSECTION OF TWO EVENTS A AND B** (p. 210). Denoted as $A \cap B$ or as "A and B."
- **MUTUALLY EXCLUSIVE (DISJOINT) EVENTS** (p. 213)
- **PROBABILITIES FOR COMPLEMENTS** (p. 210).
$P(A) + P(A^C) = 1$, $P(A) = 1 - P(A^C)$, and $P(A^C) = 1 - P(A)$
- **UNION OF TWO EVENTS A AND B** (p. 210). Denoted as $A \cup B$ or as "A or B."

Section 5.3
- **CONDITIONAL PROBABILITY** (p. 218).

$$P(B|A) = \frac{P(A \cap B)}{P(A)} = \frac{N(A \cap B)}{N(A)}$$

- **INDEPENDENT EVENTS** (p. 220). Events A and B are *independent* if $P(A|B) = P(A)$ or if $P(B|A) = P(B)$.

- **MULTIPLICATION RULE** (p. 221). $P(A \cap B) = P(B)$ $P(A|B)$ or, equivalently, $P(A \cap B) = P(A) P(B|A)$.
- **MULTIPLICATION RULE FOR INDEPENDENT EVENTS** (p. 222). If events A and B are independent, then $P(A \cap B) = P(A) P(B)$.
- **MULTIPLICATION RULE FOR n INDEPENDENT EVENTS** (p. 227). If $A, B, C, \ldots$ are independent events, then $P(A \cap B \cap C \cap \ldots) = P(A) P(B) P(C) \ldots$
- **SAMPLING WITH REPLACEMENT** (p. 223)
- **SAMPLING WITHOUT REPLACEMENT** (p. 223)

Section 5.4
- **ACCEPTANCE SAMPLING** (p. 242)
- **COMBINATION** (p. 239).

$$_nC_r = \frac{n!}{r!(n-r)!}$$

- **FACTORIAL SYMBOL n!** (p. 237). $0! = 1$; $1! = 1$; $n! = n(n-1)(n-2) \ldots 3 \cdot 2 \cdot 1$
- **MULTIPLICATION RULE FOR COUNTING** (p. 235)
- **PERMUTATION** (p. 237).

$$_nP_r = \frac{n!}{(n-r)!}$$

- **PERMUTATIONS OF NONDISTINCT ITEMS** (p. 241).

$$\frac{n!}{n_1! \cdot n_2! \cdot \ldots \cdot n_k!}$$

CHAPTER 5	**Review Exercises**

Section 5.1
For Exercises 1–5, consider the experiment of tossing a fair coin three times and find the probabilities.
1. 2 heads
2. At least 2 heads
3. 4 heads
4. 2 tails
5. At most 1 tail
6. A NEW SONNET. Literature researchers have unearthed a sonnet that they know to be by either William Shakespeare or Christopher Marlowe. The probability that the sonnet is by Marlowe is 25%.
  a. What is the probability that the sonnet is by Shakespeare?
  b. What method of assigning probability do you think was used here? Why was this method used, and not the others?

Section 5.2
7. FARMWORKERS' EDUCATIONAL LEVEL. The U.S. Department of Agriculture reports on the demographics of hired farmworkers.[6] An excerpt of the results is provided in the table, showing the percentage of noncitizen and citizen farmworkers who attained various educational levels. The educational levels are mutually exclusive. Find the following probabilities.
  a. The probability that a noncitizen farmworker is a high school graduate or has some college

  b. The probability that a citizen farmworker is a high school graduate or has some college
  c. The probability that a noncitizen farmworker has less than a ninth-grade education and has some college
  d. The probability that a farmworker is not a citizen.

	Noncitizens	Citizens
Less than 9th grade	238,008	61,776
9th–12th grade (no diploma)	57,904	152,880
High school graduate	59,784	222,144
Some college	20,304	187,200

Section 5.3
8. DRUG RESEARCH STUDIES. The *Annals of Internal Medicine* reported that 39 of the 40 research studies sponsored by a drug company had outcomes favoring a certain drug. Find the following probabilities, assuming independence.
  a. Three randomly selected research studies all favor this drug.
  b. None of the three randomly selected research studies favors this drug.
  c. At least one of three randomly selected research studies favors this drug.

9. DRUG RESEARCH STUDIES. Use the information in Exercise 8. Suppose we sample two research studies without replacement. Find the probability that the second study does not favor this drug given that the first study does not favor this drug.

GENDER AND PET PREFERENCE. Do you think your gender affects what type of pet you own? For Exercises 10–13, use the following table, showing preferences for various pets by owner gender.

Gender of owner	Cats	Dogs	Other pets	Total
Female	100	50	30	180
Male	50	50	20	120
Total	150	100	50	300

10. Find the probability that a randomly chosen person has the following characteristics.
   a. Owns a cat, $P(C)$ **b.** Owns a dog, $P(D)$

11. Find the probability that a randomly chosen person has the following characteristics.
   a. Is female and owns a dog, $P(F \cap D)$
   b. Is male and owns a dog, $P(M \cap D)$

12. Find the following conditional probabilities for a randomly chosen person.
   a. Owns a dog, given that the person is female, $P(D|F)$
   b. Owns a dog, given that the person is male, $P(D|M)$

13. If you were a dog-food manufacturer, would you advertise more on a men's TV channel or a women's TV channel? Why? Cite your evidence.

Section 5.4

14. How many distinguishable strings of letters can be made using all the letters in the word MISSISSIPPI?

15. STATISTICS QUIZ. On a statistics quiz, there are five true/false questions, four fill-in-the-blank questions, and three short-answer questions. How many different ways are there of taking this quiz?

16. INSPECTION TIME. A U.S. Army drill instructor will perform inspection on 2 soldiers in a squad of 18 soldiers. If both soldiers fail the inspection because their rifles are not clean, the entire squad will have to run a five-mile course in full gear. Three of the 18 soldiers have rifles that are not clean.
   a. Explain whether the drill instructor is using a permutation or a combination.
   b. Find the number of ways that both soldiers will fail the inspection.
   c. Find the probability that the entire squad will have to run a five-mile course in full gear.

CHAPTER 5	Quiz

True or False

1. True or false: An outcome is a collection of a series of events from the sample space of an experiment.

2. True or false: For any event A (even events like A: the moon is made of green cheese) the probability of A plus the probability of A^C always add up to 1.

Fill in the Blank

3. The minimum value that a probability can take is _____ and the maximum value is _____.

4. The union of two events is associated with the English word _____, and the intersection of two events is associated with the English word _____.

5. Someone has told you that there is a 50-50 chance of rain tomorrow. This means that the probability of rain tomorrow equals _____.

Short Answer

6. For any experiment, what is the sum of all the outcome probabilities in the sample space?

7. For which type of sampling are consecutive draws independent?

8. For two events A and B, what do we call the event containing only those outcomes that belong to both A and B?

Calculations and Interpretations

9. Consider the experiment of rolling a fair die twice. Find the following probabilities.
   a. Sum of the two dice equals 5.
   b. Sum of the two dice does not equal 5.
   c. One of the dice shows 2.
   d. Sum of the two dice equals 5 and one of the dice shows 2.
   e. Sum of the two dice equals 5 or one of the dice shows 2.

10. Suppose that A and B are any two events, with $P(B) = 0.75$ and $P(A \cap B) = 0.15$. Find $P(A|B)$.

11. Suppose that A and B are any two events, with $P(B) = 0.85$ and $P(A|B) = 0.25$. Find $P(A \cap B)$.

12. PICK A CARD. Consider the experiment of drawing a single card from a deck of 52 cards. Find the probability of observing the following events.
   a. Heart
   b. Face card (king, queen, or jack)
   c. Seven
   d. Red card
   e. Seven of hearts
   f. Red queen

HAPPINESS IN MARRIAGE. The General Social Survey tracks trends in American society through annual surveys. The married respondents were asked to characterize their feelings about being married. The results, crosstabulated with gender, are shown in the following figure. Use this information for Exercises 13–15.

		HAPPINESS OF MARRIAGE			
		VERY HAPPY	PRETTY HAPPY	NOT TOO HAPPY	Total
SEX	MALE	242	115	9	366
	FEMALE	257	149	17	423
Total		499	264	26	789

13. Find the probabilities that a randomly chosen person has the following characteristics.

   a. Is female, $P(F)$

   b. Is male, $P(M)$

   c. Is not too happily married, $P(\text{Not})$

14. Find the probabilities that a randomly chosen person has the following characteristics.

   a. Is female and not too happily married, $P(F \cap \text{Not})$

   b. Is male and not too happily married, $P(M \cap \text{Not})$

15. Are gender and being not too happily married independent? Why or why not?

16. FOOTBALL TEAMS. The four teams in the AFC South division of the National Football League are Indianapolis Colts, Jacksonville Jaguars, Tennessee Titans, and Houston Texans. Suppose the top three teams in the division this year will make the playoffs. How many different sets of teams making the playoffs are there?

17. STATE LOTTERY. In a state lottery, balls numbered 1 to 20 are placed in an urn. To win, you must choose numbers that match the three balls chosen in the order that they're chosen.

   a. Explain whether a permutation or a combination is being used.

   b. How many possible outcomes are there?

   c. Find the probability of winning this lottery if your ticket contains a single ordering of three numbers.

6 Probability Distributions

Michael Newman/Photo Edit

CASE STUDY

Text Messaging

Do you prefer receiving text messages or phone calls on your cell phone? The Chapter 6 Case Study, "Text Messaging," explores this and other questions, using data collected by the Pew Internet and American Life Project.[1] For example, their survey showed that 31% of adult Americans prefer receiving text messages to phone calls on their cell phones, compared to 53% who prefer phone calls. In Section 6.2 we determine whether it would be unusual to find 45 out of a sample of 100 American adults who prefer receiving text messages. Then in Section 6.5 we learn how to be careful of what we assume. ■

The Big Picture

Where we are coming from, and where we are headed . . .

- In Chapter 5, we learned about *probability*, which allows us to quantify the uncertainty involved in performing statistical inference in later chapters.

- However, we first need a new set of tools in our probability toolbox: *random variables* and *probability distributions*. Here, in Chapter 6, we learn these new tools, including the binomial distribution and the normal distribution.

- Chapter 7, "Sampling Distributions," is a pivotal chapter where we learn that statistics have predictable behavior, which allows us to perform the statistical inference we learn in the remainder of the book.

6.1 DISCRETE RANDOM VARIABLES

OBJECTIVES By the end of this section, I will be able to . . .

1. Identify random variables.

2. Explain what a discrete probability distribution is and construct probability distribution tables and graphs.

3. Calculate the mean, variance, and standard deviation of a discrete random variable.

1 RANDOM VARIABLES

In Chapter 5, we calculated the probabilities of outcomes from experiments. If the experiment is tossing a fair coin twice, the outcomes are HH, HT, TH, and TT. The probability of observing exactly one head in two tosses is the probability of the event $A = \{HT, TH\}$. Since the outcomes are equally likely, we used the classical method of assigning probability. The probability of $\{HT, TH\}$ is $N(A)/N(S) = 2/4 = 0.5$, where S is the sample space.

In this chapter, we develop a different approach that analyzes probability problems more efficiently. Recall from Chapter 1 that a *variable* is a characteristic that can assume different values. Suppose we define a variable $X =$ number of heads observed when 2 fair coins are tossed. In this experiment we may observe zero heads, one head, or two heads, so that the possible values of X are 0, 1, and 2. Clearly, before we conduct our experiment, we do not know how many heads we will observe. Thus, randomness plays a role in the value of the variable X, and so we call X a *random variable*.

A **random variable** is a variable that takes on quantitative values representing the results of a probability experiment, and thus its values are determined by chance. We denote random variables using capital letters such as X, Y, or Z.

In Chapter 5 (page 199), we found that the probability of observing exactly $X = $ one head was 0.5. We denote this probability using the notation

$$P(X = 1) = 0.5$$

Similarly, the probability of observing zero heads is $P(X = 0) = 0.25$, and the probability of two heads is $P(X = 2) = 0.25$.

Developing Your Statistical Sense

Random Variables Must Be Random!

The role of chance in the definition of a random variable is crucial. For example, is your age a random variable? If we are just talking about you and no one else, and we know your age, then there is no chance involved. In that case, your age is not a random variable. On the other hand, what if we select students *at random* by picking names from a hat? Then the age of the person drawn is a random variable because its value depends at least partly on chance (on which name is drawn at random).

Let's start with an example aimed at helping you move from the language of probability (experiments and outcomes) to the language of random variables.

EXAMPLE 6.1

NOTATION FOR RANDOM VARIABLES

Comstock/Jupiter Images

Suppose our experiment is to toss a single fair die, and we are interested in the number rolled. We define our random variable X to be the outcome of a single die roll.
a. Why is the variable X a random variable?
b. What are the possible values that the random variable X can take?
c. What is the notation used for rolling a 5?
d. Use random variable notation to express the probability of rolling a 5.

Solution

a. We don't know the value of X before we toss the die, which introduces an element of chance into the experiment, thereby making X a random variable.
b. The possible values for X are 1, 2, 3, 4, 5, and 6.
c. When a 5 is rolled, then X equals the outcome 5, and we write $X = 5$.
d. Recall from Section 5.1 that the probability of rolling a 5 for a fair die is 1/6. In random variable notation, we denote this as $P(X = 5) = 1/6$.

There are two main types of random variables: **discrete random variables** and **continuous random variables.** The difference between the two types relates to the possible values that each type of random variable can assume.

> **Discrete and Continuous Random Variables**
>
> • A **discrete random variable** can take either a finite or a countable number of values. Since these values may be written as a list of numbers, each value can be graphed as a separate point on a number line, with space between each point. (See Figure 6.1a.)
>
>
>
> **FIGURE 6.1a**
>
> • A **continuous random variable** can take infinitely many values. Because there are infinitely many values, the values of a continuous random variable form an interval on the number line. (See Figure 6.1b.)
>
>
>
> **FIGURE 6.1b**

Discrete random variables usually need to be counted, like 1, 2, 3, and so forth. Continuous random variables usually need to be measured, not counted, such as measuring the amount of gasoline purchased.

Examples of discrete random variables include the number of children a randomly selected person has and the number of times a randomly chosen student has been pulled over for speeding on the interstate. Continuous random variables often need to be measured, not counted. For example, the temperature in Atlanta, Georgia, at noon today may be reported as 77 degrees, but this value represents actual temperatures that may lie anywhere between 76.5 degrees and 77.5 degrees.

EXAMPLE 6.2

IDENTIFYING DISCRETE AND CONTINUOUS RANDOM VARIABLES

For the following random variables, (i) determine whether they are discrete or continuous, and (ii) indicate the possible values they can take.
a. The number of automobiles owned by a family
b. The width of your desk in this classroom
c. The number of games played in the next World Series
d. The weight of model year 2011 SUVs

Solution

a. Since the possible number of automobiles owned by a family is finite and may be written as a list of numbers, it represents a discrete random variable. The possible values are $\{0, 1, 2, 3, 4, \ldots\}$.
b. Width is something that must be measured, not counted. Width can take infinitely many different possible values, with these values forming an interval on the number line. Thus, the width of your desk is a continuous random variable. The possible values might be $1 \text{ ft} \leq W \leq 10 \text{ ft}$.
c. The number of games played in the next World Series can be counted and thus represents a discrete random variable. The possible values are finite and may be written as a list of numbers: $\{4, 5, 6, 7\}$.
d. The weight of model year 2011 SUVs must be measured, not counted, and so represents a continuous random variable. Weight can take infinitely many different possible values, with these values forming an interval on the number line: $2500 \text{ lb} \leq Y \leq 7000 \text{ lb}$.

**Now You Can Do
Exercises 7–16.**

We will return to continuous random variables in Section 6.3; Sections 6.1 and 6.2 concentrate on discrete random variables.

2 DISCRETE PROBABILITY DISTRIBUTIONS

For every random variable, there is a probability distribution that allows us to view all possible values of the random variable at a glance. Discrete probability distributions show the probabilities associated with the various values that the discrete random variable can take.

> A **probability distribution of a discrete random variable** provides all the possible values that the random variable can assume, together with the probability associated with each value. The probability distribution can take the form of a table, graph, or formula. Probability distributions describe populations, not samples.

When constructing the tabular form of a **probability distribution of a discrete random variable**, create a table with two rows:

- The top row will contain all the possible values of X.
- The bottom row will contain the probability associated with each value of X.

EXAMPLE 6.3

PROBABILITY DISTRIBUTION TABLE

Construct the probability distribution table of the number of heads observed when tossing a fair coin twice.

Solution

The probability distribution table given in Table 6.1 uses the probabilities we found on page 199.

The probabilities in Table 6.1 were assigned using the classical method, since we assumed that tossing a fair coin would result in equally likely outcomes.

Now You Can Do Exercises 17–20.

Table 6.1 Probability distribution table of the number of heads on two fair coin tosses

X = number of heads observed	0	1	2
$P(X)$ = probability of observing that many heads	1/4	1/2	1/4

Note that the probabilities in the bottom row of Table 6.1 add up to 1. Also, note that since each value in the bottom row is a probability, each value must be between 0 and 1, inclusive, that is, $0 \leq P(X) \leq 1$. We can generalize this as follows.

This first rule derives from the Law of Total Probability from Section 5.1 (page 196).

> **Rules for a Discrete Probability Distribution**
> - The sum of the probabilities of all the possible values of a discrete random variable must equal 1. That is, $\sum P(X) = 1$.
> - The probability of each value of X must be between 0 and 1, inclusive. That is, $0 \leq P(X) \leq 1$.

EXAMPLE 6.4

RECOGNIZING VALID DISCRETE PROBABILITY DISTRIBUTIONS

Identify which of the following is a valid discrete probability distribution.

a.

X	1	10	100	1000
$P(X)$	0.2	0.4	0.3	0.2

b.

X	-10	0	10	20
$P(X)$	0.5	0.3	0.4	-0.2

c.

X	Red	Green	Blue	Yellow
$P(X)$	0.1	0.3	0.4	0.2

d.

X	-5	0	5	10
$P(X)$	0.1	0.3	0.4	0.2

Solution

a. This is not a valid probability distribution, because the probabilities add up to 1.1, which is greater than 1.

b. This is not a valid probability distribution, because $P(X = 20)$ is negative.

c. This is not a valid probability distribution for a discrete random variable because the values of X are not quantitative.

Now You Can Do Exercises 21–24.

d. This is a valid probability distribution, since the probabilities sum to 1, and each probability $P(X)$ takes a value between 0 and 1.

Probability distributions can also take the form of a probability distribution graph.

EXAMPLE 6.5

DISCRETE PROBABILITY DISTRIBUTION AS A GRAPH

The number of points a soccer team gets for a game is a random variable, because it is not certain, prior to the game, how many points the team will get.

The probabilities were assigned to the random variable X using the relative frequency (empirical) method.

In Major League Soccer (MLS), teams are awarded 3 points in the standings for a win, 1 point for a tie, and 0 points for a loss. In the 30-game 2010 MLS season, the New York Red Bulls had 15 wins, 9 losses, and 6 ties.

a. Construct a probability distribution table of the number of points per game, based on the team's performance during the 2010 MLS season.

b. Construct a probability distribution graph of the number of points per game.

Solution

a. Let X = points awarded. Then the probability distribution table is given in Table 6.2.

Table 6.2 Probability distribution table of points awarded for New York Red Bulls

X = points	0	1	3
$P(X)$	9/30 = 0.3	6/30 = 0.2	15/30 = 0.5

b. The probability distribution graph is given in Figure 6.2.

- The horizontal axis is the usual x axis (the number line), and shows all the possible values that the random variable X can take, such as $X = 0$, 1, or 3. The horizontal axis gives the same information as the top row of the table.

- The vertical axis represents probability, and is the information in the bottom row in the table. A vertical bar is drawn at each value of X, with the height representing the probability of that value of X. For example, the bar of probability at $X = 0$ goes up to 0.3 and represents the probability that the New York Red Bulls will lose a game.

Given a graph of a probability distribution, you should know how to construct the probability distribution table, and vice versa.

FIGURE 6.2
Probability distribution graph of points awarded for New York Red Bulls.

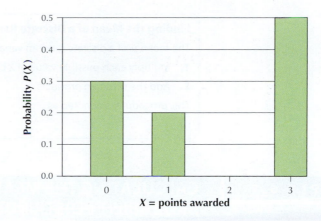

**Now You Can Do
Exercises 25–28.**

We may use probability distributions to calculate probabilities for multiple values of X. In discrete probability distributions, the outcomes are always mutually exclusive. For example, it is not possible to observe both zero heads ($X = 0$) and two heads ($X = 2$) when tossing two fair coins. Thus, we always use the Addition Rule for Mutually Exclusive Events to find the probability of two or more outcomes for a discrete random variable. For example, $P(X = 0 \text{ or } 2) = P(X = 0) + P(X = 2)$.

EXAMPLE 6.6

CALCULATING PROBABILITIES FOR MULTIPLE VALUES OF X

Use the probability distribution from Example 6.5 to find the following probabilities.
a. Probability that the New York Red Bulls are awarded either 0 or 3 points in a game
b. Probability that the New York Red Bulls are awarded both 0 and 3 points in a game
c. Probability that the New York Red Bulls are awarded at least 1 point in a game
d. Probability that the New York Red Bulls are awarded at most 1 point in a game

Solution
a. $P(X = 0 \text{ points } or \text{ 3 points}) = P(X = 0) + P(X = 3) = 0.3 + 0.5 = 0.8$. For a randomly selected game, the probability that the Red Bulls either lose the game or win the game is 0.8.
b. The outcomes $X = 0$ and $X = 3$ are mutually exclusive. Therefore, $P(X = 0 \text{ points } and \text{ 3 points}) = 0$.
c. The phrase *at least* means "that many or more." Thus we need to find: $P(X \geq 1) = P(X = 1 \text{ point } or \text{ 3 points}) = P(X = 1) + P(X = 3) = 0.2 + 0.5 = 0.7$.
d. The phrase *at most* means "that many or fewer." Hence: $P(X \leq 1) = P(X = 1 \text{ point } or \text{ 0 points}) = P(X = 1) + P(X = 0) = 0.2 + 0.3 = 0.5$.

**Now You Can Do
Exercises 29–32.**

---

3 MEAN AND VARIABILITY OF A DISCRETE RANDOM VARIABLE

Just as we can compute the mean and standard deviation of quantitative data, we can calculate the mean and standard deviation of a random variable X.

The **mean μ of a discrete random variable** X represents the mean result when the experiment is repeated an indefinitely large number of times.

> **Finding the Mean of a Discrete Random Variable X**
>
> The mean μ of a discrete random variable X is found as follows:
>
> 1. Multiply each possible value of X by its probability.
> 2. Add the resulting products.
>
> This procedure is denoted as
>
> $$\mu = \sum [X \cdot P(X)]$$

EXAMPLE 6.7

CALCULATING THE MEAN OF A DISCRETE PROBABILITY DISTRIBUTION

Note: These 250,000 teenagers constitute a population, not a sample, so the mean is μ, not $\bar{x}$.

The U.S. Department of Health and Human Services reports that there were 250,000 babies born to teenagers aged 15–18 in 2004. Of these 250,000 births, 7% were to 15-year-olds, 17% were to 16-year-olds, 29% were to 17-year-olds, and 47% were to 18-year-olds.

a. Construct the probability distribution table for $X =$ age.
b. Calculate the mean age μ.

Solution

a. The following table contains the probability distribution of the random variable $X =$ age.

$X =$ age	$P(X)$
15	0.07
16	0.17
17	0.29
18	0.47

b. To find the mean μ, we first need to multiply each possible outcome (value of X) by its probability $P(X)$. We multiply the value $X = 15$ by its probability $P(X) = 0.07$, the value $X = 16$ by its probability $P(X) = 0.17$, and so on. Then we add these four products to find the mean:

$$\mu = 15(0.07) + 16(0.17) + 17(0.29) + 18(0.47) = 17.16$$

The mean age of the mother for the babies born to teenagers aged 15–18 is 17.16 years.

What Does This Number Mean?

What does it mean to say that $\mu = 17.16$ is the mean of the random variable $X =$ age? First of all, the mean of the random variable X is definitely not the same as the mean of a sample of teenage mothers. The latter is a sample mean. For example, suppose that, for a certain hospital, the teenage mothers' ages for the last 5 such births were 16, 18, 18, 17, 18. The mean of this sample of 5 births is $\bar{x} = 17.4$. However, if we were to consider an *infinite number* of births to mothers aged 15–18, then the mean of this very large sample would converge to $\mu = 17.16$. So the mean μ of a discrete random variable is interpreted as the mean of the results from the *population* of all possible repetitions of the experiment, which is why we denote the mean of a random variable as μ.

Note: The population mean μ need not equal any values of X, nor need it be an integer.

Developing Your Statistical Sense

Why Does This Formula Work?

The formula for the mean of a discrete random variable works because it is a special case of the weighted mean (page 115). Of the population of 250,000 babies, 7%, or 17,500, were born to 15-year-olds. Thus, $w_1 = 17,500$. Similarly, we can find, $w_2 = (0.17)(250,000) = 42,500$, $w_3 = (0.29)(250,000) = 72,500$, and $w_4 = (0.47)(250,000) = 117,500$. Thus, the population weighted mean is

$$\mu = \frac{\sum w_i x_i}{\sum w_i} = \frac{(17,500)(15) + (42,500)(16) + (72,500)(17) + (117,500)(18)}{250,000}$$

Dividing through and rearranging terms give us

$$\mu = (15)(0.07) + (16)(0.17) + (17)(0.29) + (18)(0.47) = \sum [X \cdot P(X)]$$

We may also interpret the mean μ as the *balance point* of the distribution.

EXAMPLE 6.8   MEAN μ AS BALANCE POINT OF THE DISTRIBUTION

Graph the probability distribution of the random variable $X =$ age, and insert a fulcrum at the value of the mean, $\mu = 17.16$.

Solution

The probability distribution graph of $X =$ age is given in Figure 6.3. Note that the distribution is balanced at the point $\mu = 17.16$.

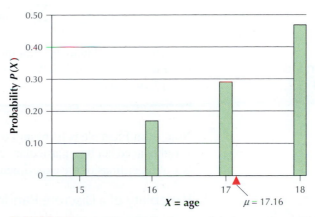

FIGURE 6.3 Probability distribution graph balances at $\mu = 17.16$.

In certain situations, we may need to identify the most likely value of the random variable X.

EXAMPLE 6.9   IDENTIFYING THE MOST LIKELY VALUE OF A DISCRETE RANDOM VARIABLE

If one of the teenagers represented in the table in Example 6.7 is chosen at random, what is the most likely age of that teenager when her baby was born?

Solution

Now You Can Do Exercises 45–47.

Since the largest probability in the probability table is $P(X = 18)$, and the longest bar in the probability graph is for $X = 18$, then 18 is the most likely age.

The mean μ of a random variable is also called the **expected value** or the **expectation of the random variable X.** It does not necessarily follow that the expected value of X is the most likely value of X. However, the expected value of X (that is, the mean μ) is often a good indication of the center of the distribution of the random variable.

> The **expected value, or expectation, of a random variable X** is the mean μ of X. It is denoted as $E(X)$. This definition holds for both discrete and continuous random variables.

EXAMPLE 6.10

EXPECTED VALUE OF A DISCRETE RANDOM VARIABLE X

Find the expected value $E(X)$ of the following discrete random variables.
a. X = number of heads in Example 6.3.
b. X = number of points awarded in Example 6.5.
c. X = age of teenage birth mothers in Example 6.7.

Solution
a. Using the probabilities in Table 6.1, we have

$$E(X) = \mu = \sum[X \cdot P(X)] = 0(0.25) + 1(0.5) + 2(0.25) = 1$$

The expected number of heads is 1
b. Using Table 6.2, we have

$$E(X) = \mu = \sum[X \cdot P(X)] = 0(0.3) + 1(0.2) + 3(0.5) = 1.7$$

The expected number of points is 1.7.

Now You Can Do Exercises 49–52.

c. From Example 6.7, $E(X) = \mu = 17.16$. The expected age of teenage mothers is 17.16 years.

Note from Example 6.10(b) and 6.10(c) that the mean or expected value of a random variable need not be a particular value of X. Rather, it is the mean of a very large number of repetitions of the experiment.

Variability of a Discrete Random Variable

Since a discrete random variable takes on quantitative values, we use the **variance** or **standard deviation of a random variable X** to help us determine whether a particular value of that random variable is unusual. Just as a random variable X has a mean (μ), which is a measure of center, so a random variable X also has a standard deviation (σ) and variance (σ^2), which are measures of spread. The variance of a discrete random variable is given by

$$\sigma^2 = \sum[(X - \mu)^2 \cdot P(X)]$$

Notice that this formula includes μ as one of its terms, so that you must first find the mean of a discrete random variable before you find the variance (or standard deviation). Recall from Chapter 3 that the standard deviation is simply the square root of the variance. The definition formula for the variance can sometimes be tedious since you must find each of the deviations $(X - \mu)$. The computational formulas below are equivalent to the definition formulas but are easier to work out.

> **Formulas for the Variance and Standard Deviation of a Discrete Random Variable**
>
> | Definition Formulas | Computational Formulas |
> |---|---|
> | $\sigma^2 = \sum[(X - \mu)^2 \cdot P(X)]$ | $\sigma^2 = \sum[X^2 \cdot P(X)] - \mu^2$ |
> | $\sigma = \sqrt{\sum[(X - \mu)^2 \cdot P(X)]}$ | $\sigma = \sqrt{\sum[X^2 \cdot P(X)] - \mu^2}$ |

EXAMPLE 6.11

CALCULATING THE VARIANCE AND STANDARD DEVIATION OF A DISCRETE RANDOM VARIABLE

X = number of credits taken	$P(X)$
12	0.1
13	0.1
14	0.1
15	0.5
16	0.1
20	0.1

 credits

Carla has 10 friends in school. She took a census of all 10 friends, asking each how many credits they had registered for that semester. Five of her friends were taking 15 credits, with one each taking 12, 13, 14, 16, and 20 credits. The resulting probability distribution table is shown to the left.

a. Find the mean μ number of credits taken.
b. Calculate the variance and standard deviation using the definition formula.
c. Calculate the variance and standard deviation using the computational formula.

Solution

a. $\mu = \sum[X \cdot P(X)] = 12(0.1) + 13(0.1) + 14(0.1) + 15(0.5) + 16(0.1) + 20(0.1) = 15$. The mean number of credits taken this semester among Carla's friends is $\mu = 15$.

b. Refer to Table 6.3. The first two columns correspond to the probability distribution of X = number of credits taken. The third column represents the calculations needed to find $(X - \mu)^2 \cdot P(X)$. Summing the values in the rightmost column provides the variance $\sigma^2 = 4$. Taking the square root of the variance gives us the standard deviation $\sigma = \sqrt{\sigma^2} = \sqrt{4} = 2$ credits.

Table 6.3 Calculating σ^2 and σ using the definition formula

X	$P(X)$	$(X - \mu)^2 \cdot P(X)$
12	0.1	$(12 - 15)^2 \cdot 0.1 = 0.9$
13	0.1	$(13 - 15)^2 \cdot 0.1 = 0.4$
14	0.1	$(14 - 15)^2 \cdot 0.1 = 0.1$
15	0.5	$(15 - 15)^2 \cdot 0.5 = 0.0$
16	0.1	$(16 - 15)^2 \cdot 0.1 = 0.1$
20	0.1	$(20 - 15)^2 \cdot 0.1 = 2.5$
		$\sigma^2 = \sum(X - \mu)^2 \cdot P(X) = 4$

c. Refer to Table 6.4 on the next page. The rightmost column contains the values $X^2 \cdot P(X)$. Summing the values in the rightmost column provides $\sum[X^2 \cdot P(X)] = 229$. To find the variance σ^2, we must subtract the square of the mean μ^2:

$$\sigma^2 = \sum[X^2 \cdot P(X)] - \mu^2 = 229 - 15^2 = 4$$

Taking the square root of the variance gives us the standard deviation $\sigma = \sqrt{\sigma^2} = \sqrt{4} = 2$ credits.

Table 6.4 Calculating σ^2 and σ using the computational formula

X	P(X)	X² · P(X)
12	0.1	$(12)^2 \cdot 0.1 = 14.4$
13	0.1	$(13)^2 \cdot 0.1 = 16.9$
14	0.1	$(14)^2 \cdot 0.1 = 19.6$
15	0.5	$(15)^2 \cdot 0.5 = 112.5$
16	0.1	$(16)^2 \cdot 0.1 = 25.6$
20	0.1	$(20)^2 \cdot 0.1 = 40$
		$\sum X^2 \cdot P(X) = 229$

Now You Can Do Exercises 53–56.

Now that we have calculated the standard deviation σ, we may use it along with the mean to determine whether values of X are outliers or moderately unusual, using the Z-score method.

EXAMPLE 6.12

Z-SCORE METHOD FOR DETERMINING AN UNUSUAL RESULT

a. Using the information from Example 6.11, determine whether X = 20 is an unusual number of credits to register for this semester.
b. Construct a probability distribution graph of X.

Solution

a. Recall from Section 3.4 (page 124) that a data value with a Z-score between 2 and 3 may be considered moderately unusual. The Z-score for X = 20 credits is

$$Z = \frac{X - \mu}{\sigma} = \frac{20 - 15}{2} = 2.5$$

Thus, among Carla's friends, it would be considered moderately unusual to take 20 credits this semester.

b. Figure 6.4 shows the probability distribution graph of X = number of credits. The mean $\mu = 15$ is indicated, along with the distances $\mu \pm 1\sigma$, $\mu \pm 2\sigma$, and $\mu \pm 3\sigma$.

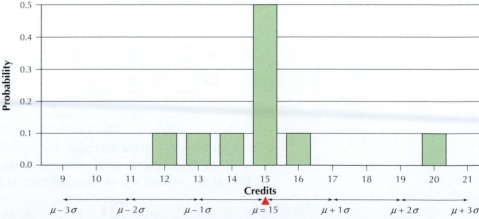

Now You Can Do Exercises 57–60.

FIGURE 6.4 X = 20 credits is moderately unusual because it lies Z = 2.5 standard deviations above the mean.

EXAMPLE 6.13 **COMPUTE THE MEAN AND STANDARD DEVIATION OF A DISCRETE RANDOM VARIABLE USING TECHNOLOGY**

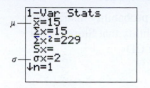

FIGURE 6.5 TI-83/84 results for mean and standard deviation of a discrete random variable.

Compute the mean and standard deviation of the probability distribution given in Example 6.11 using the TI-83/84 graphing calculator.

Solution

We use the instructions provided in the following Step-by-Step Technology Guide. The results are shown in Figure 6.5. Be careful! The calculator indicates that the mean is $\bar{x}$. It is not $\bar{x}$ but μ.

STEP-BY-STEP TECHNOLOGY GUIDE: Mean and Standard Deviation of a Discrete Random Variable

We illustrate using the data from Example 6.11.

TI-83/84

Step 1 Enter the X values in list **L1**, and the corresponding $P(X)$ values in list **L2**. See Figure 6.6a.
Step 2 Press **STAT**, highlight **CALC**, and select **1-Var Stats**.
Step 3 Type **L1** followed by a comma, followed by **L2**, as shown in Figure 6.6b. Press **ENTER**. The results are shown in Figure 6.5 above.

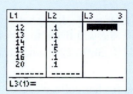

FIGURE 6.6a

FIGURE 6.6b

SECTION 6.1   Summary

1. Section 6.1 introduces the idea of random variables, which are variables whose value is determined at least partly by chance. Discrete random variables take values that are either finite or countable and may be put in a list. Continuous random variables take an infinite number of possible values, represented by an interval on the number line.

2. Discrete random variables can be described using a probability distribution, which specifies the probability

of observing each value of the random variable. Such a distribution can take the form of a table, graph, or formula. Probability distributions describe populations, not samples.

3. We can calculate and interpret the mean μ, standard deviation σ, and variance σ^2 of a discrete random variable using formulas.

SECTION 6.1   Exercises

Clarifying the Concepts

1. Explain in your own words what a random variable is. Give an example of a random variable from your own life experience.

2. Is your height a random variable? Under what circumstances would your height be considered a random variable? Under what circumstance would your height not be considered a random variable?

3. What is the difference between a discrete random variable and a continuous random variable?

4. What is the difference between a discrete random variable and a discrete probability distribution?

5. What are the two rules for a discrete probability distribution?

6. Explain the difference between $\bar{x}$ from Section 3.1 and the mean of a discrete random variable.

Practicing the Techniques

For Exercises 7–12, indicate whether the variable is a discrete or continuous random variable.

7. Number of siblings a randomly chosen person has

8. How long you will wait in your next checkout line

9. How much coffee there is in your next cup of coffee

10. How hot it will be the next time you visit the beach

11. The number of correct answers on your next multiple-choice quiz

12. How many songs you download this month

For Exercises 13–16, write down the possible values of the discrete random variables.

13. The number of students in a classroom where the maximum class size is 15

14. How many different fingers you will get paper cuts on next week

15. The number of games that the California Angels will win the next time they are in the World Series (maximum = 4)

16. The number of Donald Duck's three nephews, Huey, Dewey, and Louie, who will get into trouble in their next cartoon adventure

For Exercises 17–20, use the given information to construct a probability distribution table.

17. Shirelle enjoys listening to CDs while doing her homework. The probabilities that she will listen to $X = 0, 1, 2, 3,$ or 4 CDs tonight are 6%, 24%, 38%, 22%, and 10%, respectively.

18. Josefina loves to score goals for her college soccer team. The probabilities that she will score $X = 0, 1, 2,$ or 3 goals tonight are 0.25, 0.35, 0.25, and 0.15.

19. Joshua is going to make it big on Wall Street, if only he can graduate from college first. Joshua has invested money in a high-risk mutual fund, and has figured his probability of losing \$10,000 to be one-third, his probability of gaining \$10,000 to be one-half, and his probability of gaining \$50,000 to be one-sixth. Let X = money gained.

20. Chelsea is looking for a roommate, and would prefer a roommate who had either one or two pets. Of the 10 possible roommates who answered Chelsea's ad, 5 have no pets, 3 have one pet, 1 has two pets, and 1 has three pets.

For Exercises 21–24, determine whether the distribution represents a valid probability distribution. If it does not, explain why not.

21.

X	-10	0	10
$P(X)$	1/5	1/2	1/5

22.

X	15	16	17	20
$P(X)$	0.98	0.005	0.005	0.01

23.

X	1	2	3	4	5
$P(X)$	-0.5	0.5	0.7	0.1	0.2

24.

X	$-100,000$	50,000	100,000
$P(X)$	0.5	0.1	1.1

For Exercises 25–28, construct a probability distribution graph for the indicated discrete random variable X.

25. The number of CDs from Exercise 17

26. The number of goals from Exercise 18

27. The amount of money gained from Exercise 19

28. The number of pets from Exercise 20

For Exercises 29–32, refer to the probability distribution from Exercise 17. Find the probability that Shirelle will listen to the indicated numbers of CDs.

29. At least 3 CDs

30. At most 1 CD

31. Exactly 5 CDs

32. At least 1 CD

For Exercises 33–36, refer to the probability distribution from Exercise 18. Find the probability that Josefina will score the following numbers of goals.

33. At least 2 goals

34. At most 1 goal

35. Exactly 4 goals

36. At least 1 goal

For Exercises 37–40, refer to the probability distribution from Exercise 19. Calculate the following probabilities.

37. That Joshua will gain money on his investment

38. That Joshua will lose money

39. That Joshua will neither gain nor lose money

40. That Joshua will gain \$100,000

For Exercises 41–44, refer to the probability distribution from Exercise 20. Calculate the following probabilities.

41. That the roommate has at least 1 pet

42. That the roommate has at most 1 pet

43. That the roommate has at least 2 pets

44. That the roommate has the number of pets that Chelsea prefers

For Exercises 45–48, identify the most likely value of X, for the indicated random variables.

45. The number of CDs from Exercise 17

46. The number of goals from Exercise 18

47. The amount of money gained from Exercise 19

48. The number of pets from Exercise 20

For Exercises 49–52, find the expected value of the indicated random variable X.

49. The number of CDs from Exercise 17

50. The number of goals from Exercise 18

51. The amount of money gained from Exercise 19

52. The number of pets from Exercise 20

For Exercises 53–56, compute the variance and standard deviation of the indicated random variable X.

53. The number of CDs from Exercise 17

54. The number of goals from Exercise 18

55. The amount of money gained from Exercise 19

56. The number of pets from Exercise 20

For Exercises 57–60, use the *Z*-score method to determine whether there are any outliers or unusual data values.

57. The number of CDs from Exercise 17

58. The number of goals from Exercise 18

59. The amount of money gained from Exercise 19

60. The number of pets from Exercise 20

Applying the Concepts

61. Stanley Cup Finals. The National Hockey League championship is decided by a best-of-seven playoff called the Stanley Cup Finals. The following table shows the possible values of *X* = number of games in the series, and the frequency of each value of *X*, for the Stanley Cup Finals between 1990 and 2010.

🏒 stanleycup

X = games	Frequency
4	5
5	4
6	5
7	6

a. Explain why the number of games is a random variable.
b. Explain why the number of games is a discrete and not a continuous random variable.
c. Construct a probability distribution table for *X*.
d. Construct a probability distribution graph for *X*.
e. Find $P(X \leq 5)$.
f. Identify the most likely value of *X*.

62. Number of Courses Taught. The table provides the probability distribution for *X* = number of courses taught by faculty at all degree-granting institutions of higher learning in the United States in the fall 2010 semester.[2]

🏫 coursestaught

X = courses taught	P(X)
1	0.23
2	0.34
3	0.24
4	0.12
5	0.07

a. Explain why the number of courses taught is a random variable.
b. Explain why the number of courses taught is a discrete, and not a continuous, random variable.

c. Construct a probability distribution graph for *X*.
d. Find $P(X \geq 3)$.
e. Compute $P(X > 3)$.
f. Identify the most likely value of *X*.

63. Teenage Smokers. The National Survey on Drug Use and Health (2005) reported that 5 million young people aged 12–18 had tried tobacco products in the previous month. The table contains the proportions of the 5 million who had done so, at each age level. Let *X* = age of the person who had tried tobacco products in the previous month.

 teensmoker

X = age	P(X)
12	0.01
13	0.04
14	0.07
15	0.13
16	0.18
17	0.23
18	0.34

a. Construct a probability distribution graph for *X*.
b. Find $P(X \leq 16)$.
c. Compute $P(X < 16)$.
d. What is the difference between your answers to (**b**) and (**c**)?
e. Identify the most likely value of *X*.

64. Stanley Cup Finals. Refer to Exercise 61.
a. Calculate and interpret the mean number of games.
b. Compute the variance and standard deviation of the number of games.
c. Use the *Z*-score method to determine whether it is unusual for the Stanley Cup Finals to be a sweep (*X* = 4 games).

65. Number of Courses Taught. Refer to Exercise 62.
a. Find and interpret the expected number of courses.
b. Calculate the variance and standard deviation of the number of courses taught.
c. Use the *Z*-score method to determine whether it is unusual to teach 5 courses.

66. Teenage Smokers. Refer to Exercise 63.
a. Calculate and interpret the expected value of the variable *age*.
b. Calculate the variance and standard deviation of the variable *age*.
c. Determine whether a 12-year-old who had tried tobacco products in the previous month would be considered unusual. How about a 13-year-old?

Bringing It All Together

The Two-Dice Experiment. Use the following information for Exercises 67–70. Your experiment is to toss a pair of fair dice and find *X* = sum of the two dice.

67. Recall the sample space for the two-dice experiment from Figure 5.3 in Section 5.1 (page 200).
  a. Construct the probability distribution table of X.
  b. Graph the probability distribution of X, estimating the mean μ using the balance point method.
  c. Calculate the mean μ, and compare the result with your estimate from part **(b)**. Interpret the value of μ so that a nonspecialist would understand it.
  d. Compute the standard deviation σ of X.
  e. In your probability distribution graph from part **(b)**, label the mean μ, and indicate the size of the standard deviation σ, similar to Figure 6.4 on page 262.

68. Determine whether snake eyes ($X = 2$) is an unusual result. By symmetry, apply your finding to another value of X.

69. Note that the mean of X also happens to be the most likely value of X.
  a. Does it always happen that the mean of a discrete random variable is the same as the most likely value of that variable? If not, give a counterexample.
  b. Specify the conditions when it is true that the mean of X equals the most likely value of X.

? **70. Linear Transformation.** *What if* we add the same unknown amount k to each value of X. Describe what would happen to the following, and why.
  a. The mean of X
  b. The standard deviation of X

6.2   BINOMIAL PROBABILITY DISTRIBUTION

OBJECTIVES   By the end of this section, I will be able to . . .

1   Explain what constitutes a binomial experiment.

2   Compute probabilities using the binomial probability formula.

3   Find probabilities using the binomial tables.

4   Calculate the mean, variance, and standard deviation of the binomial random variable and find the mode of the distribution.

1   BINOMIAL EXPERIMENT

There are many different types of discrete probability distributions. Perhaps the most important is the *binomial* distribution, which we will learn about in this section. Life is full of situations where there are *only two possible outcomes* to a process.

- A baby is about to be born. Will it be a boy or a girl?

- A basketball player is about to attempt a free throw. Will she make it or miss?

- A friend of yours is also taking statistics. Will he pass or fail?

Because situations where there are only two possible outcomes are so widespread, methods have been developed to make it more convenient to analyze them. These methods begin with the definition of a **binomial experiment.**

> **Binomial Experiment**
>
> A probability experiment that satisfies the following four requirements is said to be a **binomial experiment:**
>
> 1. Each trial of the experiment has only *two possible mutually exclusive outcomes* (or is defined in such a way that the number of outcomes is reduced to two). One outcome is denoted a *success* and the other a *failure*.
>
> 2. There is a *fixed number of trials*, known in advance of the experiment.
>
> 3. The experimental outcomes are *independent* of each other.
>
> 4. The *probability* of observing a success remains the same from trial to trial.

Let's take a moment to discuss what these requirements really mean.

1. A *success* denotes simply the outcome we are interested in, without necessarily implying that the outcome is desirable. For example, for a researcher investigating college dropout rates, a dropout would be considered a success in the context of a binomial experiment.

2. Tossing a coin 10 times is a binomial experiment because we know the *fixed number of trials*. A salesman contacting customers one-by-one until he makes a sale is not a binomial experiment because he doesn't know how many customers he will have to contact.

3. Sampling without replacement would technically violate the *independence* requirement. However, recall that we may apply the 1% Guideline from Section 5.3, so that when the sample is small compared to the population, successive trials can be considered to be independent.

4. Suppose four friends are wondering how many of them will get an A in statistics. This is not a binomial experiment because the four friends presumably do not all have the same probability of success.

> Many experiments having more than two outcomes can often be defined so that there are only two outcomes. For example, the answer to a multiple-choice question that has five answer choices may be recorded as either correct or incorrect.

The outcomes of a binomial experiment, together with their probabilities, generate a special discrete probability distribution called the **binomial probability distribution.** For binomial probability distributions, there are always only two outcomes, and each outcome has a probability associated with it. The *binomial random variable,* denoted by X, represents the number of successes observed in the n trials. Note that $0 \leq X \leq n$.

EXAMPLE 6.14 **RECOGNIZING BINOMIAL EXPERIMENTS**

Determine whether each of the following experiments fulfills the conditions for a binomial experiment. If the experiment is binomial, identify the random variable X, the number of trials, the probability of success, and the probability of failure. If the experiment is not binomial, explain why not.

a. A fisherman is going fishing and will continue to fish until he catches a rainbow trout.

b. We flip a fair coin three times and observe the number of heads.

c. A market researcher at a shopping mall is asking consumers whether they use Fib detergent. She asks a sample of 4 men, one of whom is clearly the employer of the other 3.

d. The National Burglar and Fire Alarm Association reports that 34% of burglars get in through the front door. A random sample of 36 burglaries is taken, and the number of entries through the front door is noted.

Solution

a. This is not a binomial experiment because since you don't know how many fish he will catch before the rainbow trout shows up, there is not a fixed number of trials known in advance.

b. This is a binomial experiment because it fulfills the requirements:
   i. There are only two possible outcomes on each trial, with heads defined as success and tails as failure.
   ii. We know in advance that we are tossing the coin three times.
   iii. The coin doesn't remember its result from toss to toss, and so the trials are independent.
   iv. The coin is fair on each toss, and so the probability of observing heads is the same on each toss.

The binomial random variable X is the number of heads observed on the three trials; since the coin is fair, the probability of success is 0.5 and the probability of failure is 0.5. The possible values for X are 0, 1, 2, or 3.

c. This is not a binomial experiment, because the responses are not independent. The response given by the employer is likely to affect the employees' responses.

d. This is a binomial experiment because it fulfills the requirements:
   i. There are only two possible outcomes on each trial: entering through the front door or not entering through the front door.
   ii. We know in advance that the size of the random sample is 36 burglaries.
   iii. Since the sample is random, the trials are independent.
   iv. Since the sample is quite small compared to the size of the population, the probability of entering through the front door remains the same from burglary to burglary.

Now You Can Do Exercises 5–14.

The binomial random variable X is the number of front-door-entry burglaries noted for the 36 break-ins; the probability of success is 0.34 and the probability of failure is $1 - 0.34 = 0.66$.

Table 6.5 gives some notation regarding binomial experiments and the binomial distribution.

Table 6.5 Notation for binomial experiments and the binomial distribution

Symbol	Meaning
S	The outcome denoted as a success
F	The outcome denoted as a failure
$P(\text{Success}) = P(S) = p$	The probability of observing a success
$P(\text{Failure}) = P(F) = 1 - p = q$	The probability of observing a failure
n	The number of trials

Using this notation in the experiment in Example 6.14(d), we have

S = burglary through front door, and F = burglary not through front door

$$P(S) = p = 0.34, \text{ and } P(F) = 1 - p = 1 - 0.34 = 0.66 = q$$

2 BINOMIAL PROBABILITY DISTRIBUTION FORMULA

Before we examine the binomial probability distribution formula, let us recall from Section 5.4 (page 239) the formula for the **number of combinations.**

Note: In Section 5.4, we used $_nC_r$ to indicate the number of combinations. Now that we have learned about random variables, which can be denoted X, we use $_nC_X$ to represent the number of combinations.

> The **number of combinations** of X items chosen from n different items is given by
>
> $$_nC_X = \frac{n!}{X!\,(n-X)!}$$
>
> where $n!$ represents n **factorial,** which equals $n(n-1)(n-2)\cdots(2)(1)$, and $0!$ is defined to be 1.

EXAMPLE 6.15

HOW MANY TEAM COMBINATIONS IN THE INTRAMURAL VOLLEYBALL LEAGUE?

Jeffrey is in charge of drawing up a schedule for his college's intramural volleyball league. This year five teams have been fielded, and they must play each other once. How many games will be held?

Solution

The number of combinations of $n = 5$ volleyball teams taken $x = 2$ at a time is

$$_5C_2 = \frac{5!}{2!(5-2)!} = \frac{5 \cdot 4 \cdot 3 \cdot 2 \cdot 1}{(2 \cdot 1)(3 \cdot 2 \cdot 1)} = \frac{120}{(2)(6)} = 10$$

Ten games will be held.

Note: You may find the following special combinations useful. For any integer n:

$$_nC_n = 1 \qquad _nC_0 = 1$$
$$_nC_1 = n \qquad _nC_{n-1} = n$$

We are often interested in finding probabilities associated with a binomial experiment.

EXAMPLE 6.16

CONSTRUCTING A BINOMIAL PROBABILITY DISTRIBUTION

A recent study reported that about 40% of online dating-survey respondents are "hoping to start a long-term relationship" (LTR).[2] Consider the experiment of choosing three online daters at random, and let

$$X = \text{the number of "LTRers"}$$

so that a success is defined as choosing someone hoping to start a long-term relationship.
a. Construct a tree diagram for this experiment.
b. Suppose that we are interested in finding the probability that exactly two of the three online daters would be LTRers, $P(X = 2)$. In the tree diagram, highlight in blue the outcomes where exactly two of the three online daters are LTRers. Find the probability for each outcome, and use these to find $P(X = 2)$.
c. Suppose that we are interested in finding $P(X = 1)$. In the tree diagram, highlight in red the outcomes where exactly one of the three online daters is an LTRer. Find the probability for each outcome, and use these to find $P(X = 1)$.

Solution

a. Figure 6.7 shows the tree diagram for this experiment.

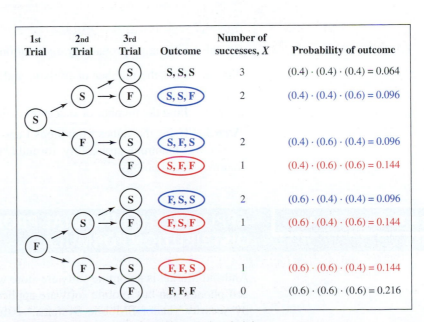

1st Trial	2nd Trial	3rd Trial	Outcome	Number of successes, X	Probability of outcome
		S	S, S, S	3	$(0.4) \cdot (0.4) \cdot (0.4) = 0.064$
	S	F	S, S, F	2	$(0.4) \cdot (0.4) \cdot (0.6) = 0.096$
S		S	S, F, S	2	$(0.4) \cdot (0.6) \cdot (0.4) = 0.096$
	F	F	S, F, F	1	$(0.4) \cdot (0.6) \cdot (0.6) = 0.144$
		S	F, S, S	2	$(0.6) \cdot (0.4) \cdot (0.4) = 0.096$
	S	F	F, S, F	1	$(0.6) \cdot (0.4) \cdot (0.6) = 0.144$
F		S	F, F, S	1	$(0.6) \cdot (0.6) \cdot (0.4) = 0.144$
	F	F	F, F, F	0	$(0.6) \cdot (0.6) \cdot (0.6) = 0.216$

FIGURE 6.7 Tree diagram and binomial probabilities.

b. As we can see from Figure 6.7, there are $(_nC_X) = (_3C_2) = 3$ different ways that exactly two of the three online daters could be LTRers (highlighted in blue).

For each of these three outcomes, the probability that $X = 2$ is $(0.4)^2(0.6) = 0.096$.

Remember: $P(S) = p$ and $P(F) = q$.

- The outcome S, S, F (second row in Figure 6.7) has probability $(p)(p)(q) = (0.4)(0.4)(0.6) = 0.096$.
- The outcome S, F, S has probability $(p)(q)(p) = (0.4)(0.6)(0.4) = 0.096$.
- The outcome F, S, S has probability $(q)(p)(p) = (0.6)(0.4)(0.4) = 0.096$.

Note that each of these products equals $(p)^2 \cdot q$, with p having exponent $X = 2$, and (q) having exponent $n - X = 3 - 2 = 1$. Thus,

$$P(X = 2) = (_3C_2)(0.4)^2(0.6)$$
$$= 3(0.096) = 0.288$$

c. Similarly, suppose that we are interested in whether exactly one $(X = 1)$ of the three online daters is an LTRer. Then, Figure 6.7 shows us, highlighted in red, that there are $(_nC_X) = (_3C_1) = 3$ different ways this could happen. Each of these outcomes has probability $(p) \cdot (q)^2 = (0.4)(0.6)^2 = 0.144$, where p has exponent $X = 1$, and q has exponent $n - X = 3 - 1 = 2$. Thus,

$$P(X = 1) = (_3C_1)(0.4)(0.6)^2$$
$$= 3(0.144) = 0.432$$

We can generalize these procedures and use the **binomial probability distribution formula** to find probabilities for the number of successes for any binomial experiment.

The Binomial Probability Distribution Formula

The probability of observing exactly X successes in n trials of a binomial experiment is

$$P(X) = (_nC_X)\, p^X\, (q)^{1-X}$$

That is,

$$P(X) = (_nC_X)\, [P(success)^{number\ of\ success} \cdot P(failure)^{number\ of\ failures}].$$

We often call this the binomial probability formula.

Developing Your Statistical Sense

Steps for Solving Binomial Probability Problems

To solve a binomial probability distribution problem, follow these steps:

Step 1. Find the number of trials n, and the probability of success on a given trial p.

Step 2. Find the number of successes X that the question is asking about.

Step 3. Using the values from Steps 1 and 2, find the required probabilities using either the binomial probability formula, the binomial tables (which we learn below), or technology.

EXAMPLE 6.17

APPLYING THE BINOMIAL PROBABILITY DISTRIBUTION FORMULA

Android Market is an online software store where owners of Android devices, such as cell phones, can buy mobile software applications called apps. According to a report by security vendor SMobile Systems,[3] 20% of Android apps available at Android Market threaten user privacy. Joshua received a random sample of 4 apps from the

Android Market when he bought his cell phone. Find the probability that the number of these apps that threaten user privacy equaled the following:

a. None
b. At least 1
c. Between 1 and 3, inclusive
d. 5

Solution

We apply the steps for solving binomial probability problems.

STEP 1 We have a random sample of four apps, so the number of trials is $n = 4$. "Success" is denoted as a particular app threatening user privacy. The report states that 20% of such apps from Android Market do so, so $p = 0.2$ and $q = 1 - 0.2 = 0.8$.

STEP 2 For (a), $X = 0$. For (b), $X \geq 1$, that is, $X = 1, 2, 3$, or 4. For (c), $1 \leq X \leq 3$, that is, $X = 1, 2$, or 3. For (d), $X = 5$.

STEP 3 We apply Step 3 for each of (a)–(d) as follows:

a. *Step 3* To find the probability that none ($X = 0$) of the apps threaten user privacy, we use the binomial probability formula:

$$P(X = 0) = (_4C_0)(0.2)^0 (0.8)^{4-0} = (1)(1)(0.4096) = 0.4096$$

So the probability that none of the apps Joshua received threaten user privacy is 0.4096.

b. *Step 3* Note that "at least 1" includes all possible values of X except $X = 0$. In other words, the two events ($X = 0$) and ($X \geq 1$) are complements of each other. Therefore, from the formula for the probability for complements in Section 5.2 (page 210), we have

$$P(X \geq 1) = 1 - P(X = 0) = 1 - 0.4096 = 0.5904$$

The probability that at least one of the apps will threaten user privacy is 0.5904.

c. *Step 3* We need to find the probability that either $X = 1$ or $X = 2$ or $X = 3$ of the apps that threaten user privacy. Since these three values of X are mutually exclusive, we find the required probability by using the Addition Rule for Mutually Exclusive Events.

$$P(1 \leq X \leq 3) = P(X = 1 \text{ or } X = 2 \text{ or } X = 3)$$
$$= P(X = 1) + P(X = 2) + P(X = 3)$$

So we calculate the following:

$$P(X = 1) = (_4C_1)(0.2)^1 (0.8)^{4-1} = (4)(0.2)(0.512) = 0.4096$$
$$P(X = 2) = (_4C_2)(0.2)^2 (0.8)^{4-2} = (6)(0.04)(0.64) = 0.1536$$
$$P(X = 3) = (_4C_3)(0.2)^3 (0.8)^{4-3} = (4)(0.008)(0.8) = 0.0256$$

Hence, $P(1 \leq X \leq 3) = 0.4096 + 0.1536 + 0.0256 = 0.5888$. The probability is 0.5888 that between 1 and 3, inclusive, of Joshua's apps will threaten user privacy.

d. *Step 3* In a binomial experiment, the number of successes X can never exceed the number of trials n. In other words, $X \leq n$, always. So, if Joshua has only $n = 4$ apps, $P(X = 5) = 0$. It is not possible that Joshua has 5 apps that threaten user privacy.

Now You Can Do
Exercises 15–22.

3 BINOMIAL DISTRIBUTION TABLES

As you can imagine, calculations involving binomial probabilities can sometimes get tedious. For example, to find the probability of observing at least 60 heads on 100 tosses of a fair coin, we would have to use the binomial formula for $X = 60$, $X = 61$, $X = 62$, and so on, right up to $X = 100$. For this type of problem, you can use Table B, Binomial Distribution, in the Appendix. If you are trying to answer a question involving unusual values of n, such as 103, or unusual values of p, such as 0.47, then you can use technology instead.

EXAMPLE 6.18

FINDING PROBABILITIES USING THE BINOMIAL TABLE

Use the binomial table and the binomial distribution from Example 6.17 to find the following probabilities:
a. None of Joshua's apps will threaten user privacy.
b. At least one of Joshua's apps will threaten user privacy.

Solution

a. From Example 6.17, we have a binomial distribution with $n = 4$ and $p = 0.2$. We next find n and p in the binomial table. In Figure 6.8:
   - Look under the n column until you find $n = 4$. That is the portion of the table you will use.
   - Then go across the top of the table until you get to $p = 0.2$.
   - For part (a), $X = 0$, so go down the X column until you see 0 under the X column on the left (and in the subgroup with $n = 4$).
   - The number in the p column is 0.4096 (see Figure 6.8), which is the same answer we calculated in Example 6.17(a).

						p
n	X	0.10	0.15	0.20	0.25	0.30
2	0	0.8100	0.7225	0.6400	0.5625	0.4900
	1	0.1800	0.2550	0.3200	0.3750	0.4200
	2	0.0100	0.0225	0.0400	0.0625	0.0900
3	0	0.7290	0.6141	0.5120	0.4219	0.3430
	1	0.2430	0.3251	0.3840	0.4219	0.4410
	2	0.0270	0.0574	0.0960	0.1406	0.1890
	3	0.0010	0.0034	0.0080	0.0156	0.0270
4	0	0.6561	0.5220	0.4096	0.3164	0.2401
	1	0.2916	0.3685	0.4096	0.4219	0.4116
	2	0.0486	0.0975	0.1536	0.2109	0.2646
	3	0.0036	0.0115	0.0256	0.0469	0.0756
	4	0.0001	0.0005	0.0016	0.0039	0.0081

$X = 1$
$X = 2$
$X = 3$
$X = 4$

FIGURE 6.8 Excerpt from the binomial tables.

b. In this case, "at least 1" means 1 or 2 or 3 or 4. So, by the Addition Rule for Mutually Exclusive Events, find the probabilities for $X = 1$, $X = 2$, $X = 3$, and

$X = 4$, and add them up. Using the same column with column head 0.20 in the table as in part (a), we add up the four probabilities.

$$P(X \geq 1) = P(X = 1) + P(X = 2) + P(X = 3) + P(X = 4)$$
$$= 0.4096 + 0.1536 + 0.0256 + 0.0016 = 0.5904$$

Now You Can Do Exercises 23–28.

This is the same answer we calculated in Example 6.17(b), but arrived at in a different way.

Next, a word about *cumulative probability*. Cumulative probability refers to the probability of *at most* a particular value of X. For example, what is the probability that at most $X = 2$ of Joshua's apps threaten user privacy? This is the cumulative probability that $X = 0$, $X = 1$, or $X = 2$. Statistical software and the TI-83/84 graphing calculator each have a function that will find cumulative binomial probabilities for you.

EXAMPLE 6.19

USING TECHNOLOGY TO FIND BINOMIAL PROBABILITIES

Using the binomial distribution from Example 6.17, use the TI-83/84 to find the following probabilities:
a. $P(X = 4)$, the probability that all 4 apps will threaten user privacy
b. $P(X \leq 2)$, the (cumulative) probability that *at most* 2 apps will threaten user privacy

Solution
We use the instructions in the Step-by-Step Technology Guide at the end of this section (page 275).
a. Figure 6.9 shows that we use the function **binompdf** with $n = 4$, $p = 0.2$, and $X = 4$. Figure 6.10 shows the result, $P(X = 4) = 0.0016$.

FIGURE 6.9 **FIGURE 6.10**

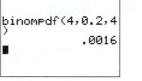

FIGURE 6.11

b. We use the function **binomcdf** with $n = 4$, $p = 0.2$, and $X = 2$. Figure 6.11 shows the result, $P(X \leq 2) = 0.9728$.

---

4 BINOMIAL MEAN, VARIANCE, STANDARD DEVIATION, AND MODE

In Section 6.1, we examined the mean, variance, and standard deviation of a discrete random variable. Since the binomial random variable X is discrete, it also has a **mean, variance,** and **standard deviation,** shown here.

Caution: These formulas work only for a binomial random variable.

> **Mean, Variance, and Standard Deviation of a Binomial Random Variable X**
>
> - Mean (or expected value): $\mu = n \cdot p$
> - Variance: $\sigma^2 = n \cdot p \cdot q$
> - Standard deviation: $\sigma = \sqrt{n \cdot p \cdot q}$

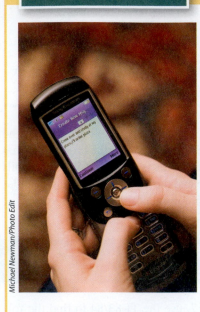

Michael Newman/Photo Edit

CASE STUDY · Text Messaging

According to the Pew Internet and American Life Project, 31% of American adults prefer to receive text messages rather than phone calls on their cell phones. Suppose we take a sample of 100 American adults.

a.   Find the mean or expected number who prefer to receive text messages.

b.   Calculate the variance σ and standard deviation σ of the number of who prefer to receive text messages.

c.   In our sample of 100, would it be unusual to observe 45 who prefer to receive text messages?

Solution

The binomial random variable here is X = the number of American adults who prefer to receive text messages rather than phone calls on their cell phones, with sample size $n = 100$, probability of success $p = 0.31$, and probability failure $q = 1 - p = 1 - 0.31 = 0.69$.

a.   The mean or expected number who prefer to receive text messages is $\mu = E(X) = n \cdot p = (100)(0.31) = 31$. American adults.

b. $\sigma^2 = n \cdot p \cdot q = (100)(0.31)(0.69) = 21.39$, expressed in "American adults squared." Then $\sigma = \sqrt{\sigma^2} = \sqrt{21.39} = 4.624932432$. (We retain so many decimal places because we need to use σ for a calculation in part **(c)**).

c.   We use the Z-score method (Section 6.1, page 262) to determine whether 45 American adults out of 100 preferring to receive text messages would be unusual. The Z-score for 45 is:

$$Z = \frac{X - \mu}{\sigma} = \frac{45 - 31}{4.624932432} \approx 3.0271$$

Now You Can Do
Exercises 49–52.

According to the Z-score method of identifying outliers, $X = 45$ American adults preferring to receive text messages rather than phone calls in a sample of 100 would be considered unusual, because it is an outlier, with $Z \geq 3$. ∎

What Do μ and σ Mean?

The value $\mu = 31$ is the "long-run" mean and the value $\sigma \approx 4.6$ is the "long-run" standard deviation. That is, if we repeat this experiment an infinite number of times, identify the number of American adults preferring to receive text messages rather than phone calls in each sample, and take the mean and standard deviation of each of these samples, they will equal $\mu = 31$ and $\sigma \approx 4.6$.

Next we consider the mode of a binomial distribution.

> The *mode* of a binomial distribution is the most likely outcome of the binomial experiment for the given values of n, p, and X, that is, the outcome with the largest probability.

The next example shows how to find the mode for a binomial distribution.

EXAMPLE 6.20

FINDING THE MOST LIKELY OUTCOME OF A BINOMIAL EXPERIMENT

Sixty percent of American adults access the Internet wirelessly, according to a 2010 report by the Pew Research Center's Internet and American Life Project. Suppose we take a random sample of $n = 3$ American adults.

a. Calculate the mean number μ of American adults who access the Internet wirelessly.

b. Use the binomial table to construct a probability distribution graph of the random variable X = the number of Americans who access the Internet wirelessly.

c. Use the binomial table or the probability distribution graph to find the most likely number of American adults who access the Internet wirelessly. Note that this represents the *mode* of the distribution.

Solution

Example 6.5 (pages 256–257) shows how to construct a probability distribution graph.

a. $\mu = n \cdot p = (3)(0.6) = 1.8$.

b. Figure 6.12 is an excerpt from the binomial table, highlighting the probabilities for $X = 0, 1, 2$, and 3, for $n = 3$ and $p = 0.6$. We use these probabilities to construct the probability distribution graph shown in Figure 6.13.

n	x	**0.55**	**0.60**
2	0	0.2025	0.1600
	1	0.4950	0.4800
	2	0.3025	0.3600
③	0	0.0911	0.0640
	1	0.3341	0.2880
	2	0.4084	0.4320
	3	0.1664	0.2160

FIGURE 6.12 Probabilities for $X = 0, 1, 2, 3$.

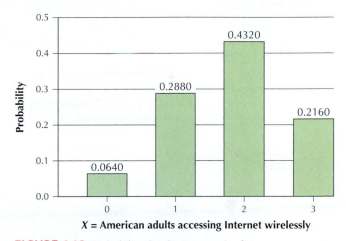

FIGURE 6.13 Probability distribution graph of X.

c. The most likely number of Americans accessing the Internet is associated with the largest probability in the highlighted section of Figure 6.12, 0.4320, which is $P(X = 2)$. Note from Figure 6.13 that $X = 2$ has the longest bar of probability. Thus, $X = 2$ is the most likely number of American adults accessing the Internet wirelessly. We say that $X = 2$ is the *mode* of the distribution of X.

Now You Can Do Exercises 53–56.

STEP-BY-STEP TECHNOLOGY GUIDE: Finding Binomial Probabilities

For Example 6.19 (page 273).

TI-83/84

Step 1 Press **2nd > DISTR** (the **VARS** key).
Step 2 Do one of **(a)** or **(b)**:
a. For individual binomial probabilities, highlight **binompdf(** and press **ENTER**. (See Figure 6.9 on page 273.)

b. For cumulative binomial probabilities, highlight **binomcdf(** and press **ENTER**.
Step 3 Enter the values for *n*, *p*, and *K*, separated by commas.
Step 4 Press **ENTER**. (See Figures 6.10 and 6.11 on page 273.)

EXCEL

Step 1 Select cell **A1**. Click the **Insert Function** icon f_x.
Step 2 In the **Search for a function** area, type **BINOMDIST**, and click **OK**.
Step 3 For **Number_s**, enter the number of successes, K. For **Trials**, enter the sample size, n. For **Probability_s**, enter the probability of success, **p**.

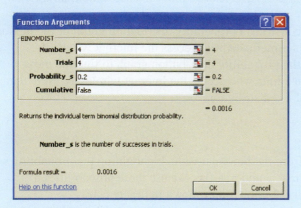

Step 4 Do one of (**a**) or (**b**):
a. For individual binomial probabilities, next to **Cumulative**, enter **false**.
b. For cumulative binomial probabilities, next to **Cumulative**, enter **true**.
Step 5 Click **OK**. See Figures 6.14 and 6.15 for illustrations using Example 6.19.

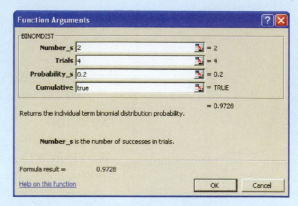

FIGURE 6.14 Example 6.19(a) using Excel.

FIGURE 6.15 Example 6.19(b) using Excel.

MINITAB

Step 1 Click **Calc > Probability Distributions > Binomial**.
Step 2 Do one of (**a**) or (**b**):
a. For individual binomial probabilities, select **Probability** and enter the number of trials **n** and probability of success **p**.

b. For cumulative binomial probabilities, select **Cumulative Probability** and enter the number of trials **n** and probability of success **p**.
Step 3 Select **Input Constant**, enter K and click **OK**.

CRUNCHIT!

We will use the data from Example 6.18.

Step 1 Click **Distribution calculator . . . Binomial**.
Step 2 For **n** enter **4**. For **p** enter **0.2**.

Step 3 For part (**a**) select = and enter **4**. For part (**b**) select ≤ and enter **2**.
Step 4 Click **Calculate**.

SECTION 6.2 **Summary**

1. The most important discrete distribution is the binomial distribution, where there are two possible outcomes, each with probability of success p, and n independent trials.

2. The probability of observing a particular number of successes can be calculated using the binomial probability distribution formula.

3. Binomial probabilities can also be found using the binomial tables or using technology.

4. There are formulas for finding the mean, variance, and standard deviation of a binomial random variable, X. The mode is the value of X with the largest probability.

SECTION 6.2 **Exercises**

Clarifying the Concepts

1. State the four requirements for a binomial experiment.

2. What is meant by a "success" in a binomial experiment? Is a success always a good thing?

3. In a binomial experiment, explain why it is not possible for X to exceed n.

4. Restate the binomial probability distribution formula using the following terms: $(_nC_x)$, the probability of success, the number of successes, the probability of failure, the number of failures.

Practicing the Techniques

For Exercises 5–14, determine whether the experiment is binomial or not. If the experiment is binomial, identify the random variable X, the number of trials n, the probability of success p, and the probability of failure q. If the experiment is not binomial, explain why not.

5. Ask ten of your friends to come to your party (remember the independence assumption).

6. Toss a fair die three times, and note the total number of spots.

7. Answer a random sample of 8 multiple-choice questions either correctly or incorrectly by random guessing. There are 4 choices, (a)–(d), for each question.

8. Toss a fair die three times, and note the number of 6s.

9. Select a student at random in the class until you come across a left-handed student.

10. Four cards are selected at random with replacement from a deck of cards, and the number of queens is observed.

11. Four cards are selected at random without replacement from a deck of cards, and the number of queens is observed.

12. Four cards are selected at random with replacement from a deck of cards, and the total number of blackjack-style points (number cards = number of points; face cards = 10 points; aces = either 1 or 11) is calculated.

13. Bob has paid to play two games at a carnival. The probability that he wins a particular game is 0.25.

14. Bob is playing a game at a carnival where he gets to play until he loses. The probability that he wins a particular game is 0.25.

For Exercises 15–28, calculate the probability of X successes for the binomial experiments with the following characteristics.

15. $n = 5, p = 0.25, X = 1$

16. $n = 5, p = 0.25, X = 0$

17. $n = 10, p = 0.5, X = 7$

18. $n = 10, p = 0.5, X = 8$

19. $n = 12, p = 0.9, X = 10$

20. $n = 12, p = 0.9, X = 11$

21. $n = 5, p = 0.25, X \le 1$

22. $n = 5, p = 0.25, X \ge 1$

23. $n = 10, p = 0.5, X = 7$ or $X = 8$

24. $n = 10, p = 0.5, X = 7$ and $X = 8$

25. $n = 12, p = 0.9, X \ge 10$

26. $n = 12, p = 0.9, X < 10$ (*Hint:* Use the result from Exercise 25.)

27. $n = 12, p = 0.9, 9 \le X \le 12$

28. $n = 12, p = 0.9, 8 \le X \le 12$

For Exercises 29–34, the binomial experiment is to toss a fair coin three times. Find the indicated probabilities.

29. Observe no heads

30. Observe one head

31. Observe two heads

32. Observe at most two heads

33. Observe at least one head

34. Observe between zero and two heads, inclusive

For Exercises 35–40, the binomial experiment is to roll a pair of dice four times, and observe the number of doubles that you roll. (*Hint:* $P(\text{Doubles}) = 1/6$.) Find the following probabilities.

35. Observe doubles on three of the rolls

36. Observe doubles on at least three of the rolls

37. Observe no doubles

38. Observe doubles on at most one of the rolls

39. Observe between one and four doubles, inclusive

40. Observe five doubles

For Exercises 41–44, the binomial experiment is to take a random sample of 5 vehicles on the interstate highway, and observe the number of vehicles obeying the speed limit. Assume that the probability that a vehicle obeys the speed limit is 0.4. Find the indicated probabilities.

41. None of the vehicles obey the speed limit.

42. At least 1 of the vehicles obeys the speed limit.

43. At most 2 of the vehicles obey the speed limit.

44. Between 1 and 3 of the vehicles obey the speed limit, inclusive.

For Exercises 45–48, conduct a survey of a random sample of 6 voters, asking each voter whether they would support an Independent for president in the next election. Assume that 15% of voters would support an Independent for president. Find the following probabilities of voters who would support an Independent for president.

45. All of the voters

46. At most 5 voters

47. At least 4 voters

48. Between 3 and 5 voters, inclusive

For each of the following binomial experiments, do the following.
  a. Find and interpret the mean μ of X.
  b. Calculate the variance σ^2 of X.
  c. Compute the standard deviation σ of X.

49. The binomial experiment in Exercises 29–34

50. The binomial experiment in Exercises 35–40

51. The binomial experiment in Exercises 41–44

52. The binomial experiment in Exercises 45–48

For each of the following binomial experiments, do the following.

    a. Construct the probability distribution graph of X.
    b. Identify the mode of X.

53. The binomial experiment in Exercises 29–34

54. The binomial experiment in Exercises 35–40

55. The binomial experiment in Exercises 41–44

56. The binomial experiment in Exercises 45–48

Applying the Concepts

57. Random Guessing on a Quiz. Suppose that you are taking a quiz of 5 multiple-choice questions (the instructor chose the questions randomly), each question having 4 possible responses. You did not study at all for the quiz and will randomly guess the correct response for each question. The random variable X is the number of correct responses.

    a. If there are 4 possible responses to each question, why is this a valid binomial experiment?
    b. State the values of n and p.
    c. Calculate the probability that you will pass this quiz by correctly responding to at least 3 of the 5 questions. Is this good news for you?
    d. Use your answer to **(c)** to find the probability that you will not pass the quiz.

58. Abandoning Landlines. The National Health Interview Survey reports that 25% of telephone users no longer use landlines, and have switched completely to cell phone use.[4] Suppose we take a random sample of 12 telephone users.

    a. Find the probability that the sample contains exactly 3 users who have abandoned their landlines.
    b. Find the probability that the sample contains at most 3 users who have abandoned their landlines.
    c. Use either the binomial table or technology to determine the most likely number of users in the sample who have abandoned their landlines.
    d. Compute the probability that the sample contains the mode number of users who have abandoned their landlines.

59. Vowels. Did you know that 37.8% of the letters in the written English language are vowels? Suppose we select 15 letters at random.

    a. Explain why we cannot use the binomial table to solve probability problems for this binomial experiment.

    b. Find the probability that the sample contains exactly 2 vowels.
    c. Find the probability that the sample contains at most 2 vowels.

60. Women in Management. According to the U.S. Government Accountability Office, women hold 40% of the management positions in the United States.[5] Suppose we take a random sample of 20 people in management positions.

    a. Find the probability that the sample contains exactly 10 women.
    b. Find the probability that the sample contains at most 1 woman.
    c. Find the probability that the sample contains between 8 and 10 women, inclusive.

61. Random Guessing on a Quiz. Refer to Exercise 57.
    a. Compute the mean, variance, and standard deviation of X. Interpret the mean.
    b. Use the Z-score method to determine which numbers of correct responses should be considered outliers.
    c. Use the binomial table to construct a probability distribution graph of X. Then state the mode of X, that is, the most likely number of correct responses.

62. Abandoning Landlines. Refer to Exercise 58.
    a. Calculate the mean, variance, and standard deviation of the number of users in the sample who have abandoned their landlines. Interpret the mean.
    b. Suppose the sample contains no users who have abandoned their landlines. Is this outcome unusual or an outlier? Use the Z-score method to find out.

63. Vowels. Refer to Exercise 59.
    a. Find the mean, variance, and standard deviation of the number of vowels.
    b. Suppose that the sample contains only 3 vowels. Use the Z-score method to determine whether this outcome is unusual or not.

64. Women in Management. Refer to Exercise 60.
    a. Find the mean, variance, and standard deviation of the number of women in management positions.
    b. Suppose that the sample contains 6 women in management positions. Use the Z-score method to determine whether this outcome is unusual or not.

65. Mean, Median, Mode. For a binomial distribution, if the mean $\mu = n \cdot p$ is a whole number, then

$$\text{mean of } X = \text{median of } X = \text{mode of } X$$

Use this equation to answer the following questions.
    a. Find the median of X for the binomial distribution in Example 6.17.
    b. Find the mode of X for the binomial distribution in Example 6.17.

c. What is the most likely value of X for the binomial distribution in Example 6.17?

66. Geometric Probability Distribution. Refer to Example 6.14(a), where a fisherman is going fishing and will continue to fish until he catches a rainbow trout. This is an example of the *geometric probability distribution*, which has the same requirements as the binomial distribution, except that there is not a fixed number of trials n. Instead, the geometric random variable X represents the number of trials until a success is observed. The geometric probability distribution formula is

$$P(X) = p(1 - p)^{X-1}$$

where p represents the probability of success. The possible values of X are $X = 1, 2, 3, \ldots$. The U.S. Census Bureau reported in 2010 that 30% of U.S. households have no access at all to the Internet. A random sample is taken of U.S. households. Let the random variable X represent the number of trials until a household is found that has access to the Internet.

a. Find the probability that $X = 1$, that is, the first household sampled has access to the Internet.
b. Find the probability that $X = 2$, that is, the first household sampled does not have access but the second household sampled does have access to the Internet.
c. Find the probability that $X = 3$, that is, the first two households sampled do not have access but the third household sampled does have access to the Internet.

67. Hypergeometric Probability Distribution. If samples are drawn from a relatively small finite population, and the sample size is larger than 1% of the population, so that the 1% Guideline (page 225) does not apply, we should not use the binomial distribution, because the samples are not independent. Instead, if we are sampling without replacement, and there are two mutually exclusive categories, then you should use the *hypergeometric probability distribution*. Suppose that N_1 objects belong to the first category ("successes"), and N_2 objects belong to the second category ("failures"). Then

the probability of getting X successes and $n - X$ failures is given by the hypergeometric probability distribution formula:

$$P(X) = \frac{(_{N_1}C_X)(_{N_2}C_{n-X})}{(_NC_n)}$$

where $N_1 + N_2 = N$, N is the population size, and n is the sample size. You are dealt 5 cards at random from a deck of 52 cards.

a. Find the probability that all 5 cards are spades.
b. Find the probability that exactly 4 cards are spades.
c. Find the probability that at least 4 cards are spades.
d. Find the probability that exactly 3 cards are spades.
e. Find the probability that at most 2 cards are spades.

68. Multinomial Distribution. The *multinomial probability distribution* is similar to the binomial distribution, except that the binomial involves only two categories, while the multinomial involves more than two categories. Suppose we have three mutually exclusive outcomes, A, B, and C, where and $p_A = P(A)$, $p_B = P(B)$, and $p_C = P(C)$. If we have a sample of n independent trials, then the probability that we get X_A outcomes of category A, X_B outcomes of category B, and X_C outcomes of category C is given by the following formula:

$$P(X_A, X_B, X_C) = \frac{n!}{X_A! \, X_B! \, X_C!} \cdot p_A^{X_A} \cdot p_B^{X_B} \cdot p_C^{X_C}$$

Suppose that 30% of students on a particular college campus are Democrats, 30% are Republicans, and 40% are Independents. Suppose we take a random sample of 10 students.

a. Find the probability that 3 are Democrat, 3 are Republican, and 4 are Independent.
b. Find the probability that 3 are Democrat, 4 are Republican, and 3 are Independent.
c. Find the probability that 4 are Democrat, 3 are Republican, and 3 are Independent.

6.3 CONTINUOUS RANDOM VARIABLES AND THE NORMAL PROBABILITY DISTRIBUTION

OBJECTIVES By the end of this section, I will be able to . . .

1 Identify a continuous probability distribution and state the requirements.
2 Calculate probabilities for the uniform probability distribution.
3 Explain the properties of the normal probability distribution.

Sections 6.1–6.2 dealt with discrete random variables, such as the binomial random variable. Next we turn to continuous random variables.

1 CONTINUOUS PROBABILITY DISTRIBUTIONS

Continuous random variables assume infinitely many possible values, with no gap between the values. For example, the height of a randomly chosen classmate of yours is a continuous random variable because it can take an infinite number of possible values.

For a given continuous random variable X, we are not interested in whether X equals any particular value. Rather, we are interested in whether X is

- greater than a particular value, or

- less than a particular value, or

- between two particular values.

That is, we are interested in whether X is located in an *interval*.

We are not interested in the probability that X equals some particular value, because this probability *always equals zero*. If this sounds crazy, then consider the following example. How much soda does a "12-ounce can" of soda actually contain? Are you sure it's 12 ounces and not 11.99999999 ounces? Or could it contain 12.00000001 ounces? In fact, the can could contain any of the infinite number of possible amounts of soda, say between 11.9 and 12.1 ounces (see Figure 6.16). Thus, any given weight of soda in the can is so unlikely that the probability that you will get exactly 12.00000000 ounces of soda in your 12-ounce can is zero.

In contrast to the graph for a discrete distribution, the graph for a continuous probability distribution is "smooth" because it represents probability at infinitely many points along an interval.

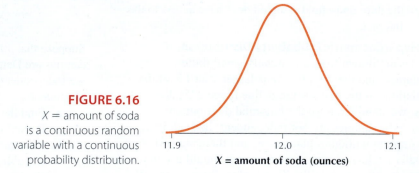

FIGURE 6.16
X = amount of soda is a continuous random variable with a continuous probability distribution.

11.9   12.0   12.1
X = amount of soda (ounces)

The graph in Figure 6.16 is called a **continuous probability distribution,** defined as follows.

Continuous Probability Distribution

A **continuous probability distribution** is represented by a graph that indicates on the horizontal axis the range of values that the continuous random variable X can take, and above which is drawn a curve, called the **density curve.** A continuous probability distribution must meet the following requirements.

Requirements for a Continuous Probability Distribution

1. The total area under the density curve must equal 1 (this is the **Law of Total Probability for Continuous Random Variables**).

2. The vertical height of the density curve can never be negative. That is, the density curve never goes below the horizontal axis.

2   CALCULATING PROBABILITIES FOR THE UNIFORM PROBABILITY DISTRIBUTION

To learn how to calculate probabilities for continuous random variables, we turn to the **uniform probability distribution.**

> The **uniform probability distribution** is a continuous distribution that has constant probability from left endpoint a to right endpoint b. Its curve is a flat, straight line, so that the shape of the uniform distribution is a rectangle.

For example, suppose the waiting time X for the campus shuttle bus follows a uniform distribution, with waiting times ranging from $a = 0$ minutes to $b = 10$ minutes. Then the uniform probability distribution is given in Figure 6.17.

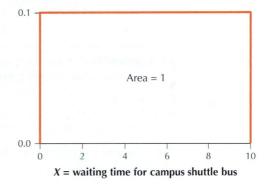

FIGURE 6.17
Waiting time X has a rectangular shape.

X = waiting time for campus shuttle bus

Note that the width of the rectangle in Figure 6.17 is $b - a = 10 - 0 = 10$. Since the total area under the density curve must equal 1 by the Law of Total Probability for Continuous Distributions, the height of the rectangle must therefore equal $1/10 = 0.1$.

So how do we represent probability for the uniform distribution, or for continuous distributions in general?

> **Probability for Continuous Distributions**
>
> The probability that a continuous random variable X takes a value in an interval is equal to the *area under the density curve above that interval.*

EXAMPLE 6.21

PROBABILITY IS REPRESENTED BY AREA

Find the probability that you will wait between 2 and 4 minutes for the campus shuttle bus.

Solution

We are interested in the interval between $X = 2$ and $X = 4$ minutes. The area above this interval forms a rectangle, shown in Figure 6.18. This area of this green rectangle represents the probability that X is between 2 and 4 minutes. The base of the rectangle equals $b - a = 4 - 2 = 2$. Since the height of the rectangle equals 0.1, we find that the area of this rectangle is

$$\text{area} = \text{base} \times \text{height} = 2 \times 0.1 = 0.2$$

Since area represents probability, we conclude that the probability is 0.2 that you will wait between 2 and 4 minutes for the campus shuttle bus.

**Now You Can Do
Exercises 7–16.**

FIGURE 6.18
Probability X between
2 and 4 equals the area
of the green rectangle.

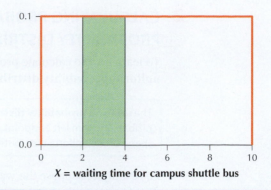

X = waiting time for campus shuttle bus

Notice from Example 6.21 that $0.2 = \dfrac{4-2}{10-0}$. We generalize this as follows:

> The probability that a uniform random variable with left endpoint a and right endpoint b takes a value in the interval [c, d] is given by
>
> $$P(c \le X \le d) = \frac{d-c}{b-a}$$

For example, the probability that you would wait between $c = 0$ and $d = 5$ minutes for the campus shuttle bus is

$$P(0 \le X \le 5) = \frac{5-0}{10-0} = 0.5$$

Now, because X is a continuous random variable, $P(X = 0) = 0$ and $P(X = 5) = 0$. Thus, $P(0 \le X \le 5) = P(0 < X < 5)$. In fact, for any continuous random variable, the inequalities $\le$ and $<$ are interchangeable, as are $\ge$ and $>$.

3 INTRODUCTION TO NORMAL PROBABILITY DISTRIBUTION

We now turn to what is considered to be the most important probability distribution in the world: the **normal probability distribution.** Sometimes referred to as the bell-shaped curve (Chapter 3), the normal distribution is a continuous distribution that has been found to model accurately such phenomena as

- the amount of rainfall in Imperial Valley, California;
- the heights and weights of high-risk infants in New York City; and
- the errors in manufacturing machine bolts in a Pennsylvania factory.

Like a discrete random variable, a continuous random variable has a mean and a standard deviation. The parameters of the normal distribution are the mean μ, which determines the center of the distribution on the number line, and the standard deviation σ, which determines the spread or shape of the distribution curve. The mean μ can be positive, negative, or zero; the standard deviation σ can never be negative.

Remember that, like all probability distributions, we are dealing with a population *of data values.*

From Figure 6.19 we can see that the normal distribution curve is symmetric about μ. If you slice the curve neatly in half at the mean μ, the result will be two pieces that are perfect mirror images of each other, as in Figure 6.19.

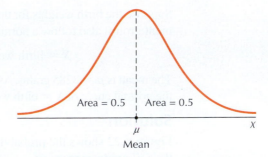

FIGURE 6.19
The normal distribution is symmetric about its mean μ.

> **Properties of the Normal Density Curve (Normal Curve)**
>
> 1. It is symmetric about the mean μ.
> 2. The highest point occurs at $X = \mu$, because symmetry implies that the mean equals the median, which equals the mode of the distribution.
> 3. The total area under the curve equals 1.
> 4. Symmetry also implies that the area under the curve to the left of μ and the area under the curve to the right of μ are both equal to 0.5 (Figure 6.19).
> 5. The normal distribution is defined for values of X extending indefinitely in both the positive and negative directions. As X moves farther from the mean, the curve approaches but never quite touches the horizontal axis.
> 6. Values of X are always found on the horizontal axis. Probabilities are represented by areas under the curve.

Figure 6.20 shows two normal density curves, with different means but the same standard deviation. Note that the two curves have precisely the same spread or shape, because each distribution has the same standard deviation, $\sigma = 2$. However, because the mean of the curve on the right is $\mu = 6$ while the mean of the curve on the left in $\mu = 2$, the curve on the right is shifted four units to the right.

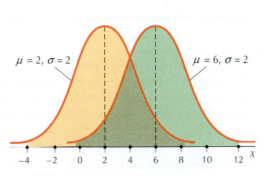

FIGURE 6.20 Different μ, same σ.

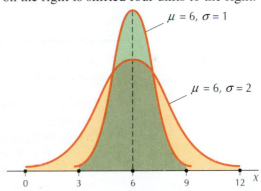

FIGURE 6.21 Same μ, different σ.

**Now You Can Do
Exercises 33 and 34.**

Since σ is a measure of spread, the larger the value of σ, the more spread out the distribution of X will be. This is illustrated in Figure 6.21. The normal distribution with the smaller standard deviation ($\sigma = 1$) has a curve with a higher peak in the center and thinner "tails" than the distribution with a larger standard deviation ($\sigma = 2$).

EXAMPLE 6.22

PROPERTIES OF THE NORMAL CURVE

A statistical study found that when nurses made home visits to pregnant teenagers to provide support services, discourage smoking, and otherwise provide care, the mean birth weight of the babies was higher for this treatment group (3285 grams) than for a control group of teenagers who were not visited (2922 grams), when the visits began before midgestation.[6] The birth weights of babies are known to follow a normal distribution.[7]

Suppose the birth weights for the babies whose mothers were visited by the nurses (treatment group) also follow a normal distribution. Then our random variable is

$$X = \text{birth weight of babies in the treatment group}$$

The mean is $\mu = 3285$ grams. Assume that the standard deviation is $\sigma = 500$ grams. Graph the normal curve of $X =$ birth weights and describe some properties of this distribution.

Solution

Figure 6.22 shows the probability graph of $X =$ birth weights. Note that the curve has the following properties:

1. It is symmetric about the mean $\mu = 3285$ grams.
2. The highest point occurs at $\mu = 3285$ grams, which is also the median and the mode.
3. The total area under the curve equals 1.
4. The area under the curve to the left of $\mu = 3285$ equals 0.5, as does the area under the curve to the right of $\mu = 3285$.

Hint: Draw a bell-shaped curve with center at $\mu = 3285$. Label the horizontal axis in increments equal to the standard deviation $\sigma = 500$. Make sure the areas to the left and right of μ are equal.

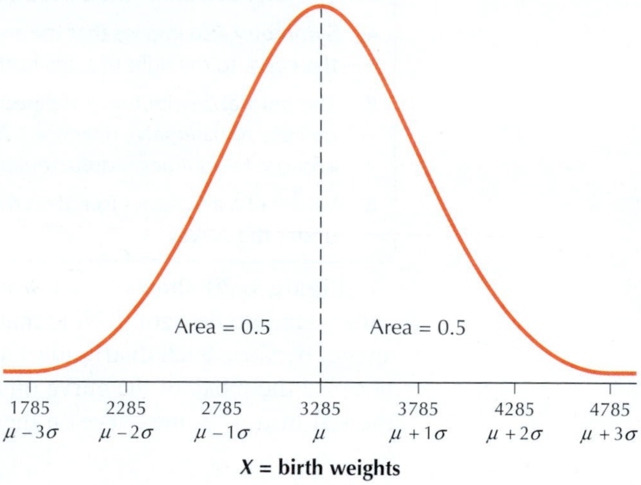

FIGURE 6.22
The normal curve of $X =$ birth weights is symmetric about its mean $\mu = 3285$.

In Chapter 3, we learned that according to the Empirical Rule the area under the normal curve has the following properties (see Figure 6.23).

1. About 68% of the area under the curve lies within 1 standard deviation of the mean.
2. About 95% of the area under the curve lies within 2 standard deviations of the mean.
3. About 99.7% of the area under the curve lies within 3 standard deviations of the mean.

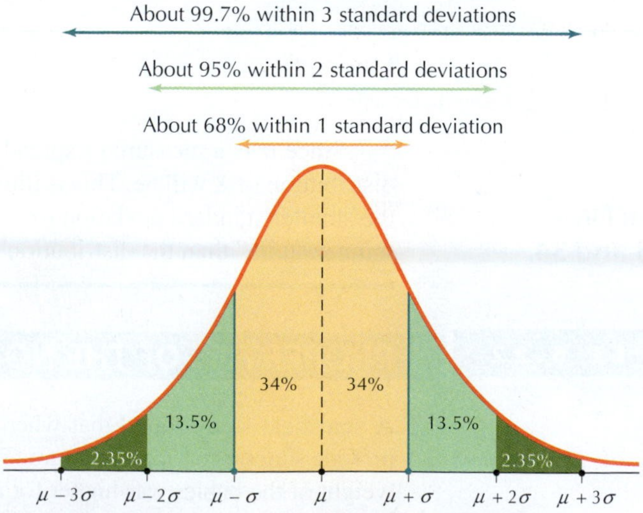

FIGURE 6.23
The Empirical Rule.

We will verify the Empirical Rule in Section 6.4.

EXAMPLE 6.23 | **EMPIRICAL RULE**

Recall the distribution of birth weights from Example 6.22.
a. What is the probability that a randomly chosen baby from the treatment group has a birth weight between 3785 grams and 4285 grams?
b. Find the probability that a randomly chosen baby from the treatment group has a birth weight greater than 425 grams.

Solution

a. Figure 6.24 shows the distribution of X = birth weights of babies from the treatment group. The area under the curve between 3785 and 4285 represents the area between $\mu + \sigma$ and $\mu + 2\sigma$. Courtesy of the Empirical Rule, Figure 6.24 tells us that the area between $\mu + \sigma$ and $\mu + 2\sigma$ is about 13.5% of the area under the curve.

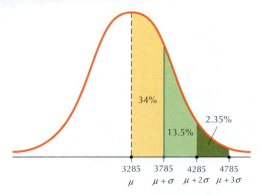

FIGURE 6.24
Some Empirical Rule probabilities for X = birth weights.

Therefore, the probability that a randomly chosen baby from the treatment group has a birth weight between 3785 grams and 4285 grams is about 0.135.

b. The area to the right of μ = 3285 equals 0.5, or 50%, of the area under the curve. To find the area to the right of X = 4285, we need to subtract the yellow area (34%) and the light green area (13.5%) from 50%: 50% − 34% − 13.5% = 2.5%. Therefore, the probability that a randomly chosen baby from the treatment group has a birth weight greater than 4285 grams is about 0.025.

We may use the probability of a birth weight greater than 4285 grams to represent the percentage or the proportion of birth weights greater than 4285.

Now You Can Do
Exercises 25–32.

SECTION 6.3 | **Summary**

1. Continuous random variables assume infinitely many possible values, with no gap between the values. Probability for continuous random variables consists of the area above an interval on the number line and under the distribution curve.

2. The uniform probability distribution has constant probability from its left to its right endpoints and is therefore shaped like a rectangle.

3. The normal distribution is the most important continuous probability distribution. It is symmetric about its mean μ and has standard deviation σ.

SECTION 6.3 | **Exercises**

Clarifying the Concepts

1. For a continuous random variable X, why are we not interested in whether X equals some particular value?

2. In the graph of a probability distribution, what is represented on the number line?

3. How is probability represented in the graph of a continuous probability distribution?

4. What are the possible values for the mean of a normal distribution? For the standard deviation?

5. True or false: The graph of the uniform distribution is always shaped like a square.

6. For continuous probability distributions, what is the difference between $P(X > 1)$ and $P(X \geq 1)$?

Practicing the Techniques

For Exercises 7–12, assume that X is a uniform random variable, with left endpoint 0 and right endpoint 100. Find the following probabilities.

7. $P(50 < X < 100)$

10. $P(15 \leq X \leq 35)$

8. $P(50 \leq X \leq 100)$

11. $P(24 < X < 25)$

9. $P(25 < X < 90)$

12. $P(25 < X < 25)$

For Exercises 13–16, assume that X is a uniform random variable, with left endpoint -5 and right endpoint 5. Compute the following probabilities.

13. $P(0 \leq X \leq 5)$

15. $P(-5 \leq X \leq -4)$

14. $P(-5 \leq X \leq 5)$

16. $P(-1 \leq X \leq 5)$

For Exercises 17–20, assume that X is a normal random variable, with mean $\mu = 4$ and standard deviation $\sigma = 2$. Use the Empirical Rule to approximate the following probabilities.

17. $P(2 \leq X \leq 6)$

19. $P(X \leq 0)$

18. $P(0 \leq X \leq 8)$

20. $P(X \geq 6)$

For Exercises 21–24, assume that X is a normal random variable, with mean $\mu = 100$ and standard deviation $\sigma = 15$. Approximate the following probabilities.

21. $P(55 \leq X \leq 145)$

23. $P(55 \leq X \leq 70)$

22. $P(0 \leq X \leq 130)$

24. $P(X \leq 85)$

Use the normal distribution from Example 6.22 for Exercises 25–32. Birth weights are normally distributed with a mean weight of $\mu = 3285$ grams and a standard deviation of $\sigma = 500$ grams.

25. What is the probability of a birth weight equal to 3285 grams?

26. What is the probability of a birth weight more than 3285 grams?

27. What is the probability of a birth weight of at least 3285 grams?

28. Is the area to the right of $X = 4285$ grams greater than or less than 0.5? How do you know this?

29. Is the area to the left of $X = 4285$ grams greater than or less than 0.5? How do you know this?

30. What is the probability of a birth weight between 2785 and 3785 grams?

31. What is the probability of a birth weight between 1785 and 4785 grams?

32. What is the probability of a birth weight between 785 and 5785 grams?

33. The two normal distributions in the accompanying figure have the same standard deviation of 5 but different means. Which normal distribution has mean 10 and which has mean 25? Explain how you know this.

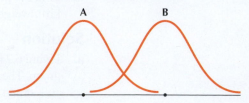

34. The two normal distributions in the figure below have the same mean of 100 but different standard deviations. Which normal distribution has standard deviation 3 and which has standard deviation 6? Explain how you know this.

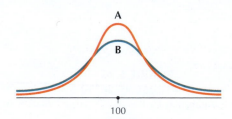

For Exercises 35–38, use the graph of the normal distribution to determine the mean and standard deviation. (*Hint:* The distance between dotted lines in the figures represents 1 standard deviation.)

35.

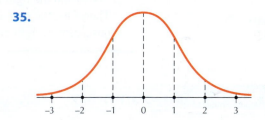

36.

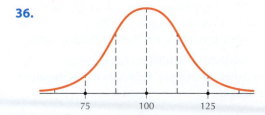

37.

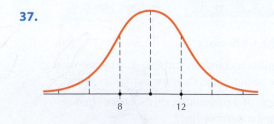

38.

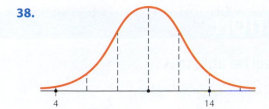

Applying the Concepts

39. Uniform Distribution: Web Page Loading Time.
Suppose that the Web page loading time for a particular home network is uniform, with left endpoint 1 second and right endpoint 5 seconds.
  a. What is the probability that a randomly selected Web page will take between 3 seconds and 4 seconds to load?
  b. Find the probability that a randomly selected Web page will take between 1 second and 2 seconds to load.
  c. How often does it take less than 1 second for a Web page to load?

40. Uniform Distribution: Random Number Generation.
Computers and calculators use the uniform distribution to generate random numbers. Suppose we have a calculator that randomly generates numbers between 0 and 1 so that they form a uniform distribution.
  a. What is the probability that a random number is generated which is less than 0.3?
  b. Find the probability that a random number is generated that is between 0.27 and 0.92.
  c. What is the probability that a random number greater than 1 is generated?

For Exercises 41–44, sketch the distribution, showing μ, $\mu + \sigma$, $\mu + 2\sigma$, $\mu + 3\sigma$, $\mu - \sigma$, $\mu - 2\sigma$, and $\mu - 3\sigma$. Then answer the questions.

41. Windy Frisco. The average wind speed in San Francisco in July is 13.6 miles per hour (mph), according to the U.S. National Oceanic and Atmospheric Administration. Suppose that the distribution of the wind speed in July in San Francisco is normal with mean $\mu = 13.6$ mph and standard deviation $\sigma = 4$ mph.
  a. Shade the region that represents wind speeds between 9.6 and 17.6 mph.
  b. What is the proportion of wind speeds between 9.6 and 17.6? (*Hint:* See Figure 6.23 (page 284).)

42. Viewers of *60 Minutes*. Nielsen Media Research reported that, for the week of October 18, 2010, 16 million viewers watched the television show *60 Minutes*. Suppose that the distribution of viewers of *60 Minutes* is normal with mean $\mu = 16$ million and standard deviation $\sigma = 4$ million.
  a. Shade the region that represents fewer than 8 million viewers.

  b. What is the probability that fewer than 8 million viewers will watch *60 Minutes*?

43. Hospital Patient Length of Stays. A study of Pennsylvania hospitals showed that the mean patient length of stay was 4.87 days with a standard deviation of 0.97 day.[8] Assume that the distribution of patient length of stays is normal. Find the probability that a randomly selected patient has a length of stay of less than 3.9 days.

44. Tobacco-Related Deaths. The World Health Organization states that tobacco is the second leading cause of death in the world. Every year, an average of 5 million people die of tobacco-related causes. Assume that the distribution is normal with mean $\mu = 5$ (in millions) and standard deviation $\sigma = 1$ (in millions).
  a. What is the probability of between 4 million and 7 million deaths?
  b. What is the probability of more than 6 million deaths?

45. Median Household Income. The Census Bureau reports that the median household income was $48,201 in 2006. Assume that the distribution of income is normal with mean $\mu = \$48,201$ and standard deviation $\sigma = \$16,000$.
  a. Find the probability that a randomly selected household has an income of greater than $80,201.
  b. What proportion of household incomes lie between $32,201 and $64,201?

46. Refer to Exercise 45. *What if* the mean μ was not $48,201 but some unknown value greater than $48,201? Describe whether the following probabilities you calculated in Exercise 45 would increase or decrease. Explain your reasoning.
  a. $P(X > \$80,201)$
  b. $P(X < \$32,201)$
  c. **Challenge Exercise.** $P(\$32,201 < X < \$64,201)$

47. Percentiles of the Uniform Distribution. The *p*th *percentile* of a continuous distribution is the value of X that is greater than or equal to $p\%$ of the values of X. Find the following percentiles of the uniform distribution in Example 6.21.
  a. 95th   **d.** 5th
  b. 90th   **e.** 10th
  c. 97.5th   **f.** 2.5th

48. Mean of the Uniform Distribution. Explain two ways that you could find the mean of the uniform distribution.
  a. Use the balance point method.
  b. Find the median (50th percentile), and argue that, since the distribution is rectangle shaped, the mean equals the median.

6.4 STANDARD NORMAL DISTRIBUTION

OBJECTIVES By the end of this section, I will be able to ...

1 Find areas under the standard normal curve, given a Z-value.

2 Find the standard normal Z-value, given an area.

1 FINDING AREAS UNDER THE STANDARD NORMAL CURVE FOR A GIVEN Z-VALUE

Note: Understanding the techniques explained in this section will allow you to analyze a whole world of data sets, even those that are not themselves normally distributed (see the Central Limit Theorem in the next chapter). Beyond this chapter, these techniques help you to calculate and understand *p*-values in Chapters 9–13.

There are many populations in the world that are normally distributed, from test scores to student heights. But there is one very special normal distribution called the **standard normal distribution.** The mean and standard deviation of the standard normal distribution make it unique.

> The **standard normal (Z) distribution** is a normal distribution with
> - mean $\mu = 0$ and
> - standard deviation $\sigma = 1$.

Because of its importance, the standard normal random variable is always denoted as a capital Z. The graph of the standard normal random variable Z is given in Figure 6.25. The standard normal curve is symmetric about its mean $\mu = 0$.

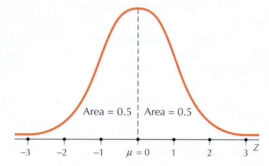

FIGURE 6.25
Z is symmetric about its mean $\mu = 0$.

We will discuss two methods for finding probabilities associated with Z, using (a) the table for finding standard normal probabilities, called the **Z table,** and (b) technology. For the Z table, see Table C in the Appendix. *The Z table provides areas under the standard normal curve to the left of a specified value of Z, denoted as Z_1* (see Figure 6.26).

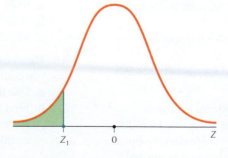

FIGURE 6.26
The Z table provides areas under the curve to the left of a specified value Z_1.

Note: Although your Z table contains only values between $Z = -3.49$ and $Z = 3.49$, there is no upper or lower limit to the values that Z may take. The curve essentially goes on forever in both the positive and the negative directions, always getting closer and closer to the horizontal axis but never quite touching it (there's a great plot for a love story in there somewhere).

Let's get acquainted with the Z table (see excerpt in Figure 6.28). Along the left side and across the top of the Z table are possible values of Z. These numbers, which in the table run from -3.49 to 3.49, are the values of Z found on the number line when you draw a graph. Down the left are the ones and tenths digits of the Z-value, and across the top is the hundredths digit. The body of the Z table contains areas (probabilities). These numbers, which run from 0.0002 to 0.9998, are areas under the standard normal curve that represent probabilities *to the left* of the specified value of Z. Table 6.6 shows the steps for finding areas under the standard normal curve, that is, for finding probabilities for specified values of Z.

Table 6.6 Steps for finding areas under the standard normal curve

Case 1	Case 2	Case 3
Find the area to the left of Z_1.	**Find the area to the right of Z_1.**	**Find the area between Z_1 and Z_2.**
Step 1 Draw the standard normal curve. Label the Z-value Z_1.	***Step 1*** Draw the standard normal curve. Label the Z-value Z_1.	***Step 1*** Draw the standard normal curve. Label the Z-values Z_1 and Z_2.
Step 2 Shade in the area to the left of Z_1.	***Step 2*** Shade in the area to the right of Z_1.	***Step 2*** Shade in the area between Z_1 and Z_2.
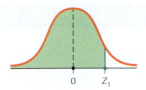	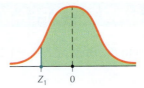	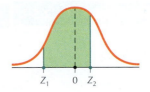
Step 3 Use the Z table to find the area to the left of Z_1.	***Step 3*** Use the Z table to find the area to the left of Z_1. The area to the right of Z_1 is then equal to 1 − (area to the left of Z_1).	***Step 3*** Use the Z table to find the area to the left of Z_1 and the area to the left of Z_2. The area between Z_1 and Z_2 is then equal to (area to the left of Z_2) − (area to the left of Z_1).

EXAMPLE 6.24   **CASE 1: FIND THE AREA TO THE LEFT OF A VALUE OF Z**

Find the area to the left of Z = 0.57.

Solution

STEP 1 First draw the standard normal curve and label Z = 0.57.

STEP 2 Shade the area to the left of 0.57, as shown in Figure 6.27.

STEP 3 In the Z table, excerpted on the next page as Figure 6.28, go down the left-hand column to 0.5 and select that row. Then go across the top row (representing the hundredth's digit) to 0.07 and select that column. The quantity at the intersection of this row and column represents the area to the left of Z = 0.57. That is, the area to the left of Z = 0.57 is 0.7157.

FIGURE 6.27
Finding the area to the left of Z.

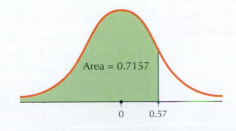

Now You Can Do
Exercises 11–18.

Standard Normal Distribution

Z	0.00	0.01	0.02	0.03	0.04	0.05	0.06	0.07	0.08	0.09
0.0	0.5000	0.5040	0.5080	0.5120	0.5160	0.5199	0.5239	0.5279	0.5319	0.5359
0.1	0.5398	0.5438	0.5478	0.5517	0.5557	0.5596	0.5636	0.5675	0.5714	0.5753
0.2	0.5793	0.5832	0.5871	0.5910	0.5948	0.5987	0.6026	0.6064	0.6103	0.6141
0.3	0.6179	0.6217	0.6255	0.6293	0.6331	0.6368	0.6406	0.6443	0.6480	0.6517
0.4	0.6554	0.6591	0.6628	0.6664	0.6700	0.6736	0.6772	0.6808	0.6844	0.6879
0.5	0.6915	0.6950	0.6985	0.7019	0.7054	0.7088	0.7123	0.7157	0.7190	0.7224
0.6	0.7257	0.7291	0.7324	0.7357	0.7389	0.7422	0.7454	0.7486	0.7517	0.7549

FIGURE 6.28 Using the Z table to find the area to the left of Z.

EXAMPLE 6.25

CASE 2: FIND THE AREA TO THE RIGHT OF A VALUE OF Z

Find the area to the right of $Z = -1.25$.

Solution

STEP 1 First draw the standard normal curve and label $Z = -1.25$.

STEP 2 Shade the area to the right of -1.25, as shown in Figure 6.29.

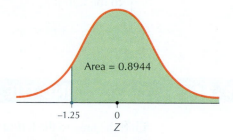

FIGURE 6.29
Finding the area to
the right of Z.

STEP 3 In the Z table, excerpted on the next page as Figure 6.30, go down the left-hand column to -1.2 and select that row. Then go across the top row to 0.05 and select that column. The area to the left of $Z = -1.25$ is therefore 0.1056. From Case 2 in Table 6.6, the area to the right of -1.25 is then

$$1 - (\text{area to the left of } -1.25) = 1 - 0.1056 = 0.8944$$

Now You Can Do Exercises 19–22.

CAUTION Remember that, although values of Z can be negative, *probabilities (or areas) can never be negative.*

Developing Your Statistical Sense

Checking That Your Answer Makes Sense

As you are finding probabilities for values of Z, you should always be checking to see that your answer makes sense. For instance, in Example 6.25, what if we had added the table area to 1 rather than subtracted the table area from 1? We would know that this answer is incorrect because the resulting probability would then have exceeded 1, and no probability can ever exceed 1.

Standard Normal Distribution

Z	0.00	0.01	0.02	0.03	0.04	0.05	0.06	0.07	0.08	0.09
−3.4	0.0003	0.0003	0.0003	0.0003	0.0003	0.0003	0.0003	0.0003	0.0003	0.0002
−3.3	0.0005	0.0005	0.0005	0.0004	0.0004	0.0004	0.0004	0.0004	0.0004	0.0003
−3.2	0.0007	0.0007	0.0006	0.0006	0.0006	0.0006	0.0006	0.0005	0.0005	0.0005
−3.1	0.0010	0.0009	0.0009	0.0009	0.0008	0.0008	0.0008	0.0008	0.0007	0.0007
−3.0	0.0013	0.0013	0.0013	0.0012	0.0012	0.0011	0.0011	0.0011	0.0010	0.0010
−1.4	0.0808	0.0793	0.0778	0.0764	0.0749	0.0735	0.0721	0.0708	0.0694	0.0681
−1.3	0.0968	0.0951	0.0934	0.0918	0.0901	0.0885	0.0869	0.0853	0.0838	0.0823
−1.2	0.1151	0.1131	0.1112	0.1093	0.1075	0.1056	0.1038	0.1020	0.1003	0.0985
−1.1	0.1357	0.1335	0.1314	0.1292	0.1271	0.1251	0.1230	0.1210	0.1190	0.1170
−1.0	0.1587	0.1562	0.1539	0.1515	0.1492	0.1469	0.1446	0.1423	0.1401	0.1379

FIGURE 6.30 Using the Z table to find the area to the right of Z.

EXAMPLE 6.26

CASE 3: FIND THE AREA BETWEEN TWO Z-VALUES (CHECKING THE ACCURACY OF THE EMPIRICAL RULE)

Recall that the Empirical Rule (page 284) states that about 68% of the area under the curve lies within 1 standard deviation of the mean, that is, between $\mu - \sigma$ and $\mu + \sigma$. Check this result for the standard normal distribution by using the Z table.

Solution

For the standard normal random variable Z, $\mu = 0$ and $\sigma = 1$, so that $\mu - \sigma = 0 - 1 = -1$ and $\mu + \sigma = 0 + 1 = 1$. Thus, using Case 3, we have $Z_1 = -1$ and $Z_2 = 1$.

STEP 1 Draw the standard normal curve. Label the Z-values $Z_1 = -1$ and $Z_2 = 1$.

STEP 2 Shade the area between -1 and 1, as shown in Figure 6.31a.

STEP 3 Find the area to the left of $Z_1 = -1$ and the area to the left of $Z_2 = 1$. The Z table gives these areas as follows: area to the left of $Z_1 = -1$ is 0.1587, and area to the left of $Z_2 = 1$ is 0.8413. We subtract the smaller area from the larger to give us the area between -1 and 1, as shown in Figures 6.31a–6.31c.

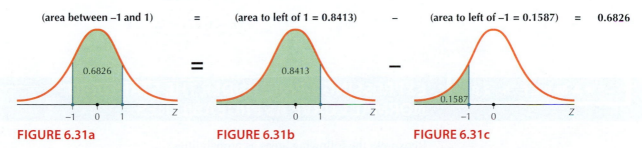

| (area between −1 and 1) | = | (area to left of 1 = 0.8413) | − | (area to left of −1 = 0.1587) | = | 0.6826 |

FIGURE 6.31a   FIGURE 6.31b   FIGURE 6.31c

Thus, the area under the Z curve within 1 standard deviation of the mean equals 0.6826. The Empirical Rule does very well for an approximation, missing the actual area by only 0.0026. Checking the accuracy of the Empirical Rule for other values of Z is left as an exercise.

Now You Can Do
Exercises 23–32.

EXAMPLE 6.27

USING TECHNOLOGY TO FIND THE AREA UNDER A STANDARD NORMAL CURVE

In Example 6.24, we found the area under the standard normal curve to the left of $Z = 0.57$ to be 0.7157. Confirm this result using technology.

Solution

We follow the instructions in the Step-by-Step Technology Guide at the end of Section 6.5 (pages 307–308). Figures 6.32a– 6.32c show the results from TI-83/84, Excel, and Minitab, respectively.

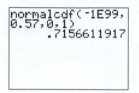

```
normalcdf(-1E99,
0.57,0,1)
      .7156611917
```

FIGURE 6.32a TI-83/84 results.

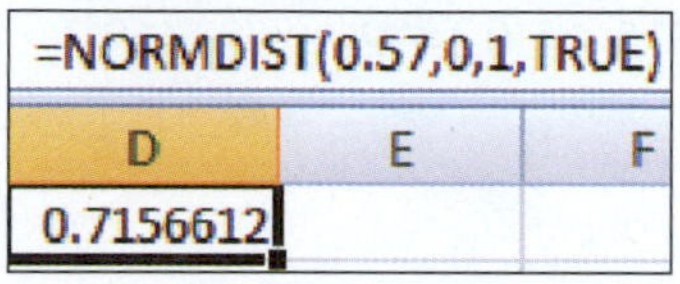

=NORMDIST(0.57,0,1,TRUE)		
D	**E**	**F**
0.7156612		

FIGURE 6.32b Excel results.

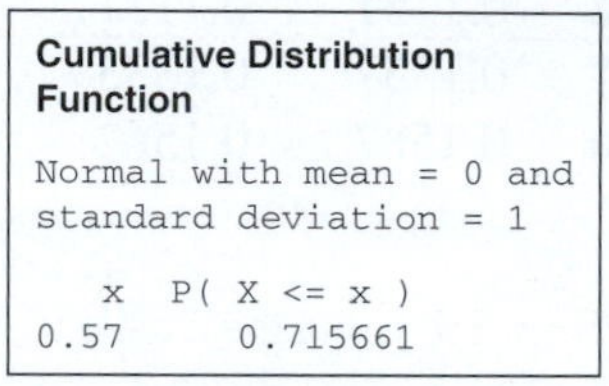

Cumulative Distribution Function

```
Normal with mean = 0 and
standard deviation = 1

    x    P( X <= x )
 0.57       0.715661
```

FIGURE 6.32c Minitab results.

The word "cumulative" in the Minitab output means "less than or equal to." Each of these results provides the area under the standard normal curve for values of Z that are less than or equal to 0.57. Each technology rounds to a different number of decimal places.

Note that the areas we have been finding in this section may also be expressed as probabilities. For continuous distributions probabilities are represented by areas under the curve above an interval. Specifically, for the standard normal distribution, probability is represented as the area above an interval under the standard normal curve. For instance, in Example 6.24, we found that the area under the standard normal curve to the left of $Z = 0.57$ is 0.7157. This may be reexpressed as follows:

"The probability that Z is less than 0.57 is 0.7157"

or

$$P(Z < 0.57) = 0.7157$$

EXAMPLE 6.28

EXPRESSING AREAS UNDER THE STANDARD NORMAL CURVE AS PROBABILITIES

Reexpress the following areas as probabilities.
a. In Example 6.25, we found the area under the standard normal curve to the right of $Z = -1.25$ to be 0.8944.
b. In Example 6.26, we found the area under the standard normal curve between $Z = -1$ and $Z = 1$ to be 0.6826.

Solution

a.   The probability that Z is greater than -1.25 is 0.8944. That is, $P(Z > -1.25) = 0.8944$.

b.   The probability that Z is between -1 and 1 is 0.6826. That is, $P(-1 < Z < 1) = 0.6826$.

Now You Can Do
Exercises 33–44.

The *Normal Density Curve* applet allows you to find areas associated with various values of Z.

2   FINDING STANDARD NORMAL Z-VALUES FOR A GIVEN AREA

In previous examples, we were given a Z-value and asked to find an area or probability. What if we turned this around, so that we are given an area, and asked to find its associated Z-value? We may call these "backwards" problems because we would need to use the Z table in reverse (unless we are using technology to solve the problem). Let's check out an example.

EXAMPLE 6.29   FINDING THE Z-VALUE WITH GIVEN AREA TO ITS LEFT

Find the Z-value with area 0.90 to its left.

Solution

STEP 1   Draw the standard normal curve. Label the Z-value Z_1.

STEP 2   Shade the area to the left of Z_1. Remember that *we are given an area and are looking for a value of Z*. Label the area to the left of Z_1 with the given area (0.90), as shown in Figure 6.33.

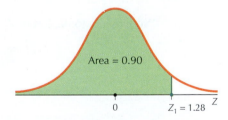

FIGURE 6.33 $Z_1 = 1.28$ is the value of Z with area 0.90 to the left of it.

Recall that the *r*th *percentile* is the value in the data set such that *r* percent of the data values fall at or below that value. Thus, $Z = 1.28$ represents the 90th percentile of the Z distribution, since it is greater than 90% of Z-values.

STEP 3   Look for 0.90 on the inside of the Z table (that is, in the body of the table), since the values *inside the table* represent areas. Because there is no 0.90 inside the table, by convention we take the area that is closest to 0.90, which is 0.8997. Next comes the trick of the backward problems, and the reason for that name. Move from 0.8997 to the left until you reach 1.2 in the first column, and then move up from 0.8997 until you get to 0.08 (see Figure 6.34). Putting these values together, we get $Z = 1.2 + 0.08 = 1.28$.

Now You Can Do
Exercises 45–52.

Standard Normal Distribution

Z	0.00	0.01	0.02	0.03	0.04	0.05	0.06	0.07	0.08	0.09
0.0	0.5000	0.5040	0.5080	0.5120	0.5160	0.5199	0.5239	0.5279	0.5319	0.5359
0.1	0.5398	0.5438	0.5478	0.5517	0.5557	0.5596	0.5636	0.5675	0.5714	0.5753
0.2	0.5793	0.5832	0.5871	0.5910	0.5948	0.5987	0.6026	0.6064	0.6103	0.6141
0.3	0.6179	0.6217	0.6255	0.6293	0.6331	0.6368	0.6406	0.6443	0.6480	0.6517
0.4	0.6554	0.6591	0.6628	0.6664	0.6700	0.6736	0.6772	0.6808	0.6844	0.6879
0.5	0.6915	0.6950	0.6985	0.7019	0.7054	0.7088	0.7123	0.7157	0.7190	0.7224
0.6	0.7257	0.7291	0.7324	0.7357	0.7389	0.7422	0.7454	0.7486	0.7517	0.7549
0.7	0.7580	0.7611	0.7642	0.7673	0.7704	0.7734	0.7764	0.7794	0.7823	0.7852
0.8	0.7881	0.7910	0.7939	0.7967	0.7995	0.8023	0.8051	0.8078	0.8106	0.8133
0.9	0.8159	0.8186	0.8212	0.8238	0.8264	0.8289	0.8315	0.8340	0.8365	0.8389
1.0	0.8413	0.8438	0.8461	0.8485	0.8508	0.8531	0.8554	0.8577	0.8599	0.8621
1.1	0.8643	0.8665	0.8686	0.8708	0.8729	0.8749	0.8770	0.8790	0.8810	0.8830
1.2	0.8849	0.8869	0.8888	0.8907	0.8925	0.8944	0.8962	0.8980	0.8997	0.9015
1.3	0.9032	0.9049	0.9066	0.9082	0.9099	0.9115	0.9131	0.9147	0.9162	0.9177

FIGURE 6.34 Using the Z table to find a value of Z for a given area.

EXAMPLE 6.30

FIND THE *Z*-VALUE WITH GIVEN AREA TO ITS RIGHT

Find the standard normal Z-value that has area 0.03 to the right of it.

Solution

STEP 1 Draw the standard normal curve. Label the Z-value Z_1. Shade the area to the right of it with the given area, as shown in Figure 6.35.

STEP 2 Since the Z table contains areas to the left of values of Z, we must find the area to the left of the specific value Z_1, as follows:

$$\text{area to left of } Z_1 = 1 - \text{area to right of } Z_1$$

So the area to the left of Z_1 is $1 - 0.03 = 0.97$.

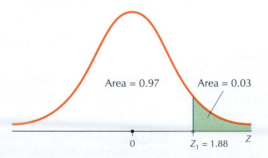

FIGURE 6.35 $Z_1 = 1.88$ has an area 0.03 to the right of it.

STEP 3 Look up 0.97 on the inside of the Z table. The closest area is 0.9699. Move from 0.9699 to the left until you reach 1.8, and then move up from 0.9699 until you get to 0.08 (see Figure 6.36). Putting these values together, we get $Z = 1.8 + 0.08 = 1.88$. In other words, the Z-value with area 0.03 to its right is $Z = 1.88$.

Now You Can Do Exercises 53–60.

Standard Normal Distribution									
Z	0.00	0.01	0.02	0.03	0.04	0.05	0.06	0.07	0.08	0.09
0.0	0.5000	0.5040	0.5080	0.5120	0.5160	0.5199	0.5239	0.5279	0.5319	0.5359
0.1	0.5398	0.5438	0.5478	0.5517	0.5557	0.5596	0.5636	0.5675	0.5714	0.5753
0.2	0.5793	0.5832	0.5871	0.5910	0.5948	0.5987	0.6026	0.6064	0.6103	0.6141
1.6	0.9452	0.9463	0.9474	0.9484	0.9495	0.9505	0.9515	0.9525	0.9535	0.9545
1.7	0.9554	0.9564	0.9573	0.9582	0.9591	0.9599	0.9608	0.9616	0.9625	0.9633
1.8	0.9641	0.9649	0.9656	0.9664	0.9671	0.9678	0.9686	0.9693	0.9699	0.9706
1.9	0.9713	0.9719	0.9726	0.9732	0.9738	0.9744	0.9750	0.9756	0.9761	0.9767

FIGURE 6.36 Using the Z table to find a value of Z for a given area.

When we learn statistical inference in later chapters, we will need to identify which Z-values divide the middle 90%, 95%, or 99% of the area under the standard normal curve from the tail area.

EXAMPLE 6.31

FIND THE VALUES OF Z THAT MARK THE BOUNDARIES OF THE MIDDLE 95% OF THE AREA

Find the two values of Z that mark the boundaries of the middle 95% of the area under the standard normal curve.

Solution

STEP 1 Draw the standard normal curve, showing the desired middle area (95%) with boundaries labeled as Z_1 and Z_2, as shown in Figure 6.37. By symmetry, there is area $= (1 - 0.95)/2 = 0.025$ in each tail.

STEP 2 Look up 0.025 on the inside of the Z table. Find Z_1 by moving to the left and up from 0.025 in the Z table, giving us $Z_1 = -1.96$.

STEP 3 Since the area in the right tail is 0.025 as well, the area to the left of Z_2 is $1 - 0.025 = 0.975$. Looking up 0.975 in the Z table gives us $Z_2 = 1.96$.

Note: Is it a coincidence that the two values of Z that determine the middle 95% of the area under the standard normal curve are 1.96 and -1.96? Not at all. Since the standard normal curve is symmetric about the mean 0, the values -1.96 and 1.96 that form the boundaries of the middle 95% must be *equidistant* from zero.

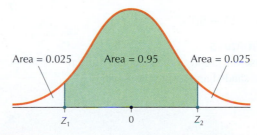

FIGURE 6.37 Z_1 and Z_2 mark the middle 95% of the Z distribution.

Area = 0.025   Area = 0.95   Area = 0.025

Z_1 0 Z_2

Now You Can Do Exercises 61–64.

Thus, the two Z-values that mark the boundaries of the middle 95% of the area under the standard normal curve are -1.96 and 1.96. This is a more precise result which states that about 95% lies between -2 and 2.

EXAMPLE 6.32 **USING TECHNOLOGY TO FIND VALUES OF Z, GIVEN AN AREA**

In Example 6.29, we found that the value of Z with area 0.90 to its left is $Z = 1.28$. Confirm this result using technology.

Solution

We follow the instructions in the Step-by-Step Technology Guide at the end of Section 6.5 (pages 307–308). Figures 6.38a–6.38c show the results from TI-83/84, Excel, and Minitab, respectively.

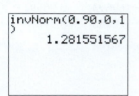

FIGURE 6.38a TI-83/84 results.

FIGURE 6.38b Excel results.

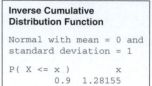

FIGURE 6.38c Minitab results.

SECTION 6.4 **Summary**

1. The standard normal distribution has mean $\mu = 0$ and standard deviation $\sigma = 1$. This distribution is often called the Z distribution. The Z table and technology can be used to find areas under the standard normal curve. In the Z table, the numbers on the outside are values of Z, and the numbers inside are areas to the left of values of Z.

2. The Z table and technology can also be used to find a value of Z, given a probability or an area under the curve.

SECTION 6.4 **Exercises**

Clarifying the Concepts

1. What is the value for the mean of the standard normal distribution?

2. What is the value for the standard deviation of the standard normal distribution?

3. True or false: The area under the Z curve to the right of $Z = 0$ is 0.5.

4. True or false: $P(Z = 0) = 0$.

Practicing the Techniques

For Exercises 5–16, use the graph of the standard normal distribution to find the shaded area using the Z table or technology.

5.

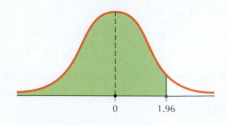

6.

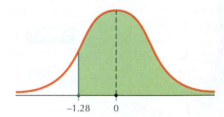

7.

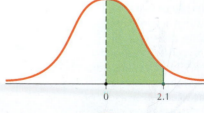

8.

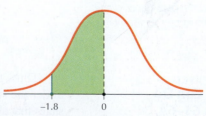

9.

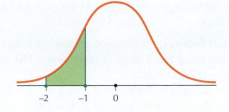

10.

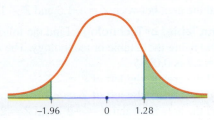

For Exercises 11–32,
   a. draw the graph.
   b. find the area using the Z table or technology.

Find the area under the standard normal curve that lies to the left of the following.

11. $Z = 1$

12. $Z = 2$

13. $Z = 3$

14. $Z = 0.5$

15. $Z = -2.7$

16. $Z = -0.9$

17. $Z = -0.2$

18. $Z = -1.2$

Find the area under the standard normal curve that lies to the right of the following.

19. $Z = 1.27$

20. $Z = 2.12$

21. $Z = -3.01$

22. $Z = -0.69$

Find the area under the standard normal curve that lies between the following.

23. $Z = 0$ and $Z = 1$

24. $Z = 1$ and $Z = 2$

25. $Z = 2$ and $Z = 3$

26. $Z = 1.28$ and $Z = 1.96$

27. $Z = -1$ and $Z = 0$

28. $Z = -2$ and $Z = -1$

29. $Z = -3$ and $Z = -2$

30. $Z = -1.96$ and $Z = -1.28$

31. $Z = -1.28$ and $Z = 1.28$

32. $Z = -2.01$ and $Z = 2.37$

For Exercises 39–50, find the indicated probability for the standard normal Z.
   a. Draw the graph.
   b. Find the area using the Z table or technology.

33. $P(Z = 0)$

34. $P(Z < 0)$

35. $P(Z < 10)$

36. $P(Z > 1.29)$

37. $P(Z < -2.17)$

38. $P(Z < 0.57)$

39. $P(-1.96 < Z < 1.96)$

40. $P(-2.07 < Z < 0.46)$

41. $P(-3.05 < Z < -0.94)$

42. $P(1.54 < Z < 2.20)$

43. $P(-100 < Z < 0)$

44. $P(-1.72 < Z < -1.57)$

For Exercises 45–52, find the Z-value with the following areas under the standard normal curve to its left. Draw the graph, then find the Z-value.

45. 0.3336   **49.** 0.95

46. 0.4602   **50.** 0.975

47. 0.3264   **51.** 0.98

48. 0.4247   **52.** 0.99

For Exercises 53–60, find the Z-value with the following areas under the standard normal curve to its right. Draw the graph, then find the Z-value.

53. 0.8078   **57.** 0.90

54. 0.3085   **58.** 0.975

55. 0.9788   **59.** 0.9988

56. 0.5120   **60.** 0.9998

For Exercises 61–65, find the values of Z that mark the boundaries of the indicated areas.

61. The middle 80%

62. The middle 95%

63. The middle 98%

64. The middle 85%

65. Find the 50th percentile of the Z distribution. (*Hint:* See margin note on page 293.)

66. Find the 75th percentile of the Z distribution.

67. Find the value of Z that is larger than 99.5% of all values of Z.

68. Find the value of Z that is smaller than 99.5% of all values of Z.

Applying the Concepts

69. Standardized Test Scores. Nicholas took a standardized test and was informed that the Z-value of his test score was 1.0. Find the percentages of test takers that Nicholas scored higher than.

70. Standardized Test Scores. Samantha's Z-value for her standardized test performance was 1.5. Calculate the proportion of test takers that Samantha scored higher than.

71. High Jump. Brandon's score in the high jump at a track-and-field event showed that he was able to jump higher than 45% of the competitors. Find the Z-value for Brandon's high-jump score.

72. Body Temperature. The body temperatures of all the students in Kayla's class were measured. Kayla's body temperature was lower than 90% of her classmates. Find the Z-value corresponding to Kayla's body temperature.

73. Checking the Empirical Rule. Check the accuracy of the Empirical Rule for $Z = 2$. That is, find the area between $Z = -2$ and $Z = 2$ using the techniques of this section. Then compare your finding with the results for $Z = 2$ using the Empirical Rule.

74. Checking the Empirical Rule. Check the accuracy of the Empirical Rule for $Z = 3$. That is, find the area between $Z = -3$ and $Z = 3$ using the techniques of this section. Then compare your finding with the results for $Z = 3$ using the Empirical Rule.

75. Without Tables or Technology. Find the following areas *without* using the Z table or technology. The area to the left of $Z = -1.5$ is 0.0668.
  a. Find the area to the right of $Z = 1.5$.
  b. Find the area to the right of $Z = -1.5$.
  c. Find the area between $Z = -1.5$ and $Z = 1.5$.

76. Without Tables or Technology. Find the following areas *without* using the Z table or technology. The area to the right of $Z = 2.7$ is 0.0035.
  a. Find the area to the left of $Z = 2.7$.
  b. Find the area to the left of $Z = -2.7$.
  c. Find the area between $Z = -2.7$ and $Z = 2.7$.

77. Values of Z That Mark the Middle 99%. Find the two values of Z that contain the middle 99% of the area under the standard normal curve.

78. Values of Z That Mark the Middle 90%. Find the two values of Z that contain the middle 90% of the area under the standard normal curve.

 Use the *Normal Density Curve* applet for Exercises 79.

79. Find the quartiles of the standard normal distribution. That is, find the 25th, 50th, and 75th percentiles of the standard normal distribution.

6.5 APPLICATIONS OF THE NORMAL DISTRIBUTION

OBJECTIVES By the end of this section, I will be able to . . .

1 Compute probabilities for a given value of any normal random variable.

2 Find the appropriate value of any normal random variable, given an area or probability.

1 FINDING PROBABILITIES FOR ANY NORMAL DISTRIBUTION

The data in problems that we face in the real world do not usually follow the standard normal distribution, Z. Instead, a problem may be stated in terms of some normal random variable X that has a mean other than 0 or a standard deviation other than 1. In cases like these, X needs to be standardized to Z so that we can use the Section 6.4 techniques.

To *standardize* things means to make them all the same. For example, college applicants take standardized tests so that the admissions officers can compare students according to a uniform assessment tool. Here, we standardize many different normal random variables X into the same standard normal Z.

Standardizing X to Z

To standardize a normal random variable X, we *transform* that normal random variable X into the standard normal random variable Z.

Suppose that X is a normal random variable with population mean μ and population standard deviation σ. We standardize X by subtracting the mean μ and dividing by the standard deviation σ. The result of this transformation is the familiar **standard normal random variable Z**.

> **Standardizing a Normal Random Variable**
>
> Any normal random variable X can be transformed into the standard normal random variable Z by *standardizing X* using the formula
>
> $$Z = \frac{X - \mu}{\sigma}$$

The key here is the following: *for a given area of interest for a normal random variable X, the corresponding area after the transformation to Z is exactly the same.* For any normal random variable X

the area between a and b

is exactly the same as

the area between $Z_a = \dfrac{(a - \mu)}{\sigma}$ and $Z_b = \dfrac{(b - \mu)}{\sigma}$ (see Figure 6.39)

So we can solve problems about areas under the nonstandard normal X curve by using the corresponding area under the Z curve.

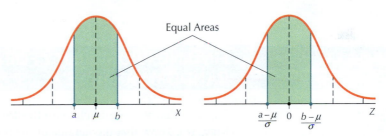

FIGURE 6.39 Corresponding areas are equal.

EXAMPLE 6.33

APRIL IN GEORGIA

The state of Georgia reports that the average temperature statewide for the month of April from 1949 to 2010 was $\mu = 61.5°F$. Assume that the standard deviation is $\sigma = 8°F$ and that temperature in Georgia in April is normally distributed. Draw the normal curve for temperatures between 45.5°F and 77.5°F, and the corresponding Z curve. Find the probability that the temperature is between 45.5°F and 77.5°F in April in Georgia.

Solution

Here we have $a = 45.5$ and $b = 77.5$, giving us

$$Z_a = \frac{a - \mu}{\sigma} = \frac{45.5 - 61.5}{8} = -2 \quad \text{and} \quad Z_b = \frac{b - \mu}{\sigma} = \frac{77.5 - 61.5}{8} = 2$$

In Figure 6.40, the area between 45.5°F and 77.5°F is the same as between $Z = -2$ and $Z = 2$. In other words

$$P(45.5 < X < 77.5) = P(-2 < Z < 2)$$

This is a Case 3 problem from Table 6.6. The Z table tells us that the area to the left of $Z_1 = -2$ is 0.0228, and the area to the left of $Z_2 = 2$ is 0.9772. The area between -2 and 2 is then equal to $0.9772 - 0.0228 = 0.9544$. The probability that temperature is between 45.5°F and 77.5°F in April in Georgia is 0.9544.

FIGURE 6.40
Find the area under
the *Z* curve and we
have found the area
under the
X curve.

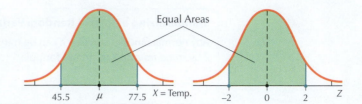

Equal Areas

45.5   μ   77.5   *X* = Temp.   −2   0   2   *Z*

Finding Probabilities for Any Normal Distribution

STEP 1   Determine the random variable *X*, the mean μ, and the standard deviation σ. Draw the normal curve for *X*, and shade the desired area.

STEP 2   Standardize by using the formula $Z = (X - \mu)/\sigma$ to find the values of *Z* corresponding to the *X*-values.

STEP 3   Draw the standard normal curve and shade the area corresponding to the shaded area in the graph of *X*.

STEP 4   Find the area under the standard normal curve using either the *Z* table or technology. This area is equal to the area under the normal curve for *X* drawn in Step 1.

EXAMPLE 6.34

FINDING PROBABILITY FOR A NORMAL RANDOM VARIABLE *X*

Suppose that you are in charge of ordering caps and gowns for senior graduation in May. You know that the heights of the population of students at your college are normally distributed with a mean of 68 inches (5 feet 8 inches) and a standard deviation of 3 inches. The gown manufacturer wants to know how many students will need to special-order their gowns because they are very tall. Find the proportion of students who are above 74 inches tall (6 feet 2 inches).

Solution

STEP 1   **Determine *X*, μ, and σ.**
We are given that the normal random variable *X* = heights of students has mean μ = 68 inches and standard deviation σ = 3 inches. In the center of the number line, mark the mean μ. Also mark on the number line the value of *X* that the problem is asking about. Figure 6.41 shows the graph of *X* (the heights of students) with the mean of 68 inches and the height of 74 inches marked.

Since you need to know the proportion of students taller than 74 inches, shade the area under the curve to the right of 74 inches. We can express this proportion as a probability, the probability that a randomly chosen student will be taller than 74 inches, or $P(X > 74)$. Just by looking at Figure 6.41 you should be able to get a rough idea of what the proportion of students taller than 74 inches will be. Certainly this proportion will be less than 50%, and probably pretty small. If you get an answer like "60%" for your proportion, you would surely know that it is wrong.

STEP 2   **Standardize.**
Now standardize the random variable *X* to the standard normal *Z*:

$$Z = \frac{X - \mu}{\sigma} = \frac{X - 68}{3}$$

Remember that you may solve problems asking for proportions or percentages by finding the appropriate probability.

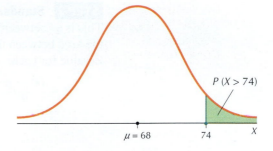

FIGURE 6.41

Graph of proportion of college students taller than 74 inches.

Find the Z-value corresponding to the height of 74 inches:

$$\frac{74 - \mu}{\sigma} = \frac{74 - 68}{3} = 2$$

So the Z-value associated with 74 inches is 2, which indicates that the height of 74 inches is 2 standard deviations above the mean of 68 inches.

STEP 3 **Draw the standard normal curve.**
Heights above 74 inches are more than 2 standard deviations above the mean, so shade the area to the right of 2 in Figure 6.42. Now find the area to the right of $Z = 2$ using the methods of Section 6.4.

STEP 4 **Find the area under the standard normal curve.**
Figure 6.42 represents a Case 2 problem from Table 6.6 (page 289). The Z table tells us that the area to the left of $Z = 2.00$ is 0.9772. Thus, the area to the right is

$$P(Z > 2) = 1 - 0.9772 = 0.0228$$

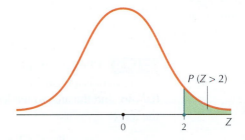

FIGURE 6.42

Graph of $P(Z > 2)$.

Now You Can Do
Exercises 3–9.

The proportion of students taller than 74 inches is 0.0228, or 2.28%. Note that this value for $P(X > 74)$ agrees with our earlier intuition that the proportion was surely less than 50% and most likely very small.

EXAMPLE 6.35

FINDING PROBABILITY THAT X LIES BETWEEN TWO GIVEN VALUES

Continuing the cap-and-gown problem, what percentage of students are between 60 and 70 inches tall?

Solution

STEP 1 **Determine X, μ, and σ.**
We have already seen that $X =$ heights of students, $\mu = 68$ inches, and $\sigma = 3$ inches. Once again, draw a graph of the distribution of heights X, with the mean 68 inches in the middle, the height 60 inches to the left of the mean, and the height 70 inches to the right of the mean, as in Figure 6.43.

STEP 2 **Standardize.**

This is a "between" example, where two values of X are given, and we are asked to find the area between them. In this case, just standardize both of these values of X to get a Z-value for each:

$$Z = \frac{60 - \mu}{\sigma} = \frac{60 - 68}{3} \approx -2.67 \quad \text{and} \quad Z = \frac{70 - \mu}{\sigma} = \frac{70 - 68}{3} \approx 0.67$$

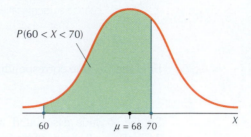

FIGURE 6.43
Graph of percentage of students between 60 and 70 inches tall.

STEP 3 **Draw the standard normal curve.**

Draw a graph of Z, shading the area between $Z = -2.67$ and $Z = 0.67$, as shown in Figure 6.44. Again, the key is that the area between $Z = -2.67$ and $Z = 0.67$ is exactly the same as the area between $X = 60$ inches and $X = 70$ inches.

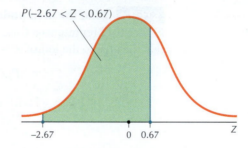

FIGURE 6.44
Graph of percentage of Z-values between -2.67 and 0.67.

STEP 4 **Find area under the standard normal curve.**

Figure 6.44 is a Case 3 problem from Table 6.6. Find the area to the left of 0.67, which is 0.7486, and the area to the left of -2.67, which is 0.0038. Subtracting the smaller from the larger gives us

$$P(-2.67 < Z < 0.67) = 0.7486 - 0.0038 = 0.7448$$

Thus, the percentage of students who are between 60 and 70 inches tall is 74.48%.

Now You Can Do
Exercises 10–14.

Check Your Answer! According to the Empirical Rule, almost all Z-values lie between -3 and 3, so it is unlikely that a randomly selected value of Z lies outside this range. You should remember this when you are doing your calculations. If you are standardizing a normal random variable X and get a very large Z-value (such as, say, 50), you should recheck your calculations because the probability that Z takes such a large value is very small.

2 FINDING A NORMAL DATA VALUE FOR A GIVEN AREA OR PROBABILITY

Sometimes we are given a probability (or proportion or area), and we are asked to find the associated value of X. Questions like these are similar to the "backward" problems of Section 6.4, so called because we must use the Z table backward or inside out. Since the formula for standardizing X gives the value for Z, we need to use our algebra skills to find the equation for X: Start with the standard normal formula $Z = (X - \mu)/\sigma$. Multiply both sides by σ to get $Z\sigma = X - \mu$. Then add μ to both sides, giving us $X = Z\sigma + \mu$.

> **Finding Normal Data Values for a Given Area or Probability**
>
> **Step 1** **Determine X, μ, and σ, and draw the normal curve for X.** Shade the desired area. Mark the position of X_1, the unknown value of X.
>
> **Step 2** **Find the Z-value corresponding to the desired area.** Look up the area you identified in Step 1 on the *inside* of the Z table. If you do not find the exact value of your area, by convention choose the area that is closest.
>
> **Step 3** **Transform this value of Z into a value of X, which is the solution.** Use the formula $X_1 = Z\sigma + \mu$.

EXAMPLE 6.36

FINDING A NORMAL DATA VALUE FOR A GIVEN AREA

Suppose that we wanted only the tallest 1% of our students to have to special-order gowns. What is the height at which tall students will have to special-order their gowns?

Solution

Notice that we are not asked to find a probability (or proportion or area). Instead, we are given a percentage (1%) and asked to find the value of X (the height) that is associated with this 1%.

STEP 1 **Determine X, μ, and σ, and draw the normal curve for X.**
We already know that X = heights of students, μ = 68 inches, and σ = 3 inches. The value of X we are interested in refers to very tall students, so that X_1 will be at the far right of the distribution of X. Only 1% of students will be taller than this height, so the area to the right of X_1 is 0.01, as shown in Figure 6.45.

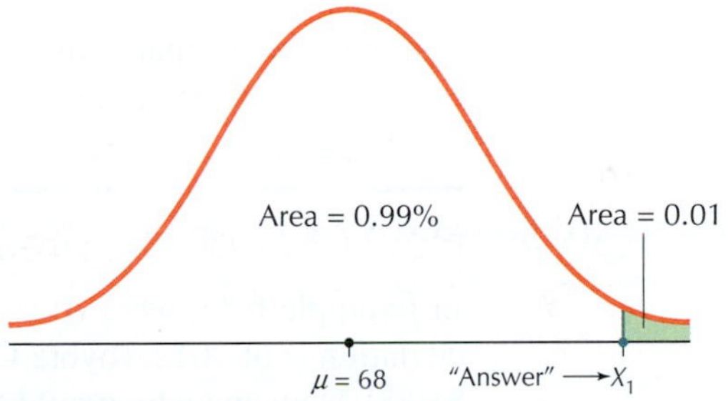

FIGURE 6.45
X_1 is the cutoff value (or critical value) of X, at which graduates will need to special-order their gowns.

Area = 0.99%   Area = 0.01

$\mu = 68$ "Answer" $\longrightarrow X_1$

STEP 2 **Find the Z-value corresponding to the desired area.**
The area to the right of X_1 equals 0.01, so that the area to the left of X_1 equals $1 - 0.01 = 0.99$. Looking up 0.99 on the inside of the Z table gives us $Z = 2.33$.

STEP 3 **Transform using the formula $X_1 = Z\sigma + \mu$.**
We calculate

$$X_1 = Z\sigma + \mu = (2.33)(3) + 68 = 74.99$$

Now You Can Do Exercises 15–22.

If we want only the tallest 1% of our students to have to special-order their gowns, the height at which tall students will have to special-order their gowns is 74.99 inches.

EXAMPLE 6.37

FINDING THE *X*-VALUES THAT MARK THE BOUNDARIES OF THE MIDDLE 95% OF *X*-VALUES

Edmunds.com reported that the average amount that people were paying for a 2012 Toyota Camry XLE was 24,725. Let X = price, and assume that price follows a normal distribution with $\mu = 24{,}725$, and $\sigma = \$1000$. Find the prices that separate the middle 95% of 2012 Toyota Camry XLE prices from the bottom 2.5% and the top 2.5%.

Solution

STEP 1 **Determine X, μ, and σ, and draw the normal curve for X.**
Let $X =$ price, $\mu = \$24{,}725$, and $\sigma = \$1000$. The middle 95% of prices are between X_1 and X_2, as shown in Figure 6.46.

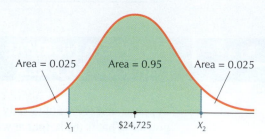

Area = 0.025   Area = 0.95   Area = 0.025

X_1 $\$24{,}725$ X_2

FIGURE 6.46
X_1 and X_2 mark the middle
95% of Camry prices.

STEP 2 **Find the Z-values corresponding to the desired area.**
The area to the left of X_1 equals 0.025, and the area to the left of X_2 equals 0.975. Looking up area 0.025 on the inside of the Z table gives us $Z_1 = -1.96$. Looking up area 0.975 on the inside of the Z table gives us $Z_2 = 1.96$.

STEP 3 **Transform using the formula $X_1 = Z\sigma + \mu$.**
We calculate

$$X_1 = Z_1\sigma + \mu = (-1.96)(1000) + 24{,}725 = 22{,}765$$

$$X_2 = Z_2\sigma + \mu = (1.96)(1000) + 24{,}725 = 26{,}685$$

**Now You Can Do
Exercises 23–26.**

The prices that separate the middle 95% of 2012 Toyota Camry XLE prices from the bottom 2.5% of prices and the top 2.5% of prices are $22,765 and $26,685.

How Change in Spread Affects Camry Prices

In Example 6.37, *what if* we ask the same question again, but this time the standard deviation σ of 2012 Toyota Camry XLE prices is not $1000 but some value less than $1000. How and why would this affect the following?

   a. The values Z_1 and Z_2 found in Step 2

   b. The value X_1 separating the middle 95% of prices from the bottom 2.5%

   c. The value X_2 separating the middle 95% of prices from the top 2.5%

Solution
Figure 6.47 illustrates the distribution of 2012 Toyota Camry XLE prices, where everything is the same as in Figure 6.46 except that the standard deviation of the prices is smaller by an unknown amount. Thus, the spread of the distribution is smaller.

   a. Since we are still asking for the middle 95% of prices, the Z-values remain the same, -1.96 and 1.96.

   b. Reexpress the formula $X_1 = Z_1\sigma + \mu$ as $X_1 = \$24{,}725 - 1.96 \cdot \sigma$. If σ is smaller than $1000, then the quantity $1.96 \cdot \sigma$, which represents the difference between the mean price and X_1, will also be smaller.

Since X_1 is less than the mean $\mu = \$24{,}725$, the smaller difference between the mean price and X_1 leads us to conclude that X_1 will be *larger* than in Example 6.38.

For example, if the new standard deviation is $\sigma = \$500$, then $X_1 = \$24{,}725 - 1.96 \cdot 500 = \$23{,}745$, which is larger than the $22,765 in Example 6.38.

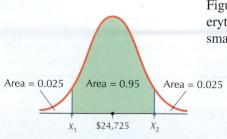

Area = 0.025   Area = 0.95   Area = 0.025

X_1 $\$24{,}725$ X_2

FIGURE 6.47 The middle 95% of prices
now has less spread.

c. Similarly, a smaller σ means a smaller quantity $1.96 \cdot \sigma$, which means that $X_2 = \$24,725 + 1.96 \cdot \sigma$ will be closer to the mean $\mu = \$24,725$. Since X_2 is larger than the mean, the new value for X_2 will be *smaller* than in Example 6.38.

The *Normal Density Curve* applet allows you to find areas associated with various values of any normal random variable.

EXAMPLE 6.38

NORMAL PROBABILITIES AND PERCENTILES USING TECHNOLOGY

Applying the information on Toyota Camry prices from Example 6.37, use the TI-83/84, Excel, or Minitab to find the following.
a. The proportion of 2007 Camry XLEs costing between \$22,000 and \$24,000, $P(22,000 \leq X \leq 24,000)$
b. The 99th percentile of Camry XLE prices, that is, find the value of X, namely, X_1, such that $P(X \leq X_1) = 0.99$

Solution

The instructions for finding these quantities are given in the Step-by-Step Technology Guide at the end of this section (page 307).

TI-83/84

a. Figure 6.48 shows that $P(22,000 \leq X \leq 24,000) = 0.6449902243 \approx 0.6450$.
b. Figure 6.49 shows that the value for X_1 such that $P(X \leq X_1) = 0.99$ is given by $X_1 = \$25,726.34788 \approx \$25,726.35$.

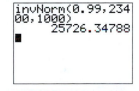

FIGURE 6.48 TI-83/84: Finding a probability.

FIGURE 6.49 TI-83/84: Finding a value of X.

Excel

a. Excel provides the cumulative probabilities $P(X \leq 22,000)$ in Figure 6.50 and $P(X \leq 24,000)$ in Figure 6.51. To find $P(22,000 \leq X \leq 24,000)$, we subtract $P(X \leq 22,000)$ from $P(X \leq 24,000)$:

$$P(22,000 \leq X \leq 24,000) = 0.725746882 - 0.080756659 = 0.644990223$$

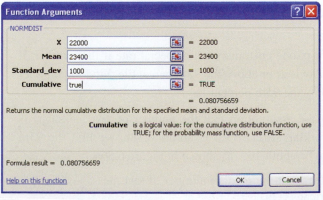

FIGURE 6.50 Excel: $P(x \leq 22,000)$.

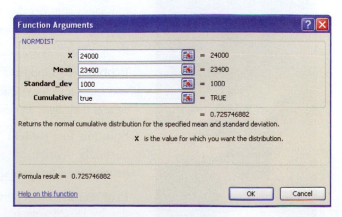

FIGURE 6.51 Excel: $P(x \leq 24,000)$.

b. Excel provides the result shown in Figure 6.52, $X_1 = \$25{,}726.34787 \approx \$25{,}726.35$.

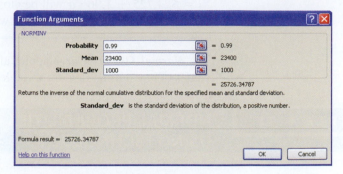

FIGURE 6.52 Excel: Finding a value of X.

Minitab

a. Like Excel, Minitab asks you to take the difference of two cumulative probabilities, $P(X \le 22{,}000)$ in Figure 6.53 and $P(X \le 24{,}000)$ in Figure 6.54:

$$P(22{,}000 \le X \le 24{,}000) = 0.725747 - 0.0807567 = 0.6449903 \approx 0.6450$$

```
Cumulative Distribution Function
Normal with mean = 23400 and standard deviation = 1000

    x   P( X <= x )
22000    0.0807567
```

FIGURE 6.53 Minitab: $P(x \le 22{,}000)$.

```
Cumulative Distribution Function
Normal with mean = 23400 and standard deviation = 1000

    x   P( X <= x )
24000    0.725747
```

FIGURE 6.54 Minitab: $P(x \le 24{,}000)$.

b. The results are given in Figure 6.55; $X_1 = \$25{,}726.30$.

```
Inverse Cumulative Distribution Function
Normal with mean = 23400 and standard deviation = 1000

P( X <= x )         x
       0.99   25726.3
```

FIGURE 6.55 Minitab: Finding a value of X.

CASE STUDY   Text Messaging: Be Careful What You Assume

Michael Newman/Photo Edit

The Pew Internet and American Life Project reports that the mean number of text messages sent per day by 18–24 year-old Americans is 109.5. Assume that the distribution of the number of text messages is normal, with $\mu = 109.5$ and standard deviation $\sigma = 35$.

Problem 1. Suppose that cell phone customers get a special rate if the number of text messages they send per day is at or above the 95th percentile. Find the number of text messages represented by the 95th percentile.

Solution to Problem 1. On the assumption that the number of text messages is normally distributed, and working similarly to Example 6.38b, we find the 95th percentile of text messages to be about 167, as shown in Figure 6.56a.

Problem 2. Pew reports further that the median number of text messages sent per day by 18–24-year-old Americans is 50.

```
invNorm(0.95,109
.5,35)
        167.0698769
```

FIGURE 6.56a 95th percentile of text messages.

a. What does this say about our assumption of normality for the distribution of text messages?

b. What shape does the distribution of the number of text messages actually take?

c. Is the actual 95th percentile of text messages greater or less than 167, and why?

Solution to Problem 2

a. In Chapter 3, we learned that, for symmetric distributions (like the normal distribution), the mean and the median were about equal (see Figure 3.4 on page 90). Since the mean number of text message 109.5 is much larger than the median of 50 text messages, then the distribution of text messages is not symmetric, and thus cannot be normal.

b. Figure 3.4 on page 90. Thus, the distribution of the number of text messages is actually right-skewed.

c. Figure 6.56b shows the (wrongly) assumed normal distribution in green and the actual right-skewed distribution in orange. Both distributions have the same mean, $\mu = 109.5$. The 95th percentile for each distribution is shown. Because the right-tail of the right-skewed distribution is extended, the 95th percentile of the right-skewed distribution is greater than the 95th percentile of the normal distribution. Thus, the actual 95th percentile of the number of text messages sent per day by 18–24-year-old Americans is greater than 167. ■

FIGURE 6.56b
Incorrect assumption of normality led us to underestimate the 95th percentile of the number of text messages.

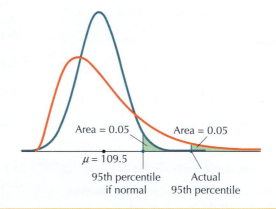

STEP-BY-STEP TECHNOLOGY GUIDE: Finding Areas, Probabilities, and Percentiles for Any Normal Distribution

TI-83/84

Finding Areas or Probabilities for Any Normal Distribution

Step 1 Press **2ⁿᵈ**, then **DISTR** (the **VARS** key).

Step 2 Press **2** to choose **normalcdf(**.

Step 3 On the home screen, enter the smaller value of X, comma, the larger value of X, comma, the mean of X, comma, the standard deviation of X, then close parenthesis. See Figure 6.48 (page 305).

Step 4 Press **ENTER**.

Note: When finding the area to the right of a value of X, use **1E99** as the larger value. When finding the area to the left of a value of X, use **-1E99** as the smaller value. Also, the shortcut for using the standard normal distribution is to specify only the lower and higher values of X. If you enter only two values, the calculator assumes you want the standard normal distribution.

Finding Percentiles for Any Normal Distribution
Step 1 Press **2nd**, then **DISTR** (the **VARS** key).
Step 2 Press **3** to choose **invNorm(**.
Step 3 On the home screen, enter the probability value or area, then the mean of *X*, then the standard deviation of *X*, then close parenthesis. See Figure 6.49 (see page 305).

Step 4 Press **ENTER**.
Note: A shortcut for finding standard normal percentiles is to enter only the value of *X* for this function, in which case the calculator assumes you want the standard normal distribution.

Note: Not all TI-83/84's have the **invNorm** function.

EXCEL

Finding Areas or Probabilities for Any Normal Distribution
Step 1 Select cell **A1** and click the **Insert Function** icon F_x.
Step 2 In the **Search for a function**, type **NORMDIST**, click **GO**, then **OK**.
Step 3 For **X**, enter the *X*-value that you want to find the probability for. For **Mean**, enter the value of μ. For **Standard_dev**, enter the value of σ. For **Cumulative**, always enter **true**. Click **OK**. See Figure 6.50 (page 305).
Step 4 Excel provides the cumulative probability, $P(X \le X_1)$ (see Example 6.38). If you need to find $P(X > X_1)$, subtract the result from 1. If you need to find $P(X_1 \le X \le X_2)$, find the two

cumulative probabilities, and subtract the lesser from the greater, as in Example 6.38.

Finding Percentiles for Any Normal Distribution
Step 1 Select cell **A1** and click the **Insert Function** icon f_x.
Step 2 In the **Search for a function**, type **NORMINV**, click **GO**, then **OK**.
Step 3 For **Probability**, enter the desired percentile in decimal form (for example, 0.99). For **Mean**, enter the value of μ. For **Standard_dev**, enter the value of σ. Click **OK**. See Figure 6.52 (page 306).

MINITAB

Finding Areas or Probabilities for Any Normal Distribution
Step 1 Click **Calc > Probability Distributions > Normal**.
Step 2 Select **Cumulative Probability**, enter the mean μ and standard deviation σ.
Step 3 Select **Input Constant**, enter the *X*-value that you want to find the probability for.
Step 4 Minitab provides the cumulative probability, $P(X \le X_1)$ (see Figure 6.53 on page 306). If you need to find $P(X > X_1)$, subtract the result from 1. If you need to find $P(X_1 \le X \le X_2)$, find

the two cumulative probabilities, and subtract the lesser from the greater, as in Example 6.38.

Finding Percentiles for Any Normal Distribution
Step 1 Click **Calc > Probability Distributions > Normal**.
Step 2 Select **Inverse Cumulative Probability,** and enter the mean μ and standard deviation σ.
Step 3 Select **Input Constant**. For the constant, enter the desired percentile in decimal form (for example, 0.99). See Figure 6.55 (page 306).

CRUNCHIT!

We will use the Toyota Camry data from Example 6.38 (page 306).

Finding Areas or Probabilities for Any Normal Distribution
Step 1 Click **Distribution calculator . . . Normal**.
Step 2 For **mean** enter **23400**. For **sd** enter the standard deviation **1000**.
Step 3 Select ≤ and enter **24000**. Click **Calculate**. The result shown is $P(X \le 24{,}000) = 0.7257468822$.
Step 4 Delete **24000** and enter **22000**. Click **Calculate**.

The result shown is $P(X \le 22{,}000) = 0.0807566592$. The answer will then be $0.7257468822 - 0.0807566592 = 0.644990223$.

Finding Percentiles for Any Normal Distribution
Step 1 Click **Distribution calculator . . . Normal**.
Step 2 For **mean** enter **23400**. For **sd** enter the standard deviation **1000**.
Step 3 Select **Quantile**. Enter the desired percentile in decimal form. For the 99th percentile, enter **0.99** and click **Calculate**.

| SECTION 6.5 | Summary |

1. Section 6.5 showed how to solve normal probability problems for any conceivable normal random variable by first standardizing *X* into *Z* and then using the methods of Section 6.4. Methods for finding probabilities for a given value of the normal random variable *X* were discussed.

2. For any normal probability distribution, values of *X* can be found for given probabilities using the formula $X_1 = Z\sigma + \mu$.

Exercises

Clarifying the Concepts

1. What does the word *standardize* mean? Explain how we use standardization in solving normal probability problems.

2. When finding a data value for a specified probability, explain why we can't just report the Z-value but must transform back to the original normal distribution.

Practicing the Techniques

For Exercises 3–14, assume that the random variable X is normally distributed with mean $\mu = 70$ and standard deviation $\sigma = 10$. Draw a graph of the normal curve with the desired probability and value of X indicated. Find the indicated probabilities by standardizing X to Z.

3. $P(X > 70)$

4. $P(X > 80)$

5. $P(X < 80)$

6. $P(X > 95)$

7. $P(X \geq 95)$

8. $P(X \geq 60)$

9. $P(X \geq 55)$

10. $P(60 < X < 100)$

11. $P(60 \leq X \leq 100)$

12. $P(90 \leq X \leq 100)$

13. $P(90 \leq X \leq 91)$

14. $P(60 \leq X \leq 70)$

For Exercises 15–26, assume that the random variable X is normally distributed with mean $\mu = 70$ and standard deviation $\sigma = 10$. Draw a graph of the normal curve with the desired probability and value of X indicated. Find the indicated values of X using the formula $X_1 = Z\sigma + \mu$.

15. The value of X larger than 95% of all X-values (that is, the 95th percentile)

16. The value of X smaller than 95% of all X-values

17. The value of X larger than 97.5% of all X-values

18. The value of X smaller than 97.5% of all X-values

19. The 1st percentile

20. The 99th percentile (note that the 1st and 99th percentiles are symmetric values of X that contain the central 98% of the area under the curve between them)

21. The 0.5th percentile

22. The 99.5th percentile

23. The two symmetric values of X that contain the central 90% of X-values between them

24. The two symmetric values of X that contain the central 95% of X-values between them

25. The two symmetric values of X that contain the central 98% of X-values between them

26. The two symmetric values of X that contain the central 99% of X-values between them

Applying the Concepts

27. Hungry Babies. Six-week-old babies consume a mean of $\mu = 15$ ounces of milk per day, with a standard deviation σ of 2 ounces. Assume that the distribution is normal. Find the probability that a randomly chosen baby consumes the following amounts of milk per day.
    a. Less than 15 ounces
    b. More than 17 ounces
    c. Between 17 and 19 ounces

28. Trading Volume. The Associated Press reports that the mean trading volume for equity and index options contracts was 3.6 million in July 2007. Assume that the distribution is normal with mean $\mu = 3.6$ (in millions) and standard deviation $\sigma = 0.5$ (in millions). Find the probability that a randomly selected day of trading has the following volume.
    a. More than 4.1 million contracts
    b. Less than 4.1 million contracts
    c. Between 3.6 million and 4.1 million contracts

29. Windy Frisco. The mean wind speed in San Francisco is 13.6 mph in July, according to the U.S. National Oceanic and Atmospheric Administration. Suppose that the distribution of the wind speed in July in San Francisco is normal with mean $\mu = 13.6$ mph and standard deviation $\sigma = 6$ mph. Find the probability that a randomly chosen day in July has the following wind speeds.
    a. 7.2 mph or less
    b. Greater than 20 mph
    c. Between 15 and 20 mph
    d. Tours to Alcatraz Island are canceled if the day is too windy, specifically if the wind speed is higher than 99% of all other wind speeds in July. Find the cutoff wind speed.
    e. Suppose that a particular day in July has no wind at all. Should this be considered unusual? On what do you base your answer?

30. Viewers of *60 Minutes*. Nielsen Media Research reported that, for the week ending October 18, 2010, 16 million viewers watched the television show *60 Minutes*. Suppose that the distribution of viewers of *60 Minutes* is normal with mean $\mu = 12$ million and standard deviation $\sigma = 4$ million. Find the probability that the following numbers of people will watch *60 Minutes*.
    a. Fewer than 10 million people
    b. Between 10 million and 11 million people

   c. More than 11 million people

   d. Find the number of viewers that represents the 75th percentile.

   e. On one particular night, 28 million people watched *60 Minutes*. Is this unusual? On what do you base your answer?

31. Hungry Babies. Refer to Exercise 27.

   a. Find the amount of milk X greater than 95% of all values of X.

   b. Find the amount of milk X less than 95% of all values of X.

   c. Compute the two symmetric amounts of milk X_1 and X_2 that contain the central 90% of X-values between them.

32. Trading Volume. Refer to Exercise 28.

   a. Calculate the trading volume X greater than 99% of all values of X.

   b. Compute the trading volume X less than 99% of all values of X.

   c. Find the two symmetric trading volumes X_1 and X_2 that contain the central 98% of X-values between them.

33. Windy Frisco. Refer to Exercise 29.

   a. Find the 90% percentile of wind speed 5.

   b. Find the 10th percentile of wind speeds.

   c. Calculate the two symmetric wind speeds X_1 and X_2 that contain the central 80% of X-values between them.

   d. Suppose that a particular day in July has no wind at all. Should this be considered unusual? Use the Z-score method for outliers to determine the answer.

34. Viewers of *60 Minutes*. Refer to Exercise 30.

   a. A sponsor will withdraw its support if the number of viewers falls below the 5th percentile. Find the 5th percentile.

   b. Suppose the network can charge more for advertising if the number of viewers is greater than the 95th percentile. Compute the 95th percentile.

   c. Calculate the two symmetric numbers of viewers X_1 and X_2 that contain the central 90% of X-values between them.

   d. On one particular night, 24 million people watched *60 Minutes*. Is this unusual? Use the Z-score method for outliers to determine the answer.

35. Hospital Patient Length of Stays. A study of Pennsylvania hospitals showed that the mean patient length of stay in 2001 was 4.87 days with a standard deviation of 0.97 day. Assume that the distribution of patient length of stays is normal.

   a. Find the probability that a randomly selected patient has a length of stay of greater than 7 days.

   b. What proportion of patient lengths of stay are between 3 and 5 days?

   c. Find the 50th percentile of patient lengths of stay. What is the relationship between the mean and the median for normal distributions?

   d. A particular patient had a length of stay of 8 days. Determine whether this is unusual.

36. Tobacco-Related Deaths. The World Health Organization states that tobacco is the second leading cause of death in the world. Every year, a mean of 5 million people die of tobacco-related causes. Assume that the distribution is normal with $\mu = 5$ million and $\sigma = 2$ million.

   a. Find the probability that more than 4 million people will die of tobacco-related causes in a particular year.

   b. Find the 25th percentile of the distribution of tobacco-related deaths.

   c. Is there a way you can use symmetry and your answer to part (c) to find the 75th percentile of the distribution of tobacco-related deaths?

   d. In one particular year, 8 million people died from tobacco-related causes. Determine whether this is unusual.

37. Stock Shares Traded. The mean number of shares traded on the New York Stock Exchange in March 2010 was 2.1 billion per day. Assume that the distribution of shares traded is normal with $\mu = 2.1$ and $\sigma = 0.6$ (both in billions of shares).

   a. Find the probability that the number of shares traded on a randomly selected day falls below 0.3 billion.

   b. What proportion of days finds the volume of shares traded between 1 billion and 2 billion?

   c. A slow trading day has fewer shares traded than 99% of all other days. Find the number of shares traded that represents this amount.

   d. Determine whether 27 billion shares traded is unusual.

38. Calories per Gram. The histogram shows the number of calories per gram for 961 food items. Assume that the population mean calories per gram is 2.25 with a standard deviation of 2.

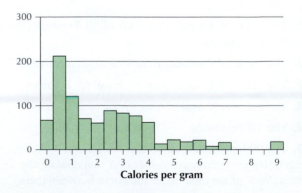

   a. Assuming that the data follow a normal distribution, what is the 5th percentile of calories per gram?

   b. Comment on whether your answer from (a) makes any sense.

c. The actual 5th percentile for this data set is 0.2 calorie per gram. Looking at the histogram, does this make more sense than your answer from **(a)**?

d. Why is your answer in **(a)** wrong?

 Use the *Normal Density Curve* applet for Exercises 39.

39. Use the applet to find the answers to the following exercises from this section.
   a. Exercise 35(a)
   b. Exercise 35(b)
   c. Exercise 36(a)
   d. Exercise 37(a)
   e. Exercise 37(c)

6.6 NORMAL APPROXIMATION TO THE BINOMIAL PROBABILITY DISTRIBUTION

OBJECTIVE By the end of this section, I will be able to . . .

1 Use the normal distribution to approximate probabilities of the binomial distribution.

1 USING THE NORMAL DISTRIBUTION TO APPROXIMATE PROBABILITIES OF THE BINOMIAL DISTRIBUTION

Recall from Section 6.2 that a binomial experiment satisfies the following four requirements: (1) Each trial must have two possible outcomes. (2) There is a fixed number of trials, n. (3) The experimental outcomes are independent. (4) The probability of observing a success is the same from trial to trial.

For certain values of n and p, it may be inconvenient to calculate probabilities for the binomial distribution. For example, if we are flipping a fair coin 100 times, so that $n = 100$ and $p = 0.5$, it may be tedious to calculate $P(X \geq 57)$, which, in the absence of technology, would involve 44 applications of the binomial probability formula. Fortunately, if the requirements are met, we may use the normal distribution to approximate such probabilities.

The binomial random variable X represents the number of successes in n trials and thus depends on the sample size n and the probability of success p. For a given probability of success p, if the sample size n gets large enough, the binomial distribution begins to resemble the normal distribution. Figure 6.57 shows the binomial probability distribution for Example 6.17 (page 270), where 20% ($p = 0.2$) of apps at Android Market threatened user privacy and Joshua received $n = 4$ apps with his new cell phone. The distribution of $X =$ the number of apps that threaten user privacy in Figure 6.57 is clearly not normal.

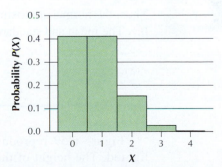

FIGURE 6.57
Binomial distribution:
$n = 4, p = 0.2$.

If we increase the sample size to $n = 64$ (Figure 6.58), the binomial distribution of X for $n = 64$ and $p = 0.2$, which is discrete, looks like it can be nicely approximated by the normal distribution, which is continuous.

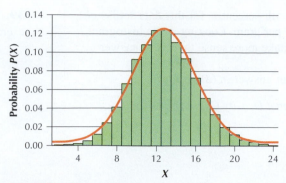

FIGURE 6.58 Binomial distribution: $n = 64, p = 0.2$.

We generalize this behavior as follows.

These values for μ_x and σ_x are the same as the values for μ and σ for a binomial random variable that we learned on page 266.

> **The Normal Approximation to the Binomial Probability Distribution**
>
> For the binomial random variable X with probability of success p and number of trials n:
> if $n \cdot p \geq 5$ and $n \cdot q \geq 5$, the binomial distribution may be approximated by a normal distribution with mean $\mu_x = n \cdot p$ and standard deviation $\sigma_x = \sqrt{n \cdot p \cdot q}$.

EXAMPLE 6.39

THE NORMAL APPROXIMATION TO THE BINOMIAL DISTRIBUTION

Punchstock/Blend

The Centers for Disease Control and Prevention reported that 20% of preschool children lack required immunizations, thereby putting themselves and their classmates at risk. For a group of $n = 64$ children with $p = 0.2$, the binomial probability distribution is shown in Figure 6.58.
a. Verify that this distribution can be approximated by a normal distribution.
b. Find the mean and standard deviation of this normal distribution.

Solution

a. The normal approximation is valid if $n \cdot p \geq 5$ and $n \cdot q \geq 5$. Substituting $n = 64$ and $p = 0.2$, we get

$$n \cdot p = (64)(0.2) = 12.8 \geq 5 \quad \text{and} \quad n \cdot q = (64)(0.8) = 51.2 \geq 5$$

Thus, the normal approximation is valid.
b. The mean and standard deviation of the normal distribution are

$$\mu_x = n \cdot p = (64)(0.2) = 12.8$$
$$\sigma_x = \sqrt{n \cdot p \cdot q} = \sqrt{(64)(0.2)(0.8)} = 3.2$$

Now You Can Do Exercises 3–8.

Figure 6.59 reproduces Figure 6.58, with the rectangle for $X = 12$ highlighted. The height of the rectangle represents the binomial probability that exactly 12 of the 64 children lack the required immunizations, that is, $P(X = 12)$. Since the width of the rectangle equals $12.5 - 11.5 = 1$, it follows that the area of the rectangle also represents the binomial probability that $X = 12$. Now, the area under the normal curve between 11.5 and 12.5 is approximately equal to

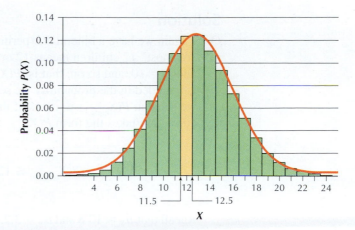

FIGURE 6.59
Normal curve approximates
binomial distribution.

this rectangle, which is $P(X = 12)$ for the binomial random variable X, with $n = 64$ and $p = 0.2$. That is

$$P(X_{\text{binomial}} = 12) \approx P(11.5 \leq Y_{\text{normal}} \leq 12.5)$$

where Y_{normal} is the normal random variable from Example 6.40(b), with mean $\mu_X = 12.8$ and standard deviation $\sigma_X = 3.2$.

The 0.5 that we add and subtract from 12 when approximating the binomial distribution with the normal distribution is called the *continuity correction,* since it is an adjustment for approximating a discrete probability with a continuous one. When using the normal approximation to the binomial, the analyst must determine which binomial rectangles are included and apply the continuity correction accordingly. This is shown in Table 6.7, which provides a listing of several types of binomial probabilities and their normal probability approximations.

Table 6.7 Binomial probabilities and approximate normal probabilities

Exact binomial probability	**Approximate normal probability**
$P(X_{\text{binomial}} = a)$	$P(a - 0.5 \leq Y_{\text{normal}} \leq a + 0.5)$
$P(X_{\text{binomial}} \leq a)$	$P(Y_{\text{normal}} \leq a + 0.5)$
$P(X_{\text{binomial}} \geq a)$	$P(Y_{\text{normal}} \geq a - 0.5)$
$P(X_{\text{binomial}} < a)$	$P(Y_{\text{normal}} < a - 0.5)$
$P(X_{\text{binomial}} > a)$	$P(Y_{\text{normal}} > a + 0.5)$
$P(a \leq X_{\text{binomial}} \leq b)$	$P(a - 0.5 < Y_{\text{normal}} \leq b + 0.5)$
$P(a < X_{\text{binomial}} < b)$	$P(a + 0.5 < Y_{\text{normal}} \leq b - 0.5)$

EXAMPLE 6.40

PERFORMING THE NORMAL APPROXIMATION TO THE BINOMIAL PROBABILITY DISTRIBUTION

For a group of $n = 64$ pre-schoolchildren with probability of lack of immunization $p = 0.2$, perform the following approximations.

a. Approximate the probability that there are at most 12 children without immunization.
b. Approximate the probability that there are more than 12 children without immunization.

Solution

Once again we have a binomial experiment with $n = 64$ and $p = 0.2$.

a. "At most" 12 children means 12 or fewer children. That is, $X = 12$ and $X = 11$ and $X = 10$, and so on; that is, $P(X_{binomial} \leq 12)$. In this case, we see that $X = 12$ is included in the probability we seek, as shown in Figure 6.60. From Table 6.7, we see that $P(X_{binomial} \leq 12)$ is of the form $P(X_{binomial} \leq a)$. Thus, our continuity correction takes the form $P(Y_{normal} \leq a + 0.5)$, where we add 0.5 to 12, so that

$$P(X_{binomial} \leq 12) \approx P(Y_{normal} \leq 12.5)$$

Recall that $\mu_X = 12.8$ and $\sigma_X = 3.2$. We use the TI-83/84, as shown in Figures 6.61 and 6.62, and find that the probability that at most 12 children lack immunization is $0.4626221269 \approx 0.4626$.

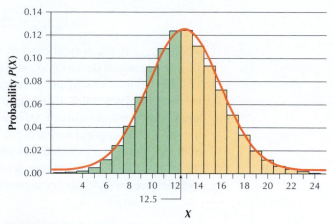

FIGURE 6.60 Approximately a binomial probability with a normal probability.

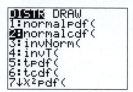

FIGURE 6.61 TI-83/84.

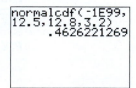

FIGURE 6.62 TI-83/84 results.

b. "More than" 12 children means $X = 13$ and $X = 14$, and so on. In other words, $X = 12$ is not included. That is, we want $P(X_{binomial} > 12)$. From Table 6.7, we see that $P(X_{binomial} > 12)$ is of the form $P(X_{binomial} > a)$. Thus, our continuity correction takes the form $P(Y_{normal} > a + 0.5)$, where we add 0.5 to 12, so that

$$P(X_{binomial} > 12) \approx P(Y_{normal} > 12.5)$$

Since the desired area is the complement of the green area in Figure 6.60, we can find the answer like this:

$$P(X_{binomial} > 12) \approx P(Y_{normal} > 12.5) = 1 - P(Y_{normal} \leq 12.5) = 1 - 0.4626 = 0.5374$$

Now You Can Do
Exercises 9–24.

The probability that more than 12 pre-schoolchildren will not have the required immunizations is 0.5374.

The *Normal Approximation to the Binomial Distributions* applet allows you to choose your own values of *n* and *p* and see how changes in these values affect the normal approximation to the binomial distribution.

SECTION 6.6	Summary

1. For certain values of *n*, *p*, and *X*, it may be inconvenient to calculate probabilities for the binomial distribution. The normal distribution can be used to approximate binomial probabilities when $n \cdot p \geq 5$ and $n \cdot q \geq 5$.

SECTION 6.6	Exercises

Clarifying the Concepts

1. Provide an example of why we would need to use the normal approximation to the binomial distribution.

2. What are the requirements for using the normal approximation to the binomial distribution?

Practicing the Techniques

For Exercises 3–8, determine whether the requirements are met for using the normal approximation to the binomial probability distribution.

3. *X* is a binomial random variable with $n = 10$ and $p = 0.5$

4. *X* is a binomial random variable with $n = 8$ and $p = 0.5$

5. *X* is a binomial random variable with $n = 10$ and $p = 0.4$

6. *X* is a binomial random variable with $n = 13$ and $p = 0.4$

7. *X* is a binomial random variable with $n = 45$ and $p = 0.1$

8. *X* is a binomial random variable with $n = 50$ and $p = 0.1$

For Exercises 9–16, let *X* be a binomial random variable with $n = 40$ and $p = 0.5$. Use the normal approximation to find the following probabilities.

9. $P(X = 20)$

10. $P(X \geq 20)$

11. $P(X > 20)$

12. $P(X \leq 20)$

13. $P(X < 20)$

14. $P(18 \leq X \leq 22)$

15. $P(18 < X < 22)$

16. $P(18 \leq X < 22)$

For Exercises 17–24, let *X* be a binomial random variable with $n = 120$ and $p = 0.1$. Use the normal approximation to find the following probabilities.

17. $P(X = 10)$

18. $P(X \geq 10)$

19. $P(X > 10)$

20. $P(X \leq 8)$

21. $P(X < 8)$

22. $P(9 \leq X \leq 11)$

23. $P(9 < X < 11)$

24. $P(9 < X \leq 11)$

Applying the Concepts

25. Gas Tax. A *New York Times*/CBS News Poll conducted in April 2007 reported that 64% of Americans would favor an increased federal tax on gasoline if it would reduce dependence on foreign oil. For a sample of 200 Americans, approximate the following probabilities.
 a. Exactly 128 would favor such a tax.
 b. At least 128 would favor such a tax.

26. Dress Casual. A survey conducted in 2007 by the Society for Human Resource Management found that the number of businesses allowing employees to "dress casually" every day dropped from 53% in 2002 to 38% in 2007. Suppose we have a sample of 50 businesses from 2002 and a sample of 50 businesses from 2007. Approximate the following probabilities.
 a. At least 25 businesses allowed casual dress every day in 2002.
 b. At least 25 businesses allowed casual dress every day in 2007.

27. Hurricane Response. A survey found that 19% of respondents in New Orleans rated the overall response by government and volunteer agencies to major hurricanes in the past three years as good or excellent, while 57% of those living in other areas did so.[9] Suppose that we have a sample of 100 people living in New Orleans and 100 people living in other areas. Approximate the following probabilities.
 a. At least 30 of the respondents living in New Orleans rated the response as good or excellent.
 b. At least 30 of the respondents living in other areas rated the response as good or excellent.

28. Disease Outbreak. A survey found that only 9% of Americans were "very confident" that the U.S. government is prepared to handle a major outbreak of an infectious disease.[10] Suppose that we have a sample of 100 Americans. Approximate the following probabilities.

a. Exactly 9 Americans are very confident.

b. At least 9 Americans are very confident.

29. Gas Tax. Refer to Exercise 25. Approximate the following probabilities.

    a. More than 128 would favor such a tax.

    b. Between 120 and 130 would favor such a tax.

30. Dress Casual. Refer to Exercise 26. Approximate the following probabilities.

    a. Fewer than 15 businesses allowed casual dress every day in 2002.

    b. Fewer than 15 businesses allowed casual dress every day in 2007.

31. Hurricane Response. Refer to Exercise 27. Approximate the following probabilities.

    a. Fewer than 20 of the respondents living in New Orleans rated the response as good or excellent.

    b. Fewer than 20 of the respondents living in other areas rated the response as good or excellent.

32. Disease Outbreak. Refer to Exercise 28. Approximate the following probabilities.

    a. More than 9 Americans are very confident.

    b. At most 9 Americans are very confident.

    c. Fewer than 9 Americans are very confident.

 Use the *Normal Approximation to the Binomial Distributions* applet for Exercise 33.

33. Select n (trials) = 10 and p (probability) = 0.2. The rectangles represent the binomial probabilities and the area under the curve represents the normal probabilities.

    a. For $n = 10$ and $p = 0.2$, is there a tight fit between the rectangles and the curve?

    b. What does this mean for whether the normal approximation should be used for a binomial distribution with $n = 10$ and $p = 0.2$?

    c. Verify whether the conditions are met for applying the normal approximation.

CHAPTER 6   **Formulas and Vocabulary**

Section 6.1

- **CONTINUOUS RANDOM VARIABLE** (p. 254)
- **DISCRETE RANDOM VARIABLE** (p. 254)
- **EXPECTED VALUE, OR EXPECTATION, OF A RANDOM VARIABLE** X (p. 260). Denoted $E(X)$
- **MEAN** μ **OF A DISCRETE RANDOM VARIABLE** (p. 258). $\mu = \sum[X \cdot P(X)]$
- **PROBABILITY DISTRIBUTION OF A DISCRETE RANDOM VARIABLE** (p. 255)
- **RANDOM VARIABLE** (p. 253)
- **RULES FOR A DISCRETE PROBABILITY DISTRIBUTION** (p. 255). $\sum P(X) = 1$ and $0 \leq P(X) \leq 1$
- **VARIANCE AND STANDARD DEVIATION OF A DISCRETE RANDOM VARIABLE** (p. 261).

Definition formulas:

$$\sigma^2 = \sum[(X - \mu)^2 \cdot P(X)]$$

$$\sigma = \sqrt{\sum[(X - \mu)^2 \cdot P(X)]}$$

Computational formulas:

$$\sigma^2 = \sum[X^2 \cdot P(X)] - \mu^2$$

$$\sigma = \sqrt{\sum[X^2 \cdot P(X)] - \mu^2}$$

Section 6.2

- **BINOMIAL EXPERIMENT** (p. 266)
- **BINOMIAL PROBABILITY DISTRIBUTION FORMULA** (p. 270). $P(X) = (_nC_X)\, p^X\, (1 - p)^{n-X}$

- **FACTORIAL (!)** (p. 268). $n! = n(n - 1)(n - 2) \ldots (2)(1)$
- **MEAN, VARIANCE, AND STANDARD DEVIATION OF A BINOMIAL RANDOM VARIABLE** X (p. 273).

Mean (or expected value):	$\mu = n \cdot p$
Variance:	$\sigma^2 = n \cdot p \cdot (1 - p)$
Standard deviation:	$\sigma = \sqrt{n \cdot p \cdot (1 - p)}$

- **NUMBER OF COMBINATIONS** (p. 268).
$$_nC_X = \frac{n!}{X!(n - X)!}$$

Section 6.3

- **CONTINUOUS PROBABILITY DISTRIBUTION** (p. 280)
- **DENSITY CURVE** (p. 280)
- **LAW OF TOTAL PROBABILITY FOR CONTINUOUS RANDOM VARIABLES** (p. 280)
- **PROBABILITY FOR CONTINUOUS DISTRIBUTIONS** (p. 281)
- **PROPERTIES OF THE NORMAL DENSITY CURVE (NORMAL CURVE)** (p. 283)
- **REQUIREMENTS FOR A CONTINUOUS PROBABILITY DISTRIBUTION** (p. 280)
- **UNIFORM PROBABILITY DISTRIBUTION** (p. 281)

Section 6.4

- **STANDARD NORMAL (Z) DISTRIBUTION** (p. 288)
- **Z TABLE FOR THE STANDARD NORMAL RANDOM VARIABLE Z** (p. 288)

Section 6.5

- **FINDING NORMAL DATA VALUES FOR SPECIFIED PROBABILITIES** (p. 303)

- **FINDING PROBABILITIES FOR ANY NORMAL DISTRIBUTION** (p. 300)
- **STANDARDIZING A NORMAL RANDOM VARIABLE** (p. 299).

$$Z = \frac{X - \mu}{\sigma}$$

Section 6.6
- **NORMAL APPROXIMATION TO THE BINOMIAL PROBABILITY DISTRIBUTION** (p. 312)

CHAPTER 6 Review Exercises

Section 6.1

1. EARLY LUNCH. Chad has gotten to lunch early and is waiting for his friends to catch up. He figures that the probability that one friend shows up is 25%; two friends, 35%; three friends, 20%; and more than three friends, 5%.

a. What is the probability that no friends show up?
b. What is the probability that more than one friend shows up?

2. CONNECTICUT LOTTO. The Connecticut Lottery Corporation runs a game called Connecticut Classic Lotto. The player picks six different numbers from 1 to 44 and pays $1 to play.

- If your picks match 3 of the 6 numbers chosen (probability 0.02381), you win $2.
- If you match 4 out of 6 (probability 0.001495), you win $50.
- If you match 5 out of 6 (probability 0.0000323), you win $2000.
- If you match all 6 numbers (probability 0.0000001417), you win the jackpot. The jackpot on July 23, 2004, was $2,600,000.
  a. Construct the probability distribution of your winnings. Make sure to include the probability of not winning anything (in effect, losing $1), which equals 1 minus the four probabilities specified above.
  b. Find the expected winnings. Compare this with the price to play.

Section 6.2

3. GESTATIONAL DIABETES. Gestational diabetes occurs in 8% of pregnancies, according to the American Diabetes Association. A random sample of 20 pregnancies is taken.

a. Find the probability that none of the pregnancies results in gestational diabetes.
b. Find the probability that at least 1 of the pregnancies results in gestational diabetes.
c. Find the probability that at most 2 of the pregnancies result in gestational diabetes.

4. PRICE OF GAS. The Pew Research Center reports that, in March 2011, 90% of Americans said they were bearing mostly bad news about the price of gasoline. A random sample of 15 Americans is taken.

a. Find the probability that 12 Americans said that they were bearing mostly bad news about price of gas.

b. Find the probability that between 12 and 14 Americans said that they were bearing mostly bad news about the price of gas.
c. Find the mean, variance, and standard deviation. Interpret the mean.

Section 6.3

SYSTOLIC BLOOD PRESSURE. Use the following information for Exercises 5–8. A study found that the mean systolic blood pressure was 106 mm Hg.[11] Assume that systolic blood pressure follows a normal distribution with mean $\mu = 106$ and standard deviation $\sigma = 8$.

5. What is the probability that a randomly chosen systolic blood pressure is equal to 106 mm Hg?

6. What is the probability that a randomly selected systolic blood pressure is more than 106 mm Hg?

7. Is the area to the right of $X = 110$ mm Hg greater than or less than 0.5? How do you know this?

8. What is the probability that a randomly selected systolic blood pressure is between 98 and 114 mm Hg?

Section 6.4

For Exercises 9–14, (a) draw the graph, and (b) find the indicated area using the Z table or technology.

Find the area under the standard normal curve that lies to the left of the following.
9. $Z = 2.1$
10. $Z = 2.9$

Find the area under the standard normal curve that lies to the right of the following.
11. $Z = -2.2$
12. $Z = -2.9$

Find the area under the standard normal curve that lies between the following.
13. $Z = -1.28$ and $Z = 1.28$
14. $Z = -1.04$ and $Z = 1.51$

15. SOUTH DAKOTA SPEEDS. The National Motorists Association reports that, in South Dakota, the mean speed on interstate highways is 68.3 mph. Denote the mean to be $\mu = 68.3$ mph, and assume that the distribution is normal and $\sigma = 4$ mph.

a. Find the probability that a randomly chosen vehicle is traveling faster than the 65 mph speed limit.
b. What percentage of vehicles travel slower than 60 mph?
c. What proportion of vehicles travel at speeds between 65 and 68.3 mph?

d. The National Motorists Association asserts that speeding tickets should be issued only for drivers whose speeds exceed the 85th percentile. If the police in South Dakota followed this rule, then at what speed would they start handing out speeding tickets?

e. Suppose that someone from South Dakota never drives faster than 55 mph on the interstate. Is this unusual? On what do you base your answer?

Section 6.5

16. DRUNK-DRIVING DEATHS. In the United States, a mean of 48 people *per day* are killed in vehicle accidents involving a drunk driver. Assume that the distribution of drunk-driving accident deaths per day is normal, $\mu = 48$, and $\sigma = 12$.

a. Find the probability that at most 12 people will be killed in drunk-driving accidents today.

b. Find the probability that between 50 and 80 people will be killed in drunk-driving accidents today.

c. Find the 99.5th percentile of the number of drunk-driving accident deaths.

d. Suppose on one particular day in the United States, 60 people are killed in drunk-driving accidents. Is this unusual? On what do you base your answer?

17. MATH SCORES. The National Center for Education Statistics reports that mean scores on the standardized math test for eighth-graders in 2009 increased slightly from those for previous years. The mean score in 2009 was $\mu = 283$. Assume $\sigma = 10$.

a. Find the probability that the test score of a randomly selected eighth-grader was greater than 290.

b. What proportion of test scores was between 295 and 300?

c. Suppose students who scored at the 5th percentile or lower could not graduate. Find the 5th percentile test score.

d. Suppose you know someone who scored 258 on the test. Is this unusual? On what do you base your answer?

Section 6.6

18. REINSTATE THE DRAFT? A *New York Times*/CBS News Poll found that 97% of young Americans (aged 17–29) oppose reinstating the military draft. Suppose we take a random sample of 400 young Americans. Use the normal approximation to approximate the following probabilities.

a. Exactly 388 oppose reinstating the military draft.

b. All 400 oppose reinstating the military draft.

c. More than 388 oppose reinstating the military draft.

d. At most 388 oppose reinstating the military draft.

e. Fewer than 388 oppose reinstating the military draft.

f. Between 385 and 390 (inclusive) oppose reinstating the military draft.

CHAPTER 6   Quiz

True or False

1. True or false: The following is a continuous and not a discrete random variable: How much coffee there is in your next cup of coffee.

2. True or false: The following is an example of a binomial experiment: Rolling a pair of dice 3 times and observing the sum of the two dice.

3. True or false: Our distributions for continuous random variables are for samples and not for populations.

Fill in the Blank

4. The probability that a randomly chosen value of a normally distributed random variable will be greater than the mean is _____.

5. The probability that a randomly chosen value of a normally distributed random variable will be equal to the mean is _____.

6. The standard deviation of a normal random variable can never take a value that is less than _____.

Short Answer

7. Is the following a discrete or continuous random variable: The number of goals your college soccer team will score in its next game.

8. Recording the gender of the next 20 babies born at City Hospital is an example of what kind of experiment?

9. What are the values for the mean and standard deviation of the standard normal distribution?

Calculations and Interpretations

10. CHURCH BAZAAR. Lenny has gone down to the church bazaar with his family. There is a game there where if you roll two dice and get a sum of at least 9, you will win $5; otherwise, you don't win anything.

a. Construct the probability distribution for the amount you win playing this game.

b. What are the expected winnings?

c. What would be a fair (break-even) price for the church to ask you to pay to play this game?

11. CEOS DRIVING LUXURY CARS. According to CareerBuilder.com, 19% of company CEOs drive luxury cars. Suppose a random sample is taken of 100 company CEOs.

a. Find the probability that the sample contains 20 CEOs who drive luxury cars.

b. What is the most likely number of CEOs who drive luxury cars?

c. Find the mean, variance, and standard deviation. Interpret the mean.

d. Suppose the sample contains 40 CEOs who drive luxury cars. Is this unusual? Explain how you determine this.

12. GAMBLING LOSSES. Treatment providers for problem gamblers report that men who approached them for intervention had lost a mean of $2849 in the preceding four weeks, according to a 2002 report.[12] Assume that the distribution of gambling losses is normally distributed with mean $\mu = \$2849$ and standard deviation $\sigma = \$900$.

a. Find the probability that a randomly selected male had lost more than $4000.

b. What percentage of males lost between $3000 and $4000?

c. Suppose that a gambling support group is trying to identify those who lose the most, as measured by the 95th percentile. How much money in gambling losses does this represent?

d. Suppose you know of a male problem gambler who lost $1000 in four weeks and then approached a treatment provider. Is this amount unusual? On what do you base your answer?

7 Sampling Distributions

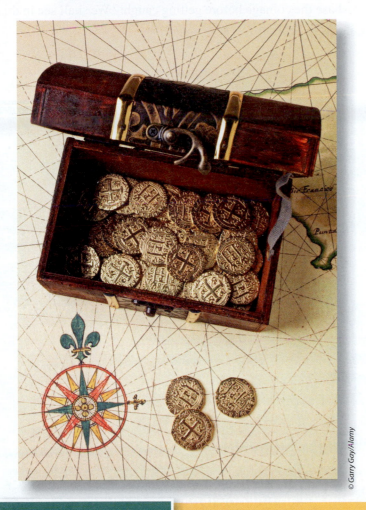

© Garry Gay/Alamy

CASE STUDY

Trial of the Pyx: How Much Gold Is in Your Gold Coins?

The kings of bygone England had a problem: How much gold should they put into their gold coins? After all, the very commerce of the kingdom depended on the purity of the currency. How did the lords of the realm ensure that the coins floating around the kingdom contained reliable amounts of gold?

From the year 1282, the Trial of the Pyx has been held annually in London to ensure that newly minted coins adhere to the standards of the realm. It is the responsibility of the presiding judge to ensure that the trial proceeds lawfully and to inform Her Majesty's Treasury of the verdict. Six members of the Company of Goldsmiths compose the jury, who are given two months to test the coins. It works like this: A ceremonial boxwood chest, called the Pyx, is brought forth, and a sample of 100 of the coins cast that year at the mint is put into it. The Pyx is then weighed. In times past, each gold coin, called a guinea, had an expected weight of 128 grams, so the total weight of the guineas in the Pyx was expected to be 12,800 grams.

(continues)

If the weight of the coins in the Pyx was much less than 12,800 grams, the jury concluded that the Master of the Mint was cheating the crown by pocketing the excess gold, and he was severely punished. On the other hand, if the coins in the Pyx weighed much more than 12,800 grams, that wasn't good either, since it cut down on the profits produced by the kings' coin-minting monopoly.

By how much could the Master of the Mint debase the coinage before getting caught? We shall see in this chapter's Case Study, Trial of the Pyx, which unfolds in Section 7.2. ■

The Big Picture

Where we are coming from, and where we are headed . . .

- In Chapters 1–4 we learned ways to describe data sets using numbers, tables, and graphs. Then in Chapters 5 and 6 we learned the tools of probability and probability distributions that allow us to quantify uncertainty.

- Here, in Chapter 7, "Sampling Distributions," we will discover that seemingly random statistics, such as the sample mean $\bar{x}$, have *predictable behaviors*. The special type of distribution we use to describe these behaviors is called the *sampling distribution of the sample mean*. This leads us to perhaps the most important result in statistical inference: the Central Limit Theorem.

- The sampling distributions we learn in this chapter form the basis for most of the statistical inference we perform in the remainder of the book. For example, in Chapter 8, "Confidence Intervals," we will learn how to estimate an unknown parameter with a certain level of confidence.

7.1 INTRODUCTION TO SAMPLING DISTRIBUTIONS

OBJECTIVES By the end of this section, I will be able to . . .

1 Explain the sampling distribution of the sample mean $\bar{x}$.

2 Describe the sampling distribution of the sample mean $\bar{x}$ when the population is normal.

3 Find probabilities and percentiles for the sample mean when the population is normal.

In Chapter 6 we dealt with probability distributions, which describe populations. Here, in Chapter 7, we return to the use of sample data, in order to show how populations and their samples are connected.

1   SAMPLING DISTRIBUTION OF THE SAMPLE MEAN $\bar{x}$

In this chapter we will develop methods that will allow us to quantify the *behavior* of statistics like $\bar{x}$. For the sample mean $\bar{x}$, this behavior is expressed in the **sampling distribution of the sample mean.**

> The **sampling distribution of the sample mean** $\bar{x}$ for a given sample size n consists of the collection of the means of all possible samples of size n from the population.

First, we illustrate the collection of the sample means into the sampling distribution of the sample mean $\bar{x}$.

EXAMPLE 7.1

CONSTRUCTING THE SAMPLING DISTRIBUTION OF THE SAMPLE MEAN $\bar{x}$

Suppose we are interested in how long it takes the five members of the student government to commute to school. The times (in minutes) are given in Table 7.1 Since these five people are *all* the members of the student government, they represent a *population*.

Table 7.1 Commuting times for the five members of the student government

Amber	Brandon	Chantal	Dave	Emma
10	20	5	30	15

a. Calculate the population mean commuting time μ.
b. Take all possible samples of size $n = 3$ and calculate the mean $\bar{x}$ of each sample.

Solution

a. The mean commuting time of this population is

$$\mu = \frac{\sum x}{n} = \frac{10 + 20 + 5 + 30 + 15}{5} = 16 \text{ minutes}$$

b. Table 7.2 shows all possible samples of size $n = 3$ from the five students, along with the respective sample means.

Table 7.2 All possible samples of size 3 from population of student government members

Sample	Amber Brandon Chantal	Amber Brandon Dave	Amber Brandon Emma	Amber Chantal Dave	Amber Chantal Emma	Amber Dave Emma	Brandon Chantal Dave	Brandon Chantal Emma	Brandon Dave Emma	Chantal Dave Emma
Data	10	10	10	10	10	10	20	20	20	5
	20	20	20	5	5	30	5	5	30	30
	5	30	15	30	15	15	30	15	15	15
$\bar{x}$	11.67	20	15	15	10	18.33	18.33	13.33	21.67	16.67

The bottom row in Table 7.2 contains the sample means for all possible samples of size $n = 3$. That is, this row represents the sampling distribution of the sample mean $\bar{x}$ for $n = 3$.

Note from Table 7.2 that the value for the sample mean $\bar{x}$ varies from sample to sample. Thus, $\bar{x}$ is a *random variable*. This random variable exhibits *sampling variability* because its value changes from sample to sample. Fortunately, there are patterns (predictable behaviors) in how the sample mean $\bar{x}$ varies. Like any distribution, the sampling distribution of the sample mean has a balance point, and therefore a mean.

Figure 7.1 provides a dotplot of the sample means in Table 7.2, along with the mean of these sample means, indicated at the balance point $\mu = 16$. Figure 7.1 represents the sampling distribution of the sample mean for this example.

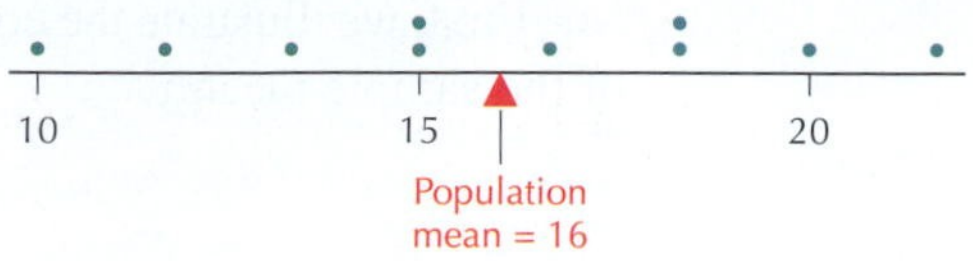

FIGURE 7.1

Sampling distribution of the sample mean for Example 7.1.

Population
mean = 16

The mean value of the ten sample means is

$$\frac{11.67 + 20 + 15 + 15 + 10 + 18.33 + 18.33 + 13.33 + 21.67 + 16.67}{10} = 16$$

Note that this value is exactly equal to the population mean $\mu = 16$. That is, the sampling distribution of $\bar{x}$ is centered at μ. We generalize this result as follows.

Note: It is convenient to number a set of important *facts*, as we build toward the Central Limit Theorem for Means and the Central Limit Theorem for Proportions.

> **Fact 1: Mean of the Sampling Distribution of the Sample Mean $\bar{x}$**
>
> The mean of the sampling distribution of the sample mean $\bar{x}$ is the value of the population mean μ. It can be denoted as $\mu_{\bar{x}} = \mu$ and read as "the mean of the sampling distribution of $\bar{x}$ is μ."

Note: In this example, the precise relationship between the two standard deviations is

$$\sigma_{\bar{x}} = \sqrt{\frac{N - n}{N - 1}} \cdot \frac{\sigma}{\sqrt{n}}$$

where N is the population size and n is the sample size. This gives

$$\sigma_{\bar{x}} = \sqrt{\frac{5 - 3}{5 - 1}} \cdot \frac{8.6023}{\sqrt{3}} \approx 3.5119$$

However, the coefficient

$$\sqrt{\frac{N - n}{N - 1}}$$

called the *finite population correction factor* is required only for special cases (like this textbook example) where the population is not much larger than the sample. This finite population correction factor does not apply when sampling with replacement, and its value tends to zero as the sample size approaches the population size. However, for most real-world problems, and for the remainder of this book, we dispense with this coefficient and assume that the population size is very large compared to the sample size.

Next we would like to uncover information regarding the spread of the sampling distribution of $\bar{x}$. The population standard deviation of the original commute times in Table 7.1 is

$$\sigma = \sqrt{\frac{\sum(x - \mu)^2}{N}}$$

$$= \sqrt{\frac{[(10 - 16)^2 + (20 - 16)^2 + (5 - 16)^2 + (30 - 16)^2 + (15 - 16)^2]}{5}}$$

$$\approx 8.6023$$

And the population standard deviation of the sampling distribution of $\bar{x}$ is

$$\sigma_{\bar{x}} = \sqrt{\frac{\sum(\bar{x} - \mu)^2}{10}} = \sqrt{\frac{[(11.67 - 16)^2 + (20 - 16)^2 + \cdots + (16.67 - 16)^2]}{10}}$$

$$\approx 3.5119$$

Note that the standard deviation of the sample means is smaller than the original standard deviation.

> **Fact 2: Standard Deviation of the Sampling Distribution of the Sample Mean $\bar{x}$**
>
> The standard deviation of the sampling distribution of the sample mean $\bar{x}$ is $\sigma_{\bar{x}} = \sigma/\sqrt{n}$, where σ is the population standard deviation and n is the sample size. $\sigma_{\bar{x}}$ is called the **standard error of the mean.**

Note the $\sqrt{n}$ in the denominator of the formula. Because of this factor, the larger the sample size, the tighter the resulting sampling distribution. Larger sample sizes lead to smaller variability, which results in more precise estimation.

EXAMPLE 7.2

FINDING THE MEAN AND STANDARD DEVIATION OF THE SAMPLING DISTRIBUTION OF $\bar{x}$

According to CanEquity Mortgage Company, the mean age of mortgage applicants in the city of Toronto is 37 years old. Assume that the population standard deviation is $\sigma = 6$ years. Find the mean and standard deviation for the sampling distribution of $\bar{x}$ for the following sample sizes: (a) 4, (b) 100, (c) 225.

Solution

We have $\mu_{\bar{x}} = \mu = 37$. Note that this value for $\mu_{\bar{x}}$ does not depend on the sample size, so the value is true for any sample size. We also have $\sigma = 6$.

a. $n = 4$. Then $\sigma_{\bar{x}} = \dfrac{\sigma}{\sqrt{n}} = \dfrac{6}{\sqrt{4}} = 3$. The standard error of the mean for $n = 4$ is $\sigma_{\bar{x}} = 3$.

b. $n = 100$. Then standard error $\sigma_{\bar{x}} = \dfrac{\sigma}{\sqrt{n}} = \dfrac{6}{\sqrt{100}} = 0.6$.

Now You Can Do Exercises 5–10.

c. $n = 225$. Then standard error $\sigma_{\bar{x}} = \dfrac{\sigma}{\sqrt{n}} = \dfrac{6}{\sqrt{225}} = 0.4$.

What Does This Number Mean?

Consider $\sigma_{\bar{x}} = 0.6$ for $n = 100$. This is a measure of the variability of the sampling distribution of $\bar{x}$ for this sample size. That is, if we take samples of size 100, our estimation of the population mean age μ of all mortgage applicants in Toronto will be within 0.6 year of the true population mean most of the time.

2 SAMPLING DISTRIBUTION OF $\bar{x}$ FOR A NORMAL POPULATION

To find out what form the sampling distribution of $\bar{x}$ takes when the population is normal, we consider the following example, in which we examine a small, normally distributed population and find the sample means for all possible samples of a certain size.

EXAMPLE 7.3

SAMPLING DISTRIBUTION OF $\bar{x}$ FOR A NORMAL POPULATION

In Example 6.37 (pages 303–304), the average statewide temperature in Georgia in the month of April was normally distributed with a mean of $\mu = 61.5°F$ and a standard deviation of $\sigma = 8°F$. Using Minitab, 1000 samples of size $n = 2$ were generated from this normal distribution, and the sample means $\bar{x}$ were calculated for each sample. Construct a histogram and observe the shape of the distribution.

Solution

Figure 7.2 shows the histogram of the means from the 1000 samples of size $n = 2$. As you may have expected, the histogram looks quite normal, even with this tiny sample size of $n = 2$.

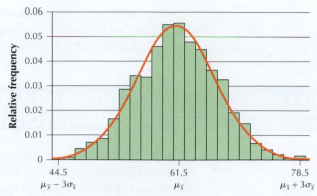

FIGURE 7.2 Sampling distribution of the sample means of size $n = 2$ from a normal population.

Example 7.3 considered a very small sample size, $n = 2$. In fact, this outcome is true for all normal populations. Using Facts 1 and 2, we can summarize this sampling distribution as follows.

Note: Let the notation

$$\text{normal } (\mu, \sigma/\sqrt{n})$$

denote a normal distribution with mean of μ and standard deviation of $\sigma/\sqrt{n}$.

> **Fact 3: Sampling Distribution of the Sample Mean for a Normal Population**
>
> For a normal population, the sampling distribution of the sample mean $\bar{x}$ is distributed as normal $(\mu, \sigma/\sqrt{n})$, where μ is the population mean and σ is the population standard deviation.

Once we know that the sample mean is normally distributed, we can use the method we learned in Section 6.5 (page 298) to standardize and produce Z, just as we would for any normal random variable.

> **Fact 4: Standardizing a Normal Sampling Distribution for Means**
>
> When the sampling distribution of $\bar{x}$ is normal, we may standardize to produce the standard normal random variable Z as follows:
>
> $$Z = \frac{\bar{x} - \mu_{\bar{x}}}{\sigma_{\bar{x}}} = \frac{\bar{x} - \mu}{\sigma/\sqrt{n}}$$
>
> where μ is the population mean, σ is the population standard deviation, and n is the sample size.

3 FINDING PROBABILITIES AND PERCENTILES USING A SAMPLING DISTRIBUTION

Since we know that the sampling distribution of the sample mean $\bar{x}$ is normal when the population is normally distributed, we can use the techniques of Section 6.5 to answer questions about the means of samples taken from normal populations.

EXAMPLE 7.4

FINDING PROBABILITIES USING THE SAMPLE MEAN: COMPARING $P(x > 80)$ WITH $P(\bar{x} > 80)$

Suppose that statistics quiz scores for a certain instructor are normally distributed with mean 70 and standard deviation 10.
 a. Find the probability that a randomly chosen student's score will be above 80.
 b. Suppose that, over the years, the instructor has had many sections of size $n = 25$. Describe the sampling distribution of the sample mean.

c.   Find the probability that a sample of 25 quiz scores will have a mean score greater than 80.

Solution

a.   This is a normal probability problem, which we learned how to do in Section 6.5.

$$P(X > 80) = P\left(\frac{X - 70}{10} > \frac{80 - 70}{10}\right)$$

$$= P(Z > 1) = 1 - 0.8413 = 0.1587$$

Case 2 shows how to find area to the right of a Z-value.

using Case 2 from Table 6.6 (page 289). Therefore, there is a 15.87% probability that a randomly chosen student will have a quiz score above 80 (Figure 7.3a).

b.   We are given $\mu = 70$ and $\sigma = 10$. So by Fact 1, $\mu_{\bar{x}} = \mu = 70$. And by Fact 2,

$$\sigma_{\bar{x}} = \frac{\sigma}{\sqrt{n}} = \frac{10}{\sqrt{25}} = 2.$$

Next, we are given that the population of quiz scores is normal. Therefore, by Fact 3, the sampling distribution of $\bar{x}$ is distributed as normal (70, 2) (see Figure 7.3b).

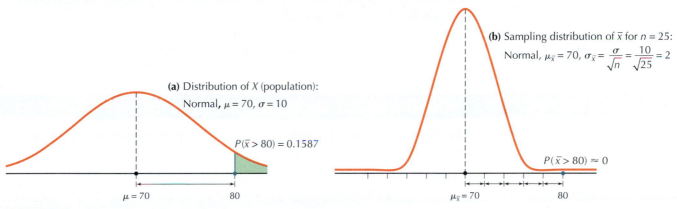

(a) Distribution of X (population):
Normal, $\mu = 70$, $\sigma = 10$

$P(\bar{x} > 80) = 0.1587$

$\mu = 70$ 80

(b) Sampling distribution of $\bar{x}$ for $n = 25$:
Normal, $\mu_{\bar{x}} = 70$, $\sigma_{\bar{x}} = \dfrac{\sigma}{\sqrt{n}} = \dfrac{10}{\sqrt{25}} = 2$

$P(\bar{x} > 80) \approx 0$

$\mu_{\bar{x}} = 70$ 80

FIGURE 7.3 Distribution of X and sampling distribution of $\bar{x}$ for Example 7.4.

c.   Once we know that the sample mean is normally distributed, we can standardize the quiz score, as we have for other normal random variables. Just be sure to use $\sigma_{\bar{x}} - 2$, the standard error of the mean, and not $\sigma = 10$, the standard deviation for the population. $\sigma_{\bar{x}}$ is always smaller.

Applying Fact 4,

$$Z = \frac{\bar{x} - \mu_{\bar{x}}}{\sigma_{\bar{x}}} = \frac{\bar{x} - \mu}{\sigma/\sqrt{n}} = \frac{\bar{x} - 70}{10/\sqrt{25}} = \frac{\bar{x} - 70}{2}$$

We need to standardize the score of 80 as well.

$$Z = \frac{80 - \mu_{\bar{x}}}{\sigma_{\bar{x}}} = \frac{80 - \mu}{\sigma/\sqrt{n}} = \frac{80 - 70}{10/\sqrt{25}} = \frac{80 - 70}{2} = 5$$

Hence,

$$P(\bar{x} > 80) = P\left(\frac{\bar{x} - 70}{10/\sqrt{25}} > \frac{80 - 70}{10/\sqrt{25}}\right) = P\left(\frac{\bar{x} - 70}{2} > \frac{80 - 70}{2}\right) = P(Z > 5) \approx 0$$

as shown in Figure 7.3b. Since Z is standard normal, nearly all observations lie between -3 and 3. Thus, the Z table does not go up to 5 since the probabilities

**Now You Can Do
Exercises 11–16.**

are so close to zero. The TI-83 provides the more precise probability of $P(Z > 5) = 0.000000287$, or about 3 in 10 million. This instructor just does not give easy quizzes!

What Does This Probability Mean?

There is essentially no chance that the sample mean $\bar{x}$ on one of the quizzes will be greater than 80. Compare this to the nearly 16% chance that a *particular* student's score would be above 80. Figure 7.3 shows the graphs of the distributions of quiz scores and class means. Both distributions are centered at $\mu_{\bar{x}} = \mu = 70$, but the standard deviations differ. The arrow in Figure 7.3a represents the standard deviation of X, $\sigma = 10$, and it shows that $x = 80$ is only 1 standard deviation above the mean $\mu = 70$. The arrows in Figure 7.3b represent the standard error of the mean, $\sigma_{\bar{x}} = 2$, and they illustrate that $\bar{x} = 80$ lies 5 standard errors above the mean $\mu_{\bar{x}} = 70$. Thus, class means are less variable than individual student scores.

In Chapter 6, we found the percentiles of normally distributed random variables. Since the sampling distribution of $\bar{x}$ is normal, we are able to find the percentiles of the $\bar{x}$'s as well. Once the appropriate Z-value is found, we use the following equation to transform the Z-value into an $\bar{x}$-value.

$$\bar{x} = Z \cdot \sigma_{\bar{x}} + \mu = Z \cdot \frac{\sigma}{\sqrt{n}} + \mu$$

EXAMPLE 7.5

FINDING A VALUE OF $\bar{x}$, GIVEN A PROBABILITY OR AREA

Using the information in Example 7.4, find the 95th percentile of the class mean quiz scores.

Solution

The 95th percentile of the class means is the value of $\bar{x}$ with area 0.95 to the left of it.

Since we want the 95th percentile, we seek 0.95 on the *inside* of the Z table. Since 0.95 is not in the Z table, we take the closest value. Since the two closest values, 0.9495 and 0.9505, are equally close, we split the difference. Working backward from 0.9495, we find $Z = 1.64$, and for 0.9505 we find $Z = 1.65$. Splitting the difference, we get $Z = 1.645$. This value of $Z = 1.645$ is the 95th percentile of the standard normal distribution.

Since we are looking for a sample mean quiz score, 1.645 cannot be the answer. We need to "unstandardize" by transforming this value of Z to an $\bar{x}$-value:

$$\bar{x} = Z \cdot \sigma_{\bar{x}} + \mu = 1.645(2) + 70 = 73.29$$

**Now You Can Do
Exercises 17–20.**

Thus, the 95th percentile of the sample means for the statistics quizzes is 73.29.

EXAMPLE 7.6

FINDING PROBABILITIES AND PERCENTILES USING SAMPLE MEANS

Use the information from Examples 7.4 and 7.5.
a. Find the 5th percentile of the class mean quiz scores.
b. What two symmetric values for the sample mean quiz score contain the middle 90% of all sample means between them?
c. Verify that $P(66.71 \leq \bar{x} \leq 73.29) = 0.90$.

Solution

a. Since the sampling distribution is normal, it is also symmetric. Thus, the 95th percentile and the 5th percentile are the same distance away from the mean. Since the 95th percentile is $(73.29 - 70) = 3.29$ above the mean, the 5th percentile must be 3.29 below the mean, or $(70 - 3.29) = 66.71$.

b. This is just another way of asking for the 5th and 95th percentiles, which we found in Example 7.5 and here in part (**a**). (See Figure 7.4.) The answer is 66.71 and 73.29.

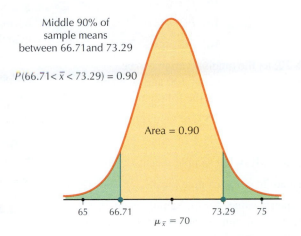

FIGURE 7.4
Middle 90% of the sample means.

c. We seek $P(66.71 < \bar{x} < 73.29)$, as shown in Figure 7.4. Proceeding with the calculations, we have, as expected,

$$P(66.71 < \bar{x} < 73.29) = P\left(\frac{66.71 - 70}{2} < \frac{\bar{x} - 70}{2} < \frac{73.29 - 70}{2}\right)$$
$$= P(-1.645 < Z < 1.645) = 0.95 - 0.05 = 0.90$$

**Now You Can Do
Exercises 21–28.**

In Section 7.2, we tackle the more challenging problem of finding the sampling distribution of the sample mean for non-normal populations.

SECTION 7.1   ## Summary

1. The sampling distribution of the sample mean $\bar{x}$ for a given sample size n consists of the collection of the means of all possible samples of size n from the population. The mean of the sampling distribution of $\bar{x}$ is the value of the population mean μ (Fact 1). The standard error is $\sigma_{\bar{x}} = \sigma/\sqrt{n}$, where σ is the population standard deviation (Fact 2).

2. For a normal population, the sampling distribution of $\bar{x}$ is distributed as normal $(\mu, \sigma/\sqrt{n})$, where μ is the population mean and σ is the population standard deviation (Fact 3).

3. We can use Facts 3 and 4 to find probabilities and percentiles using sample means.

SECTION 7.1   ## Exercises

Clarifying the Concepts

1. Explain in your own words what statistical inference means.

2. Explain what a sampling distribution is. Why are sampling distributions so important?

3. For a normal population, what can we say about the sampling distribution of the sample mean?

4. True or false: $\mu_{\bar{x}} = \mu$ and $\sigma_{\bar{x}} = \sigma/\sqrt{n}$ regardless of whether or not the sampling distribution of $\bar{x}$ is normal.

Practicing the Techniques

For Exercises 5–10, find $\mu_{\bar{x}}$ and $\sigma_{\bar{x}}$, the mean and standard deviation of the sampling distribution of $\bar{x}$.

5. $\mu = 100, \sigma = 20, n = 25$

6. $\mu = 100, \sigma = 20, n = 100$

7. $\mu = 0, \sigma = 10, n = 9$

8. $\mu = 0, \sigma = 10, n = 25$

9. $\mu = -10, \sigma = 5, n = 100$

10. $\mu = -10, \sigma = 5, n = 400$

For Exercises 11–16, let the random variable X be normally distributed, with mean $\mu = 5$ and standard deviation $\sigma = 3$. Let $n = 9$. Find the following probabilities.

11. $P(\bar{x} > 6)$ **14.** $P(\bar{x} < 7)$

12. $P(\bar{x} < 4)$ **15.** $P(\bar{x} > 3)$

13. $P(4 < \bar{x} < 6)$ **16.** $P(3 < \bar{x} < 7)$

For Exercises 17–22, let the random variable X be normally distributed, with mean $\mu = 100$ and standard deviation $\sigma = 15$. Let $n = 4$. Find the following values of $\bar{x}$.

17. The value of $\bar{x}$ greater than 95% of values of $\bar{x}$

18. The value of $\bar{x}$ smaller than 95% of values of $\bar{x}$

19. The 97.5th percentile of the sample means

20. The 2.5th percentile of the sample means

21. The two symmetric values for the sample mean that contain the middle 90% of sample means. (*Hint:* Use your answers to Exercises 17 and 18.)

22. The two symmetric values for $\bar{x}$ that contain the middle 95% of $\bar{x}$ values

For Exercises 23–30, assume that the random variable X follows a normal distribution, with mean $\mu = 50$ and standard deviation $\sigma = 10$. That is, X is normal (50, 10). Let $n = 25$.

23. Describe the sampling distribution of $\bar{x}$ for $n = 25$.

24. Calculate the probability that $\bar{x}$ is less than 48.

25. Without using your calculator, use the symmetry of the normal distribution to calculate the probability that $\bar{x}$ is greater than 52.

26. Without using your calculator, use your answers from Exercises 24 and 25 to compute the probability that $\bar{x}$ lies between 48 and 52.

27. Find the value of $\bar{x}$ that is greater than 95% of all values of $\bar{x}$.

28. Without using your calculator, use the symmetry of the normal distribution to calculate the value of $\bar{x}$ that is smaller than 95% of all values of $\bar{x}$.

29. What are the two symmetric values for the sample mean that contain the middle 90% of sample means?

30. What proportion of $\bar{x}$ values lies outside the values you found in the previous exercise?

For Exercises 31–38, assume that the random variable X is normally distributed, with mean $\mu = 10$ and standard deviation $\sigma = 4$. Let $n = 16$.

31. Describe the sampling distribution of $\bar{x}$ for $n = 16$.

32. Calculate the probability that $\bar{x}$ exceeds 11.

33. Without using your calculator, use the symmetry of the normal distribution to calculate the probability that $\bar{x}$ is less than 9.

34. Without using your calculator, use your answers from Exercises 32 and 33 to compute the probability that $\bar{x}$ lies between 9 and 11.

35. Find the value of $\bar{x}$ that is greater than 97.5% of all values of $\bar{x}$.

36. Without using your calculator, use the symmetry of the normal distribution to calculate the value of $\bar{x}$ that is smaller than 97.5% of all values of $\bar{x}$.

37. What are the two symmetric values for the sample mean that contain the middle 95% of sample means?

38. What proportion of $\bar{x}$ values lies outside the values you found in the previous exercise?

Applying the Concepts

39. Lab Rat Reaction Time. A laboratory rat's mean reaction time to a stimulus is $\mu = 1.7$ seconds, with a standard deviation of $\sigma = 0.3$ second. Let the sample size be $n = 9$.

  a. Find the mean of the sampling distribution $\mu_{\bar{x}}$ and the standard error $\sigma_{\bar{x}}$.

  b. Calculate the probability that the sample mean reaction time will be less than 1.6 seconds.

  c. Compute $P(\bar{x} > 1.8)$.

40. Student Heights. The heights of a population of students have a mean of 68 inches (5 feet 8 inches) and a standard deviation of 4 inches. Suppose we take a sample of $n = 16$ students.

  a. Find the mean of the sampling distribution $\mu_{\bar{x}}$ and the standard error $\sigma_{\bar{x}}$.

  b. Calculate the probability that the sample mean student height will be greater than 70 inches.

  c. Compute $P(\bar{x} < 65)$.

41. Initial Public Offerings. A monitor of initial public offerings (IPOs) reports that the mean amount of stock offered was $100 million with a standard deviation of $40 million. Suppose that IPO amounts are normally distributed and that we take a sample of 4 IPOs. Find the probability that the sample mean IPO amount will have the following values.

  a. Greater than $125 million

  b. Between $120 million and $140 million

  c. Less than $95 million

42. Teacher Salaries. Suppose the salaries of teachers in your city are normally distributed with a mean of $50,000 and a standard deviation of $5000. Suppose we take samples of size 25 teachers. Find the probability that the sample mean salary will have the following values.

  a. More than $52,000

  b. Less than $47,000

  c. Between $52,000 and $53,000

43. Lab Rat Reaction Time. Refer to Exercise 39.
   a. Find the sample mean reaction time greater than 95% of all sample mean reaction times.
   b. Find the value of $\bar{x}$ smaller than 95% of all sample mean reaction times.
   c. What are the two symmetric values of $\bar{x}$ that contain the middle 90% of sample means?

44. Student Heights. Refer to Exercise 40.
   a. Find the sample mean student height greater than 97.5% of all sample mean student heights.
   b. Find the value of $\bar{x}$ smaller than 97.5% of all sample mean student heights.
   c. What are the two symmetric values of $\bar{x}$ that contain the middle 95% of sample means?

45. Initial Public Offerings. Refer to Exercise 41.
   a. What are the two symmetric values of $\bar{x}$ that contain the middle 99% of sample means?
   b. Draw a graph of the sampling distribution of $\bar{x}$, showing $\mu_{\bar{x}}$, along with the two symmetric values of $\bar{x}$ from part (c). Shade the area under the curve between these two values of $\bar{x}$, and indicate the amount of area this represents.

46. Teacher Salaries. Refer to Exercise 42.
   a. What are the two symmetric values of $\bar{x}$ that contain the middle 80% of sample means?
   b. Draw a graph of the sampling distribution of $\bar{x}$, showing $\mu_{\bar{x}}$, along with the two symmetric values of $\bar{x}$ from part (c). Shade the area under the curve between these two values of $\bar{x}$, and indicate the amount of area this represents.

Japan Earthquakes. Use the following information for Exercises 47–49. On March 11, 2011, a magnitude 9.0 earthquake struck off the shore of Honshu, Japan. The quake and the resulting tsunami led to massive destruction and the deaths of thousands of people. Shown here are the magnitudes of a set of 5 aftershocks that occurred later that same day.[1]

Consider these magnitudes to be a population.

   Aftershock magnitudes: 7.9   7.7   6.5   6.3   6.1

47. Answer the following.
   a. How many samples of size $n = 2$ can we generate from this tiny population of size 5?
   b. Compute the population mean μ.
   c. Calculate the population standard deviation σ.

48. Take every possible sample of size $n = 2$ from this population.
   a. Find the mean magnitude $\bar{x}$ of each sample.
   b. Construct a dotplot of the sample mean magnitudes, using Figure 7.1 as a guide.
   c. Where would the balance point be located in the dotplot from (b)? Indicate it on the plot.
   d. Recall that the balance point represents the mean. What is your estimate of the mean $\mu_{\bar{x}}$ using this balance point?

49. Refer to your work in Exercises 47 and 48.
   a. Find the mean of all the sample mean magnitudes in Exercise 48(a).
   b. Does the value for the mean from Exercise 49(a) agree with the value of the population mean from Exercise 47(b)? Which fact from this section does this reflect?

50. A Fair Die. Consider a fair six-sided die. Suppose we take samples of size 16 and are interested in the population mean of the die rolls.
   a. Find $\mu_{\bar{x}}$.
   b. Find $\sigma_{\bar{x}}$. (*Hint:* First find the standard deviation of a fair die roll using a frequency distribution.)

Bringing It All Together

SAT Math Scores. Use this information for Exercises 51–55. The College Board (**www.collegeboard.com**) reports that the nationwide mean math SAT score is 5.5. Assume that the standard deviation is 116 and that the scores are normally distributed.

51. What is the probability that a randomly selected SAT math score will be less than 500?

52. As a researcher, you are looking at samples of SAT math scores of size 16.
   a. Find $\mu_{\bar{x}}$.
   b. Find $\sigma_{\bar{x}}$.
   c. What can you say about the sampling distribution of the sample mean? How do you know this?

53. Refer to Exercise 52.
   a. What is the probability that a sample of 16 students will have a sample mean math SAT score below 500?
   b. Why is the probability so much lower for the sample mean than for a particular student?

54. Refer to Exercise 53. What if the population standard deviation was greater than 116. Explain how this would affect the following, if at all.
   a. Probability that a randomly selected SAT math score will be less than 500
   b. $\mu_{\bar{x}}$
   c. $\sigma_{\bar{x}}$
   d. Sampling distribution of the sample mean

55. *What if* the population standard deviation was greater than $116. Explain how this would affect the following, if at all.
   a. Probability that a sample of 16 SAT math scores will have a mean less than 500
   b. 99.5th percentile of the sample mean SAT math scores
   c. 0.5th percentile of the sample mean SAT math scores

7.2   CENTRAL LIMIT THEOREM FOR MEANS

OBJECTIVES   By the end of this section, I will be able to . . .

1   Use normal probability plots to assess normality.

2   Describe the sampling distribution of $\bar{x}$ for skewed and symmetric populations as the sample size increases.

3   Apply the Central Limit Theorem for Means to solve probability questions about the sample mean.

1   ASSESSING NORMALITY USING NORMAL PROBABILITY PLOTS

Much of the analysis we carry out in this text requires that the sample data come from a population that is normally distributed. But how do we assess whether a data set is normally distributed? Histograms, dotplots, and stem-and-leaf displays may be used. But a more precise graphical tool for assessing normality is the **normal probability plot.** A normal probability plot is a scatterplot of the estimated cumulative normal probabilities (expressed as percents) against the corresponding data values in the data set.

Figure 7.5 shows the normal probability plot for a sample of normally distributed data. The points are arrayed nicely along the straight line, and all the points lie within the curved bounds. Figure 7.6 shows the normal probability plot for a sample of right-skewed data. The points do not line up in a straight line, and many points lie outside the curved bounds, indicating that the data set is not normal.

> **Analyzing Normal Probability Plots**
>
> If the points in the normal probability plot either cluster around a straight line or nearly all fall within the curved bounds, then it is likely that the data set is normal. Systematic deviations off the straight line are evidence against the claim that the data set is normal.

Now You Can Do Exercises 7–10.

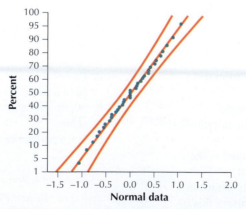

FIGURE 7.5 Normal probability plot of normal data.

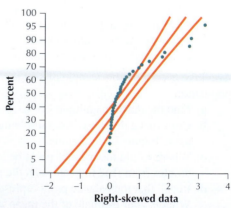

FIGURE 7.6 Normal probability plot of right-skewed data.

2 SAMPLING DISTRIBUTION OF $\bar{x}$ FOR SKEWED POPULATIONS

In Section 7.1, we discovered that the sampling distribution for the sample mean for a normal population is also normal. What if the population is not normal? In this section, we use a simulation study to learn how the sampling distribution of the sample mean $\bar{x}$ for non-normal populations becomes approximately normal as the sample size increases.

EXAMPLE 7.7

SIMULATION STUDY: SAMPLE MEANS FROM STRONGLY SKEWED POPULATION

Nutrition

The data set **Nutrition** on your CD and the companion Web site contains nutrition information on a population of 961 foods.

a. Construct a histogram of the potassium content of these 961 foods, and describe the shape of the population distribution.

b. Using Minitab, take 500 random samples of sizes $n = 10, 20,$ and 30 from the population. Assess the normality of the resulting sampling distributions of $\bar{x}$ using histograms and normal probability plots.

Solution

a. A histogram of the potassium content of these foods is shown in Figure 7.7, revealing a strongly right-skewed, non-normal data set.

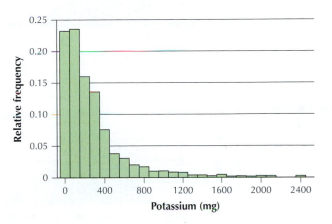

FIGURE 7.7
Potassium content is strongly right-skewed, not normal.

b. Using Minitab, we take 500 random samples of size $n = 10$ from the population. We find the means of the 500 samples shown in the graphs in Figure 7.8 (on page 334).
   - $n = 10$: The sampling distribution of $\bar{x}$ is skewed (Figure 7.8a).
   - $n = 20$: The sampling distribution of $-x$ is still somewhat skewed (Figure 7.8b).
   - $n = 30$: Despite a few outliers, the sampling distribution of $\bar{x}$ is approximately normal (Figure 7.8c).

For a skewed population, we have seen that the sampling distribution of the sample mean becomes approximately normal as the sample size reaches 30. For a less skewed population, we can expect that the sampling distribution of $\bar{x}$ approximates a normal distribution for smaller sample sizes.

3 APPLYING THE CENTRAL LIMIT THEOREM FOR MEANS

Based on our simulation study, we may conclude that *regardless of the population, the sampling distribution of the sample mean becomes approximately normal as the sample size gets larger*. We can then combine this statement with Fact 3 (page 326) to form the **Central Limit Theorem for Means.**

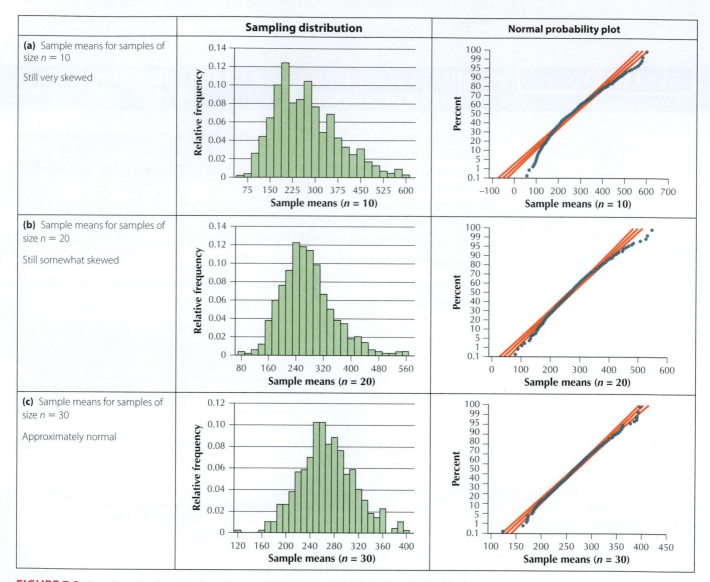

FIGURE 7.8 Sampling distribution of $\bar{x}$ and normal probability plots for $n = 10$, 20, and 30.

Central Limit Theorem for Means

Given a population with mean μ and standard deviation σ, the sampling distribution of the sample mean $\bar{x}$ becomes approximately normal $(\mu, \sigma/\sqrt{n})$ as the sample size gets larger, regardless of the shape of the population.

How large does the sample size have to be before the Central Limit Theorem for Means takes effect? In general, it depends on the degree of symmetry, or skewness, of the population. In the simulation study (Figure 7.8), we saw that the sampling distribution of $\bar{x}$ was approximately normal even for a skewed population when $n = 30$. Thus, we shall abide by the following rule of thumb.

> **Rule of Thumb for When to Use the Central Limit Theorem for Means**
>
> We consider $n \geq 30$ as large enough to apply the Central Limit Theorem for Means for any population.

Developing Your Statistical Sense

The Central Limit Theorem

The Central Limit Theorem (CLT) is one of the most important results in statistics. Worldwide, much statistical inference is based on the CLT. It actually makes fairly intuitive sense, doesn't it? If we find the mean of a sample of data values, in many cases the extreme values will tend to balance out. However, remember that the mean is very sensitive to outliers. In a small sample, there may not be enough nonextreme values to balance the influence of the outliers. This is what was happening early in the potassium simulation (for example, Figure 7.8a). However, as the sample sizes increase, the influence of extreme values diminishes and the resulting sample means start to migrate toward the center.

Combining Fact 3 and the Central Limit Theorem for Means, we can identify three possible situations for the sampling distribution of $\bar{x}$.

> **Three Possible Situations for the Sampling Distribution of the Sample Mean $\bar{x}$**
>
> 1. **The population is normal.** Therefore the sampling distribution of $\bar{x}$ is *normal* (Fact 3, page 326).
>
> 2. **The population is either non-normal or of unknown distribution *and* the sample size is at least 30.** Therefore the sampling distribution of $\bar{x}$ is *approximately normal* (Central Limit Theorem for Means).
>
> 3. **The population is either non-normal or of unknown distribution *and* the sample size is less than 30.** Therefore we have *insufficient information* to conclude that the sampling distribution of the sample mean $\bar{x}$ is either normal or approximately normal.

Of course, in the real world, no one will tell you which of the three situations applies. You need to investigate the assumptions of each of the situations to determine for yourself which one applies.

EXAMPLE 7.8

APPLICATION OF THE CENTRAL LIMIT THEOREM FOR THE MEAN

The U.S. Small Business Administration (SBA) provides information on the number of small businesses for each metropolitan area in the United States.[2] Figure 7.9 shows a histogram of our population for this example, the number of small businesses in each of the 328 cities nationwide. (For example, Austin, Texas, has 22,305 small businesses, while Pensacola, Florida, has 6020.) The mean is $\mu = 12,485$ and the standard deviation is $\sigma = 21,973$.

a. Find the probability that a random sample of size $n = 36$ cities will have a mean number of small businesses greater than 17,000.

b. Find the 90th percentile of sample means.

Solution

a. Clearly, the population is not normal, but the sample size $n = 36$ is large enough, so the Central Limit Theorem applies. The sampling distribution of the sample mean $\bar{x}$ is approximately normal. Next we need to find $\mu_{\bar{x}}$ and $\sigma_{\bar{x}}$. Facts 1 and 2 tell us that

$$\mu_{\bar{x}} = \mu = 12,485 \qquad \text{and} \qquad \sigma_{\bar{x}} = \frac{\sigma}{\sqrt{n}} = \frac{21,973}{\sqrt{36}} \approx 3662.1667$$

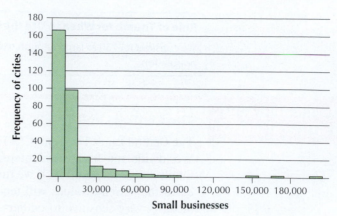

FIGURE 7.9 Population is skewed, so a large sample is needed to apply the Central Limit Theorem.

Therefore, as the CLT indicates, the sampling distribution of $\bar{x}$ is approximately normal ($\mu_{\bar{x}} = 12{,}485$, $\sigma_{\bar{x}} = 3662.1667$). We are then left to solve a normal probability problem using the methods of Sections 6.4 and 6.5. Figure 7.10 shows the sampling distribution of $\bar{x}$ and the probability we are interested in, $P(\bar{x} > 17{,}000)$. Using Fact 4, we standardize:

Since the CLT has shown the distribution is approximately normal, we can use normal distribution methods to solve the problem.

$$Z = \frac{17{,}000 - \mu_{\bar{x}}}{\sigma_{\bar{x}}} = \frac{17{,}000 - 12{,}485}{3662.1667} \approx 1.2329 \approx 1.23$$

Thus, $P(\bar{x} > 17{,}000) \approx P(Z > 1.23)$, as shown in Figure 7.11. We therefore look up $Z = 1.23$ in the Z table and subtract this table area (0.8907) from 1 to get the desired tail area:

We round Z to 2 decimal places to allow use of the Z table in finding $P(Z > 1.23)$.

$$P(Z > 1.23) = 1 - 0.8907 = 0.1093$$

Now You Can Do Exercises 17–22.

The probability is 0.1093 that a random sample of 36 cities will have a mean number of small businesses greater than 17,000.

b. We proceed just as we did for Example 7.5 in Section 7.1 (page 328). We seek the area 0.90 on the inside of the Z table. We find the closest area = 0.8997, which gives us $Z = 1.28$. Transforming Z to a sample mean value, we calculate:

$$\bar{x} = Z \cdot \sigma_{\bar{x}} + \mu = 1.28(3662.1667) + 12{,}485 \approx 17{,}173$$

Now You Can Do Exercises 23–28.

The value of $\bar{x} \approx 17{,}173$ is the 90th percentile of sample means. That is, of all possible sample means for $n = 36$, $\bar{x} \approx 17{,}173$ lies at or above 90% of them.

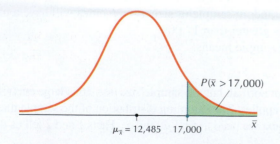

FIGURE 7.10 Area to the right of $\bar{x} = 17{,}000$ equals. . . .

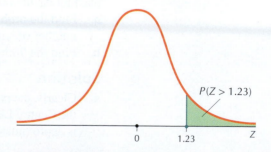

FIGURE 7.11 Area to the right of $Z = 1.23$.

EXAMPLE 7.9

SOMETIMES THERE IS INSUFFICIENT INFORMATION TO SOLVE THE PROBLEM

Using the same data set as in Example 7.8, suppose the sample size is only $n = 10$. Now try again to find the probability that a random sample of size $n = 10$ will have a mean number of small businesses greater than 17,000.

Solution

The population is skewed (not normal) and the sample size $n = 10$ is less than the minimum $n = 30$ required to apply the Central Limit Theorem. Therefore, we have insufficient information to conclude that the sampling distribution of the sample mean $\bar{x}$ is either normal or approximately normal. Unfortunately, we cannot find the probability that a random sample of $n = 10$ cities will have a mean number of small businesses greater than 17,000.

CASE STUDY

Trial of the Pyx: How Much Gold Is in Your Gold Coins?

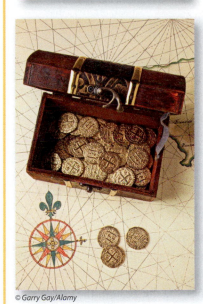

© Garry Gay/Alamy

Medieval English kings devised a procedure to ensure that the coins of the realm contained the proper amount of gold. A sample of 100 of the gold coins that were cast each year was placed in a ceremonial box called the Pyx. At the chosen time, the Company of Goldsmiths jury weighed the gold coins. The mean weight of the entire sample of coins was supposed to be 128 grams. If the mean weight was much less than 128 grams, the jury concluded that the Master of the Mint was cheating the crown by pocketing the excess gold, and he was severely punished. If the mean weight of the coins was within 3.2 grams of the expected 128 grams, the jury accepted the year's gold as pure. Thus, the mean weight had to lie between 127.68 grams and 128.32 grams.

Problem 1. Can we estimate what the jury used for a standard deviation?

Solution to Problem 1. Let's assume that "much less than" indicated a measurement that is 2 or more standard deviations below average. For the sampling distribution of $\bar{x}$, then, this would indicate a range of $0.32 = 2\sigma_{\bar{x}}$ between 127.68 and the mean 128. Therefore, $\sigma_{\bar{x}} = 0.16$. And therefore, by the Empirical Rule, for instance, approximately 95% of the sample mean observations for the Trial of the Pyx would have been between 127.68 and 128.32. Since $\sigma_{\bar{x}} = \sigma/\sqrt{n}$, it follows that $\sigma = \sqrt{100} \cdot 0.16 = 1.6$ grams.

Problem 2. What were the chances that the Master of the Mint would have been caught and punished if he were in fact cheating the throne?

Solution to Problem 2. What if the Master of the Mint set the mean amount of gold in the population of all coins to be $\mu = 127.9$ grams instead of the required 128, shortchanging the crown by a tenth of a gram of gold per coin? The jury would never have noticed this, would they?

Let's calculate the probability that the Master of the Mint would have passed the Trial of the Pyx if the mean amount of gold in the coins had been only 127.9 grams. We've seen that the Master of the Mint would have passed the Trial of the Pyx if $127.68 < \bar{x} < 128.32$. Now, because 100 is a large sample size, the Central Limit

(continues)

Theorem tells us that the sampling distribution of $\bar{x}$ is approximately normal, with $\mu_{\bar{x}} = \mu = 127.9$ and $\sigma_{\bar{x}} = \dfrac{\sigma}{\sqrt{n}} = \dfrac{1.6}{\sqrt{100}} = 0.16$.

Standardizing using Fact 5:

$$Z = \frac{127.68 - \mu_{\bar{x}}}{\sigma_{\bar{x}}} = \frac{127.68 - 127.9}{0.16} \approx -1.38 \qquad \text{and}$$

$$Z = \frac{128.32 - \mu_{\bar{x}}}{\sigma_{\bar{x}}} = \frac{128.32 - 127.9}{0.16} \approx 2.63$$

Solving using Table 6.6 in Section 6.4 (page 289):

$$P(-1.38 < Z < 2.63) = 0.9957 - 0.0838 = 0.9119$$

That is, the chances of the crown accepting the coins as pure, even if the Master of the Mint had been shortchanging by a tenth of a gram per coin, were over 91% (Figure 7.12).

Clipart.com

Note: Sir William Sharington, 1493–1553, Master of the Mint during the turbulent Tudor era in England. He debased the currency, issued worthless coinage, and diverted the real gold to fund Thomas Seymour's conspiracy to topple the government and seize young King Edward VI. Sharington was arrested in 1548 or 1549, but he later received pardon and became Sheriff of Wiltshire for a short time before he died.

FIGURE 7.12
Sampling distribution if population mean gold weight is reduced to 127.9 grams.

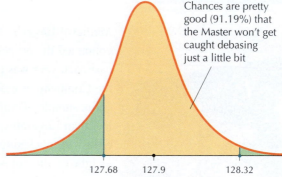

Chances are pretty good (91.19%) that the Master won't get caught debasing just a little bit

127.68   127.9   128.32

Problem 3. Would the Master of the Mint have been satisfied with this small amount of debasement? Would he have quit while he was ahead?

Solution to Problem 3. No way! The following year the Master of the Mint decided to debase the currency even further, setting the mean amount of gold in the coins to be $\mu = 127.3$ grams per coin.

We need to find the probability of the Master passing the Trial of the Pyx if the mean amount of gold in a coin was 127.3 grams instead of the required 128 grams per coin. We use the same calculations, with $\mu_{\bar{x}} = 127.3$ grams. Standardizing:

$$Z = \frac{127.68 - \mu_{\bar{x}}}{\sigma_{\bar{x}}} = \frac{127.68 - 127.3}{0.16} \approx 2.38 \qquad \text{and}$$

$$Z = \frac{128.32 - \mu_{\bar{x}}}{\sigma_{\bar{x}}} = \frac{128.32 - 127.3}{0.16} \approx 6.38$$

Then $P(2.38 < Z < 6.38) \approx 1 - 0.9913 = 0.0087$.

In other words, the Master of the Mint actually would have stood very little chance—less than 1% probability—of passing the Trial of the Pyx if he cheated by this much (Figure 7.13).

England is a great country for retaining fine old traditions. Today England's Company of Goldsmiths still operates the London Assay Office where the purity of the kingdom's coin is tested at the annual Trial of the Pyx. ■

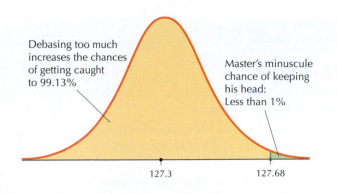

FIGURE 7.13
Sampling distribution if
population mean gold weight
is reduced to 127.3 grams.

 The *Central Limit Theorem* applet allows you to experiment with various sample sizes and see how the Central Limit Theorem for Means behaves in action.

STEP-BY-STEP TECHNOLOGY GUIDE: Constructing Normal Probability Plots

TI-83/84

Assume that the data set is in list L1.
Step 1   Access STAT PLOTS by pressing **2nd Y**.
Step 2   Select **1:Plot1**. Press **ENTER**.
Step 3   Move the cursor over **On** and press **ENTER**.
Step 4   Select the normal probability plot type by moving the cursor to the lower-right plot among the choices for Type. Press **ENTER**.

Step 5   For **Data List**, enter **L1**.
Step 6   For **Data Axis**, choose **X**.
Step 7   Press **ZOOM**, then **9: ZoomStat**.

MINITAB

Assume that the data set is in column **C1**.
Step 1   From the menu, select **Graph**, then click **Probability Plot**.
Step 2   Select **Single** and click **OK**.

Step 3   In the Probability Plot dialog box, select **C1**, and click **OK**. The normal probability plot for the data set in **C1** is then generated.

CRUNCHIT!

We will use the data from Example 7.8, Small Businesses.
Step 1   Click **File . . .** then highlight **Load from Larose2e . . .** Chapter 7 . . . and click on **Example 7.8**.

Step 2   Click **Graphics** and select **QQ Plot**. For **Sample** select **Businesses**. Then click **Calculate**.

SECTION 7.2 Summary

1. Normal probability plots are used to assess the normality of a data set.

2. A simulation study showed that the sampling distribution of $\bar{x}$ for a skewed population achieved approximate normality when n reached 30.

3. The Central Limit Theorem is one of the most important results in statistics and is stated as follows: given a population with mean μ and standard deviation σ, the sampling distribution of the sample mean $\bar{x}$ becomes approximately normal (μ, $\sigma/\sqrt{n}$) as the sample size gets larger, regardless of the shape of the population.

Clarifying the Concepts

1. Explain what we use a normal probability plot for. What should we look for in a normal probability plot?

2. Use the Central Limit Theorem to explain what happens to the sampling distribution of $\bar{x}$ as the sample size gets larger.

3. According to our rule of thumb, what is the minimum sample size for approximate normality of the sampling distribution of $\bar{x}$?

4. State the three possible situations for the sampling distribution of $\bar{x}$.

5. Suppose we would like to decrease the size of the standard error to half its original size. How much do we have to increase the sample size?

6. State the conditions when the sampling distribution of $\bar{x}$ is neither normal nor approximately normal.

Practicing the Techniques

For Exercises 7–10, determine whether the normal probability plots indicate acceptable normality of the data set.

7.

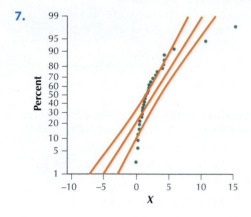

8.

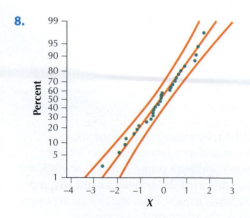

9.

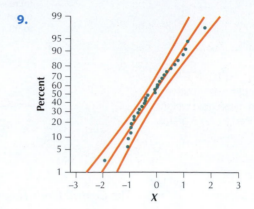

10.

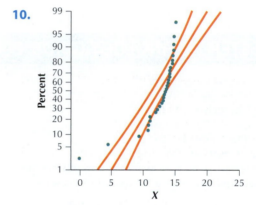

For Exercises 11–16, provide **(a)** $\mu_{\bar{x}}$ and **(b)** $\sigma_{\bar{x}}$, and determine whether the sampling distribution of $\bar{x}$ is normal, approximately normal, or unknown. (*Hint:* See the three possible situations on page 335.)

11. SAT scores are normally distributed, with $\mu = 516$ and $\sigma = 116$. A sample of size $n = 9$ is taken.

12. SAT scores are not normally distributed, with $\mu = 516$ and $\sigma = 116$. A sample of size $n = 36$ is taken.

13. Systolic blood pressure readings are not normally distributed, with $\mu = 80$ and $\sigma = 8$. A sample of size $n = 64$ is taken.

14. Systolic blood pressure readings are not normally distributed, with $\mu = 80$ and $\sigma = 8$. A sample of size $n = 25$ is taken.

15. The gas mileage for 2010 Toyota Prius hybrid vehicles is not normally distributed, with $\mu = 50$ miles per gallon and $\sigma = 6$. A sample of size $n = 16$ is taken.

16. The gas mileage for 2010 Toyota Prius hybrid vehicles is not normally distributed, with $\mu = 50$ miles per gallon and $\sigma = 6$. A sample of size $n = 64$ is taken.

For the situations in Exercises 17–22, if possible find the indicated probability. If not possible, explain why not.

17. The situation in Exercise 11—find $P(\bar{x} > 540)$.

18. The situation in Exercise 12—find $P(\bar{x} < 500)$.

19. The situation in Exercise 13—find $P(\bar{x} < 82)$.

20. The situation in Exercise 14—find $P(\bar{x} < 78)$.

21. The situation in Exercise 15—find $P(\bar{x} < 48)$.

22. The situation in Exercise 16—find $P(\bar{x} < 52)$.

For the situations in Exercises 23–28, if possible find the indicated value of $\bar{x}$. If not possible, explain why not.

23. The pollen count distribution for Los Angeles in September is not normally distributed, with $\mu = 8$ and $\sigma = 1$. A sample of size 64 is taken. Find the sample mean pollen count larger than 75% of all sample means.

24. The pollen count distribution for Los Angeles in September is not normally distributed, with $\mu = 8$ and $\sigma = 1$. A sample of size 16 is taken. Find the sample mean pollen count larger than 75% of all sample means.

25. Prices for boned trout are normally distributed, with $\mu = \$3.10$ per pound and $\sigma = \$0.30$. A sample of size 16 is taken. Find the sample mean price that is smaller than 90% of sample means.

26. Prices for boned trout are not normally distributed, with $\mu = \$3.10$ per pound and $\sigma = \$0.30$. A sample of size 16 is taken. Find the sample mean price that is smaller than 90% of sample means.

27. Accountant incomes are not normally distributed, with $\mu = \$60,000$ per year and $\sigma = \$10,000$. A sample of 100 is taken. Find the 5th percentile of sample mean incomes.

28. Accountant incomes are normally distributed, with $\mu = \$60,000$ per year and $\sigma = \$10,000$. A sample of 100 is taken. Find the 95th percentile of sample mean incomes.

Applying the Concepts

29. Cholesterol Levels. The Centers for Disease Control and Prevention reports that the mean serum cholesterol level in Americans is 202. Assume that the standard deviation is 45. There is no information about the distribution. We take a sample of 36 Americans.

   a. Find $P(\bar{x} > 212)$.
   b. Calculate $P(192 < \bar{x} < 212)$.

30. Tennessee Temperatures. According to the National Oceanic and Atmospheric Administration, the mean temperature for Nashville, Tennessee, in the month of January between 1872 and 2011 was 38.6°F. Assume that the standard deviation is 10°F, but the distribution is unknown. If we take a sample of $n = 36$, find the following probabilities.

   a. $P(\bar{x} < 40)$
   b. $P(40 < \bar{x} < 41)$

31. Computers per School. The National Center for Educational Statistics (http://nces.ed.gov) reported that the mean number of instructional computers per public school nationwide was 124. Assume that the standard deviation is 50 computers and that there is no information about the shape of the distribution. Suppose we take a sample of size 100 public schools. Compute the following probabilities.

   a. $P(\bar{x} < 110)$
   b. $P(110 < \bar{x} < 124)$
   c. How do we know the distribution of the sample mean?

32. Stock Prices. A stockbroker was examining her track record. The mean net gain in stock price for all her clients' portfolios was $4, with a standard deviation of $6. She has no information about the distribution.

   a. She takes a sample of 16 stocks. If possible, find the probability that the sample will have a mean net loss in stock price (i.e., $P(-x < 0)$). If not possible, explain why not.
   b. Now she takes a sample of 36 stocks. Calculate $P(\bar{x} < 0)$.

33. Cholesterol Levels. Refer to Exercise 29.

   a. Find the sample mean serum cholesterol level that is larger than 95% of all such sample means.
   b. Calculate the sample mean serum cholesterol level that is smaller than 95% of all such sample means.

34. Tennessee Temperatures. Refer to Exercise 30.

   a. Find the sample mean temperature that is larger than 97.5% of all such sample means.
   b. Calculate the sample mean temperature that is smaller than 97.5% of all such sample means.
   c. Draw a graph of the sampling distribution of $\bar{x}$. Indicate $\mu_{\bar{x}}$, the two $\bar{x}$ values from **(a)** and **(b)**, and the area between them.

35. Computers per School. Refer to Exercise 31.

   a. Find the 0.5th percentile of sample mean numbers of computers.
   b. Compute the 99.5th percentile of sample mean numbers of computers.
   c. Draw a graph of the sampling distribution of $\bar{x}$. Indicate $\mu_{\bar{x}}$, the two $\bar{x}$ values from **(a)** and **(b)**, and the area between them.

36. Stock Prices. Refer to Exercise 32 for $n = 36$.

   a. Find the 90th percentile of sample mean net gains.
   b. Compute the 10th percentile of sample mean net gains.
   c. Draw a graph of the sampling distribution of $\bar{x}$. Indicate $\mu_{\bar{x}}$, the two $\bar{x}$ values from **(a)** and **(b)**, and the area between them.

Bringing It All Together

Adjusted Gross Income. Use the following information for Exercises 37–40. The population mean adjusted gross income for instructors at a certain college is $\mu = \$50,000$ with standard deviation $\sigma = \$30,000$. Here is the normal probability plot for the population of instructors.

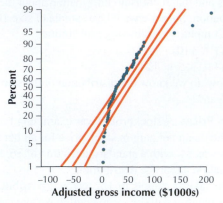

Normal probability plot of adjusted gross income.

37. Does the normal probability plot show evidence in favor of normality or against normality? What characteristics of the plot illustrate this evidence?

38. If possible, find the probability that a random sample of $n = 16$ instructors will have a mean adjusted gross income between $40,000 and $60,000. If not possible, explain why not.

39. If possible, find the probability that a random sample of $n = 36$ instructors will have a mean adjusted gross income between $40,000 and $60,000. If not possible, explain why not.

40. Refer to Exercise 39. What if the sample size used was some unspecified value greater than 36? Describe how and why this change would have affected the following, if at all. Would the quantities increase, decrease, remain unchanged? Or is there insufficient information to tell what would happen? Explain your answers.

    a. $\mu_{\bar{x}}$

    b. $\sigma_{\bar{x}}$

    c. $Z = \dfrac{\bar{x} - \mu_{\bar{x}}}{\sigma_{\bar{x}}}$

    d. $P(\$40,000 < \bar{x} < \$60,000)$

Use the *Central Limit Theorem* applet for Exercises 41 and 42.

41. Describe the shape of the sampling distribution of $\bar{x}$ for the following sample sizes.

    a. 2   **b.** 5   **c.** 30

42. At what sample size would you say the sampling distribution of $\bar{x}$ becomes approximately normal?

7.3   CENTRAL LIMIT THEOREM FOR PROPORTIONS

OBJECTIVES   By the end of this section, I will be able to . . .

1   Explain the sampling distribution of the sample proportion $\hat{p}$.

2   Apply the Central Limit Theorem for Proportions to solve probability questions about the sample proportion.

1   SAMPLING DISTRIBUTION OF THE SAMPLE PROPORTION $\hat{p}$

The sample mean is not the only statistic that can have a sampling distribution. Every sample statistic has a sampling distribution. One of the most important is the sampling distribution of the sample proportion $\hat{p}$.

> Suppose each individual in a population either has or does not have a particular characteristic. If we take a sample of size n from this population, the **sample proportion** $\hat{p}$ (read "p-hat") is
>
> $$\hat{p} = \frac{X}{n}$$
>
> where X represents the number of individuals in the sample that have the particular characteristic. We use $\hat{p}$ to estimate the unknown value of the population proportion p. In Section 6.2, we were introduced to $\hat{p}$ as the sample proportion of successes in a binomial experiment.

EXAMPLE 7.10

CALCULATING THE SAMPLE PROPORTION $\hat{p}$

In 2010, the Pew Internet and American Life Project surveyed 3000 Americans, and found 1410 who owned an MP3 player (such as an iPod). Calculate the sample proportion of Americans who own an MP3 player.

Solution

The survey sample size is $n = 30$, and the number of successes is $X = 1410$. We calculate

$$\hat{p} = \frac{X}{n} = \frac{1410}{3000} = 0.47$$

Thus, the sample proportion of Americans who own an MP3 player is 0.47. That is, $\hat{p} = 0.47$, or 47%, of Americans in the sample own an MP3 player.

Like $\bar{x}$, the sample proportion $\hat{p}$ varies from sample to sample. And since we do not know its value prior to taking the sample, $\hat{p}$ is a random variable. Just as we learned the Central Limit Theorem for Means in Section 7.2, here in Section 7.3, we develop a Central Limit Theorem for Proportions, where the **sampling distribution of the sample proportion** becomes approximately normal if the right conditions are satisfied.

> The **sampling distribution of the sample proportion** $\hat{p}$ for a given sample size n consists of the collection of the sample proportions of all possible samples of size n from the population.
>
> In general, the **sampling distribution of any particular statistic** for a given sample size n consists of the collection of the values of that sample statistic across all possible samples of size n.

Recall that in Section 7.1 we found that the mean of the sampling distribution of the sample mean $\bar{x}$ is $\mu_{\bar{x}} = \mu$ and the standard error of the mean is $\sigma_{\bar{x}} = \sigma/\sqrt{n}$. We now learn the mean and standard error of the sampling distribution of the sample proportion $\hat{p}$.

> **Fact 5: Mean of the Sampling Distribution of the Sample Proportion** $\hat{p}$
>
> The mean of the sampling distribution of the sample proportion $\hat{p}$ is the value of the population proportion p. This may be denoted as $\mu_{\hat{p}} = p$ and read as "the mean of the sampling distribution of $\hat{p}$ is p."

Fact 5 provides a measure of center for the sampling distribution of the sample proportion $\hat{p}$, and Fact 6 provides a measure of spread.

Note: Just as for $\sigma_{\bar{x}}$ (see page 324), the finite population correction factor

$$\sqrt{\frac{N-n}{N-1}}$$

should be used when the population is not much larger than the sample.

> **Fact 6: Standard Deviation of the Sampling Distribution of the Sample Proportion** p
>
> The standard deviation of the sampling distribution of the sample proportion $\hat{p}$ is
>
> $\sigma_{\hat{p}} = \sqrt{\frac{p \cdot q}{n}}$, where p is the population proportion and n is the sample size. $\sigma_{\hat{p}}$ is called the **standard error of the proportion.**

EXAMPLE 7.11

MEAN AND STANDARD ERROR OF $\hat{p}$

The National Institutes of Health reported that color blindness linked to the X chromosome afflicts 8% of men. Suppose we take a random sample of 100 men and let p denote the proportion of men in the population who have color blindness linked to the X chromosome. Find $\mu_{\hat{p}}$ and $\sigma_{\hat{p}}$.

Solution

First, we note that this is a binomial experiment with $p = 0.08$ and $n = 100$. Fact 5 tells us that $\mu_{\hat{p}} = p$, that is, the sampling distribution of the sample proportion $\hat{p}$ has a mean of $p = 0.08$. Fact 6 states that the standard error is

$$\sigma_{\hat{p}} = \sqrt{\frac{p \cdot q}{n}} = \sqrt{\frac{0.08 \cdot (1 - 0.08)}{100}} = \sqrt{0.000736} \approx 0.02713$$

What Do These Numbers Mean?

Imagine that we repeatedly draw random samples of 100 men and observe the proportion of men $\hat{p}$ in each sample who have color blindness linked to the X chromosome. Each sample provides us with a value for $\hat{p}$. Eventually, the values of $\hat{p}$, when graphed, form the sampling distribution shown in Figure 7.14.

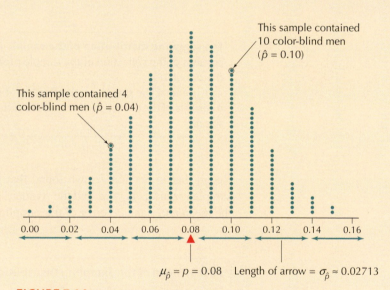

FIGURE 7.14 Sampling distribution of sample proportion $\hat{p}$.

Note that $\mu_{\hat{p}} = p = 0.08$ is located at the balance point of this distribution, which we should expect since the mean proportion of these samples is $\mu_{\hat{p}} = p = 0.08$. Each arrow represents 1 standard error $\sigma_{\hat{p}} = 0.02713$. Note that nearly all the sample proportions lie within 3 standard errors of the mean.

Unfortunately, the sampling distribution of $\hat{p}$ is not always normal. Recall from Section 7.2 that the approximate normality provided by the Central Limit Theorem for Means was a useful tool for solving probability problems for the sample mean $\bar{x}$. Similarly, in order to solve probability problems for the sample proportion $\hat{p}$, we need a way to achieve approximate normality for the sampling distribution of $\hat{p}$. Conditions for the approximate normality of the sampling distribution of $\hat{p}$ are as follows.

Fact 7: Conditions for Approximate Normality of the Sampling Distribution of the Sample Proportion $\hat{p}$

The sampling distribution of the sample proportion $\hat{p}$ may be considered approximately normal only if both the following conditions hold:

$$n \cdot p \geq 5 \quad \text{and} \quad n \cdot q \geq 5$$

The **minimum sample size** required to produce approximate normality in the sampling distribution of $\hat{p}$ is the *larger* of either

$$n_1 = \frac{5}{p} \quad \text{or} \quad n_2 = \frac{5}{q}$$

(rounded up to the next integer).

2 APPLYING THE CENTRAL LIMIT THEOREM FOR PROPORTIONS

Using information from Facts 5, 6, and 7, we express the Central Limit Theorem for Proportions.

Central Limit Theorem for Proportions

The sampling distribution of the sample proportion $\hat{p}$ follows an approximately normal distribution with mean $\mu_{\hat{p}} = p$ and standard deviation $\sigma_{\hat{p}} = \sqrt{\frac{p \cdot q}{n}}$ when both the following conditions are satisfied: $n \cdot p \geq 5$ and $n \cdot q \geq 5$.

EXAMPLE 7.12

APPLYING THE CENTRAL LIMIT THEOREM FOR PROPORTIONS

In Example 7.11, we learned that color blindness linked to the X chromosome afflicts 8% of men. Determine the approximate normality of the sampling distribution of $\hat{p}$, the proportion of men who have color blindness linked to the X chromosome, for samples of size (a) 50 and (b) 100.

Solution

We need to check both conditions to find whether the sampling distribution of $\hat{p}$ is approximately normal.

a. We are given that $p = 0.08$ and $n = 50$.

$$n \cdot p = 50 \cdot 0.08 = 4 \quad \text{and} \quad n \cdot q = 50 \cdot (0.92) = 46$$

Since 4 is not ≥ 5, the first condition is not satisfied. The Central Limit Theorem for Proportions cannot be used. We cannot conclude that the sampling distribution of $\hat{p}$ is approximately normal.

b. Here $p = 0.08$ and $n = 100$.

$$n \cdot p = 100 \cdot 0.08 = 8 \quad \text{and} \quad n \cdot q = 100 \cdot (0.92) = 92$$

Since both 8 and 92 are ≥ 5, both conditions are satisfied. The Central Limit Theorem for Proportions applies, and we can conclude that the sampling distribution of $\hat{p}$ is approximately normal. From Example 7.11 we have $\mu_{\hat{p}} = 0.08$ and $\sigma_{\hat{p}} = 0.02713$. Thus, the sampling distribution of $\hat{p}$ is approximately normal with $\mu_{\hat{p}} = 0.08$ and $\sigma_{\hat{p}} = 0.02713$.

Now You Can Do Exercises 7–18.

EXAMPLE 7.13 **MINIMUM SAMPLE SIZE FOR APPROXIMATE NORMALITY**

The Texas Workforce Commission reported that the state unemployment rate in March 2007 was 4.3%. Let $p = 0.043$ represent the population proportion of unemployed workers in Texas.

a. Find the minimum size of the samples that produces a sampling distribution of $\hat{p}$ that is approximately normal.

b. Describe the sampling distribution of $\hat{p}$ if we use this minimum sample size.

Solution

a. Using Fact 7, the minimum sample size required is the *larger* of either

$$n_1 = \frac{5}{p} \quad \text{or} \quad n_2 = \frac{5}{q}$$

Here

$$n_1 = \frac{5}{p} = \frac{5}{0.043} \approx 116.3 \quad \text{and} \quad n_2 = \frac{5}{q} = \frac{5}{0.957} \approx 5.2$$

The larger of n_1 and n_2 is $n_1 = 116.3$. However, it is unclear what "0.3" of a worker means. So we round up to the next integer: $n = 117$. Therefore, the minimum sample size required to produce a sampling distribution of $\hat{p}$ that is approximately normal is $n = 117$ Texas workers. We confirm that this satisfies our conditions:

$$n \cdot p = (117)(0.043) = 5.031 \geq 5 \quad \text{and} \quad n \cdot q = (117)(0.957) = 111.969 \geq 5$$

b. We have $\mu_{\hat{p}} = 0.043$ and

$$\sigma_{\hat{p}} = \sqrt{\frac{pq}{n}} = \sqrt{\frac{0.043(0.957)}{117}} \approx \sqrt{0.00035172} \approx 0.01875$$

Now You Can Do Exercises 19–24.

Since the conditions are met, the Central Limit Theorem for Proportions applies. The sampling distribution of $\hat{p}$ is approximately normal ($\mu_{\hat{p}} = 0.043$, $\sigma_{\hat{p}} = 0.01875$).

In those cases where we determine that the sampling distribution of $\hat{p}$ is approximately normal, we can then proceed to determine probabilities or find percentiles using the normal distribution methods we learned in Chapter 6. Fact 8 is similar to Fact 4.

> **Fact 8: Standardizing a Normal Sampling Distribution for Proportions**
>
> When the sampling distribution of $\hat{p}$ is approximately normal, we can standardize to produce the standard normal Z:
>
> $$Z = \frac{\hat{p} - \mu_{\hat{p}}}{\sigma_{\hat{p}}} = \frac{\hat{p} - p}{\sqrt{\frac{pq}{n}}}$$
>
> where p is the population proportion of successes and n is the sample size.

EXAMPLE 7.14 **APPLYING THE CENTRAL LIMIT THEOREM FOR PROPORTIONS**

Using the information in Example 7.13, find the probability that a sample of Texas workers will have a proportion unemployed greater than 9% for samples of size (a) 30 respondents and (b) 117 respondents.

Solution

a. We found in Example 7.13(a) that this sample size of $n = 30$ does not meet the minimum sample size required for the sampling distribution of $\hat{p}$ to be approximately normal, so we cannot conclude that the sampling distribution of $\hat{p}$ is approximately normal. Thus, we cannot solve this problem.

b. From Example 7.13(b), the sampling distribution of $\hat{p}$ is approximately normal with mean $\mu_{\hat{p}} = 0.043$ and standard deviation $\sigma_{\hat{p}} = 0.01875$. We are then faced with a normal probability problem similar to those in Section 6.5. Figure 7.15 shows the sampling distribution of $\hat{p}$ and the probability we are interested in, $P(\hat{p} > 0.09)$. Using Fact 8, we standardize as follows:

> *Again we can use our normal distribution methods since the CLT for proportions gives us approximate normality.*

$$Z = \frac{0.09 - \mu_{\hat{p}}}{\sigma_{\hat{p}}} = \frac{0.09 - 0.043}{0.01875} \approx 2.51$$

Thus, $P(\hat{p} > 0.09) = P(Z > 2.51)$, as shown in Figure 7.16.

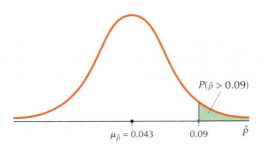

$P(\hat{p} > 0.09)$

$\mu_{\hat{p}} = 0.043$ 0.09 $\hat{P}$

FIGURE 7.15 Area to the right of $\hat{p} = 0.09$ equals. . . .

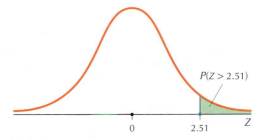

$P(Z > 2.51)$

0   2.51 Z

FIGURE 7.16 Area to the right of $Z = 2.51$.

Following Table 6.6 (page 289), we look up $Z = 2.51$ in the Z table and subtract this table area (0.9940) from 1 to get the desired tail area. That is,

$$P(Z > 2.51) = 1 - 0.9940 = 0.0060$$

Now You Can Do Exercises 25–32.

So the probability that the sample proportion of unemployed Texas workers will exceed 0.09 is 0.0060.

EXAMPLE 7.15 APPLYING THE CLT FOR PROPORTIONS TO FIND A PERCENTILE

Using the information from Example 7.13, find the 99th percentile of sample proportions for $n = 117$.

Solution

The 99th percentile shown in Figure 7.17 separates the top 1% of sample proportions from the lower 99%. Thus, the area to the left of the 99th percentile is 0.99. We look up $Z = 0.99$ on the inside of the Z table, and the closest value we can find is 0.9901. The Z-value associated with 0.9901 is 2.33. We need to transform this Z-value back to the scale of sample proportions. Use

$$\hat{p} = Z \cdot \sigma_{\hat{p}} + \mu_{\hat{p}} = (2.33)(0.01875) + 0.043 \approx 0.0867$$

The 99th percentile of the sampling distribution of $\hat{p}$ is 0.0867.

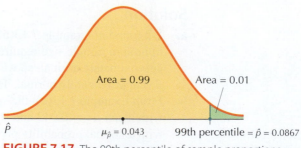

FIGURE 7.17 The 99th percentile of sample proportions.

**Now You Can Do
Exercises 33–38.**

| EXAMPLE 7.16 | PITFALLS OF USING AN APPROXIMATION |

Use symmetry and the results from Example 7.15 to find the 1st percentile of the sampling distribution of $\hat{p}$ for $n = 117$.

Solution

Note: What can we do to estimate the 1st percentile? One way is to use simulation. Generate samples of size $n = 117$ from the population of the original survey respondents, record the sample proportion from each, and simply choose the 1st percentile. Proceeding in this manner, we estimate the 1st percentile as 0.0128.

By symmetry, the 1st percentile will be the same distance below the mean that the 99th percentile is above the mean. The 99th percentile, 0.0867, lies $(0.0867 - 0.043) = 0.0437$ above the mean. Therefore, the 1st percentile lies 0.0437 below the mean:

$$\hat{p} = (0.043 - 0.0437) = -0.0007$$

However, this value of -0.0007 is negative and cannot represent a sample proportion. This negative result is obtained because the normality of the sampling distribution of $\hat{p}$ is only approximate and not exact.

| SECTION 7.3 | Summary |

1. The sampling distribution of the sample proportion $\hat{p}$ for a given sample size n consists of the collection of the sample proportions of all possible samples of size n from the population.

2. According to the Central Limit Theorem for Proportions, the sampling distribution of the sample proportion $\hat{p}$ follows

an approximately normal distribution with mean $\mu_{\hat{p}} = p$ and standard deviation $\sigma_{\hat{p}} = \sqrt{pq/n}$ when both the following conditions are satisfied: (1) $n \cdot p \geq 5$ and (2) $n \cdot q \geq 5$.

| SECTION 7.3 | Exercises |

Clarifying the Concepts

1. Explain what a sample proportion is, using as an example the courses for which you got an A last semester.

2. What is the mean of the sampling distribution of $\hat{p}$?

3. Give the formula for the standard error of the proportion.

4. What are the requirements for the sampling distribution of $\hat{p}$ to be approximately normal?

5. Suppose you double the sample size. What happens to the standard error of the proportion?

6. For the following values of X and n, calculate the sample proportion $\hat{p}$.

a. $X = 10$, $n = 40$
b. $X = 25$, $n = 75$
c. Number of successes $= 27$, number of trials $= 54$
d. Number of successes $= 1000$, number of trials $= 1$ million

Practicing the Techniques

In Exercises 7–18, samples are taken. Find **(a)** $\mu_{\hat{p}}$ and **(b)** $\sigma_{\hat{p}}$, and **(c)** determine whether the sampling distribution of $\hat{p}$ is approximately normal or unknown.

7. $p = 0.5$, $n = 100$

8. $p = 0.5$, $n = 5$

9. $p = 0.01, n = 100$

10. $p = 0.01, n = 500$

11. $p = 0.9, n = 40$

12. $p = 0.9, n = 50$

13. $p = 0.02, n = 200$

14. $p = 0.02, n = 250$

15. $p = 0.98, n = 250$

16. $p = 0.98, n = 200$

17. $p = 0.99, n = 500$

18. $p = 0.99, n = 100$

In Exercises 19–24, find the minimum sample size that produces a sampling distribution of $\hat{p}$ that is approximately normal.

19. $p = 0.5$

20. $p = 0.25$

21. $p = 0.1$

22. $p = 0.05$

23. $p = 0.01$

24. $p = 0.001$

For Exercises 25–32, if possible find the indicated probability. If it is not possible, explain why not.

25. $p = 0.5, n = 100, P(\hat{p} > 0.55)$

26. $p = 0.5, n = 5, P(\hat{p} > 0.55)$

27. $p = 0.01, n = 100, P(\hat{p} > 0.011)$

28. $p = 0.01, n = 500, P(\hat{p} > 0.011)$

29. $p = 0.9, n = 40, P(0.88 < \hat{p} < 0.91)$

30. $p = 0.9, n = 50, P(0.88 < \hat{p} < 0.91)$

31. $p = 0.02, n = 200, P(\hat{p} < 0.021)$

32. $p = 0.02, n = 250, P(\hat{p} < 0.021)$

For Exercises 33–38, find the indicated value of $\hat{p}$. If it is not possible, explain why not.

33. $p = 0.5, n = 100$, value of $\hat{p}$ larger than 90% of all values of $\hat{p}$

34. $p = 0.5, n = 400$, value of $\hat{p}$ larger than 90% of all values of $\hat{p}$

35. $p = 0.9, n = 64$, 95th percentile of values of $\hat{p}$

36. $p = 0.9, n = 144$, 95th percentile of values of $\hat{p}$

37. $p = 0.1, n = 64$, 10th percentile of values of $\hat{p}$

38. $p = 0.1, n = 144$, 10th percentile of values of $\hat{p}$

Applying the Concepts

39. Abandoning Landlines. The National Health Interview Survey reports that 25% of telephone users no longer use landlines, and have switched completely to cell phone use. Suppose we take samples of size 36.
  a. Find the mean and standard error of the sampling distribution of $\hat{p}$, the sample proportion of telephone users who no longer use landlines.
  b. Describe the sampling distribution of $\hat{p}$.
  c. Compute the probability that $\hat{p}$ exceeds 0.26.

40. LeBron James. During the 2009–2010 National Basketball Association season, 50.3% of LeBron James's shots from the floor were successful. Suppose we take a sample of 50 of LeBron's shots.
  a. Find $\mu_{\hat{p}}$ and $\sigma_{\hat{p}}$ for the sample proportion of LeBron's shots that were good.
  b. Describe the sampling distribution of $\hat{p}$.
  c. Calculate $P(\hat{p} > 0.60)$.

41. Small Business Jobs. According to the U.S. Small Business Administration, small businesses provide 75% of the new jobs added to the economy. Suppose we take samples of 20 new jobs.
  a. Find $\mu_{\hat{p}}$ and $\sigma_{\hat{p}}$ for the sample proportion of new jobs added to the economy that are provided by small businesses.
  b. Calculate $P(\hat{p} > 0.69)$.
  c. Compute $P(0.775 < \hat{p} < 0.8)$.

42. AIDS and Drug Use. The Centers for Disease Control and Prevention reported that, in 2008, 13% of males living with AIDS contracted it through intravenous drug use. Suppose we take samples of 49 males living with AIDS.
  a. Find $\mu_{\hat{p}}$ and $\sigma_{\hat{p}}$ for the sample proportion of males living with AIDS who contracted it through intravenous drug use.
  b. Calculate $P(\hat{p} < 0.04)$.
  c. Compute $P(0.10 < \hat{p} < 0.15)$.

43. Abandoning Landlines. Refer to Exercise 39.
  a. Find the 5th and 95th percentiles of the sample proportions.
  b. Draw a graph showing the sampling distribution of $\hat{p}$, centered at p, with the 5th and 95th percentiles, and the area of 0.90 under the curve between them shaded.
  c. Suppose only 2 of 36 phone users abandoned their landlines. Would this be considered an outlier? Explain your reasoning. (*Hint*: Use the Z-score method.)
  d. Determine which sample proportions would be considered outliers.

44. LeBron James. Refer to Exercise 40.
  a. Find the 2.5th and 97.5th percentiles of the sample proportions.
  b. Draw a graph showing the sampling distribution of $\hat{p}$, centered at p, with the 2.5th and 97.5th percentiles, and the area of 0.95 under the curve between them shaded.
  c. Suppose LeBron James was shooting at 65% accuracy in a particular game. Would that be considered "hot shooting" by his standards? Explain your reasoning. (*Hint*: Use the Z-score method.)

d. Suppose LeBron James was shooting at 35% accuracy in a particular game. Would that be considered "poor shooting" by his standards? Explain your reasoning.

45. Small Business Jobs. Refer to Exercise 41.
  a. Find the 0.5th and 99.5th percentiles of the sample proportions.
  b. Draw a graph showing the sampling distribution of $\hat{p}$, with the area between the 0.5th and 99.5th percentiles shaded.
  c. Suppose 14 of 20 new jobs added to the economy were provided by small business. Would this be considered unusual? Explain your reasoning.

46. AIDS and Drug Use. Refer to Exercise 42.
  a. Find the 2.5th and 97.5th percentiles of the sample proportions.
  b. Draw a graph showing the sampling distribution of $\hat{p}$, with the area between the 2.5th and 97.5th percentiles shaded.
  c. Calculate $P(\hat{p} < 0.12)$.
  d. Suppose someone claimed that the proportion of all males living with AIDS who contracted it through intravenous drug use was less than 0.12. Based on the probability you calculated in **(c)**, do you think there is strong evidence against this claim?

? 47. AIDS and Drug Use. Refer to Exercises 42 and 46. *What if* we increased the sample size to some unspecified larger number. Describe how and why the following quantities would change, if at all.
  a. $\mu_{\hat{p}}$
  b. $\sigma_{\hat{p}}$
  c. $P(\hat{p} < 0.04)$
  d. $P(0.10 < \hat{p} < 0.15)$
  e. $P(0.45 < \hat{p} < 0.49)$

  f. 2.5th percentile of the sample proportions
  g. 97.5th percentile of the sample proportions

Bringing It All Together

Partners Checking Up On Each Other. Use the following information for Exercises 48–51. According to a study in the journal *Computers in Human Behavior*,[3] 65% of the college women surveyed checked the call histories on the cell phones of their partners, while 41% of the males did so.

48. Suppose we take a sample of 100 college females and 100 college males.
  a. Find $\mu_{\hat{p}}$ and $\sigma_{\hat{p}}$ for the sample proportion of females checking the call histories of their partners.
  b. Find $\mu_{\hat{p}}$ and $\sigma_{\hat{p}}$ for the sample proportion of males checking the call histories of their partners.

49. Refer to Exercise 48. Calculate the following probabilities.
  a. That more than 65% of the females checked the call histories of their partners
  b. That more than 65% of the males checked the call histories of their partners
  c. That less than 41% of the females checked the call histories of their partners
  d. That less than 41% of the males checked the call histories of their partners

50. Refer to Exercise 48.
  a. Find the 2.5th and 97.5th percentiles of the sample proportions of females checking the call histories of their partners.
  b. Find the 2.5th and 97.5th percentiles of the sample proportions of males checking the call histories of their partners.

51. Suppose someone claimed that there really was no difference in the proportions of females and males who check the call histories on their partners' cell phones. How would you use the results from Exercises 49 and 50 to address this claim?

CHAPTER 7 — Formulas and Vocabulary

Section 7.1
- **MEAN OF THE SAMPLING DISTRIBUTION OF THE SAMPLE MEAN $\bar{x}$, FACT 1** (p. 324). Denoted as $\mu_{\bar{x}} = \mu$.
- **SAMPLING DISTRIBUTION OF THE SAMPLE MEAN $\bar{x}$** (p. 323)
- **SAMPLING DISTRIBUTION OF $\bar{x}$ FOR A NORMAL POPULATION, FACT 3** (p. 326)
- **STANDARD ERROR OF THE MEAN, FACT 2** (p. 324). $\sigma_{\bar{x}} = \sigma/\sqrt{n}$.
- **STANDARDIZING A NORMAL SAMPLING DISTRIBUTION FOR MEANS, FACT 4** (p. 326).
$$Z = \frac{\bar{x} - \mu_{\bar{x}}}{\sigma_{\bar{x}}} = \frac{\bar{x} - \mu}{\sigma/\sqrt{n}}$$

Section 7.2
- **CENTRAL LIMIT THEOREM FOR MEANS** (p. 334)
- **NORMAL PROBABILITY PLOT** (p. 332)

Section 7.3
- **CENTRAL LIMIT THEOREM FOR PROPORTIONS** (p. 345)

- **CONDITIONS FOR APPROXIMATE NORMALITY, FACT 7** (p. 345)
- **MEAN OF THE SAMPLING DISTRIBUTION OF THE SAMPLE PROPORTION $\hat{p}$, FACT 5** (p. 343). Denoted as $\mu_{\hat{p}} = p$.
- **MINIMUM SAMPLE SIZE REQUIRED** (p. 345)
- **SAMPLE PROPORTION $\hat{p}$** (p. 342). $\hat{p} = x/n$.
- **SAMPLING DISTRIBUTION FOR ANY STATISTIC** (p. 343)
- **SAMPLING DISTRIBUTION OF THE SAMPLE PROPORTION $\hat{p}$** (p. 343)
- **STANDARD ERROR OF THE PROPORTION, FACT 6** (p. 343). $\sigma_{\hat{p}} = \sqrt{pq/n}$
- **STANDARDIZING A NORMAL SAMPLING DISTRIBUTION FOR PROPORTIONS, FACT 8** (p. 346).

$$Z = \frac{\hat{p} - \mu_{\hat{p}}}{\sigma_{\hat{p}}} = \frac{\hat{p} - p}{\sqrt{pq/n}}$$

CHAPTER 7 **Review Exercises**

Section 7.1

For Exercises 1–5, find $\mu_{\bar{x}}$ and $\sigma_{\bar{x}}$, the mean and standard deviation of the sampling distribution of $\bar{x}$.

1. $\mu = 10$, $\sigma = 5$, $n = 25$
2. $\mu = 10$, $\sigma = 5$, $n = 36$
3. $\mu = 10$, $\sigma = 5$, $n = 49$
4. $\mu = 50$, $\sigma = 40$, $n = 4$
5. $\mu = 50$, $\sigma = 40$, $n = 16$

For Exercises 6–9, assume that X is normal ($\mu = 10$, $\sigma = 4$) and $n = 25$.

6. Find the sampling distribution of $\bar{x}$ for $n = 25$.
7. Find the probability that $\bar{x}$ exceeds 11.
8. Without using your calculator, find the probability that $\bar{x}$ is less than 9.
9. Without using your calculator, find the probability that $\bar{x}$ lies between 9 and 11.

Section 7.2

For Exercises 10 and 11, if possible find the indicated probability. If it is not possible, explain why not.

10. Scores on a psychological test are not normally distributed, with $\mu = 100$ and $\sigma = 15$. A sample of size 25 is taken. Find $P(94 < \bar{x} < 103)$.
11. Scores on a psychological test are normally distributed, with $\mu = 100$ and $\sigma = 15$. A sample of size 25 is taken. Find $P(94 < \bar{x} < 103)$.

For Exercises 12 and 13, find the indicated value of $\bar{x}$. If it is not possible, explain why not.

12. Scores on a psychological test are not normally distributed, with $\mu = 100$ and $\sigma = 15$. A sample of size 25 is taken. Find the 50th percentile of sample means.
13. Scores on a psychological test are normally distributed, with $\mu = 100$ and $\sigma = 15$. A sample of size 25 is taken. Find the 50th percentile of sample means.
14. **COCAINE AND HEART ATTACKS.** The American Medical Association reported: "During the first hour after using cocaine, the user's risk of heart attack increases nearly 24 times. The average age of people in the study who suffered heart attacks

soon after using cocaine was only 44. That's about 17 years younger than the average heart attack patient. Of the 38 cocaine users who had heart attacks, 29 had no prior symptoms of heart disease."[4] Assume that the standard deviation of the age of people who suffered heart attacks soon after using cocaine was 10 years and we take a sample of size 38.

    a. Find the 97.5th percentile of the mean age at heart attack after using cocaine.
    b. Find the 2.5th percentile of the mean age at heart attack after using cocaine.
    c. Between which two sample mean ages that are symmetric about the population mean lie 95% of mean ages of all people who suffered heart attacks soon after using cocaine?
    d. By hand, sketch a plot of how this would look.

Section 7.3

For Exercises 15 and 16, if possible find the indicated probability. If it is not possible, explain why not.

15. $p = 0.1$, $n = 40$, $P(\hat{p} < 0.12)$
16. $p = 0.1$, $n = 50$, $P(\hat{p} < 0.12)$

For Exercises 17 and 18, find the indicated value of $\hat{p}$. If it is not possible, explain why not.

17. $p = 0.02$, $n = 400$, the value of $\hat{p}$ smaller than 75% of all p values
18. $p = 0.02$, $n = 625$, the value of $\hat{p}$ smaller than 75% of all p values
19. **WOMEN AND MEN AND DEPRESSION.** According to the National Institute for Mental Health, 12% of women are affected by a depressive disorder each year. Suppose we take samples of 49 women. Answer the following.

    a. Find $P(\hat{p} > 0.15)$, where $\hat{p}$ represents the sample proportion of women who are affected by a depressive disorder each year.
    b. Calculate $P(0.12 < \hat{p} < 0.15)$.
    c. Use your answer to **(a)** to calculate $P(\hat{p} < 0.15)$.
    d. Find the 5th and 95th percentiles of the sample proportion.

CHAPTER 7 **Quiz**

True or False

1. True or false: For a normal population, the sampling distribution of the sample mean is always normal.
2. True or false: Since the Central Limit Theorem takes effect at $n = 30$, it doesn't make sense to get larger samples.

Fill in the Blank

3. The distance between the point estimate and its target parameter is called the _____ _____ [two words].

4. If the population is either non-normal or of unknown distribution and the sample size is large, then the sampling distribution of $\bar{x}$ is _____ _____ (two words).

Short Answer

5. If the population is either non-normal or of unknown distribution and the sample size is small, then do we know the sampling distribution of $\bar{x}$?

6. The sampling distribution of the sample proportion $\hat{p}$ may be considered approximately normal only if *both* the following conditions hold: (1) _____ and (2) _____.

Calculations and Interpretations

SOYBEAN CROP. Protein content in a particular farmer's soybean crop is normally distributed, with a mean of 40 grams and a standard deviation of 20 grams. Suppose we take samples of size 100 soy plants. Use this information for Exercises 7 and 8.

7. a. Find the probability that the sample mean protein content will be less than 38 grams.
   b. Find the probability that the sample mean protein content will be between 36.08 and 43.92 grams.
   c. Find the probability that the sample mean protein content will be greater than 42.5 grams.

8. Refer to Exercise 7.
   a. Find the sample mean protein content higher than 99.5% of all such sample means.
   b. Find the sample mean protein content lower than 99.5% of all such sample means.
   c. Between which two values does the middle 99% of sample mean protein content lie?

STUDENT HEIGHTS. Use this information for Exercises 9 and 10. The heights of the population of students at a college are normally distributed with a mean of 68 inches (5 feet 8 inches) and a standard deviation of 3 inches. Suppose we take samples of 100 students.

9. a. Find the probability that the sample mean height will exceed 68.6 inches.
   b. Find the probability that the sample mean height will be less than 67.4 inches.
   c. Find the probability that the sample mean height will be between 67.4 and 68.6 inches.

10. a. Find the 99.5th percentile of sample mean heights.
   b. Find the 0.5th percentile of sample mean heights.
   c. Between which two values do the middle 99% of sample mean heights lie?

11. MEN AND DEPRESSION. According to the National Institute for Mental Health, 6.6% of men are affected by a depressive disorder each year.
   a. If we take samples of 100 men, find $P(\hat{p} < 0.066)$.
   b. If we take samples of 100 men, find $P(0.05 < \hat{p} < 0.066)$.
   c. If we take samples of 100 men, find the 2.5th and 97.5th percentiles of the sample proportion.

8 Confidence Intervals

AP Photo/Tertius Pickard

CASE STUDY

Health Effects of the *Deepwater Horizon* Oil Spill

On April 20, 2010, an explosion occurred on the *Deepwater Horizon* oil drilling rig 48 miles off the coast of Louisiana, causing a fireball visible 35 miles away. The *Deepwater Horizon* sank, leaving oil gushing from the seafloor into the Gulf of Mexico and creating the largest oil spill in United States history.

An army of cleanup workers fanned out across the states bordering the Gulf of Mexico in an effort to rescue wildlife, protect beaches, and save wetlands. Many of these workers were exposed to oil, chemical dispersants, cleaners, and other chemicals. The National Institute for Occupational Safety and Health was concerned about the health effects on the workers of exposure to the oil and chemicals. A survey was taken of the exposed workers to determine the extent of their injuries or symptoms. We shall use the new statistical tools that we learn in Chapter 8 to examine the results of this survey in Section 8.3 in the Case Study, Health Effects of the *Deepwater Horizon* Oil Spill. ■

The Big Picture

Where we are coming from, and where we are headed . . .

- We stand on the threshold of the two most important statistical inference methods: confidence intervals and hypothesis testing. From descriptive statistics in Chapters 1–4 through probability and probability distributions in Chapters 5–6 and sampling distributions in Chapter 7, everything that we have studied thus far has been in preparation for this moment.

- Here in Chapter 8, we learn about *confidence interval estimation*, where we can infer with a certain level of confidence that our target parameter lies within a particular interval.

- Every chapter from here to the end of the book will uncover a new and different topic in statistical inference. In Chapter 9, "Hypothesis Testing," we will learn about the most prevalent method of statistical inference.

8.1 Z INTERVAL FOR THE POPULATION MEAN

OBJECTIVES By the end of this section, I will be able to . . .

1 Calculate a point estimate of the population mean.

2 Calculate and interpret a Z interval for the population mean when the population is normal and when the sample size is large.

3 Find ways to reduce the margin of error.

4 Calculate the sample size needed to estimate the population mean.

1 CALCULATE A POINT ESTIMATE OF THE POPULATION MEAN

Recall from Section 1.2 that characteristics of a sample, such as the sample mean $\bar{x}$, are called **statistics,** while characteristics of a population, like the population mean μ, are called **parameters. Statistical inference** consists of methods for estimating and drawing conclusions about parameters, based on the corresponding statistic. For example, we use the known value of $\bar{x}$ to estimate the unknown value of μ.

Suppose a random sample of 30 male students at your school produced a sample mean height of $\bar{x} = 70$ inches. We could then use this statistic $\bar{x} = 70$ to *infer* that the population mean height μ of all male students at your school was close to 70 inches. This value of $\bar{x} = 70$ is called a **point estimate** of the population mean μ.

> **Point estimation** is the process of estimating unknown population parameters by known sample statistics. The value of each sample statistic used as an estimate is called a **point estimate.**

| EXAMPLE 8.1 | CALCULATING A POINT ESTIMATE |

© Hollyjauch/Dreamstime.com

Suppose we are interested in estimating the population mean price for pumpkins across all 50 states. Shown here is the mean 2008 price per state for pumpkins for a sample of 5 states, in cents per pound, as published by the United States Department of Agriculture.

a. Find the sample mean price $\bar{x}$.
b. Express $\bar{x}$ as the point estimate of μ, the unknown population mean price for pumpkins.

State	Price per pound (in cents)
California	15
Michigan	16
New York	36
Ohio	24
Pennsylvania	16

Solution

a. The sample mean price per pound is calculated as

$$\bar{x} = \frac{\sum x}{n} = \frac{15 + 16 + 36 + 24 + 16}{5} = 21.4$$

Now You Can Do Exercises 11–14.

b. The point estimate of μ, the unknown nationwide mean price per pound of pumpkins, is the sample mean $\bar{x} = 21.4$ cents per pound.

However, since a sample is only a small subset of the population, generalizing from a sample to the population carries the risk that the point estimate may not be very accurate. For example, do you think that the population mean price of pumpkins μ exactly equals our point estimate of 21.4 cents per pound? It's not likely, since we learned in Example 7.1 (page 323) that different samples will produce different sample means, and thus different point estimates of μ. Our point estimate $\bar{x} = 21.4$ may be close to μ or it may be far from μ. In other words, *we have no measure of confidence* that our point estimate is close to μ. There has to be a better way, and there is: *confidence intervals*, the subject of this chapter.

2 THE *Z* INTERVAL FOR THE POPULATION MEAN

Although we cannot measure how confident we are of $\bar{x}$ as a point estimate for μ, we can use the point estimate $\bar{x}$ to find an interval that is likely to contain μ. Suppose we are interested in estimating the mean height of the students at your school. Since the students in your class are a sample of the population of students at your school, suppose we calculate the sample mean height of the students in your class to be $\bar{x} = 67.5$ inches (5 feet 7½ inches tall).

We may then use $\bar{x} = 67.5$ inches as a point estimate of the unknown population mean height of all students at your school. However, this estimate is not likely to be exactly correct. To address this uncertainty in our estimate, we can use a range of heights instead, such as 67.5 inches, give or take an inch, which we write

67.5 inches $\pm$ 1 inch

and would equal the interval

$$(66.5 \text{ inches}, 68.5 \text{ inches})$$

The "1 inch" is called the *margin of error*. We might then say that

> we are 90% confident that the mean height of all students at our school lies in the interval 67.5 inches $\pm$ 1 inch (see the figure in the margin).

To increase the confidence in our estimate, we increase the margin of error, so that we might say

> we are 95% confident that the mean height of all students at our school lies in the interval 67.5 inches $\pm$ 2 inches

or the interval (65.5 inches, 69.5 inches). These two intervals are examples of what are called **confidence intervals.**

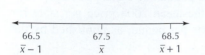

We are 90% confident that μ lies between 66.5 inches and 68.5 inches.

> A **confidence interval** is an estimate of a parameter consisting of an interval of numbers based on a point estimate, together with a **confidence level** specifying the probability that the interval contains the parameter.

For example, our estimate that the mean height of all students at our school would lie in the interval (66.5 inches, 68.5 inches) was reported with confidence level

$$90\% = (1 - 0.10) \cdot 100\%$$

Confidence intervals are often reported in the format:

$$(\text{lower bound, upper bound})$$

In the 90% confidence interval above, we have lower bound = 66.5 and upper bound = 68.5.

A confidence level of 90% for a confidence interval means that the probability is 0.9 that the population parameter lies between the lower bound and the upper bound. Recall that in previous chapters we calculated probabilities for normal distributions using the standard normal Z. We can use Z to develop the formula for the Z confidence intervals for the population mean.

But before we do so, we need to define some notation.

- Let α (alpha) be some small constant, usually ($0 < \alpha \leq 0.10$).

- Define $Z_{\alpha/2}$ to be the value of (standard normal) Z that has area $\alpha/2$ to the right of it (see Figure 8.1). For example, for $\alpha = 0.05$, $\alpha/2 = 0.25$ and $Z_{\alpha/2} = Z_{0.025} = 1.96$, as we know from Example 6.32 in Section 6.4.

- Since the Z distribution is symmetric, the area to the left of $-Z_{\alpha/2}$ is also $\alpha/2$.

- Thus, area $1 - \alpha$ lies in the interval of values of Z between $-Z_{\alpha/2}$ and $Z_{\alpha/2}$. That is, the area $1 - \alpha$ lies in the interval $-Z_{\alpha/2} < Z < Z_{\alpha/2}$ (see Figure 8.1).

FIGURE 8.1
$Z_{\alpha/2}$ is the value of Z that has area $\alpha/2$ to the right of it.

Area = $\alpha/2$ Area = $1 - \alpha$ Area = $\alpha/2$

$-Z_{\alpha/2}$ 0 $Z_{\alpha/2}$

Next, we use the facts we learned in Chapter 7 about the sampling distribution of the sample mean to develop the formula for the confidence interval for the mean.

- Fact 1: $\mu_{\bar{x}} = \mu$.
- Fact 2: $\sigma_{\bar{x}} = \sigma/\sqrt{n}$ (standard error of the mean).
- Fact 3: Sampling distribution is normal when the population is normal.
- Fact 4: Standardize $\bar{x}$ to get

$$Z = \frac{\bar{x} - \mu}{\sigma/\sqrt{n}}$$

Plugging this formula for Z back into the earlier inequality, $-Z_{\alpha/2} < Z < Z_{\alpha/2}$, gives

$$-Z_{\alpha/2} < \frac{\bar{x} - \mu}{\sigma/\sqrt{n}} < Z_{\alpha/2}$$

We then use algebra to isolate μ as the middle term:

$$\bar{x} - Z_{\alpha/2}(\sigma/\sqrt{n}) < \mu < \bar{x} + Z_{\alpha/2}(\sigma/\sqrt{n})$$

Therefore, since areas represent probabilities, we can write

$$P\big(\bar{x} - Z_{\alpha/2}(\sigma/\sqrt{n}) < \mu < \bar{x} + Z_{\alpha/2}(\sigma/\sqrt{n})\big) = 1 - \alpha$$

The quantities on either side of μ in this inequality represent the lower bound and the upper bound for a $100(1 - \alpha)\%$ confidence interval for μ. Since this confidence interval for μ is based on the standard normal Z distribution, it is called the **Z interval for the population mean μ.**

Z Interval for the Population Mean μ

The Z interval for μ may be constructed only when either of the following two conditions are met:

- The population is normally distributed, and the value of σ is known.
- The sample size is large (≥ 30), and the value of σ is known.

When a random sample of size n is taken from a population, a $100(1 - \alpha)\%$ confidence interval for μ is given by

$$\text{lower bound} = \bar{x} - Z_{\alpha/2}(\sigma/\sqrt{n})$$
$$\text{upper bound} = \bar{x} + Z_{\alpha/2}(\sigma/\sqrt{n})$$

where $1 - \alpha$ is the confidence level. The Z interval can also be written as

$$\bar{x} \pm Z_{\alpha/2}(\sigma/\sqrt{n})$$

and is denoted

$$(\text{lower bound, upper bound})$$

To use the Z interval for μ, the value of σ must be known.

Now You Can Do Exercises 15–20.

Two important results from Chapter 7 form the conditions that allow us to construct the Z interval for μ:

- The first condition comes from Fact 3 in Section 7.1: if the population is normal, then the sampling distribution of $\bar{x}$ is also normal.

- The second condition is a result of the Central Limit Theorem for Means (from Section 7.2): if the sample size is large, then the sampling distribution of $\bar{x}$ is approximately normal.

Table 8.1 provides a listing of $Z_{\alpha/2}$ values for the most common confidence levels.

Table 8.1 $Z_{\alpha/2}$ values for common confidence levels

Confidence level $(1 - \alpha)100\%$	α	$\alpha/2$	$Z_{\alpha/2}$
$100(1 - 0.10)\% = 90\%$	0.10	0.05	1.645
$100(1 - 0.05)\% = 95\%$	0.05	0.025	1.96
$100(1 - 0.01)\% = 99\%$	0.01	0.005	2.576

**Now You Can Do
Exercises 21–26.**

EXAMPLE 8.2

CONSTRUCTING A CONFIDENCE INTERVAL FOR THE MEAN OF A NORMAL POPULATION

The College Board reports that the scores on the 2010 SAT Math test were normally distributed. A sample of 25 SAT scores had a mean of $\bar{x} = 510$. Assume that the population standard deviation of such scores is $\sigma = 100$. Construct a 90% confidence interval for the population mean SAT score on the 2010 SAT Math test.

Solution

Because the population is normal and the population standard deviation σ is known, the requirements for the Z interval are met:

$$\text{lower bound} = \bar{x} - Z_{\alpha/2}(\sigma/\sqrt{n}) \qquad \text{upper bound} = \bar{x} + Z_{\alpha/2}(\sigma/\sqrt{n})$$

We are given $\bar{x} = 510$, $\sigma = 100$, and $n = 25$. From Table 8.1 we have $Z_{\alpha/2} = 1.645$. Thus

$$\text{lower bound} = 510 - 1.645(100/\sqrt{25}) = 477.1$$
$$\text{upper bound} = 510 + 1.645(100/\sqrt{25}) = 542.9$$

**Now You Can Do
Exercises 27–29.**

We are 90% confident that the population mean SAT score on the 2010 Mathematics SAT test lies between 477.1 and 542.9.

**What Does This
Confidence Interval
Mean?**

What does the 90% mean in the phrase *90% confidence interval*? If we take sample after sample for a very long time, then in the long run, the proportion of intervals that will contain the population mean μ will equal 90%.

Interpreting Confidence Intervals

You may use the following generic interpretation for the confidence intervals that you construct: "We are 90% (or 95% or 99% and so on) confident that the population mean _____ (for example, SAT Math score) lies between _____ (lower bound) and _____ (upper bound)."

The *Z* interval for the population mean μ takes the form

$$\text{point estimate} \pm \text{margin of error } E$$

where the point estimate equals $\bar{x}$ and the margin of error E equals $Z_{\alpha/2}(\sigma/\sqrt{n})$.

> The **margin of error *E*** is a measure of the precision of the confidence interval estimate. For the *Z* interval, the margin of error takes the form $E = Z_{\alpha/2}(\sigma/\sqrt{n})$.

For example, the confidence interval from Example 8.2 has the form

$$\text{point estimate} \pm \text{margin of error } E$$
$$= \bar{x} \pm E$$
$$= \bar{x} \pm Z_{\alpha/2}(\sigma/\sqrt{n})$$
$$= 510 \pm 32.9$$

Later in this section we learn ways to reduce the margin of error.

Developing Your Statistical Sense

What Is Random Here?

It is important to understand that *it is the interval that is random, not the population mean μ.* The interval is formed by sample statistics like $\bar{x}$, and for each different sample we get different values for the statistics. So the interval is random because it is constructed using $\bar{x}$, which is also random. The population mean μ, though unknown, is nevertheless *constant*.

Examine Figure 8.2, which shows a set of 10 90% confidence intervals in the form $\bar{x} \pm E$, along with the population mean μ. Note that the intervals are random while μ is constant. It turns out that 9 out of 10 of the samples (90%) produced confidence intervals that contained μ. But it did not have to turn out this way. The 90% refers to the proportion of intervals that will contain μ after a great many samples are taken.

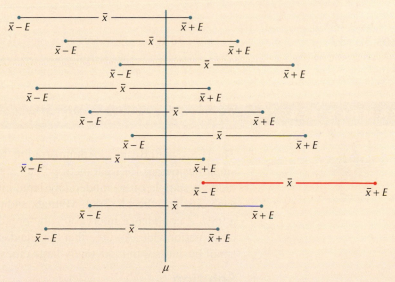

FIGURE 8.2 The intervals are random; μ is constant.

Exactostock/Superstock

EXAMPLE 8.3

CONSTRUCTING A Z INTERVAL FOR THE POPULATION MEAN FOR A LARGE SAMPLE SIZE

The Washington State Department of Ecology reported that the mean lead contamination in trout in the Spokane River is 1 part per million (ppm), with a standard deviation of 0.5 ppm.[1] Suppose a sample of $n = 100$ trout has a mean lead contamination of $\bar{x} = 1$ ppm. Assume that $\sigma = 0.5$ ppm.
a. Determine whether the requirements are met for constructing the Z interval for μ.
b. Construct a 95% confidence interval for μ, the population mean lead contamination in all trout in the Spokane River.
c. Interpret the confidence interval.

Solution

a. We are not given any information about the distribution of the population, so we don't know if the population is normally distributed. However, the sample size $n = 100$ is greater than 30 and the value of $\sigma = 0.5$ is known; therefore we can proceed to construct the confidence interval.
b. The formula for the confidence interval is given by

$$\text{lower bound} = \bar{x} - Z_{\alpha/2}(\sigma/\sqrt{n})$$
$$\text{upper bound} = \bar{x} + Z_{\alpha/2}(\sigma/\sqrt{n})$$

We are given $n = 100$, $\bar{x} = 1$, and $\sigma = 0.5$. For a confidence level of 95%, Table 8.1 provides the value of $Z_{\alpha/2} = Z_{0.025} = 1.96$. Plugging into the formula:

$$\text{lower bound} = 1 - 1.96(0.5/\sqrt{100}) = 1 - 1.96(0.05) = 1 - 0.098 = 0.902$$
$$\text{upper bound} = 1 + 1.96(0.5/\sqrt{100}) = 1 + 1.96(0.05) = 1 + 0.098 = 1.098$$

Note: As a check on your arithmetic, make sure that
$$\frac{(\text{lower bound} + \text{upper bound})}{2} = \bar{x}.$$

c. We are 95% confident that μ, the population mean lead contamination for all trout on the Spokane River, lies between 0.902 ppm and 1.098 ppm. (See Figure 8.3.)

Now You Can Do Exercises 30–32.

0.902 $\bar{x} = 1$ 1.098

FIGURE 8.3 95% Confidence interval for the population mean lead contamination.

EXAMPLE 8.4

smallbiz30

Z INTERVALS FOR μ USING TECHNOLOGY

The U.S. Small Business Administration (SBA) provides information on the number of small businesses for each metropolitan area in the United States. Table 8.2 contains a random sample of 30 moderately large cities and the number of small businesses in each city. Use the TI-83/84, Minitab, and the WHFStat Add-ins for Excel to construct a 95% Z confidence interval for the population mean number of small businesses in cities nationwide. Assume that the standard deviation is $\sigma = 4300$ for the number of small businesses in moderately large cities.

Solution

We shall use the instructions provided in the Step-by-Step Technology Guide at the end of this section (page 365). Since the sample size $n = 30$ is large (≥ 30), it is not necessary to check for normality.

Table 8.2 Small businesses in a sample of 30 cities

City	Small businesses	City	Small businesses	City	Small businesses
Orlando, FL	32,751	Cincinnati, OH	25,618	Nashville, TN	21,736
Kansas City, MO	32,750	Salt Lake City, UT	25,107	New Orleans, LA	21,565
San Jose, CA	30,921	Las Vegas, NV	24,867	Oklahoma City, OK	21,102
West Palm Beach, FL	30,226	Monmouth, NJ	24,255	Hartford, CT	20,677
Charlotte, NC	28,739	Columbus, OH	23,786	Jacksonville, FL	20,168
Indianapolis, IN	27,397	Raleigh, NC	23,566	Grand Rapids, MI	18,636
Sacramento, CA	27,189	Providence, RI	23,205	Buffalo, NY	18,285
Milwaukee, WI	26,456	Norfolk, VA	22,844	Richmond, VA	18,015
Fort Worth, TX	25,735	Greensboro, NC	22,359	Louisville, KY	17,754
Middlesex, NJ	25,726	Austin, TX	22,305	Greenville, SC	16,791

The results for the TI-83/84 in Figure 8.4 show that the 95% *Z* confidence interval for the population mean number of small businesses per city is

$$\text{lower bound} = 22{,}479, \text{ upper bound} = 25{,}556$$

Figure 8.4 also shows the sample mean $\bar{x} = 24{,}017.7$, the sample standard deviation $s = 4322.473886$, and the sample size $n = 30$.

The Minitab results are provided in Figure 8.5. The "assumed standard deviation" is indicated to be $\sigma = 4300$. Then the sample size $n = 30$, the sample mean $\bar{x} = 24{,}018$ (rounded), and the sample standard deviation $s = 4322$ (rounded) are displayed. "SE Mean" refers to the standard error of the mean, but we don't need it here. Finally, the 95% confidence interval is given as (lower bound = 22,479, upper bound = 25,556).

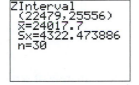

```
ZInterval
 (22479,25556)
 x̄=24017.7
 Sx=4322.473886
 n=30
```

FIGURE 8.4 TI-83/84 results.

One-Sample Z: Small Businesses

```
The assumed standard deviation = 4300

Variable           N    Mean   StDev   SE Mean      95% CI
Small Businesses   30   24018   4322       785   (22479, 25556)
```

FIGURE 8.5 Minitab results.

The results from the WHFStat Add-ins for Excel are shown in Figure 8.6.

SUMMARY STATISTICS		
Sample Mean	**Sample Size**	**Standard Deviation**
24017.7	30	4300

ONE SAMPLE Z CONFIDENCE INTERVAL	
Confidence Level	**Critical Z Value**
95 %	1.96

Confidence Interval

24017.7	+/-	1538.735
22478.96	to	25556.43

FIGURE 8.6 WHFStats Add-ins results.

The confidence level 95% is shown, along with the critical Z value, $Z_{\alpha/2} = 1.96$. The confidence interval is then shown:

$$\text{lower bound} = 22{,}478.96, \text{upper bound} = 25{,}556.43$$

This 95% confidence interval can also be expressed as (22,478.96, 25,556.43).

3   WAYS TO REDUCE THE MARGIN OF ERROR

Recall that the Z interval for μ takes the form

$$\text{point estimate} \pm \text{margin of error} = \bar{x} \pm E$$

Remember that the "±" notation *always* represents a pair of numbers.

where $E = Z_{\alpha/2}(\sigma/\sqrt{n})$. We interpret the margin of error E for a $(1 - \alpha)100\%$ confidence interval for μ as follows:

"We can estimate μ to within E units with $(1 - \alpha)100\%$ confidence."

EXAMPLE 8.5

FINDING AND INTERPRETING THE MARGIN OF ERROR

In Example 8.3, the Z interval for the population mean lead contamination (in ppm) for all trout on the Spokane River is

$$\text{lower bound} = 1 - 1.96(0.5/\sqrt{100}) = 1 - 1.96(0.05) = 1 - 0.098 = 0.902$$
$$\text{upper bound} = 1 + 1.96(0.5/\sqrt{100}) = 1 + 1.96(0.05) = 1 + 0.098 = 1.098$$

a. Find the margin of error E.
b. Express the confidence interval in the form "point estimate ± margin of error."
c. Interpret the margin of error E.

Solution

a. We find the margin of error as follows:

$$E = Z_{\alpha/2}(\sigma/\sqrt{n}) = 1.96(0.5/\sqrt{100}) = 1.96(0.05) = 0.098$$

b. The point estimate is $\bar{x} = 1$. Thus, the 95% confidence interval for the population mean lead contamination (in ppm) for all trout on the Spokane River takes the following form:

$$\text{point estimate} \pm \text{margin of error}$$
$$= \bar{x} \pm Z_{\alpha/2}(\sigma/\sqrt{n})$$
$$= 1 \pm 0.098$$

**Now You Can Do
Exercises 33–38.**

c. We interpret the margin of error E by saying that we can estimate the population mean lead contamination for all trout in the Spokane River to *within 0.098 ppm with 95% confidence.*

Of course, we would like our confidence interval estimates to be as precise as possible. Therefore, we would like the margin of error to be as small as possible, which would in turn result in a tighter confidence interval. Tighter confidence intervals are better, since the likely maximum difference between the sample mean and the population mean is reduced.

Note: When it comes to the margin of error *E*, smaller is better!

So how do we reduce the size of the margin of error? Let's look at the margin of error for the *Z* interval:

$$E = Z_{\alpha/2}(\sigma/\sqrt{n})$$

Since the population standard deviation σ is fixed, only $Z_{\alpha/2}$ and n can vary. There are therefore two strategies for decreasing the margin of error:

- *Decrease the confidence level*, which would decrease the value of $Z_{\alpha/2}$ (see Table 8.1), and

- *Increase the sample size n*, since dividing by a larger $\sqrt{n}$ will reduce *E*.

EXAMPLE 8.6

DECREASING THE MARGIN OF ERROR BY DECREASING THE CONFIDENCE LEVEL

For the confidence interval for the population mean lead contamination in Example 8.3, suppose we reduce the confidence level from 95% to 90% and leave everything else unchanged. Find the new margin of error. Describe how the margin of error has changed.

Solution

For confidence level 90%, $Z_{\alpha/2} = 1.645$, giving the following margin of error:

$$E = Z_{\alpha/2}(\sigma/\sqrt{n}) = 1.645\,(0.5/\sqrt{100}) \approx 0.082$$

Decreasing the confidence level from 95% to 90% decreases the margin of error from 0.098 to 0.082 ppm.

Developing Your Statistical Sense

There's No Free Lunch

The margin of error in Example 8.6 is smaller than the one in Example 8.3, which is good because it gives a more precise estimate of μ. However, this smaller margin of error is due entirely to the decrease in the confidence level, which is not good. In statistical data analysis, there is rarely a free lunch. The trade-off here is that, while the margin of error went down, so did the confidence level, from 95% to 90%. On the other hand, confidence intervals that are too wide can be useless. For example, we can be 99.9999% confident that the population mean age of college students in Florida lies between 15 and 75 years old. But, so what? The interval is too wide to be of practical use. More useful would be a 95% confidence interval that the population mean age of college students in Florida lies between 20 and 27.

This leads us to Strategy 2 for reducing the margin of error: increase the sample size. *The only way to have both high confidence and a tight interval is to boost the sample size.*

EXAMPLE 8.7

DECREASING THE MARGIN OF ERROR BY INCREASING THE SAMPLE SIZE

For the confidence interval for the population mean lead contamination in Example 8.3, suppose the results were based on a sample of size $n = 400$ rather than $n = 100$. Leaving everything else unchanged, find the new margin of error, and describe how the margin of error has changed.

Solution

For $n = 400$, the margin of error is

$$E = Z_{\alpha/2}(\sigma/\sqrt{n}) = 1.96(0.5/\sqrt{400}) = 0.049$$

Increasing the sample size from $n = 100$ to $n = 400$ has decreased the margin of error from 0.098 to 0.049 ppm.

"More data" is a familiar refrain in statistical analysis. Of course, increasing the sample size often raises pocketbook issues, since large samples can get very expensive ("We would like a large-sample estimate of the amount of damage sustained by Corvettes hitting a wall at 90 mph"). Sometimes obtaining large samples is simply impossible. Suppose an astronomer has developed a new technique for predicting corona effects during solar eclipses; she will have to wait a while (say, a few hundred years) to build up a large sample. So, take samples as large as realistically possible to keep the width of the confidence interval as narrow as possible.

4 SAMPLE SIZE FOR ESTIMATING THE POPULATION MEAN

When samples are plentiful and cheap, *arbitrarily precise confidence intervals with arbitrarily high confidence are possible simply by taking sufficiently large samples.*

Therefore, the question arises: *How large a sample size do I need* to get a tight confidence interval with a high confidence level?

EXAMPLE 8.8

SAMPLE SIZE FOR ESTIMATING THE POPULATION MEAN

Suppose we want to estimate to within $1000 the mean salary μ of all college graduates who were business majors. How many business majors would we sample to estimate the mean salary to within $1000 with 95% confidence?

Solution

"Within $1000" means that the margin of error E is $1000. Recall that the margin of error for 95% confidence is given by

$$E = 1.96\,(\sigma/\sqrt{n})$$

where 1.96 is the $Z_{\alpha/2}$ value associated with 95% confidence. Since the desired margin of error is 1000,

$$E = 1000 = 1.96\,(\sigma/\sqrt{n})$$

Solving for n gives us

$$n = \left(\frac{1.96\sigma}{1000}\right)^2$$

Suppose we know that $\sigma = \$5000$. Then:

$$n = \left(\frac{1.96 \cdot 5000}{1000}\right)^2 = 96.04$$

When finding the required sample size, if the formula results in a decimal, we always round up to the next whole number. Thus, we need a sample size of $n = 97$ for a confidence level of 95%.

Note: We solve for n as follows:
$$1000 = 1.96\,(\sigma/\sqrt{n})$$
Multiply both sides by $\sqrt{n}$:
$$1000\sqrt{n} = 1.96\sigma$$
Divide both sides by 1000:
$$\sqrt{n} = \left(\frac{1.96\sigma}{1000}\right)$$
Square both sides to get the formula for n:
$$n = \left(\frac{1.96\sigma}{1000}\right)^2$$

Now You Can Do Exercises 41–48.

We generalize the result from Example 8.8 as follows.

> **Sample Size for Estimating the Population Mean**
>
> The sample size for a *Z* interval that estimates the population mean μ to within a margin of error *E* with confidence $100(1 - \alpha)\%$ is given by
>
> $$n = \left[\frac{(Z_{\alpha/2})\sigma}{E} \right]^2$$
>
> where $Z_{\alpha/2}$ is the value associated with the desired confidence level (Table 8.1), *E* is the desired margin of error, and σ is the population standard deviation. By convention, whenever this formula yields a sample size with a decimal, *always round up to the next whole number.*

We round up because (a) the sample size *n* must be a whole number and (b) rounding down will lead to a value of *n* with less than the desired confidence level.

The *Normal Density Curve* applet may be used to find $Z_{\alpha/2}$ critical values for confidence levels not listed in Table 8.1.

The *Confidence Interval* applet allows you to see for yourself how individual samples generate intervals that either do or do not contain the population mean.

STEP-BY-STEP TECHNOLOGY GUIDE: *Z* Confidence Intervals

We illustrate how to construct the confidence interval for Example 8.4 (page 360).

TI-83/84

If you have the data values:
Step 1   Enter the data into list **L1** (Figure 8.7).
Step 2   Press **STAT**, highlight **TESTS**.
Step 3   Press **7** (for **ZInterval**).
Step 4   For input (**Inpt**), highlight **Data** and press **ENTER** (Figure 8.8).
a. For σ, enter the assumed value of **4300**.
b. For **List**, press **2nd** then **L1**.
c. For **Freq**, enter **1**.
d. For **C-Level** (confidence level), enter the appropriate confidence level (e.g., **0.95**), and press **ENTER**.
e. Highlight **Calculate** and press **ENTER**. The results are shown in Figure 8.4 in Example 8.4.

If you have the summary statistics:
Step 1   Press **STAT**, highlight **TESTS**.
Step 2   Press **7** (for **ZInterval**).
Step 3   For input (**Inpt**), highlight **Stats** and press **ENTER** (Figure 8.9).
a. For σ, enter the assumed value of **4300**.
b. For $\bar{x}$, enter the sample mean **24017.7**.
c. For *n*, enter the sample size **30**.
d. For **C-Level** (confidence level), enter the appropriate confidence level (e.g., **0.95**), and press **ENTER**.
e. Highlight **Calculate** and press **ENTER**. The results are shown in Figure 8.4 in Example 8.4.

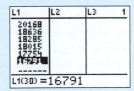

FIGURE 8.7

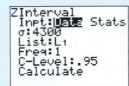

FIGURE 8.8

FIGURE 8.9

EXCEL

If you have the data values:
Step 1   Enter the data into column A.
Step 2   Load the **WHFStat Add-ins**.
Step 3   Select **Add-ins > Macros > Estimating a Mean > Z Confidence Interval**.
Step 4   Click **Select Dataset Range**, highlight A1–A30, and click **OK**.
Step 5   Input **4300** for the **Population Standard Deviation**, select the **95%** confidence level, and click **OK**.
The results are displayed in Figure 8.6 in Example 8.4.

If you have the summary statistics:
Step 1   Load the **WHFStat Add-ins**.
Step 2   Select **Add-ins > Macros > Estimating a Mean > Z Confidence Interval**.
Step 3   Click **Input Summary Statistics**, enter 24017.7 for the **Sample Mean**, enter **30** for the **Sample Size**, and click **OK**.
Step 4   Input **4300** for the **Population Standard Deviation**, select the **95%** confidence level, and click **OK**.
The results are displayed in Figure 8.6 In Example 8.4.

MINITAB

If you have the data values:

Step 1 Enter the data into column C1.
Step 2 Click **Stat > Basic Statistics > 1-Sample Z.**
Step 3 Click **Samples in Columns** and select **C1.**
Step 4 Click **Options**, enter **95** as the **Confidence Level**, and click **OK.**
Step 5 Enter **4300** for **Sigma** and click **OK.**
The results are displayed in Figure 8.5 in Example 8.4.

If you have the summary statistics:

Step 1 Click **Stat > Basic Statistics > 1-Sample Z.**
Step 2 Click **Summarized Data.**
Step 3 Enter the **Sample Size 30** and the **Sample Mean** 24017.7.
Step 4 Enter **4300** for the **Standard Deviation.**
Step 5 Click **Options**, enter **95** as the **Confidence Level**, click **OK**, and click **OK** again.
The results are displayed in Figure 8.5 in Example 8.4.

CRUNCHIT!

If you have the data values:

Step 1 Click **File** . . . then highlight **Load from Larose2e . . . Chapter 8** . . . and click on Example 8.4.
Step 2 Click **Statistics . . . Z** and select **1-sample.**
Step 3 With the **Columns** tab chosen, for **Sample** select **Businesses.** For **Standard Deviation**, enter **4300.**
Step 4 Select the **Confidence Interval** tab, and enter **95** for the **Confidence Interval Level.**
Then click **Calculate.**

If you have the summary statistics:

Step 1 Click **Statistics . . . Z** and select **1-sample.**
Step 2 Choose the **Summarized** tab. For **n** enter the sample size 30; for **Sample Mean** enter 24017.7. For **Standard Deviation**, enter **4300.**
Step 3 Select the **Confidence Interval** tab, and enter **95** for the **Confidence Interval Level.**
Then click **Calculate.**

SECTION 8.1 Summary

1. Using a single statistic only, such as $\bar{x}$, to estimate a population parameter is called point estimation. The value of the statistic is called the point estimate.

2. A confidence interval estimate of a parameter consists of an interval of numbers generated by a point estimate, together with an associated confidence level specifying the probability that the interval contains the parameter. The $100(1 - \alpha)\%$ Z confidence interval for μ is given by the interval

$$\text{lower bound} = \bar{x} - Z_{\alpha/2}(\sigma/\sqrt{n})$$
$$\text{upper bound} = \bar{x} + Z_{\alpha/2}(\sigma/\sqrt{n})$$

where $1 - \alpha$ is the confidence level. If σ is not known, then the Z interval cannot be used.

3. The margin of error E is a measure of the precision of the confidence interval estimate. For the Z interval, the margin

of error takes the form

$$E = Z_{\alpha/2}(\sigma/\sqrt{n})$$

Usually, our confidence intervals take the form

$$\text{point estimate} \pm \text{margin of error}$$

4. To use a Z interval to estimate the population mean μ to within a margin of error E with confidence $100(1 - \alpha)\%$, the required sample size is given by

$$n = \left[\frac{(Z_{\alpha/2})\sigma}{E}\right]^2$$

where $Z_{\alpha/2}$ is associated with the desired confidence level (Table 8.1), E is the desired margin of error, and σ is the population standard deviation. Round up to the next integer if there is a decimal.

SECTION 8.1 Exercises

Clarifying the Concepts

1. Explain why a point estimate, together with a margin of error, is more likely to capture the value of a population parameter than a point estimate alone.

2. What are two ways of presenting a confidence interval?

3. Suppose that a 95% confidence interval for the population mean football score is (15, 25). Interpret this confidence interval.

4. True or false: It is the confidence interval that is random, not the population mean μ.

5. Let E represent the margin of error. Explain what the "$\pm$" notation means in $\bar{x} \pm E$.

6. What is the difference between *confidence interval* and *confidence level*?

7. Assume that the confidence level increases.
   a. What happens to the value of $Z_{\alpha/2}$?
   b. Explain why this happens. Draw a sketch to help you.

8. Suppose your supervisor wants to (a) increase the confidence level from 95% to 99% and (b) keep the width of the confidence interval small. What is the only way to accomplish this?

9. What happens to the required sample size for estimating the population mean as the confidence level is increased? Decreased?

10. What happens to the required sample size for estimating the population mean as the margin of error is increased? Decreased?

Practicing the Techniques

For the data sets shown in Exercises 11–14, calculate the point estimate of the population mean μ.

11.

2	3	1	3	1

12.

8	4	6	4	8

13.

11	17	14	17	11

14.

96	104	100	96	104

For Exercises 15–20, random samples are drawn. Indicate whether or not we can use the Z confidence interval for μ.

15. The sample size is large ($n \geq 30$) and σ is unknown.

16. The original population is normal and σ is known.

17. The sample size is large ($n \geq 30$) and σ is known.

18. The sample size is small ($n < 30$), the original population is normal, and σ is known.

19. The sample size is large ($n \geq 30$), the original population is not normal, and σ is known.

20. The original population is not normal, and σ is not known.

For Exercises 21–26, find the value of $Z_{\alpha/2}$.

21. Confidence level = 99%

22. $\alpha = 0.05$

23. Confidence level = 95%

24. $\alpha/2 = 0.025$

25. Confidence level = 90%

26. $\alpha = 0.01$

For Exercises 27–32, answer the following questions.
   a. Calculate $\sigma/\sqrt{n}$.
   b. Find $Z_{\alpha/2}$ for a confidence interval for μ with 95% confidence.
   c. Construct and interpret a 95% confidence interval for μ.

27. A random sample of $n = 16$ with sample mean $\bar{x} = 35$ is drawn from a normal population in which $\sigma = 2$.

28. A random sample of $n = 25$ with sample mean $\bar{x} = 50$ is drawn from a normal population in which $\sigma = 5$.

29. A random sample of $n = 9$ with sample mean $\bar{x} = 15$ is drawn from a normal population in which $\sigma = 6$.

30. A random sample of $n = 64$ with sample mean $\bar{x} = 10$ is drawn from a population in which $\sigma = 4$.

31. A random sample of $n = 49$ with sample mean $\bar{x} = 20$ is drawn from a population in which $\sigma = 7$.

32. A random sample of $n = 81$ with sample mean $\bar{x} = 100$ is drawn from a population in which $\sigma = 18$.

For Exercises 33–38, do the following.
   a. Compute the margin of error for the confidence interval constructed in the indicated exercise.
   b. Interpret this value for the margin of error.

33. Confidence interval from Exercise 27

34. Confidence interval from Exercise 28

35. Confidence interval from Exercise 29

36. Confidence interval from Exercise 30

37. Confidence interval from Exercise 31

38. Confidence interval from Exercise 32

39. A random sample of $n = 25$ is drawn from a normal population in which $\sigma = 2$. The sample mean is $\bar{x} = 10$. For (a)–(c), construct and interpret confidence intervals for μ with the indicated confidence levels. Then answer the question in (d).
   a. 90%
   b. 95%
   c. 99%
   d. What can you conclude about the width of the interval as the confidence level increases?

40. A random sample of $n = 100$ is drawn from a population in which $\sigma = 5$. The sample mean is $\bar{x} = 50$. For parts (a)–(c), construct and interpret confidence intervals for μ with the indicated confidence levels. Then answer the question in (d).
   a. 99%
   b. 95%
   c. 90%
   d. What can you conclude about the width of the interval as the confidence level decreases?

Suppose we are estimating μ. For Exercises 41–43, find the required sample size.

41. $\sigma = 10$, confidence level 90%, margin of error 32

42. $\sigma = 10$, confidence level 90%, margin of error 16

43. $\sigma = 10$, confidence level 90%, margin of error 8

44. What happens to the required sample size when the margin of error is halved and σ and the confidence level stay the same?

Suppose we are estimating μ. For Exercises 45–47, find the required sample size.

45. $\sigma = 10$, confidence level 90%, margin of error 8

46. $\sigma = 10$, confidence level 95%, margin of error 8

47. $\sigma = 10$, confidence level 99%, margin of error 8

48. What happens to the required sample size when the confidence level increases and the margin of error and σ stay the same?

Applying the Concepts

For each of Exercises 49–52, do the following.
  a. Find the point estimate of the population mean.
  b. Calculate $\sigma/\sqrt{n}$.
  c. Find $Z_{\alpha/2}$ for a confidence interval for the indicated confidence level.
  d. Construct and interpret a confidence interval with the indicated confidence level for the population mean.

49. Consumption of Carbonated Beverages. The U.S. Department of Agriculture reports that the mean American consumption of carbonated beverages per year is greater than 52 gallons. A random sample of 30 Americans yielded a sample mean of 69 gallons. Assume that the population standard deviation is 20 gallons. Let the confidence level be 95%.

50. Stock Shares Traded. The *Statistical Abstract of the United States* reports that the mean daily number of shares traded on the New York Stock Exchange (NYSE) in March 2010 was 2129 million. Assume that the population standard deviation equals 500 million shares. Suppose that, in a random sample of 36 days from the present year, the mean daily number of shares traded equals 2 billion. Let the confidence level be 95%.

51. Engaging with Science. A psychological study found that the mean length of time that boys remained engaged with a science exhibit at a museum was 107 seconds with a standard deviation of 117 seconds.[2] Assume that the 117 seconds represents the population standard deviation. The sample size is 36 and let the confidence level be 95%.

52. Latino Tobacco Consumption. The Bureau of Labor Statistics reported that the mean amount spent by all American citizens on tobacco products and smoking supplies is $308; the mean for American Latinos is $177. Assume that σ, the standard deviation for American Latinos, equals $150. Assume that the data on American Latinos represents a sample of size 36. Let the confidence level be 90%.

53. Consumption of Carbonated Beverages. Refer to Exercise 49.
  a. Compute and interpret the margin of error.
  b. How large a sample size is needed to estimate μ to within 25 gallons with 95% confidence?
  c. How large a sample size is needed to estimate μ to within 5 gallons with 95% confidence?

54. Stock Shares Traded. Refer to Exercise 50.
  a. Calculate and interpret the margin of error.
  b. How large a sample size (trading days) is needed to estimate the population mean number of shares traded per day to within 100 million with 95% confidence?
  c. How large a sample size (trading days) is needed to estimate the population mean number of shares traded per day to within 10 million with 95% confidence? How many years does this number of days translate into?

55. Engaging with Science. Refer to Exercise 51.
  a. Find and interpret the margin of error.
  b. How large a sample size is needed to estimate μ to within 30 seconds with 95% confidence?
  c. How large a sample size is needed to estimate μ to within 3 seconds with 95% confidence?

56. Latino Tobacco Consumption. Refer to Exercise 52.
  a. Compute and interpret the margin of error.
  b. How large a sample size would have been required if the BLS had wanted to estimate the population mean amount spent by American Latinos to within $50 with 95% confidence?
  c. How large a sample size would have been required if the BLS had wanted to estimate the population mean amount spent by American Latinos to within $10 with 95% confidence?

57. Carbon Emissions. The following table represents the carbon emissions (in millions of tons) from consumption of fossil fuels, for a random sample of 5 nations.[3] Assume $\sigma = 200$ million tons.

 carbon

Nation	Emissions
Brazil	361
Germany	844
Mexico	398
Great Britain	577
Canada	631

  a. Assess the normality of the data, using a normal probability plot. (*Hint:* See page 360.)
  b. Assuming that carbon emissions are normally distributed, construct and interpret a 90% confidence interval for the population mean carbon emissions.
  c. Calculate and interpret the margin of error for the confidence interval in part (**b**).
  d. How large a sample size do we need to estimate μ to within 50 million tons with 90% confidence?

58. *Deepwater Horizon* Cleanup Costs. The following table represents the amount of money distributed by BP to a random sample of 6 Florida counties, for cleanup of the

Deepwater Horizon oil spill, in millions of dollars.[4] Assume $\sigma = \$350{,}000$.

 deepwaterclean

County	Cleanup costs ($ millions)
Broward	0.85
Escambia	0.70
Franklin	0.50
Pinellas	1.15
Santa Rosa	0.50
Walton	1.35

a. Assess the normality of the data, using a normal probability plot.
b. Assuming that the cleanup costs are normally distributed, construct and interpret a 95% confidence interval for the population mean cleanup cost.
c. Calculate and interpret the margin of error for the confidence interval in part (**b**).
d. How large a sample size do we need to estimate μ to within $50,000 with 95% confidence?

59. Wii Game Sales. The following table represents the number of units sold in the United states for the week ending March 26, 2011, for a random sample of 8 Wii games.[5] Assume $\sigma = 30{,}000$.

 wiisales

Game	Units (1000s)	Game	Units (1000s)
Wii Sports Resort	65	Zumba Fitness	56
Super Mario All Stars	40	Wii Fit Plus	36
Just Dance 2	74	Michael Jackson	42
New Super Mario Bros.	16	Lego Star Wars	110

a. Assess the normality of the data, using a normal probability plot.
b. Assuming that the game sales are normally distributed, construct and interpret a 99% confidence interval for the population mean number of units sold.
c. Calculate and interpret the margin of error for the confidence interval in part (**b**).
d. How large a sample size do we need to estimate μ to within 5000 units with 99% confidence?

60. A Rainy Month in Georgia? The following table represents the total rainfall (in inches) for the month of February 2011 for a random sample of ten locations in Georgia.[6] Assume $\sigma = 0.64$ inches.

 georgiarain

Location	Rainfall (inches)	Location	Rainfall (inches)
Athens	4.72	Atlanta	4.25
Augusta	4.31	Cartersville	3.03
Dekalb	2.96	Fulton	4.36
Gainesville	4.06	Lafayette	3.75
Marietta	3.20	Rome	3.26

a. Assess the normality of the data, using a normal probability plot.
b. Assuming that the rainfall amounts are normally distributed, construct and interpret a 95% confidence interval for the population mean rainfall in inches.
c. Calculate and interpret the margin of error for the confidence interval in part (**b**).
d. How large a sample size do we need to estimate μ to within 0.1 inch with 95% confidence?

61. Short-Term Memory. In a famous research paper in the psychology literature, George Miller found that the amount of information humans could process in short-term memory was 7 bits (pieces of information), plus or minus 2 bits.[7] Let us assume that the title of Miller's paper ("The Magical Number Seven, Plus or Minus Two") refers to a confidence interval. Assume that $\sigma = 10$ bits.

a. What is the point estimate for the amount of information all humans can process in short-term memory?
b. What is the margin of error? Interpret this number.
c. The most common confidence level in the psychological literature is 95%. Which value for $Z_{\alpha/2}$ is associated with 95% confidence?
d. How large a sample size did Miller use to find the confidence interval in the title, assuming that he used 95% confidence?
e. Suppose he had wanted the title to read "The Magical Number Seven, Plus or Minus One"? How large a sample size would he have needed?

62. Commuting Distances. A university is trying to attract more commuting students from the local community. As part of the research into the modes of transportation students use to commute to the university, a survey was conducted asking how far commuting students commuted from home to school each day. A random sample of 30 students provided the distances (in miles) shown in the table below. Assume that the standard deviation is $\sigma = 3$ miles.

 commutedist

14	10	14	12	12	11	5	6	9	14	9	9	4	7	15
9	7	7	12	10	15	10	6	11	9	11	10	11	7	12

a. Compute and interpret the margin of error for a confidence interval with 95% confidence.
b. Construct and interpret a 95% confidence interval for the population mean commuting distance.

Small Businesses. Use this information for Exercise 71. The United States Small Business Administration publishes data on the number of small businesses in each of 327 metropolitan areas. This data is in the data file **Small Businesses.**

 smallbusinesses

63. Follow steps (a)–(e).
a. Find the sample mean number of small firms per metropolitan area.
b. Generate a histogram of the number of small firms per metropolitan area.
c. Generate a normal probability plot of the number of small firms in each metropolitan area. What is your conclusion regarding the normality of the distribution of the number of firms?
d. Construct and interpret a 95% confidence interval for the population number of small firms per metropolitan area. Assume that the standard deviation is 25,000 firms.
e. On the histogram, indicate the location of the confidence interval.

 Use the *Confidence Interval* applet for Exercises 64.

64. Set the confidence level to 90%. Click "Sample 50" to produce 50 simple random samples (SRSs) and display the resulting 90% confidence intervals for μ.
a. What is the percent hit, that is, the proportion of the confidence intervals that actually contain the true value of μ?
b. Keep clicking "Sample 50" until 1000 confidence intervals are generated. What is the percent hit?
c. It is not likely (though it is possible) that the percent hit in (b) exactly equals 90%. Explain why the percent hit is not equal to 90% when we asked for a confidence level of 90%.

Use the *Normal Density Curve* applet for Exercises 65 and 66.

65. Use the applet to find $Z_{\alpha/2}$ critical values for unusual confidence levels. Select **2-Tail,** and click and drag the flags so that the central area and not the tail area is highlighted. Verify that the $Z_{\alpha/2}$ critical value for 95% confidence is 1.96.

66. Use the applet to find $Z_{\alpha/2}$ critical values for the following confidence levels.
a. 80%
b. 85%
c. 98%

8.2 *t* INTERVAL FOR THE POPULATION MEAN

OBJECTIVES By the end of this section, I will be able to . . .

1 Describe the characteristics of the *t* distribution.

2 Calculate and interpret a *t* interval for the population mean.

1 INTRODUCING THE *t* DISTRIBUTION

In Section 8.1 we constructed confidence intervals for the population mean μ assuming that the population standard deviation σ was known. This assumption may be valid for certain fields such as quality control. However, in many real-world problems, we do not know the value of σ, and thus cannot use a Z interval to estimate the mean. When σ is unknown, we use the sample standard deviation s to construct a confidence interval that is likely to contain the population mean.

Fact 4 from Chapter 7 showed us that we could standardize $\bar{x}$ to derive the standard normal random variable:

$$Z = \frac{\bar{x} - \mu}{\sigma/\sqrt{n}}$$

Unfortunately, however, if we replace the unknown σ in this equation with the known s, we can no longer obtain the standard normal Z because s, being a statistic, is itself a random variable. Instead, $\dfrac{\bar{x} - \mu}{s/\sqrt{n}}$ follows an entirely new and different distribution, called the *t* **distribution.**

> ### *t* Distribution
>
> For a normal population, the distribution of
>
> $$t = \frac{\bar{x} - \mu}{s/\sqrt{n}}$$
>
> follows a *t* distribution, with $n - 1$ degrees of freedom, where $\bar{x}$ is the sample mean, μ is the unknown population mean, s is the sample standard deviation, and n is the sample size.

Developing Your Statistical Sense

Degrees of Freedom

Notice that the definition of the *t* distribution includes a new concept called *degrees of freedom*. Degrees of freedom is a measure that determines how the *t* distribution changes as the sample size changes. The idea of degrees of freedom is that, in a sum of *n* numbers, you need to know only the first $n - 1$ of these numbers to find the *n*th number because you already know the sum. For example, suppose you know that the sum of $n = 3$ numbers is 10 and are told that the first two numbers are 5 and 1. Then you can deduce that the last number is $10 - (5 + 1) = 4$. The first two numbers have the *freedom* to take on any values, but the third number must take a particular value. Thus, there are only $n - 1$ independent pieces of information. The concept is similar for the *t* distribution. Since we use the sample standard deviation *s* to estimate the unknown σ and since *s* is known, only $n - 1$ independent pieces of information are needed to find the value of *t*. Thus, we say that $t = \dfrac{\bar{x} - \mu}{s/\sqrt{n}}$ follows a *t* distribution with $n - 1$ degrees of freedom.

Figure 8.10 displays a comparison of some *t* curves with the *Z* curve. Note that there is only one *Z* distribution (or curve), but there is a different *t* curve for every different degrees of freedom (df), that is, for every different sample size. The degrees of freedom, df $= n - 1$, determines the shape of the *t* distribution, just as the mean and variance uniquely determine the shape of the normal distribution. All *t* curves have several characteristics in common.

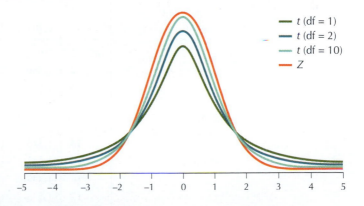

FIGURE 8.10
Different *t* curve for different degrees of freedom (df $= n - 1$).

> ### Characteristics of the *t* Distribution
>
> - Centered at zero. The mean of *t* is zero, just as with *Z*.
> - Symmetric about its mean zero, just as with *Z*.
> - As df decreases, the *t* curve gets flatter, and the area under the *t* curve decreases in the center and increases in the tails. That is, the *t* curve has heavier tails than the *Z* curve.
> - As df increases toward infinity, the *t* curve approaches the *Z* curve, and the area under the *t* curve increases in the center and decreases in the tails.

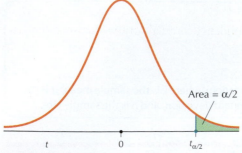

FIGURE 8.11 $t_{\alpha/2}$ has area to the right of it.

Similar to the definition of $Z_{\alpha/2}$ in Section 8.1, we can define $t_{\alpha/2}$ to be the value of the t distribution with area $a/2$ to the right of it, as seen in Figure 8.11. Table 8.1 in Section 8.1 provides the $Z_{\alpha/2}$ values for certain common confidence levels. Unfortunately, because there is a different t curve for each sample size, there are many possible $t_{\alpha/2}$ values. You will need to use the t table (Table D in the Appendix) to find the value of $t_{\alpha/2}$, as follows.

Procedure for Finding $t_{\alpha/2}$

Step 1   Go across the row marked "Confidence level" in the t table (Table D in the Appendix) until you find the column with the desired confidence level at the top. The $t_{\alpha/2}$ value is in this column somewhere.

Step 2   Go down the column until you see the correct number of degrees of freedom on the left. The number in that row and column is the desired value of $t_{\alpha/2}$.

EXAMPLE 8.9   FINDING $t_{\alpha/2}$

Find the value of $t_{\alpha/2}$ that will produce a 95% confidence interval for μ if the sample size is $n = 20$.

Solution

Note: For the newer TI-84s

1. Press **2nd DISTR** and select **4:invT.**

2. Enter the area to the *left* of the t value, then **comma,** then df = $n - 1$.

3. Press **ENTER**.

For example, **invT(0.975,19)** gives 2.093024022. The TI-83 does not have this function.

STEP 1   We go across the row labeled "Confidence level" in the t table (Figure 8.12) until we see the 95% confidence level. Our $t_{\alpha/2}$ is somewhere in this column.

STEP 2   The degrees of freedom are df = $n - 1 = 20 - 1 = 19$. We go down the column until we see 19 on the left. The number in that row is our $t_{\alpha/2}$, 2.093.

t-Distribution

				Confidence level		
		80%	90%	95%	98%	99%
				Area in one tail		
		0.10	0.05	0.025	0.01	0.005
				Area in two tails		
		0.20	0.10	0.05	0.02	0.01
df	1	3.078	6.314	12.706	31.821	63.657
	2	1.886	2.920	4.303	6.965	9.925
	3	1.638	2.353	3.182	4.541	5.841
	14	1.345	1.761	2.145	2.624	2.977
	15	1.341	1.753	2.131	2.602	2.947
	16	1.337	1.746	2.120	2.583	2.921
	17	1.333	1.740	2.110	2.567	2.898
	18	1.330	1.734	2.101	2.552	2.878
	19	1.328	1.729	2.093	2.539	2.861
	20	1.325	1.725	2.086	2.528	2.845
	21	1.323	1.721	2.080	2.518	2.831

Now You Can Do Exercises 5–8.

FIGURE 8.12 Use the confidence level and the degrees of freedom to find $t_{\alpha/2}$.

2 *t* INTERVAL FOR THE POPULATION MEAN

The *t* distribution provides the following confidence interval for the unknown population mean μ, called the ***t* interval.**

---

t Interval for μ

The *t* interval for μ may be constructed whenever *either* of the following conditions is met:

- The population is normal.
- The sample size is large ($n \geq 30$).

Suppose a random sample of size n is taken from a population with unknown mean μ. A $100(1 - \alpha)\%$ confidence interval for μ is given by the interval

$$\text{lower bound} = \bar{x} - t_{\alpha/2}(s/\sqrt{n}), \text{ upper bound} = \bar{x} + t_{\alpha/2}(s/\sqrt{n})$$

where $\bar{x}$ is the sample mean, $t_{\alpha/2}$ is associated with the confidence level and $n - 1$ degrees of freedom, and s is the sample standard deviation. The *t* interval may also be written as

$$\bar{x} \pm t_{\alpha/2}(s/\sqrt{n})$$

and is denoted

$$(\text{lower bound, upper bound})$$

Note: Suppose that σ is unknown, *and* the population is either non-normal or of unknown distribution, *and* the sample size is not large. Then we should not use the *t* interval. Rather, we need to turn to nonparametric methods, for example, the *sign interval* or the *Wilcoxon interval*. (See Nonparametric Statistics chapter, available online.)

EXAMPLE 8.10

fourthfeet

INTERVAL FOR μ

Suppose a children's shoe manufacturer is interested in estimating the population mean length of fourth-graders' feet. A random sample of 20 fourth-graders' feet yielded the following foot lengths, in centimeters.[8]

| 22.4 | 23.4 | 22.5 | 23.2 | 23.1 | 23.7 | 24.1 | 21.0 | 21.6 | 20.9 |
| 25.5 | 22.8 | 24.1 | 25.0 | 24.0 | 21.7 | 22.0 | 22.7 | 24.7 | 23.5 |

Construct a 95% confidence interval for μ, the population mean length of all fourth-graders' feet.

Solution

We do not know the population standard deviation σ, so we cannot use the *Z* interval. We can construct a *t* interval whenever either the population is normal or the sample size is large. The sample size here is 20, which is not large ($n \geq 30$), so we must check for normality. Figure 8.13 shows the normal probability plot of the foot lengths. The points generally line up along the line, so the assumption of normality is validated for this data set. We can then proceed to construct the *t* interval for μ.

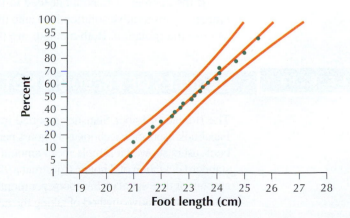

FIGURE 8.13
Fourth-grade foot lengths are normally distributed.

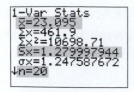

The TI-83/84 provides the summary statistics shown here, giving $n = 20$, $\bar{x} = 23.095$, and $s \approx 1.280$. All that is left is to find $t_{\alpha/2}$. In Example 8.9, we found the value of $t_{\alpha/2}$ for confidence level = 95% and $n = 20$ to be $t_{\alpha/2} = 2.093$. The 95% confidence interval then becomes

$$\text{lower bound} = \bar{x} - t_{\alpha/2}(s/\sqrt{n})$$
$$= 23.095 - 2.093(1.280/\sqrt{20}) \approx 23.095 - 0.599 = 22.496$$

$$\text{upper bound} = \bar{x} + t_{\alpha/2}(s/\sqrt{n})$$
$$= 23.095 + 2.093(1.280/\sqrt{20}) \approx 23.095 + 0.599 = 23.694$$

This interval is denoted

$$(22.496, 23.694)$$

We are 95% confident that the population mean length of fourth-graders' feet lies between 22.496 and 23.694 cm. (See Figure 8.14.)

Now You Can Do
Exercises 9–24.

FIGURE 8.14 95% *t* Confidence interval for population mean foot length.

Developing Your Statistical Sense

_t_ Intervals May Offer More Peace of Mind than _Z_ Intervals

In Example 8.10, if we had assumed that the population standard deviation σ was known ($\sigma = 1.280$), then the 95% Z interval for the population mean length of fourth-grade feet would have been

$$\text{lower bound} = \bar{x} - Z_{\alpha/2}(\sigma/\sqrt{n})$$
$$= 23.095 - 1.96(1.280/\sqrt{20}) \approx 23.095 - 0.561 = 22.534$$
$$\text{upper bound} = \bar{x} + Z_{\alpha/2}(\sigma/\sqrt{n})$$
$$= 23.095 + 1.96(1.280/\sqrt{20}) \approx 23.095 + 0.561 = 23.656$$

Note that this Z interval (22.534, 23.656) is only slightly more precise than the t interval (22.496, 23.694). However, the Z interval depends on prior knowledge of the value of σ. If the value of σ is inaccurate, then the Z interval will be misleading and overly optimistic. With even moderate sample sizes, reporting the t interval rather than the Z interval may offer peace of mind to the data analyst.

If the degrees of freedom needed to find $t_{\alpha/2}$ do not appear in the df column of the t table, a conservative solution is to take the next row with smaller df. Alternatively, we can use interpolation. Both methods are illustrated in Example 8.11.

EXAMPLE 8.11   **DEGREES OF FREEDOM NOT IN THE _t_ TABLE**

The Bureau of Labor Statistics reported in 2010 that the mean amount of time spent by Facebook users on Facebook is 11 hours per month. Suppose a random sample of 49 Facebook users showed a sample mean amount of time of 11 hours per month with a sample standard deviation of 7 hours. Construct a 99% confidence interval for the population mean amount of time spent on Facebook per month for all Facebook users. When finding $t_{\alpha/2}$, use (a) the conservative method of taking the next row with smaller *df*, and (b) interpolation.

Solution

Since σ is unknown and the sample size is large, we proceed to construct the *t* interval for μ. We have $n = 49$, $\bar{x} = 11$, and $s = 7$. Now we must find $t_{\alpha/2}$. The confidence level is 99% and the degrees of freedom are $n - 1 = 49 - 1 = 48$. *Unfortunately, the value of 48 for the df does not appear in the df column.*

a.   The next row with df smaller than 48 would be df $= 40$. Thus, the "conservative" $t_{\alpha/2}$ is 2.704. We then proceed to construct the 99% confidence interval:

$$\bar{x} \pm t_{\alpha/2}\,(s/\sqrt{n}) = 11 \pm 2.704(7/\sqrt{49}) = (8.296, 13.704)$$

b.   Alternatively, you could interpolate as follows. Since df $= 48$ is 8/10 of the distance between 40 and 50, we can estimate $t_{\alpha/2}$ by taking 8/10 of the distance from the *t*-value for df $= 40$ to the *t*-value for df $= 50$, and subtracting the result from the *t*-value for df $= 40$:

$$\frac{8}{10}\left[(t_{\alpha/2}\text{ for df} = 40) - (t_{\alpha/2}\text{ for df} = 50)\right] = \frac{8}{10}(2.704 - 2.678) = 0.0208$$

> Using a smaller degrees of freedom is *conservative*, that is, cautious. This means that the resulting confidence interval will not be more precise than is warranted by the data.

Thus, $t_{\alpha/2}$ for df $= 48$ would be $2.704 - 0.0208 = 2.6832$, using interpolation. The 99% confidence interval using interpolation is thus

$$\bar{x} \pm t_{\alpha/2}(s/\sqrt{n}) = 11 \pm 2.6832(7/\sqrt{49}) = (8.3168, 13.6832)$$

**Now You Can Do
Exercises 25–32.**

Note that the confidence interval using the conservative method is somewhat wider, reflecting the conservative choice of $t_{\alpha/2}$.

Recall that the margin of error for the *Z* interval equals $Z_{\alpha/2} \cdot (\sigma/\sqrt{n})$. For the *t* interval, since σ is unknown, the margin of error is given as follows.

> **Margin of Error for the *t* Interval**
>
> $$E = t_{\alpha/2} \cdot \left(\frac{s}{\sqrt{n}}\right)$$
>
> The margin of error *E* for a $(1 - \alpha)$100% *t* interval for μ can be interpreted as follows: "We can estimate μ to within *E* units with $(1 - \alpha)$100% confidence."

EXAMPLE 8.12

FINDING AND INTERPRETING THE MARGIN OF ERROR FOR THE FOURTH-GRADER FOOT LENGTHS

Use the statistics observed in Example 8.10.
a.   Find the margin of error for the 95% confidence interval for mean foot lengths.
b.   Interpret the margin of error.

Solution

a.   From Example 8.10, $n = 20$ and $s = 1.280$. Also, for a confidence level of 95%, $t_{\alpha/2} = 2.093$. Therefore, the margin of error of fourth-grade foot length is

$$E = t_{\alpha/2} \cdot \left(\frac{s}{\sqrt{n}}\right) = (2.093) \cdot \frac{1.280}{\sqrt{20}} \approx 0.599$$

**Now You Can Do
Exercises 44–48.**

b.   We can estimate the population mean of fourth-grade foot lengths to within 0.599 centimeter with 95% confidence.

What Does the Margin of Error Mean?

The margin of error $E = 0.599$ provides an indication of the accuracy of the confidence interval estimate for confidence level = 95%. That is, if we repeatedly take many samples of size 20 fourth-graders, our sample mean $\bar{x}$ will be within $E = 0.599$ centimeter of the unknown population mean μ in 95% of those samples.

EXAMPLE 8.13 *t* INTERVALS FOR μ USING TECHNOLOGY

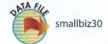

DATA FILE smallbiz30

In Example 8.4, we considered a sample of 30 randomly selected moderately large cities and counted the number of small businesses in each city (see Table 8.2, page 361). We found that the sample mean $\bar{x} = 24{,}017.7$ and the sample standard deviation $s = 4322.473886$. However, this time we are not assuming that we know the value of the population standard deviation, σ. Use the TI-83/84, Minitab, and the WHFStat Add-ins for Excel to construct a 95% *t* confidence interval for the population mean number of small businesses in moderately sized cities nationwide.

Solution

We use the instructions provided in the Step-by-Step Technology Guide on page 377. Since the sample size $n = 30$ is large (≥ 30), it is not necessary to check for normality.

The results for the TI-83/84 in Figure 8.15 display the 95% *t* confidence interval for the population mean number of small businesses per city to be

(lower bound = 22,404, upper bound = 25,632)

```
TInterval
 (22404,25632)
 x̄=24017.7
 Sx=4322.473886
 n=30
```

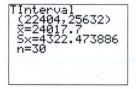

FIGURE 8.15 TI-83/84 results.

They also show the sample mean $\bar{x} = 24{,}017.7$, the sample standard deviation $s = 4322.473886$, and the sample size $n = 30$.

The Minitab results are shown in Figure 8.16, providing the sample size $n = 30$, the sample mean $\bar{x} = 24{,}017.7$, the sample standard deviation $s = 4322.5$, the standard error (SE mean) $s_{\bar{x}} = \frac{s}{\sqrt{n}} = \frac{4322.5}{\sqrt{30}} = 789.2$, and the 95% *t* confidence interval (22,403.7, 25,631.7).

One-Sample T: Small Business

Variable	N	Mean	StDev	SE Mean	95% CI
Small Business	30	24017.7	4322.5	789.2	(22403.7, 25631.7)

FIGURE 8.16 Minitab results.

The results from the WHFStat Add-ins for Excel are shown in Figure 8.17. Displayed are the sample mean $x = 24{,}017.7$, the sample size $n = 30$, the degrees of freedom $df = n - 1 = 29$, the sample standard deviation $s = 4322.474$, and the standard error

$$s_{\bar{x}} = \frac{s}{\sqrt{n}} = \frac{4322.474}{\sqrt{30}} \approx 789.1722.$$

SUMMARY STATISTICS		
Sample Mean	**Sample Size**	**Sample Standard Deviation**
24017.7	30	4322.474
	Deg of Freedom	**Standard Error (SE)**
	29	789.1722

ONE SAMPLE T CONFIDENCE INTERVAL			
Confidence Level			**Critical T Value**
95 %			2.045231
Confidence Interval			
24017.7	+/-	1614.039	
22403.66	to	25631.74	

FIGURE 8.17 Results from WHFStat Add-ins for Excel.

The confidence level 95% is shown, along with the critical *t* value, $t_{\alpha/2} = 2.045231$. The confidence interval is then shown in the form

$$\text{point estimate} \pm \text{margin of error}$$
$$= 24{,}017.7 \pm 1614.039$$

so the margin of error is

$$E = t_{\alpha/2}\,(s/\sqrt{n}) = 1614.039$$

The confidence interval is also shown as "22,403.66 to 25,631.74."

STEP-BY-STEP TECHNOLOGY GUIDE: *t* Confidence Intervals

We illustrate how to construct the *t* confidence interval for Example 8.13 (page 376).

TI-83/84

If you have the data values:
Step 1 Enter the data into list **L1**.
Step 2 Press **STAT**, highlight **TESTS**.
Step 3 Press **8** (for **TInterval**, see Figure 8.18).
Step 4 For input (**Inpt**), highlight **Data** and press **ENTER** (Figure 8.19).
a. For **List**, press **2nd** then **L1**.
b. For **Freq**, enter **1**.
c. For **C-Level** (confidence level), enter the appropriate confidence level (for example, **0.95**), and press **ENTER**.
d. Highlight **Calculate** and press **ENTER**. The results are shown in Figure 8.15 in Example 8.13.

If you have the summary statistics:
Step 1 Press **STAT**, highlight **TESTS**.
Step 2 Press **8** (for **TInterval**, see Figure 8.18).
Step 3 For input (**Inpt**), highlight **Stats** and press **ENTER** (Figure 8.20).
a. For $\bar{x}$, enter the sample mean 24017.7.
b. For **Sx**, enter the sample standard deviation 4322.473886.
c. For *n*, enter the sample size **30**.
d. For **C-Level** (confidence level), enter the appropriate confidence level (for example, **0.95**), and press **ENTER**.
e. Highlight **Calculate** and press **ENTER**. The results are shown in Figure 8.15 in Example 8.13.

FIGURE 8.18

FIGURE 8.19

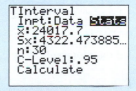

FIGURE 8.20

EXCEL

If you have the data values:
Step 1 Enter the data into column A.
Step 2 Load the **WHFStat Add-ins**.
Step 3 Select **Add-ins > Macros > Estimating a Mean > t Confidence Interval**.
Step 4 Click **Select Dataset Range**, highlight **A1–A30**, and click **OK**.
Step 5 Select the **95%** confidence level, and click **OK**. The results are shown in Figure 8.17 in Example 8.13.

If you have the summary statistics:
Step 1 Load the **WHFStat Add-ins**.
Step 2 Select **Add-ins > Macros > Estimating a Mean > Z Confidence Interval**.
Step 3 Click **Input Summary Statistics**, enter 24017.7 for the **Sample Mean**, enter **30** for the **Sample Size**, enter **4322.473886** for the **Sample Standard Deviation**, and click **OK**.
Step 4 Select the **95%** confidence level and click **OK**. The results are shown in Figure 8.17 in Example 8.13.

MINITAB

If you have the data values:
Step 1 Enter the data into column C1.
Step 2 Click **Stat > Basic Statistics > 1-Sample t**.
Step 3 Click **Samples in Columns** and select **C1**.
Step 4 Click **Options**, enter **95** as the **Confidence Level**, click **OK**, and click **OK** again.
The results are shown in Figure 8.16 in Example 8.13.

If you have the summary statistics:
Step 1 Click **Stat > Basic Statistics > 1-Sample t**.
Step 2 Click **Summarized Data**.
Step 3 Enter the **Sample Size 30**, the **Sample Mean 24017.7**, and **4322.473886** for the **Standard Deviation**.
Step 4 Click **Options**, enter **95** as the **Confidence Level**, click **OK**, and click **OK** again.
The results are shown in Figure 8.16 in Example 8.13.

CRUNCHIT!

If you have the data values:
Step 1 Click **File** . . . then highlight **Load from Larose2e** . . .
Chapter 8 . . . and click on **Example 8.13**.
Step 2 Click **Statistics** . . . **t** and select **1-sample**.
Step 3 With the **Columns** tab chosen, for **Sample** select
Businesses.
Step 4 Select the **Confidence Interval** tab, and enter **95** for the
Confidence Interval Level. Then click **Calculate**.

If you have the summary statistics:
Step 1 Click **Statistics** . . . **t** and select **1-sample**.
Step 2 Choose the **Summarized** tab. For **n** enter the sample
size **30**; for **Sample Mean** enter 24017.7. For **Standard Deviation**,
enter 4322.473886.
Step 3 Select the **Confidence Interval** tab, and enter **95** for the
Confidence Interval Level. Then click **Calculate**.

SECTION 8.2 Summary

1. For a normal population, the distribution of

$$t = \frac{\bar{x} - \mu}{s/\sqrt{n}}$$

follows a t distribution, with $n - 1$ degrees of freedom, where $\bar{x}$ is the sample mean, μ is the unknown population mean, s is the sample standard deviation, and n is the sample size. The t distribution is symmetric about its mean 0, just like the Z distribution. However, the t distribution is flatter.

2. A $100(1 - \alpha)\%$ confidence interval for μ is given by the interval

$$\bar{x} \pm t_{\alpha/2} \, (s/\sqrt{n})$$

where $\bar{x}$ is the sample mean, $t_{\alpha/2}$ is associated with the confidence level and $n - 1$ degrees of freedom, s is the sample standard deviation, and n is the sample size. We can construct a t interval whenever *either* of the following conditions is met: the population is normal, or the sample size is large ($n \geq 30$).

SECTION 8.2 Exercises

Clarifying the Concepts

1. Why do we need the t interval? Why can't we always use Z intervals?

2. Suppose that σ is known. Can we still use a t interval?

3. As the sample size gets larger and larger, what happens to the t curve?

4. State the formula for the margin of error for the t interval.

Practicing the Techniques

5. For the following scenarios, we are taking a random sample from a normal population with σ unknown. Find $t_{\alpha/2}$.
  a. Confidence level 90%, sample size 10
  b. Confidence level 95%, sample size 10
  c. Confidence level 99%, sample size 10

6. For the following scenarios we are taking a random sample from a normal population with σ unknown. Find $t_{\alpha/2}$.
  a. Confidence level 95%, sample size 10
  b. Confidence level 95%, sample size 15
  c. Confidence level 95%, sample size 20

7. Refer to Exercise 5.
  a. Describe what happens to the value of $t_{\alpha/2}$, as the confidence level increases, for a given sample size.

  b. Draw a sketch of the t curve for sample size $n = 10$, and explain why the value of $t_{\alpha/2}$ changes as it does.

8. Refer to Exercise 6.
  a. Describe what happens to the value of $t_{\alpha/2}$, as the sample size increases, for a given confidence level.
  b. Draw a sketch of the t curve for a confidence level of 95%, and explain why the value of $t_{\alpha/2}$ changes as it does.

For the data sets shown in Exercises 9–12, do the following.
  a. Calculate $\bar{x}$ and s.
  b. Find $t_{\alpha/2}$.
  c. Construct and interpret a 95% confidence interval for μ.

9.

2	3	1	3	1

10.

8	4	6	4	8

11.

11	17	14	17	11

12.

96	104	100	96	104

For Exercises 13–18, we are taking a random sample from a normal population with σ unknown.
  a. Find $t_{\alpha/2}$.
  b. Construct the confidence interval for μ with the indicated confidence level.
  c. Sketch the confidence interval on a number line.

13. Confidence level 95%, sample size 25, sample mean 10, sample standard deviation 5

14. Confidence level 90%, sample size 9, sample mean 22, sample standard deviation 3

15. Confidence level 95%, $n = 4, \bar{x} = 50, s = 6$

16. Confidence level 99%, $n = 16, \bar{x} = 0, s = 8$

17. Confidence level 90%, $n = 9, \bar{x} = -20, s = 6$

18. Confidence level 95%, $n = 25, \bar{x} = 0, s = 15$

For Exercises 19–24, we are taking a random sample from a population with σ unknown. However, do not assume that the population is normally distributed.
  a. Find $t_{\alpha/2}$.
  b. Construct the confidence interval for μ with the indicated confidence level.
  c. Sketch the confidence interval on a number line.

19. Confidence level 95%, sample size 100, sample mean 100, sample standard deviation 10.

20. Confidence level 90%, sample size 64, sample mean 250, sample standard deviation 20.

21. Confidence level 99%, $n = 64, \bar{x} = 35, s = 8$

22. Confidence level 95%, $n = 400, \bar{x} = 42, s = 10$

23. Confidence level 90%, $n = 81, \bar{x} = -20, s = 6$

24. Confidence level 95%, $n = 225, \bar{x} = 0, s = 15$

For Exercises 25–28, find the value of $t_{\alpha/2}$ using the following methods.
  a. The conservative approach
  b. Interpolation

25. Confidence level 95%, $n = 55, \bar{x} = 100, s = 15$

26. Confidence level 99%, $n = 117, \bar{x} = 100, s = 15$

27. Confidence level 90%, $n = 46, \bar{x} = 10, s = 2$

28. Confidence level 95%, $n = 46, \bar{x} = 10, s = 2$

For Exercises 29–32, assume the data come from a normal distribution. Calculate and interpret the confidence interval for μ using the value of $t_{\alpha/2}$ you found in the indicated exercises.

29. Exercise 25
  a. Part (a)   **b.** Part (b)

30. Exercise 26
  a. Part (a)   **b.** Part (b)

31. Exercise 27
  a. Part (a)   **b.** Part (b)

32. Exercise 28
  a. Part (a)   **b.** Part (b)

For each of Exercises 33–40, we are taking a random sample from a population with σ unknown. If the conditions are met, construct the indicated t interval for μ. If not, explain why not.

33. Confidence level 95%, $n = 25, \bar{x} = 100, s = 10$

34. Confidence level 90%, $n = 16, \bar{x} = 250, s = 20$

35. Confidence level 95%, $n = 225, \bar{x} = 10, s = 5$, normal population

36. Confidence level 90%, $n = 81, \bar{x} = 22, s = 3$

37. Confidence level 99%, $n = 16, \bar{x} = 35, s = 8$

38. Confidence level 95%, $n = 25, \bar{x} = 42, s = 10$, normal population

39. Confidence level 95%, $n = 36, \bar{x} = 50, s = 6$

40. Confidence level 99%, $n = 64, \bar{x} = 0, s = 8$

For Exercises 41–48, calculate and interpret the margin of error for the confidence interval from the indicated exercise.

41. Exercise 9   **45.** Exercise 13

42. Exercise 10   **46.** Exercise 14

43. Exercise 11   **47.** Exercise 15

44. Exercise 12   **48.** Exercise 16

Applying the Concepts

49. Sickle-Cell Anemia. The U.S. Department of Health and Human Services reports that the mean length of stay in hospital for sickle-cell anemia patients in 2008 was $\bar{x} = 5.3$ days with a standard deviation of $s = 7.6$ days. For a sample of 100 patients, do the following.
  a. Find $t_{\alpha/2}$ for a confidence interval with 95% confidence.
  b. Construct and interpret a 95% confidence interval for the population mean length of stay for all sickle-cell anemia patients.

50. Student Loans. The Pew Research Center (pewresearch.org) reports that the mean student loan amount in 2008 was $15,425 for students obtaining a bachelor's degree. Suppose a sample of 400 students had a sample mean loan amount of $15,425 and a sample standard deviation student loan amount of $20,000. Do the following.
  a. Find $t_{\alpha/2}$ for a confidence interval with 90% confidence.
  b. Construct and interpret a 90% confidence interval for the population mean student loan amount for all students obtaining a bachelor's degree.

51. Parking Meters. A tried-and-true revenue stream for large cities has been the funds collected from parking meters. A random sample of 75 parking meters yielded a mean of $120 per meter with a standard deviation of $30.
  a. Find $t_{\alpha/2}$ for a confidence interval with 95% confidence.

b. Construct and interpret a 95% confidence interval for the population mean revenue collected from all parking meters.

52. Teachers Graded. A 2007 study reported in *Science* magazine stated that fifth-grade teachers scored a mean of 3.4 (out of 7) points for "providing evaluative feedback to students on their work."[9] Assume that the sample size was 36 and the sample standard deviation was 1.5.

   a. Find $t_{\alpha/2}$ for a confidence interval with 90% confidence.

   b. Construct and interpret a 90% confidence interval for the population mean points scored by fifth-grade teachers for providing evaluative feedback.

53. Sickle-Cell Anemia. Refer to Exercise 49.

   a. Calculate and interpret the margin of error.

   b. If the sample size is increased to 400, describe what will happen to the margin of error.

54. Student Loans. Refer to Exercise 50.

   a. Calculate and interpret the margin of error.

   b. If the sample size is decreased to 100, describe what will happen to the margin of error.

55. Parking Meters. Refer to Exercise 51.

   a. Compute the margin of error and interpret it.

   b. Describe two ways of reducing this margin of error. Which method is more desirable, and why?

56. Teachers Graded. Refer to Exercise 52.

   a. Compute the margin of error and interpret it.

   b. Describe two ways of reducing this margin of error. Which method is more desirable, and why?

For Exercises 57–60, the normality of the data was confirmed in the Section 8.1 exercises.

57. Carbon Emissions. The following table represents the carbon emissions (in millions of tons) from consumption of fossil fuels, for a random sample of 5 nations.[10]

 carbon

Nation	Emissions
Brazil	361
Germany	844
Mexico	398
Great Britain	577
Canada	631

   a. Construct and interpret a 90% *t* confidence interval for the population mean carbon emissions.

   b. Calculate and interpret the margin of error for the confidence interval in part **(a)**.

   c. Explain two ways we could decrease the margin of error. Which method is preferable, and why?

58. *Deepwater Horizon* Cleanup Costs. The following table represents the amount of money disbursed by BP to a random sample of 6 Florida counties, for cleanup of the *Deepwater Horizon* oil spill, in millions of dollars.[11]

deepwaterclean

County	Cleanup costs ($ millions)
Broward	0.85
Escambia	0.70
Franklin	0.50
Pinellas	1.15
Santa Rosa	0.50
Walton	1.35

   a. Construct and interpret a 95% *t* confidence interval for the population mean cleanup cost.

   b. Calculate and interpret the margin of error for the confidence interval in part **(a)**.

   c. Explain two ways we could decrease the margin of error. Which method is preferable, and why?

59. Wii Game Sales. The following table represents the number of units sold in the United States for the week ending March 26, 2011, for a random sample of 8 Wii games.[12]

wiisales

Game	Units (1000s)	Game	Units (1000s)
Wii Sports Resort	65	Zumba Fitness	56
Super Mario All Stars	40	Wii Fit Plus	36
Just Dance 2	74	Michael Jackson	42
New Super Mario Bros.	16	Lego Star Wars	110

   a. Construct and interpret a 99% confidence interval for the population mean number of units sold.

   b. Calculate and interpret the margin of error for the confidence interval in part **(b)**.

   c. How could we increase the precision of our confidence interval without decreasing the confidence level?

60. A Rainy Month in Georgia? The following table represents the total rainfall (in inches) for the month of February 2011 for a random sample of 10 locations in Georgia.[13]

georgiarain

Location	Rainfall (inches)	Location	Rainfall (inches)
Athens	4.72	Atlanta	4.25
Augusta	4.31	Cartersville	3.03
Dekalb	2.96	Fulton	4.36
Gainesville	4.06	Lafayette	3.75
Marietta	3.20	Rome	3.26

a. Construct and interpret a 95% confidence interval for the population mean rainfall in inches.
b. Calculate and interpret the margin of error for the confidence interval in part (b).
c. How could we increase the precision of our confidence interval without decreasing the confidence level?

61. Hybrid Car Gas Mileage. The accompanying table shows the city gas mileage for 6 hybrid cars, as reported by the Environmental Protection Agency and **www.hybridcars.com** in 2007.

hybridmiles

Vehicle	Mileage (mpg)
Honda Accord	30
Ford Escape (2wd)	36
Toyota Highlander	33
Saturn VUE Green Line	27
Lexus RX 400h	31
Lexus GS 450h	25

a. Use technology to construct a normal probability plot of the gas mileages. Confirm that the distribution appears to be normal.
b. Find $t_{\alpha/2}$ for a confidence interval with 90% confidence.
c. Compute and interpret the margin of error E for a confidence interval with 90% confidence.
d. Construct and interpret a 90% confidence interval (t interval) for the population mean mileage.

62. Hybrid Car Gas Mileage II. The table contains the complete listing of 12 hybrid vehicle gas mileages shown on **www.hybridcars.com** in 2007.

hybridmiles2

Vehicle	Mileage (mpg)
Honda Insight	61
Toyota Prius	60
Honda Civic	50
Toyota Camry	43
Honda Accord	30
Ford Escape (2wd)	36
Ford Escape	33
Mercury Mariner	33
Toyota Highlander	33
Saturn VUE Green Line	27
Lexus RX 400h	31
Lexus GS 450h	25

a. Use technology to construct a normal probability plot of the gas mileages.
b. Is there evidence that the distribution is not normal?
c. Can you proceed to construct a t interval? Why or why not?

63. Calories in Breakfast Cereals. What is the mean number of calories in a bowl of breakfast cereal? A random sample of 6 well-known breakfast cereals yielded the following calorie data.

cerealcalories

Cereal	Calories
Apple Jacks	110
Cocoa Puffs	110
Mueslix	160
Cheerios	110
Corn Flakes	100
Shredded Wheat	80

a. Use technology to construct a normal probability plot of the number of calories.
b. Is there evidence that the distribution is not normal?
c. Can we proceed to construct a t interval? Why or why not?

64. Commuting Distances. A university is trying to attract more commuting students from the local community. As part of the research into the modes of transportation students use to commute to the university, a survey was conducted asking how far commuting students commuted from home to school each day. A random sample of 30 students provided the distances (in miles) shown.

commutedist

14	10	14	12	12	11	5	6	9	14	9	9	4	7	15
9	7	7	12	10	15	10	6	11	9	11	10	11	7	12

a. Find $t_{\alpha/2}$ for a confidence interval with 95% confidence.
b. Compute and interpret the margin of error for a confidence interval with 95% confidence.
c. Construct and interpret a 95% t confidence interval for the population mean commuting distance.

65. Consider the confidence interval we found for the fourth-graders' foot lengths in Example 8.10. *What if* we increased the sample size to some unspecified value but everything else stayed the same. Describe what, if anything, would happen to each of the following measures and why.
a. $t_{\alpha/2}$
b. Margin of error E
c. Width of the confidence interval

Bringing It All Together

Cigarette Consumption. Use the following information for Exercises 66–71. Health officials are interested in estimating the population mean number of cigarettes smoked annually per capita in order to evaluate the efficacy of their antismoking campaign. A random sample of 8 U.S. counties yielded the following numbers of cigarettes smoked per capita: 2206, 2391, 2540, 2116, 2010, 2791, 2392, 2692.

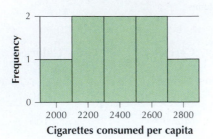

66. Evaluate the normality assumption using the accompanying histogram. Is it appropriate to construct a t interval using this data set? Why or why not? What is it about the histogram that tells you one way or the other?

67. Find the point estimate of μ, the population mean number of cigarettes smoked per capita.

68. Compute the sample standard deviation s.

69. Find $t_{\alpha/2}$ for a confidence interval with 90% confidence.

70. Compute and interpret the margin of error E for a confidence interval with 90% confidence. What is the meaning of this number?

71. Construct and interpret a 90% confidence interval for the population mean number of cigarettes smoked per capita.

8.3   *Z* INTERVAL FOR THE POPULATION PROPORTION

OBJECTIVES   By the end of this section, I will be able to . . .

1. Calculate the point estimate $\hat{p}$ of the population proportion p.
2. Construct and interpret a Z interval for the population proportion p.
3. Compute and interpret the margin of error for the Z interval for p.
4. Determine the sample size needed to estimate the population proportion.

1 POINT ESTIMATE $\hat{p}$ OF THE POPULATION PROPORTION p

So far we have dealt with interval estimates of the population mean μ only. However, we may also be interested in an interval estimate for the population proportion of successes, p. Recall from Section 7.3 that the sample proportion of successes

$$\hat{p} = \frac{x}{n} = \frac{\text{number of successes}}{\text{sample size}}$$

is a point estimate of the population proportion p.

EXAMPLE 8.14   **COMMUNITY COLLEGE SURVEY OF STUDENT ENGAGEMENT**

Collaborative learning in college helps students prepare for life in the business world, where employees are required to work together in teams. The Community College Survey of Student Engagement reports on the proportion of students who have worked with classmates outside class to prepare a group assignment during the current academic year.[14] Suppose that a random sample of 300 students is polled, and 174 students respond that they did indeed work on a group project this year. Calculate the point estimate $\hat{p}$ of the population proportion p.

Solution

We have $n = 300$ students and $x = 174$. Thus

$$\hat{p} = \frac{x}{n} = \frac{174}{300} = 0.58$$

The point estimate of the population proportion p of community college students who have worked with classmates outside class to prepare a group assignment during the current academic year is $\hat{p} = 0.58$.

**Now You Can Do
Exercises 3–6.**

Of course, different samples of community college students may turn up different sample proportions $\hat{p}$. These are point estimates, and thus they carry no measure of confidence in their accuracy. The point estimates are probably close to the true values, but it's possible that they are not. They may be far from the true values. Only by using confidence intervals can we make probability statements about the accuracy of the estimates.

2   Z INTERVAL FOR THE POPULATION PROPORTION p

Recall the Central Limit Theorem for Proportions in Section 7.3.

> **Central Limit Theorem for Proportions**
>
> The sampling distribution of the sample proportion $\hat{p}$ follows an approximately normal distribution with mean $\mu_{\hat{p}} = p$ and standard deviation $\sigma_{\hat{p}} = \sqrt{\frac{p \cdot q}{n}}$ when *both* the following conditions are satisfied: (1) $n \cdot p \geq 5$ and (2) $n \cdot q \geq 5$ where $q = 1 - p$.

We can use the Central Limit Theorem for Proportions to construct confidence intervals for the population proportion p. Because the confidence interval for p is based on the standard normal Z distribution, it is called the **Z interval for the population proportion p.** Because p is unknown, the conditions and the formula for $\sigma_{\hat{p}}$ substitute $\hat{p}$ for p.

> **Z Interval for p**
>
> The Z interval for p may be performed only if *both* the following conditions are met: $n \cdot \hat{p} \geq 5$ and $n \cdot \hat{q} \geq 5$. When a random sample of size n is taken from a binomial population with unknown population proportion p, the $100(1 - \alpha)\%$ confidence interval for p is given by
>
> $$\text{lower bound} = \hat{p} - Z_{\alpha/2} \sqrt{\frac{\hat{p} \cdot \hat{q}}{n}}$$
>
> $$\text{upper bound} = \hat{p} + Z_{\alpha/2} \sqrt{\frac{\hat{p} \cdot \hat{q}}{n}}$$
>
> Alternatively,
>
> $$\hat{p} \pm Z_{\alpha/2} \sqrt{\frac{\hat{p} \cdot \hat{q}}{n}}$$
>
> where $\hat{p}$ is the sample proportion of successes, $\hat{q} = 1 - \hat{p}$, n is the sample size, and $Z_{\alpha/2}$ depends on the confidence level.

For convenience, we repeat Table 8.1 here, showing the $Z_{\alpha/2}$ values for the most common confidence levels.

Table 8.1 $Z_{\alpha/2}$ values for common confidence levels

Confidence level	α	$\alpha/2$	$Z_{\alpha/2}$
90%	0.10	0.05	1.645
95%	0.05	0.025	1.96
99%	0.01	0.005	2.576

EXAMPLE 8.15

Z INTERVAL FOR THE POPULATION PROPORTION p

Note that the population is binomial because each student either (a) has worked with classmates in this way or (b) has not.

Using the survey data from Example 8.14, (a) verify that the conditions for constructing the Z interval for p have been met, and (b) construct a 95% confidence interval for the population proportion of community college students who have worked with classmates outside class to prepare a group assignment during the current academic year.

Solution

a. We have $n = 300$ students and $x = 174$. We check the conditions for the confidence interval:

$$n \cdot \hat{p} = (300) \cdot (0.58) = 174 \geq 5 \quad \text{and} \quad n \cdot \hat{q} = (300) \cdot (0.42) = 126 \geq 5.$$

The conditions for constructing the Z interval for p have been met.

b. From Table 8.1, the confidence level of 95% gives $Z_{\alpha/2} = 1.96$. Thus, the confidence interval is

$$\text{lower bound} = \hat{p} - Z_{\alpha/2} \sqrt{\frac{\hat{p} \cdot \hat{q}}{n}} = 0.58 - 1.96 \sqrt{\frac{0.58(0.42)}{300}}$$

$$= 0.58 - 1.96(0.0284956137) \approx 0.58 - 0.05585 = 0.52415$$

$$\text{upper bound} = \hat{p} + Z_{\alpha/2} \sqrt{\frac{\hat{p} \cdot \hat{q}}{n}} = 0.58 + 1.96 \sqrt{\frac{0.58(0.42)}{300}}$$

$$= 0.58 + 1.96(0.0284956137) \approx 0.58 + 0.05585 = 0.63585$$

We are 95% confident that the population proportion of community college students who have worked with classmates outside class to prepare a group assignment during the current academic year lies between 0.52415 and 0.63585. (See Figure 8.21.)

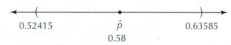

| 0.52415 | $\hat{p}$ | 0.63585 |
| | 0.58 | |

Now You Can Do
Exercises 7–22.

FIGURE 8.21 95% Confidence interval for the population proportion of community college students who have worked with classmates outside class to prepare a group assignment.

EXAMPLE 8.16

Z INTERVALS FOR p USING TECHNOLOGY

A 2005 poll by the Center for Social Research at Stony Brook University asked, "Should high school athletes who test positive for steroids or other performance-enhancing drugs be banned from high school athletic teams, or not?" Of the 830 randomly selected respondents, 631 responded, "Yes, they should be banned." Use technology to find a 95% confidence interval for the population proportion of all Americans who think such athletes should be banned.

Solution

We use the instructions provided in the Step-by-Step Technology Guide at the end of this section (page 389). The results for the TI-83/84 in Figure 8.22 display the 95% confidence interval for the population proportion of Americans who think such athletes should be banned to be

$$(\text{lower bound} = 0.7312, \text{upper bound} = 0.78929)$$

FIGURE 8.22 TI-83/84 results.

They also show the sample proportion $\hat{p} = 0.7602409639$ and the sample size $n = 830$.

The results for Minitab are shown in Figure 8.23. At this point, we consider only the statistics in **blue**. The remaining material will be explained in Chapter 9. Minitab provides the sample number of successes $X = 631$, the sample size $n = 830$, the sample proportion $\hat{p} = 0.7602409639$ (rounded to 0.760241), and the 95% confidence interval for p (0.731196, 0.789286).

```
Test and CI for One Proportion
Test of p = 0.5 vs p not = 0.5

Sample    X    N   Sample p        95% CI          Z-Value  P-Value
1        631  830  0.760241  (0.731196, 0.789286)   14.99    0.000
```

FIGURE 8.23 Minitab results for the Z interval for p.

3   MARGIN OF ERROR FOR THE *Z* INTERVAL FOR *p*

For the Z interval for the population proportion p, the margin of error is given as follows.

> **Margin of Error for the Z Interval for p**
> $$E = Z_{\alpha/2} \cdot \sqrt{\frac{\hat{p} \cdot \hat{q}}{n}}$$
> The margin of error E for a $(1 - \alpha)100\%$ Z interval for p can be interpreted as follows:
> "We can estimate p to within E with $(1 - \alpha)100\%$ confidence."

Note that, just like the confidence interval for μ, the Z interval for p takes the form

$$\text{point estimate} \pm \text{margin of error}$$

$$= \hat{p} \pm Z_{\alpha/2} \sqrt{\frac{\hat{p} \cdot \hat{q}}{n}}$$

$$= \hat{p} \pm E$$

EXAMPLE 8.17

POLLS AND THE FAMOUS "PLUS OR MINUS 3 PERCENTAGE POINTS"

There is hardly a day that goes by without some new poll coming out. Especially during election campaigns, polls influence the choice of candidates and the direction of their policies. For example, the Gallup Organization polled 1012 American adults, asking them, "Do you think there should or should not be a law that would ban the possession of handguns, except by the police and other authorized persons?" Of the 1012 randomly chosen respondents, 638 said that there should NOT be such a law.
a.   Check that the conditions for the Z interval for p have been met.
b.   Find and interpret the margin of error E.
c.   Construct and interpret a 95% confidence interval for the population proportion of all American adults who think there should not be such a law.

Solution

The sample size is $n = 1012$. The observed proportion is $\hat{p} = \dfrac{638}{1012} \approx 0.63$, so $\hat{q}\,(1 - \hat{p}) = 0.37$.
a.   We next check the conditions for the confidence interval:

$$n \cdot \hat{p} = (1012) \cdot (0.63) = 637.56 \geq 5 \quad \text{and} \quad n \cdot \hat{q} = (1012) \cdot (0.37) = 374.44 \geq 5$$

b. The confidence level of 95% implies that our $Z_{\alpha/2}$ equals 1.96 (from Table 8.1). Thus, the margin of error equals

$$E = Z_{\alpha/2} \cdot \sqrt{\frac{\hat{p} \cdot \hat{q}}{n}} = 1.96 \cdot \sqrt{\frac{0.63(0.37)}{1012}} \approx 0.02975 \approx 0.03$$

c. The 95% confidence interval is

point estimate $\pm$ margin of error

$$= \hat{p} \pm Z_{\alpha/2} \sqrt{\frac{\hat{p} \cdot \hat{q}}{n}}$$

$$= \hat{p} \pm E$$

$$\approx 0.63 \pm 0.03$$

$$= (\text{lower bound} = 0.60, \text{ upper bound} = 0.66)$$

**Now You Can Do
Exercises 23–34.**

Thus, we are 95% confident that the population proportion of all American adults who think that there should not be such a law lies between 60% and 66%.

**Developing Your
Statistical Sense**

Famous "Plus or Minus 3 Points"

Note that this confidence interval was obtained by adding and subtracting 3% from the 63% point estimate. That is, the poll has a margin of error of $E = 3$ percentage points $= 0.03$. This is the famous "plus or minus 3 percentage points" used in many news reports. However, newscasters rarely announce the confidence level of the poll. National pollsters almost always use 95% as their confidence level and usually try to select the sample size necessary to create a margin of error of about 3%. We learn how they do this next.

4 SAMPLE SIZE FOR ESTIMATING THE POPULATION PROPORTION

Next we consider the question: How large a sample size do I need to estimate the population proportion p to within margin of error E with $100(1 - \alpha)\%$ confidence? The margin of error of the confidence interval for proportions equals

$$E = Z_{\alpha/2} \cdot \sqrt{\frac{\hat{p} \cdot \hat{q}}{n}}$$

Solving for n gives us

$$n = \hat{p} \cdot \hat{q} \left(\frac{Z_{\alpha/2}}{E}\right)^2 \qquad \text{(Equation 8.1)}$$

Unfortunately, Equation 8.1 depends on prior knowledge of $\hat{p}$. So, if we have such information about $\hat{p}$ available from some earlier sample, then we use Equation 8.1 to determine the required sample. However, what if we do not know the value of $\hat{p}$?

Figure 8.24 plots the sample size requirements for a 95% confidence interval for p, with a desired margin of error of 0.03, for values of $\hat{p}$ ranging from 0.01 to 0.99, representing all sample proportions from 1% to 99%. Note that the plot is symmetric, and therefore the largest required sample size occurs at the midpoint $\hat{p} = 0.5$. Thus, $\hat{p} = 0.5$ is the most conservative value for $\hat{p}$. When the actual value of $\hat{p}$ is not known, we use the following formula:

$$n = \left(\frac{0.5 \cdot Z_{\alpha/2}}{E}\right)^2$$

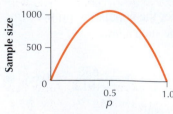

FIGURE 8.24 Sample size required for the range of values p.

> **Sample Size for Estimating a Population Proportion**
>
> When $\hat{p}$ is known, the sample size needed to estimate the population proportion p to within a margin of error E with confidence $100(1 - \alpha)\%$ is given by
>
> $$n = \hat{p} \cdot \hat{q} \left[\frac{Z_{\alpha/2}}{E} \right]^2$$
>
> where $Z_{\alpha/2}$ is the value associated with the desired confidence level, E is the desired margin of error, and $\hat{p}$ is the sample proportion of successes available from some earlier sample and $\hat{q} = 1 - \hat{p}$. Round up to the next integer.
>
> When $\hat{p}$ is unknown, we use
>
> $$n = \left[\frac{0.5 \cdot Z_{\alpha/2}}{E} \right]^2$$

These formulas are illustrated using the following two examples.

EXAMPLE 8.18

SAMPLE SIZE FOR ESTIMATING A SAMPLE PROPORTION WHEN $\hat{p}$ IS KNOWN

Refer to Example 8.17. Suppose that the Gallup Organization now wanted to estimate the population proportion of those who think there should not be a law that would ban the possession of handguns to within a margin of error of $E = 0.01$ with 95% confidence. How large a sample size is needed?

Solution

From Example 8.17, we have the sample proportion $\hat{p} = 0.63$. The confidence level of 95% implies that our $Z_{\alpha/2} = 1.96$, and the desired margin of error is $E = 0.01$. Thus, the required sample size is

$$n = \hat{p} \cdot \hat{q} \left(\frac{Z_{\alpha/2}}{E} \right)^2 = 0.63(0.37) \left(\frac{1.96}{0.01} \right)^2 \approx 8954.77$$

Now You Can Do Exercises 35–40.

Rounding up, this gives us a required sample size of 8955. The smaller margin of error requires a larger sample size.

EXAMPLE 8.19

REQUIRED SAMPLE SIZE FOR POLLS

Suppose the Dimes-Newspeak organization would like to take a poll on the proportion of Americans who will vote Republican in the next presidential election. How large a sample size does the Dimes-Newspeak organization need to estimate the proportion to within plus or minus 3 percentage points ($E = 0.03$) with 95% confidence?

Solution

The 95% confidence implies that the value for $Z_{\alpha/2}$ is 1.96. Since there is no information available about the value of the population proportion of all Americans who will vote Republican in the next election, we use 0.5 as our most conservative value of p:

$$n = \left[\frac{0.5 \cdot Z_{\alpha/2}}{E} \right]^2 = \left[\frac{(0.5)(1.96)}{0.03} \right]^2 \approx 1067.11$$

Now You Can Do Exercises 41–48.

So if the pollsters would like to estimate the population proportion of all American voters who will vote Republican in the upcoming election to within 3% with 95% confidence, they will need a sample of 1068 voters (don't forget to round up!).

AP Photo/Tertius Pickard

AP Photo/U.S. Coast Guard

Health Effects of the *Deepwater Horizon* Oil Spill

The *Deepwater Horizon* oil drilling platform exploded on April 20, 2010, killing 11 workers and causing the largest oil spill in American history. Many Americans participated in the cleanup of coastal property and wildlife habitat, including the cleaning of the wildlife. The National Institute for Occupational Safety and Health (NIOSH) conducted a randomly sampled survey of 54 of these workers who were exposed to oil, dispersant, cleaners, and other chemicals. Of these 54 workers, 25 reported skin problems, such as itchy skin or rash, as a result of exposure to these chemicals. Suppose we are interested in constructing a 95% confidence interval for the population proportion of all wildlife workers who reported such skin problems.

a. What is the point estimate of p, the population proportion of workers reporting skin problems?

b. Are the conditions met for constructing the desired confidence interval?

c. What is the critical value $Z_{\alpha/2}$?

d. Calculate the margin of error $E = Z_{\alpha/2} \cdot \sqrt{\dfrac{\hat{p} \cdot \hat{q}}{n}}$. Interpret the margin of error.

e. Express the confidence interval for p in terms of the values for the point estimate $\pm$ the margin of error.

f. Calculate the lower and upper bounds for the confidence interval. Interpret the confidence interval.

g. How large a sample size would be needed to estimate the population proportion of all wildlife workers who reported such skin problems to within 0.1330 with 95% confidence? Comment on your answer.

h. Suppose we now want the estimate to be within 0.1330 with 99% confidence rather than 95%. Will the required sample size be larger or smaller and why? Verify your statement by finding the required sample size.

Solution

a. Of the 54 workers, 25 reported skin problems, so the point estimate of p, the population proportion of workers reporting skin problems, is $\hat{p} = 25/54 \approx 0.4630$.

b. The conditions for constructing the confidence interval for p have been met, since

$$n \cdot \hat{p} = (54)(0.4630) \approx 25 \geq 5 \qquad \text{and} \qquad n \cdot \hat{q} = (54)(0.5370) \approx 29 \geq 5.$$

c. For confidence level 95%, we have from Table 8.1 (page 358), $Z_{\alpha/2} = 1.96$.

d. The margin of error is:

$$E = Z_{\alpha/2} \cdot \sqrt{\frac{\hat{p} \cdot \hat{q}}{n}} \approx (1.96) \cdot \sqrt{\frac{0.4630(0.5370)}{54}} \approx 0.1330.$$

We interpret this as follows: "We can estimate the population proportion of workers reporting skin problems to within 0.1330 with 95% confidence."

e. We may express our confidence interval in terms of point estimate $\pm$ the margin of error, as follows: 0.4630 ± 0.1330.

f. Lower bound $= 0.4630 - 0.1330 = 0.3300$,   Upper bound $= 0.4630 + 0.1330 = 0.5960$

We are 95% confident that the population proportion of all wildlife cleanup workers who suffered from skin problems lies between 0.3300 and 0.5960.

g. We have

$$n = \hat{p} \cdot \hat{q}\left(\frac{Z_{\alpha/2}}{E}\right) = 0.4630(0.5370)\left(\frac{1.96}{0.1330}\right)^2 \approx 53.9963$$

which rounds up to $n = 54$. This is precisely the sample size that we originally had, which did in fact give us precisely this margin of error of $E = 0.1330$.

h. Because the confidence level has increased while all other quantities have stayed the same, the required sample size will also increase. We now have

$$n = \hat{p} \cdot \hat{q}\left(\frac{Z_{\alpha/2}}{E}\right) = 0.4630(0.5370)\left(\frac{2.576}{0.1330}\right)^2 \approx 93.2704$$

which rounds up to $n = 94$, a larger required sample size for a larger desired confidence level. ■

STEP-BY-STEP TECHNOLOGY GUIDE: *Z* Confidence Intervals for *p*

We illustrate how to construct the *Z* confidence interval for *p* from Example 8.16 (page 384).

TI-83/84

Step 1   Press **STAT** and highlight **TESTS**.
Step 2   Scroll down to **A** (for **1-PropZInt**, see Figure 8.25), and press **ENTER**.
Step 3   For **x**, enter the number of success, **631**.
Step 4   For **n**, enter the sample size **830**.
Step 5   For **C-Level** (confidence level), enter the appropriate confidence level (e.g., **0.95**), and press **ENTER** (Figure 8.26).
Step 6   Highlight **Calculate** and press **ENTER**. The results are shown in Figure 8.22 in Example 8.16.

FIGURE 8.25

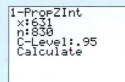

FIGURE 8.26

MINITAB

Step 1   Click **Stat > Basic Statistics > 1-Proportion**.
Step 2   Click **Summarized Data**.
Step 3   Enter the **Number of Trials** (n) **830** and the **Number of Events** (X) **631**.

Step 4   Click on **Options**, enter **95** as the **Confidence Level**, select **Use test and interval based on normal distribution**, and click **OK**. Then click **OK** again.
The results are shown in Figure 8.23 in Example 8.16.

CRUNCHIT!

Step 1   Click **Statistics . . . Proportion** and select **1-sample**.
Step 2   Choose the **Summarized** tab. For **n** enter the number of trials **830**; for **Successes** enter **631**.

Step 3   Select the **Confidence Interval** tab, and enter **95** for the **Confidence Interval Level**. Then click **Calculate**.

SECTION 8.3   Summary

1. The sample proportion of successes

$$\hat{p} = \frac{x}{n} = \frac{\text{number of successes}}{\text{sample size}}$$

is a point estimate of the population proportion *p*.

2. The $100(1 - \alpha)\%$ confidence interval for the population proportion *p* is given by

$$\hat{p} \pm Z_{\alpha/2}\sqrt{\frac{\hat{p} \cdot \hat{q}}{n}}$$

where $\hat{p}$ is the sample proportion of successes $\hat{q} = 1 - \hat{p}$, *n* is the sample size, and $Z_{\alpha/2}$ depends on the confidence level. The *Z* interval for *p* may be constructed only if *both* the following conditions apply: $n \cdot \hat{p} \geq 5$ and $n \cdot \hat{q} \geq 5$.

3. Note that the confidence interval for *p* takes on the form

point estimate $\pm$ margin of error

where $\hat{p}$ is the point estimate of p and $E = Z_{\alpha/2} \sqrt{\hat{p} \cdot \hat{q}/n}$ is the margin of error.

4. Suppose we would like to estimate the population proportion p to within a margin of error E with confidence $100(1 - \alpha)\%$. If $\hat{p}$ is known, then the required sample size needed is given by

$$n = \hat{p} \cdot \hat{q} \left(\frac{Z_{\alpha/2}}{E} \right)^2$$

If $\hat{p}$ is not known, then the required sample size needed is given by

$$n = \left(\frac{0.5 \cdot Z_{\alpha/2}}{E} \right)^2$$

SECTION 8.3 Exercises

Clarifying the Concepts

1. Suppose the population proportion of successes p is known. Is it useful to construct a confidence interval for p?

2. A news broadcast mentions that the sample size of a poll is about 1000 and that the margin of error is plus or minus 3 percentage points. How do we know that the pollsters are using a 95% confidence level?

Practicing the Techniques

For Exercises 3–6, calculate the point estimate $\hat{p}$ of the population proportion p.

3. Sample size = 100, number of successes = 40

4. Sample size = 500, number of successes = 100

5. $n = 1000$, $x = 560$

6. $n = 10,000$, $x = 2057$

For Exercises 7–22, do the following:
   a. Find $Z_{\alpha/2}$.
   b. Determine whether the conditions for constructing a confidence interval for p are met.
   c. If the conditions are met, construct a confidence interval for p with the indicated confidence level.
   d. If the conditions are met, sketch the confidence interval using a graph similar to Figure 8.21.

7. Confidence level 95%, sample size 100, sample proportion 0.2

8. Confidence level 95%, sample size 100, sample proportion 0.1

9. Confidence level 95%, sample size 100, sample proportion 0.05

10. Confidence level 95%, sample size 100, sample proportion 0.04

11. Confidence level 90%, $n = 25$, $\hat{p} = 0.2$

12. Confidence level 95%, $n = 25$, $\hat{p} = 0.2$

13. Confidence level 99%, $n = 25$, $\hat{p} = 0.2$

14. Confidence level 95%, $n = 25$, $\hat{p} = 0.16$

15. Confidence level 95%, sample size 25, number of successes 12

16. Confidence level 90%, sample size 81, number of successes 8

17. Confidence level 99%, sample size 100, number of successes 50

18. Confidence level 99%, sample size 20, number of successes 1

19. Confidence level 95%, $n = 64$, $x = 26$

20. Confidence level 99%, $n = 144$, $x = 80$

21. Confidence level 90%, $n = 49$, $x = 18$

22. Confidence level 95%, $n = 15$, $x = 26$

For Exercises 23–34, calculate the margin of error for the confidence interval from the indicated exercise.

23. Exercise 7

24. Exercise 8

25. Exercise 9

26. Refer to Exercises 23–25.
   a. Write a sentence describing what happens to the margin of error as the sample proportion decreases, while the sample size and confidence level stay the same.
   b. What effect does the behavior you observed in (a) have on the width of the confidence interval?

27. Exercise 11

28. Exercise 12

29. Exercise 13

30. Refer to Exercises 27–29.
   a. Write a sentence describing what happens to the margin of error as the confidence level increases, while the sample size and the sample proportion stay the same.
   b. What effect does the behavior you observed in (a) have on the width of the confidence interval?

31. For the following samples, find the margin of error E for a 95% confidence interval for p.
   a. 5 successes in 10 trials
   b. 50 successes in 100 trials
   c. 500 successes in 1000 trials
   d. 5000 successes in 10,000 trials

32. For the following samples, find the margin of error E for a 95% confidence interval for p.
   a. 10 successes in 100 trials
   b. 20 successes in 100 trials
   c. 30 successes in 100 trials
   d. 40 successes in 100 trials
   e. 50 successes in 100 trials

33. Refer to Exercise 31.
   a. Write a sentence describing what happens to the margin of error as the sample size increases while $\hat{p}$ remains constant.
   b. What effect will the behavior you observed in **(a)** have on the width of the confidence interval?

34. Refer to Exercise 32.
   a. Write a sentence describing what happens to the margin of error as the sample proportion approaches 0.5 while the sample size remains constant.
   b. What effect will the behavior you observed in **(a)** have on the width of the confidence interval?

For Exercises 35–39, we are estimating p and we know the value of $\hat{p}$. Find the required sample size.

35. Confidence level 95%, margin of error 0.03, $\hat{p} = 0.3$

36. Confidence level 95%, margin of error 0.03, $\hat{p} = 0.7$

37. Confidence level 95%, margin of error 0.03, $\hat{p} = 0.1$

38. Confidence level 95%, margin of error 0.03, $\hat{p} = 0.01$

39. Confidence level 95%, margin of error 0.03, $\hat{p} = 0.001$

40. Using Exercises 37–39, describe what happens to the required sample size when $\hat{p}$ gets very small.

For Exercises 41–46, we are estimating p and we do not know the value of $\hat{p}$. Find the required sample size.

41. Confidence level 90%, margin of error 0.03

42. Confidence level 95%, margin of error 0.03

43. Confidence level 99%, margin of error 0.03

44. Confidence level 95%, margin of error 0.015

45. Confidence level 95%, margin of error 0.0075

46. Confidence level 95%, margin of error 0.00375

47. Using Exercises 41–43, describe what happens to the required sample size as the confidence level increases.

48. Using Exercises 44–46, describe what happens to the required sample size when the margin of error is halved and the confidence level stays constant.

Applying the Concepts

For Exercises 49–52, do the following.
   a. Find $Z_{\alpha/2}$.
   b. Determine whether the conditions are met for constructing a confidence interval for p.
   c. If the conditions are met, construct and interpret a confidence interval for p with the indicated confidence level, and sketch the confidence interval on the number line. If the conditions are not met, state why not.

49. Married Millennials. *Millennials* refers to the generation of young people aged 18–29 in 2010, because they are the first generation to come of age in the new millennium. A 2010 Pew Research Center study found that 183 of a sample of 830 American millennials were married. Use a 99% confidence level.

50. Rather Be Fishing? A study found that Minnesota, at 38%, leads the nation in the proportion of people who go fishing.[15] Assume that the study sample size was 100 and use a 95% confidence level.

51. Spring Break and Drinking. A study released by the American Medical Association found that 83% of college female respondents agreed that heavier drinking occurs on spring break trips than is typically found on campus. Assume that the sample size was 25 and use a 90% confidence level.

52. NASCAR Fans and Pickup Trucks. *American Demographics* magazine reported that 40% of a sample of NASCAR racing attendees said they owned a pickup truck. Suppose the sample size was 1000. Construct a 95% confidence interval for the population proportion of NASCAR racing attendees who own a pickup truck.

For Exercises 53–56, do the following for the confidence interval from the indicated exercises.
   a. Calculate the margin of error.
   b. Explain what this value for the margin of error means.

53. Married Millennials. Exercise 49

54. Rather Be Fishing? Exercise 50

55. Spring Break and Drinking. Exercise 51

56. NASCAR Fans and Pickup Trucks. Exercise 52

57. Hawaii Residents Thriving. The Gallup Organization collects data on the well-being of residents in the 50 states. In 2011, the highest proportion of residents that are reported to be "thriving" is in Hawaii, with 65.5% thriving. (Gallup categorizes respondents as thriving who report fewer health problems, fewer sick days, lower levels of stress, sadness and anger, and higher levels of happiness and respect.) Suppose the poll is based on 1000 Hawaii residents.[16]
   a. Find the margin of error using a 95% confidence level. What does this number mean?
   b. Construct and interpret a 95% confidence interval for the population proportion of all Hawaiians who are thriving.

58. Does Heavy Debt Lead to Ulcers? An AP–AOL Poll reported on June 9, 2008, that 27% of respondents carrying heavy mortgage or credit card debt also said that they had stomach ulcers.[17] How large a sample size is needed to estimate the population proportion of respondents carrying heavy debt who also have stomach ulcers to within 1% with 99% confidence?

59. Mozart Effect. Harvard University's Project Zero (**pzweb.harvard.edu**) found that listening to certain kinds of music, including Mozart, improved spatial-temporal reasoning abilities in children. Suppose that, in a sample of 100 randomly chosen fifth-graders, 65 performed better on a spatial-temporal achievement test after listening to a Mozart sonata. If appropriate, find a 95% confidence interval for the population proportion of all fifth-graders who performed better after listening to a Mozart sonata.

60. Mozart Effect. Refer to Exercise 59. *What if* we increase the confidence level to 99% while changing nothing else. Explain what would happen to the following statistics and why.
   a. $Z_{\alpha/2}$
   b. Margin of error
   c. Width of the confidence interval

The Famous ± 3 Percentage Points. Use the information from Example 8.17 for Exercises 61 and 62.

61. *What if* the sample size is higher than 1012, but otherwise everything else is the same as in the example. How would this affect the following?
   a. Margin of error
   b. $Z_{\alpha/2}$
   c. Width of the confidence interval

62. *What if* the confidence level is lower than 95%, but otherwise everything else is the same as in the example. How would this affect the following?

   a. Margin of error
   b. $Z_{\alpha/2}$
   c. Width of the confidence interval

Bringing It All Together

Drug Companies and Research Studies. Use this information for Exercises 63–65. The *Annals of Internal Medicine* reported that 39 of the 40 research studies with acknowledged sponsorship by a drug company had outcomes favoring the drug under investigation.[18]

63. If appropriate, construct and interpret a 90% confidence interval for the population proportion of all studies sponsored by drug companies that have outcomes favoring the drug. If not appropriate, clearly state why not.

64. The article in the *Annals of Internal Medicine* found that 89 of the 112 studies *without* acknowledged drug company support had outcomes favoring the drug. If appropriate, construct a 95% confidence interval for the population proportion of all studies without acknowledged drug company support which have outcomes favoring the drug. If not appropriate, clearly state why not.

65. Refer to Exercise 64. *What if* we decrease the confidence level to 90%, while changing nothing else. Explain precisely what would happen to the following statistics and why.
   a. $Z_{\alpha/2}$
   b. Margin of error
   c. Width of the confidence interval

8.4 CONFIDENCE INTERVALS FOR THE POPULATION VARIANCE AND STANDARD DEVIATION

OBJECTIVES By the end of this section, I will be able to . . .

1 Describe the properties of the χ^2 (chi-square) distribution, and find critical values for the χ^2 distribution.

2 Construct and interpret confidence intervals for the population variance and standard deviation.

We have seen how confidence intervals can be used to estimate the unknown value of a population mean or a population proportion. However, the variability of a population is also important. As we have learned, less variability is usually better. For example, a tool manufacturer relies on a quality control technician (who has a strong background in statistics) to make sure that the tools the company is making do not vary appreciably from the required specifications. Otherwise, the tools may be too large or too small. Data analysts therefore construct confidence intervals to estimate the unknown value of the population parameters that measure variability: the population variance σ^2 and the population standard deviation σ.

We first need to become acquainted with the χ^2 **(chi-square) distribution,** which is used to construct these confidence intervals.

1   PROPERTIES OF THE χ^2 (CHI-SQUARE) DISTRIBUTION

The χ^2 (pronounced *ky-square*, to rhyme with "my square") distribution was discovered in 1875 by the German physicist Friedrich Helmert and further developed in 1900 by the English statistician Karl Pearson.

The χ^2 random variable is continuous. Just as we did with the normal and t distributions, we can find probabilities associated with values of χ^2, and vice versa. Like any continuous distribution, probability is represented by area below the curve above an interval. We examine the properties of the χ^2 distribution and then learn how to use the χ^2 table to find the critical values of the χ^2 distribution.

Properties of the χ^2 Distribution

- Just as for any continuous random variable, the total area under the χ^2 curve equals 1.
- The value of the χ^2 random variable is never negative, so the χ^2 curve starts at 0. However, it extends indefinitely to the right, with no upper bound.
- Because of the characteristics just described, the χ^2 curve is right-skewed.
- There is a different curve for every different degrees of freedom, $n - 1$. As the number of degrees of freedom increases, the χ^2 curve begins to look more symmetric (Figure 8.27).

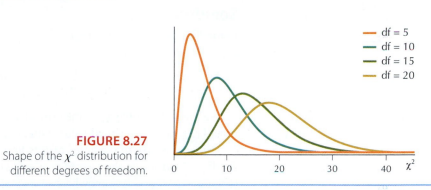

**Now You Can Do
Exercises 5–8.**

FIGURE 8.27
Shape of the χ^2 distribution for different degrees of freedom.

To construct the confidence intervals in this section, we shall need to find the critical values of a χ^2 distribution for the given confidence level $100(1 - \alpha)\%$, using either the χ^2 table (Table E in the Appendix) or technology. The χ^2 table is somewhat similar to the t table (Table D in the Appendix); both tables show the degrees of freedom in the left column. The area to the right of the χ^2 critical value is given across the top of the table.

Since the χ^2 distribution is not symmetric, we cannot construct the confidence interval for σ^2 using the "point estimate $\pm$ margin of error" method. Rather, the lower bound and upper bound for the confidence interval are determined using two χ^2 critical values:

$\chi^2_{1-\alpha/2}$ = the value of the χ^2 distribution with area $1 - \alpha/2$ to its right (Figure 8.28)

$\chi^2_{\alpha/2}$ = the value of the χ^2 distribution with area $\alpha/2$ to its right (Figure 8.28).

For instance, for a 95% confidence interval $(1 - \alpha) = 0.95$, $\alpha/2 = 0.025$ and $1 - \alpha/2 = 0.975$. Thus, $\chi^2_{0.975}$ represents the value of the χ^2 distribution with area $1 - \alpha/2 = 0.975$ to the right of the χ^2 critical value. The second critical value $\chi^2_{0.025}$ represents the value of the χ^2 distribution with area $\alpha/2 = 0.025$ to the right of the χ^2 critical value.

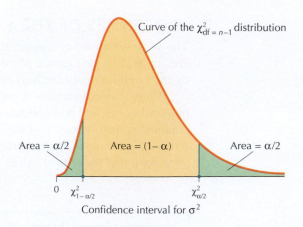

FIGURE 8.28
χ^2 critical values.

Confidence interval for σ^2

EXAMPLE 8.20

FINDING THE χ^2 CRITICAL VALUES

Find χ^2 critical values for a 90% confidence interval, where we have a sample size of size $n = 10$.

Solution

For a 90% confidence interval

$$(1 - \alpha) = 0.90 \qquad \frac{\alpha}{2} = \frac{0.10}{2} = 0.05 \qquad 1 - \frac{\alpha}{2} = 1 - 0.05 = 0.95$$

Note: If the appropriate degrees of freedom are not given in the χ^2 table, the conservative solution is to take the next row with the smaller df.

Now You Can Do Exercises 9–16.

So we are seeking (1) $\chi^2_{0.95}$, the critical value with area $1 - \alpha/2 = 0.95$ to the right of it, and (2) $\chi^2_{0.05}$, the critical value with area $\alpha/2 = 0.05$ to the right of it.

Since $n = 10$, the degrees of freedom is df $= n - 1 = 10 - 1 = 9$. To find $\chi^2_{0.95}$ for df $= 9$, go across the top of the χ^2 table (Table E in the Appendix) until you see 0.95 (Figure 8.29). $\chi^2_{0.95}$ is somewhere in that column. Now go down that column until you see your number of degrees of freedom df $= 9$. Thus, for df $= 9$, $\chi^2_{0.95} = 3.325$. For a χ^2 distribution with 9 degrees of freedom, there is area $= 0.95$ to the right of 3.325. Similarly, $\chi^2_{0.05}$ is found in the column labeled "0.05" and the row corresponding to df $= 9$. We find that $\chi^2_{0.05} = 16.919$, as shown in Figure 8.30.

Chi-Square (χ^2) Distribution
Area to the Right of Critical Value

Degrees of Freedom	0.995	0.99	0.975	0.95	0.90	0.10	0.05	0.025	0.01	0.005
1	—	—	0.001	0.004	0.016	2.706	3.841	5.024	6.635	7.879
2	0.010	0.020	0.051	0.103	0.211	4.605	5.991	7.378	9.210	10.597
3	0.072	0.115	0.216	0.352	0.584	6.251	7.815	9.348	11.345	12.838
4	0.207	0.297	0.484	0.711	1.064	7.779	9.488	11.143	13.277	14.860
5	0.412	0.554	0.831	1.145	1.610	9.236	11.071	12.833	15.086	16.750
6	0.676	0.872	1.237	1.635	2.204	10.645	12.592	14.449	16.812	18.548
7	0.989	1.239	1.690	2.167	2.833	12.017	14.067	16.013	18.475	20.278
8	1.344	1.646	2.180	2.733	3.490	13.362	15.507	17.535	20.090	21.955
9	1.735	2.088	2.700	3.325	4.168	14.684	16.919	19.023	21.666	23.589
10	2.156	2.558	3.247	3.940	4.865	15.987	18.307	20.483	23.209	25.188

FIGURE 8.29 Finding $\chi^2_{0.95}$ and $\chi^2_{0.05}$ using the χ^2 table.

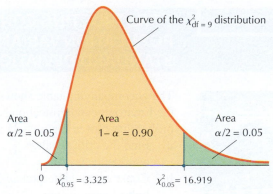

FIGURE 8.30 χ^2 critical values for the χ^2 distribution with df = 9.

2 CONSTRUCTING CONFIDENCE INTERVALS FOR THE POPULATION VARIANCE AND STANDARD DEVIATION

We derive the formula for a $100(1 - \alpha)\%$ confidence interval for the population variance σ^2. Suppose we take a random sample of size n from a normal population with mean μ and standard deviation σ. Then the statistic

$$\chi^2 = \frac{(n-1)s^2}{\sigma^2}$$

follows a χ^2 distribution with $n - 1$ degrees of freedom, where s^2 represents the sample variance. From Figure 8.28, we see that $100(1 - \alpha)\%$ of the values of χ^2 lie between $\chi^2_{1-\alpha/2}$ and $\chi^2_{\alpha/2}$. These values are described as

$$\chi^2_{1-\alpha/2} < \frac{(n-1)s^2}{\sigma^2} < \chi^2_{\alpha/2}$$

Rearranging this inequality so that σ^2 is in the numerator gives us the formula for the $100(1 - \alpha)\%$ confidence interval for σ^2:

$$\frac{(n-1)s^2}{\chi^2_{\alpha/2}} < \sigma^2 < \frac{(n-1)s^2}{\chi^2_{1-\alpha/2}}$$

Thus the lower bound of the confidence interval for σ^2 is $\dfrac{(n-1)s^2}{\chi^2_{\alpha/2}}$, and the upper bound is $\dfrac{(n-1)s^2}{\chi^2_{1-\alpha/2}}$. Taking the square root of each gives us the lower and upper bounds for the confidence interval for σ.

Confidence Interval for the Population Variance σ^2

Suppose we take a sample of size n from a normal population with mean μ and standard deviation σ. Then a $100(1 - \alpha)\%$ confidence interval for the population variance σ^2 is given by

$$\text{lower bound} = \frac{(n-1)s^2}{\chi^2_{\alpha/2}}, \text{upper bound} = \frac{(n-1)s^2}{\chi^2_{1-\alpha/2}}$$

where s^2 represents the sample variance and $\chi^2_{1-\alpha/2}$ and $\chi^2_{\alpha/2}$ are the critical values for a χ^2 distribution with $n - 1$ degrees of freedom.

Confidence Interval for the Population Standard Deviation σ

A $100(1 - \alpha)\%$ confidence interval for the population standard deviation σ is then given by

$$\text{lower bound} = \sqrt{\frac{(n-1)s^2}{\chi^2_{\alpha/2}}}, \text{upper bound} = \sqrt{\frac{(n-1)s^2}{\chi^2_{1-\alpha/2}}}$$

EXAMPLE 8.21

CONSTRUCTING CONFIDENCE INTERVALS FOR THE POPULATION VARIANCE σ^2 AND POPULATION STANDARD DEVIATION σ

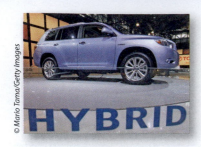

hybridmiles

The accompanying table shows the city gas mileage for 6 hybrid cars, as reported by the Environmental Protection Agency and **www.hybridcars.com** in 2007. The normal probability plot in Figure 8.31 indicates that the data are normally distributed.

a. Find the critical values $\chi^2_{1-\alpha/2}$ and $\chi^2_{\alpha/2}$ for a confidence interval with a 95% confidence level.
b. Construct and interpret a 95% confidence interval for the population variance of hybrid gas mileage.
c. Construct and interpret a 95% confidence interval for the population standard deviation of hybrid gas mileage.

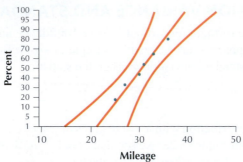

FIGURE 8.31 Normal probability plot of mileage.

Vehicle	Mileage (mpg)
Honda Accord	30
Ford Escape (2wd)	36
Toyota Highlander	33
Saturn VUE Green Line	27
Lexus RX 400h	31
Lexus GS 450h	25

Solution

a. There are $n = 6$ hybrid cars in our sample, so the degrees of freedom equal $n - 1 = 5$. For a 95% confidence interval,

$$(1 - \alpha) = 0.95 \qquad \alpha/2 = 0.025 \qquad 1 - \alpha/2 = 0.975$$

From the χ^2 table (Table E in the Appendix), therefore,

$$\chi^2_{1-\alpha/2} = \chi^2_{0.975} = 0.831 \qquad \chi^2_{\alpha/2} = \chi^2_{0.025} = 12.833$$

Figures 8.32 and 8.33 show these results using Excel and Minitab.

A1		fx	=CHIINV(0.975,5)

	A	B	C	D	E
1	0.831212				
2	12.8325				

A2		fx	=CHIINV(0.025, 5)

	A	B	C	D	E
1	0.831212				
2	12.8325				

FIGURE 8.32 Excel results.

```
Inverse Cumulative Distribution Function
Chi-Square with 5 DF

P( X <= x )          x
     0.025   0.831212
```

```
Inverse Cumulative Distribution Function
Chi-Square with 5 DF

P( X <= x )          x
     0.975   12.8325
```

(a)   (b)

FIGURE 8.33 Minitab results.

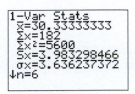

FIGURE 8.34 TI-83/84 results.

b. Figure 8.34 shows the descriptive statistics for the hybrid car gas mileages, as obtained by the TI-83/84. The sample standard deviation is $s = 3.983298466$. Thus, our 95% confidence interval for σ^2 is given by

$$\text{lower bound} = \frac{(n-1)s^2}{\chi^2_{\alpha/2}} = \frac{(5)3.983298466^2}{12.833} \approx 6.181978754 \approx 6.18$$

$$\text{upper bound} = \frac{(n-1)s^2}{\chi^2_{1-\alpha/2}} = \frac{(5)3.983298466^2}{0.831} \approx 95.46730848 \approx 95.47$$

We are 95% confident that the population variance σ^2 lies between 6.18 and 95.47 miles per gallon squared, that is, (mpg)^2. (Recall that the variance is measured in *units squared*.) Since it is unclear what miles per gallon squared means, we prefer to construct a confidence interval for the population standard deviation σ.

c. Using the results from part (b),

$$\text{lower bound} = \sqrt{\frac{(n-1)s^2}{\chi^2_{\alpha/2}}} = \sqrt{6.181978754} \approx 2.486358533 \approx 2.49$$

$$\text{upper bound} = \sqrt{\frac{(n-1)s^2}{\chi^2_{1-\alpha/2}}} = \sqrt{95.46730848} \approx 9.770737356 \approx 9.77$$

We are 95% confident that the population standard deviation σ lies between 2.49 and 9.77 miles per gallon. Figure 8.35 shows the two confidence intervals obtained using Minitab.

Now You Can Do
Exercises 17–24 and 33.

```
                              CI for        CI for
Variable  Method      StDev           Variance
Mileage   Standard    (2.49, 9.77)    (6.2, 95.4)
```

FIGURE 8.35 Minitab results showing the confidence intervals.

STEP-BY-STEP TECHNOLOGY GUIDE: χ^2 Distribution

EXCEL

Finding the Critical Values $\chi^2_{1-\alpha/2}$ and $\chi^2_{\alpha/2}$
Step 1 Select cell **A1**. Click the **Insert Function** icon f_x.
Step 2 For **Search for a Function**, type **chiinv**, click **GO**, then click **OK**.
Step 3 To find $\chi^2_{1-\alpha/2}$: For **Probability**, enter $1 - \alpha/2$ (such as 0.975 for a 95% confidence interval), and for **Deg_ freedom** enter

the degrees of freedom. Excel displays the value of $\chi^2_{1-\alpha/2}$ in the cell.
Step 4 To find $\chi^2_{\alpha/2}$: Repeat Steps 1–2. For **Probability**, enter $\alpha/2$ (such as **0.025** for a 95% confidence interval), and for **Deg_freedom** enter the degrees of freedom. Excel displays the value of $\chi^2_{\alpha/2}$ in the cell.

MINITAB

Finding the Critical Values $\chi^2_{1-\alpha/2}$ and $\chi^2_{\alpha/2}$
Step 1 Click **Calc > Probability Distributions > Chi-Square**.
Step 2 Select **Inverse cumulative probability**, and enter the **Degrees of freedom**.
Step 3 To find $\chi^2_{1-\alpha/2}$: For **Input constant**, enter the area to the *left* of the desired critical value. For $\chi^2_{1-\alpha/2}$, this will be $\alpha/2$ (such as 0.025). Click **OK**.
Step 4 To find $\chi^2_{\alpha/2}$: Repeat Steps 1 and 2. For **Input constant**, enter the area to the *left* of the desired critical value. For $\chi^2_{\alpha/2}$, this will be $1 - \alpha/2$ (such as **0.975**). Click **OK**.

Step 5 Minitab displays the values of $\chi^2_{1-\alpha/2}$ and $\chi^2_{\alpha/2}$ in the session window.

Finding a 100(1 − α)% Confidence Interval for σ
Step 1 Enter the data into column C1.
Step 2 Select **Stat > Basic Statistics > Variance . . .**
Step 3 For **Samples in columns**, select **C1**.
Step 4 Click **Options**, choose the confidence level, and click **OK**. The confidence interval for σ is reported in the output, as shown in Figure 8.35.

CRUNCHIT!

Finding the Critical Values $\chi^2_{1-\alpha/2}$ and $\chi^2_{\alpha/2}$
Step 1 Click **Distribution Calculator** and select **Chi-square**.
Step 2 For **df** enter the degrees of freedom.
Step 3 Select the **Quantile** tab. Enter the area $\alpha/2$ (such as **0.025** for a 95% confidence interval). Click **Calculate**. CrunchIt! displays the value of $\chi^2_{1-\alpha/2}$.

Step 4 Enter the area $1 - \alpha/2$ (such as **0.975** for a 95% confidence interval). Click **Calculate**. CrunchIt! displays the value of $\chi^2_{\alpha/2}$.

SECTION 8.4 Summary

1. The χ^2 continuous random variable takes values that are never negative, so the χ^2 distribution curve starts at 0 and extends indefinitely to the right. Thus, the χ^2 curve is right-skewed and not symmetric. There is a different curve for every different degrees of freedom, $n - 1$. To find χ^2 critical values, we can use either the χ^2 table or technology.

2. If the population is normally distributed, we use the χ^2 distribution to construct a $100(1 - \alpha)\%$ confidence

interval for the population variance σ^2, which is given by

$$\text{lower bound} = \frac{(n-1)s^2}{\chi^2_{\alpha/2}}, \quad \text{upper bound} = \frac{(n-1)s^2}{\chi^2_{1-\alpha/2}}$$

where s^2 represents the sample variance and $\chi^2_{1-\alpha/2}$ and $\chi^2_{\alpha/2}$ are the critical values for a χ^2 distribution with $n - 1$ degrees of freedom. The confidence interval for σ is found by taking the square root of these lower and upper bounds.

SECTION 8.4 Exercises

Clarifying the Concepts

1. To construct a confidence interval for σ^2 or σ, what must be true about the population?

2. Explain the difference between σ^2 and s^2.

3. Explain why we need to find two different critical values to construct the confidence intervals in this section. Why can't we just use the "point estimate ± margin of error" method we used earlier in this chapter?

4. Provide an example from the real world where it would be important to estimate the variability of a data set.

Determine whether each proposition in Exercises 5–8 is true or false. If it is false, restate the proposition correctly.

5. The χ^2 curve is symmetric.

6. The value of the χ^2 random variable is never negative.

7. The χ^2 curve is right-skewed.

8. The total area under the χ^2 curve equals 1.

Practicing the Techniques

For Exercises 9–14, find the critical values $\chi^2_{1-\alpha/2}$ and $\chi^2_{\alpha/2}$ for the given confidence level and sample size.

9. Confidence level 90%, $n = 25$

10. Confidence level 95%, $n = 25$

11. Confidence level 99%, $n = 25$

12. Confidence level 95%, $n = 10$

13. Confidence level 95%, $n = 15$

14. Confidence level 95%, $n = 20$

15. Consider the critical values you calculated in Exercises 9–11. Describe what happens to the critical values for a given sample size as the confidence level increases.

16. Consider the critical values you calculated in Exercises 12–14. Describe what happens to the critical values for a given confidence level as the sample size increases.

In Exercises 17–22, a random sample is drawn from a normal population. The sample of size $n = 25$ has a sample variance of $s^2 = 10$. Construct the specified confidence interval.

17. 90% confidence interval for the population variance σ^2

18. 95% confidence interval for the population variance σ^2

19. 99% confidence interval for the population variance σ^2

20. 90% confidence interval for the population standard deviation σ

21. 95% confidence interval for the population standard deviation σ

22. 99% confidence interval for the population standard deviation σ

23. Consider the confidence intervals you constructed in Exercises 17–19. Describe what happens to the lower bound and upper bound of a confidence interval for σ^2 as the confidence level increases but the sample size stays the same.

24. Consider the confidence intervals you constructed in Exercises 20–22. Describe what happens to the lower bound and upper bound of a confidence interval for σ as the confidence level increases but the sample size stays the same.

In Exercises 25–30, a random sample is drawn from a normal population. The sample variance is $s^2 = 10$. Construct the specified confidence interval.

25. 95% confidence interval for the population variance σ^2 for a sample of size $n = 10$

26. 95% confidence interval for the population variance σ^2 for a sample of size $n = 15$

27. 95% confidence interval for the population variance σ^2 for a sample of size $n = 20$

28. 95% confidence interval for the population standard deviation σ for a sample of size $n = 10$

29. 95% confidence interval for the population standard deviation σ for a sample of size $n = 15$

30. 95% confidence interval for the population standard deviation σ for a sample of size $n = 20$

31. Consider the confidence intervals you constructed in Exercises 25–27. Describe what happens to the lower bound and upper bound of a confidence interval for σ^2 as the sample size increases but the confidence level stays the same.

32. Consider the confidence intervals you constructed in Exercises 28–30. Describe what happens to the lower bound and upper bound of a confidence interval for σ as the sample size increases but the confidence level stays the same.

33. Biomass Power Plants. Power plants around the country are retooling in order to consume biomass instead of or in addition to coal. The table contains a random sample of 10 such power plants and the amount of biomass they consumed in 2006 in trillions of Btu (British thermal units). The normal probability plot indicates acceptable normality.

 biomass

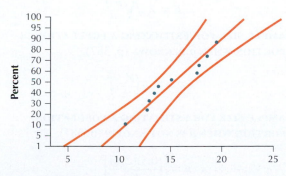

Normal probability plot of biomass consumed (trillions of Btu).

Power plant	Location	Biomass consumed (trillions of Btu)
Georgia Pacific	Choctaw, AL	13.4
Jefferson Smurfit	Nassau, FL	12.9
International Paper	Richmond, GA	17.8
Gaylord Container	Washington, LA	15.1
Escanaba Paper	Delta, MI	19.5
Weyerhaeuser	Martin, NC	18.6
International Paper	Georgetown, SC	13.8
Bowater Newsprint	McMinn, TN	10.6
Covington Facility	Covington, VA	12.7
Mosinee Paper	Marathon, WI	17.6

Sources: Energy Information Administration, Form EIA-860, "Annual Electric Generator Report," and Form EIA-906, "Power Plant Report."

a. Find the critical values $\chi^2_{1-\alpha/2}$ and $\chi^2_{\alpha/2}$ for a 95% confidence interval for σ^2.

b. Construct and interpret a 95% confidence interval for the population variance σ^2 of the amount of biomass consumed.

c. Construct and interpret a 95% confidence interval for the population standard deviation σ of the amount of biomass consumed.

34. Most Active Stocks. The table shows the ten most traded stocks on the New York Stock Exchange on October 3, 2007, together with their closing prices and net change in price, in dollars. Use only the net change data for this analysis. Assume that the net change data are normally distributed.

activestock

Stock	Closing price	Net change
Micron Technology	$10.74	−1.05
Ford Motor Company	$ 8.43	−0.14
Citigroup	$47.89	0.03
Advanced Micro Devices	$13.23	0.03
EMC Corporation	$21.13	−0.24
Commerce Bancorp	$38.84	−0.63
General Electric Company	$41.55	−0.57
Avaya	$16.95	−0.07
Sprint Nextel Corporation	$18.76	−0.24
iShares:Taiwan	$17.18	−0.18

Source: USA Today. http://markets.usatoday.com.

a. Find the critical values $\chi^2_{1-\alpha/2}$ and $\chi^2_{\alpha/2}$ for a 95% confidence interval for σ^2.

b. Construct and interpret a 95% confidence interval for the population variance σ^2 of net price changes.

35. Biomass Power Plants. Refer to Exercise 33.
  a. What are the units you used to interpret your confidence interval in (**b**)?
  b. What are the units you used to interpret your confidence interval in (**c**)?
  c. Which units are more easily understood by most people?

36. Most Active Stocks. Refer to Exercise 34.
  a. What are the units you used to interpret your confidence interval in (**b**)?
  b. Do you think that those units would be easily understood by most people?
  c. What would the units be for a confidence interval for the population standard deviation σ?
  d. Construct and interpret a 95% confidence interval for σ.

37. *Deepwater Horizon* **Cleanup Costs.** The following table represents the amount of money disbursed by BP to a random sample of 6 Florida counties, for cleanup of the *Deepwater Horizon* oil spill, in millions of dollars.[19] The normality of the data was confirmed in the Section 8.1 exercises. Construct and interpret a 95% confidence interval for σ.

 deepwaterclean

County	Cleanup costs ($ millions)
Broward	0.85
Escambia	0.70
Franklin	0.50
Pinellas	1.15
Santa Rosa	0.50
Walton	1.35

38. Wii Game Sales. The following table represents the number of units sold in the United States for the week ending March 26, 2011, for a random sample of 8 Wii games.[20] The normality of the data was confirmed in the Section 8.1 exercises. Construct and interpret a 95% confidence interval for σ.

🔴 wiisales

Game	Units (1000s)	Game	Units (1000s)
Wii Sports Resort	65	Zumba Fitness	56
Super Mario All Stars	40	Wii Fit Plus	36
Just Dance 2	74	Michael Jackson	42
New Super Mario Bros.	16	Lego Star Wars	110

CHAPTER 8   **Formulas and Vocabulary**

Section 8.1
- **CONFIDENCE INTERVAL** (p. 356)
- **CONFIDENCE LEVEL** (p. 356)
- **MARGIN OF ERROR E FOR THE Z INTERVAL FOR μ** (p. 359).

$$E = Z_{\alpha/2}(\sigma/\sqrt{n})$$

- **POINT ESTIMATE** (p. 354)
- **POINT ESTIMATION** (p. 354)
- **SAMPLE SIZE FOR ESTIMATING THE POPULATION MEAN** (p. 364).

$$n = \left[\frac{(Z_{\alpha/2})\sigma}{E}\right]^2$$

- **Z INTERVAL FOR μ** (p. 357).

$$\text{lower bound} = \bar{x} - Z_{\alpha/2}(\sigma/\sqrt{n})$$
$$\text{upper bound} = \bar{x} + Z_{\alpha/2}(\sigma/\sqrt{n})$$

Section 8.2
- **DEGREES OF FREEDOM** (p. 371)
- **MARGIN OF ERROR E FOR THE t INTERVAL FOR μ** (p. 375).

$$E = t_{\alpha/2}(s/\sqrt{n})$$

- **t DISTRIBUTION** (p. 371)
- **t DISTRIBUTION CHARACTERISTICS** (p. 371)
- **t INTERVAL FOR μ** (p. 373).

$$\text{lower bound} = \bar{x} - t_{\alpha/2}(s/\sqrt{n})$$
$$\text{upper bound} = \bar{x} + t_{\alpha/2}(s/\sqrt{n})$$

Section 8.3
- **CENTRAL LIMIT THEOREM FOR PROPORTIONS** (p. 383)
- **MARGIN OF ERROR E FOR THE Z INTERVAL FOR p** (p. 385).

$$E = Z_{\alpha/2}\sqrt{\frac{\hat{p}\,\hat{q}}{n}}$$

- **SAMPLE SIZE FOR ESTIMATING A POPULATION PROPORTION WHEN $\hat{p}$ IS KNOWN** (p. 387).

$$n = \hat{p}\,\hat{q}\left(\frac{Z_{\alpha/2}}{E}\right)^2$$

- **SAMPLE SIZE FOR ESTIMATING A POPULATION PROPORTION WHEN $\hat{p}$ IS NOT KNOWN** (p. 387).

$$n = \left[\frac{(0.5)(Z_{\alpha/2})}{E}\right]^2$$

- **Z INTERVAL FOR p** (p. 383).

$$\text{lower bound} = \hat{p} - Z_{\alpha/2}\sqrt{\frac{\hat{p}\,\hat{q}}{n}}$$

$$\text{upper bound} = \hat{p} + Z_{\alpha/2}\sqrt{\frac{\hat{p}\,\hat{q}}{n}}$$

Section 8.4

- **χ^2 (CHI SQUARE) DISTRIBUTION PROPERTIES** (p. 393)

- **CONFIDENCE INTERVAL FOR THE POPULATION STANDARD DEVIATION σ** (p. 395).

$$\text{lower bound} = \sqrt{\frac{(n-1)s^2}{\chi^2_{\alpha/2}}} \qquad \text{upper bound} = \sqrt{\frac{(n-1)s^2}{\chi^2_{1-\alpha/2}}}$$

- **CONFIDENCE INTERVAL FOR THE POPULATION VARIANCE σ^2** (p. 395).

$$\text{lower bound} = \frac{(n-1)s^2}{\chi^2_{\alpha/2}} \qquad \text{upper bound} = \frac{(n-1)s^2}{\chi^2_{1-\alpha/2}}$$

CHAPTER 8 Review Exercises

Section 8.1

For Exercises 1 and 2, answer the following questions.
  a. Calculate $\sigma/\sqrt{n}$.
  b. Find $Z_{\alpha/2}$ for a confidence interval for μ with 95% confidence.
  c. Compute and interpret E, the margin of error for a confidence interval μ with 95% confidence.
  d. Construct and interpret a 95% confidence interval for μ.

1. A sample of $n = 25$ with sample mean $\bar{x} = 50$ is drawn from a normal population in which $\sigma = 10$.

2. A sample of $n = 100$ with sample mean $\bar{x} = 50$ is drawn from a population in which $\sigma = 10$.

3. THE MOZART EFFECT. A random sample of 45 children showed a mean increase of 7 IQ points after listening to a Mozart piano sonata for about 10 minutes. The distribution of such increases is unknown, but the standard deviation is assumed to be 2 IQ points.
  a. Find the point estimate of the increase in IQ points for all children after listening to Mozart.
  b. Calculate $\sigma/\sqrt{n}$.
  c. Find $Z_{\alpha/2}$ for a confidence interval with 90% confidence.
  d. Compute and interpret the margin of error for a confidence interval with 90% confidence.
  e. Construct and interpret a 90% confidence interval for the mean increase in IQ points for all children after listening to a Mozart piano sonata for about 10 minutes.

Suppose we are estimating μ. For Exercises 4–6, find the required sample size.

4. $\sigma = 50$, confidence level 95%, margin of error 10
5. $\sigma = 30$, confidence level 95%, margin of error 10
6. $\sigma = 10$, confidence level 95%, margin of error 10

7. CLINICAL PSYCHOLOGY. A clinical psychologist would like to estimate the population mean number of episodes her patients have suffered in the past year. Assume that the standard deviation is 10 episodes. How many patients will she have to examine if she wants her estimate to be within 2 episodes with 90% confidence?

Section 8.2

For Exercises 8–10, construct the indicated confidence interval if appropriate. If it is not appropriate, explain why not.

8. Confidence level 90%, $n = 25$, $\bar{x} = 22$, $s = 5$, non-normal population
9. Confidence level 90%, $n = 25$, $\bar{x} = 22$, $s = 5$, normal population
10. Confidence level 90%, $n = 100$, $\bar{x} = 22$, $s = 5$, non-normal population

11. CIGARETTE CONSUMPTION. Health officials are interested in estimating the population mean number of cigarettes smoked per capita in order to evaluate the efficacy of the antismoking campaign. A random sample of 8 U.S. counties yielded the following numbers of cigarettes smoked annually per capita: 2206, 2391, 2540, 2116, 2010, 2791, 2392, 2692. Assume the data are normally distributed.
  a. Construct a 95% confidence interval for the population mean per capita number of cigarettes smoked in all U.S. counties.
  b. Construct a 99% confidence interval for the population mean per capita number of cigarettes smoked in all U.S. counties.

Section 8.3

For Exercises 12 and 13, follow steps **(a)**–**(d)**.
  a. Find $Z_{\alpha/2}$.
  b. Determine whether the conditions are met.
  c. Calculate and interpret the margin of error, $E = Z_{\alpha/2}\sqrt{\hat{p}\,\hat{q}/n}$
  d. Construct a confidence interval for p with the indicated confidence level, and sketch the confidence interval on the number line.

12. Confidence level 95%, $n = 100$, $\hat{p} = 0.1$
13. Confidence level 95%, $n = 500$, $\hat{p} = 0.99$
14. ECSTASY AND EMERGENCY ROOM VISITS. According to the National Institute on Drug Abuse (**www.drugabuse.gov**), 77% of the emergency room patients who mentioned MDMA (Ecstasy) as a factor in their admission were age 25 and under. Assume that the sample size is 200.
  a. Calculate and interpret the margin of error for confidence level 95%.

b. Construct and interpret a 95% confidence interval for the population proportion of all emergency room patients mentioning MDMA (Ecstasy) as a factor in their admission who are age 25 and under.

For Exercises 15–17, we are estimating p and we know the value of $\hat{p}$. Find the required sample size.
15. Confidence level 99%, margin of error 0.03, $\hat{p} = 0.9$
16. Confidence level 95%, margin of error 0.03, $\hat{p} = 0.99$
17. Confidence level 95%, margin of error 0.03, $\hat{p} = 0.999$

For Exercises 18–20, we are estimating p and we do not know the value of $\hat{p}$. Find the required sample size.
18. Confidence level 90%, margin of error 0.05
19. Confidence level 90%, margin of error 0.03
20. Confidence level 90%, margin of error 0.01

Section 8.4
For Exercises 21–24, a random sample is drawn from a normal population. The sample of size $n = 36$ has a sample variance of $s^2 = 100$. Construct the specified confidence interval.
21. 90% confidence interval for the population variance σ^2
22. 95% confidence interval for the population variance σ^2
23. 90% confidence interval for the population standard deviation σ

24. 95% confidence interval for the population standard deviation σ
25. UNION MEMBERSHIP. The table contains the total union membership for seven randomly selected states. Construct and interpret a 95% confidence interval for σ. Assume the data are normally distributed.

🔴 unionmember

State	Union membership (1000s)
Florida	397
Indiana	334
Maryland	342
Massachusetts	414
Minnesota	395
Texas	476
Wisconsin	386

Source: U.S. Bureau of Labor Statistics.

| CHAPTER 8 | Quiz |

True or False
1. True or false: In Figure 8.2 (page 359), since the confidence level is 90%, then 90% of the intervals must contain μ. Explain your answer.
2. True or false: The t curve is symmetric about 0, just like the Z curve is. Therefore we can use all our symmetry techniques with the t curve as well.

Fill in the Blank
3. Suppose we cut a margin of error in half. The sample size requirement then becomes _____ times larger.
4. Our estimate of μ is _____ precise using the t curve rather than the Z curve.

Short Answer
5. α is used to find the value of $Z_{\alpha/2}$. Is α a probability or a value of x or a value of Z?
6. What are the conditions for constructing a t interval?

Calculations and Interpretations
7. COLLEGE EDUCATION COSTS. A random sample of 49 colleges yielded a mean cost of college education of $30,500 per year. Assume that the population standard deviation is $3000.
    a. Compute and interpret the margin of error for a confidence interval with 90% confidence.
    b. Construct and interpret a 90% confidence interval for the population mean cost of college education.

8. CRASH TEST DATA. The National Highway Traffic Safety Administration collects data on crash tests for new motor vehicles. They reported that the mean femur load (force applied to the femur) in a frontal crash for the passenger in a Ford Equinox SUV was 1003 pounds. Assume that the population standard deviation was 210 pounds and the sample size was 49.
    a. Compute and interpret the margin of error for a confidence interval with 90% confidence.
    b. Construct and interpret a 90% confidence interval for the population mean femur load in a frontal crash for the passenger in a Ford Equinox SUV.

9. 9/11 AND RELIGIOUS ATTENDANCE. The Pew Research Center reported that, in a survey of 3733 randomly selected respondents, 991 had attended a religious service in response to the attacks on the World Trade Center and the Pentagon.
    a. If appropriate, find the margin of error for confidence level 95%. What does this number mean?
    b. Construct, if appropriate, a 95% confidence interval for the population proportion of Americans who attended a religious service in response to the attacks on the World Trade Center and the Pentagon.

10. INDEPENDENCE FOR QUEBEC? A poll conducted by the newspaper *La Presse* reported that 340 of 1000 randomly chosen Quebec adults surveyed would vote "Yes" in a referendum for independence from Canada.

a. If appropriate, find the margin of error for confidence level 99%. What does this number mean?

b. If appropriate, find a 99% confidence interval for the population proportion of all Quebec residents who favor independence for the province of Quebec.

11. TAX RETURNS. Recall from Section 3.2 that Ashley and Brandon work at an accounting firm preparing tax returns. Their Chief Accountant kept careful track of the amount of time (in hours) for all the tax returns that they prepared in the last week of March, shown in the accompanying table. Assume both data sets are normally distributed.

taxreturn

Ashley	5	7	8	9	11
Brandon	3	5	7	11	14

a. Construct and interpret a 95% confidence interval for the population standard deviation of Ashley's preparation time.

b. Construct and interpret a 95% confidence interval for the population standard deviation of Brandon's preparation time.

12. QUALITY OF EDUCATION IN AMERICA. The National Assessment of Educational Progress (NAEP) administers exams to a nationwide sampling of students to assess the quality of education in America. Suppose NAEP would like to estimate the population proportion of American schoolchildren who would answer a given question correctly. Find a sample size which would give a margin of error of 0.03 with 90% confidence.

9 Hypothesis Testing

William R. McIver Collection, American Heritage Center, University of Wyoming.

CASE STUDY

The Golden Ratio

What do Euclid's *Elements*, the Parthenon of ancient Greece, the *Mona Lisa*, and the beadwork of the Shoshone tribe of Native Americans have in common? An appreciation for the *golden ratio*. Suppose we have two quantities A and B, with $A > B > 0$. Then, A/B is called the golden ratio if

$$\frac{A + B}{A} = \frac{A}{B}$$

that is, if the ratio of the sum of the quantities to the larger quantity equals the ratio of the larger to the smaller.

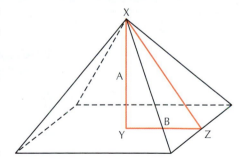

FIGURE 9.1

The golden ratio permeates ancient, medieval, Renaissance, and modern art and architecture. For example, the Egyptians constructed their great pyramids using the golden ratio. (Specifically, in Figure 9.1, if $A = \overline{XY}$ is the height from the top vertex to the base, and $B = \overline{YZ}$ is the distance from the center of the base to the edge, then $(A + B)/A = A/B$.) Some mathematicians have said that the golden ratio may be intrinsically pleasing to the human species. Support for this conjecture would be especially strong if evidence was found for the use of the golden ratio in non-Western artistic traditions. In the Case Study on page 445, we use hypothesis testing to determine whether the decorative beaded rectangles sewn by the Shoshone tribe of Native Americans follow the golden ratio. ◼

The Big Picture

Where we are coming from, and where we are headed . . .

- Chapter 8's topic, confidence intervals, represents only the first of a large family of topics in statistical inference.

- Hypothesis testing is the most widely used method for statistical inference, forming the bedrock of the scientific method, and touching nearly every field of scientific endeavor, from medicine to business to psychology. It is also the basis for business-oriented decision-making methods. Here, in Chapter 9, we learn how to perform hypothesis tests for the population mean, the population proportion, and the population standard deviation.

- In Chapter 10, "Two-Sample Inference," we will learn confidence intervals and hypothesis tests for comparing parameters from two populations.

9.1 INTRODUCTION TO HYPOTHESIS TESTING

OBJECTIVES By the end of this section, I will be able to . . .

1 Construct the null hypothesis and the alternative hypothesis from the statement of the problem.

2 State the two types of errors made in hypothesis tests: the Type I error, made with probability α, and the Type II error, made with probability β.

Researchers are interested in investigating many different types of questions, such as the following:

- An accountant may wish to examine whether evidence exists for corporate tax fraud.

- A Department of Homeland Security executive may want to test whether a new surveillance method will uncover terrorist activity.

- A sociologist may want to examine whether the mayor's economic policy is increasing poverty in the city.

Questions such as these can be tackled using statistical **hypothesis testing,** which is a statistical inference process for using sample data to render a decision about claims regarding the unknown value of a population parameter. In this section we will learn how to make decisions about the values of a population mean.

1 CONSTRUCTING THE HYPOTHESES

The basic idea of hypothesis testing is the following:

1. We need to make a *decision* about the value of a population parameter, such as the population mean μ or the population proportion p.

2. Unfortunately, the true value of that parameter is *unknown*.

3. Therefore, there may be different *hypotheses* about the true value of this parameter.

Statistical hypothesis testing is a way of formalizing the decision-making process so that a decision can be rendered about the value of the parameter. We craft *two competing statements (hypotheses)* about the value of the population parameter (either μ, p, or σ) and gather evidence to conclude that one of the hypotheses is likely to be true.

> **The Hypotheses**
>
> - The status quo hypothesis represents what has been tentatively assumed about the value of the parameter and is called the **null hypothesis,** denoted as H_0.
> - The **alternative hypothesis,** or **research hypothesis,** denoted as H_a, represents an alternative claim about the value of the parameter.

Hypothesis testing is like conducting a criminal trial. In a trial in the United States, the defendant is innocent until proven guilty, and the jury must evaluate the truth of two competing hypotheses:

$$H_0 : \text{defendant is not guilty} \quad \text{versus} \quad H_a : \text{defendant is guilty}$$

The not-guilty hypothesis is considered the **null hypothesis H_0** because the jurors must assume it is true until proven otherwise. The **alternative hypothesis H_a,** that the defendant is guilty, *must be demonstrated* to be true, beyond a reasonable doubt. How does a court of law determine whether the defendant is convicted or acquitted? This judgment is based upon the *evidence,* the hard facts heard in court. Similarly, in hypothesis testing, the researcher draws a conclusion based on the evidence provided by the sample data.

In Sections 9.1–9.4, we will examine hypotheses for the unknown mean μ. The null hypothesis will be a claim about a certain specified value for μ denoted μ_0, and the alternative hypothesis will be a claim about other values for μ. The hypotheses have one of the three possible forms shown in Table 9.1. The right-tailed test and the left-tailed test are called one-tailed tests. In Section 9.2 we will find out why we use this terminology.

Table 9.1 The three possible forms for the hypotheses for a test for μ

Form	Null and alternative hypotheses
Right-tailed test	$H_0 : \mu = \mu_0$ versus $H_a : \mu > \mu_0$
Left-tailed test	$H_0 : \mu = \mu_0$ versus $H_a : \mu < \mu_0$
Two-tailed test	$H_0 : \mu = \mu_0$ versus $H_a : \mu \neq \mu_0$

EXAMPLE 9.1

CONSTRUCTING A HYPOTHESIS TEST

The medical information Web site **MayoClinic.com** reports that a 16-ounce Starbucks Park Place brewed coffee contains 350 milligrams (mg) of caffeine. Suppose a local health organization is interested in whether the mean amount of caffeine in this coffee

D Hurst/Alamy

is greater than 350 mg. They intend to take a random sample of Starbucks Park Place brewed coffees, and measure the amount of caffeine in each one. Construct the appropriate hypothesis test for this situation.

Solution

The local health organization is interested in whether the mean amount of caffeine is *greater than* 350 mg. The only form of the hypothesis test that contains the ">" symbol is the right-tailed test. Thus, we write a null hypothesis and an alternative hypothesis for a right-tailed test:

$$H_0 : \mu = 350 \quad \text{versus} \quad H_a : \mu > 350$$

The null hypothesis H_0 states that the population mean μ equals 350 mg. The alternative hypothesis $H_a : \mu > 350$ states that the population mean amount of caffeine is *greater than* 350 mg. Here, $\mu_0 = 350$, which is the possible value of μ specified in the example. (By the way, the National Institutes of Health recommend that caffeine intake be limited to 250 mg per day. I wonder if they will make an exception for finals week.)

The first task in hypothesis testing is to form hypotheses. To convert a word problem into two hypotheses, look for certain key words that can be expressed mathematically. Table 9.2 shows how to convert words typically found in word problems into symbols.

Table 9.2 Key English words, with mathematical symbols and synonyms

English words	Symbol	Synonyms
Equal	=	Is; has stayed the same
Not equal	≠	Is different from; has changed from; differs from
Greater than	>	Is more than; exceeds; has increased
Less than	<	Is below; is smaller than; has decreased

Once you have identified the key words, use the associated mathematical symbol to write the two hypotheses. The following strategy can be used to write the hypotheses.

> **Strategy for Constructing the Hypotheses About μ**
>
> **Step 1** Search the word problem for certain key English words and select the associated symbol from Table 9.2.
>
> **Step 2** Determine the form of the hypotheses listed in Table 9.1 that uses this symbol.
>
> **Step 3** Find the value of μ_0 (the number that answers the question: "greater than what?" or "less than what?") and write your hypotheses in the appropriate forms.

EXAMPLE 9.2

APPLYING THE STRATEGY FOR CONSTRUCTING THE HYPOTHESES ABOUT μ

The mean annual rainfall in Arizona has been 8 inches per year, according to the *World Almanac*. But weather researchers are interested in whether this already small amount of rain will decrease, leading to drought conditions in the state. Write a null hypothesis and an alternative hypothesis that describe this situation.

Solution

Let's use our strategy to construct the hypotheses needed to test this claim.

STEP 1 **Search the word problem for certain key English words and select the appropriate symbol.**

The problem uses the word "decrease," which means, "less than." Thus we will write a hypothesis that contains the $<$ symbol.

STEP 2 **Determine the form of the hypotheses.**

From Table 9.1, we see that the symbol $<$ means that we use a left-tailed test:

$$H_0 : \mu = \mu_0 \quad \text{versus} \quad H_a : \mu < \mu_0$$

STEP 3 **Find the value for μ_0 and write your hypotheses.**

The alternative hypothesis H_a states that the mean annual rainfall μ is less than some value μ_0. Less than what? Eight inches per year. Write the two hypotheses with $\mu_0 = 8$.

$$H_0 : \mu = 8 \quad \text{versus} \quad H_a : \mu < 8$$

Now You Can Do Exercises 9–14.

CAUTION

Do not blindly apply this strategy without thinking about what you are doing. Rather, use the strategy to help formulate your own hypotheses. *There is no substitute for thinking through the problem!*

Now that we know how to construct hypotheses, we next consider when sufficient evidence exists to reject the null hypothesis.

> **Statistical Significance**
>
> A result is said to be **statistically significant** if it is unlikely to have occurred due to chance.

Suppose that you are a researcher for a pharmaceutical research company. You are investigating the side effects of a new cholesterol-lowering medication and would like to determine whether the medication will decrease the population mean systolic blood pressure level from the current mean of 110. If so, then a warning will have to be given not to prescribe the new medication to patients whose blood pressure is already low. The appropriate hypotheses are

$$H_0 : \mu = 110 \quad \text{versus} \quad H_a : \mu < 110$$

where μ represents the population mean systolic blood pressure and $\mu_0 = 110$. To determine which of these hypotheses is correct, we take a sample of randomly selected patients who are taking the medication. We record their systolic blood pressure levels and calculate the sample mean $\bar{x}$ and sample standard deviation s. Most likely, the mean of this sample of patients' systolic blood pressure levels will not be exactly equal to 110, even if the null hypothesis is true.

Now, suppose that the sample mean blood pressure $\bar{x}$ is less than the hypothesized population mean of 110. *Is the difference due simply to chance variation, or is it evidence of a real side effect of the cholesterol medication?* Let's consider some possible values for $\bar{x}$:

- $\bar{x} = 109$: The difference between $\bar{x}$ and $\mu = 110$ is only 1. Depending on the variability present in the sample, the researcher would likely not reject the null hypothesis because this small difference is probably due to chance variation. The result is not statistically significant.

- $\bar{x} = 90$: The difference between $\bar{x}$ and $\mu = 110$ is 20. Depending on the variability present in the sample, the researcher would probably conclude that this difference is so large that it is unlikely that it is due to chance variation. Thus, the researcher would reject the null hypothesis H_0 in favor of the alternative hypothesis H_a. The result is statistically significant.

To summarize: *in a hypothesis test, we compare the sample mean $\bar{x}$ with the value μ_0 of the population mean used in the H_0 hypothesis. If the difference is large, then H_0 is rejected. If the difference is not large, then H_0 is not rejected.* The question is, "Where do you draw the line?" Just how large a difference is large enough? The hypothesis-testing procedure will show us.

2 TYPE I AND TYPE II ERRORS

Next, we take a closer look at some of the thorny issues involved in performing a hypothesis test. Let's return to the example of a criminal trial. The jury will convict the defendant if they find evidence compelling enough to reject the null hypothesis of "not guilty" *beyond a reasonable doubt*. However, jurors are only human; sometimes their decisions are correct and sometimes they are not. Thus, the jury's verdict will be one of the following outcomes:

1. An innocent defendant is wrongfully convicted.
2. A guilty defendant is convicted.
3. A guilty defendant is wrongfully acquitted.
4. An innocent defendant is acquitted.

Recall that we can write the two hypotheses for a criminal trial as

$$H_0: \text{defendant is not guilty} \quad \text{versus} \quad H_a: \text{defendant is guilty}$$

Table 9.3 shows the possible verdicts on the left and the two hypotheses across the top.

Table 9.3 Four possible outcomes of a criminal trial

	Reality	
Jury's decision	H_0 true: Defendant did not commit the crime	H_0 false: Defendant did commit the crime
Reject H_0: Find defendant guilty	Type I error	Correct decision
Do not reject H_0: Find defendant not guilty	Correct decision	Type II error

Let's look at the two possible decisions the jury can make. It can find the defendant guilty: the jury *rejects the claim* in the null hypothesis H_0. Alternatively, the jury can find the defendant not guilty: the jury *does not reject* the null hypothesis H_0. There are two ways for the jury to render the *correct decision*.

> **Two Ways of Making the Correct Decision**
> - To not reject H_0 when H_0 is true.
>   Example: To find the defendant not guilty when in reality he did not commit the crime.
> - To reject H_0 when H_0 is false.
>   Example: To find the defendant guilty when in reality he did commit the crime.

Unfortunately, there are also two ways for the jury to render an incorrect decision. In statistics, the two incorrect decisions are called **Type I** and **Type II** errors.

> **Two Types of Errors**
> - **Type I error:** To reject H_0 when H_0 is true.
>   Example: To find the defendant guilty when in reality he did not commit the crime.
> - **Type II error:** To not reject H_0 when H_0 is false.
>   Example: To find the defendant not guilty when in reality he did commit the crime.

Now You Can Do Exercises 15–24.

Developing Your Statistical Sense

A Decision Is Not Proof

It is important to understand that the decision to reject or not reject H_0 does not prove anything. The decision represents whether or not there is sufficient evidence against the null hypothesis. This is our best judgment given the data available. You cannot claim to have *proven* anything about the value of a population parameter unless you elicit information from the entire population, which is usually not possible.

We can make decisions about population parameters using the limited information available in a sample because we base our decisions on *probability*. When the difference between the sample mean $\bar{x}$ and the hypothesized population mean μ_0 is large, then the null hypothesis is *probably* not correct. When the difference is small, then the data are *probably* consistent with the null hypothesis. But we don't know for sure.

> The probability of a Type I error is denoted as α **(alpha)**. We set the value of α to be some small constant, such as 0.01, 0.05, or 0.10, so that there is only a small probability of rejecting a true null hypothesis.

To say that $\alpha = 0.05$ means that, if this hypothesis test were repeated over and over again, the long-term probability of rejecting a true null hypothesis would be 5%. The **level of significance** of a hypothesis test is another name for α, the probability of rejecting H_0 when H_0 is true. A smaller α makes it harder to wrongfully reject H_0 just by chance. If the consequences of making a Type I error are serious, then the level of significance should be small, such as $\alpha = 0.01$. If the consequences of making a Type I error are not so serious, then one may choose a larger value for the level of significance, such as $\alpha = 0.05$ or $\alpha = 0.10$.

The probability of a Type II error is denoted as β **(beta).** This is the probability of not rejecting H_0 when H_0 is false, such as acquitting someone who is really guilty. Making α smaller inevitably makes β larger (for a fixed sample size). Of course, our goal is to simultaneously minimize both α and β. Unfortunately, the only way to do this is to increase the sample size.

There are only two possible hypothesis-testing conclusions:

Note: When we reject H_0, we say that the results are statistically significant. If we do not reject H_0, the results are not statistically significant.

- Reject H_0, or
- Do not reject H_0.

SECTION 9.1   **Summary**

1. Statistical hypothesis testing is a way of formalizing the decision-making process so that a decision can be rendered about the unknown value of the parameter. The status quo hypothesis that represents what has been tentatively assumed about the value of the parameter is called the null hypothesis and is denoted as H_0. The alternative hypothesis, or research hypothesis, denoted as H_a, represents an alternative conjecture about the value of the parameter.

2. When performing a hypothesis test, there are two ways of making a correct decision: to not reject H_0 when H_0 is true and to reject H_0 when H_0 is false. Also, there are two types of error: a Type I error is to reject H_0 when H_0 is true, and a Type II error is to not reject H_0 when H_0 is false. The probability of a Type I error is denoted as α (alpha). The probability of a Type II error is denoted as β (beta).

Clarifying the Concepts

1. What are some characteristics of the null hypothesis? The alternative hypothesis?

2. Explain what is meant by μ_0.

3. In the hypothesis test for the population mean, how many forms of the hypotheses are there? Write out these forms.

4. In a criminal trial, what are the two possible decision errors? What do statisticians call these errors?

5. When does a Type I error occur? A Type II error?

6. What are the two correct decisions that can be made?

7. Say we want to test whether the population mean is less than 100, and the sample we take yields a sample mean of 90. Is this sufficient evidence that the population mean is less than 10? Explain why or why not.

8. True or false: If the consequences of making a Type I error are serious, then the data analyst should choose a larger level of significance.

Practicing the Techniques

For Exercises 9–14, provide the null and alternative hypotheses.

9. Test whether μ is greater than 10.

10. Test whether μ is less than 100.

11. Test whether μ is different from 0.

12. Test whether or not μ equals 4.0.

13. Test whether μ has changed from 36.

14. Test whether μ exceeds -4.

For Exercises 15–18, do the following.
  a. Provide the null and alternative hypotheses.
  b. Determine if a correct decision has been made. If an error has been made, indicate which type of error.

15. Child Abuse. The U.S. Administration for Children and Families reported that the national rate for child abuse referrals was 43.9 per 1000 children in 2005. A hypothesis test was carried out that tested whether the population mean referral rate had increased this year from the 2005 level. The null hypothesis was not rejected. Suppose that, in actuality, the population mean child abuse referral rate for this year is 45 per 1000 children.

16. Travel Costs. A motorists' guide reported that travel costs were greater than 15 cents per mile. Suppose that this report was based on a hypothesis test and that in actuality the population mean travel costs were lower than 15 cents per mile.

17. Eating Trends. According to the NPD Group, higher gasoline prices are causing consumers to go out to eat less and eat at home more.[1] Suppose that this report found that the mean number of meals prepared and eaten at home is less than 700 per year, and that in actuality the population mean number of such meals is 600.

18. Hybrid Vehicles. A study by **Edmunds.com** showed that owners of hybrid vehicles can recoup their initial increased cost through reduced fuel consumption in less than three years. Suppose that the report was based on a hypothesis test and that in actuality the population mean number of years it takes to recoup their initial cost is two years.

Applying the Concepts

For Exercises 19–24, do the following.
  a. Provide the null and alternative hypotheses.
  b. Describe the two ways a correct decision could be made in the context of the problem.
  c. Describe what a Type I error would mean in the context of the problem.
  d. Describe what a Type II error would mean in the context of the problem.

19. Shares Traded on the Stock Market. The *Statistical Abstract of the United States* reports that the mean daily number of shares traded on the New York Stock Exchange in 2005 was 1.602 billion. Based on a sample of this year's trading results, a financial analyst would like to test whether the mean number of shares traded will be larger than the 2005 level.

20. Traffic Light Cameras. The Ministry of Transportation in the province of Ontario reported that the installation of cameras that take pictures at traffic lights has decreased the mean number of fatal and injury collisions to 339.1 per year. A hypothesis test was performed to determine whether the population mean number of such collisions has changed.

21. Price of Milk. The Bureau of Labor Statistics reports that the mean price for a gallon of milk in January 2011 was $3.34. Suppose that we conduct a hypothesis test to investigate if the population mean price of milk this year has increased.

22. Americans' Height. Americans used to be on average the tallest people in the world. That is no longer the case, according to a study by Dr. Richard Steckel, professor of economics and anthropology at The Ohio State University. The Norwegians and Dutch are now the tallest, at 178 centimeters, followed by the Swedes at 177, and then the Americans, with a mean height of 175 centimeters (approximately 5 feet 9 inches). According to Dr. Steckel, "The average height of Americans has been pretty much stagnant for 25 years."[2] Suppose that we conduct a hypothesis test to investigate whether the population mean height of Americans this year has changed from 175 centimeters.

23. Credit Score in Florida. According to **CreditReport .com**, the mean credit score in Florida in 2006 was 673. Suppose that a hypothesis test was conducted to determine if the mean credit score in Florida has decreased since that time.

24. Salary of College Grads. According to the U.S. Census Bureau, the mean salary of college graduates in 2002 was $52,200. Suppose that a hypothesis test was carried out to determine whether the population mean salary of college graduates has increased.

9.2   *Z* TEST FOR THE POPULATION MEAN: CRITICAL-VALUE METHOD

OBJECTIVES   By the end of this section, I will be able to . . .

1   Explain the essential idea about hypothesis testing for the population mean.

2   Perform the *Z* test for the mean, using the critical-value method.

1   THE ESSENTIAL IDEA ABOUT HYPOTHESIS TESTING FOR THE MEAN

Recall that in Section 9.1 we wanted to determine whether the population mean systolic blood pressure μ was less than 110 and we considered the hypotheses

$$H_0 : \ \mu = 110 \quad \text{versus} \quad H_a : \ \mu < 110$$

We stated that a large difference between the observed sample mean $\bar{x}$ and the hypothesized mean $\mu_0 = 110$ would result in the rejection of the null hypothesis H_0. The question is, "How large is large?"

The *Z* test for the mean tells us when our results are statistically significant. To learn how this test works, consider the following. A sample of $n = 25$ patients who are taking the medication shows a sample mean systolic blood pressure level of $\bar{x} = 104$; further assume that the population standard deviation systolic blood pressure reading is $\sigma = 10$, and that the population of such readings is normal. Would this value $\bar{x} = 104$ represent sufficient evidence to reject H_0 and conclude that $\mu < 110$?

Recall from Chapter 7 that the sampling distribution of the sample mean $\bar{x}$ is the collection of sample means of all possible samples of size n. When the population is normal, or the sample size is large, the sampling distribution of $\bar{x}$ is approximately normal, with mean $\mu_{\bar{x}} = \mu$ and standard error $\sigma_{\bar{x}} = \sigma/\sqrt{n}$. The idea behind the *Z* test is to determine where our sample mean $\bar{x} = 104$ falls within the sampling distribution. Is $\bar{x} = 104$ somewhere near the middle of the sampling distribution, or is it an outlier? Now, if H_0 is true, then $\mu = \mu_0 = 110$ and we may standardize $\bar{x}$ to get

Note: Here we are using Facts 1–4 and the Central Limit Theorem from Chapter 7.

$$Z = \frac{\bar{x} - \mu_0}{\sigma/\sqrt{n}}$$

Substituting, we get

$$Z = \frac{\bar{x} - \mu_0}{\sigma/\sqrt{n}} = \frac{104 - 110}{10/\sqrt{25}} = -3$$

In other words, $\bar{x} = 104$ lies 3 standard errors below the hypothesized mean $\mu_0 = 110$. Thus, if we accept that the null hypothesis is true, then $\bar{x} = 104$ is an outlier, an extreme value (see Figure 9.2). That is, if H_0 is true, then the probability of observing $\bar{x} \leq 104$ is very small ($P(Z < -3) = 0.0013$), since the corresponding *Z*-value lies in the tail of the distribution, and nearly all the values of $\bar{x}$ are greater than 104.

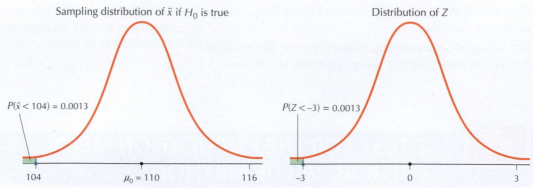

FIGURE 9.2 An extreme value of $\bar{x}$ calls for rejection of H_0.

Thus we must choose one of the following two scenarios:

1. H_0 is true, the value of μ_0 is accurate, and our observation of this extreme value of $\bar{x}$ is an amazingly unlikely event.

2. H_0 is not correct, and the true value of μ is closer to $\bar{x}$.

Developing Your Statistical Sense

The Data Prevail!

When faced with the above situation, since we don't want to base our decisions on "amazingly unlikely events," we therefore would conclude that H_0 is not correct. Remember that the null hypothesis is just a conjecture, but the sample mean $\bar{x}$ represents directly observable "hard data." The scientific method states that, when there is a conflict between a conjecture and the observed data, the data prevail, and we need to rethink our null hypothesis.

This conclusion illustrates the **essential idea about hypothesis testing for the mean.**

> **The Essential Idea About Hypothesis Testing for the Mean**
>
> When the observed value of $\bar{x}$ is unusual or extreme in the sampling distribution of $\bar{x}$ that assumes H_0 is true, we should reject H_0. Otherwise, we should not reject H_0.

All the remaining parts of Sections 9.2–9.4, all the steps and all the calculations, are really just ways to implement this essential idea.

Note that our Z statistic

$$Z = \frac{\bar{x} - \mu_0}{\sigma/\sqrt{n}}$$

We are developing the Z test using a left-tailed test, but the essential idea applies to right-tailed tests and two-tailed tests as well.

contains four quantities, three of which are taken from data. The sample mean $\bar{x}$ and the sample size n are characteristics of the sample data, and the population standard deviation σ represents the population data. Thus, we call this statistic Z_{data}.

> **The Test Statistic Z_{data}**
>
> The test statistic used for the Z test for the mean is
>
> $$Z_{\text{data}} = \frac{\bar{x} - \mu_0}{\sigma/\sqrt{n}}$$

For the blood pressure data, we have

$$Z_{\text{data}} = \frac{\bar{x} - \mu_0}{\sigma/\sqrt{n}} = \frac{104 - 110}{10/\sqrt{25}} = -3$$

Z_{data} is an example of a **test statistic,** a statistic generated from a data set for the purposes of testing a statistical hypothesis. We will meet several other test statistics throughout the remainder of the text. The hypothesis test in this section and Section 9.3 is called the *Z test* because the test statistic Z_{data} comes from the standard normal *Z* distribution.

EXAMPLE 9.3

© Maria Teijeiro/Getty Images

CALCULATING Z_{data}

Do you have a debit card? How often do you use it? ATM network operator Star System of San Diego reported that active users of debit cards used them an average of 11 times per month.[3] Suppose a random sample of 36 people used debit cards last month an average of $\bar{x} = 11.5$ times. Assume the population standard deviation $\sigma = 3$. We would like to test whether people use debit cards on average more than 11 times per month.

Solution

Using our strategy for constructing the hypotheses from Section 9.1, the key words "more than" mean ">," and the ">" symbol occurs only in the right-tailed test. Answering the question "More than what?" is $\mu_0 = 11$. Thus our hypotheses are

$$H_0 : \mu = 11 \quad \text{versus} \quad H_a : \mu > 11$$

so that $\mu_0 = 11$. The sample size is $n = 36$, with a sample mean of $\bar{x} = 11.5$, and $\sigma = 3$. Thus

$$Z_{data} = \frac{\bar{x} - \mu_0}{\sigma/\sqrt{n}} = \frac{11.5 - 11}{3/\sqrt{36}} = 1$$

**Now You Can Do
Exercises 9–16.**

---

2 PERFORMING THE *Z* TEST FOR THE MEAN, USING THE CRITICAL-VALUE METHOD

In the critical-value method for the *Z* test, we compare Z_{data} with a *threshold value,* or **critical value** of *Z*, called $\mathbf{Z_{crit}}$. The value of Z_{crit} separates *Z* into two regions (see Figure 9.3):

- **Critical region:** the values of Z_{data} for which we reject H_0
- **Noncritical region:** the values of Z_{data} for which we do not reject H_0

> - The **critical region** consists of the range of values of the test statistic Z_{data} for which we reject the null hypothesis.
> - The **noncritical region** consists of the range of values of the test statistic Z_{data} for which we do not reject the null hypothesis.
> - The value of *Z* that separates the critical region from the noncritical region is called the **critical value Z_{crit}**.

Z_{crit} represents the *boundary* between values of Z_{data} which are statistically significant and those which are not statistically significant. The value of Z_{crit} depends on the value of α, the probability of wrongly rejecting H_0. A smaller value of α will make it harder to reject H_0, that is, harder to find statistical significance. Thus, α is called the **level of significance** of the hypothesis test.

The value of Z_{crit} depends on (a) the form of the hypothesis test, and (b) the level of significance α. Table 9.4 on the next page shows values of Z_{crit} for the most commonly used levels of significance α. It also shows the location of the critical region.

Table 9.4 Table of critical values Z_{crit} for common values of the level of significance α

Level of significance α	Form of Hypothesis Test		
	Right-tailed $H_0: \mu = \mu_0$ $H_a: \mu > \mu_0$	Left-tailed $H_0: \mu = \mu_0$ $H_a: \mu < \mu_0$	Two-tailed $H_0: \mu = \mu_0$ $H_a: \mu \neq \mu_0$
0.10	$Z_{crit} = 1.28$	$Z_{crit} = -1.28$	$Z_{crit} = 1.645$
0.05	$Z_{crit} = 1.645$	$Z_{crit} = -1.645$	$Z_{crit} = 1.96$
0.01	$Z_{crit} = 2.33$	$Z_{crit} = -2.33$	$Z_{crit} = 2.58$
Critical region	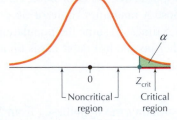	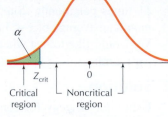	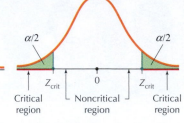
Rejection rule:	Reject H_0 if $Z_{data} \geq Z_{crit}$	Reject H_0 if $Z_{data} \leq Z_{crit}$	Reject H_0 if $Z_{data} \leq -Z_{crit}$ or $Z_{data} \geq Z_{crit}$

EXAMPLE 9.4

FINDING Z_{crit} AND THE CRITICAL REGION

For the hypotheses

$$H_0: \mu = 110 \quad \text{versus} \quad H_a: \mu < 110$$

where μ represents the population mean systolic blood pressure, let the level of significance $\alpha = 0.05$.
a. Find the critical value Z_{crit}.
b. Graph the distribution of Z, showing the critical region.

Solution

We have a left-tailed test and level of significance $\alpha = 0.05$, so Table 9.4 tells us that the critical value is $Z_{crit} = -1.645$. The graph showing the critical region is provided in Figure 9.3. We would reject H_0 for values of Z_{data} that are $\leq Z_{crit} = -1.645$.

FIGURE 9.3
Critical region for a left-tailed test lies in the left (lower) tail.

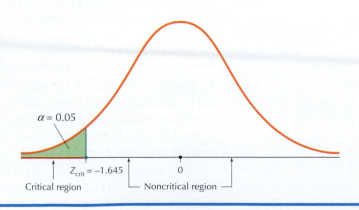

**Now You Can Do
Exercises 17–24.**

What Does the Left-Tailed Test Mean?

A hypothesis test of the form

$$H_0 : \mu = \mu_0 \quad \text{versus} \quad H_a : \mu < \mu_0$$

is called a *left-tailed* test because the critical region lies in the left (lower) tail. Similarly, the critical region for a right-tailed test lies in the right (upper) tail. The critical region for a two-tailed test lies in both the lower and upper tails.

We are now ready to learn the steps for performing the *Z* test for the population mean using the critical-value method.

Z Test for the Population Mean μ: Critical-Value Method

When a random sample of size *n* is taken from a population where the population standard deviation σ is known, you can use the *Z* test if (a) the population is normal, or (b) the sample size is large ($n \geq 30$).

Step 1   **State the hypotheses.**
Use one of the forms from Table 9.4. State the meaning of μ.

Step 2   **Find Z_{crit} and state the rejection rule.**
Use Table 9.4 and the given level of significance α.

Step 3   **Calculate Z_{data}.**

$$Z_{data} = \frac{\bar{x} - \mu_0}{\sigma/\sqrt{n}}$$

Step 4   **State the conclusion and the interpretation.**
If Z_{data} falls in the critical region, then reject H_0; otherwise, do not reject H_0. Interpret your conclusion so that a nonspecialist (that is, someone who has not had a course in statistics) can understand.

What Does This Conclusion Mean?

Interpreting Your Conclusion for Nonspecialists

Recall that a data analyst needs to interpret the results so that nonspecialists can understand them. You can use the following generic interpretation for the two possible conclusions. Just remember that generic interpretations are no substitute for thinking clearly about the problem and the implications of the conclusion.

Interpreting the Conclusion

- If you reject H_0, the interpretation is: *There is evidence at level of significance a that [whatever H_a says].*

- If you do not reject H_0, the interpretation is: *There is insufficient evidence at level of significance α that [whatever H_a says].*

Next, we illustrate the critical-value method of performing a right-tailed *Z* test, a left-tailed *Z* test, and a two-tailed *Z* test for μ.

EXAMPLE 9.5   **Z TEST FOR μ, CRITICAL-VALUE METHOD, RIGHT-TAILED TEST**

Using the debit card sample described in Example 9.3, test at level of significance $\alpha = 0.01$ whether people use debit cards on average more than 11 times per month.

Solution

We may apply the Z test because the sample is large ($n \geq 30$), and the population standard deviation σ is known.

STEP 1 **State the hypotheses.**

From Example 9.3, our hypotheses are

$$H_0 : \mu = 11 \quad \text{versus} \quad H_a : \mu > 11$$

where μ represents the population mean number of times people use their debit cards per month.

STEP 2 **Find Z_{crit} and state the rejection rule.**

We have a right-tailed test and level of significance $\alpha = 0.01$, which, from Table 9.4, tell us that $Z_{crit} = 2.33$. Because we have a right-tailed test, the rejection rule will be "Reject H_0 if $Z_{data} \geq Z_{crit}$," that is, "Reject H_0 if $Z_{data} \geq 2.33$" (see Figure 9.4).

STEP 3 **Find Z_{data}.**

From Example 9.3, we have $Z_{data} = 1$.

STEP 4 **State the conclusion and interpretation.**

Our rejection rule states that we will reject H_0 if $Z_{data} \geq 2.33$. Since $Z_{data} = 1$, which is *not* ≥ 2.33, the conclusion is to *not* reject H_0 (Figure 9.4). Even though the sample mean of 11.5 exceeds 11, it does not do so by a wide enough margin to overcome the reasonable doubt that the difference between this sample mean $\bar{x} = 11.5$ and the hypothesized value $\mu_0 = 11$ may have been due to chance. We interpret our conclusion as follows: "There is insufficient evidence at the 0.01 level of significance that the population mean monthly debit card use is greater than 11 times per month."

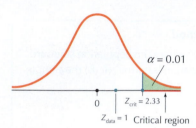

FIGURE 9.4 Critical region for a right-tailed test.

$\alpha = 0.01$

$Z_{crit} = 2.33$

$Z_{data} = 1$ Critical region

Now You Can Do
Exercises 27–29.

EXAMPLE 9.6   **Z TEST FOR μ, CRITICAL-VALUE METHOD, LEFT-TAILED TEST**

For the hypotheses in Example 9.4, perform the Z test for the population mean, using level of significance $\alpha = 0.05$.

Solution

STEP 1 **State the hypotheses.**

From Example 9.4, we have

$$H_0 : \mu = 110 \quad \text{versus} \quad H_a : \mu < 110$$

where μ represents the population mean systolic blood pressure reading.

STEP 2 **Find Z_{crit} and state the rejection rule.**

Example 9.4 gives us the critical value $Z_{crit} = -1.645$, and Table 9.4 tells us that, for level of significance $\alpha = 0.05$, we will reject H_0 if $Z_{data} \leq Z_{crit}$, that is, if $Z_{data} \leq -1.645$ (Figure 9.5).

STEP 3 **Calculate Z_{data}.**

From page 413, we know that

$$Z_{data} = \frac{\bar{x} - \mu_0}{\sigma/\sqrt{n}} = \frac{104 - 110}{10/\sqrt{25}} = -3$$

STEP 4 **State the conclusion and the interpretation.**

In Step 2 we stated that we would reject H_0 if $Z_{data} \leq -1.645$. Since our $Z_{data} = -3 \leq -1.645$, we therefore reject H_0. Our interpretation is: "There is evidence at level of significance $\alpha = 0.05$ that the population mean systolic blood pressure reading is less than 110."

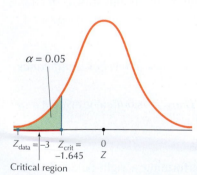

$\alpha = 0.05$

$Z_{data} = -3$ $Z_{crit} = -1.645$ 0 Z

Critical region

FIGURE 9.5 Critical region for a left-tailed test.

Now You Can Do
Exercises 30–32.

EXAMPLE 9.7   ## Z TEST FOR μ, CRITICAL-VALUE METHOD, TWO-TAILED TEST

When the level of hemoglobin in the blood is too low, a person is anemic. Unusually high levels of hemoglobin are undesirable as well and can be associated with dehydration. The optimal hemoglobin level is 13.8 grams per deciliter (g/dl). Suppose a random sample of $n = 25$ women at a certain college showed a sample mean hemoglobin of $\bar{x} = 11.8$ g/dl, the population standard deviation of hemoglobin level is $\sigma = 5$ g/dl, and hemoglobin level is normally distributed. We are interested in testing whether the population mean hemoglobin level differs from 13.8 g/dl. Perform the appropriate hypothesis test, using level of significance $\alpha = 0.10$.

Solution

We may use the Z test, since the population of hemoglobin levels is normally distributed, and the population standard deviation σ is known.

STEP 1   **State the hypotheses.**
The key words "differs from" indicate a two-tailed test, with $\mu_0 = 13.8$. Thus, our hypotheses are

$$H_0 : \mu = 13.8 \quad \text{versus} \quad H_a : \mu \neq 13.8$$

where μ represents the population mean hemoglobin level.

STEP 2   **Find Z_{crit} and state the rejection rule.**
We have a two-tailed test and level of significance $\alpha = 0.10$. Using this information, Table 9.4 tells us that the critical value $Z_{crit} = 1.645$ and that we will reject H_0 if $Z_{data} \leq -1.645$ or if $Z_{data} \geq 1.645$ (Figure 9.6).

STEP 3   **Calculate Z_{data}.**
We have $\bar{x} = 11.8$, $n = 25$, $\sigma = 5$, and $\mu_0 = 13.8$. Substituting:

$$Z_{data} = \frac{\bar{x} - \mu_0}{\sigma/\sqrt{n}} = \frac{11.8 - 13.8}{5/\sqrt{25}} = -2$$

STEP 4   **State the conclusion and the interpretation.**
$Z_{data} = -2$, which is ≤ -1.645. Therefore we reject H_0. There is evidence at level of significance $\alpha = 0.10$ that the population mean hemoglobin level differs from 13.8.

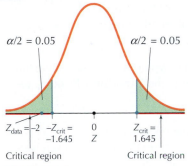

$\alpha/2 = 0.05$ $\alpha/2 = 0.05$

$Z_{data} = -2$ $-Z_{crit} = -1.645$ 0 Z $Z_{crit} = 1.645$

Critical region   Critical region

FIGURE 9.6 Critical region for a two-tailed test.

Now You Can Do Exercises 33 and 34.

STEP-BY-STEP TECHNOLOGY GUIDE: Z Test for μ

To learn how to use technology to perform the Z test for the mean, see the Step-by-Step Technology Guide on page 432.

SECTION 9.2   ## Summary

1. The essential idea about hypothesis testing for the mean is as follows: When the observed value of $\bar{x}$ is unusual or extreme in the sampling distribution of $\bar{x}$ that assumes H_0 is true, we should reject H_0. Otherwise, we should not reject H_0.

2. The critical region consists of the range of values of the test statistic Z_{data} for which we reject the null hypothesis. The value of Z that separates the critical region from the noncritical region is called the critical value Z_{crit}. In the critical-value method for the Z test for the mean, we compare Z_{data} with Z_{crit}.

Clarifying the Concepts

1. What is the essential idea about hypothesis testing for the mean?

2. What does Z_{data} represent?

3. Explain what a test statistic is.

4. Describe the difference between the critical region and the noncritical region.

5. Clearly describe what Z_{crit} is.

6. Suppose we reject H_0 for the hypothesis test $H_0: \mu = 5$ versus $H_a: \mu < 5$. Provide the generic interpretation.

7. How did the right-tailed test get its name?

8. True or false: The value of Z_{crit} does not depend at all on the sample data.

Practicing the Techniques

For Exercises 9–34, assume that the conditions for performing the Z test are met.

For Exercises 9–16, calculate Z_{data}.

9. $H_0: \mu = 10$ vs. $H_a: \mu > 10, \bar{x} = 11, \sigma = 5, n = 25$

10. $H_0: \mu = 10$ vs. $H_a: \mu > 10, \bar{x} = 12, \sigma = 5, n = 25$

11. $H_0: \mu = 10$ vs. $H_a: \mu > 10, \bar{x} = 12.5, \sigma = 5, n = 25$

12. $H_0: \mu = 7$ vs. $H_a: \mu < 7, \bar{x} = 6, \sigma = 4, n = 16$

13. $H_0: \mu = 7$ vs. $H_a: \mu < 7, \bar{x} = 5.5, \sigma = 4, n = 16$

14. $H_0: \mu = 7$ vs. $H_a: \mu < 7, \bar{x} = 4, \sigma = 4, n = 16$

15. $H_0: \mu = 100$ vs. $H_a: \mu \neq 100, \bar{x} = 90, \sigma = 10, n = 25$

16. $H_0: \mu = -50$ vs. $H_a: \mu \neq -50, \bar{x} = -55, \sigma = 5, n = 9$

For Exercises 17–24, do the following:
   a. Find the critical value Z_{crit}.
   b. Sketch the critical region, using the figures in Table 9.4 as a guide.
   c. State the rejection rule.

17. $H_0: \mu = 10$ vs. $H_a: \mu > 10$, level of significance $\alpha = 0.10$

18. $H_0: \mu = 10$ vs. $H_a: \mu > 10$, level of significance $\alpha = 0.05$

19. $H_0: \mu = 10$ vs. $H_a: \mu > 10$, level of significance $\alpha = 0.01$

20. $H_0: \mu = 7$ vs. $H_a: \mu < 7$, level of significance $\alpha = 0.10$

21. $H_0: \mu = 7$ vs. $H_a: \mu < 7$, level of significance $\alpha = 0.05$

22. $H_0: \mu = 7$ vs. $H_a: \mu < 7$, level of significance $\alpha = 0.01$

23. $H_0: \mu = 100$ vs. $H_a: \mu \neq 100$, level of significance $\alpha = 0.05$

24. $H_0: \mu = -50$ vs. $H_a: \mu \neq -50$, level of significance $\alpha = 0.01$

25. Consider your results from Exercises 17–19. Describe what happens to **(a)** Z_{crit} and **(b)** the critical region, for a right-tailed test when the only change is the decrease in the level of significance α.

26. Consider your results from Exercises 20–22. Explain what happens to **(a)** Z_{crit} and **(b)** the critical region, for a left-tailed test as the level of significance α decreases but everything else stays the same.

For Exercises 27–34, use the hypotheses and data from the indicated exercises to perform the Z test for μ by doing the following steps.
   a. State the hypotheses.
   b. Find Z_{crit} and state the rejection rule.
   c. State the value of Z_{data} from the indicated exercise.
   d. State the conclusion and the interpretation.

27. Use Z_{data} from Exercise 9 and Z_{crit} from Exercise 17.

28. Use Z_{data} from Exercise 10 and Z_{crit} from Exercise 18.

29. Use Z_{data} from Exercise 11 and Z_{crit} from Exercise 19.

30. Use Z_{data} from Exercise 12 and Z_{crit} from Exercise 20.

31. Use Z_{data} from Exercise 13 and Z_{crit} from Exercise 21.

32. Use Z_{data} from Exercise 14 and Z_{crit} from Exercise 22.

33. Use Z_{data} from Exercise 15 and Z_{crit} from Exercise 23.

34. Use Z_{data} from Exercise 16 and Z_{crit} from Exercise 24.

Applying the Concepts

For Exercises 35–42, do the following.
   a. State the hypotheses.
   b. Find Z_{crit} and the critical region.
   c. Find Z_{data}. Also, draw a standard normal Z curve showing Z_{crit}, the critical region, and Z_{data}.
   d. State the conclusion and the interpretation.

35. Facebook Connections. According to **Facebook.com**, the mean number of community pages, groups, and events that users are connected to is 80. A random sample of 64 Facebook users showed a mean of 86 connections to community pages, groups, and events. Assume $\sigma = 48$. Test using level of significance $\alpha = 0.05$ whether the population mean number of connections to community pages, groups, and events is greater than 80.

36. Marketing Manager Salaries. The Web site **salary.com** reports that the mean salary for marketing managers is $80,000. A random sample of 25 marketing managers taken during the recession showed a mean salary of $75,000. Assume normality and $\sigma = \$10,000$. Test using level of significance $\alpha = 0.01$ whether the population mean salary for marketing managers fell during the recession.

37. Text Messages. The Pew Internet and American Life Project reports that American adults send a mean of 10 text messages per day. A random sample of 100 American adults showed a mean of 12 text messages per day. Assume $\sigma = 20$. Test using level of significance $\alpha = 0.01$ whether the population mean number of text messages per day differs from 10.

38. Video Gamers. Can't pry the PlayStation away from your dad? The Entertainment Software Association reports that the mean age of video gamers is 37 years old. A random sample of 36 video gamers had a mean age of 36. Assume $\sigma = 6$. Test using level of significance $\alpha = 0.05$ whether the population mean age of video gamers is less than 37.

39. Gas Prices. The American Automobile Association reported in June 2011 that the mean price for a gallon of regular gasoline was $3.70. A random sample of 25 gas stations had a mean price of $3.90. Assume normality and $\sigma = \$0.50$. Test using level of significance $\alpha = 0.05$ whether the population mean price for a gallon of regular gasoline has risen since June 2011.

40. Household Size. The U.S. Census Bureau reports that the mean household size equals 2.58 persons. A random sample of 900 households provides a mean size of 2.56 persons. Assume $\sigma = 0.6$. Conduct a hypothesis test using level of significance $\alpha = 0.10$ to determine whether the population mean household size this year is less than 2.58.

41. Americans' Height. A random sample of 400 Americans yields a mean height of 176 centimeters. Assume $\sigma = 2.5$. Conduct a hypothesis test to investigate whether the population mean height of Americans has changed from 175 centimeters, using level of significance $\alpha = 0.10$.

42. Price of Milk. The U.S. Bureau of Labor Statistics reported that the mean price for a gallon of milk in 2011 was $3.34. A random sample of 100 retail establishments this year provides a mean price of $3.39. Assume $\sigma = \$0.25$. Perform a hypothesis test using level of significance $\alpha = 0.05$ to investigate whether the population mean price of milk this year has increased from the 2011 value.

43. Accountants' Salaries. According to the *Wall Street Journal*, the mean salary for accountants in Texas in 2007 was $50,529. A random sample of 16 Texas accountants this year showed a mean salary of $52,000. We assume that the population standard deviation equals $4000. The histogram of the salary (in $1000s) is shown here. If it is appropriate to apply the Z test, then do so, using the critical-value method and level of significance $\alpha = 0.05$. If not, then explain clearly why not.

Salaries of 16 accountants.

Bringing It All Together

44. Honda Civic Gas Mileage. Cars.com reported in 2007 that the mean city gas mileage for the Honda Civic was 30 mpg. This year, a random sample of 20 Honda Civics had a mean gas mileage of 36 mpg. Assume $\sigma = 5$ mpg. A Minitab histogram of the data is shown here.

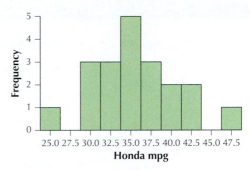

Miles per gallon of 20 imported Hondas.

    a. Is it appropriate to apply the Z test? Explain clearly why or why not.
    b. Test at level of significance $\alpha = 0.10$ whether the population mean city gas mileage has increased since 2007.
    c. What if we now performed the same test on the same data but used $\alpha = 0.05$ instead? Without carrying out the hypothesis test, state whether this would affect our conclusion. Why or why not?

45. Honda Civic Gas Mileage. Refer to Exercise 44. Try to answer the following questions by thinking about the relationship between the statistics rather than by redoing all the calculations. *What if* the 36 mpg is a typo. We are not sure what the actual sample mean is, but it is less than 36 mpg.
    a. How does this affect Z_{data}?
    b. How does this affect Z_{crit}?
    c. How does this affect the conclusion?

46. Automobile Operation Cost. The Bureau of Transportation Statistics reports that the mean cost of operating an automobile in the United States, including gas and oil, maintenance and tires, is 5.9 cents per mile. Suppose that a sample taken this year of 100 automobiles shows a mean operating cost of 6.2 cents per mile, and assume that the population standard deviation is 1.5 cents per mile. Test whether the population mean cost is greater than 5.9 cents per mile, using level of significance $\alpha = 0.05$.

  a. Is it appropriate to apply the Z test? Why or why not?

  b. We have a sample mean that is greater than the mean in the null hypothesis of 5.9 cents. Isn't this enough by itself to reject the null hypothesis? Explain why or why not.

  c. How many standard deviations above the mean is the 6.2 cents per mile? Do you think this is extreme?

47. Automobile Operation Cost. Refer to Exercise 46.

  a. Construct the hypotheses.

  b. Find the Z critical value and state the rejection rule.

  c. Calculate the value of the test statistic Z_{data}.

  d. State the conclusion and the interpretation.

48. Sodium. Work with the **Nutrition** data set.

 Nutrition

  a. Use technology to explore the variable *sodium*.

  b. Use technology to test at level of significance $\alpha = 0.05$ whether the population mean amount of sodium is greater than 280 mg. Let $\sigma = 625$ mg.

  c. Use technology to test at level of significance $\alpha = 0.05$ whether the population mean amount of sodium is greater than 290 mg. Let $\sigma = 625$ mg.

9.3   *Z* TEST FOR THE POPULATION MEAN: *p*-VALUE METHOD

OBJECTIVES   By the end of this section, I will be able to . . .

  1   Perform the Z test for the mean, using the p-value method.

  2   Assess the strength of evidence against the null hypothesis.

  3   Describe the relationship between the p-value method and the critical-value method.

  4   Use the Z confidence interval for the mean to perform the two-tailed Z test for the mean.

1   THE *p*-VALUE METHOD OF PERFORMING THE *Z* TEST FOR THE MEAN

In Section 9.2 we considered the critical-value method for performing the Z test, which works by comparing one Z-value (Z_{data}) with another Z-value (Z_{crit}). In this section we introduce the p-value method, which works by comparing one probability (the p-value) to another probability (α). The two methods are equivalent for the same level of significance α, giving you the same conclusion.

The ***p*-value** is a measure of how well (or how poorly) the data fit the null hypothesis.

> **p-Value**
>
> The **p-value** is the probability of observing a sample statistic (such as $\bar{x}$ or Z_{data}) at least as extreme as the statistic actually observed if we assume that the null hypothesis is true.
>
> Roughly speaking, the p-value represents the probability of observing the sample statistic if the null hypothesis is true. Since the term *p-value* mean "probability value," its value must always lie between 0 and 1.

A p-value is a probability associated with Z_{data} and tells us whether or not Z_{data} is an extreme value. The method for calculating p-values depends on the form of the hypothesis test (Table 9.5).

- For a right-tailed test, the p-value is in the right (or upper) tail area.
- For a left-tailed test, the p-value is in the left (or lower) tail area.
- For a two-tailed test, the p-value lies in both tails.

Table 9.5 Finding the p-value depends on the form of the hypothesis test

	Right-tailed test	**Left-tailed test**	**Two-tailed test**
Type of hypothesis test	$H_0: \mu = \mu_0$ $H_a: \mu > \mu_0$	$H_0: \mu = \mu_0$ $H_a: \mu < \mu_0$	$H_0: \mu = \mu_0$ $H_a: \mu \neq \mu_0$
p-Value is tail area associated with Z_{data}	p-value = $P(Z > Z_{\text{data}})$ Area to right of Z_{data}	p-value = $P(Z < Z_{\text{data}})$ Area to left of Z_{data}	p-value = $P(Z > \lvert Z_{\text{data}} \rvert)$ $+ P(Z < -\lvert Z_{\text{data}} \rvert)$ $= 2 \cdot P(Z > \lvert Z_{\text{data}} \rvert)$ Sum of the two tail areas.

EXAMPLE 9.8

FINDING THE p-VALUE

For each of the following hypothesis tests, calculate and graph the p-value.

a. $H_0: \mu = 3.0$ versus $H_a: \mu > 3.0, Z_{\text{data}} = 1$
b. $H_0: \mu = 10$ versus $H_a: \mu < 10, Z_{\text{data}} = -1.5$
c. $H_0: \mu = 100$ versus $H_a: \mu \neq 100, Z_{\text{data}} = -2$

Solution

a. We have a right-tailed test, so that the p-value equals the area in the right tail:

$$\text{p-value} = P(Z > Z_{\text{data}}) = P(Z > 1)$$

To review how to calculate these probabilities, see Table 6.6 on page 289.

The Z table gives the probability for $P(Z < 1)$, and thus

$$\text{p-value} = P(Z > 1) = 1 - P(Z < 1) = 1 - 0.8413 = 0.1587 \text{ (Figure 9.7a)}.$$

b. We have a left-tailed test, so that the p-value equals the area in the left tail:

$$\text{p-value} = P(Z < Z_{\text{data}}) = P(Z < -1.5) = 0.0668 \text{ (Figure 9.7b)}.$$

Remember that probability is represented by area under the curve.

c. Here we have a left-tailed test, so that the p-value equals the sum of the areas in the two tails:

$$\begin{aligned}\text{p-value} &= P(Z > \lvert Z_{\text{data}} \rvert) + (Z < -\lvert Z_{\text{data}} \rvert)\\ &= P(Z > \lvert -2 \rvert) + (Z < -\lvert -2 \rvert)\\ &= P(Z > 2) + (Z < -2)\\ &= 0.0228 + 0.0228 = 0.0456 \text{ (Figure 9.7c)}\end{aligned}$$

Now You Can Do
Exercises 7–14.

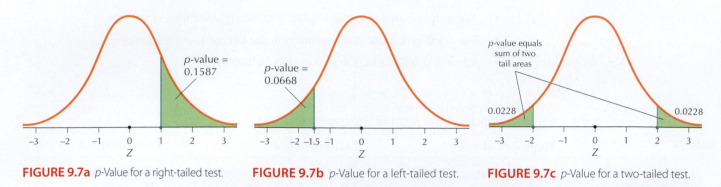

FIGURE 9.7a p-Value for a right-tailed test.   **FIGURE 9.7b** p-Value for a left-tailed test.   **FIGURE 9.7c** p-Value for a two-tailed test.

Since a p-value is based on the value of Z_{data}, the p-value tells us whether or not Z_{data} is an extreme value. Unusual and extreme values of $\bar{x}$, and therefore of Z_{data}, will have a small p-value, while values of $\bar{x}$ and Z_{data} nearer to the center of the distribution will have a large p-value.

> **Assuming H_0 is true:**
>
> Unusual and extreme values of $\bar{x}$ and Z_{data} ⟷ Small p-value
> (close to 0; see Figure 9.7c)
>
> Values of $\bar{x}$ and Z_{data} near center ⟷ Large p-value
> (greater than, say, 0.15; see Figure 9.7a)

A small p-value indicates a conflict between your sample data and the null hypothesis, and will thus *lead us to reject H_0*. However, how small is small? We learned in Section 9.1 that the probability of Type I error α is chosen by the researcher to be small, usually 0.01, 0.05, or 0.10. Thus, a p-value is small if it is $\leq \alpha$. This leads us to the **rejection rule** that tells us when we may reject the null hypothesis.

This rejection rule can be applied to any type of hypothesis test we perform in Chapters 9–11 using the p-value method.

> The rejection rule for performing a hypothesis test using the p-value method is:
> **Reject H_0 when the p-value $\leq \alpha$. Otherwise, do not reject H_0.**

The value of α represents the boundary between results that are statistically significant (where we reject H_0) and results that are not statistically significant (where we do not reject H_0). Thus, α is called the *level of significance* of the hypothesis test.

Here are the steps for performing the Z test for μ using the p-value method.

> ### Z Test for the Population Mean μ: p-Value Method
>
> When a random sample of size n is taken from a population where the standard deviation σ is known, you can use the Z test if either (a) the population is normal, or (b) the sample size is large ($n \geq 30$).
>
> **Step 1   State the hypotheses and the rejection rule.**
>
> Use one of the forms from Table 9.5 to write the hypotheses. State the meaning of μ. The rejection rule is "Reject H_0 if the p-value $\leq \alpha$."
>
> **Step 2   Calculate Z_{data}.**
>
> $$Z_{\text{data}} = \frac{\bar{x} - \mu_0}{\sigma/\sqrt{n}}$$
>
> where the sample mean $\bar{x}$ and the sample size n represent the sample data, and the population standard deviation σ represents the population data.

Step 3   **Find the *p*-value.**
Either use technology to find the *p*-value, or calculate it using the form in Table 9.5 that corresponds to your hypotheses.

Step 4   **State the conclusion and interpretation.**
If the *p*-value $\leq \alpha$, then reject H_0. Otherwise do not reject H_0. Interpret your conclusion so that a nonspecialist (someone who has not had a course in statistics) can understand, as follows:
- Interpretation when you reject H_0 : *There is evidence at level of significance α that [whatever H_a says].*
- Interpretation when you do not reject H_0 : *There is insufficient evidence at level of significance α that [whatever H_a says].*

EXAMPLE 9.9

THE *Z* TEST FOR THE MEAN USING THE *p*-VALUE METHOD: LEFT-TAILED TEST

The technology Web site **www.cnet.com** publishes user reviews of computers, software, and other electronic gadgetry. The mean user rating, on a scale of 1–10, for the Dell XPS 410 desktop computer as of September 10, 2007, was 7.2. Assume that the population standard deviation of user ratings is known to be $\sigma = 0.9$. A random sample taken this year of $n = 81$ user ratings for the Dell XPS 410 showed a mean of $\bar{x} = 7.05$. Using level of significance $\alpha = 0.05$, test whether the population mean user rating for this computer has fallen since 2007.

Solution

The sample size $n = 81$ is large, and the population standard deviation σ is known. We may therefore perform the *Z* test for the mean.

STEP 1   **State the hypotheses and the rejection rule.**
The key words here are "has fallen," which means "is less than." The answer to the question "Less than what?" gives us $\mu_0 = 7.2$. Thus, our hypotheses are

$$H_0: \ \mu = 7.2 \quad \text{versus} \quad H_a: \ \mu < 7.2$$

where μ refers to the population mean user rating for the Dell XPS 410 computer. We will reject H_0 if the *p*-value $\leq \alpha = 0.05$.

STEP 2   **Calculate Z_{data}.**
We have $\bar{x} = 7.05$, $\mu_0 = 7.2$, $n = 81$, and $\sigma = 0.9$. Thus, our test statistic is

$$Z_{data} = \frac{\bar{x} - \mu_0}{\sigma/\sqrt{n}} = \frac{7.05 - 7.2}{0.9/\sqrt{81}} = -1.5$$

STEP 3   **Find the *p*-value.**
Our hypotheses represent a left-tailed test from Table 9.5. Thus

$$p\text{-value} = P(Z < Z_{data}) = P(Z < -1.5)$$

This is a Case 1 problem from Table 6.6 (page 289). The *Z* table (Appendix Table C) provides us with the area to the left of $Z = -1.5$ (Figure 9.8):

$$P(Z < -1.5) = 0.0668$$

Thus, the *p*-value is 0.0668.

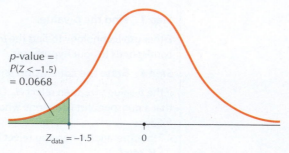

p-value $=$
$P(Z < -1.5)$
$= 0.0668$

$Z_{data} = -1.5$ 0

FIGURE 9.8 The p-value 0.0668 is not ≤ 0.05, so do not reject H_0.

STEP 4 **State the conclusion and interpretation.**
Our level of significance is $\alpha = 0.05$ (from Step 1). Since the p-value $= 0.0668$ is *not* ≤ 0.05, we therefore do *not* reject H_0. There is insufficient evidence at level of significance $\alpha = 0.05$ that the population mean user rating for a Dell XPS 410 computer is less than 7.2.

**Now You Can Do
Exercises 17–19.**

EXAMPLE 9.10   **THE p-VALUE METHOD USING TECHNOLOGY: TWO-TAILED TEST**

Brisbane

The birth weights, in grams (1000 grams = 1 kilogram $\approx$ 2.2 pounds), of a random sample of 44 babies from Brisbane, Australia, have a sample mean weight $\bar{x} = 3276$ grams. Formerly, the mean birth weight of babies in Brisbane was 3200 grams. Assume that the population standard deviation $\sigma = 528$ grams. Is there evidence that the population mean birth weight of Brisbane babies now differs from 3200 grams? Use technology to perform the appropriate hypothesis test, with level of significance $\alpha = 0.10$.

**What Results
Might We Expect?**

Note from Figure 9.9 that the sample mean birth weight $\bar{x} = 3276$ grams is close to the hypothesized mean birth weight of $\mu_0 = 3200$ grams. This value of $\bar{x}$ is not extreme and thus does not seem to offer strong evidence that the hypothesized mean birth weight is wrong. Therefore, we might expect to *not reject* the hypothesis that $\mu_0 = 3200$ grams.

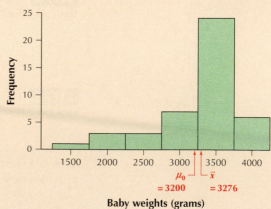

FIGURE 9.9 Sample mean, $\bar{x} = 3276$, is close to hypothesized mean, $\mu_0 = 3200$.

Solution

Since the sample size $n = 44$ is large and $\sigma = 528$ is known, we may proceed with the *Z* test for μ.

STEP 1 **State the hypotheses and the rejection rule.**
The key words "differs from" mean that we have a two-tailed test:

$$H_0 : \mu = 3200 \quad \text{versus} \quad H_a : \mu \neq 3200$$

where μ refers to the population mean birth weight of Brisbane babies. We will reject H_0 if the *p*-value $\leq \alpha = 0.10$.

STEP 2 **Calculate** Z_{data}.
We will use the instructions provided in the Step-by-Step Technology Guide at the end of this section (page 432). Figure 9.10 shows the TI-83/84 results from the *Z* test for μ:

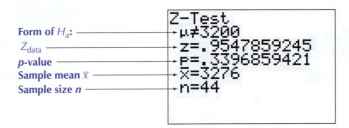

Form of H_a:
Z_{data}
p-value
Sample mean $\bar{x}$
Sample size n

FIGURE 9.10
TI-83/84 results.

$$Z_{\text{data}} = \frac{\bar{x} - \mu_0}{\sigma / \sqrt{n}} = \frac{3276 - 3200}{528 / \sqrt{44}} = 0.9547859245 \approx 0.9548$$

Figure 9.11 shows the Minitab results, where

- "Test of $\mu = 3200$ versus not 3200" refers to the hypotheses being tested, $H_0 : \mu = 3200$ versus $H_a : \mu \neq 3200$.

- "The assumed standard deviation $= 528$" refers to our assumption that $\sigma = 528$.

- SE Mean refers to the standard error of the mean, that is, $\sigma / \sqrt{n}$. You can see that $528 / \sqrt{44} \approx 79.60$.

- 90% CI represents a 90% *Z* confidence interval for μ.

- *Z* refers to our test statistic:

Different software rounds the results to different numbers of decimal places.

$$Z_{\text{data}} = \frac{\bar{x} - \mu_0}{\sigma / \sqrt{n}} = (3276 - 3200)/(528 / \sqrt{44}) = 0.9547859245 \approx 0.9548$$

- *P* represents our *p*-value of 0.340.

FIGURE 9.11
Minitab results.

```
One-Sample Z

Test of mu = 3200 vs not = 3200
The assumed standard deviation = 528


 N     Mean   SE Mean        90% CI          Z      P
44   3276.0      79.6   (3145.1, 3406.9)   0.95  0.340
```

STEP 3 **Find the *p*-value.**
We have a two-tailed test from Step 1, so that from Table 9.5 our *p*-value is (Figure 9.12)

$$p\text{-Value} = 2 \cdot P(Z > |Z_{\text{data}}|) = 2 \cdot P(Z > 0.9548) \approx 2 \cdot (0.1698)$$

$$= 0.3396$$

STEP 4 **State the conclusion and interpretation.**
Since 0.3396 is not ≤ 0.10, we do not reject H_0. There is insufficient evidence that the population mean birth weight differs from 3200 grams.

FIGURE 9.12 *p*-Value is sum of two tail areas: $0.1698 + 0.1698 = 0.3396$.

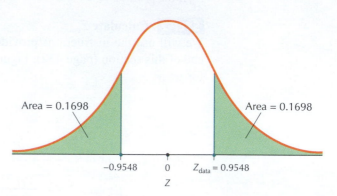

Area = 0.1698   Area = 0.1698

-0.9548 0 $Z_{\text{data}} = 0.9548$

Z

Now You Can Do Exercises 20–22.

The *p-value* applet allows you to experiment with various hypotheses, means, standard deviations, and sample sizes in order to see how changes in these values affect the *p*-value.

2 ASSESSING THE STRENGTH OF EVIDENCE AGAINST THE NULL HYPOTHESIS

The hypothesis-testing methods we have shown so far deliver a simple "yes-or-no" conclusion: either "Reject H_0," or "Do not reject H_0." There is no indication of how strong the evidence is for rejecting the null hypothesis. Was the decision close? Was it a no-brainer? On the other hand, the *p*-value itself represents the *strength of evidence against the null hypothesis*. There is extra information here, which we should not ignore.

For instance, we can directly compare the results of hypothesis tests. Suppose that we have two hypothesis tests that both result in not rejecting the null hypothesis, with level of significance $\alpha = 0.05$. However, Test A has a *p*-value of 0.06, while Test B has a *p*-value of 0.57. Clearly, Test A came very close to rejecting the null hypothesis and shows a fair amount of evidence against the null hypothesis, while Test B shows no evidence at all against the null hypothesis. A simple statement of the "yes-or-no" conclusion misses the clear distinction between these two situations.

Of course, we are free to determine whether the results are significant using whatever α level we wish. For example. Test A would have rejected H_0 for any α value 0.06 or higher. Some data analysts in fact do not think in terms of rejecting or not rejecting the null hypothesis. Rather, they think completely in terms of *assessing the strength of evidence against the null hypothesis*.

For many (though not all) data domains, Table 9.6 provides a thumbnail impression of the strength of evidence against the null hypothesis for various *p*-values. For certain domains (such as the physical sciences), however, alternative interpretations are appropriate.

The p-value provides us with the smallest level of significance at which the null hypothesis would be rejected, that is, the smallest value of α at which the results would be considered significant.

Table 9.6 Strength of evidence against the null hypothesis for various levels of *p*-value

p-Value	Strength of evidence against H_0
p-value ≤ 0.001	Extremely strong evidence
$0.001 <$ *p*-value ≤ 0.01	Very strong evidence
$0.01 <$ *p*-value ≤ 0.05	Solid evidence
$0.05 <$ *p*-value ≤ 0.10	Moderate evidence
$0.10 <$ *p*-value ≤ 0.15	Slight evidence
$0.15 <$ *p*-value	No evidence

Note: Use Table 9.6 for all exercises that ask for an assessment of the strength of evidence against the null hypothesis.

EXAMPLE 9.11

ASSESSING THE STRENGTH OF EVIDENCE AGAINST H_0

Assess the strength of evidence against H_0 shown by the *p*-values in (a) Example 9.9 and (b) Example 9.10.

Solution

a. In Example 9.9, we tested H_0: $\mu = 7.2$ versus H_a: $\mu < 7.2$, where μ refers to the population mean user rating for the Dell XPS 410 computer. Our *p*-value of 0.0668 implies that there is *moderate* evidence against the null hypothesis that the population mean user rating for the Dell XPS 410 computer is 7.2 or higher.

b. In Example 9.10, we tested H_0: $\mu = 3200$ versus H_a: $\mu \neq 3200$, where μ refers to the population mean birth weight of Brisbane babies (in grams). Our *p*-value of 0.3397 implies that there is *no* evidence against the null hypothesis that the population mean birth weight of Brisbane babies equals 3200 grams.

**Now You Can Do
Exercises 23–28.**

**Developing Your
Statistical Sense**

The Role of the Level of Significance α

Suppose that in Example 9.9, our level of significance α was 0.10 rather than 0.05. Would this have changed anything? Certainly. Since our *p*-value of 0.0668 is less than the new $\alpha = 0.10$, we would reject H_0. Think about that for a moment. *The data haven't changed at all, but our conclusion is reversed simply by changing α.* What is a data analyst to make of a situation like this? There are two alternatives.

1. Since we don't want the choice of α to dictate our conclusion, then perhaps we should turn to a direct assessment of the strength of evidence against the null hypothesis, as provided in Table 9.6. In this case, the *p*-value of about 0.0668 would offer moderate evidence against the null hypothesis, *regardless of the value of α.*
2. Obtain more data, perhaps through a call for further research.

3 THE RELATIONSHIP BETWEEN THE *p*-VALUE METHOD AND THE CRITICAL-VALUE METHOD

Figure 9.13 shows the relationships between the *p*-value method and the critical-value method. The top half represents values of *Z* and the critical-value method that we studied in Section 9.2. The bottom half represents probabilities and the *p*-value method that we studied in this section. The left half represents statistics associated with the observed sample data. The right half represents critical-value thresholds for significance that these statistics are compared against.

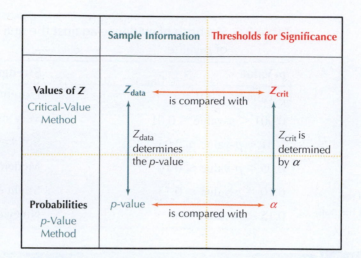

FIGURE 9.13
Critical-value method
and p-value method
are equivalent.

Since Z_{data} helps us to determine the p-value, these two values are related. Similarly, since the level of significance α helps to determine the value of Z_{crit}, these two values are related. Moreover, just as we compare Z_{data} with the threshold Z_{crit}, we compare the p-value statistic with the α threshold to determine significance. Thus, the two methods for carrying out hypothesis tests are equivalent and, in fact, are quite thoroughly interwoven.

Figures 9.14a and 9.14b illustrate this equivalence for a right-tailed test. The rejection rule for the p-value method is to reject H_0 when the p-value $\leq \alpha$. The rejection rule for the critical-value method is to reject H_0 when $Z_{data} \geq Z_{crit}$. Note in Figures 9.14a and 9.14b how the p-value is determined by Z_{data}, and α is determined by Z_{crit}. In Figure 9.14a, when $Z_{data} < Z_{crit}$, it must also happen that the p-value $> \alpha$. In both cases we do not reject H_0. However, in Figure 9.14b, when $Z_{data} \geq Z_{crit}$, it also follows that the p-value is $\leq \alpha$. In both cases we reject H_0. Thus, the p-value method and the critical-value method are equivalent.

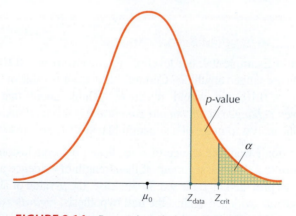

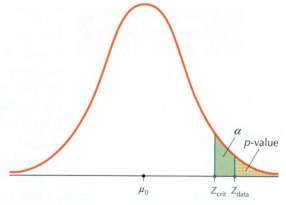

FIGURE 9.14a For a right-tailed test, $Z_{data} < Z_{crit}$ only when p-value $> \alpha$.

FIGURE 9.14b For a right-tailed test, $Z_{data} \geq Z_{crit}$ only when p-value $\leq \alpha$.

4 USING CONFIDENCE INTERVALS FOR μ TO PERFORM TWO-TAILED HYPOTHESIS TESTS ABOUT μ

Consider a two-tailed hypothesis test for μ:

$$H_0 : \mu = \mu_0 \quad \text{versus} \quad H_a : \mu \neq \mu_0$$

and recall the $100(1 - \alpha)\%$ Z confidence interval for μ from Section 8.1:

$$\bar{x} \pm Z_{\alpha/2}(\sigma/\sqrt{n})$$

Both inference methods are based on the *Z* statistic:

$$Z = \frac{\bar{x} - \mu}{\sigma/\sqrt{n}}$$

so it makes sense that the two-tailed hypothesis test and the confidence interval are equivalent.

Equivalence of a Two-Tailed Hypothesis Test and a Confidence Interval

- If a certain value for μ_0 lies *outside* the corresponding $100(1 - \alpha)\%$ *Z* confidence interval for μ, then the null hypothesis specifying this value for μ_0 would be *rejected* for level of significance α (see Figure 9.15).

- Alternatively, if a certain value for μ_0 lies *inside* the $100(1 - \alpha)\%$ *Z* confidence interval for μ, then the null hypothesis specifying this value for μ_0 would *not be rejected* for level of significance α.

FIGURE 9.15

Reject H_0 for values of μ_0 that lie outside confidence interval (a, b).

Lower Bound = a   Upper Bound = b

Reject H_0 Do not reject H_0 Reject H_0

Table 9.7 shows the confidence levels and associated α levels of significance that will produce the equivalent inference.

Table 9.7 Confidence levels for equivalent α levels of significance

Confidence level	Level of significance α
90%	0.10
95%	0.05
99%	0.01

We may thus use a single confidence interval to test as many values of μ_0 as we like.

EXAMPLE 9.12   **LEAD CONTAMINATION IN TROUT, REVISITED**

Recall Example 8.3 from Section 8.1 (page 360), where we were 95% confident using a *Z* interval that the population mean lead contamination for all trout in the Spokane River lies between 0.902 and 1.098 ppm. Once we have constructed the 95% confidence interval, we may test as many possible values for μ_0 as we like. If any values of μ_0 lie inside the confidence interval, that is, between 0.902 and 1.098, we will not reject H_0 for this value of μ_0. If any values of μ_0 lie outside the confidence interval, that is, either to the left of 0.902 or to the right of 1.098, we will reject H_0, as shown in Figure 9.16.

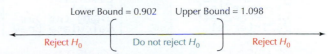

Lower Bound = 0.902   Upper Bound = 1.098

Reject H_0 Do not reject H_0 Reject H_0

FIGURE 9.16 Reject H_0 for values of μ_0 that lie outside (0.902, 1.098).

Test using level of significance $\alpha = 0.05$ whether the population mean lead contamination differs from these values: **(a)** 0.900, **(b)** 0.910, **(c)** 1.100.

Solution

We set up the three two-tailed hypothesis tests as follows:

- **a.** $H_0 : \mu = 0.900$ versus $H_a : \mu \neq 0.900$
- **b.** $H_0 : \mu = 0.910$ versus $H_a : \mu \neq 0.910$
- **c.** $H_0 : \mu = 1.100$ versus $H_a : \mu \neq 1.100$

To perform each hypothesis test, simply observe where each value of μ_0 falls on the number line shown in Figure 9.16. For example, in the first hypothesis test, the hypothesized value $\mu_0 = 0.900$ lies outside the interval (0.902, 1.098). Thus, we reject H_0. The three hypothesis tests are summarized here.

Now You Can Do Exercises 29–34.

Value of μ_0	Form of hypothesis test, with $\alpha = 0.01$	Where μ_0 lies in relation to 95% confidence interval	Conclusion of hypothesis test
a. 0.900	$H_0: \mu = 0.900$ vs. $H_a: \mu \neq 0.900$	Outside	Reject H_0
b. 0.910	$H_0: \mu = 0.910$ vs. $H_a: \mu \neq 0.910$	Inside	Do not reject H_0
c. 1.100	$H_0: \mu = 1.100$ vs. $H_a: \mu \neq 1.100$	Outside	Reject H_0

STEP-BY-STEP GUIDE TO TECHNOLOGY: Z test for μ

We will use the birth weight data from Example 9.10 (page 426).

TI-83/84

If you have the data values:

Step 1   Enter the data into list **L1**.
Step 2   Press **STAT**, highlight **TESTS**, and press **ENTER**.
Step 3   Press **1** (for **Z-Test**; see Figure 9.17).
Step 4   For input (**Inpt**), highlight **Data** and press **ENTER** (Figure 9.18).
a. For μ_0, enter the value of μ_0, **3200**.
b. For σ, enter the value of σ, **528**.
c. For **List**, press **2nd**, then **L1**.
d. For **Freq**, enter **1**.
e. For μ, select the form of H_a. Here we have a right-tailed test, so highlight $> \mu_0$ and press **ENTER**.
f. Highlight **Calculate** and press **ENTER**. The results are shown in Figure 9.10 in Example 9.10.

If you have the summary statistics:

Step 1   Press **STAT**, highlight **TESTS**, and press **ENTER**.
Step 2   Press **1** (for **Z-Test**; see Figure 9.18).
Step 3   For input (**Inpt**), highlight **Stats** and press **ENTER** (Figure 9.19).
a. For μ_0, enter the value of μ_0, **3200**.
b. For σ, enter the value of σ, **528**.
c. For $\bar{x}$, enter the sample mean **3276**.
d. For **n**, enter the sample size **44**.
e. For μ, select the form of H_a. Here we have a right-tailed test, so highlight $> \mu_0$ and press **ENTER**.
f. Highlight **Calculate** and press **ENTER**. The results are shown in Figure 9.10 in Example 9.10.

FIGURE 9.17

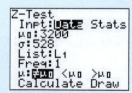

FIGURE 9.18

FIGURE 9.19

EXCEL

WHFStat Macros

Step 1   Enter the data into column A. (If you have only the summary statistics, go to Step 2.)
Step 2   Load the **WHFStat Macros**.
Step 3   Select **Add-Ins > Macros > Testing a Mean > Z Test − Confidence Interval − One Sample**.

Step 4   Select cells A1 to A44 as the **Dataset Range**. (Alternatively, you may enter the summary statistics.)
Step 5   Select your **Confidence level**, which should be $1 - \alpha$. Here, because $\alpha = 0.10$, we select **90%**.
Step 6   Enter the **Population Standard Deviation**, $\sigma = 528$.
Step 7   Enter the **Null Hypothesis Value**, $\mu 0 = 3200$, and click **OK**.

MINITAB

If you have the data values:
Step 1 Enter the data into column C1.
Step 2 Click **Stat > Basic Statistics > 1-Sample Z**.
Step 3 Click **Samples in Columns** and select **C1**.
Step 4 Enter **528** as **Standard Deviation**.
Step 5 For **Test Mean**, enter **3200**.
Step 6 Click **Options**.
a. Choose your **Confidence Level** as $100(1 - \alpha)$. Our level of significance α here is 0.10, so the confidence level is 90.0.
b. Select **Greater Than** to symbolize the right-tailed test.
Step 7 Click **OK** and click **OK** again. The results are shown in Figure 9.11 in Example 9.10.

If you have the summary statistics:
Step 1 Click **Stat > Basic Statistics > 1-Sample Z**.
Step 2 Click **Summarized Data**.
Step 3 Enter the **Sample Size 44** and the **Sample Mean 3276**.
Step 4 Click **Options**.
a. Choose your **Confidence Level** as $100(1 - \alpha)$. Our level of significance α here is 0.10, so the confidence level is 90.0.
b. Select **Greater Than** to symbolize the right-tailed test.
Step 5 Click **OK** and click **OK** again. The results are shown in Figure 9.11 in Example 9.10.

CRUNCHIT!

If you have the data values:
Step 1 Click **File** ... then highlight **Load from Larose2e** ... **Chapter 9** ... and click on **Example 9.10**.
Step 2 Click **Statistics ... Z** and select **1-sample**.
Step 3 With the **Columns** tab chosen, for **Sample** select **Weight**. For **Standard Deviation**, enter **528**.
Step 4 Select the **Hypothesis Test** tab. For **Mean under null hypothesis**, enter **3200**. For **Alternative** select **Greater than**. Then click **Calculate**.

If you have the summary statistics:
Step 1 Click **File** ... then highlight **Load from Larose2e** ... **Chapter 9** ... and click on **Example 9.10**.
Step 2 Click **Statistics ... Z** and select **1-sample**.
Step 3 Choose the **Summarized** tab. For **n** enter the sample size **44**; for **Sample Mean** enter **3276**. For **Standard Deviation**, enter **528**.
Step 4 Select the **Hypothesis Test** tab. For **Mean under null hypothesis**, enter **3200**. For **Alternative** select **Greater than**. Then click **Calculate**.

| **SECTION 9.3** | **Summary** |

1. The p-value can be thought of as the probability of observing a sample statistic at least as extreme as the statistic in your sample if we assume that the null hypothesis is true. The rejection rule for the p-value method is to reject H_0 when the p-value $\leq \alpha$, the level of significance.

2. The p-value can be used to assess the strength of evidence against the null hypothesis.

3. The critical-value method and the p-value method are equivalent, and related in several ways.

4. We can use a single confidence interval for μ to help us perform any number of corresponding two-tailed hypothesis tests about μ.

| **SECTION 9.3** | **Exercises** |

Clarifying the Concepts

1. True or false: It is possible to get a p-value equal to 1.5.

2. State the rejection rule for the p-value method for performing the Z test for μ.

3. Explain why we might want to assess the strength of evidence against the null hypothesis, rather than delivering a simple "reject H_0 or do not reject H_0" conclusion.

4. What is the criterion for rejecting H_0 when using a confidence interval to perform a two-tailed hypothesis test for μ?

5. True or false: For a right-tailed test, when $Z_{data} \leq Z_{crit}$, the p-value is always $\leq \alpha$.

6. For (a)–(c), indicate whether or not the quantity represents a probability.
 a. Z_{data}
 b. p-value
 c. α

Practicing the Techniques

For Exercises 7–34, assume that the conditions for performing the Z test are met.

For Exercises 7–14, find the p-value.

7. $H_0 : \mu = 5$ vs. $H_a : \mu > 5, Z_{data} = 1$

8. $H_0 : \mu = 5$ vs. $H_a : \mu > 5, Z_{data} = 2$

9. $H_0 : \mu = 5$ vs. $H_a : \mu > 5, Z_{data} = 3$

10. $H_0 : \mu = 20$ vs. $H_a : \mu < 20, Z_{data} = -2.5$

11. $H_0 : \mu = 42$ vs. $H_a : \mu < 42, Z_{data} = -2.5$

12. $H_0 : \mu = 50$ vs. $H_a : \mu \neq 50, Z_{data} = 2.9$

13. $H_0 : \mu = 50$ vs. $H_a : \mu \neq 50, Z_{data} = -2.9$

14. $H_0 : \mu = 100$ vs. $H_a : \mu \neq 100, Z_{data} = -1.27$

15. Refer to Exercises 7–9. Explain what happens to the p-value for a right-tailed test as Z_{data} increases.

16. Refer to Exercises 12 and 13. What can we say about the p-values of two two-tailed tests whose values of Z_{data} have the same absolute value?

For Exercises 17–22, perform the Z test for μ using level of significance $\alpha = 0.05$ by doing the following steps.
  a. State the hypotheses and the rejection rule.
  b. Calculate Z_{data}.
  c. Find the p-value.
  d. State the conclusion and the interpretation.

17. $H_0 : \mu = 98.6$ vs. $H_a : \mu > 98.6, \bar{x} = 99.1, \sigma = 10, n = 100$

18. $H_0 : \mu = 32$ vs. $H_a : \mu < 32, \bar{x} = 27, \sigma = 20, n = 25$

19. $H_0 : \mu = -0.1$ vs. $H_a : \mu > -0.1, \bar{x} = 0, \sigma = 1, n = 400$

20. $H_0 : \mu = 100$ vs. $H_a : \mu \neq 100, \bar{x} = 102.3, \sigma = 15, n = 100$

21. $H_0 : \mu = -50$ vs. $H_a : \mu \neq -50, \bar{x} = -46, \sigma = 15, n = 100$

22. $H_0 : \mu = 0$ vs. $H_a : \mu \neq 0, \bar{x} = -1.7, \sigma = 4.5, n = 81$

For Exercises 23–28, use the indicated p-value to assess the strength of evidence against the null hypothesis, using Table 9.6.

23. p-Value from Exercise 17

24. p-Value from Exercise 18

25. p-Value from Exercise 19

26. p-Value from Exercise 20

27. p-Value from Exercise 21

28. p-Value from Exercise 22

For Exercises 29–34, a $100(1 - \alpha)\%$ confidence interval is given. Use the confidence interval to test using level of significance α whether μ differs from each of the indicated hypothesized values.

29. A 95% Z confidence interval for μ is $(-2.7, 6.9)$. Hypothesized values μ_0 are
  a. -3 b. -2 c. 0
  d. 5 e. 7

30. A 99% Z confidence interval for μ is $(45, 55)$. Hypothesized values μ_0 are
  a. 0 b. 44 c. 50
  d. 54 e. 56

31. A 90% Z confidence interval for μ is $(-10, -5)$. Hypothesized values μ_0 are
  a. -3 b. -8 c. -11
  d. 0 e. 7

32. A 95% Z confidence interval for μ is $(1024, 2056)$. Hypothesized values μ_0 are
  a. 1000 b. 2000 c. 3000
  d. 0 e. 1025

33. A 95% Z confidence interval for μ is $(0, 1)$. Hypothesized values μ_0 are
  a. 1.5 b. -1 c. 0.5
  d. 0.9 e. 1.2

34. A 95% Z confidence interval for μ is $(1.3275, 1.4339)$. Hypothesized values μ_0 are
  a. 1.3 b. 1.35 c. 1.4
  d. 1.45 e. 1.3275

Applying the concepts

For Exercises 35–40, do the following.
  a. State the hypotheses and the rejection rule.
  b. Calculate Z_{data}.
  c. Find the p-value.
  d. State the conclusion and the interpretation.

35. Child Abuse. The U.S. Administration for Children and Families reports that the national rate for child abuse referrals is 43.9 per 1000 children. Suppose that a random sample of 1000 children shows 47 child abuse referrals. Assume $\sigma = 5$. Test whether the population mean referral rate has increased, using level of significance $\alpha = 0.10$.

36. California Warming. A 2007 report found that the mean temperature in California increased from 1950 to 2000 by 2 degrees Fahrenheit (°F). Suppose that a random sample of 36 California locations showed a mean increase of 4°F over 1950 levels. Assume $\sigma = 0.5$. Test whether the population mean temperature increase in California is greater than 2°F, at level of significance $\alpha = 0.05$.

37. Eating Trends. According to an NPD Group report, the mean number of meals prepared and eaten at home is less than 700 per year. Suppose that a random sample of 100 households showed a sample mean number of meals prepared and eaten at home of 650. Assume $\sigma = 25$. Test whether the population mean number of such meals is less than 700, using level of significance $\alpha = 0.10$.

38. DDT in Breast Milk. Researchers compared the amount of DDT in the breast milk of 12 Latina women in the Yakima Valley of Washington State with the amount of DDT in breast milk in the general U.S. population.[4] They measured the mean DDT level in the general population to be 47.2 parts per billion (ppb) and the mean DDT level in the 12 Latina women to be 219.7 ppb. Assume $\sigma = 36$ and a normally distributed population. Test whether the population mean DDT level in the breast milk of Latina women in the Yakima Valley is greater than that of the general population, using level of significance $\alpha = 0.01$.

39. Stock Market. The *Statistical Abstract of the United States* reports that the mean daily number of shares traded on the New York Stock Exchange in 2010 was 2 billion. Let this value represent the hypothesized population mean, and assume that the population standard deviation equals 0.5 billion shares. Suppose that, in a random sample of 36 days from the present year, the mean daily number of shares traded equals 2.1 billion. We are interested in testing whether the population mean daily number of shares traded differs from 2 billion using level of significance $\alpha = 0.05$.

40. Tree Rings. Do trees grow more quickly when they are young? The International Tree Ring Data Base collected data on a particular 440-year-old Douglas fir tree.[5] The mean annual ring growth in the tree's first 80 years of life was 1.4261 millimeters (mm). A random sample of size 100 taken from the tree's later years showed a sample mean growth of 0.56 mm per year. Assume $\sigma = 0.5$ mm and a normally distributed population. Test whether the population mean annual ring growth in the tree's later years is less than 1.4261 mm, using level of significance $\alpha = 0.05$.

41. Hybrid Vehicles. A study by **Edmunds.com** examined the time it takes for owners of hybrid vehicles to recoup their additional initial cost through reduced fuel consumption. Suppose that a random sample of 9 hybrid cars showed a sample mean time of 2.1 years. Assume that the population is normal with $\sigma = 0.2$. Test using level of significance $\alpha = 0.01$ whether the population mean time it takes owners of hybrid cars to recoup their initial cost is less than three years.

42. Americans' Height. Americans used to be on average the tallest people in the world. That is no longer the case, according to a study by Dr. Richard Steckel, professor of economics and anthropology at The Ohio State University. The Norwegians and Dutch are now the tallest, at 178 centimeters, followed by the Swedes at 177, and then the Americans, with a mean height of 175 centimeters (approximately 5 feet 9 inches). According to Dr. Steckel, "The average height of Americans has been pretty much stagnant for 25 years."[6] Suppose a random sample of 100 Americans taken this year shows a mean height of 174 centimeters, and we assume $\sigma = 10$ centimeters. Test using level of significance $\alpha = 0.01$ whether the population mean height of Americans this year has changed from 175 centimeters.

43. Cost of Education. The College Board reports that the mean annual cost of education at a private four-year college was $22,218 for the 2006–2007 school year. Suppose that a random sample of 49 private four-year colleges this year gives a mean cost of $24,000 per year. Assume the population standard deviation is $3000.

a. Construct a 95% confidence interval for the population mean annual cost.
b. Use the confidence interval to test at level of significance $\alpha = 0.05$ whether the population mean annual cost differs from the following amounts.

   i. $24,000   **iii.** $23,200
  ii. $23,000   **iv.** $25,000

Health Care Premiums. Use the following information for Exercises 44–46. According to the National Coalition on Health Care, the mean annual premium for an employer health plan covering a family of four cost $13,100 in 2010. A random sample of 100 families of four showed a mean annual premium of $13,700. Assume $\sigma = 3000.

44. Test whether the population mean annual premium is >$13,100, using level of significance $\alpha = 0.05$.

45. *What if* the sample mean premium equaled some value larger than $13,700, while everything else stayed the same. Explain how this change would affect the following, if at all.
a. The hypotheses
b. Z_{crit}
c. The critical region
d. Z_{data}
e. The conclusion

46. Test whether the population mean annual premium is >$13,100 using level of significance $\alpha = 0.01$. Compare your conclusion with the conclusion in Exercise 44. Comment.

Mean Family Size. Use the following information for Exercises 47–49. According to the *Statistical Abstract of the United States,* the mean family size in 2010 was 3.14 persons, reflecting a slow decrease since 1980, when the mean family size was 3.29 persons. Has this trend continued to the present day? Suppose a random sample of 225 families taken this year yields a sample mean size of 3.05 persons, and suppose we assume that the population standard deviation of family sizes is 1 person.

47. Test whether the population mean family size in America has decreased since 2010, using the *p*-value method and level of significance $\alpha = 0.05$. (Try using the *p-value* applet to help you solve this problem.)

48. Refer to Exercise 47
a. What is the smallest *p*-value for which you will reject H_0?
b. Which type of error is it possible that we are making, a Type I error or a Type II error? Which type of error are we certain we are not making?
c. Suppose a newspaper headline referring to the study was "Mean Family Size Decreasing." Is the headline supported or not supported by the data and the hypothesis test?

49. Refer to Exercises 47 and 48, *What if* the 3.05 persons had been a typo, and the actual sample mean was 3.00 persons. How would this have affected the following?

   a. Z_{data}

   b. The *p*-value

   c. The conclusion

50. Women's Heart Rates. A random sample of 15 women produced the normal probability plot for their heart rates shown here. The sample mean was 75.6 beats per minute. Suppose the population standard deviation is known to be 9.

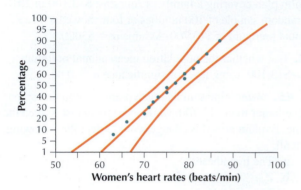

   a. Discuss the evidence for or against the normality assumption. Should we use the Z test? Why or why not?

   b. Assume that the plot does not contradict the normality assumption; test whether the population mean heart rate for all women *is less than* 78, using level of significance $\alpha = 0.05$.

   c. Test whether the population mean heart rate for all women *differs from* 78, using $\alpha = 0.05$.

51. Challenge Exercise. Refer to the previous exercise.

   a. Compare your conclusions from Exercises 50(**b**) and 50(**c**). Note that the conclusions differ but the meanings of the hypotheses tested also differ. Combine the two conclusions into a single sentence. Do you find this sentence difficult to explain?

   b. Explain in your own words the difference between the hypotheses in Exercises 50(**b**) and 50(**c**). Also, explain how there could be evidence that the population mean heart rate is *less than 78 but not different from 78.*

   c. Assess the strength of the evidence against the null hypothesis for the hypothesis tests in Exercises 50(**b**) and 50(**c**).

Bringing It All together

Sodium in Breakfast Cereal. Use the following information for Exercises 52–55. A random sample of 23 breakfast cereals containing sodium had a mean sodium content per serving of 192.39 grams. Assume that the population standard deviation equals 50 grams. We are interested in whether the population mean sodium content per serving is less than 210 grams.

52. a. The normal probability plot of the sodium content is shown here. Should we proceed to apply the Z test? Why or why not?

   b. Test whether the population mean sodium content per serving is less than 210 grams, using level of significance $\alpha = 0.01$.

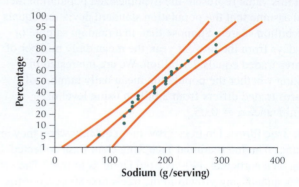

53. *What if* the population standard deviation of 50 grams had been a typo, and the actual population standard deviation was smaller. How would this have affected the following?

   a. The standard deviation of the sampling distribution

   b. Z_{data}

   c. *p*-value

   d. The conclusion

54. *What if* our level of significance α equaled 0.05 instead of 0.01.

   a. Perform the appropriate hypothesis test using the *p*-value method, but this time using level of significance $\alpha = 0.05$.

   b. Note that your conclusion differs from that obtained using level of significance $\alpha = 0.01$. Have the data changed? Why did your conclusion change?

   c. Suggest two alternatives for addressing the contradiction between Exercise 52(**b**) and Exercise 54(**a**).

55. Assess the strength of the evidence against the null hypothesis.

56. Texas Towns. Work with the **Texas** data set for the following.

 Texas

   a. How many observations are in the data set? How many variables?

   b. Use technology to explore the variable *tot_occ*, which lists the total occupied housing units for each county in Texas. Generate numerical summary statistics and graphs for the total occupied housing units. What is the sample mean? The sample standard deviation? Comment on the symmetry or skewness of the data set.

   c. Suppose we are using the data in this data set as a sample of the total occupied housing units of all the counties in the southwestern United States and let $\sigma = 88,400$. Use technology to test at level of significance $\alpha = 0.05$ whether the population mean total occupied housing units for these counties differs from 40,000.

t **TEST FOR THE POPULATION MEAN**

OBJECTIVES   By the end of this section, I will be able to . . .

1   Perform the *t* test for the mean using the critical-value method.

2   Carry out the *t* test for the mean using the *p*-value method.

3   Use confidence intervals to perform two-tailed hypothesis tests.

1 *t* **TEST FOR** μ **USING THE CRITICAL-VALUE METHOD**

Note: Students may wish to review the characteristics of the *t* distribution on page 371.

In many real-world scenarios, the value of the population standard deviation σ is unknown. When this occurs, we should use neither the Z interval nor the Z test. Recall that in Section 8.2 we used the *t* distribution to find a confidence interval for the mean when σ was not known. The situation is similar for hypothesis testing.

Let $\bar{x}$ be the sample mean, μ be the unknown population mean, s be the sample standard deviation, and n be the sample size. The *t* statistic

$$t = \frac{\bar{x} - \mu}{s/\sqrt{n}}$$

with $n - 1$ degrees of freedom may be used when either the population is normal or the sample size is large. We call this *t* statistic t_{data} because its value depends largely on the sample data and the population data.

> The test statistic used for the *t* test for the mean is
> $$t_{data} = \frac{\bar{x} - \mu_0}{s/\sqrt{n}}$$

t_{data} represents the number of standard errors $\bar{x}$ lies above or below μ_0.

Extreme values of $\bar{x}$, that is, values of $\bar{x}$ that are significantly far from the hypothesized μ, will translate into extreme values of t_{data}. In other words, just as with Z_{data}, when $\bar{x}$ is far from μ_0, t_{data} will be far from 0. We answer the question "How extreme is extreme?" using the critical-value method by finding a *critical value* of *t*, called t_{crit}. This threshold value t_{crit} separates the values of t_{data} for which we reject H_0 (the *critical region*) from the values of t_{data} for which we will not reject H_0 (the *noncritical region*). Because there is a different *t* curve for every different sample size, you need to know the following to find the value of t_{crit}:

The degrees of freedom is a measure of how the *t* distribution changes as the sample size changes.

- the form of the hypothesis test (one-tailed or two-tailed)
- the degrees of freedom (df $= n - 1$)
- the level of significance α

> *t* **Test for the Population Mean** μ**: Critical-Value Method**
>
> When a random sample of size *n* is taken from a population, you can use the *t* test if either the population is normal or the sample size is large ($n \geq 30$).
>
> *Step 1*   **State the hypotheses.**
> Use one of the forms from Table 9.8. State the meaning of μ.
>
> *Step 2*   **Find** t_{crit} **and state the rejection rule.**
> Use Table D in the Appendix and Table 9.8.
>
> *Step 3*   **Calculate** t_{data}**.**
> $$t_{data} = \frac{\bar{x} - \mu_0}{s/\sqrt{n}}$$
>
> *Step 4*   **State the conclusion and the interpretation.**
> If t_{data} falls within the critical region, then reject H_0. Otherwise, do not reject H_0. Interpret your conclusion so that a nonspecialist can understand.

Table 9.8 contains the critical regions and rejection rules for the t test.

Table 9.8 Critical regions and rejection rules for various forms of the t test for μ

	Right-tailed test	Left-tailed test	Two-tailed test
Form of test	$H_0: \mu = \mu_0$ $H_a: \mu > \mu_0$ level of significance α	$H_0: \mu = \mu_0$ $H_a: \mu < \mu_0$ level of significance α	$H_0: \mu = \mu_0$ $H_a: \mu \neq \mu_0$ level of significance α
Critical region	0 t_{crit} Noncritical region — Critical region	$-t_{crit}$ 0 Critical region — Noncritical region	$-t_{crit}$ 0 t_{crit} Critical region — Noncritical region — Critical region
Rejection rule	Reject H_0 if $t_{data} \geq t_{crit}$	Reject H_0 if $t_{data} \leq -t_{crit}$	Reject H_0 if $t_{data} \geq t_{crit}$ or $t_{data} \leq -t_{crit}$

EXAMPLE 9.13

t TEST FOR μ USING CRITICAL-VALUE METHOD: LEFT-TAILED TEST

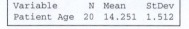

```
Variable      N  Mean   StDev
Patient Age  20  14.251  1.512
```

Minitab description statistics.

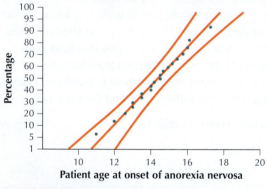

FIGURE 9.20 Normal probability plot for age at onset of anorexia nervosa.

We are interested in testing, using level of significance $\alpha = 0.05$, whether the mean age at onset of anorexia nervosa in young women has been decreasing. Assume that the previous mean age at onset was 15 years old. Data were gathered for a study of the onset age for this disorder.[7] From these data, a random sample was taken of $n = 20$ young women who were admitted under this diagnosis to the Toronto Hospital for Sick Children. The Minitab descriptive statistics shown here indicate a sample mean age of $\bar{x} = 14.251$ years and a sample standard deviation of $s = 1.512$ years. If appropriate, perform the t test.

Solution

Since the sample size $n = 20$ is not large, we need to verify normality. The normal probability plot of the ages at onset in Figure 9.20 indicates that the ages in the sample are normally distributed. We may proceed to perform the t test for the mean.

STEP 1 **State the hypotheses.**
The key word "decreasing" guides us to state our hypotheses as follows:

$$H_0: \mu = 15 \quad \text{versus} \quad H_a: \mu < 15$$

where μ refers to the population mean age at onset.

STEP 2 **Find t_{crit} and state the rejection rule.**
Our hypotheses from Step 1 indicate that we have a left-tailed test, meaning that the critical region represents an area in the left tail (see Figure 9.22). To find t_{crit}, we turn to the t table, an excerpt of which is shown in Figure 9.21. Since we have a one-tailed test, under "Area in one tail," select the column with our α value 0.05. Then choose the row with our df $= n - 1 = 20 - 1 = 19$, so that we get $t_{crit} = 1.729$. Because we have a left-tailed test, the rejection rule from Table 9.8 is "Reject H_0 if $t_{data} \leq -t_{crit}$"; that is, we will reject H_0 if $t_{data} \leq -1.729$.

				Area in one tail
		0.10	0.05	0.025
				Area in two tails
		0.20	0.10	0.05
df	1	3.078	6.314	12.706
	2	1.886	2.920	4.303
	3	1.638	2.353	3.182
	4	1.533	2.132	2.776
	5	1.476	2.015	2.571
	6	1.440	1.943	2.447
	7	1.415	1.895	2.365
	8	1.397	1.860	2.306
	9	1.383	1.833	2.262
	10	1.372	1.812	2.228
	11	1.363	1.796	2.201
	12	1.356	1.782	2.179
	13	1.350	1.771	2.160
	14	1.345	1.761	2.145
	15	1.341	1.753	2.131
	16	1.337	1.746	2.120
	17	1.333	1.740	2.110
	18	1.330	1.734	2.101
	19	1.328	1.729	2.093
	20	1.325	1.725	2.086

FIGURE 9.21 Finding t_{crit} for a one-tailed test. For a two-tailed test, use "Area in two tails."

STEP 3 Calculate t_{data}.
We have $n = 20$, $\bar{x} = 14.251$, and $s = 1.512$ years. Also, $\mu_0 = 15$, since this is the hypothesized value of μ stated in H_0. Therefore, our test statistic is

$$t_{data} = \frac{\bar{x} - \mu_0}{s/\sqrt{n}} = \frac{14.251 - 15}{1.512/\sqrt{20}} \approx -2.2154$$

STEP 4 State the conclusion and interpretation.
The rejection rule from Step 2 says to reject H_0 if $t_{data} \leq -1.729$. From Step 3, we have $t_{data} = -2.2154$. Since -2.2154 is less than -1.729, our conclusion is to reject H_0. If you prefer the graphical approach, consider Figure 9.22, which shows where t_{data} falls in relation to the critical region. Since $t_{data} = -2.2154$ falls within the critical region, our conclusion is to reject H_0. There is evidence at level of significance $\alpha = 0.05$ that the population mean age of onset has decreased from its previous level of 15 years.

FIGURE 9.22
Our $t_{data} = -2.2154$ falls in the critical region.

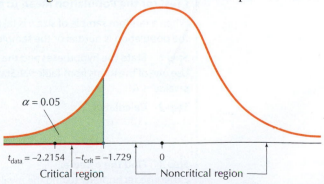

$\alpha = 0.05$

$t_{data} = -2.2154$ $-t_{crit} = -1.729$ 0

Critical region   Noncritical region

Now You Can Do
Exercises 3–8.

EXAMPLE 9.14 *t* TEST FOR μ USING CRITICAL VALUE METHOD: TWO-TAILED TEST

The Pew Internet and American Life Project reported in 2010 that the mean number of text messages sent and received daily by teenagers is 50. Suppose another researcher disputes this finding and is interested in testing whether the population mean number of text messages differs from 50. A random sample of $n = 100$ teenagers yields a sample mean of $\bar{x} = 47.75$ text messages, with a sample standard deviation of $s = 15$ messages. If the conditions are met, perform the appropriate hypothesis test using level of significance $\alpha = 0.10$.

Solution

Since $n = 36 > 30$, we may proceed with the *t* test.

See Example 8.11 (pages 374–375) for why we are taking the next higher df.

STEP 1 **State the hypotheses.**
The key words "differs from" indicate a two-tailed test, with $\mu_0 = 50$, because we are testing whether μ differs from 50. So our hypotheses are

$$H_0 : \ \mu = 50 \quad \text{versus} \quad H_a : \ \mu \neq 50$$

where μ represents the population mean number of text messages sent and received by teenagers daily.

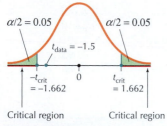

FIGURE 9.23 Critical region for two-tailed test.

STEP 2 **Find t_{crit} and state the rejection rule.**
To find t_{crit} for a two-tailed test with level of significance $\alpha = 0.10$, we look in the 0.10 column in the "Area in two tails" section of Table D in the Appendix. The degrees of freedom df $= n - 1 = 99$ are not listed, so we take the next higher degrees of freedom, df $= 90$, giving us $t_{\text{crit}} = 1.662$. From Table 9.8, the rejection rule is: "Reject H_0 if $t_{\text{data}} \geq 1.662$ or $t_{\text{data}} \leq -1.662$.

STEP 3 **Calculate t_{data}:**

$$t_{\text{data}} = \frac{\bar{x} - \mu_0}{s/\sqrt{n}} = \frac{47.75 - 50}{15/\sqrt{100}} = -1.5$$

Now You Can Do Exercises 9–14.

STEP 4 **State the conclusion and the interpretation.**
See Figure 9.23. $t_{\text{data}} = -1.5$ is not ≥ 1.662 and it is not ≤ -1.662; therefore, we do not reject H_0. There is insufficient evidence at level of significance $\alpha = 0.10$ that the population mean number of text messages differs from 50.

2 *t* TEST FOR μ USING THE *p*-VALUE METHOD

We may also use the *p*-value method for performing the *t* test for μ. The critical-value method and the *p*-value are equivalent, so they will provide identical conclusions.

---

t Test for the Population Mean μ: p-Value Method

When a random sample of size *n* is taken from a population, you can use the *t* test if either the population is normal or the sample size is large ($n \geq 30$).

Step 1 **State the hypotheses and the rejection rule.**
Use one of the forms from Table 9.9. State the meaning of μ. The rejection rule is "Reject H_0 if the *p*-value $\leq \alpha$."

Step 2 **Calculate t_{data}.**

$$t_{\text{data}} = \frac{\bar{x} - \mu_0}{s/\sqrt{n}}$$

Step 3 **Find the *p*-value.**
Either use technology to find the *p*-value or estimate the *p*-value using Table D, *t* Distribution, in the Appendix.

Step 4   **State the conclusion and the interpretation.**
If the *p*-value $\le \alpha$, then reject H_0. Otherwise, do not reject H_0. Interpret your conclusion so that a nonspecialist can understand.

The definition of a *p*-value for a *t* test is similar to the *p*-value for a *Z* test. Unusual and extreme values of $\bar{x}$, and therefore of t_{data}, will have a small *p*-value, while values of $\bar{x}$ and t_{data} nearer to the center of the distribution will have a large *p*-value. Table 9.9 summarizes the definition of the *p*-value for *t* tests. Note that we will not be finding these *p*-values manually but will either (a) use a computer or calculator or (b) estimate them using the *t* table.

Table 9.9 *p*-Values for *t* tests

	Right-tailed test	**Left-tailed test**	**Two-tailed test**
Form of test	$H_0 : \mu = \mu_0$ $H_a : \mu > \mu_0$ level of significance α	$H_0 : \mu = \mu_0$ $H_a : \mu < \mu_0$ level of significance α	$H_0 : \mu = \mu_0$ $H_a : \mu \ne \mu_0$ level of significance α
p-Value is tail area associated with t_{data}	*p*-value = $P(t > t_{data})$ Area to the right of t_{data}	*p*-value = $P(t < t_{data})$ Area to the left of t_{data}	*p*-value = $P(t > \lvert t_{data}\rvert) + P(t < -\lvert t_{data}\rvert)$ $= 2 \cdot P(t > \lvert t_{data}\rvert)$ Sum of the two tail areas

---

EXAMPLE 9.15

t TEST USING THE *p*-VALUE METHOD: RIGHT-TAILED TEST

City	Price
Baltimore	$3.75
Chicago	$3.00
Detroit	$2.70
Hartford	$3.66
Houston	$3.36
Los Angeles	$3.32
Miami	$3.80
New York	$3.92
Philadelphia	$3.91
St. Louis	$3.58

milkprice

FIGURE 9.24 Normal probability plot of milk prices.

The U.S. Bureau of Labor Statistics reports that the mean price for a gallon of milk in January 2011 was $3.34. Gallons of milk were bought in a sample of $n = 10$ different cities, with the prices shown in the accompanying table. Test using level of significance $\alpha = 0.10$ whether the population mean price for a gallon of milk is greater than $3.34.

Solution

We first check whether the conditions for performing the *t* test are met. Because our sample size is small, we must check for normality. The normal probability plot in Figure 9.24 shows acceptable normality, allowing us to proceed with the *t* test.

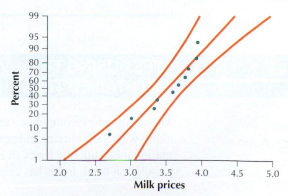

STEP 1   **State the hypotheses and the rejection rule.**

The key words "is greater than" means that we have a right-tailed test. Answering the question "Greater than what?" gives us $\mu_0 = 3.34$.

$$H_0 : \mu = 3.34 \quad \text{versus} \quad H_a : \mu > 3.34$$

where μ represents the population mean price of milk. We will reject H_0 if the p-value $\leq \alpha = 0.10$.

STEP 2   **Calculate t_{data}.**

We use the instructions from the Step-by-Step Technology Guide on page 447. Figure 9.25 shows the TI-83/84 results from the t test for μ.

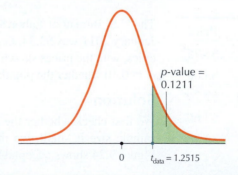

FIGURE 9.25
TI-83/84 results for right-tailed t test.

Form of H_a:
t_{data}
p-value
Sample mean $\bar{x}$
Sample standard deviation s
Sample size n

```
T-Test
μ>3.34
t=1.251511662
p=.1211489193
x̄=3.5
Sx=.4042826294
n=10
```

For a more accurate calculation of the p-value, we retain 9 decimal places for the value of t_{data}.

Using the statistics from Figure 9.25 we have the test statistic

$$t_{\text{data}} = \frac{\bar{x} - \mu_0}{s/\sqrt{n}} = \frac{3.5 - 3.34}{0.4042826294/\sqrt{10}} = 1.251511662 \approx 1.2515$$

STEP 3   **Find the p-value.**

From Figures 9.25 and 9.26, we have

$$p\text{-value} = P(t \geq 1.251511662) = 0.1211489193 \approx 0.1211$$

STEP 4   **State the conclusion and the interpretation.**

The p-value ≈ 0.1211 is not less than the level of significance $\alpha = 0.10$, so therefore do not reject H_0. There is insufficient evidence at level of significance $\alpha = 0.10$ that the population mean price of milk is greater than \$3.34.

FIGURE 9.26
The p-value for a right-tailed t test.

p-value = 0.1211

0 $t_{\text{data}} = 1.2515$

**Now You Can Do
Exercises 15–20.**

EXAMPLE 9.16

t TEST USING THE p-VALUE METHOD: TWO-TAILED TEST

cancercare

The table below contains a random sample of 10 highly rated cancer care facilities, along with their nursing index (nurse-to-patient ratio), in 2007.[8] Suppose that the population mean nursing index in 2005 was 1.6 nurses per cancer patient. Test whether the population mean index has changed using level of significance $\alpha = 0.05$.

Hospital	Index
Memorial Sloan Kettering Cancer Center	1.5
M. D. Anderson Cancer Center	2.0
Johns Hopkins Hospital	2.3
Mayo Clinic	2.8
Dana Farber Cancer Institute	0.8
Univ. of Washington Medical Center	2.2
Duke University Medical Center	1.8
Univ. of Chicago Hospitals	2.3
UCLA Medical Center	2.2
UC San Francisco Medical Center	2.3

Solution

Since the sample size is small, we check normality. The normal probability plot (Figure 9.27) is not perfectly linear, but there are no points outside the bounds, and it is difficult to determine normality for such small sample sizes. We proceed to perform the *t* test, with the caveat that the normality assumption could be better supported and that more data would be helpful.

FIGURE 9.27

Normal probability plot of nursing index

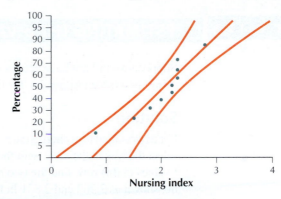

STEP 1 **State the hypotheses and the rejection rule.**
The key words "has changed" means that we have a two-tailed test:

$$H_0: \mu = 1.6 \quad \text{versus} \quad H_a: \mu \neq 1.6$$

where μ represents the population mean nursing index. We will reject H_0 if the *p*-value $\leq \alpha = 0.05$.

STEP 2 **Calculate t_{data}.**
We use the instructions supplied in the Step-by-Step Technology Guide at the end of this section. Figure 9.28 shows the TI-83/84 results from the *t* test for μ.

FIGURE 9.28
TI-83/84 results.

Form of H_a: —
t_{data} —
p-value —
Sample mean $\bar{x}$ —
Sample standard deviation s —
Sample size n —

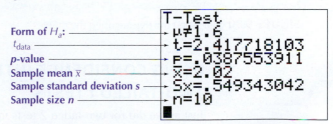

```
T-Test
μ≠1.6
t=2.417718103
p=.0387553911
x=2.02
Sx=.549343042
n=10
```

Using the statistics from Figure 9.28, we have the test statistic

We are retaining decimal places for more accurate calculation of the *p*-value.

$$t_{\text{data}} = \frac{\bar{x} - \mu_0}{s/\sqrt{n}} = \frac{2.02 - 1.6}{0.549343042/\sqrt{10}} \approx 2.417718103 \approx 2.4177$$

STEP 3 Find the *p*-value.
From Figures 9.28 and 9.29, we have

$$p\text{-value} = P(t > |2.417718103|) + P(t < -|2.417718103|) \approx 0.03876$$

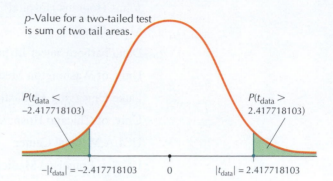

p-Value for a two-tailed test is sum of two tail areas.

$P(t_{\text{data}} < -2.417718103)$

$P(t_{\text{data}} > 2.417718103)$

$-|t_{\text{data}}| = -2.417718103$ 0 $|t_{\text{data}}| = 2.417718103$

FIGURE 9.29
The *p*-value for a two-tailed test.

STEP 4 State the conclusion and interpretation.
The *p*-value of 0.03876 is less than $\alpha = 0.05$. We therefore reject H_0. There is evidence at level of significance $\alpha = 0.05$ that the population mean nurse-to-patient ratio differs from 1.6.

Now You Can Do
Exercises 21–26.

EXAMPLE 9.17

ESTIMATING THE *p*-VALUE USING THE *t* TABLE

Suppose we did not have access to technology. Estimate the *p*-value from Example 9.16 using the *t* table (Appendix Table D).

Solution

For a two-tailed test, choose the row of the *t* table with the heading "Area in two tails." Then select the row in the table with the appropriate degrees of freedom df = $n - 1 = 9$. Of the *t*-values in this row, find the two *t*-values between which the value of $t_{\text{data}} = 2.4177$ would lie, shown as 2.262 and 2.821 in Figure 9.30. The *p*-value must, therefore, lie between the corresponding *p*-values, 0.05 and 0.02. Thus we estimate the *p*-value for Example 9.16 to lie between 0.02 and 0.05, which of course it does: *p*-value ≈ 0.03876.

Now You Can Do
Exercises 27–30.

		Area in two tails				
		0.20	0.10	0.05	0.02	0.01
df	9	1.383	1.833	2.262	2.821	3.250

t_{data} lies between 2.262 and 2.821, so the *p*-value lies between 0.05 and 0.02

FIGURE 9.30 Estimating the *p*-value using the *t* table.

3 USING CONFIDENCE INTERVALS TO PERFORM TWO-TAILED *t* TESTS

Just as we did for two-tailed *Z* tests in Section 9.3, we may use a $100(1 - \alpha)\%$ *t* confidence interval to perform a two-tailed *t* test with level of significance α for various hypothesized values of μ_0. The strategy is the same: if a certain value for μ_0 lies outside the $100(1 - \alpha)\%$ *t* confidence interval for μ, then the null hypothesis specifying this value for μ_0 would be rejected. Otherwise it would not be rejected.

EXAMPLE 9.18

USING A CONFIDENCE INTERVAL TO PERFORM TWO-TAILED *t* TESTS

Example 8.10 (pages 373–374) provided a 95% confidence interval for the population mean length (in centimeters) of fourth-graders' feet as (22.496, 23.694). Test using level of significance $\alpha = 0.05$ whether the population mean length of fourth-graders' feet differs from these values: (a) 22 cm, (b) 23 cm, (c) 24 cm.

Solution

The key words "differs from" mean that we are using two-tailed tests. Then, for each hypothesized value of μ_0, we determine whether it falls inside or outside the given confidence interval.

a. $H_0 : \mu = 22$ versus $H_a : \mu \neq 22$
The confidence interval is (22.496, 23.694), and since $\mu_0 = 22$ lies outside the interval (see Figure 9.31), we reject H_0.

b. $H_0 : \mu = 23$ versus $H_a : \mu \neq 23$
$\mu_0 = 23$ lies inside the interval, so we do not reject H_0.

c. $H_0 : \mu = 24$ versus $H_a : \mu \neq 24$
$\mu_0 = 24$ lies outside the interval, so we reject H_0.

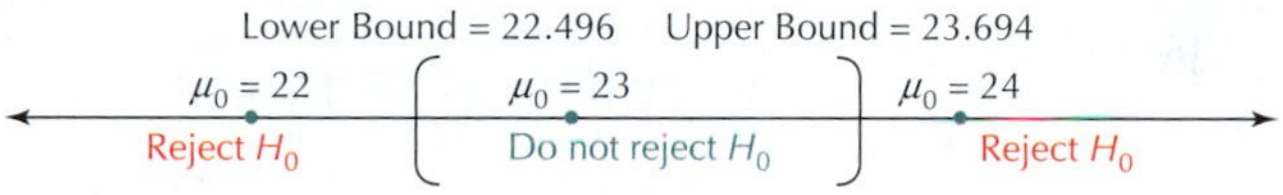

**Now You Can Do
Exercises 31–36.**

FIGURE 9.31 Reject H_0 for values of μ_0 that lie outside (22.496, 23.694).

CASE STUDY   ## The Golden Ratio

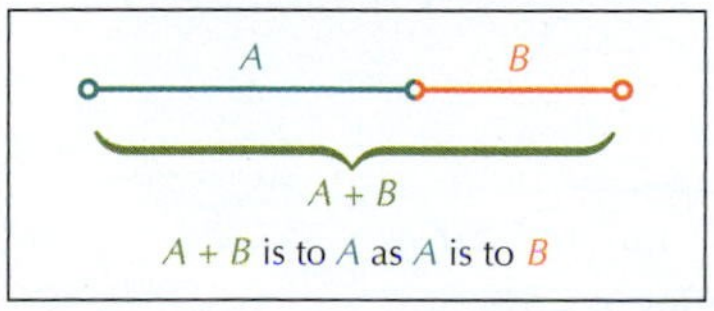

© Purestock

FIGURE 9.32 The golden ratio.

Euclid's *Elements*, the Parthenon, the *Mona Lisa*, and the beadwork of the Shoshone tribe all have in common an appreciation for the *golden ratio*.

> Suppose we have two quantities *A* and *B*, with *A* > *B* > 0. Then *A/B* is called the *golden ratio* if
>
> $$\frac{A + B}{A} = \frac{A}{B}$$
>
> that is, if the ratio of the sum of the quantities to the larger quantity equals the ratio of the larger to the smaller (see Figure 9.32).

Euclid wrote about the golden ratio in his *Elements,* calling it the "extreme and mean ratio." The ratio of the width *A* and height *B* of the Parthenon, one of the most famous temples in ancient Greece, equals the golden ratio (Figure 9.32). If you enclose the face of Leonardo da Vinci's *Mona Lisa* in a rectangle, the resulting ratio of the long side to the short side follows the golden ratio (Figure 9.33 on the next page). The golden ratio has a value of approximately 1.618.

Now we will test whether there is evidence for the use of the golden ratio in the artistic traditions of the Shoshone, a Native American tribe from the American West.

(continues)

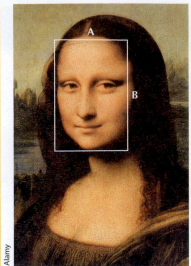

FIGURE 9.33 Mona Lisa's face follows the golden ratio.

Figure 9.34 shows a detail of a nineteenth-century Shoshone beaded dress that belonged to Nahtoma, the daughter of Chief Washakie of the Eastern Shoshone.[9] It is intriguing to consider whether Shoshone beaded rectangles such as those on this dress follow the golden ratio.

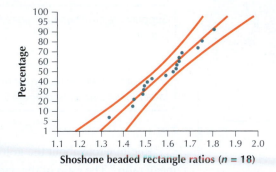

William R. McIver Collection, American Heritage Center, University of Wyoming.

FIGURE 9.34 Beaded dress of Nahtoma, daughter of Chief Washakie, showing rectangles that may follow the golden ratio.

Table 9.10 contains the ratios of lengths to widths of 18 beaded rectangles made by Shoshone artisans.[10] We will perform a hypothesis test to determine whether the population mean ratio of Shoshone beaded rectangles equals the golden ratio of 1.618.

DATA FILE Shoshone

Table 9.10 Ratio of length to width of a sample of Shoshone beaded rectangles

1.44300	1.75439	1.64204	1.66389	1.63666
1.51057	1.33511	1.52905	1.73611	1.80832
1.44928	1.48810	1.62602	1.49254	
1.65017	1.59236	1.49701	1.65017	

Since the population standard deviation for such rectangles is unknown, we must use a *t* test rather than a *Z* test. Our sample size $n = 18$ is not large, so we must assess whether the data are normally distributed. Figure 9.35 shows the normal probability plot indicating acceptable support for the normality assumption. We proceed with the *t* test, using level of significance $\alpha = 0.05$.

FIGURE 9.35
Normal probability plot.

Shoshone beaded rectangle ratios ($n = 18$)

Solution

We use the TI-83/84 to perform this hypothesis test, using the Step-by-Step Technology Guide at the end of this section.

Step 1   State the hypotheses and the rejection rule.

Since we are interested in whether the population mean length-to-width ratio of Shoshone beaded rectangles *equals* the golden ratio of 1.618, we perform a two-tailed test:

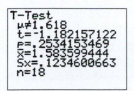

FIGURE 9.36 TI-83/84 results.

$$H_0: \mu = 1.618 \quad \text{versus} \quad H_a: \mu \neq 1.618$$

where μ represents the population mean length-to-width ratio of Shoshone beaded rectangles. We will reject H_0 if the *p*-value ≤ 0.05.

Step 2   Find t_{data}.
From Figure 9.36, we have $t_{\text{data}} \approx -1.1822$

Step 3   Find the *p*-value.
From Figure 9.36, we have *p*-value $\approx 0.1267 + 0.1267 = 0.2534$ (Figure 9.37).

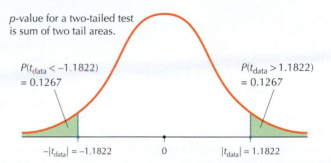

FIGURE 9.37
p-Value for *t* test.

Step 4   State the conclusion and interpretation.
Since *p*-value ≈ 0.2534 is *not* $\leq \alpha = 0.05$, we do *not* reject H_0. Thus, there is insufficient evidence at level of significance $\alpha = 0.05$ that the population mean ratio differs from 1.618. In other words, the data do not reject the claim that Shoshone beaded rectangles follow the same golden ratio exhibited by the Parthenon and the *Mona Lisa*. ∎

STEP-BY-STEP TECHNOLOGY GUIDE: *t* test for μ

We will use the nurse-to-patient ratio data from Example 9.16 (page 442).

TI-83/84

If you have the data values:
Step 1   Enter the data into list **L1**.
Step 2   Press **STAT**, highlight **TESTS**, and press **ENTER**.
Step 3   Press **2** (for **T-Test**; see Figure 9.38).
Step 4   For input (**Inpt**), highlight **Data** and press **ENTER** (Figure 9.39).
a. For μ_0, enter the value of μ_0, **1.6**.
b. For **List**, press **2nd**, then **L1**.
c. For **Freq**, enter **1**.
d. For μ, select the form of H_a. Here we have a two-tailed test, so highlight $\neq \mu_0$ and press **ENTER** (Figure 9.39).
e. Highlight **Calculate** and press **ENTER**. The results are shown in Figure 9.28 in Example 9.16.

If you have the summary statistics:
Step 1   Press **STAT**, highlight **TESTS**, and press **ENTER**.
Step 2   Press **2** (for **T-Test**; see Figure 9.38).
Step 3   For input (**Inpt**), highlight **Stats** and press **ENTER** (Figure 9.40).
a. For μ_0, enter the value of μ_0, **1.6**.
b. For **Sx**, enter the value of *s*, **0.549343042**.
c. For $\bar{x}$, enter the sample mean **2.02**.
d. For n, enter the sample size **10**.
e. For μ, select the form of H_a. Here we have a two-tailed test, so highlight $\neq \mu_0$ and press **ENTER** (Figure 9.40).
f. Highlight **Calculate** and press **ENTER**. The results are shown in Figure 9.28 in Example 9.16.

FIGURE 9.38

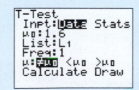

FIGURE 9.39

FIGURE 9.40

EXCEL

WHFStat Add-ins
Step 1 Enter the data into column A. (If you have only the summary statistics, go to Step 2.)
Step 2 Load the **WHFStat Add-ins**.
Step 3 Select **Add-Ins > Macros > Testing a Mean > t Test − Confidence Interval − One Sample**.

Step 4 Select cells A1 to A10 as the **Dataset Range**. (Alternatively, you may enter the summary statistics.)
Step 5 Select your **Confidence level**, which should be $1 - \alpha$. Here, because $\alpha = 0.05$, we select **95%**.
Step 6 Enter the **Null Hypothesis Value**, $\mu_0 = 1.6$, and click **OK**.

MINITAB

If you have the data values:
Step 1 Enter the data into column C1.
Step 2 Click **Stat > Basic Statistics > 1-Sample t**.
Step 3 Click **Samples in Columns** and select **C1**.
Step 4 For **Test Mean**, enter **1.6**.
Step 5 Click **Options**.
a. Choose your **Confidence Level** as $100(1 - \alpha)$. Our level of significance α here is 0.05, so the confidence level is 95.0.
b. Select **not Equal** for the **Alternative**.
Step 6 Click **OK** and click **OK** again.

If you have the summary statistics:
Step 1 Click **Stat > Basic Statistics > 1-Sample t**.
Step 2 Click **Summarized Data**.
Step 3 Enter the **Sample Size 10**, the **Sample Mean 2.02**, and the **Sample Standard Deviation 0.549343042**.
Step 4 Click **Options**.
a. Choose your **Confidence Level** as $100(1 - \alpha)$. Our level of significance α here is 0.05, so the confidence level is 95.0.
b. Select **not Equal** for the two-tailed test.
Step 5 Click **OK** and click **OK** again.

CRUNCHIT!

If you have the data values:
Step 1 Click **File . . .** then highlight **Load from Larose2e . . . Chapter 9 . . .** and click on **Example 9.16**.
Step 2 Click **Statistics . . . t** and select **1-sample**. With the **Columns** tab chosen, for **Sample** select **Index**.
Step 3 Select the **Hypothesis Test** tab. For **Mean under null hypothesis**, enter **1.6**. For **Alternative** select **Two-sided**. Then click **Calculate**.

If you have the summary statistics:
Step 1 Click **File . . .** then highlight **Load from Larose2e . . . Chapter 9 . . .** and click on **Example 9.16**.
Step 2 Click **Statistics . . . t** and select **1-sample**.
Step 3 Choose the **Summarized** tab. For **n** enter the sample size 10; for **Sample Mean** enter **2.02**. For **Standard Deviation**, enter **0.549343042**.
Step 4 Select the **Hypothesis Test** tab. For **Mean under null hypothesis**, enter **1.6**. For **Alternative** select **Two-sided**. Then click **Calculate**.

SECTION 9.4   Summary

1. The test statistic used for the *t* test for the mean is

$$t_{\text{data}} = \frac{\bar{x} - \mu_0}{s/\sqrt{n}}$$

with $n - 1$ degrees of freedom. The *t* test may be used under either of the following conditions: (a) the population is normal, or (b) the sample size is large ($n \geq 30$). For the critical-value method, we compare the values of t_{data} and t_{crit}. If t_{data} falls in the critical region, we reject H_0.

2. For the *p*-value method, we reject H_0 if the *p*-value $\leq \alpha$.

3. We may use $100(1 - \alpha)\%$ *t* confidence interval to perform two-tailed *t* tests at level of significance α for various values of μ_0.

SECTION 9.4   Exercises

Clarifying the Concepts

1. What assumption is required for performing the *Z* test that is not required for the *t* test?

2. What do we use to estimate the unknown population standard deviation σ?

Practicing the Techniques

For Exercises 3–14, do the following.
  a. State the hypotheses.
  b. Calculate the t critical value t_{crit} and state the rejection rule. Also, sketch the critical region.
  c. Find the test statistic t_{data}.
  d. State the conclusion and the interpretation.

3. $H_0: \mu = 22$ vs. $H_a: \mu < 22, \bar{x} = 20, s = 4$, $n = 31, \alpha = 0.05$

4. $H_0: \mu = 3$ vs. $H_a: \mu < 3, \bar{x} = 2, s = 1, n = 41$, $\alpha = 0.10$

5. $H_0: \mu = 11$ vs. $H_a: \mu > 11, \bar{x} = 12, s = 3$, $n = 16, \alpha = 0.01$, population is normal

6. $H_0: \mu = 80$ vs. $H_a: \mu > 80, \bar{x} = 82, s = 5, n = 9$, $\alpha = 0.05$, population is normal

7. A random sample of size 25 from a normal population yields $\bar{x} = 104$ and $s = 10$. Researchers are interested in finding whether the population mean exceeds 100, using level of significance $\alpha = 0.01$.

8. A random sample of size 100 from a population with an unknown distribution yields a sample mean of -5 and a sample standard deviation of 5. Researchers are interested in finding whether the population mean is less than -4, using level of significance $\alpha = 0.05$.

9. $H_0: \mu = 102$ vs. $H_a: \mu \neq 102, \bar{x} = 106, s = 10$, $n = 81, \alpha = 0.05$

10. $H_0: \mu = 95$ vs. $H_a: \mu \neq 95, \bar{x} = 99, s = 10$, $n = 31, \alpha = 0.01$

11. $H_0: \mu = 1000$ vs. $H_a: \mu \neq 1000, \bar{x} = 975, s = 100, n = 25, \alpha = 0.10$, population is normal

12. $H_0: \mu = -10$ vs. $H_a: \mu \neq -10, \bar{x} = -8, s = 5$, $n = 25, \alpha = 0.05$, population is normal

13. A random sample of size 36 from a population with an unknown distribution yields $\bar{x} = 10$ and $s = 3$. Researchers are interested in finding whether the population mean differs from 9, using level of significance $\alpha = 0.10$.

14. A random sample of size 16 from a normal population yields $\bar{x} = 995$ and $s = 15$. Researchers are interested in finding whether the population mean differs from 1000, using level of significance $\alpha = 0.01$.

For Exercises 15–26, do the following.
  a. State the hypotheses and the rejection rule using the p-value method.
  b. Calculate the test statistic t_{data}.
  c. Find the p-value. (Use technology or estimate the p-value.)
  d. State the conclusion and the interpretation.

15. $H_0: \mu = 10$ vs. $H_a: \mu < 10, \bar{x} = 7, s = 5, n = 81$, $\alpha = 0.01$

16. $H_0: \mu = 50$ vs. $H_a: \mu < 50, \bar{x} = 42, s = 8$, $n = 41, \alpha = 0.05$

17. $H_0: \mu = 100$ vs. $H_a: \mu > 100, \bar{x} = 120, s = 50$, $n = 25, \alpha = 0.10$, population is normal

18. $H_0: \mu = 3.0$ vs. $H_a: \mu > 3.0, \bar{x} = 3.2, s = 0.5$, $n = 25, \alpha = 0.05$, population is normal

19. A random sample of size 400 from a population with an unknown distribution yields a sample mean of 230 and a sample standard deviation of 5. Researchers are interested in finding whether the population mean is greater than 200, using level of significance $\alpha = 0.05$.

20. A random sample of size 100 from a population with an unknown distribution yields $\bar{x} = 27$ and $s = 10$. Researchers are interested in finding whether the population mean is less than 28, using level of significance $\alpha = 0.05$.

21. $H_0: \mu = 25$ vs. $H_a: \mu \neq 25, \bar{x} = 25, s = 1$, $n = 31, \alpha = 0.01$

22. $H_0: \mu = 98.6$ vs. $H_a: \mu \neq 98.6, \bar{x} = 99, s = 10$, $n = 81, \alpha = 0.05$

23. $H_0: \mu = 3.14$ vs. $H_a: \mu \neq 3.14, \bar{x} = 3.17, s = 0.5$, $n = 9, \alpha = 0.10$, population is normal

24. $H_0: \mu = 2.72$ vs. $H_a: \mu \neq 2.72, \bar{x} = 2.57, s = 0.1$, $n = 25, \alpha = 0.05$, population is normal

25. A random sample of size 9 from a normal population yields $\bar{x} = 1$ and $s = 0.5$. Researchers are interested in finding whether the population mean differs from 0, using level of significance $\alpha = 0.05$.

26. A random sample of size 16 from a normal population yields $\bar{x} = 2.2$ and $s = 0.3$. Researchers are interested in finding whether the population mean differs from 2.0, using level of significance $\alpha = 0.01$.

For Exercises 27–30, use the t table to estimate the p-value for the hypothesis tests in the indicated exercises.

27. Exercise 3

28. Exercise 4

29. Exercise 9

30. Exercise 10

For Exercises 31–36, a $100(1 - \alpha)\%$ t confidence interval is given. Use the confidence interval to test using level of significance α whether μ differs from each of the indicated hypothesized values.

31. A 95% t confidence interval for μ is $(1, 4)$. Hypothesized values μ_0 are
  a. 0   **b.** 2   **c.** 5

32. A 99% t confidence interval for μ is $(57, 58)$. Hypothesized values μ_0 are
  a. 55.5   **b.** 59.5   **c.** 57.5

33. A 90% t confidence interval for μ is $(-20, -10)$. Hypothesized values μ_0 are
  a. -21 **b.** -5 **c.** -12

34. A 95% *t* confidence interval for μ is (2010, 2015). Hypothesized values μ_0 are

    a. 2012   **b.** 2007   **c.** 2014

35. A 95% *t* confidence interval for μ is (−1, 1). Hypothesized values μ_0 are

    a. 1.5   **b.** −1.5   **c.** 0

36. A 95% *t* confidence interval for μ is (19,570, 20,105). Hypothesized values μ_0 are

    a. 20,000   **b.** 21,000   **c.** 19,571

Applying the Concepts

37. Health Care Costs. The U.S. Agency for Healthcare Research and Quality (www.ahrq.gov) reports that, in 2010, the mean cost of a stay in the hospital for American women aged 18–44 was $15,200. A random sample of 400 hospital stays of women aged 18–44 showed a mean cost of $16,000, with a standard deviation of $5000. Test whether the population mean cost has increased since 2010, using level of significance $\alpha = 0.05$.

38. iPhone Apps. According to a 2010 Nielsen survey,[11] the mean number of apps downloaded by iPhone users is 40. Suppose a sample of 36 iPhone users downloaded an average of 45 apps, with a standard deviation of 24. Test whether the population mean number of apps is greater than 40, using level of significance $\alpha = 0.10$.

39. Facebook Friends. According to **Facebook.com**, the mean number of Facebook friends is 130. Suppose a sample of 100 Facebook users has a mean number of 110 Facebook friends, with a standard deviation of 50. Test whether the population mean number of Facebook friends is less than 130, using level of significance $\alpha = 0.05$.

40. Small Business Employees. The U.S. Census Bureau reports that the average number of employees in a small business is 16.1. Suppose a sample of 49 small businesses showed a mean of 15 employees, with a standard deviation of 25. Test whether the population mean number of employees in a small business is different from 16.1, using level of significance $\alpha = 0.01$.

Internet Response Times. Use the following information for Exercises 41–42. The Web site **www.Internettrafficreport .com** monitors Internet traffic worldwide and reports on the response times of randomly selected servers.

41. On June 6, 2011, the Web site reported the following response times to Asia, in milliseconds:

165 175 2221 872 311 127 195 1801 769 225 261 249 421

We would like to test whether the population mean response time is slower than 180 milliseconds, using a *t* test and level of significance $\alpha = 0.05$. A boxplot of the data is provided.

(*Hint:* The boxplot is right-skewed and the normal distribution is symmetric.) Can we proceed with the *t* test? Explain.

42. On June 6, 2011, the Web site reported the following response times to Asia, in milliseconds:

61 32 50 73 51 42 55 65 59 57 76 77 67 71

The normal probability plot of the data is also shown. We would like to perform a *t* test.

    a. Are the conditions for performing the *t* test satisfied? Explain how.

    b. Test using level of significance $\alpha = 0.05$ whether the population mean response time is less than 60 milliseconds.

    c. Explain why we can't use a *Z* test for this problem.

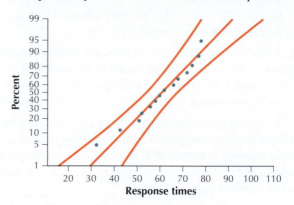

Top Gas Mileage. Use the following information for Exercises 43–45. The top ten vehicles for city gas mileage in 2007, as reported by the Environmental Protection Agency, are shown in the following table, along with the normal probability plot.

🔺 topmileage

Car	Mileage	Car	Mileage
Toyota Yaris	39	Honda Fit	38
Chevrolet Aveo	37	Nissan Versa	34
Pontiac G5	34	Dodge Caliber	32
VW Eos	32	Ford Escape	31
Saturn Sky	30	BMW 525	30

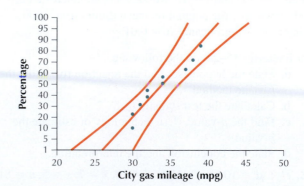

Normal probability plot.

43. We are interested in testing whether the population mean city mileage of such cars is greater than 30 mpg.

 a. Is it appropriate to apply the *t* test for the mean? Why or why not?

 b. Test, using the estimated *p*-value method at level of significance $\alpha = 0.01$, whether the population mean city mileage exceeds 30 mpg.

44. Answer the following.

 a. Repeat your test from Exercise 43(**b**), this time using level of significance $\alpha = 0.001$.

 b. How do you think we should resolve the apparent contradiction in 43(**b**) and part (**a**) of this exercise?

 c. Assess the strength of the evidence against the null hypothesis. Does this change depend on which level of α you use?

45. *What if* we changed μ_0 to some larger value (though still smaller than $\bar{x}$). Otherwise, everything else remains unchanged. Describe how this change would affect the following, if at all.

 a. t_{data}

 b. t_{crit}

 c. The *p*-value

 d. The conclusion from Exercise 43(**b**)

 e. The conclusion from Exercise 44(**a**)

 f. The strength of the evidence against the null hypothesis

Bringing It All Together

Community College Tuition. Use the following information for Exercises 46–47. The College Board reported that the mean tuition and fees at community colleges nationwide was $2272. Data were gathered on the total tuition and fees for a random sample of ten community colleges this year. The normal probability plot and Minitab *t* test output are shown here.

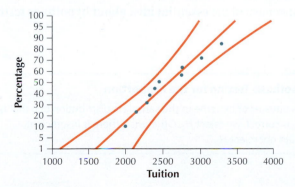

Normal probability plot.

```
Test of mu = 2272 vs not = 2272

Variable   N    Mean    StDev  SE Mean       95% CI           T     P
tuition   10  2538.92  404.75   127.99  (2249.38, 2828.46)  2.09  0.067
```

Minitab *t* test output.

46. Analysts are interested in whether the population mean tuition and fees this year have increased.

 a. Is it appropriate to apply the *t* test for the mean? Why or why not?

 b. It appears that the data analyst who produced the Minitab printout asked for the wrong hypothesis test. How can we tell?

47. Refer to your work in the previous exercise.

 a. Test whether the population mean tuition and fees have increased using level of significance $\alpha = 0.05$. How can we use the *p*-value on the Minitab printout to find the *p*-value needed for this right-tailed hypothesis test?

 b. Compare the conclusion from (**a**) with the conclusion we would have gotten had we not noticed that the data analyst performed the wrong hypothesis test. What are some of the possible consequences of making an error of this sort?

 c. Based on your experiences in these exercises, write a sentence about the importance of understanding the statistical modeling behind the "point and click" power of statistical software.

48. Challenge Exercise. Refer to your work in the previous exercise.

 a. Note that we have concluded that there is insufficient evidence that the population mean cost has changed, but that there is evidence that the population mean cost has increased. How can the mean cost have increased without changing? Explain what is going on here, in terms of either critical regions or *p*-values.

 b. Assess the strength of the evidence against the null hypothesis for the test in Exercise 50(**a**).

New York Towns. Work with the **New York** data set for Exercises 49 and 50.

 New York

49. Use technology to find the summary statistics for the variable *tot_pop*, which lists the population for each of the towns and cities in New York with at least 1000 people.

50. Suppose we are using the data in this data set as a sample of the population of all the towns and cities in the northeastern United States with at least 1000 people. Use technology to test at level of significance $\alpha = 0.05$ whether the population mean population of these towns differs from 50,000.

9.5 Z TEST FOR THE POPULATION PROPORTION

OBJECTIVES By the end of this section, I will be able to . . .

1 Perform the Z test for p using the critical-value method.

2 Carry out the Z test for p using the p-value method.

3 Use confidence intervals for p to perform two-tailed hypothesis tests about p.

1 THE Z TEST FOR p USING THE CRITICAL-VALUE METHOD

Thus far, we have dealt with testing hypotheses about the population mean μ only. In this section, we will learn how to perform the Z test for the population proportion p.

For our point estimate of the unknown population proportion p, we use the sample proportion $\hat{p} = x/n$, where x equals the number of successes.

Just as with the Z test for the mean, in the Z test for the proportion the null hypothesis will include a certain hypothesized value for the unknown parameter, which we call p_0. For example, the hypotheses for the two-tailed test have the following form:

$$H_0: p = p_0 \quad \text{versus} \quad H_a: p \neq p_0$$

where p_0 represents a particular hypothesized value of the unknown population proportion p. For instance, if a researcher is interested in determining whether the population proportion of Americans who support increased funding for higher education differs from 50%, then $p_0 = 0.50$ and $q_0 = 1 - p_0 = 0.50$.

If we assume H_0 is correct, then the population proportion of successes is p_0. Then Facts 5 and 6 from Section 7.3 tell us that the sampling distribution of p has a mean of p_0 and the standard deviation

$$\sigma_{\hat{p}} = \sqrt{\frac{p \cdot q}{n}} = \sqrt{\frac{p_0 \cdot q_0}{n}}$$

since we claim in H_0 that $p = p_0$. $\sigma_{\hat{p}}$ is called the **standard error of the proportion.** Fact 7 from Section 7.3 tells us that the sampling distribution of $\hat{p}$ is approximately normal whenever both of the following conditions are met: $n \cdot p \geq 5$ and $n \cdot q \geq 5$. This leads us to the following statement of the **essential idea about hypothesis testing for the proportion.**

> **The Essential Idea About Hypothesis Testing for the Proportion**
>
> When the sample proportion $\hat{p}$ is unusual or extreme in the sampling distribution of $\hat{p}$ that is based on the assumption that H_0 is correct, we reject H_0. Otherwise, there is insufficient evidence against H_0, and we should not reject H_0.

The remainder of this section explains the details of implementing hypothesis testing for the proportion. The critical-value method for the Z test for p is similar to that of the Z test for μ, in that we compare one Z-value (Z_{data}) with another Z-value (Z_{crit}). In this section, Z_{data} represents the number of standard errors ($\sigma_{\hat{p}}$) the sample proportion $\hat{p}$ lies above or below the hypothesized proportion p_0.

For example, if a baseball player has $x = 30$ hits in $n = 100$ at-bats, his batting average is $\hat{p} = x/n = 30/100 = 0.3$ (or .300).

> The test statistic used for the Z test for the proportion is
>
> $$Z_{data} = \frac{\hat{p} - p_0}{\sqrt{\dfrac{p_0 \cdot q_0}{n}}}$$
>
> where $\hat{p}$ is the observed sample proportion of successes, p_0 is the value of p hypothesized in H_0, $q_0 = 1 - p_0$ and n is the sample size.

Now You Can Do Exercises 7–14.

To find the Z_{crit} critical values, the critical regions, or the rejection rules, you can use Table 9.11.

Table 9.11 Table of critical values Z_{crit} for common values of the level of significance α

Level of significance α	Form of Hypothesis Test		
	Right-tailed $H_0 : p = p_0$ $H_a : p > p_0$	**Left-tailed** $H_0 : p = p_0$ $H_a : p < p_0$	**Two-tailed** $H_0 : p = p_0$ $H_a : p \neq p_0$
0.10	$Z_{crit} = 1.28$	$Z_{crit} = -1.28$	$Z_{crit} = 1.645$
0.05	$Z_{crit} = 1.645$	$Z_{crit} = -1.645$	$Z_{crit} = 1.96$
0.01	$Z_{crit} = 2.33$	$Z_{crit} = -2.33$	$Z_{crit} = 2.58$
Rejection rule	Reject H_0 if $Z_{data} \geq Z_{crit}$	Reject H_0 if $Z_{data} \leq Z_{crit}$	Reject H_0 if $Z_{data} \leq -Z_{crit}$ or $Z_{data} \geq Z_{crit}$

EXAMPLE 9.19

CALCULATING Z_{data} FOR THE Z TEST FOR PROPORTION

The Centers for Disease Control and Prevention reported in 2010 that 20% of Americans smoked tobacco.[12] A random sample of $n = 400$ Americans found 76 who smoked. We are interested in testing whether the population proportion of Americans who smoke has changed from 20%. Calculate the test statistic Z_{data}.

Solution

The key words "has changed" indicate a two-tailed test. "Changed from what?" The hypothesized proportion $p_0 = 0.20$. The hypotheses are

$$H_0 : p = 0.20 \quad \text{versus} \quad H_a : p \neq 0.20$$

The sample proportion of those who smoke is

$$\hat{p} = \frac{x}{n} = \frac{\text{number in sample who smoke}}{\text{sample size}} = \frac{76}{400} = 0.19$$

We then calculate the value of the test statistic Z_{data}:

$$Z_{data} = \frac{\hat{p} - p_0}{\sqrt{\frac{p_0 \cdot q_0}{n}}} = \frac{0.19 - 0.20}{\sqrt{\frac{0.20(0.80)}{400}}} = \frac{-0.01}{0.02} = -0.5$$

Now You Can Do Exercises 7–14.

> ### *Z* Test for the Population Proportion *p*: Critical-Value Method
>
> When a random sample of size n is taken from a population, you can use the Z test for the proportion if both of the normality conditions are satisfied:
>
> $$n \cdot p_0 \geq 5 \quad \text{and} \quad n \cdot q_0 \geq 5$$
>
> **Step 1 State the hypotheses.**
> Use one of the forms from Table 9.11. State the meaning of p.
>
> **Step 2 Find Z_{crit} and state the rejection rule.**
> Use Table 9.11.
>
> **Step 3 Calculate Z_{data}.**
>
> $$Z_{data} = \frac{\hat{p} - p_0}{\sigma_{\hat{p}}} = \frac{\hat{p} - p_0}{\sqrt{\frac{p_0 \cdot q_0}{n}}}$$
>
> **Step 4 State the conclusion and the interpretation.**
> If Z_{data} falls in the critical region, then reject H_0. Otherwise, do not reject H_0. Interpret the conclusion so that a nonspecialist can understand.

EXAMPLE 9.20

Z TEST FOR *p* USING THE CRITICAL-VALUE METHOD

Refer to Example 9.19. Test whether the population proportion of Americans who smoke has changed from 20%, using the critical-value method and level of significance $\alpha = 0.10$.

Solution

As a check on your arithmetic, the two quantities you obtain when checking the normality conditions should add up to n. Here $80 + 320 = 400$.

First we check that both of our normality conditions are met. From Example 9.19, we have $p_0 = 0.20$ and $n = 400$.

$$n \cdot p_0 = (400)(0.20) = 80 \geq 5 \quad \text{and} \quad n \cdot q_0 = (400)(0.80) = 320 \geq 5$$

The normality conditions are met and we may proceed with the hypothesis test.

STEP 1 State the hypotheses.
From Example 9.19 our hypotheses are

$$H_0: p = 0.20 \quad \text{versus} \quad H_a: p \neq 0.20$$

where p represents the population proportion of Americans who smoke tobacco.

STEP 2 Find Z_{crit} and state the rejection rule.
We have a two-tailed test, with $\alpha = 0.10$. This gives us our critical value $Z_{crit} = 1.645$ and the rejection rule from Table 9.11. Reject H_0 if $Z_{data} \geq 1.645$ or $Z_{data} \leq -1.645$ (Figure 9.41).

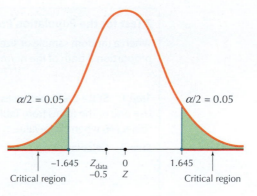

FIGURE 9.41
Z_{data} does not fall in the critical region.

STEP 3 **Calculate Z_{data}.**
From Example 9.19, we have $Z_{data} = -0.5$

STEP 4 **State the conclusion and the interpretation.**
The test statistic $Z_{data} = -0.5$ is not ≥ 1.645 and not ≤ -1.645. Thus, we do not reject H_0. There is insufficient evidence at level of significance $\alpha = 0.10$ that the population proportion of Americans who smoke tobacco differs from 20%.

Now You Can Do
Exercises 15–18.

2 *Z* TEST FOR *p*: THE *p*-VALUE METHOD

The *p*-value method for the *Z* test for *p* is equivalent to the critical-value method. The *p*-values are defined similarly to those for the *Z* test for μ, as shown in Table 9.12.

Table 9.12 Finding the *p*-value depends on the form of the hypothesis test

	Right-tailed test	**Left-tailed test**	**Two-tailed test**
Type of hypothesis test	$H_0: p = p_0$ $H_a: p > p_0$	$H_0: p = p_0$ $H_a: p < p_0$	$H_0: p = p_0$ $H_a: p \neq p_0$
_p_-Value is tail area associated with Z_{data}	$p\text{-value} = P(Z > Z_{data})$ Area to right of Z_{data}	$p\text{-value} = P(Z < Z_{data})$ Area to left of Z_{data}	$p\text{-value} = P(Z > \lvert Z_{data}\rvert)$ $\qquad + P(Z < -\lvert Z_{data}\rvert)$ $\qquad = 2 \cdot P(Z > \lvert Z_{data}\rvert)$ Sum of the two tail areas.

Note that the *p*-value has precisely the same definition and behavior as in the *Z* test for the mean. That is, the *p*-value is roughly a measure of how extreme your value of Z_{data} is and takes values between 0 and 1, with small values indicating extreme values of Z_{data}.

Developing Your Statistical Sense

The Difference Between the *p*-Value and the Population Proportion *p*

Be careful to distinguish between the *p*-value and the population proportion *p*. The latter represents the population proportion of successes for a binomial experiment and is a population parameter. The *p*-value is the probability of observing a value of Z_{data} at least as extreme as the Z_{data} actually observed. The *p*-value depends on the sample data, but the population proportion *p* does not depend on the sample data.

> **Z Test for the Population Proportion p: p-Value Method**
>
> When a random sample of size n is taken from a population, you can use the Z test for the proportion if both of the normality conditions are satisfied:
>
> $$n \cdot p_0 \geq 5 \quad \text{and} \quad n \cdot q_0 \geq 5$$
>
> **Step 1   State the hypotheses and the rejection rule.**
> Use one of the forms from Table 9.12. State the meaning of p. State the rejection rule as "Reject H_0 when the p-value $\leq \alpha$."
>
> **Step 2   Calculate Z_{data}.**
>
> $$Z_{data} = \frac{\hat{p} - p_0}{\sqrt{\dfrac{p_0 \cdot q_0}{n}}}$$
>
> **Step 3   Find the p-value.**
> Either use technology to find the p-value, or calculate it using the form in Table 9.12 that corresponds to your hypotheses.
>
> **Step 4   State the conclusion and the interpretation.**
> If the p-value $\leq \alpha$, then reject H_0. Otherwise do not reject H_0. Interpret your conclusion so that a nonspecialist can understand.

EXAMPLE 9.21

Z TEST FOR p USING THE p-VALUE METHOD

Getty Images/Stockbyte Platinum

The National Transportation Safety Board publishes statistics on the number of automobile crashes that people in various age groups have. Young people aged 18–24 have an accident rate of 12%, meaning that on average 12 out of every 100 young drivers per year had an accident. A researcher claims that the population proportion of young drivers having accidents is greater than 12%. Her study examined 1000 young drivers aged 18–24 and found that 134 had an accident this year. Perform the appropriate hypothesis test using the p-value method with level of significance $\alpha = 0.05$.

Solution

First we check that both of our normality conditions are met. Since we are interested in whether the proportion has increased from 12%, we have $p_0 = 0.12$.

$$n \cdot p_0 = (1000)(0.12) = 120 \geq 5 \quad \text{and} \quad n \cdot q_0 = (1000)(0.88) = 880 \geq 5$$

The normality conditions are met and we may proceed with the hypothesis test.

STEP 1 **State the hypotheses and the rejection rule.**
Our hypotheses are

$$H_0: p = 0.12 \quad \text{versus} \quad H_a: p > 0.12$$

where p represents the population proportion of young people aged 18–24 who had an accident. We reject the null hypothesis if the p-value $\leq \alpha = 0.05$.

STEP 2 **Calculate Z_{data}.**
Our sample proportion is $\hat{p} = 134/1000 = 0.134$. Since $p_0 = 0.12$, the standard error of $\hat{p}$ is

$$\sigma_{\hat{p}} = \sqrt{\frac{p_0 \cdot q_0}{n}} = \sqrt{\frac{(0.12)(0.88)}{1000}} \approx 0.0103$$

Thus, our test statistic is

We report Z_{data} to 2 decimal places to allow the use of the Z table to calculate the p-value.

$$Z_{data} = \frac{\hat{p} - p_0}{\sqrt{\dfrac{p_0 \cdot q_0}{n}}} = \frac{0.134 - 12}{\sqrt{\dfrac{(0.12)(0.88)}{1000}}} \approx 1.36$$

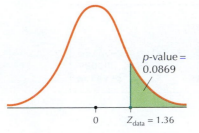

FIGURE 9.42 *p*-Value for a right-tailed test equals area to right of Z_{data}.

Now You Can Do
Exercises 19–22.

That is, the sample proportion $\hat{p} = 0.134$ lies approximately 1.36 standard errors above the hypothesized proportion $p_0 = 0.12$.

STEP 3   **Find the *p*-value.**
Since we have a right-tailed test, our *p*-value from Table 9.12 is $P(Z > Z_{data})$. This is a Case 2 problem from Table 6.6 (page 289), where we find the tail area by subtracting the *Z* table area from 1 (Figure 9.42):

$$P(Z > Z_{data}) = P(Z > 1.36) = 1 - 0.9131 = 0.0869$$

STEP 4   **State the conclusion and the interpretation.**
Since the *p*-value is not $\leq \alpha = 0.05$, we do not reject H_0. There is insufficient evidence that the population proportion of young people aged 18–24 who had an accident has increased.

EXAMPLE 9.22   **PERFORMING THE *Z* TEST FOR *p* USING TECHNOLOGY**

A study reported that 1% of American Internet users who are married or in a long-term relationship met on a blind date or through a dating service.[13] A survey of 500 American Internet users who are married or in a long-term relationship found 8 who met on a blind date or through a dating service. If appropriate, test whether the population proportion has increased. Use the *p*-value method with level of significance $\alpha = 0.05$.

Solution
We have $p_0 = 0.01$ and $n = 500$. Checking the normality conditions, we have

$$n \cdot p_0 = (500)(0.01) = 5 \geq 5 \quad \text{and} \quad n \cdot q_0 = (500)(0.99) = 495 \geq 5$$

The normality conditions are met and we may proceed with the hypothesis test.

STEP 1   **State the hypotheses and the rejection rule.**
Our hypotheses are

$$H_0: p = 0.01 \quad \text{versus} \quad H_a: p > 0.01$$

where *p* represents the population proportion of American Internet users who are married or in a long-term relationship and who met on a blind date or through a dating service. We will reject H_0 if the *p*-value ≤ 0.05.

STEP 2   **Calculate Z_{data}.**
We use the instructions supplied in the Step-by-Step Technology Guide on page 459. Figure 9.43 shows the TI-83/84 results from the *Z* test for *p*, and Figure 9.44 shows the results from Minitab.

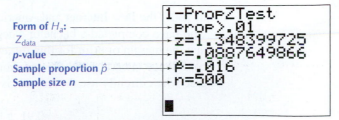

FIGURE 9.43   TI-83/84 results.

Note: Minitab and TI-83/84 round results to different numbers of decimal places.

```
Test of p = 0.01 vs p > 0.01
                                  95%
                                 Lower
 Sample  X    N  Sample p        Bound   Z-Value  P-Value
 1       8  500  0.016000     0.006770      1.35    0.089
         X    n     p̂        (not used)    Z_data   p-value
```

FIGURE 9.44   Minitab results.

We have

$$Z_{\text{data}} = \frac{\hat{p} - p_0}{\sqrt{\dfrac{p_0 \cdot q_0}{n}}} = \frac{0.016 - 0.01}{\sqrt{\dfrac{(0.01)(0.99)}{500}}} \approx 1.348399725$$

which concurs with the TI-83/84 results in Figure 9.43.

STEP 3   **Find the *p*-value.**
From Figures 9.43, 9.44, and 9.45, we have

$$p\text{-value} = P(Z > 1.348399725) = 0.0887649866 \approx 0.08876$$

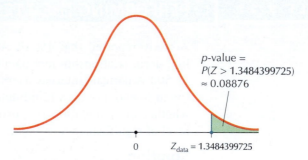

p-value =
$P(Z > 1.3484399725)$
≈ 0.08876

0          $Z_{\text{data}} = 1.3484399725$

FIGURE 9.45   *p*-Value for a right-tailed test.

STEP 4   **State the conclusion and interpretation.**
Since *p*-value ≈ 0.08876 is *not* $\leq \alpha = 0.05$, we do *not* reject H_0. There is insufficient evidence that the population proportion of American Internet users who are married or in a long-term relationship and who met on a blind date or through a dating service has increased.

3   USING CONFIDENCE INTERVALS FOR *p* TO PERFORM TWO-TAILED HYPOTHESIS TESTS ABOUT *p*

Just as for μ, we can use a $100(1 - \alpha)\%$ confidence interval for the population proportion p in order to perform a set of two-tailed hypothesis tests for p.

EXAMPLE 9.23   **USING A CONFIDENCE INTERVAL FOR *p* TO PERFORM TWO-TAILED HYPOTHESIS TESTS ABOUT *p***

In 2007, the Pew Internet and American Life Project reported that 91% of Americans who have completed a bachelor's degree currently use the Internet. Pew also reports that the margin of error for this survey (confidence level = 95%) was $\pm 3\%$. The 95% confidence interval for the population proportion of Americans with a bachelor's degree who currently use the Internet is therefore

$$0.91 \pm 0.03 = (0.88, 0.94)$$

Use the confidence interval to test, using level of significance $\alpha = 0.05$, whether the population proportion differs from
a. 0.85   **b.** 0.90   **c.** 0.95

Solution

There is equivalence between a $100(1 - \alpha)\%$ confidence interval for p and a two-tailed test for p with level of significance α. Values of p_0 that lie outside the confidence interval lead to rejection of the null hypothesis, while values of p_0 within the confidence interval lead to not rejecting the null hypothesis. Figure 9.46 illustrates the 95% confidence interval for p.

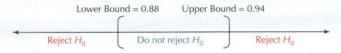

Lower Bound = 0.88   Upper Bound = 0.94

Reject H_0 Do not reject H_0 Reject H_0

FIGURE 9.46 H_0 for values p_0 that lie outside the interval (0.88, 0.94).

We would like to perform the following two-tailed hypothesis tests:

a. $H_0 : p = 0.85$ versus $H_a : p \neq 0.85$
b. $H_0 : p = 0.90$ versus $H_a : p \neq 0.90$
c. $H_0 : p = 0.95$ versus $H_a : p \neq 0.95$

To perform each hypothesis test, simply observe where each value of p_0 falls on the number line. For example, in the first hypothesis test, the hypothesized value $p_0 = 0.85$ lies outside the interval (0.88, 0.94). Thus, we reject H_0. The three hypothesis tests are summarized here.

Now You Can Do Exercises 23–26.

Value of p_0	Form of hypothesis test, with $\alpha = 0.05$		Where p_0 lies in relation to 95% confidence interval	Conclusion of hypothesis test
a. 0.85	$H_0 : p = 0.85$	$H_a : p \neq 0.85$	Outside	Reject H_0
b. 0.90	$H_0 : p = 0.90$	$H_a : p \neq 0.90$	Inside	Do not reject H_0
c. 0.95	$H_0 : p = 0.95$	$H_a : p \neq 0.95$	Outside	Reject H_0

STEP-BY-STEP TECHNOLOGY GUIDE: *Z* test for *p*

We will use the information from Example 9.22 (page 457).

TI-83/84

Step 1   Press **STAT**, highlight **TESTS**, and press **ENTER**.
Step 2   Press 5 (for **1-PropZTest**; see Figure 9.47).
Step 3   For p_0, enter the value of p_0, **0.01**.
Step 4   For **x**, enter the number of successes, **8**.
Step 5   For **n**, enter the number of trials **500**.
Step 6   For **prop**, enter the form of H_a. Here we have a right-tailed test, so highlight >p_0 and press **ENTER** (see Figure 9.48).
Step 7   Highlight **Calculate** and press **ENTER**. The results are shown in Figure 9.43 in Example 9.22.

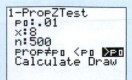

```
EDIT CALC TESTS
1:Z-Test…
2:T-Test…
3:2-SampZTest…
4:2-SampTTest…
5⯈1-PropZTest…
6:2-PropZTest…
7↓ZInterval…
```

FIGURE 9.47

```
1-PropZTest
 p0:.01
 x:8
 n:500
 prop≠p0 <p0 >p0
 Calculate Draw
```

FIGURE 9.48

EXCEL

WHFStat Add-ins

Step 1 Enter the data into column A. (If you have only the summary statistics, go to Step 2.)

Step 2 Load the **WHFStat Add-ins**.

Step 3 Select **Add-Ins > Macros > Testing a Proportion > One Sample**.

Step 4 Enter the **Number of successes 8**.

Step 5 Enter the **Sample size 500**.

Step 6 Enter the **Testing Proportion, $p_0 = 0.01$**.

Step 7 Select your **Confidence level**, which should be $1 - \alpha$. Here, because $\alpha = 0.05$, we select **95%**.

Step 8 Click **OK**.

MINITAB

If you have the summary statistics:

Step 1 Click **Stat > Basic Statistics > 1 Proportion**.

Step 2 Click **Summarized Data**.

Step 3 Enter the **Number of trials 500** and the **Number of Events 8**.

Step 4 Click **Options**.

a. Choose your **Confidence Level** as $100(1 - \alpha)$. Our level of significance α here is 0.05, so the confidence level is 95.0.

b. Enter **0.01** for the **Test Proportion**.

c. Select **Greater than** for the **Alternative**.

d. Check **Use test and interval based on normal distribution**.

Step 5 Click **OK** and click **OK** again. The results are shown in Figure 9.44 in Example 9.22.

CRUNCHIT!

Step 1 Click **File** . . . then highlight **Load from Larose2e** . . . **Chapter 9** . . . and click on **Example 9.22**.

Step 2 Click **Statistics** . . . **Proportion** and select **1-sample**.

Step 3 Choose the **Summarized** tab. For **n** enter the number of trials **500**; for **Successes** enter **8**.

Step 4 Select the **Hypothesis Test** tab. For **Proportion under null hypothesis**, enter **0.01**.

For **Alternative** select **Greater than**. Then click **Calculate**.

SECTION 9.5 Summary

1. The test statistic used for the Z test for the proportion is

$$Z_{\text{data}} = \frac{\hat{p} - p_0}{\sqrt{\dfrac{p_0 \cdot q_0}{n}}}$$

where $\hat{p}$ is the observed sample proportion of successes, p_0 is the value of p hypothesized in H_0, $q_0 = 1 - p_0$ and n is the sample size. Z_{data} represents the number of standard deviations ($\sigma_{\hat{p}}$) the sample proportion $\hat{p}$ lies above or below the hypothesized proportion p_0. Extreme values of $\hat{p}$ will

be associated with extreme values of Z_{data}. The Z test for the proportion may be performed using either the p-value method or the critical-value method. For the critical-value method, we compare the values of Z_{data} and Z_{crit}. If Z_{data} falls in the critical region, we reject H_0.

2. For the p-value method, we reject H_0 if the p-value $\leq \alpha$.

3. We can use a single $100(1 - \alpha)\%$ confidence interval for p to help us perform any number of two-tailed hypothesis tests about p with level of significance α.

SECTION 9.5 Exercises

Clarifying the Concepts

1. What is the difference between $\hat{p}$ and p?

2. What are the conditions for the Z test for p?

3. Explain the essential idea about hypothesis testing for the proportion.

4. Explain what p_0 refers to.

5. What possible values can p_0 take?

6. What is the difference between p and a p-value?

Practicing the Techniques

For Exercises 7–9, find the value of the test statistic Z_{data} for a right-tailed test with $p_0 = 0.4$.

7. A sample of size 50 yields 30 successes.

8. A sample of size 50 yields 40 successes.

9. A sample of size 50 yields 45 successes.

10. What kind of pattern do we observe in the value of Z_{data} for a right-tailed test as the number of successes becomes more extreme?

For Exercises 11–13, find the value of the test statistic Z_{data} for a two-tailed test with $p_0 = 0.5$.

11. A sample of size 80 yields 20 successes.

12. A sample of size 80 yields 30 successes.

13. A sample of size 80 yields 40 successes.

14. What kind of pattern do we observe in the value of Z_{data} as the sample proportion approaches p_0?

For Exercises 15–18, do the following.
  a. Check the normality conditions.
  b. State the hypotheses.
  c. Find Z_{crit} and the rejection rule.
  d. Calculate Z_{data}.
  e. Compare Z_{crit} with Z_{data}. State the conclusion and the interpretation.

15. Test whether the population proportion is less than 0.5. A random sample of size 225 yields 100 successes. Let level of significance $\alpha = 0.05$.

16. Test whether the population proportion differs from 0.3. A random sample of size 100 yields 25 successes. Let level of significance $\alpha = 0.01$.

17. Test whether the population proportion exceeds 0.6. A random sample of size 400 yields 260 successes. Let level of significance $\alpha = 0.05$.

18. Test whether p differs from 0.4. A random sample of size 900 yields 400 successes. Let level of significance $\alpha = 0.10$.

For Exercises 19–22, do the following.
  a. Check the normality conditions.
  b. State the hypotheses and the rejection rule for the p-value method, using level of significance $\alpha = 0.05$.
  c. Find Z_{data}.
  d. Find the p-value.
  e. Compare the p-value with level of significance $\alpha = 0.05$. State the conclusion and the interpretation.

19. Test whether the population proportion exceeds 0.4. A random sample of size 100 yields 44 successes.

20. Test whether the population proportion is less than 0.2. A random sample of size 400 yields 75 successes.

21. Test whether the population proportion differs from 0.5. A random sample of size 900 yields 475 successes.

22. Test whether the population proportion exceeds 0.9. A random sample of size 1000 yields 925 successes.

For Exercises 23–26, a $100(1 - \alpha)\%$ Z confidence interval for p is given. Use the confidence interval to test using level of significance α whether p differs from each of the indicated hypothesized values.

23. A 95% Z confidence interval for p is (0.1, 0.9). Hypothesized values p_0 are
  a. 0
  b. 1
  c. 0.5

24. A 99% Z confidence interval for p is (0.51, 0.52). Hypothesized values p_0 are
  a. 0.511
  b. 0.521
  c. 0.519

25. A 90% Z confidence interval for p is (0.1, 0.2). Hypothesized values p_0 are
  a. 0.09
  b. 0.9
  c. 0.19

26. A 95% Z confidence interval for p is (0.05, 0.95). Hypothesized values p_0 are
  a. 0.01
  b. 0.5
  c. 0.06

Applying the Concepts

27. Baptists in America. A study reported that 17.2% of Americans identified themselves as Baptists.[14] A survey of 500 randomly selected Americans showed that 85 of them were Baptists. If appropriate, test using level of significance $\alpha = 0.10$ whether the population proportion of Americans who are Baptists has changed.

28. Births to Unmarried Women. The National Center for Health Statistics reported: "Childbearing by unmarried women increased to record levels for the Nation in 2005."[15] In that year, 36.8% of all births were to unmarried women. Suppose that a random sample taken this year of 1000 births showed 380 to unmarried women. If appropriate, test whether the population proportion has increased since 2005, using level of significance $\alpha = 0.05$.

29. Twenty-Somethings. According to the U.S. Census Bureau, 7.1% of Americans were between the ages of 20 and 24. Suppose that a random sample of 400 Americans taken this year yields 35 between the ages of 20 and 24. If appropriate, test whether the population proportion of Americans aged 20–24 is different from 7.1%. Use level of significance $\alpha = 0.01$.

30. Nonmedical Pain Reliever Use. The National Survey on Drug Use and Health reported that 4.8% of persons aged 12 or older used a prescription pain reliever nonmedically.[16] Suppose that a random sample of 900 persons aged 12 or older found 54 that had used a prescription pain reliever nonmedically. If appropriate, test whether the population proportion has increased, using level of significance $\alpha = 0.01$.

31. Ethnic Asians in California. A research report states that 12.3% of California residents were of Asian ethnicity.[17] Suppose that a random sample of 400 California residents yields 52 of Asian ethnicity. We are interested in whether the population proportion of California residents of Asian ethnicity has risen.

   a. Is it appropriate to perform the Z test for the proportion? Why or why not?

   b. Is there evidence that the population proportion of California residents of Asian ethnicity has risen? Test using the p-value method at level of significance $\alpha = 0.05$.

32. Affective Disorders Among Women. What do you think is the most common nonobstetric (not related to pregnancy) reason for hospitalization among 18- to 44-year-old American women? According to the U.S. Agency for Healthcare Research and Quality (**www.ahrq.gov**), this is the category of *affective* disorders, such as depression. Of hospitalizations among 18- to 44-year-old American women, 7% were for affective disorders. Suppose that a random sample taken this year of 1000 hospitalizations of 18- to 44-year-old women showed 80 admitted for affective disorders. We are interested in whether the population proportion of hospitalizations for affective disorders has changed since 2002. Test using the p-value method and level of significance $\alpha = 0.10$.

33. Latino Household Income. The U.S. Census Bureau reported that 15.3% of Latino families had household incomes of at least $75,000. We are interested in whether the population proportion has changed, using the critical-value method and level of significance $\alpha = 0.01$. Suppose that a random sample of 100 Latino families reported 23 with household incomes of at least $75,000.

   a. Is it appropriate to perform the Z test for the proportion? Why or why not?

   b. Perform the appropriate hypothesis test.

34. Eighth-Grade Alcohol Use. The National Institute on Alcohol Abuse and Alcoholism reported that 45.6% of eighth-graders had used alcohol.[18] A random sample of 100 eighth-graders this year showed that 41 of them had used alcohol.

   a. Is it appropriate to perform the Z test for the proportion? Why or why not?

   b. Is there evidence that the population proportion of eighth-graders who used alcohol has changed? Test using the p-value method at level of significance $\alpha = 0.05$.

35. Eighth-Grade Alcohol Use. Refer to Exercise 34.

   a. Evaluate the strength of evidence against the null hypothesis.

   b. Suppose that we decide to carry out the same Z test as Exercise 34(**b**), however, this time using the critical-value method. Without actually performing the test, what would the conclusion be and why?

   c. Would a 95% Z interval for p contain $p = 0.456$? Explain.

Bringing It All Together

Children and Environmental Tobacco Smoke at Home. Use the following information for Exercises 36–39. The Environmental Protection Agency reported that 11% of children aged 6 and under were exposed to environmental tobacco smoke (ETS) at home on a regular basis (at least four times per week).[19] A random sample of 100 children aged 6 and under showed that 6% of these children had been exposed to ETS at home on a regular basis.

36. Answer the following.

   a. Is it appropriate to perform the Z test for the proportion? Why or why not?

   b. Test at level of significance $\alpha = 0.05$ whether the population proportion of children aged 6 and under exposed to ETS at home on a regular basis has decreased.

37. Refer to Exercise 36.

   a. Which is the only possible error you can be making here, a Type I or a Type II error? What are some consequences of this error?

   b. Suppose that a newspaper headline reported "Second-hand Smoke Prevalence Down." How would you respond? Does your inference support this headline?

38. Refer to your work in Exercise 36.

   a. Test at level of significance $\alpha = 0.10$ whether the population proportion of children aged 6 and under exposed to ETS at home on a regular basis has decreased.

   b. How do you explain the different conclusions you got in the two hypothesis tests above?

   c. Evaluate the strength of evidence against the null hypothesis.

39. Refer to Exercise 36. *What if* the sample proportion $\hat{p}$ decreased, but everything else stayed the same. Describe what would happen to the following, and why.

   a. $\sigma_{\hat{p}}$

   b. Z_{data}

   c. The p-value

   d. α

   e. The conclusion

40. Chapter 8 Case Study, Continued. On page 388 we calculated the 95% confidence interval for p, the population proportion of all wildlife cleanup workers who experienced skin problems, to be (0.330, 0.596). Test using level of significance $\alpha = 0.05$ whether p differs from: **(a)** 0.3, **(b)** 0.4, **(c)** 0.5, **(d)** 0.6.

Car Accidents Among Young Drivers. For Exercises 41 and 42, refer to Example 9.21.

41. Suppose that our sample size and the number of successes are doubled, so that $\hat{p}$ remains the same. Otherwise, everything else is the same as in the original example. Describe how this change would affect the following.

a. $\sigma_{\hat{p}}$
b. Z_{data}
c. The p-value
d. α
e. The conclusion

42. Suppose that the hypothesized proportion p_0 was no longer 0.12. Instead, p_0 takes some value between 0.12 and 0.134. Otherwise, everything else is the same as in the original example. Describe how this change would affect the following.

a. $\sigma_{\hat{p}}$
b. Z_{data}
c. The p-value
d. α
e. The conclusion

9.6   CHI-SQUARE TEST FOR THE POPULATION STANDARD DEVIATION

OBJECTIVES   By the end of this section, I will be able to . . .

1   Perform the χ^2 test for σ using the critical-value method.

2   Carry out the χ^2 test for σ using the p-value method.

3   Use confidence intervals for σ to perform two-tailed hypothesis tests about σ.

1 χ^2 (CHI-SQUARE) TEST FOR σ USING THE CRITICAL-VALUE METHOD

In Section 8.4 (pages 392–400) we used the χ^2 distribution to help us construct confidence intervals for the population variance and standard deviation. Here, in Section 9.6, we will use the χ^2 distribution to perform hypothesis tests about the population standard deviation σ. Why might we be interested in doing so? A pharmaceutical company that wishes to ensure the safety of a particular new drug would perform statistical tests to make sure that the drug's effect was consistent and did not vary widely from patient to patient. The biostatisticians employed by the company would therefore perform a hypothesis test to make sure that the population standard deviation σ was not too large.

Under the assumption that $H_0 : \sigma = \sigma_0$ is true, the χ^2 statistic takes the following form:

$$\chi^2_{\text{data}} = \frac{(n-1)s^2}{\sigma_0^2}$$

For the hypothesis test about σ, our test statistic is called χ^2_{data} because the values of $n - 1$ and s^2 come from the observed data. The test statistic χ^2_{data} takes a moderate value when the value of s^2 is moderate assuming H_0 is true, and χ^2_{data} takes an extreme value when the value of s^2 is extreme assuming H_0 is true. This leads us to the following.

> **The Essential Idea About Hypothesis Testing for the Standard Deviation**
> When the observed value of χ^2_{data} is unusual or extreme on the assumption that H_0 is true, we should reject H_0. Otherwise, there is insufficient evidence against H_0, and we should not reject H_0.

The remainder of Section 9.6 explains the details of implementing hypothesis testing for the standard deviation. The χ^2 test for σ may be performed using the p-value method or the critical-value method. We begin with the critical-value method.

> **χ^2 Test for σ: Critical-Value Method**
> This hypothesis test is valid only if we have a random sample from a normal population.
>
> **Step 1 State the hypotheses.**
> Use one of the forms in Table 9.13. State the meaning of σ.
>
> **Step 2 Find the χ^2 critical value or values and state the rejection rule.**
> Use Table 9.13.
>
> **Step 3 Calculate χ^2_{data}.**
> Either use technology to find the value of the test statistic χ^2_{data} or calculate the value of χ^2_{data} as follows:
>
> $$\chi^2_{\text{data}} = \frac{(n-1)s^2}{\sigma_0^2}$$
>
> which follows a χ^2 distribution with $n-1$ degrees of freedom, and where s^2 represents the sample variance.
>
> **Step 4 State the conclusion and the interpretation.**
> If χ^2_{data} falls in the critical region, then reject H_0. Otherwise do not reject H_0. Interpret your conclusion so that a nonspecialist can understand.

The χ^2 critical values in the right-tailed, left-tailed, or two-tailed tests use the following notations: χ^2_α, $\chi^2_{1-\alpha}$, $\chi^2_{\alpha/2}$, and $\chi^2_{1-\alpha/2}$ (see Table 9.13). In each case, *the subscript indicates the area to the right of the χ^2 critical value*. Find these values just as you did in Section 8.4, using either technology or Table E, Chi-Square (χ^2) Distribution, in the Appendix.

Table 9.13 Critical values and rejection rules for the χ^2 test for σ

Right-tailed test	Left-tailed test	Two-tailed test
$H_0 : \sigma = \sigma_0$ $H_a : \sigma > \sigma_0$	$H_0 : \sigma = \sigma_0$ $H_a : \sigma < \sigma_0$	$H_0 : \sigma = \sigma_0$ $H_a : \sigma \neq \sigma_0$
Critical value: χ^2_α Reject H_0 if $\chi^2_{\text{data}} \geq \chi^2_\alpha$ level of significance α	Critical value: $\chi^2_{1-\alpha}$ Reject H_0 if $\chi^2_{\text{data}} \leq \chi^2_{1-\alpha}$ level of significance α	Critical values: $\chi^2_{\alpha/2}$ and $\chi^2_{1-\alpha/2}$ Reject H_0 if $\chi^2_{\text{data}} \geq \chi^2_{\alpha/2}$ or if $\chi^2_{\text{data}} \leq \chi^2_{1-\alpha/2}$ level of significance α

| EXAMPLE 9.24 | | χ^2 TEST FOR σ USING THE CRITICAL-VALUE METHOD |

Alabama	48
Arkansas	37
Iowa	33
Massachusetts	50
Minnesota	45
Oregon	63
South Carolina	66
Utah	52

 lowincome

The table contains the numbers of children (in 1000s) living in low-income households without health insurance for a random sample of 8 states.[20] Test whether the population standard deviation σ of children living in low-income households without health insurance differs from 10,000, using level of significance $\alpha = 0.05$.

Solution

The normal probability plot indicates acceptable normality.

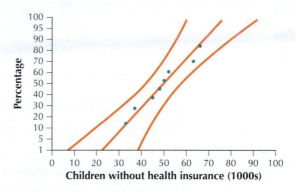

Normal probability plot for children without health insurance.

STEP 1 State the hypotheses.
The phrase "differs from" indicates that we have a two-tailed test. The value $\sigma_0 = 10$ answers the question "Differs from what?" (Note that σ_0 is 10, and not 10,000, since the data are expressed in thousands.) Thus, we have our hypotheses:

$$H_0 : \sigma = 10 \quad \text{versus} \quad H_a : \sigma \neq 10$$

where σ represents the population standard deviation of number of children living in low-income households without health insurance.

STEP 2 Find the χ^2 critical values and state the rejection rule.
We have $n = 8$, so degrees of freedom $= n - 1 = 7$. Since α is given as 0.05, $\alpha/2 = 0.025$ and $1 - \alpha/2 = 0.975$. Then, from the χ^2 table (Appendix Table E), we have $\chi^2_{\alpha/2} = \chi^2_{0.025} = 16.013$, and $\chi^2_{1-\alpha/2} = \chi^2_{0.975} = 1.690$. We will reject H_0 if χ^2_{data} is either $\geq \chi^2_{\alpha/2} = 16.013$ or $\leq \chi^2_{1-\alpha/2} = 1.690$.

FIGURE 9.49 TI-83/84 results.

STEP 3 Find χ^2_{data}.
The TI-83/84 descriptive statistics in Figure 9.49 tell us that the sample variance is

$$s^2 = 11.41114743^2$$

Thus

$$\chi^2_{\text{data}} = \frac{(n-1)s^2}{\sigma_0^2} = \frac{(8-1)11.41114743^2}{10^2} \approx 9.115$$

STEP 4 State the conclusion and the interpretation.
In Step 2 we said that we would reject H_0 if χ^2_{data} was either ≥ 16.013 or ≤ 1.690. Since $\chi^2_{\text{data}} = 9.115$ is neither ≥ 16.013 nor ≤ 1.690 (see Figure 9.50), we do not reject H_0. There is insufficient evidence at level of significance $\alpha = 0.05$ that the population standard deviation of the numbers of children living in low-income households without health insurance differs from 10,000.

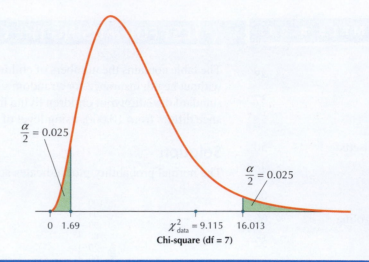

Now You Can Do
Exercises 17–28.

FIGURE 9.50
$\chi^2_{\text{data}} = 9.115$ does not fall in critical region.

2 χ^2 TEST FOR σ USING THE p-VALUE METHOD

We may also use the p-value method to perform the χ^2 test for σ.

χ^2 Test for σ: p-Value Method

This hypothesis test is valid only if we have a random sample from a normal population.

Step 1 **State the hypotheses and the rejection rule.**
Use one of the forms in Table 9.14. State the rejection rule as "Reject H_0 when the p-value $\leq \alpha$." State the meaning of σ.

Step 2 **Calculate χ^2_{data}.**
Either use technology to find the value of the test statistic χ^2_{data} or calculate the value of χ^2_{data} as follows:

$$\chi^2_{\text{data}} = \frac{(n-1)s^2}{\sigma_0^2}$$

which follows a χ^2 distribution with $n-1$ degrees of freedom, and where s^2 represents the sample variance.

Step 3 **Find the p-value.**
Use Table 9.14.

Step 4 **State the conclusion and the interpretation.**
If the p-value $\leq \alpha$, then reject H_0. Otherwise, do not reject H_0. Interpret your conclusion so that a nonspecialist can understand.

Table 9.14 p-Value method for the χ^2 test for σ

Right-tailed test	Left-tailed test	Two-tailed test
$H_0 : \sigma = \sigma_0$	$H_0 : \sigma = \sigma_0$	$H_0 : \sigma = \sigma_0$
$H_a : \sigma > \sigma_0$	$H_a : \sigma < \sigma_0$	$H_a : \sigma \neq \sigma_0$
p-value $= P(\chi^2 > \chi^2_{\text{data}})$	p-value $= P(\chi^2 < \chi^2_{\text{data}})$	If $P(\chi^2 > \chi^2_{\text{data}}) \leq 0.5$, then
Area to right of χ^2_{data}	Area to left of χ^2_{data}	

a. χ^2_{data} is on the right side of the distribution

b. p-value $= 2 \cdot P(\chi^2 > \chi^2_{\text{data}})$

If $P(\chi^2 > \chi^2_{\text{data}}) > 0.5$, then

a. χ^2_{data} is on the left side of the distribution

b. p-value $= 2 \cdot P(\chi^2 < \chi^2_{\text{data}})$

χ^2 TEST FOR σ USING THE p-VALUE METHOD AND TECHNOLOGY

Alamy

Power plants around the country are retooling in order to consume biomass instead of or in addition to coal. The following table contains a random sample of 10 such power plants and the amount of biomass they consumed in 2006, in trillions of Btu (British thermal units).[21] Test whether the population standard deviation is greater than 2 trillion Btu using level of significance $\alpha = 0.05$.

Power plant	Location	Biomass consumed (trillions of Btu)
Georgia Pacific Naheola Mill	Choctaw, AL	13.4
Jefferson Smurfit Fernandina Beach	Nassau, FL	12.9
International Paper Augusta Mill	Richmond, GA	17.8
Gaylord Container Bogalusa	Washington, LA	15.1
Escanaba Paper Company	Delta, MI	19.5
Weyerhaeuser Plymouth NC	Martin, NC	18.6
International Paper	Georgetown, SC	13.8
Bowater Newsprint	McMinn, TN	10.6
Covington Facility	Covington, VA	12.7
Mosinee Paper	Marathon, WI	17.6

Solution

The normal probability plot in Figure 9.51 indicates acceptable normality, allowing us to proceed with the hypothesis test.

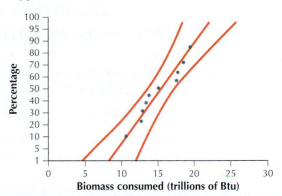

FIGURE 9.51 Normal probability of biomass.

STEP 1 State the hypotheses and the rejection rule.
The phrase "greater than" indicates that we have a right-tailed test. The question "Greater than what?" tells us that $\sigma_0 = 2$, giving us

$$H_0: \sigma = 2 \quad \text{versus} \quad H_a: \sigma > 2$$

We reject H_0 if the p-value $\leq \alpha = 0.05$.

STEP 2 Find χ^2_{data}.
We use the Step-by-Step Technology Guide on page 469. The TI-83/84 descriptive statistics in Figure 9.52 tell us that the sample variance is

$$s^2 = 2.990354866^2$$

```
1-Var Stats
x̄=15.2
Σx=152
Σx²=2390.88
Sx=2.990354866
σx=2.836899716
↓n=10
```

FIGURE 9.52 TI-83/84 results.

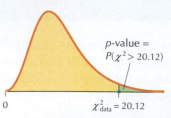

FIGURE 9.53 *p*-Value for χ^2 test.

Thus

$$\chi^2_{\text{data}} = \frac{(n-1)s^2}{\sigma_0^2} = \frac{(10-1)2.990354866^2}{2^2} \approx 20.12$$

STEP 3 Find the *p*-value.

For our right-tailed test, Table 9.14 tells us that

$$p\text{-value} = P(\chi^2 > \chi^2_{\text{data}}) = P(\chi^2 > 20.12)$$

That is, the *p*-value is the area to the right of $\chi^2_{\text{data}} = 20.12$, as shown in Figure 9.53. To find the *p*-value, we use the instructions provided in the Step-by-Step Technology Guide provided at the end of this section. The TI-83/84 results shown in Figure 9.54a tell us that *p*-value $= P(\chi^2 > 20.12) = 0.0171861114$.

The Excel and Minitab results in Figures 9.54b and 9.54c agree with this *p*-value. (Excel and Minitab do not exactly match the TI-83/84 *p*-value because they round the *p*-values to fewer decimal places.) Instead of providing the *p*-value directly, Minitab gives the area to the left of χ^2_{data}: $P(X \leq 20.12) = 0.982814$. We therefore need to subtract the given value from 1 to get the *p*-value:

$$p\text{-value} = 1 - 0.982814 = 0.017186$$

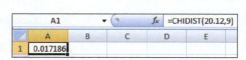

```
X²cdf(20.12,1E99
,9)
    .0171861114
```

FIGURE 9.54a TI-83/84 results.

FIGURE 9.54b Excel results.

	Cumulative Distribution Function
	Chi-Square with 9 DF
x	P(X <= x)
20.12	0.982814

FIGURE 9.54c Minitab results.

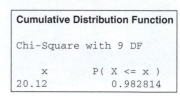

	A1		f_x =CHIDIST(20.12,9)		
	A	B	C	D	E
1	0.017186				

Now You Can Do Exercises 29–34.

STEP 4 State the conclusion and the interpretation.

Since *p*-value $= 0.0171861114 \leq \alpha = 0.05$, we reject H_0. There is evidence that the population standard deviation is greater than 2 trillion Btu.

3 USING CONFIDENCE INTERVALS FOR σ TO PERFORM TWO-TAILED HYPOTHESIS TESTS FOR σ

Suppose we have a $100(1 - \alpha)\%$ confidence interval for σ, of the form (lower bound, upper bound), and are interested in two-tailed hypothesis tests using level of significance α of the form:

$$H_0 : \sigma = \sigma_0 \quad \text{versus} \quad H_a : \sigma \neq \sigma_0$$

We will not reject H_0 for values of σ_0 that lie between the lower bound and upper bound of the confidence interval, and we will reject H_0 for values of σ_0 that lie outside this interval.

EXAMPLE 9.26

USING CONFIDENCE INTERVALS FOR σ TO CONDUCT TWO-TAILED χ^2 TESTS FOR σ

A 95% confidence interval for the population mean sodium content of breakfast cereals, in milligrams (mg) per serving, is given by

$$(44.53 \text{ mg}, 81.50 \text{ mg})$$

Assume that the data are normally distributed. Test using level of significance $\alpha = 0.05$ whether σ differs from the following.

a. 80 mg

b. 40 mg

Solution

a. For the hypothesis test $H_0 : \sigma = 80$ versus $H_a : \sigma \neq 80$, $\sigma_0 = 80$ lies between the lower bound 44.53 and the upper bound 81.50 of the confidence interval, and we therefore do not reject H_0. There is insufficient evidence that the population standard deviation of sodium content differs from 80 mg.

b. For the hypothesis test $H_0 : \sigma = 40$ versus $H_a : \sigma \neq 40$, $\sigma_0 = 40$ lies outside the confidence interval, and we therefore reject H_0. There is evidence that the population standard deviation of sodium content differs from 40 mg.

Now You Can Do
Exercises 35–38.

STEP-BY-STEP TECHNOLOGY GUIDE: Finding χ^2 *p-value*

We will use the information from Example 9.25 (page 467). The steps for finding the χ^2 critical values are given in the Step-by-Step Technology Guide at the end of Section 8.4 (pages 397–398).

TI-83/84

Step 1   Enter the data into List **L1**.
Step 2   Press **2nd > DISTR**, then χ^2 **cdf(**, and press **ENTER**.
Step 3   On the home screen, enter the value of χ^2_{data}, comma, 1E99, comma, degrees of freedom, close parenthesis, as shown

in Figure 9.54a. (Remember that this "ε" is inserted by pressing 2nd, followed by the comma key.)
Step 4   Press **ENTER**. The results for Example 9.25 are shown in Figure 9.54a.

EXCEL

Step 1   Select cell **A1**. Click the **Insert Function** icon f_x.
Step 2   For **Search for a Function**, type **chidist** and click **OK**.

Step 3   For **x**, enter the value of χ^2_{data}, and for **Deg_freedom**, enter the degrees of freedom. Excel displays the *p*-value in the cell in the dialog box, as shown in Figure 9.54b.

MINITAB

Step 1   Click on **Calc > Probability Distributions > Chi-Square**.
Step 2   Select **Cumulative probability**, and enter the **Degrees of freedom**.
Step 3   For **Input constant**, enter the value of χ^2_{data} and click **OK**.

Step 4   Minitab displays the area to the left of χ^2_{data} in the session window, as shown in Figure 9.54c. To find the *p*-value, subtract this area from 1.

CRUNCHIT!

Step 1   Click **Distribution Calculator** and select **Chi-square**.
Step 2   For **df** enter the degrees of freedom.
Step 3   Select the **Probability** tab. Enter the value of χ^2_{data} and click **Calculate**. CrunchIt! displays the cumulative probability,

that is, the area to the left of χ^2_{data}. To find the *p*-value, subtract this value from 1.

| **SECTION 9.6** | **Summary** |

1. Under the assumption that H_0 is true, the χ^2 statistic takes the following form:

$$\chi^2_{data} = \frac{(n-1)s^2}{\sigma_0^2}$$

The hypothesis test about σ may be performed using the *p*-value method or the critical-value method. Either way, the test is valid only if we have a random sample from a normal population. The critical-value method compares χ^2_{data} with one or two critical values.

2. The p-value method compares the p-value to level of significance α.

3. If we have a $100(1 - \alpha)\%$ confidence interval for σ, of the form (lower bound, upper bound), and are interested in a two-tailed hypothesis test, using α, of the form

$$H_0 : \sigma = \sigma_0 \quad \text{versus} \quad H_a : \sigma \neq \sigma_0$$

we will not reject H_0 for values of σ_0 that lie between the lower bound and upper bound of the confidence interval. We will reject H_0 for values of σ_0 that lie outside this interval.

SECTION 9.6 **Exercises**

Clarifying the Concepts

1. Think of one instance where an analyst would be interested in performing a hypothesis test about the population standard deviation σ.

2. What is the difference between σ and σ_0?

3. Does it make sense to test whether $\sigma < 0$? Explain.

4. What condition must be fulfilled for us to perform a hypothesis test about σ?

5. Explain how we can use a confidence interval to determine significance.

6. In the previous exercise, what must be the relationship between α and the confidence level?

Practicing the Techniques

For Exercises 7–10, construct the hypotheses.

7. Test whether the population standard deviation is greater than 10.

8. Test whether the population standard deviation is less than 5.

9. Test whether the population standard deviation differs from 3.

10. Test whether the population standard deviation is greater than 100.

For Exercises 11–16, a random sample is drawn from a normal population. Calculate χ^2_{data}.

11. We are testing whether $\sigma > 1$ and have a sample of size $n = 21$ with a sample variance of $s^2 = 3$.

12. We are testing whether $\sigma < 5$ and have a sample of size $n = 11$ with a sample variance of $s^2 = 25$.

13. We are testing whether $\sigma \neq 3$ and have a sample of size $n = 16$ with a standard deviation of $s = 2.5$.

14. We are testing whether $\sigma > 10$ and have a sample of size $n = 14$ with a standard deviation of $s = 12$.

15. We are testing whether $\sigma < 20$, and have a sample of size $n = 8$ and a sample variance of $s^2 = 350$.

16. We are testing whether $\sigma \neq 5$ and have a sample of size $n = 26$ with a standard deviation of $s = 5$.

For Exercises 17–22, a random sample is drawn from a normal population. The values of χ^2_{data} for these exercises

are found in Exercises 11–16. Find the critical value or values.

17. We are testing whether $\sigma > 1$, using $\alpha = 0.05$, and have a sample of size $n = 21$ and a sample variance of $s^2 = 3$. Find χ^2_α.

18. We are testing whether $\sigma < 5$, using $\alpha = 0.05$, and have a sample of size $n = 11$ and a sample variance of $s^2 = 25$. Find $\chi^2_{1-\alpha}$.

19. We are testing whether $\sigma \neq 3$, using $\alpha = 0.05$, and have a sample of size $n = 16$ and a standard deviation of $s = 2.5$. Find $\chi^2_{\alpha/2}$ and $\chi^2_{1-\alpha/2}$.

20. We are testing whether $\sigma > 10$, using $\alpha = 0.01$, and have a sample of size $n = 14$ and a standard deviation of $s = 12$. Find χ^2_α.

21. We are testing whether $\sigma < 20$, using $\alpha = 0.10$, and have a sample of size $n = 8$ and a sample variance of $s^2 = 350$. Find $\chi^2_{1-\alpha}$.

22. We are testing whether $\sigma \neq 5$, using $\alpha = 0.05$, and have a sample of size $n = 26$ with a standard deviation of $s = 5$. Find $\chi^2_{\alpha/2}$ and $\chi^2_{1-\alpha/2}$.

For Exercises 23–28, a random sample is drawn from a normal population. The values of χ^2_{data} for these exercises were found in Exercises 11–16. The critical values were found in Exercises 17–22. Do the following.
  a. State the rejection rule.
  b. Compare χ^2_{data} with the critical value or values. State the conclusion and interpretation.

23. The data in Exercise 17

24. The data in Exercise 18

25. The data in Exercise 19

26. The data in Exercise 20

27. The data in Exercise 21

28. The data in Exercise 22

For Exercises 29–34, a random sample is drawn from a normal population. Do the following.
  a. Draw a χ^2 distribution and indicate the location of χ^2_{data}.
  b. Find the p-value and indicate the p-value in your distribution in **(a)**.
  c. Compare the p-value with level of significance $\alpha = 0.05$. State the conclusion and interpretation.

29. The data in Exercise 11

30. The data in Exercise 12

31. The data in Exercise 13

32. The data in Exercise 14

33. The data in Exercise 15

34. The data in Exercise 16

For Exercises 35–38 a $100(1 - \alpha)\%$ χ^2 confidence interval for σ is given. Use the confidence interval to test using level of significance α whether σ differs from each of the indicated hypothesized values.

35. A 95% χ^2 confidence interval for σ is (1, 4). Hypothesized values σ_0 are
   a. 0
   b. 2
   c. 5

36. A 99% χ^2 confidence interval for σ is (10, 25). Hypothesized values σ_0 are
   a. 15
   b. 26
   c. 5

37. A 90% χ^2 confidence interval for σ is (100, 200). Hypothesized values σ_0 are
   a. 150
   b. 250
   c. 0

38. A 95% χ^2 confidence interval for σ is (127, 698). Hypothesized values σ_0 are
   a. 125
   b. 128
   c. 700

Applying the Concepts

39. DDT in Breast Milk. Researchers compared the amount of DDT in the breast milk of a random sample of 12 Latina women in Yakima Valley in Washington State with the amount of DDT in breast milk in the general U.S. population.[22] They measured the standard deviation of the amount of DDT in the general population to be 36.5 parts per billion (ppb). Assume that the population is normally distributed. We are interested in testing whether the population standard deviation of DDT level in the breast milk of Latina women in Yakima Valley is greater than that of the general population, using level of significance $\alpha = 0.01$.
   a. The sample variance is $s^2 = 119{,}025$. Calculate χ^2_{data}.
   b. Perform the appropriate hypothesis test.

40. Tree Rings. Does the growth of trees vary more when the trees are young? The International Tree Ring Data Base collected data on a particular 440-year-old Douglas fir tree.[23] The standard deviation of the annual ring growth in the tree's first 80 years of life was 0.8 millimeter (mm) per year. Assume that the population is normal. We are interested in testing

whether the population standard deviation of annual ring growth in the tree's later years is less than 0.8 mm per year.
   a. The sample variance for a random sample of size 100 taken from the tree's later years is $s^2 = 0.3136$. Calculate χ^2_{data}.
   b. Perform the appropriate hypothesis test.

41. Union Membership. The following table contains the total union membership (in 1000s) for 7 randomly selected states.[24] Assume that the distribution is normal. We are interested in whether the population standard deviation of union membership σ differs from 30,000, using level of significance $\alpha = 0.05$.

 unionmember

Florida	397
Indiana	334
Maryland	342
Massachusetts	414
Minnesota	395
Texas	476
Wisconsin	386

   a. The sample variance is $s^2 = 2245.67$. Calculate χ^2_{data}.
   b. Perform the appropriate hypothesis test.
   c. Would $\sigma = 30{,}000$ lie inside or outside a 95% Z interval for σ? Explain.

42. Fourth-Grade Feet. Suppose a children's shoe manufacturer is interested in estimating the variability of fourth-graders' feet. A random sample of 20 fourth-graders' feet yielded the following foot lengths, in centimeters.[25] The normality of the data was verified in Example 8.10 (page 373). Test whether the population standard deviation of foot lengths σ is less than 1 centimeter using level of significance $\alpha = 0.05$.

 fourthfeet

22.4	23.4	22.5	23.2	23.1	23.7	24.1	21.0	21.6	20.9
25.5	22.8	24.1	25.0	24.0	21.7	22.0	22.7	24.7	23.5

43. Does Score Variability Differ by Gender? Recently, researchers have been examining the evidence for whether there is greater variability in boys' scores than girls' scores on cognitive abilities tests. For example, one study found that boys were overrepresented at both the top and the bottom of nonverbal reasoning tests and quantitative reasoning tests.[26] Suppose that the standard deviation for girls' scores is known to be 50 points for a particular test and that the population of all scores is normal. A random sample of 101 boys has a sample variance of 2600. Test whether the population standard deviation for boys exceeds 50 points, using level of significance $\alpha = 0.05$.

44. Heart Rate Variability. A reduction in heart rate variability is associated with elevated levels of stress, since the body continues to pump adrenaline after high-stress situations, even when at rest.[27] Suppose the standard deviation of heartbeats in the general population is 20 beats per minute, and that the population of heart rates is normal. A random sample of 50 individuals leading high-stress lives has a sample variance of 200 beats per minute. Test using level of significance $\alpha = 0.05$ whether the population standard deviation for those leading high-stress lives is lower than that in the general population.

9.7 PROBABILITY OF TYPE II ERROR AND THE POWER OF A HYPOTHESIS TEST

OBJECTIVES By the end of this section, I will be able to . . .

1 Calculate the probability of Type II error for a Z test for μ.

2 Compute the power of a Z test for μ and construct a power curve.

1 PROBABILITY OF A TYPE II ERROR

In Section 9.1 we defined a Type II error as follows:

Type II error: not rejecting H_0 when H_0 is false

For example, the criminal trial scenario on page 407 had the following hypotheses:

H_0 : defendant is not guilty   versus H_a : defendant is guilty

In this case, a Type II error was to find the defendant not guilty (not reject H_0) when in reality he did commit the crime (H_0 is false). In this section we learn how to calculate the probability of making a Type II error for a Z test for μ, called β (beta), and to use the value of β to compute the power of a Z test for μ.

Calculating β, the Probability of a Type II Error

Use the following steps to calculate β, the probability of a Type II error.

Step 1

Recall that Z_{crit} divides the critical region from the noncritical region. Let $\bar{x}_{crit}$ be the value of the sample mean $\bar{x}$ associated with Z_{crit}. The following table shows how to calculate $\bar{x}_{crit}$ for the three forms of the hypothesis test.

Form of test			Value of $\bar{x}_{crit}$
Right-tailed	$H_0: \mu = \mu_0$	vs. $H_a: \mu > \mu_0$	$\bar{x}_{crit} = \mu_0 + Z_{crit} \cdot \dfrac{\sigma}{\sqrt{n}}$
Left-tailed	$H_0: \mu = \mu_0$	vs. $H_a: \mu < \mu_0$	$\bar{x}_{crit} = \mu_0 - Z_{crit} \cdot \dfrac{\sigma}{\sqrt{n}}$
Two-tailed	$H_0: \mu = \mu_0$	vs. $H_a: \mu \neq \mu_0$	$\bar{x}_{crit,\ lower} = \mu_0 - Z_{crit} \cdot \dfrac{\sigma}{\sqrt{n}}$ $\bar{x}_{crit,\ upper} = \mu_0 + Z_{crit} \cdot \dfrac{\sigma}{\sqrt{n}}$

Here, μ_0 is the hypothesized value of the population mean, σ is the population standard deviation, and n is the sample size.

Step 2
Let μ_a represent a particular value for the population mean μ chosen from the values indicated in the alternative hypothesis H_a. Draw a normal curve centered at μ_a, with the value or values of $\bar{x}_{\text{crit}}$ from Step 1 indicated (see Example 9.27).

Step 3
Calculate β for the particular μ_a chosen using the following table.

Form of test			β = probability of Type II error
Right-tailed	$H_0: \mu = \mu_0$	vs.	$H_a: \mu > \mu_0$	The area under the normal curve drawn in Step 2 to the left of $\bar{x}_{\text{crit}}$.
Left-tailed	$H_0: \mu = \mu_0$	vs.	$H_a: \mu < \mu_0$	The area under the normal curve drawn in Step 2 to the right of $\bar{x}_{\text{crit}}$.
Two-tailed	$H_0: \mu = \mu_0$	vs.	$H_a: \mu \neq \mu_0$	The area under the normal curve drawn in Step 2 between $\bar{x}_{\text{crit, lower}}$ and $\bar{x}_{\text{crit, upper}}$.

Let us illustrate the steps for calculating β, the probability of a Type II error, using an example.

EXAMPLE 9.27

CALCULATING β, THE PROBABILITY OF A TYPE II ERROR

In Example 9.3, we tested whether people use debit cards on average more than 11 times per month. The hypotheses are

$$H_0: \mu = 11 \quad \text{versus} \quad H_a: \mu > 11$$

where μ represents the population mean debit card usage per month. From Example 9.3 we have $n = 36$, $\bar{x} = 11.5$, and $\sigma = 3$, and from Example 9.5 we have $Z_{\text{crit}} = 2.33$.

a. State what a Type II error would be in this case.

b. Let $\mu_a = 13$. That is, suppose the population mean debit card usage is actually 13 times per month. Calculate β, the probability of making a Type II error when $\mu_a = 13$.

Solution

a. We make a Type II error when we do not reject H_0 when H_0 is false. In this case, a Type II error would be to conclude that the population mean debit card usage was 11 times per month when in actuality it was more than 11 times per month.

b. We follow the steps for calculating β.

| STEP 1 | We have a right-tailed test, so that

$$\bar{x}_{\text{crit}} = \mu_0 + Z_{\text{crit}} \cdot \frac{\sigma}{\sqrt{n}} = 11 + 2.33 \cdot \frac{3}{\sqrt{36}} = 12.165$$

| STEP 2 | Figure 9.55 shows the normal curve centered at $\mu_a = 13$, with $\bar{x}_{\text{crit}} = 12.165$ labeled.

| STEP 3 | The right-tailed test tells us that β equals the area under the normal curve drawn in Step 2 to the left of $\bar{x}_{\text{crit}} = 12.165$. This is the shaded area in Figure 9.55. Since area represents probability, we have

$$\beta = P(\bar{x} < 12.165) \text{ when } \mu_a = 13$$

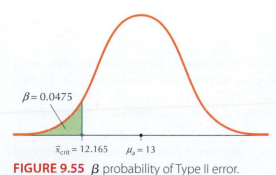

$\beta = 0.0475$

$\bar{x}_{\text{crit}} = 12.165$ $\mu_a = 13$

FIGURE 9.55 β probability of Type II error.

This is a Case 1 problem from Table 6.6 on page 289.

Standardizing with $\mu_a = 13$, $\sigma = 3$, and $n = 36$:

$$\beta = P(\bar{x} < 12.165)$$

$$= P\left(Z < \frac{12.165 - 13}{3/\sqrt{36}}\right)$$

$$= P(Z < -1.67) = 0.0475$$

Thus, $\beta = 0.0475$. This represents the probability of making a Type II error, that is, of not rejecting the hypothesis that the population mean debit card usage is 11 times per month when in actuality it is 13 times per month.

Now You Can Do
Exercises 5a, b, c–16a, b, c.

2 POWER OF A HYPOTHESIS TEST

It is a correct decision to reject the null hypothesis when the null hypothesis is false. The probability of making this type of correct decision is called the **power of the test.**

> **Power of a Hypothesis Test**
>
> The **power of a hypothesis test** is the probability of rejecting the null hypothesis when the null hypothesis is false. Power is calculated as
>
> $$\text{power} = 1 - \beta$$

EXAMPLE 9.28

POWER OF A HYPOTHESIS TEST

Calculate the power, for the particular alternative value of the mean, of the hypothesis test in Example 9.27.

Solution

The probability of a Type II error was found in Example 9.27 to be $\beta = 0.0475$. Thus, the power of the hypothesis test is

Now You Can Do
Exercises 5d–16d.

$$\text{power} = 1 - \beta = 1 - 0.0475 = 0.9525$$

The probability of correctly rejecting the null hypothesis is 0.9525.

Type II Error and Power of the Test

Suppose that we have the same hypothesis test from Example 9.27 and the same value $\bar{x}_{\text{crit}} = 12.165$. Now, *what if* we decrease μ_a such that it is less than 13 but still larger than 12.165. Describe what will happen to the following, and why.

    a. The probability of a Type II error, β

    b. The power of the test, $1 - \beta$

Solution

    a. Consider Figure 9.56. The distribution of sample means remains centered at μ_a, so that a smaller μ_a will "slide" the normal curve toward the value of $\bar{x}_{\text{crit}} = 12.165$. This results in a larger area to the left of 12.165, as you can see by comparing Figure 9.56 with Figure 9.55. Therefore, a smaller μ_a leads to an *increase* in the probability of a Type II error, β.

    b. As β increases, $1 - \beta$ decreases. Therefore, a smaller μ_a leads to a decrease in the power of the test.

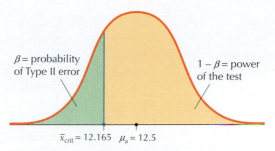

β = probability of Type II error

$1 - \beta$ = power of the test

$\bar{x}_{crit} = 12.165$ $\mu_a = 12.5$

FIGURE 9.56   Smaller μ_a leads to an increase in β.

A **power curve** plots the values for the power of the test versus the values of μ_a.

EXAMPLE 9.29

POWER CURVE

a. Calculate the power of the hypothesis test from Example 9.27 for the following values of μ_a: 11.0, 11.5, 12.0, 12.165, 12.5, 13.5.

b. Construct the power curve by graphing the values for the power of the test on the vertical axis against the values of μ_a on the horizontal axis.

Solution

a. We have $\bar{x}_{crit} = 12.165$, $\sigma = 3$, and $n = 36$. The calculations are provided in the following table.

μ_a	Probability of Type II error: β	Power of the test: $1 - \beta$
11.0	$P\left(Z < \dfrac{12.165 - 11}{3/\sqrt{36}}\right) = P(Z < 2.33) = 0.9901$	$1 - 0.9901 = 0.0099$
11.5	$P\left(Z < \dfrac{12.165 - 11.5}{3/\sqrt{36}}\right) = P(Z < 1.33) = 0.9082$	$1 - 0.9082 = 0.0918$
12.0	$P\left(Z < \dfrac{12.165 - 12}{3/\sqrt{36}}\right) = P(Z < 0.33) = 0.6293$	$1 - 0.6293 = 0.3707$
12.165	$P\left(Z < \dfrac{12.165 - 12.165}{3/\sqrt{36}}\right) = P(Z < 0.00) = 0.5$	$1 - 0.5 = 0.5$
12.5	$P\left(Z < \dfrac{12.165 - 12.5}{3/\sqrt{36}}\right) = P(Z < -0.67) = 0.2514$	$1 - 0.2514 = 0.7486$
13.5	$P\left(Z < \dfrac{12.165 - 13.5}{3/\sqrt{36}}\right) = P(Z < -2.67) = 0.0038$	$1 - 0.0038 = 0.9962$

b. Figure 9.57 represents a power curve, since it plots the values for the power of the test on the vertical axis against the values of μ_a on the horizontal axis. Note that, as μ_a moves farther away from the hypothesized mean $\mu_0 = 11$, the power of the test increases. This is because it is more likely that the null hypothesis will be correctly rejected as the actual value of the mean μ_a gets farther away from the hypothesized value μ_0.

For completeness, we include the power for $\mu_a = 13$ from Example 9.28 in this power curve.

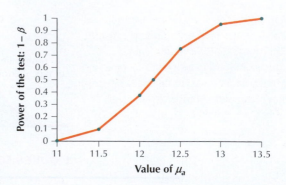

**Now You Can Do
Exercises 17 and 18.**

FIGURE 9.57 Power curve.

SECTION 9.7 Summary

1. We may calculate β, the probability of making a Type II error for a Z test for μ, given a particular alternative value for the population mean, μ_a.

2. The power of a hypothesis test is the probability of rejecting a false null hypothesis, and is calculated as *power* $= 1 - \beta$. We may then build the power curve by plotting the power against values of μ_a.

SECTION 9.7 Exercises

Clarifying the Concepts

1. Explain what a Type II error is.

2. Describe what $\bar{x}_{crit}$ is.

3. In words, what do we mean by the power of a hypothesis test?

4. How do we calculate the power of a test?

Practicing the Techniques

For Exercises 5–18, assume that the conditions for performing the Z test are met. Do the following.
  a. Calculate the value or values of $\bar{x}_{crit}$.
  b. Draw a normal curve, centered at μ_a, with the value or values of $\bar{x}_{crit}$ indicated.
  c. Calculate β, the probability of a Type II error for that value of μ_a. Shade the corresponding area under the normal curve.
  d. Calculate the power of the hypothesis test.

5. $H_0: \mu = 50$ vs. $H_a: \mu > 50$, $\alpha = 0.10$, $\sigma = 4$, $n = 25$, $\mu_a = 51$

6. $H_0: \mu = 50$ vs. $H_a: \mu > 50$, $\alpha = 0.10$, $\sigma = 4$, $n = 25$, $\mu_a = 52$

7. $H_0: \mu = 50$ vs. $H_a: \mu > 50$, $\alpha = 0.10$, $\sigma = 4$, $n = 25$, $\mu_a = 53$

8. $H_0: \mu = 50$ vs. $H_a: \mu > 50$, $\alpha = 0.10$, $\sigma = 4$, $n = 25$, $\mu_a = 54$

9. $H_0: \mu = 50$ vs. $H_a: \mu > 50$, $\alpha = 0.10$, $\sigma = 4$, $n = 25$, $\mu_a = 55$

10. $H_0: \mu = 50$ vs. $H_a: \mu > 50$, $\alpha = 0.10$, $\sigma = 4$, $n = 25$, $\mu_a = 56$

11. $H_0: \mu = 100$ vs. $H_a: \mu < 100$, $\alpha = 0.05$, $\sigma = 12$, $n = 36$, $\mu_a = 96$

12. $H_0: \mu = 100$ vs. $H_a: \mu < 100$, $\alpha = 0.05$, $\sigma = 12$, $n = 36$, $\mu_a = 94$

13. $H_0: \mu = 100$ vs. $H_a: \mu < 100$, $\alpha = 0.05$, $\sigma = 12$, $n = 36$, $\mu_a = 92$

14. $H_0: \mu = 100$ vs. $H_a: \mu < 100$, $\alpha = 0.05$, $\sigma = 12$, $n = 36$, $\mu_a = 90$

15. $H_0: \mu = 100$ vs. $H_a: \mu < 100$, $\alpha = 0.05$, $\sigma = 12$, $n = 36$, $\mu_a = 88$

16. $H_0: \mu = 100$ vs. $H_a: \mu < 100$, $\alpha = 0.05$, $\sigma = 12$, $n = 36$, $\mu_a = 86$

17. Refer to Exercises 5–10. Construct the power curve for the given values of μ_a.

18. Refer to Exercises 11–16. Construct the power curve for the given values of μ_a.

Applying the Concepts

19. Stock Market. The *Statistical Abstract of the United States* reports that the mean daily number of shares traded on the New York Stock Exchange in 2005 was 1.6 billion. Let this value represent the hypothesized population mean, and assume that the population standard deviation equals 0.5 billion shares. Suppose that we have a random sample of 36 days from the present year, and we are interested in testing whether the population mean daily number of shares traded has increased since 2005, using level of significance $\alpha = 0.05$.

  a. Describe what a Type II error would mean in the context of this problem.
  b. What is the probability of making a Type II error when the actual mean number of shares traded is 1.65 billion?
  c. What is the probability of making a Type II error when the actual mean number of shares traded is 1.70 billion?
  d. What is the probability of making a Type II error when the actual mean number of shares traded is 1.75 billion?
  e. What is the probability of making a Type II error when the actual mean number of shares traded is 1.80 billion?
  f. Calculate the power of the hypothesis test for the values of μ_a given in (b)–(e).
  g. Construct the power curve for the values of μ_a given in (b)–(e).

20. Credit Score in Florida. According to **CreditReport.com**, the mean credit score in Florida in 2006 was 673. Suppose we have a random sample of 900 credit scores in Florida, and assume that the population standard deviation is 150. We are interested in testing using level of significance $\alpha = 0.05$ whether the population mean credit score in Florida has decreased since that time.

  a. Describe what a Type II error would mean in the context of this problem.
  b. What is the probability of making a Type II error when the actual mean credit score is 670?
  c. What is the probability of making a Type II error when the actual mean credit score is 665?
  d. What is the probability of making a Type II error when the actual mean credit score is 660?
  e. What is the probability of making a Type II error when the actual mean credit score is 655?
  f. Calculate the power of the hypothesis test for the values of μ_a given in (b)–(e).
  g. Construct the power curve for the values of μ_a given in (b)–(e).

21. Accountants' Salary. According to **Salary.com**, the mean salary for entry-level accountants in 2010 was $41,560. Let this value represent the hypothesized population mean, and assume that the population standard deviation equals $5000. Suppose we have a random sample of 100 entry-level accountants and wish to test using level of significance $\alpha = 0.05$ whether the population mean salary has changed since 2010.

  a. Describe what a Type II error would mean in the context of this problem.

  b. What is the probability of making a Type II error when the actual mean salary is $42,000?
  c. What is the probability of making a Type II error when the actual mean salary is $43,000?
  d. What is the probability of making a Type II error when the actual mean salary is $44,000?
  e. What is the probability of making a Type II error when the actual mean salary is $45,000?
  f. Calculate the power of the hypothesis test for the values of μ_a given in (b)–(e).
  g. Construct the power curve for the values of μ_a given in (b)–(e).

22. Price of Milk. The U.S. Bureau of Labor Statistics reports that the mean price for a gallon of milk in 2005 was $3.24. Suppose that we have a random sample taken this year of 400 gallons of milk, and assume that the population standard deviation equals $1.00. We would like to conduct a hypothesis test using level of significance $\alpha = 0.01$ to investigate if the population mean price of milk this year has increased.

  a. Describe what a Type II error would mean in the context of this problem.
  b. What is the probability of making a Type II error when the actual mean price is $3.30?
  c. What is the probability of making a Type II error when the actual mean price is $3.50?
  d. What is the probability of making a Type II error when the actual mean price is $3.70?
  e. What is the probability of making a Type II error when the actual mean price is $3.90?
  f. Calculate the power of the hypothesis test for the values of μ_a given in (b)–(e).
  g. Construct the power curve for the values of μ_a given in (b)–(e).

23. Hybrid Vehicles. A 2006 study by **Edmunds.com** examined the time it takes for owners of hybrid vehicles to recoup their additional initial cost through reduced fuel consumption. Suppose we have a random sample of 9 hybrid cars. Assume that the population is normal with $\sigma = 0.2$. We would like to test using level of significance $\alpha = 0.01$ whether the population mean time it takes owners of hybrid cars to recoup their initial cost is less than three years.

  a. Describe what a Type II error would mean in the context of this problem.
  b. What is the probability of making a Type II error when the actual mean time is 2.5 years?
  c. What is the probability of making a Type II error when the actual mean time is 2 years?
  d. What is the probability of making a Type II error when the actual mean time is 1.5 years?
  e. What is the probability of making a Type II error when the actual mean time is 1 year?
  f. Calculate the power of the hypothesis test for the values of μ_a given in (b)–(e).
  g. Construct the power curve for the values of μ_a given in (b)–(e).

CHAPTER 9 **Formulas and Vocabulary**

Section 9.1

- α (ALPHA) (p. 411)
- ALTERNATIVE HYPOTHESIS (p. 407)
- β (BETA) (p. XXX)
- HYPOTHESIS TESTING (p. 406)
- LEVEL OF SIGNIFICANCE (p. 411)
- NULL HYPOTHESIS (p. 407)
- STATISTICAL SIGNIFICANCE (p. 409)
- TYPE I ERROR (p. 411)
- TYPE II ERROR (p. 411).

Section 9.2

- CRITICAL REGION (p. 415)
- CRITICAL VALUE, Z_{crit} (p. 415)
- ESSENTIAL IDEA ABOUT HYPOTHESIS TESTING FOR THE MEAN (p. 414)
- LEVEL OF SIGNIFICANCE α (p. 415)
- NONCRITICAL REGION (p. 415)
- TEST STATISTIC (p. 415)
- Z_{data} (p. 414).

$$Z_{\text{data}} = \frac{\bar{x} - \mu_0}{\sigma/\sqrt{n}}$$

Section 9.3

- p-VALUE (p. 422)
- REJECTION RULE FOR PERFORMING A HYPOTHESIS TEST USING THE p-VALUE METHOD (p. 424).

Section 9.4

- CRITICAL VALUE, t_{crit} (p. 437)
- t_{data} (p. 437).

$$t_{\text{data}} = \frac{\bar{x} - \mu_0}{s/\sqrt{n}}$$

Section 9.5

- ESSENTIAL IDEA ABOUT HYPOTHESIS TESTING FOR THE PROPORTION (p. XXX)
- STANDARD ERROR OF THE PROPORTION (p. 432)
- Z_{data} FOR THE HYPOTHESIS TEST FOR THE POPULATION PROPORTION (p. 452).

$$Z_{\text{data}} = \frac{\hat{p} - p_0}{\sqrt{\dfrac{p_0(1 - p_0)}{n}}}$$

Section 9.6

- χ^2_{data} (p. 464)

$$\chi^2_{\text{data}} = \frac{(n - 1)s^2}{\sigma_0^2}$$

- ESSENTIAL IDEA ABOUT HYPOTHESIS TESTING FOR THE STANDARD DEVIATION (p. 464).

Section 9.7

- POWER CURVE (p. 475)
- POWER OF A HYPOTHESIS TEST (p. 474).

CHAPTER 9 **Review Exercises**

Section 9.1

For Exercises 1–3, provide the null and alternative hypotheses.

1. Test whether $\mu < 12$.
2. Test whether $\mu > 10$.
3. Test whether μ is below zero.

For Exercises 4–6, do the following.
   a. Provide the null and alternative hypotheses.
   b. Describe the two ways a correct decision could be made.
   c. Describe what a Type I error would mean in the context of the problem.
   d. Describe what a Type II error would mean in the context of the problem.

4. **HOUSEHOLD SIZE.** The U.S. Census Bureau reported that the mean household size is 2.58 persons. We conduct a hypothesis test to determine whether the population mean household size has changed.

5. **SPEEDING-RELATED TRAFFIC FATALITIES.** The National Highway Traffic Safety Administration reports that the mean number of speeding-related traffic fatalities over the Thanksgiving holiday period is 202.7. We conduct a hypothesis test to examine whether the population mean number of such fatalities has decreased.

6. **SALARIES OF ASSISTANT PROFESSORS.** Salaries.com reports that the median salary for assistant professors in science was \$49,934. We use this median salary to estimate that the mean salary in 2005 was \$50,000. A hypothesis test was conducted to determine if the population mean salary of assistant professors in science has increased.

Section 9.2

For Exercises 7–9, find the value of Z_{data}.

7. $\bar{x} = 59$, $\sigma = 10$, $n = 100$, $\mu_0 = 60$
8. $\bar{x} = 59$, $\sigma = 5$, $n = 100$, $\mu_0 = 60$
9. $\bar{x} = 59$, $\sigma = 1$, $n = 100$, $\mu_0 = 60$

For each of the following hypothesis tests in Exercises 10–12, do the following.
   a. Find the value of Z_{crit}.
   b. Find the critical-value rejection rule.

c. Draw a standard normal curve and indicate the critical region.
d. State the conclusion and interpretation.

10. $H_0 : \mu = \mu_0$ versus $H_a : \mu \neq \mu_0$, $\alpha = 0.01$, $Z_{data} = -2.5$

11. $H_0 : \mu = \mu_0$ versus $H_a : \mu > \mu_0$, $\alpha = 0.10$, $Z_{data} = 1.5$

12. $H_0 : \mu = \mu_0$ versus $H_a : \mu > \mu_0$, $\alpha = 0.05$, $Z_{data} = -2.5$

For Exercises 13 and 14, do the following.
a. State the hypotheses.
b. Find the value of Z_{crit} and the rejection rule. Also, draw a standard normal curve, indicating the critical region.
c. Calculate Z_{data}. Draw a standard normal curve showing Z_{crit}, the critical region, and Z_{data}.
d. State the conclusion and the interpretation.

13. CREDIT SCORES IN FLORIDA. According to CreditReport.com, the mean credit score in Florida in 2006 was 673. A random sample of 144 Florida residents this year shows a mean credit score of 650. Assume $\sigma = 50$. Perform a hypothesis test using level of significance $\alpha = 0.05$ to determine if the population mean credit score in Florida has decreased.

14. SALARY OF COLLEGE GRADS. It pays to stay in school. According to the U.S. Census Bureau, the mean salary of college graduates is $52,200, whereas the mean salary of those with "some college" is $36,800. A random sample of 100 college graduates provides a sample mean salary of $55,000. Assume $\sigma = \$3000$. Perform a hypothesis test to determine whether the population mean salary of college graduates has increased, using level of significance $\alpha = 0.10$.

Section 9.3

For Exercises 15 and 16, perform the following steps.
a. State the hypotheses and the rejection rule for the p-value method.
b. Calculate Z_{data}.
c. Find the p-value. Draw the standard normal curve, with Z_{data} and the p-value indicated on it.
d. State the conclusion and the interpretation.

15. We are interested in testing at level of significance $\alpha = 0.05$ whether the population mean differs from 500. A random sample of size 100 is taken, with a mean of 520. Assume $\sigma = 50$.

16. We would like to test at level of significance $\alpha = 0.01$ whether the population mean is less than -10. A random sample of size 25 is taken from a normal population. The sample mean is -12. Assume $\sigma = 2$.

17. HEALTH CARE EXPENDITURES. We are interested in whether the population mean per capita annual expenditures on health care have increased since 2007, when the mean was $6096 per person.[28] A random sample taken this year of 100 Americans shows mean annual health care expenditures of $8000. Suppose that prior research has indicated that the population standard deviation of such expenditures is $1600. Perform the appropriate hypothesis test, using the p-value method and level of significance $\alpha = 0.01$.

18. THE OLD COFFEE MACHINE. A random sample of 36 cups of coffee dispensed from the old coffee machine in the lobby had a mean amount of coffee of 7 ounces per cup. Assume that the population standard deviation is 1 ounce.
a. Construct a 95% confidence interval for the population mean amount of coffee dispensed by the old coffee machine in the lobby.
b. Use the confidence interval to test at level of significance $\alpha = 0.05$ whether the population mean amount of coffee dispensed by the old coffee machine in the lobby differs from the following amounts, in ounces.
  i. 6.9
  ii. 7.5
  iii. 6.7
  iv. 7

Section 9.4

For Exercises 19–21, find the critical value t_{crit} and sketch the critical region. Assume normality.

19. $H_0 : \mu = 100$, $H_a : \mu > 100$, $n = 8$, $\alpha = 0.10$

20. $H_0 : \mu = 100$, $H_a : \mu > 100$, $n = 8$, $\alpha = 0.05$

21. $H_0 : \mu = 100$, $H_a : \mu > 100$, $n = 8$, $\alpha = 0.01$

22. Describe what happens to the t critical value t_{crit} for right-tailed tests as α decreases.

23. A random sample of size 16 from a normal population yields a sample mean of 10 and a sample standard deviation of 3. Test whether the population mean differs from 9, using level of significance $\alpha = 0.10$.

24. A random sample of size 144 from an unknown population yields a sample mean of 45 and a sample standard deviation of 10. Test whether the population mean differs from 45, using level of significance $\alpha = 0.10$.

Section 9.5

For Exercises 25–27, do the following.
a. Check the normality conditions.
b. State the hypotheses.
c. Find Z_{crit} and the rejection rule.
d. Calculate Z_{data}.
e. State the conclusion and the interpretation.

25. Test whether the population proportion exceeds 0.8. A random sample of size 1000 yields 830 successes. Let $\alpha = 0.10$.

26. Test whether the population proportion is below 0.2. A random sample of size 900 yields 160 successes. Let $\alpha = 0.05$.

27. Test whether the population proportion is not equal to 0.4. A random sample of size 100 yields 55 successes. Let $\alpha = 0.01$.

For Exercises 28 and 29, do the following.
a. Check the normality conditions.
b. State the hypotheses and the rejection rule for the p-value method, using level of significance $\alpha = 0.05$.
c. Calculate Z_{data}.
d. Calculate the p-value.
e. State the conclusion and the interpretation.

28. Test whether the population proportion differs from 0.7. A random sample of size 144 yields 110 successes.

29. Test whether the population proportion is less than 0.25. A random sample of size 100 yields 25 successes.

30. DSL INTERNET SERVICE. The U.S. Department of Commerce reports that 41.6% of Internet users preferred DSL as their method of service delivery.[29] A random sample of 1000 Internet users shows 350 who preferred DSL. If appropriate, test whether the population proportion who prefer DSL has decreased, using level of significance $\alpha = 0.05$.

Section 9.6

For Exercises 31 and 32, do the following.
   a. State the hypotheses.
   b. Find the χ^2 critical value or values, and state the rejection rule.
   c. Find χ^2_{data}. Also, draw a χ^2 distribution and indicate χ^2_{data} and the χ^2 critical value or values.
   d. State the conclusion and the interpretation.

31. We are testing whether $\sigma > 6$ and have a random sample of size 20 with a standard deviation of $s = 9$. Let $\alpha = 0.05$.

32. We are testing whether $\sigma \neq 10$ and have a random sample of size 26 with a sample variance of 90. Let $\alpha = 0.05$.

For Exercises 33 and 34, do the following.
   a. State the hypotheses and the p-value rejection rule for $\alpha = 0.05$.
   b. Find χ^2_{data}.
   c. Find the p-value. Also, draw a χ^2 distribution and indicate χ^2_{data} and the p-value.
   d. State the conclusion and the interpretation.

33. We are testing whether $\sigma < 35$ and have a random sample of size 8 with a sample variance of 1200.

34. We are testing whether $\sigma \neq 50$ and have a random sample of size 26 with a standard deviation of $s = 45$.

35. PRISONER DEATHS IN STATE CUSTODY. The following table contains the numbers of prisoners who died in state

custody for a random sample of 5 states.[30] Assume normality. Using $\alpha = 0.01$ and the p-value method, test whether the population standard deviation of prisoners who died in state custody differs from 50.

🔴 prisonerdeath

New York	171
Pennsylvania	149
Michigan	140
Ohio	121
Georgia	122

Section 9.7

For Exercises 36–40, assume that the conditions for performing the Z test are met. Do the following.
   a. Calculate the value or values of $\bar{x}_{crit}$.
   b. Draw a normal curve, centered at μ_a, with the value or values of $\bar{x}_{crit}$ indicated.
   c. Calculate β, the probability of a Type II error for that value of μ_a. Shade the corresponding area under the normal curve.
   d. Calculate the power of the hypothesis test.

36. $H_0: \mu = 100$ vs. $H_a: \mu \neq 100$, $\alpha = 0.01$, $\sigma = 15$, $n = 64$, $\mu_a = 103$

37. $H_0: \mu = 100$ vs. $H_a: \mu \neq 100$, $\alpha = 0.01$, $\sigma = 15$, $n = 64$, $\mu_a = 106$

38. $H_0: \mu = 100$ vs. $H_a: \mu \neq 100$, $\alpha = 0.01$, $\sigma = 15$, $n = 64$, $\mu_a = 109$

39. $H_0: \mu = 100$ vs. $H_a: \mu \neq 100$, $\alpha = 0.01$, $\sigma = 15$, $n = 64$, $\mu_a = 112$

40. $H_0: \mu = 100$ vs. $H_a: \mu \neq 100$, $\alpha = 0.01$, $\sigma = 15$, $n = 64$, $\mu_a = 115$

41. Refer to Exercises 36–40. Construct the power curve for the given values of μ_a.

CHAPTER 9   Quiz

True or False

1. True or false: It is possible that both the null and alternative hypotheses are correct at the same time.

2. True or false: The conclusion you draw from performing the critical-value method for the Z test is the same as the conclusion you draw from performing the p-value method for the Z test.

3. True or false: We do not need the estimated p-value method if we have access to a computer or calculator.

Fill in the Blank

4. To reject H_0 when H_0 is true is a Type _____ error.

5. An extreme value of $\bar{x}$ is associated with a _____ p-value.

6. The rejection rule for performing a hypothesis test using the p-value method is to reject H_0 when the p-value is less than _____.

Short Answer

7. Under what conditions may we apply the Z test for the population proportion?

8. What does a small p-value indicate with respect to the null hypothesis? A large p-value?

9. Does the value of Z_{data} change when the form of the hypothesis test changes (for example, left-tailed instead of right-tailed)?

Calculations and Interpretations

10. ATM Fees. Do you hate paying the extra fees imposed by banks when withdrawing funds from an automated teller machine (ATM) not owned by your bank? The Federal Reserve System reports that the mean such fee is $1.14. A random sample of 36 such transactions yielded a mean of $1.07 in extra fees. Suppose the population standard deviation of such extra fees is $0.25.

    a. Test using level of significance $\alpha = 0.05$ whether there has been a reduction in the population mean fee charged on such transactions.

    b. Which type of error is it possible that we are making, a Type I error or a Type II error? Which type of error are we certain we are not making?

11. Alcohol-Related Fatal Car Accidents. The National Traffic Highway Safety Commission keeps statistics on the "mean years of potential life lost" in alcohol-related fatal automobile accidents. For males the mean years of life lost is 32. That is, on average, males involved in fatal drinking-and-driving accidents had their lives cut short by 32 years. A random sample of 36 alcohol-related fatal accidents had a mean years of life lost of 33.8, with a standard deviation of 6 years.

    a. Test whether the population mean years of life lost has changed, using a t test and level of significance $\alpha = 0.10$.

    b. Assess the strength of the evidence against the null hypothesis.

12. Preterm Births. The U.S. National Center for Health Statistics reports that, in 2005, the percentage of infants delivered at less than 37 weeks of gestation was 12.7%, up from 10.6% in 1990.[31] Has this upward trend continued? A random sample taken this year of 400 births contained 57 preterm births. Test whether the population proportion of preterm births has increased from 12.7%, using the p-value method and level of significance $\alpha = 0.05$.

13. Active Stocks. On October 3, 2007, the 10 most traded stocks on the New York Stock Exchange were those shown in the following table, which gives their closing prices and net change in price, in dollars. Use only the net change data for this analysis. Assume normality. Using for level of significance $\alpha = 0.10$ and the critical-value method, test whether the population standard deviation of net price change is less than 25 cents.

 activestock

Stock	Closing price	Net change
Micron Technology, Inc.	$10.74	−1.05
Ford Motor Company	$ 8.43	−0.14
Citigroup, Inc.	$47.89	0.03
Advanced Micro Devices	$13.23	0.03
EMC Corporation	$21.13	−0.24
Commerce Bancorp, Inc.	$38.84	−0.63
General Electric	$41.55	−0.57
Avaya, Inc.	$16.95	−0.07
Sprint Nextel Corporation	$18.76	−0.24
iShares:Taiwan	$17.18	−0.18

10 Two-Sample Inference

Design Pics/Superstock

CASE STUDY

Do Prior Student Evaluations Influence Students' Ratings of Professors?

A study in 1950 reported that instructor reputation affected students' ratings of their instructors.[1] Towler and Dipboye uncovered experimental evidence in support of this phenomenon.[2] They randomly assigned to students one of two summaries of prior student evaluations, one for a "charismatic instructor" and the other for a "punitive instructor." The "charismatic" summary included such phrases as "always lively and stimulating in class" and "always approachable and treated students as individuals." The "punitive" summary included such phrases as "did not show an interest in students' progress" and "consistently seemed to grade students harder." All subjects were then shown the same 20-minute lecture video given by the same instructor. They were asked to rate the instructor using three questions, and a summary rating score was calculated.

Were students' ratings influenced by the prior student evaluations? We examine this question further in the Case Study on page 505. ■

The Big Picture

Where we are coming from, and where we are headed . . .

● Thus far, our statistical inference has been limited to one population and one sample. In Chapter 8 we learned to construct confidence intervals, and in Chapter 9 we learned how to perform hypothesis tests, but all for a single population parameter.

● Here, in Chapter 10, "Two-Sample Inference," we perform inference on the differences in the parameters of two populations. For example, we may be interested in whether there is a difference in the population proportions of women and men who post personal information on the Internet.

● In Chapter 11, we will turn to inference methods for categorical data, such as contingency tables.

10.1 INFERENCE FOR MEAN DIFFERENCE—DEPENDENT SAMPLES

OBJECTIVES By the end of this section, I will be able to . . .

1 Distinguish between independent samples and dependent samples.

2 Perform hypothesis tests for the population mean difference for dependent samples.

3 Construct and interpret confidence intervals for the population mean difference for dependent samples.

4 Use a t interval for μ_d to perform t tests about μ_d.

1 INDEPENDENT SAMPLES AND DEPENDENT SAMPLES

Chapter 10 is about two-sample inference. The type of inference we apply depends on whether the data come from **independent samples** or **dependent samples.**

> **Independent Samples and Dependent Samples**
>
> Two samples are **independent** when the subjects selected for the first sample do not determine the subjects in the second sample. Two samples are **dependent** when the subjects in the first sample determine the subjects in the second sample. The data from dependent samples are called **matched-pair** or **paired** samples.

For example, suppose we are interested in comparing the heights of girl-boy fraternal twins. Selecting the girl twin for the first sample automatically results in the boy twin's being selected for the second sample. This is an example of dependent sampling, and the boy-girl pairs are called **matched-pair** samples or **paired** samples.

However, suppose we are interested in comparing the heights of females and males in general. Then, if we took a random sample of 20 females at your school and another random sample of 20 males at your school, these samples would be **independent,** because the females selected in the first sample do not determine the males selected in the second sample.

EXAMPLE 10.1

DEPENDENT OR INDEPENDENT SAMPLING?

Indicate whether each of the following experiments uses an independent or dependent sampling method.
a.   A study wished to compare the differences in price between name-brand merchandise and store-brand merchandise. Name-brand and store-brand items of the same size were purchased from each of the following six categories: paper towels, shampoo, cereal, ice cream, peanut butter, and milk.
b.   A study wished to compare traditional acupuncture with usual clinical care for a certain type of lower-back pain.[3] The 241 subjects suffering from persistent nonspecific lower-back pain were randomly assigned to receive either traditional acupuncture or the usual clinical care. The results were measured at 12 and 24 months.

Solution

a.   For a given store, each name-brand item in the first sample is associated with exactly one store-brand item of that size in the second sample. Therefore, the items in the first sample determine the items in the second sample. This is an example of dependent sampling.
b.   The subjects were randomly assigned to receive either of the two treatments. Thus, the subjects that received acupuncture did not determine those who received clinical care, and vice versa. This is an example of independent sampling.

**Now You Can Do
Exercises 5–8.**

2   DEPENDENT SAMPLE t TEST FOR THE POPULATION MEAN OF THE DIFFERENCES

Table 10.1 shows students' scores on two statistics quizzes. The "After" row (sample 1) contains scores after the students sought help in the Math Center, and the "Before" row (sample 2) shows scores before they had help. The observations are taken from the same students before and after they had help. Thus, sample 1 and sample 2 are dependent, matched-pair data.

Table 10.1 Statistics quiz scores of seven students before and after visiting the Math Center

Student	Ashley	Brittany	Chris	Dave	Emily	Fran	Greg
After (sample 1)	66	68	74	88	89	91	100
Before (sample 2)	50	55	60	70	75	80	88

Notice that each student's score improved on the second quiz:

Ashley: $66 - 50 = 16$ Emily: $89 - 75 = 14$
Brittany: $68 - 55 = 13$ Fran: $91 - 80 = 11$
Chris: $74 - 60 = 14$ Greg: $100 - 88 = 12$
Dave: $88 - 70 = 18$

The key idea behind dependent sampling is that we consider the set of these seven differences {16, 13, 14, 18, 14, 11, 12} as a sample so that we can perform inference on these differences. In other words, we no longer have two samples. By matching the samples element by element and taking the difference, we have transformed two samples into one that is the sample of differences (Figure 10.1). We have already learned how to perform inference using a single sample, so the remainder of this section uses techniques you have used before.

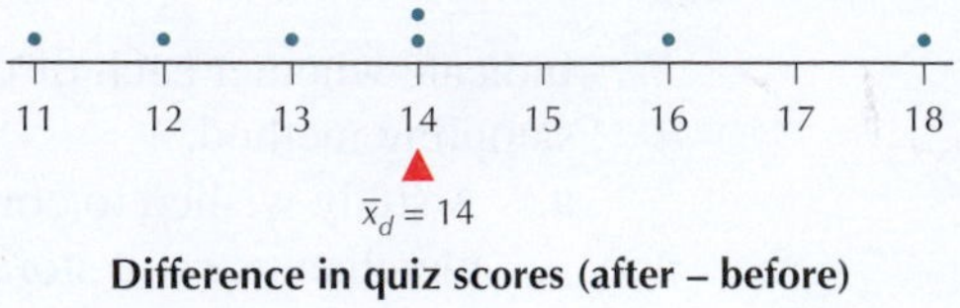

FIGURE 10.1
Taking the differences reduces a two-sample problem to a single sample of differences.

$\bar{x}_d = 14$

Difference in quiz scores (after − before)

The sample mean of the set of differences is

**Now You Can Do
Exercises 9–14.**

$$\bar{x}_d = \frac{16 + 13 + 14 + 18 + 14 + 11 + 12}{7} = 14$$

as illustrated in Figure 10.1. The sample of differences can be considered representative of the *population* of these differences, where the population represents all students who took statistics quizzes before and after visiting the Math Center. The sample mean difference $\bar{x}_d = 14$ is a point estimate of the *population mean difference* μ_d, the unknown mean difference in the (after − before) quiz scores for all students who visited the Math Center. Since μ_d is unknown, we need to perform hypothesis tests and construct confidence intervals to learn about its value.

⚠ CAUTION

Note that μ_d always refers to sample 1 minus sample 2, never sample 2 minus sample 1. For example, μ_d represents the mean difference between the students' "after" scores and the "before" scores on the statistics quizzes in Table 10.1.

Paired Sample *t* Test for the Population Mean of the Differences μ_d: Critical-Value Method

For matched-pair data taken from dependent samples of two populations, find the differences to produce a random sample of the differences between the populations. You can use the *t* test whenever *either* of the following conditions is met:

- The population of differences is normal, or
- The sample size of differences is large ($n \geq 30$).

Notice that since we have only one sample of differences this procedure is very similar to the one-sample *t* test from Section 9.4.

Step 1 State the hypotheses.
Use one of the hypothesis test forms in Table 10.2. State the meaning of μ_d.
Step 2 Find t_{crit}, and state the rejection rule.
To find t_{crit}, use the *t* table and degrees of freedom $n − 1$. To find the rejection rule, use Table 10.2.
Step 3 Calculate t_{data}.

$$t_{data} = \frac{\bar{x}_d}{s_d / \sqrt{n}}$$

which follows an approximate *t* distribution with degrees of freedom $n − 1$.
Step 4 State the conclusion and the interpretation.
Compare t_{data} with t_{crit}.

Table 10.2 Critical regions and rejection rules for dependent sample *t* test

	Right-tailed test	Left-tailed test	Two-tailed test
Form of test	$H_0 : \mu_d = \mu_0$ $H_a : \mu_d > \mu_0$ level of significance α	$H_0 : \mu_d = \mu_0$ $H_a : \mu_d < \mu_0$ level of significance α	$H_0 : \mu_d = \mu_0$ $H_a : \mu_d \neq \mu_0$ level of significance α
Critical region			
Rejection rule	Reject H_0 if $t_{\text{data}} \geq t_{\text{crit}}$	Reject H_0 if $t_{\text{data}} \leq -t_{\text{crit}}$	Reject H_0 if $t_{\text{data}} \geq t_{\text{crit}}$ or $t_{\text{data}} \leq -t_{\text{crit}}$

EXAMPLE 10.2

PAIRED *t* TEST USING THE CRITICAL-VALUE METHOD

groceries

Are name-brand groceries more expensive than store-brand groceries? A sample of 6 randomly selected grocery items yielded the price data shown in Table 10.3. Test at level of significance $\alpha = 0.05$ whether the population mean μ_d of the differences in price (name brand minus store brand) is greater than zero. Or, more informally, test whether the name-brand items at the grocery store cost more on average than the store-brand items.

Table 10.3 Prices of name-brand and store-brand grocery items

Item	Paper towels	Shampoo	Cereal	Ice cream	Peanut butter	Milk
Name brand	$1.29	$4.69	$3.59	$3.49	$2.79	$2.99
Store brand	$1.29	$3.99	$3.39	$2.69	$2.39	$3.49
Differences	$0.00	$0.70	$0.20	$0.80	$0.40	−$0.50

Solution

The normal probability plot of the differences shows acceptable normality, allowing us to proceed with the hypothesis test.

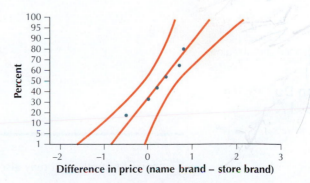

STEP 1 **State the hypotheses.**
"Greater than" implies that $\mu_d > 0$, leading to the hypotheses

$$H_0 : \mu_d = 0 \qquad H_a : \mu_d > 0$$

where μ_d represents the population mean difference in price between name-brand and store-brand merchandise.

STEP 2 **Find the critical value t_{crit} and state the rejection rule.**
Use $n - 1$ degrees of freedom. Here $n = 6$, so df $= n - 1 = 5$. Since we have a right-tailed test with $\alpha = 0.05$, we find our t-critical value by choosing the column in the t table with area 0.05 in one tail: $t_{crit} = 2.015$. The right-tailed test tells us that our rejection rule is to reject H_0 when t_{data} is greater than 2.015.

STEP 3 **Find t_{data}.**
We need to calculate $\bar{x}_d$ and s_d.

$$\bar{x}_d = \frac{\sum x}{n} = \frac{0.00 + 0.70 + 0.20 + 0.80 + 0.40 - 0.50}{6} \approx \$0.267$$

$$s_d = \sqrt{\frac{\sum (x - \bar{x}_d)^2}{n - 1}}$$

$$= \sqrt{\frac{(0.00 - 0.267)^2 + (0.70 - 0.267)^2 + (0.20 - 0.267)^2 + (0.80 - 0.267)^2 + (0.40 - 0.267)^2 + (-0.50 - 0.267)^2}{5}}$$

$$\approx \$0.48$$

This gives

$$t_{data} = \frac{\bar{x}_d}{s_d/\sqrt{n}} = \frac{0.267}{0.48/\sqrt{6}} \approx 1.36$$

STEP 4 **State the conclusion and the interpretation.**
Since $t_{data} = 1.36$ is not greater than $t_{crit} = 2.015$ (Figure 10.2), do not reject H_0. There is insufficient evidence that brand-name grocery items cost more, on average, than store-brand items at the $\alpha = 0.05$ level of significance. It appears that the brand-name milk was on sale; otherwise the conclusion may very well have been different.

FIGURE 10.2
1.360 does not fall within the critical region.

**Now You Can Do
Exercises 15–17.**

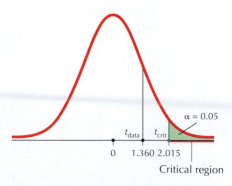

The paired sample t test may also be performed using the p-value method.

> **Paired Sample _t_ Test for the Population Mean of the Differences μ_d: _p_-Value Method**
>
> For matched-pair data taken from dependent samples of two populations, find the differences to produce a random sample of the differences between the populations. You can use the _t_ test whenever _either_ of the following conditions is met:
>
> - The population of differences is normal, or
> - The sample size of differences is large ($n \geq 30$).
>
> **Step 1 State the hypotheses and the rejection rule.**
> Use one of the hypothesis test forms from Table 10.4 for a test at level of significance α. State the meaning of μ_d. The rejection rule is reject H_0 if the _p_-value is less than α.
> **Step 2 Calculate t_{data}.**
>
> $$t_{\text{data}} = \frac{\bar{x}_d}{s_d/\sqrt{n}}$$
>
> which follows an approximate _t_ distribution with degrees of freedom $n - 1$.
> **Step 3 Find the _p_-value.**
> If you have access to technology, use it to find the _p_-value. Otherwise, calculate the _p_-value using one of the test forms in Table 10.4.
> **Step 4 State the conclusion and the interpretation.**
> Compare the _p_-value with α.

Table 10.4 _p_-Values for dependent sample _t_ tests

	Right-tailed test	Left-tailed test	Two-tailed test
Form of test	$H_0 : \mu_d = \mu_0$ $H_a : \mu_d > \mu_0$	$H_0 : \mu_d = \mu_0$ $H_a : \mu_d < \mu_0$	$H_0 : \mu_d = \mu_0$ $H_a : \mu_d \neq \mu_0$
_p_-Value	_p_-value $= P(t > t_{\text{data}})$ Area to the right of t_{data}	_p_-value $= P(t < t_{\text{data}})$ Area to the left of t_{data}	_p_-value $= P(t > \|t_{\text{data}}\|) + P(t < -\|t_{\text{data}}\|)$ $= 2 \cdot P(t > \|t_{\text{data}}\|)$ Sum of the two-tailed areas

---

EXAMPLE 10.3

PAIRED SAMPLE _t_ TEST FOR μ_d: THE _p_-VALUE METHOD

A study was carried out to determine whether Reiki touch therapy was useful in the reduction of mean pain level in chronic pain sufferers, including cancer patients.[4] The pain level reported by a random sample of 13 patients before and after Reiki touch therapy is shown in Table 10.5. Test whether there has been a mean reduction in pain level after the Reiki therapy, using level of significance $\alpha = 0.05$. In other words, test whether the population mean difference μ_d is less than zero, where μ_d is defined as the (after − before) difference in pain level.

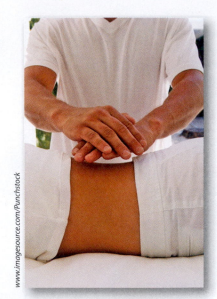

Table 10.5 Pain level reported by 13 patients before and after Reiki touch therapy

Patient	1	2	3	4	5	6	7	8	9	10	11	12	13
After	3	1	0	0	2	1	2	1	0	4	1	4	8
Before	6	2	2	3	3	4	2	5	1	6	6	4	8
Difference	−3	−1	−2	−3	−1	−3	0	−4	−1	−2	−5	0	0

Solution

For each patient, we subtract the "before" pain level from the "after" pain level to arrive at a set of $n = 13$ differences, highlighted in Table 10.5. The normal probability plot of the differences indicates acceptable normality, given the small sample size. The Minitab results from the t test are provided here.

 reki

```
Test of mu = 0 vs < 0
                                              95%
                                             Upper
Variable   N      Mean     StDev   SE Mean    Bound       T       P
[Diff]    13   -1.92308  1.60528  0.44522  -1.12956   -4.32   0.000
```

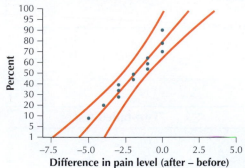

Difference in pain level (after – before)

STEP 1 **State the hypotheses and the rejection rule.**
We are interested in testing whether there was a mean reduction in pain level, which would mean that the mean pain level would be lower after the Reiki therapy than before. This implies that the population mean difference in pain level, $\mu_d =$ (after – before), is *less than* 0. Thus, from Table 10.4, the hypotheses are

$$H_0 : \mu_d = 0 \qquad H_a : \mu_d < 0$$

where μ_d represents the population mean difference in pain level. We will reject H_0 if the p-value ≤ 0.05.

STEP 2 **Find t_{data}.**
As provided in the Minitab results,

$$t_{\text{data}} = \frac{\bar{x}_d}{s_d/\sqrt{n}} = \frac{-1.92308}{1.60528/\sqrt{13}} \approx -4.32$$

which follows an approximate t distribution with degrees of freedom $n - 1 = 13 - 1 = 12$.

STEP 3 **Find the p-value.**
For a left-tailed test, the p-value is the area to the left of t_{data}. This area is essentially 0, as shown in Figure 10.3 and provided by Minitab,

$$P(t < t_{\text{data}}) = P(t < -4.32) \approx 0.000$$

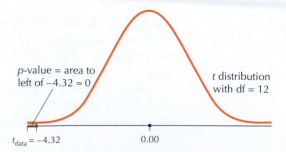

FIGURE 10.3
The p-value $= P(t < -4.32) \approx 0.000$.

STEP 4   **State the conclusion and the interpretation.**
Since p-value $\approx 0.000 \leq \alpha = 0.05$, we reject H_0. There is evidence that $\mu_d < 0$ and hence that the population mean difference in pain level (after $-$ before) has decreased. That is, there is evidence at level of significance $\alpha = 0.05$ that the Reiki touch therapy has worked to reduce the mean pain level for chronic pain sufferers.

Now You Can Do
Exercises 18–20.

3   *t* INTERVALS FOR THE POPULATION MEAN DIFFERENCE FOR DEPENDENT SAMPLES

Recall that in Section 8.2 we used the formula $\bar{x} \pm t_{\alpha/2}(s/\sqrt{n})$ to calculate the t interval for the population mean μ. Here, to estimate the population mean of the differences μ_d, we use essentially the same formula, substituting $\bar{x}_d$ for $\bar{x}$ and s_d for s.

To construct this confidence interval, we need

$\bar{x}_d$ = mean of the differences of the two samples

s_d = standard deviation of the differences of the two samples

n = sample size of differences

$t_{\alpha/2}$ = critical value associated with confidence level $1 - \alpha$ and degrees of freedom $n - 1$.

Confidence Interval for Population Mean Difference μ_d (Dependent Samples)

For matched-pair data taken from dependent samples of two populations, find the differences to produce a random sample of the differences between the populations. A $100(1 - \alpha)\%$ confidence interval for μ_d, the population mean of the differences, is given by

$$\text{lower bound} = \bar{x}_d - t_{\alpha/2}\left(\frac{s_d}{\sqrt{n}}\right) \qquad \text{upper bound} = \bar{x}_d + t_{\alpha/2}\left(\frac{s_d}{\sqrt{n}}\right)$$

where $\bar{x}_d$ and s_d represent the sample mean and sample standard deviation of the differences, respectively, of the set of n paired differences, $d_1, d_2, d_3, \ldots, d_n$, and where $t_{\alpha/2}$ is based on $n - 1$ degrees of freedom. This t interval applies whenever *either* of the following conditions is met:

- The population of differences is normal, or
- The sample size of differences is large ($n \geq 30$).

The $100(1 - \alpha)\%$ confidence interval for μ_d may also be expressed in the form

$$\bar{x}_d \pm t_{\alpha/2}\left(\frac{s_d}{\sqrt{n}}\right)$$

EXAMPLE 10.4   ***t* CONFIDENCE INTERVAL FOR μ_d**

Use the "before" and "after" quiz scores from Table 10.1 to construct a 95% t confidence interval for the population mean of the differences in the statistics quiz scores. Is there evidence that the Math Center tutoring leads to a mean improvement in the quiz scores?

Solution

The normal probability plot of the differences shows acceptable normality, allowing us to construct the confidence interval.

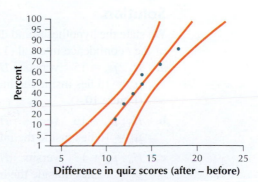

We ignore the original raw data (see Table 10.1) and concentrate only on the set of sample differences: {16, 13, 14, 18, 14, 11, 12}. For the data set of $n = 7$ differences, we find the mean and standard deviation. We found earlier that $\bar{x}_d = 14$. Now we calculate

$$s_d = \sqrt{\frac{\sum(x - \bar{x}_d)^2}{n - 1}}$$

$$= \sqrt{\frac{(16 - 14)^2 + (13 - 14)^2 + (14 - 14)^2 + (18 - 14)^2 + (14 - 14)^2 + (11 - 14)^2 + (12 - 14)^2}{7 - 1}}$$

$$\approx 2.3805$$

For 95% confidence with $n - 1 = 6$ degrees of freedom, $t_{a/2}$ equals 2.447 (see Appendix Table D). Using these values,

lower bound $= \bar{x}_d - t_{a/2}(s_d/\sqrt{n})$ upper bound $= \bar{x}_d + t_{a/2}(s_d/\sqrt{n})$

$\quad = 14 - (2.447)(2.3805/\sqrt{7})$ $\quad = 14 + (2.447)(2.3805/\sqrt{7})$

$\quad \approx 14 - 2.2017 = 11.7983$ $\quad \approx 14 + 2.2017 = 16.2017$

We are 95% confident that the population mean of the differences between quiz scores before and after visiting the Math Center lies between 11.7983 points and 16.2017 points. If there were no mean change in the quiz scores, the difference would be 0, which is not in this confidence interval. Thus, we have evidence that the Math Center tutoring leads to a significant mean improvement in the quiz scores with 95% confidence.

Now You Can Do Exercises 21–26.

4 USE A t INTERVAL FOR μ_d TO PERFORM t TESTS ABOUT μ_d

Given a $100(1 - \alpha)\%$ t confidence interval for μ_d, we may perform two-tailed t tests for various values of μ_d, just as we did for the single sample case in Section 9.4. The methodology is the same: if a certain value for μ_d lies outside the $100(1 - \alpha)\%$ t confidence interval for μ_d, then the null hypothesis specifying this value for μ_d would be rejected. Otherwise it would not be rejected.

EXAMPLE 10.5

USING A t INTERVAL FOR μ_d TO PERFORM t TESTS ABOUT μ_d

Example 10.4 provided a 95% t confidence interval for the population mean of the differences between quiz scores before and after visiting the Math Center as (11.7983, 16.2017). Test using level of significance $\alpha = 0.05$ whether the population mean of the differences between quiz scores before and after visiting the Math Center differs from these values: (a) 15 points, (b) 16 points, (c) 17 points.

Solution

We state the hypotheses and determine if each proposed value μ_0 lies inside or outside of the t confidence interval (11.7983, 16.2017).

a. $H_0 : \mu_d = 15$ versus $H_a : \mu_d \neq 15$
$\mu_0 = 15$ lies inside the interval (11.7983, 16.2017), so we do not reject H_0 (Figure 10.4).

b. $H_0 : \mu_d = 16$ versus $H_a : \mu_d \neq 16$
$\mu_0 = 16$ lies inside the interval, so we do not reject H_0.

c. $H_0 : \mu_d = 17$ versus $H_a : \mu_d \neq 17$
$\mu_0 = 17$ lies outside the interval, so we reject H_0.

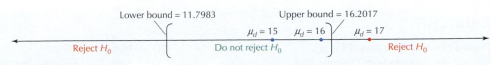

Lower bound = 11.7983   Upper bound = 16.2017

$\mu_d = 15$ $\mu_d = 16$ $\mu_d = 17$

Reject H_0 Do not reject H_0 Reject H_0

Now You Can Do Exercises 27–30.

FIGURE 10.4 Reject H_0 for values of μ_d that lie outside the t confidence interval.

STEP-BY-STEP TECHNOLOGY GUIDE: Confidence Intervals and Hypothesis Tests for μ_d

TI-83/84

Hypothesis Test
(Example 10.2 is used to illustrate the procedure.)
Step 1   Enter samples 1 and 2 in lists L1 and L2.
Step 2   Type **(L1 – L2) STO** $\rightarrow$ **L3** and press **ENTER**.
Step 3   Press **STAT** and highlight **TESTS**.
Step 4   For the hypothesis test, press **2** (for the **T-Test**). The T-Test menu appears.
Step 5   For input (**Inpt**), highlight **Data** and press **ENTER**. (If given the summary statistics for the differences, choose **STATS**.)
Step 6   For μ_0, enter the hypothesized value. For **List**, press **2nd** then **L3**. For **Freq**, enter **1**. Choose the form of the hypothesis test, and press **ENTER** (Figure 10.5).

Step 7   When the cursor is over **Calculate**, make sure all your entries are correct, and press **ENTER**. The results are shown in Figure 10.6.

FIGURE 10.5

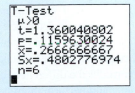

FIGURE 10.6

Confidence Interval
(Example 10.4 is used to illustrate the procedure.)
Step 1   Enter samples 1 and 2 in lists L1 and L2.
Step 2   Type **(L1 – L2) STO** $\rightarrow$ **L3** and press **ENTER** (Figure 10.7).
Step 3   Press **STAT** and highlight **TESTS**.
Step 4   Press **8** (for the **TInterval**).

Step 5   For input (**Inpt**), highlight **Data** and press **ENTER**. (If given the summary statistics for the differences, choose **STATS**.)
Step 6   For **List**, press **2nd** then **L3**. For **Freq**, enter **1**. Enter the **C-Level** (confidence level, such as **0.95** for 95%), and press **ENTER** (Figure 10.8).
Step 7   Highlight **Calculate** and press **ENTER**. The results are shown in Figure 10.9.

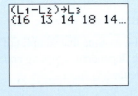

FIGURE 10.7

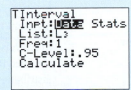

FIGURE 10.8

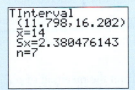

FIGURE 10.9

EXCEL

Hypothesis Test
Step 1   Enter samples 1 and 2 in columns A and B.
Step 2   Click **Data > Data Analysis > t-Test: Paired Two Sample for Means**, and click **OK**.

Step 3   For **Variable 1 Range**, highlight the cells for sample 1 in column A, and for **Variable 2 Range**, highlight the cells for sample 2 in column B.
Step 4   Enter the **Hypothesized Mean Difference** (usually 0), and enter a value for **alpha**. Then click **OK**.

MINITAB

Confidence Interval and Hypothesis Test
Step 1   Enter samples 1 and 2 in columns C1 and C2.
Step 2   Click **Stat > Basic Statistics > Paired t**.
Step 3   For **First Sample**, enter **C1**, and for **Second Sample**, enter **C2**.

Step 4   Click **Options**.
a. For the confidence interval, specify the **Confidence Level**, then click **OK** twice.
b. For the hypothesis test, specify the form of the alternative hypothesis, then click **OK** twice.

CRUNCHIT!

Paired *t* test and *t* interval for μ_d.
We will use the data from Example 10.3.
Step 1 Click **File** . . . then highlight **Load from LaroseFundamentals2e** . . . **Chapter 10** . . . and click on **Example 10.3**.
Step 2 Click **Statistics** and select **t** . . . **Paired**. For **First Variable** select **After**. For **Second Variable** select **Before**.

Hypothesis Test
Step 3 Select the **Hypothesis Test** tab, choose the correct form of the **Alternative** hypothesis, and click **Calculate**.
For the confidence interval:
Step 4 Select the **Confidence Interval** tab, enter the **Confidence Interval Level**, and click **Calculate**.

SECTION 10.1 Summary

1. Two samples are independent when the subjects selected for the first sample do not determine the subjects in the second sample. Two samples are dependent when the subjects in the first sample determine the subjects in the second sample. The data from dependent samples are called matched-pair or paired samples. The key concept in this section is that we consider the differences of matched-pair data as a single sample, and perform inference on this sample of differences.

2. The paired sample *t* test for the population mean of the differences μ_d can be used either when the population is normal or the sample size is large ($n \geq 30$). The test may be

carried out using either the critical-value method or the *p*-value method.

3. A $100(1 - \alpha)\%$ confidence interval for μ_d, the population mean of the differences, is given by $\bar{x}_d \pm t_{\alpha/2}(s_d/\sqrt{n})$, where $\bar{x}_d$ and s_d represent the sample mean and sample standard deviation of the differences, respectively, of the set of n paired differences, $d_1, d_2, d_3, \ldots, d_n$, and where $t_{\alpha/2}$ is based on $n - 1$ degrees of freedom.

4. This confidence interval may be used to conduct two-tailed hypothesis tests for μ_d.

SECTION 10.1 Exercises

Clarifying the Concepts

1. When are two samples considered independent?

2. When are two samples considered dependent?

3. What do we call the data obtained from dependent sampling?

4. How do we interpret the meaning of μ_d?

Practicing the Techniques

Determine whether the experiments in Exercises 5–8 represent an independent sampling method or a dependent sampling method. Explain your answer.

5. The Jacksonville Jaguars are interested in comparing the performance of their first-year players. For each player, a sample is taken of their games from their last year in college and compared to a sample of games taken from their first year in the pros.

6. For her senior project, an exercise science major takes a sample of females majoring in exercise science, and a sample of females from her college not majoring in exercise science. She records the body mass index for each subject.

7. Before the first lecture, an algebra instructor gives a pretest to his students to determine the students' algebra readiness. At the end of the course, the instructor gives a post-test to the same students and compares the results with the pretest.

8. The sheriff's department takes a sample of vehicle speeds on a certain stretch of road and compares the results to a sample of vehicle speeds on a certain stretch of a different road. Both roads have the same posted speed limit.

In Exercises 9–14, assume that samples of differences are obtained through dependent sampling and follow a normal distribution. Calculate $\bar{x}_d$ and s_d.

9.

Subject	1	2	3	4	5
Sample 1	3.0	2.5	3.5	3.0	4.0
Sample 2	2.5	2.5	2.0	2.0	1.5

10.

Subject	1	2	3	4	5	6
Sample 1	10	12	9	14	15	8
Sample 2	8	11	10	12	14	9

11.

Subject	1	2	3	4	5	6	7
Sample 1	20	25	15	10	20	30	15
Sample 2	30	30	20	20	25	35	25

12.

Subject	1	2	3	4	5	6	7
Sample 1	1.5	1.8	2.0	2.5	3.0	3.2	4.0
Sample 2	1.0	1.7	2.1	2.0	2.7	2.9	3.3

13.

Subject	1	2	3	4	5	6	7	8
Sample 1	0	0.5	0.75	1.25	1.9	2.5	3.2	3.3
Sample 2	0.25	0.25	0.75	1.5	1.8	2.2	3.3	3.4

14.

Subject	1	2	3	4	5	6	7	8
Sample 1	105	88	103	97	115	125	122	92
Sample 2	110	95	108	97	116	127	125	95

15. For the data in Exercise 9, test whether $\mu_d > 0$, using the critical-value method and level of significance $\alpha = 0.05$.

16. For the data in Exercise 10, test whether $\mu_d \neq 0$, using the critical-value method and level of significance $\alpha = 0.01$.

17. For the data in Exercise 11, test whether $\mu_d < 0$, using the critical-value method and level of significance $\alpha = 0.10$.

18. For the data in Exercise 12, test whether $\mu_d > 0$, using the p-value method and level of significance $\alpha = 0.01$.

19. For the data in Exercise 13, test whether $\mu_d \neq 0$, using the p-value method and level of significance $\alpha = 0.05$.

20. For the data in Exercise 14, test whether $\mu_d < 0$, using the p-value method and level of significance $\alpha = 0.10$.

21. Using the data from Exercise 9, construct a 95% confidence interval for μ_d.

22. Using the data from Exercise 10, construct a 99% confidence interval for μ_d.

23. Using the data from Exercise 11, construct a 90% confidence interval for μ_d.

24. Using the data from Exercise 12, construct a 99% confidence interval for μ_d.

25. Using the data from Exercise 13, construct a 95% confidence interval for μ_d.

26. Using the data from Exercise 14, construct a 90% confidence interval for μ_d.

For Exercises 27–30 a $100(1 - \alpha)\%$ t confidence interval for μ_d is given. Use the confidence interval to test using level of significance α whether μ_d differs from each of the indicated hypothesized values.

27. A 95% t confidence interval for μ_d is $(-5, 5)$. Hypothesized values are
  a. 0
  b. −6
  c. 4

28. A 99% t confidence interval for μ_d is $(-10, -4)$. Hypothesized values are
  a. −12
  b. 0
  c. 4

29. A 90% t confidence interval for μ_d is $(10, 20)$. Hypothesized values are
  a. −10
  b. 25
  c. 0

30. A 95% t confidence interval for μ_d is $(0, 1)$. Hypothesized values are
  a. 0.41
  b. 0.29
  c. 1.23

Applying the Concepts

31. New Car Prices. Kelley's Blue Book (**kbb.com**) publishes data on new and used cars. The following table contains the manufacturer's suggested retail price for four vehicles, model years 2006 and 2007. We are interested in the difference in price between the 2006 models and the 2007 models. Assume that the population of price differences is normally distributed.

carprice
  a. Find the mean of the differences, $\bar{x}_d$, and the standard deviation of the differences, s_d.
  b. Test whether 2007 models are on average more expensive, using level of significance $\alpha = 0.05$.

	Subaru Forester	Honda CR-V	Toyota RAV-4	Nissan Sentra
2006	$22,420	$20,990	$22,980	$13,815
2007	$21,820	$22,395	$23,630	$15,375

32. Mozart Effect? A researcher claims that listening to Mozart improves scores on math quizzes. A random sample of five students took math quizzes, first before and then after listening to Mozart.

mozart
  a. Find the mean of the differences, $\bar{x}_d$, and the standard deviation of the differences, s_d.
  b. Perform the appropriate hypothesis test for determining whether the results support the researcher's claim, using level of significance $\alpha = 0.10$. Assume normality.

Student	1	2	3	4	5
Before	75	50	80	85	95
After	85	45	85	95	95

33. High and Low Temperatures. The University of Waterloo Weather Station tracks the daily low and high temperatures in Waterloo, Ontario, Canada. Table 10.6 contains a random sample of the daily high and low temperatures for May 1–May 10, 2006, in degrees centigrade. Assume that the temperature differences are normally distributed.

 waterlootemp

a. Find the mean of the differences, $\bar{x}_d$, and the standard deviation of the differences, s_d.

b. Test using level of significance $\alpha = 0.01$ whether the population mean difference between high and low temperatures differs from zero.

Table 10.6 High and low temperatures

May date	1	2	3	4	5	6	7	8	9	10
High temp.	19.0	19.8	23.3	21.1	15.2	9.9	17.2	21.7	21.2	23.9
Low temp.	7.4	3.0	3.9	7.9	4.4	0.7	−1.1	2.3	6.6	5.8

34. Falling Home Sales Prices. A credit crunch gripped the nation in 2007–2008, leading to record numbers of mortgage foreclosures and declines in home sales prices. The following table provides the median home sales prices for four regions of the country in the first quarter (January–March) of 2007 and the first quarter of 2008. Assume that the differences are normally distributed.

 homesales

a. Find the mean of the differences, $\bar{x}_d$, and the standard deviation of the differences, s_d.

b. Test whether the population mean difference between the first quarter 2007 median price and the first quarter 2008 median price differs from zero, using level of significance $\alpha = 0.10$.

	Northeast	Midwest	South	West
Jan.–Mar. 2007	$370,300	$212,800	$222,900	$341,500
Jan.–Mar. 2008	$326,600	$201,900	$204,800	$298,900

Source: U.S. Census Bureau.

35. New Car Prices. Use the information in Exercise 31 to construct and interpret a 95% confidence interval for μ_d, the population mean difference in price.

36. Mozart Effect? Use the data from Exercise 32 to construct and interpret a 95% confidence interval for μ_d, the population mean difference in quiz scores before and after listening to Mozart.

37. High and Low Temperatures. Use the information in Exercise 33 for the following.

a. Construct and interpret a 99% confidence interval for μ_d, the population mean difference in temperature.

b. Explain how your confidence interval supports your conclusion to the hypothesis test in Exercise 33.

38. Falling Home Sales Prices. Use the information in Exercise 34 for the following.

a. Construct and interpret a 99% confidence interval for μ_d, the population mean difference in price.

b. Explain how your confidence interval supports your conclusion to the hypothesis test in Exercise 34.

39. Math Scores Worldwide. The National Center for Educational Statistics publishes the results from the Trends in International Math and Science Study (TIMSS). Table 10.7 contains the 1995 and 2007 mean mathematics scores for eighth-graders from various countries. Assume that the population of score differences is normally distributed.

 mathscore

a. Construct a 90% confidence interval for μ_d, the population mean difference in score.

b. Using level of significance $\alpha = 0.10$, test whether the 2007 scores differ from the 1995 scores, on average.

Table 10.7 Eighth-grade math scores

Country	1995	2007
Singapore	609	593
Japan	581	570
England	498	513
United States	492	508
Russia	524	512
Australia	509	496
Scotland	493	487
Cyprus	468	465
Norway	498	469
Iran	418	403

40. Collisions Before and After. The Washington Department of Transportation compared collision data on particular sections of roadway before and after a series of road improvements to determine whether road improvements lowered the number of collisions per year (Table 10.8).[5] Assume that the differences are normally distributed.

collisions

Table 10.8 Collision data

Location	Before	After
Seattle	77.5	43.8
Shoreline	63.3	33.6
Alderton	49.9	40.3
Snoqualmie	19.4	10.4
Sunnyside	12.0	11.7
Ritzville	39.0	23.7
Milton	14.5	11.2
Spokane	114.7	77.3
Kent	25.3	13.8
Vancouver	22.4	4.3

a. Find the point estimate of the mean decrease in collisions per year.

b. Find a 95% confidence interval for the population mean of the differences, μ_d.

c. Using level of significance $\alpha = 0.01$, test whether the improvements have lowered the population mean number of collisions per year.

Bringing It All Together

Home Sales. Use the following information for Exercises 41–45. The number of sales of single family residences in a random sample of towns in eastern Connecticut is provided in Table 10.9 for the time periods January–September 2006 and January–September 2007. Assume that the differences are normally distributed.

 cthomesales

41. Explain why these are dependent samples and not independent samples.

42. Calculate the following statistics.
   a. $\bar{x}_d$
   b. s_d
   c. t_{data}

43. Test using level of significance $\alpha = 0.05$ whether the population mean number of home sales μ_d differs from 2006 to 2007.

44. Construct a 95% confidence interval for μ_d.

45. *What if* we added a certain number of home sales to every entry in the table. How would this change affect the conclusion?

Table 10.9 Home sales in eastern Connecticut

Town	2006	2007
Andover	31	32
Bolton	46	39
Coventry	180	137
East Hartford	469	405
Ellington	121	121
Hebron	98	74
Manchester	475	479
Somers	60	73
South Windsor	154	161
Stafford	114	89
Suffield	114	121
Tolland	146	141
Vernon	210	213
Windsor	327	288
Windsor Locks	120	128

Source: Manchester (CT) *Journal-Inquirer,* November 7, 2007.

10.2 INFERENCE FOR TWO INDEPENDENT MEANS

OBJECTIVES By the end of this section, I will be able to . . .

1 Perform and interpret *t* tests about $\mu_1 - \mu_2$ using Welch's method.[6]

2 Compute and interpret *t* intervals for $\mu_1 - \mu_2$ using Welch's method.

3 Use confidence intervals for $\mu_1 - \mu_2$ to perform two-tailed *t* tests about $\mu_1 - \mu_2$.

4 Perform and interpret *t* tests and *t* intervals about $\mu_1 - \mu_2$ using the pooled variance method.

5 Apply *Z* tests and *Z* intervals for $\mu_1 - \mu_2$ when σ_1 and σ_2 are known.

1 INDEPENDENT SAMPLE *t* TEST FOR $\mu_1 - \mu_2$

On page 140 in Chapter 3 we used boxplots to find evidence of a difference between male and female body temperature for a sample of 65 women and a sample of 65 men. The summary statistics are shown in Table 10.10.

Table 10.10 Summary statistics for female versus male body temperatures in °F

Gender	Sample size	Sample mean body temperature	Sample standard deviation	Population mean body temperature
Females (sample 1)	$n_1 = 65$	$\bar{x}_1 = 98.394$	$s_1 = 0.743$	$\mu_1 = ?$
Males (sample 2)	$n_2 = 65$	$\bar{x}_2 = 98.105$	$s_2 = 0.699$	$\mu_2 = ?$

However, since the female subjects did not determine the male subjects, and vice versa, the 65 women and 65 men represent independent samples, so we cannot use the dependent sampling methods we learned in Section 10.1.

Note that for independent samples, we have two sample sizes, n_1 and n_2, two sample means, $\bar{x}_1$ and $\bar{x}_2$, two sample standard deviations, s_1 and s_2, and two unknown population means, μ_1 and μ_2. Since we are interested in the difference in the population means, we consider the quantity

$$\mu_1 - \mu_2$$

Developing Your Statistical Sense

The Difference Difference

There is a difference in interpretation between the quantity $\mu_1 - \mu_2$ and the quantity μ_d from Section 10.1. Here, $\mu_1 - \mu_2$ refers to the difference in population means, whereas μ_d represents the population mean of the paired differences.

In previous chapters we used the statistic $\bar{x}$ to learn about the parameter μ. Here we shall use the statistic $\bar{x}_1 - \bar{x}_2$ to perform inference about the parameter $\mu_1 - \mu_2$, whose value is unknown. Note from Table 10.10 that the value of $\bar{x}_1 - \bar{x}_2$ for these samples is

$$\bar{x}_1 - \bar{x}_2 = 98.394 - 98.105 = 0.289$$

We use $\bar{x}_1 - \bar{x}_2 = 0.289$ as a *point estimate* of $\mu_1 - \mu_2$. If we repeat the experiment an infinite number of times, then the values of $\bar{x}_1 - \bar{x}_2$ will form a distribution called the **sampling distribution of $\bar{x}_1 - \bar{x}_2$.**

It is unlikely that the experimenter will have knowledge of both population standard deviations σ_1 and σ_2. Therefore, we use the estimates of σ_1 and σ_2 provided by the sample standard deviations s_1 and s_2. Recall from Section 8.2 that, when the population standard deviation σ is unknown, and if either the population is normal or the sample size is large, the quantity

$$t = \frac{\bar{x} - \mu}{s/\sqrt{n}}$$

has a t distribution with $n - 1$ degrees of freedom. By analogy, we have the following sampling distribution.

> **Sampling Distribution of $\bar{x}_1 - \bar{x}_2$**
>
> When random samples are drawn independently from two populations with population means μ_1 and μ_2, and either (a) the two populations are normally distributed, or (b) the two sample sizes are large (at least 30), then the quantity
>
> $$t = \frac{(\bar{x}_1 - \bar{x}_2) - (\mu_1 - \mu_2)}{\sqrt{\dfrac{s_1^2}{n_1} + \dfrac{s_2^2}{n_2}}}$$
>
> approximately follows a t distribution with degrees of freedom equal to the smaller of $n_1 - 1$ and $n_2 - 1$, where $\bar{x}_1$ and s_1 represent the mean and standard deviation of the sample taken from population 1, and $\bar{x}_2$ and s_2 represent the mean and standard deviation of the sample taken from population 2.

This t statistic is called Welch's approximate t, after the twentieth-century English statistician Bernard Lewis Welch. Although there are other distributions that statisticians use to estimate the difference between two population means, we use this approximation because it is conservative and easy to calculate.

Researchers are often interested in testing whether the mean of one population is greater than, less than, or different from the mean of another population. Thus, we next learn how to perform hypothesis tests for the difference in population means $\mu_1 - \mu_2$. Usually the most important hypothesized value for $\mu_1 - \mu_2$ is 0. Consider the two-tailed hypothesis test

$$H_0 : \mu_1 - \mu_2 = 0 \quad \text{versus} \quad H_a : \mu_1 - \mu_2 \neq 0$$

which is equivalent to

$$H_0 : \mu_1 = \mu_2 \quad \text{versus} \quad H_a : \mu_1 \neq \mu_2$$

In practice, the hypothesized difference between the two population means is nearly always $(\mu_1 - \mu_2)_0 = 0$. Thus, the test statistic takes the following form:

$$t_{\text{data}} = \frac{(\bar{x}_1 - \bar{x}_2) - 0}{\sqrt{\dfrac{s_1^2}{n_1} + \dfrac{s_2^2}{n_2}}} = \frac{(\bar{x}_1 - \bar{x}_2)}{\sqrt{\dfrac{s_1^2}{n_1} + \dfrac{s_2^2}{n_2}}}$$

Just as in Section 9.4, if t_{data} is extreme, then it represents evidence against the null hypothesis. The hypothesis test may be carried out using either the critical-value method or the p-value method.

> **Welch's Hypothesis Test for the Difference in Two Population Means: Critical-Value Method**
>
> The hypothesis test applies whenever *either*
>
>   **a.** Both populations are normally distributed, or
>
>   **b.** Both samples are large, that is $n_1 \geq 30$ and $n_2 \geq 30$.
>
> *Step 1* **State the hypotheses.**
> Use one of the forms from Table 10.11 on the next page. State the meaning of μ_1 and μ_2.
> *Step 2* **Find t_{crit} and state the rejection rule.**
> To find t_{crit}, use the t table and degrees of freedom the *smaller* of $n_1 - 1$ and $n_2 - 1$. To find the rejection rule, use Table 10.11.
> *Step 3* **Calculate t_{data}.**
>
> $$t_{\text{data}} = \frac{(\bar{x}_1 - \bar{x}_2)}{\sqrt{\dfrac{s_1^2}{n_1} + \dfrac{s_2^2}{n_2}}}$$
>
> which follows an approximate t distribution with degrees of freedom the smaller of $n_1 - 1$ and $n_2 - 1$.
> *Step 4* **State the conclusion and the interpretation.**
> Compare t_{data} with t_{crit}.

Table 10.11 Critical regions and rejection rules for t test for $\mu_1 - \mu_2$

	Right-tailed test	**Left-tailed test**	**Two-tailed test**
Form of test	$H_0: \mu_1 = \mu_2$ $H_a: \mu_1 > \mu_2$ level of significance α	$H_0: \mu_1 = \mu_2$ $H_a: \mu_1 < \mu_2$ level of significance α	$H_0: \mu_1 = \mu_2$ $H_a: \mu_1 \neq \mu_2$ level of significance α
Critical region	α 0 t_{crit} Noncritical region — Critical region	α $-t_{crit}$ 0 Critical region — Noncritical region	$\alpha/2$ $\alpha/2$ $-t_{crit}$ 0 t_{crit} Critical region — Noncritical region — Critical region
Rejection rule	Reject H_0 if $t_{data} \geq t_{crit}$	Reject H_0 if $t_{data} \leq -t_{crit}$	Reject H_0 if $t_{data} \geq t_{crit}$ or $t_{data} \leq -t_{crit}$

EXAMPLE 10.6

t TEST FOR $\mu_1 - \mu_2$: CRITICAL-VALUE METHOD

Using Table 10.10, test whether women's population mean body temperature differs from that of men, using the critical-value method and $\alpha = 0.05$.

Solution

Both sample sizes are large ($n_1 = n_2 = 65 \geq 30$), so we can perform the hypothesis test.

STEP 1 State the hypotheses.
The key words "differs from" indicate a two-tail test:

$$H_0: \mu_1 = \mu_2 \quad \text{versus} \quad H_a: \mu_1 \neq \mu_2$$

where μ_1 and μ_2 represent the population mean body temperature for women and men, respectively.

STEP 2 Find t_{crit} and state the rejection rule.
The required degrees of freedom is the smaller of $n_1 - 1$ and $n_2 - 1$, which is $65 - 1 = 64$. Again df = 64, but we use the conservative df = 60 in the t table in Appendix Table D. For $\alpha = 0.05$, this gives $t_{crit} = 2.000$. We have a two-tailed test, so Table 10.11 gives us the following rejection rule:

$$\text{Reject } H_0 \text{ if } t_{data} \geq 2.000 \text{ or } t_{data} \leq -2.000$$

STEP 3 Find t_{data}.

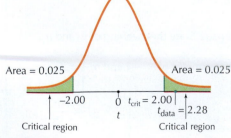

Area = 0.025   Area = 0.025

−2.00   0 t_{crit} = 2.00
 t_{data} = |2.28|
 t

Critical region   Critical region

FIGURE 10.10 $t_{data} = 2.28$ falls within the critical region.

**Now You Can Do
Exercises 3–6.**

$$t_{data} = \frac{(\bar{x}_1 - \bar{x}_2)}{\sqrt{\dfrac{s_1^2}{n_1} + \dfrac{s_2^2}{n_2}}} = \frac{(98.394 - 98.105)}{\sqrt{\dfrac{(0.743)^2}{65} + \dfrac{(0.699)^2}{65}}} \approx 2.28$$

STEP 4 State the conclusion and the interpretation.
The test statistic $t_{data} = 2.28$ is greater than $t_{crit} = 2.000$ (see Figure 10.9). We therefore reject H_0. There is evidence at level of significance $\alpha = 0.05$ that the difference in population mean body temperatures is not the same for women and men.

We may also use the *p*-value method to perform the independent sample *t* test for $\mu_1 - \mu_2$.

> **Welch's Hypothesis Test for the Difference in Two Population Means: *p*-Value Method**
>
> The hypothesis test applies whenever *either*
>
> **a.** Both populations are normally distributed, or
>
> **b.** Both samples are large, that is $n_1 \geq 30$ and $n_2 \geq 30$.
>
> *Step 1* **State the hypotheses and the rejection rule.**
> Use one of the forms from Table 10.12. State the meaning of μ_1 and μ_2. The rejection rule is *Reject H_0 if the p-value is $\leq \alpha$.*
>
> *Step 2* **Calculate t_{data}.**
>
> $$t_{data} = \frac{(\bar{x}_1 - \bar{x}_2)}{\sqrt{\dfrac{s_1^2}{n_1} + \dfrac{s_2^2}{n_2}}}$$
>
> which follows an approximate *t* distribution with degrees of freedom the smaller of $n_1 - 1$ and $n_2 - 1$.
>
> *Step 3* **Find the *p*-value.**
> Use technology or estimate using the *t* table.
>
> *Step 4* **State the conclusion and the interpretation.**
> Compare the *p*-value with α.

Table 10.12 *p*-Values for *t* test for $\mu_1 - \mu_2$

	Right-tailed test	**Left-tailed test**	**Two-tailed test**
Form of test	$H_0 : \mu_1 = \mu_2$ $H_a : \mu_1 > \mu_2$	$H_0 : \mu_1 = \mu_2$ $H_a : \mu_1 < \mu_2$	$H_0 : \mu_1 = \mu_2$ $H_a : \mu_1 \neq \mu_2$
***p*-Value**	*p*-value $= P(t > t_{data})$ Area to the right of t_{data}	*p*-value $= P(t < t_{data})$ Area to the left of t_{data}	*p*-value $= P(t > \lvert t_{data} \rvert) +$ $P(t < -\lvert t_{data} \rvert) = 2 \cdot P(t > \lvert t_{data} \rvert)$ Sum of the two tail areas

---

EXAMPLE 10.7

amleague
natleague

t TEST FOR $\mu_1 - \mu_2$ USING THE *p*-VALUE METHOD AND TECHNOLOGY

Many baseball fans hold that, because of the designated hitter rule, there are more runs scored in the American League than in the National League. Perform an independent samples *t* test to find out whether that was indeed the case in 2006. Use the TI-83/84 or Excel, the *p*-value method, and level of significance $\alpha = 0.05$. Table 10.13 contains the mean runs per game (RPG) for a random sample of six teams from each league.

Table 10.13 Major League Baseball runs scored per game, 2006 regular season

American League: Sample 1		National League: Sample 2	
Team	**RPG**	**Team**	**RPG**
New York Yankees	5.74	Philadelphia Phillies	5.34
Chicago White Sox	5.36	Atlanta Braves	5.24
Texas Rangers	5.15	Colorado Rockies	5.02
Detroit Tigers	5.07	Arizona Diamondbacks	4.77
Boston Red Sox	5.06	Florida Marlins	4.68
Los Angeles Angels	4.73	Houston Astros	4.54

Solution

Because the samples are small, we must determine whether both populations are normally distributed. The normal probability plots for RPG for each league indicate acceptable normality, so we may perform the hypothesis test.

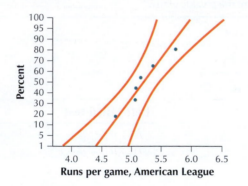

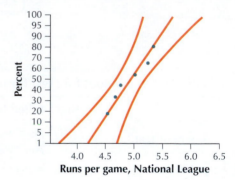

Note: Our degrees of freedom, the smaller of $n_1 - 1$ and $n_2 - 1$, is $6 - 1 = 5$. However, the TI-83/84 shows df = 9.966314697, and the Excel output rounds this to 10. Why does the technology use different degrees of freedom than we do? Recall that we are using Welch's approximation to the t distribution. The TI-83/84, Excel, Minitab, and other technology calculate the degrees of freedom as follows:[7]

$$df = \frac{\left(\frac{s_1^2}{n_1} + \frac{s_2^2}{n_2}\right)^2}{\frac{\left(\frac{s_1^2}{n_1}\right)^2}{n_1 - 1} + \frac{\left(\frac{s_2^2}{n_2}\right)^2}{n_2 - 1}}$$

This provides a more accurate determination of the degrees of freedom than our method. However, our method is a conservative estimate that is easier to calculate, and it is recommended for hand calculations.

STEP 1 **State the hypotheses and the rejection rule.**

Since the American League represents sample 1 and we are interested in whether the American League has scored *more* runs than the National League, we have the following hypotheses:

$$H_0 : \mu_1 = \mu_2 \quad \text{versus} \quad H_a : \mu_1 > \mu_2$$

where μ_1 and μ_2 represent the population mean runs per game for the American League and National League, respectively. The rejection rule is to reject H_0 if p-value ≤ 0.05.

STEP 2 **Find t_{data}.**

We use the instructions provided in the Step-by-Step Technology Guide at the end of this section. From either Figure 10.11 or Figure 10.12,

$$t_{\text{data}} = \frac{(\bar{x}_1 - \bar{x}_2)}{\sqrt{\frac{s_1^2}{n_1} + \frac{s_2^2}{n_2}}} \approx \frac{(5.185 - 4.932)}{\sqrt{\frac{0.339^2}{6} + \frac{0.320^2}{6}}} \approx 1.3301$$

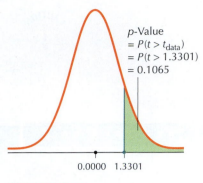

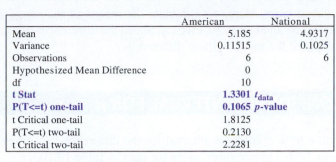

	American	National
Mean	5.185	4.9317
Variance	0.11515	0.1025
Observations	6	6
Hypothesized Mean Difference	0	
df	10	
t Stat	**1.3301** t_{data}	
P(T<=t) one-tail	**0.1065** *p*-value	
t Critical one-tail	1.8125	
P(T<=t) two-tail	0.2130	
t Critical two-tail	2.2281	

FIGURE 10.11
TI-83/84 output.

FIGURE 10.12 Excel output.

FIGURE 10.13 The *p*-value for the right-tailed *t* test.

STEP 3 Find the *p*-value.
From either Figure 10.11 or Figure 10.12,

$$p\text{-value} = P(t > t_{\text{data}}) = P(t > 1.3301) = 0.1065$$

This *p*-value is illustrated in Figure 10.13.

STEP 4 State the conclusion and the interpretation.
Our *p*-value of 0.1065 is not ≤ 0.05. Therefore, we do not reject H_0. There is insufficient evidence that the population mean runs scored per game is greater in the American League than in the National League.

Now You Can Do Exercises 7–10.

2 *t* CONFIDENCE INTERVALS FOR $\mu_1 - \mu_2$

Recall from Section 8.2 that to estimate the unknown population mean μ, we can use a *t* confidence interval:

$$\bar{x} \pm E = \bar{x} \pm t_{a/2} (s/\sqrt{n})$$

where E is the **margin of error.** By analogy, here the *t* interval for $\mu_1 - \mu_2$ takes the following form.

Welch's Confidence Interval for $\mu_1 - \mu_2$

For two independent random samples taken from two populations with population means μ_1 and μ_2, a $100(1 - \alpha)\%$ **confidence interval for $\mu_1 - \mu_2$** is given by

$$(\bar{x}_1 - \bar{x}_2) \pm t_{a/2}\sqrt{\frac{s_1^2}{n_1} + \frac{s_2^2}{n_2}}$$

where $\bar{x}_1$, s_1, and n_1 represent the mean, standard deviation, and sample size of the sample taken from population 1 and $\bar{x}_2$, s_2, and n_2 represent the mean, standard deviation, and sample size of the sample taken from population 2, and $t_{a/2}$ is associated with the confidence level and degrees of freedom of the smaller of $n_1 - 1$ and $n_2 - 1$.

The *t* interval applies whenever *either* of the following conditions is met:

- Both populations are normally distributed, or
- Both sample sizes are large.

Margin of Error E

The **margin of error** for a $100(1 - \alpha)\%$ confidence interval for $\mu_1 - \mu_2$ is given by

$$E = t_{a/2} \cdot \sqrt{\frac{s_1^2}{n_1} + \frac{s_2^2}{n_2}}$$

Thus, the confidence interval for $\mu_1 - \mu_2$ takes the form $(\bar{x}_1 - \bar{x}_2) \pm E$.

This is a confidence interval for the difference in two population means, which is not the same as in Section 10.1, which was for the population mean of the differences of matched pairs. Here, we calculate the means of the samples and then compute the difference. In Section 10.1 we calculated the differences of sample values first and then computed the mean of these differences.

EXAMPLE 10.8

CONFIDENCE INTERVAL FOR $\mu_1 - \mu_2$

Find a 95% confidence interval for the difference in women's and men's population mean body temperatures, using the data in Table 10.10.

Solution

Both sample sizes are large ($n_1 = n_2 = 65 \geq 30$), so we may construct the interval. For $t_{\alpha/2}$, the required degrees of freedom is the smaller of $n_1 - 1$ and $n_2 - 1$, which is $65 - 1 = 64$. We again use the conservative df = 60. For 95% confidence, then, $t_{\alpha/2} = 2.00$. The margin of error is

$$E = t_{\alpha/2} \cdot \sqrt{\frac{s_1^2}{n_1} + \frac{s_2^2}{n_2}} \approx (2.00) \cdot \sqrt{\frac{(0.743)^2}{65} + \frac{(0.699)^2}{65}} = 0.253$$

The 95% confidence interval is then

$$(\bar{x}_1 - \bar{x}_2) \pm E = (98.394 - 98.105) \pm 0.253 = 0.289 \pm 0.253 = (0.036, 0.542).$$

Now You Can Do Exercises 11–16.

We are 95% confident that the difference in population means $\mu_1 - \mu_2$ lies between 0.036°F and 0.542°F. Since 0 is not contained in this interval, we may conclude that $\mu_1 \neq \mu_2$, just as we did in Example 10.6.

3 USING CONFIDENCE INTERVALS TO PERFORM HYPOTHESIS TESTS

As in earlier sections, we may use a $100(1 - \alpha)\%$ t confidence interval for $\mu_1 - \mu_2$ to perform two-tailed t tests about $\mu_1 - \mu_2$.

> **Equivalence of a Two-Tailed t Test About $\mu_1 - \mu_2$ and a t Confidence Interval for $\mu_1 - \mu_2$**
>
> - If a certain value for $\mu_1 - \mu_2$ lies *outside* the corresponding $100(1 - \alpha)\%$ t confidence interval for $\mu_1 - \mu_2$, then the null hypothesis specifying this value would be *rejected* for level of significance α.
>
> - Alternatively, if a certain value for $\mu_1 - \mu_2$ lies *inside* the $100(1 - \alpha)\%$ t confidence interval for $\mu_1 - \mu_2$, then the null hypothesis specifying this value would *not be rejected* for level of significance α.

EXAMPLE 10.9

USING A t CONFIDENCE INTERVAL TO PERFORM A TWO-TAILED t TEST ABOUT $\mu_1 - \mu_2$

amleague
natleague

a. Construct a 95% confidence interval for the difference in runs scored per game in the American League and National League, using the data from Example 10.7.
b. Test using level of significance $\alpha = 0.05$ whether the population mean number of runs scored per game in the American League differs from the population mean number of runs scored in the National League.

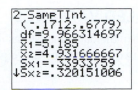

FIGURE 10.14 TI-83/84 results.

Now You Can Do
Exercises 17–20.

Solution

a.   Figure 10.14 shows the 95% confidence interval for $\mu_1 - \mu_2$, where μ_1 and μ_2 represent the population mean runs per game in the American and National Leagues, respectively:

$$(-0.1712, 0.6779)$$

b.   The confidence interval in part (a) does contain 0. That is, 0 lies between -0.1712 and 0.6779. Therefore, with level of significance $\alpha = 0.05$, we do not reject the hypothesis that there is no difference between population mean runs scored per game in the American and National Leagues.

CASE STUDY

Do Prior Student Evaluations Influence Students' Ratings of Professors?

In this case study the students in one sample were shown positive evaluations of an instructor and the students in a second sample were shown negative evaluations of the instructor. Then all subjects were shown the same 20-minute lecture video given by the same instructor. They were then asked to rate the instructor using three questions, and a summary rating score was calculated. Were students' ratings influenced by the prior student evaluations?

We investigate this question by constructing a 95% confidence interval for the difference in population mean ratings $\mu_1 - \mu_2$. Assume that both populations are normally distributed and that the samples are drawn independently.

Reputation	Subjects	Sample mean rating	Sample standard deviation
Charismatic (sample 1)	$n_1 = 25$	$\bar{x}_1 = 2.613$	$s_1 = 0.533$
Punitive (sample 2)	$n_2 = 24$	$\bar{x}_2 = 2.236$	$s_2 = 0.543$

The degrees of freedom is the smaller of $n_1 - 1 = 25 - 1 = 24$ and $n_2 - 1 = 24 - 1 = 23$. Thus, df $= 23$. Then, for 95% confidence, from the t table, $t_{\text{crit}} = 2.069$. Then the 95% confidence interval for $\mu_1 - \mu_2$ is

$$(\bar{x}_1 - \bar{x}_2) \pm t_{a/2} \sqrt{\frac{s_1^2}{n_1} + \frac{s_2^2}{n_2}}$$

$$= (2.613 - 2.236) \pm 2.069 \sqrt{\frac{0.533^2}{25} + \frac{0.543^2}{24}}$$

$$\approx 0.377 \pm 2.069 \cdot (0.1538)$$

$$\approx (0.059, 0.695)$$

We are 95% confident that the difference in population mean instructor ratings $\mu_1 - \mu_2$ among the two groups of students lies between 0.059 and 0.695. Since this interval does not contain 0, we can conclude that the difference in population mean ratings is significant at level of significance $\alpha = 0.05$. ∎

4 t INFERENCE FOR $\mu_1 - \mu_2$ USING POOLED VARIANCE

Recall that the variance equals the square of the standard deviation.

An alternative method for t inference may be applied when the data analyst has reason to believe that $\sigma_1^2 = \sigma_2^2$, that is, the variances of the two populations are equal. A *pooled estimate* s_{pooled}^2 of the common variance $\sigma_1^2 = \sigma_2^2 = \sigma^2$ is used.

> **Pooled Estimate of the Common Variance σ^2**
>
> $$s_{pooled}^2 = \frac{(n_1 - 1)s_1^2 + (n_2 - 1)s_2^2}{n_1 + n_2 - 2}$$

Some statisticians think that the pooled variance method should be used sparingly.[8]

The conditions for performing t inference using pooled variance are the same as for Welch's method (page 499), with the additional condition that $\sigma_1^2 = \sigma_2^2$. The test statistic t_{data} for the pooled variance t test is then given by

$$t_{data} = \frac{(\bar{x}_1 - \bar{x}_2)}{\sqrt{s_{pooled}^2 \left(\frac{1}{n_1} + \frac{1}{n_2}\right)}}$$

We illustrate the pooled variance t test and the pooled variance t confidence interval using the following two examples.

EXAMPLE 10.10

POOLED VARIANCE t TEST

The University of Michigan Consumer Sentiment Index measures consumer optimism, with a baseline of 100 equal to the level of consumer optimism in December 1964. Summary statistics for August 2008 are provided in the following table, for families with incomes above and below $75,000. Use the critical-value method for the pooled variance t test to test whether the population mean consumer sentiment index of families with income above $75,000 is greater than that of families with income below $75,000. Assume $\sigma_1^2 = \sigma_2^2$ and use level of significance $\alpha = 0.01$.

Income	Sample size	Sample mean index	Sample standard deviation
Above $75,000	$n_1 = 31$	$\bar{x}_1 = 67.5$	$s_1 = 11.6$
Below $75,000	$n_2 = 31$	$\bar{x}_2 = 60.2$	$s_2 = 11.2$

Solution

STEP 1 **State the hypotheses.**

$$H_0 : \mu_1 = \mu_2 \quad \text{versus} \quad H_a : \mu_1 > \mu_2$$

where μ_1 and μ_2 represent the population mean consumer sentiment index for families with incomes above and below $75,000, respectively.

STEP 2 **Find t_{crit}.**

The degrees of freedom for the pooled variance t test equals $n_1 + n_2 - 2 = 31 + 31 - 2 = 60$. From the t table we obtain the critical value $t_{crit} = 2.390$. Reject H_0 if $t_{data} \geq 2.390$.

STEP 3   Calculate s^2_{pooled} and t_{data}.

$$s^2_{pooled} = \frac{(n_1 - 1)s_1^2 + (n_2 - 1)s_2^2}{n_1 + n_2 - 2} = \frac{(31 - 1)11.6^2 + (31 - 1)11.2^2}{31 + 31 - 2} = 130$$

Plugging this value into the following formula for the test statistic, we obtain

$$t_{data} = \frac{\bar{x}_1 - \bar{x}_2}{\sqrt{s^2_{pooled}\left(\frac{1}{n_1} + \frac{1}{n_2}\right)}} = \frac{67.5 - 60.2}{\sqrt{130\left(\frac{1}{31} + \frac{1}{31}\right)}} \approx 2.521$$

STEP 4   **Conclusion and interpretation.**
The test statistic $t_{data} \approx 2.521$ is greater than the critical value $t_{crit} = 2.390$. Therefore we reject H_0. There is evidence that the population mean consumer sentiment index of families with incomes above \$75,000 is greater than that of families with incomes below \$75,000.

Now You Can Do
Exercises 21 and 22.

The pooled variance method may also be used to construct a t confidence interval for $\mu_1 - \mu_2$.

EXAMPLE 10.11   **POOLED VARIANCE t CONFIDENCE INTERVAL FOR $\mu_1 - \mu_2$**

Use the data from Example 10.10 to construct a 99% confidence interval for the difference in population mean optimism indices. Use the pooled variance method.

Solution

The $100(1 - \alpha)\%$ confidence interval for $\mu_1 - \mu_2$ using the pooled variance method is given by the following formula:

$$\bar{x}_1 - \bar{x}_2 \pm t_{a/2}\sqrt{s^2_{pooled}\left(\frac{1}{n_1} + \frac{1}{n_2}\right)}$$

where $t_{a/2}$ is found using $n_1 + n_2 - 2$ degrees of freedom. For degrees of freedom $n_1 + n_2 - 2 = 60$, we have $t_{a/2} = 2.660$. Thus, our 99% confidence interval is:

$$67.5 - 60.2 \pm (2.660)\sqrt{130\left(\frac{1}{31} + \frac{1}{31}\right)} = 7.3 \pm 7.703 = (-0.403, 15.003)$$

Now You Can Do
Exercises 23 and 24.

We are 95% confident that the difference in population mean optimism indices lies between -0.403 and 15.003.

Developing Your Statistical Sense

Easier to Reject H_0 Using a One-Tailed Test

Note that the 99% confidence interval for $\mu_1 - \mu_2$ contains zero, so that a two-tailed test for the difference in population means would not have rejected $H_0 : \mu_1 = \mu_2$ for level of significance $\alpha = 0.01$. Contrast this with our rejection of the null hypothesis for the right-tailed (one-tailed) test with level of significance $\alpha = 0.01$ in Example 10.10. We can therefore observe that it is easier to reject the null hypothesis for a one-tailed test than for a two-tailed test with the same level of significance.

5 Z INFERENCE FOR $\mu_1 - \mu_2$ WHEN σ_1 AND σ_2 ARE KNOWN

When the population standard deviations σ_1 and σ_2 are known, the data analyst may prefer to use Z inference for $\mu_1 - \mu_2$, since the margin of error for Z inference is smaller than for t inference. The conditions for performing Z inference for $\mu_1 - \mu_2$ are similar to Welch's method (page 499), with the additional condition that σ_1 and σ_2 are known. We illustrate the two-sample Z test and the Z confidence interval for $\mu_1 - \mu_2$ using the following two examples.

Do not use Z inference for $\mu_1 - \mu_2$ unless both σ_1 and σ_2 are known.

| EXAMPLE 10.12 | TWO-SAMPLE Z TEST |

A Kaiser Family Foundation report found that the mean amount of time that young people aged 8–18 spend talking on their cell phones is $\bar{x}_1 = 33$ minutes per day, while the mean amount of time spent watching TV shows on their cell phones is $\bar{x}_2 = 49$ minutes per day.[9] Assume that the sample sizes are $n_1 = 50$ and $n_2 = 40$, and that the population standard deviations are known to be $\sigma_1 = 15$ minutes and $\sigma_2 = 20$ minutes. Test using the critical-value method and level of significance $\alpha = 0.05$ whether the population mean amount of time young people spending talking on their cell phones is less than the population mean amount of time they spend watching TV shows on their cell phones.

Solution

STEP 1   State the hypotheses.

$$H_0 : \mu_1 = \mu_2 \quad \text{versus} \quad H_a : \mu_1 < \mu_2$$

where μ_1 and μ_2 represent the population mean amount of time young people spend talking and watching TV shows on their cell phones, respectively.

STEP 2   Find Z_{crit}.
From Table 9.4 (page 416), we have $Z_{\text{crit}} = -1.645$. Reject H_0 if $Z_{\text{data}} \leq -1.645$.

STEP 3   Calculate Z_{data}.
The test statistic for the Z test for $\mu_1 - \mu_2$ takes the form

$$Z_{\text{data}} = \frac{\bar{x}_1 - \bar{x}_2}{\sqrt{\dfrac{\sigma_1^2}{n_1} + \dfrac{\sigma_2^2}{n_2}}} = \frac{33 - 49}{\sqrt{\dfrac{15^2}{50} + \dfrac{20^2}{40}}} \approx -4.202$$

STEP 4   Conclusion and interpretation.
The test statistic $Z_{\text{data}} \approx -4.202$ is less than the critical value $Z_{\text{crit}} = -1.645$. Therefore we reject H_0. There is evidence that the population mean amount of time young people spending talking on their cell phones is less than the population mean amount of time they spend watching TV shows on their cell phones.

**Now You Can Do
Exercises 25 and 26.**

When σ_1 and σ_2 are known, we can also construct a Z confidence interval for $\mu_1 - \mu_2$.

| EXAMPLE 10.13 | Z CONFIDENCE INTERVAL FOR $\mu_1 - \mu_2$ |

Use the data from Example 10.12 to construct a 95% Z confidence interval for the difference in population mean amount of time spent using cell phones.

Solution

The $100(1 - \alpha)\%$ Z confidence interval for $\mu_1 - \mu_2$ is as follows:

$$\bar{x}_1 - \bar{x}_2 \pm Z_{a/2}\sqrt{\frac{\sigma_1^2}{n_1} + \frac{\sigma_2^2}{n_2}}$$

From Table 8.1 (page 358) we have $Z_{a/2} = 1.96$. Thus, our 95% confidence interval is

$$33 - 49 \pm (1.96)\sqrt{\frac{15^2}{50} + \frac{20^2}{40}} \approx -16 \pm 7.463 = (-23.463, -8.537)$$

**Now You Can Do
Exercises 27 and 28.**

We are 95% confident that the difference in population mean amounts of time spent on cell phones talking and watching TV shows lies between -23.463 minutes and -8.537 minutes.

STEP-BY-STEP TECHNOLOGY GUIDE: Two-Sample *t* Test and Confidence Interval for $\mu_1 - \mu_2$

TI-83/84

Welch's *t* Test and Confidence Interval for $\mu_1 - \mu_2$
We use two different examples to illustrate the two different options for performing a two-sample *t* test or confidence interval for $\mu_1 - \mu_2$ using the TI-83/84.

Data Option. (Example 10.7 is used to illustrate this method.)
Step 1 Enter the American League data into List **L1** and the National League data into List **L2**.
Step 2 Press **STAT** and highlight **TESTS**.
Step 3 Press **4** (for the **2-Samp TTest**). The **2-Samp TTest** menu appears.
Step 4 For input (**INPT**), move the cursor over **Data** and press **ENTER**.
Step 5 For **List1** and **List2**, enter **L1** and **L2**.
Step 6 For **Freq1** and **Freq2**, enter **1**.
Step 7 For μ_1, choose the form of H_a. For Example 10.7, choose "$> \mu_2$" and press **ENTER**.
Step 8 For **Pooled**, select **No** because we are not assuming the variances are equal, and do not need an estimate of the common variance (Figure 10.15).
Step 9 Press **Calculate**. The results for Example 10.7 are shown in Figure 10.11 on page 503.

Stats Option. (Example 10.8 is used to illustrate this method.)
Here you enter the summary statistics.
Step 1 Press **STAT** and highlight **TESTS**.
Step 2 Press **4** (for the **2-Samp TTest**). The **2-Samp TTest** menu appears.
Step 3 For input (**Inpt**), move the cursor over **Stats** and press **ENTER**.
Step 4 For $\bar{x}_1$, enter **98.394**.
Step 5 For Sx_1, enter **0.743**.
Step 6 For n_1, enter **65.**
Step 7 For $\bar{x}_2$, enter **98.105**.
Step 8 For Sx_2, enter **0.699**.
Step 9 For n_2, enter **65** (Figure 10.16).
Step 10 For μ_1, choose the form of H_a. For Example 10.8, choose "$\neq \mu_2$" and press **ENTER**.
Step 11 For **Pooled**, press **No** (Figure 10.17).
Step 12 Press **Calculate**.

Welch's Two-Sample *t* interval for $\mu 1 - \mu 2$
Follow the same steps as for the *t* test, except select **0: 2-SampTInt**. Also, to select confidence level (**C-Level**), enter **0.95** for 95%, at Step 2 for example.

FIGURE 10.15

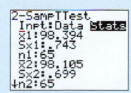

FIGURE 10.16

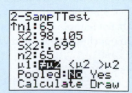

FIGURE 10.17

Pooled Variance t Test and Confidence Interval for $\mu_1 - \mu_2$
Follow the same steps as for Welch's method, except select **Yes** for **Pooled** in Step 8.

Z Test for $\mu_1 - \mu_2$
Data Option
Step 1   Enter the data into Lists **L1** and **L2**.
Step 2   Press **STAT** and highlight **TESTS**.
Step 3   Press **3** (for the **2-Samp Z Test**). The **2-Samp Z Test** menu appears.
Step 4   For input (**INPT**), move the cursor over **Data** and press **ENTER**.
Step 5   Enter the values for σ_1 and σ_2.
Step 6   For **List1** and **List2**, enter **L1** and **L2**.
Step 7   For **Freq1** and **Freq2**, enter **1**.

Step 8   For μ_1, choose the form of H_a and press **ENTER**.
Step 9   Press **Calculate**.

Stats Option
Step 1   Press **STAT** and highlight **TESTS**.
Step 2   Press **3** (for the **2-Samp Z Test**). The **2-Samp Z Test** menu appears.
Step 3   For input (**Inpt**), move the cursor over **Stats** and press **ENTER**.
Step 4   Enter the values for σ_1, σ_2, $\bar{x}_1$, n_1, $\bar{x}_2$, and n_2.
Step 5   For μ_1, choose the form of H_a and press **ENTER**.
Step 6   Press **Calculate**.

Z Confidence Interval for $\mu_1 - \mu_2$
Follow the same steps as for the Z test, except select **9: 2-SampleZInt** at Step 2.

EXCEL

Welch's t Test for $\mu_1 - \mu_2$
Step 1   Enter Sample 1 and Sample 2 data into columns A and B, respectively.
Step 2   Select **Data > Data Analysis > t-Test: Two-Sample Assuming Unequal Variances**, and click **OK**.
Step 3   For the **Dataset Range**, select the cells in column A for the **Variable 1 range** and the cells in column B for the **Variable 2 range**.

For the hypothesized mean difference, enter **0**, enter your value for **Alpha**, and click **OK**.

Pooled Variance t Test for $\mu_1 - \mu_2$
Follow the same steps as for Welch's method, except select **t-test: Two-Sample Assuming Equal Variances** in Step 2.

Z Test for $\mu_1 - \mu_2$
Step 1   Enter Sample 1 and Sample 2 data into columns A and B, respectively.
Step 2   Select **Data > Data Analysis > Z-Test: Two-Sample for Means**, and click **OK**.
Step 3   For the **Dataset Range**, select the cells in column A for the **Variable 1 range** and the cells in column B for the **Variable 2 range**. For the hypothesized mean difference, enter **0**.
Step 4   Enter the values for σ_1^2 and σ_2^2, and the value α for **ALPHA** and click **OK**.

MINITAB

Welch's t Test and Confidence Interval for $\mu_1 - \mu_2$
Step 1   Enter Sample 1 and Sample 2 data into columns C1 and C2, respectively.
Step 2   Click **Stat > Basic Statistics > 2-Sample t**.
Step 3   **a.** If you have the data values, select **Samples in different columns**, and select **C1** and **C2** as your two columns.
b. If you have the summary statistics, select **summarized data** and enter the **sample size**, **mean**, and **standard deviation** for each of the **first** and **second** samples.

Step 4   Click **Options** and select the form of the **Alternative** hypothesis.
Step 5   Click **OK** and click **OK** again.

Pooled Variance t Test and Confidence Interval for $\mu_1 - \mu_2$
Follow the same steps as for Welch's method, except select **Assume equal variances** at the end of Step 3.

CRUNCHIT!

Welch's t Test and t Interval for $\mu_1 - \mu_2$
We will use the data from Example 10.7.
Step 1   Click **File . . .** then highlight **Load from Larose2e . . . Chapter 10 . . .** and click on **Example 10.7**.
Step 2   Click **Statistics** and select **t . . . 2-sample**. Select the **Columns** tab. For **Sample 1** select **American**. For **Sample 2** select **National**. Do not check the **Pooled Variance** option.

For the hypothesis test:
Step 3   Select the **Hypothesis Test** tab, choose the correct form of the **Alternative** hypothesis, and click **Calculate**.

For the confidence interval:
Step 3   Select the **Confidence Interval** tab, enter the **Confidence Interval Level**, and click **Calculate**.

Pooled Variance t Test and t Interval for $\mu_1 - \mu_2$
Use the same steps as for Welch's t test and t interval, except make sure to check the **Pooled Variance** option in Step 2.

Z test for $\mu_1 - \mu_2$
We will use the data from Example 10.12.
Step 1   Click **File . . .** then highlight **Load from LaroseFundamentals2e . . . Chapter 10 . . .** and click on **Example 10.12**.
Step 2   Click **Statistics** and select **Z . . . 2-sample**. Select the **Summarized** tab. For **Sample 1** enter $n_1 = 100$ and $\bar{x}_1 = 33$, and for **Sample 2** enter $n_2 = 81$ and $\bar{x}_2 = 49$.
Step 3   Enter the population standard deviations, $\sigma_1 = 15$ and $\sigma_2 = 20$. Choose the correct form of the **Alternative** hypothesis, and click **Calculate**.

1. Section 10.2 examines inferential methods for $\mu_1 - \mu_2$, the difference between the means of two independent populations. Two-sample t tests may be carried out using either the p-value method or the critical-value method.

2. $100(1 - \alpha)\%$ t confidence intervals for $\mu_1 - \mu_2$ are developed and illustrated.

3. The use of t confidence intervals for $\mu_1 - \mu_2$ to perform two-tailed t tests is illustrated.

4. The pooled variance method for t inference may be applied when the data analyst has reason to believe that the variances of the two populations are equal.

5. When the population standard deviations σ_1 and σ_2 are known, the data analyst may prefer to use Z inference for $\mu_1 - \mu_2$.

Clarifying the Concepts

1. What are the conditions that permit us to perform the two-sample t test?

2. If a $100(1 - \alpha)\%$ confidence interval for $\mu_1 - \mu_2$ contains 0, then with level of significance α what is our conclusion regarding the hypothesis that there is no difference in the population means?

Practicing the Techniques

For Exercises 3–6, perform the indicated Welch's hypothesis test using the critical-value method. The summary statistics were taken from random samples that were drawn independently. For each exercise follow these steps.
   a. State the hypotheses.
   b. Find the critical value t_{crit} and the rejection rule for this test.
   c. Calculate t_{data}.
   d. Compare t_{data} with t_{crit}. State and interpret your conclusion.

3. Test at level of significance $\alpha = 0.10$ whether $\mu_1 \neq \mu_2$.

Sample 1	$n_1 = 36$	$\bar{x}_1 = 10$	$s_1 = 2$
Sample 2	$n_2 = 36$	$\bar{x}_2 = 8$	$s_2 = 2$

4. Test at level of significance $\alpha = 0.05$ whether $\mu_1 > \mu_2$.

Sample 1	$n_1 = 64$	$\bar{x}_1 = 20$	$s_1 = 3$
Sample 2	$n_1 = 64$	$\bar{x}_1 = 18$	$s_1 = 2$

5. Test at level of significance $\alpha = 0.01$ whether $\mu_1 < \mu_2$.

Sample 1	$n_1 = 100$	$\bar{x}_1 = 70$	$s_1 = 10$
Sample 2	$n_1 = 50$	$\bar{x}_1 = 80$	$s_1 = 12$

6. Test at level of significance $\alpha = 0.05$ whether $\mu_1 > \mu_2$.

Sample 1	$n_1 = 60$	$\bar{x}_1 = 100$	$s_1 = 20$
Sample 2	$n_2 = 40$	$\bar{x}_2 = 90$	$s_2 = 10$

For Exercises 7–10, perform the indicated Welch's hypothesis test using the p-value method. The summary statistics were taken from random samples that were drawn independently. For each exercise follow these steps.
   a. State the hypotheses and the rejection rule.
   b. Calculate t_{data}.
   c. Find the p-value.
   d. Compare the p-value with level of significance α. State and interpret your conclusion.

7. Test at level of significance $\alpha = 0.10$ whether $\mu_1 \neq \mu_2$.

Sample 1	$n_1 = 64$	$\bar{x}_1 = 0$	$s_1 = 3$
Sample 2	$n_2 = 49$	$\bar{x}_2 = 1$	$s_2 = 1$

8. Test at level of significance $\alpha = 0.05$ whether $\mu_1 > \mu_2$.

Sample 1	$n_1 = 255$	$\bar{x}_1 = 103$	$s_1 = 17$
Sample 2	$n_2 = 400$	$\bar{x}_2 = 95$	$s_2 = 11$

9. Test at level of significance $\alpha = 0.05$ whether $\mu_1 < \mu_2$.

Sample 1	$n_1 = 100$	$\bar{x}_1 = 50$	$s_1 = 10$
Sample 2	$n_2 = 100$	$\bar{x}_2 = 75$	$s_2 = 15$

10. Test at level of significance $\alpha = 0.01$ whether $\mu_1 \neq \mu_2$.

Sample 1	$n_1 = 30$	$\bar{x}_1 = -10$	$s_1 = 5$
Sample 2	$n_2 = 30$	$\bar{x}_2 = -5$	$s_2 = 2$

For Exercises 11–16, do the following for the designated data:

    a. Provide the point estimate of $\mu_1 - \mu_2$.

    b. Calculate the margin of error for the confidence level indicated.

    c. Construct and interpret a t confidence interval for $\mu_1 - \mu_2$ with the confidence level indicated.

11. Data in Exercise 3, confidence level = 90%

12. Data in Exercise 4, confidence level = 95%

13. Data in Exercise 5, confidence level = 99%

14. Data in Exercise 6, confidence level = 95%

15. Data in Exercise 7, confidence level = 95%

16. Data in Exercise 8, confidence level = 90%

For Exercises 17–20 a $100(1 - \alpha)\%$ t confidence interval for $\mu_1 - \mu_2$ is given. Use the confidence interval to test using level of significance α whether $\mu_1 - \mu_2$ differs from each of the designated hypothesized values.

17. A 95% t confidence interval for $\mu_1 - \mu_2$ is (10, 15). Hypothesized values are

    a. 0   **b.** 12   **c.** 16

18. A 99% t confidence interval for $\mu_1 - \mu_2$ is (0, 100). Hypothesized values are

    a. 1   **b.** 99   **c.** 101

19. A 90% t confidence interval for $\mu_1 - \mu_2$ is (−10, 10). Hypothesized values are

    a. −10.1   **b.** −9.9   **c.** 0

20. A 95% t confidence interval for $\mu_1 - \mu_2$ is (−25, −15). Hypothesized values are

    a. −16   **b.** −26   **c.** 0

For Exercises 21–22, perform the indicated hypothesis test using the pooled variance method. The summary statistics were taken from random samples that were drawn independently. Assume $\sigma_1^2 = \sigma_2^2$.

21. Test at level of significance $\alpha = 0.10$ whether $\mu_1 > \mu_2$.

Sample 1	$n_1 = 36$	$\bar{x}_1 = 54$	$s_1 = 10$
Sample 2	$n_2 = 36$	$\bar{x}_2 = 52$	$s_2 = 11$

22. Test at level of significance $\alpha = 0.05$ whether $\mu_1 < \mu_2$.

Sample 1	$n_1 = 250$	$\bar{x}_1 = 3.0$	$s_1 = 0.25$
Sample 2	$n_2 = 150$	$\bar{x}_2 = 3.2$	$s_2 = 0.30$

For Exercises 23–24, construct a 95% confidence interval for $\mu_1 - \mu_2$ for the indicated data using the pooled variance method.

23. The data in Exercise 21

24. The data in Exercise 22

For Exercises 25 and 26, perform the indicated hypothesis test using the Z test. The summary statistics were taken from random samples that were drawn independently. Assume that σ_1 and σ_2 are known.

25. Test at level of significance $\alpha = 0.05$ whether $\mu_1 > \mu_2$.

Sample 1	$n_1 = 49$	$\bar{x}_1 = 100$	$\sigma_1 = 1$
Sample 2	$n_2 = 36$	$\bar{x}_2 = 99$	$\sigma_2 = 2$

26. Test at level of significance $\alpha = 0.10$ whether $\mu_1 < \mu_2$.

Sample 1	$n_1 = 64$	$\bar{x}_1 = 72$	$\sigma_1 = 3$
Sample 2	$n_2 = 100$	$\bar{x}_2 = 76$	$\sigma_2 = 5$

For Exercises 27 and 28, construct a 95% Z confidence interval for $\mu_1 - \mu_2$ for the indicated data.

27. The data in Exercise 25

28. The data in Exercise 26

Applying the Concepts

For Exercises 29–48, use Welch's t test and t interval unless otherwise indicated.

29. PC Sales. A personal computer company launched an advertising campaign in the hopes of boosting sales. A random sample (sample 1) of 16 days before the advertising blitz showed mean sales of 120 computers per day with a standard deviation of 30. A random sample of 15 days after the advertisements appeared showed mean sales of 125 computers per day with a standard deviation of 35. If it is appropriate, test whether $\mu_1 < \mu_2$. If not, explain why not.

30. Foreclosures. A random sample (sample 1) of 20 counties in 2007 had a mean number of foreclosures on single-family residences of 50 and a standard deviation of 25. A random sample (sample 2) of 25 counties in 2008 had a mean number of foreclosures of 70 and a standard deviation of 35. Assume that the number of foreclosures per county is normally distributed in both 2007 and 2008. If it is appropriate, test whether $\mu_1 < \mu_2$. If not, explain why not.

31. Income in California and Los Angeles. According to random samples taken by the Bureau of Economic Analysis, the mean income for Sacramento County and Los Angeles County, California, was $31,987 and $33,179, respectively. Suppose the samples had the following sample statistics.

Sacramento County	$n_1 = 36$	$\bar{x}_1 = \$31{,}987$	$s_1 = \$5000$
Los Angeles County	$n_2 = 49$	$\bar{x}_2 = \$33{,}179$	$s_2 = \$6000$

    a. Provide the point estimate of the difference in population means $\mu_1 - \mu_2$.

b. Calculate the margin of error for a confidence level of 95%.

c. Construct and interpret a 95% confidence interval for $\mu_1 - \mu_2$.

d. Test at level of significance $\alpha = 0.05$ whether $\mu_1 < \mu_2$.

e. Explain whether the confidence interval in (c) could have been used to perform the hypothesis test in (d). Why or why not?

32. Math Scores. The Institute of Educational Sciences published the results of the Trends in International Math and Science Study. The sample mean mathematics scores for students from the United States and Hong Kong were 518 and 575, respectively. Suppose independent random samples are drawn from each population, and assume that the populations are normally distributed with the following summary statistics.

USA	$n_1 = 10$	$\bar{x}_1 = 518$	$s_1 = 80$
Hong Kong	$n_2 = 12$	$\bar{x}_2 = 575$	$s_2 = 70$

a. Provide the point estimate of the difference in population means $\mu_1 - \mu_2$.

b. Calculate the margin of error for a confidence level of 90%.

c. Construct and interpret a 90% confidence interval for $\mu_1 - \mu_2$.

d. Test at level of significance $\alpha = 0.01$ whether $\mu_1 < \mu_2$.

e. Provide two reasons why the confidence interval in (c) could not have been used to perform the hypothesis test in (d).

33. Children per Classroom. According to **www.localschooldirectory.com**, the sample mean number of children per teacher in the towns of Cupertino, California, and Santa Rosa, California, are 20.9 and 19.3, respectively. Suppose random samples of classrooms are taken from each county, with the following sample statistics.

Cupertino	$n_1 = 36$	$\bar{x}_1 = 20.9$	$s_1 = 5$
Santa Rosa	$n_2 = 64$	$\bar{x}_2 = 19.3$	$s_2 = 4$

a. Construct and interpret a 99% confidence interval for $\mu_1 - \mu_2$.

b. Use the confidence interval in (a) to test at level of significance $\alpha = 0.01$ whether μ_1 differs from μ_2.

34. Property Taxes. Suppose you want to move to either a small town in Ohio (sample 1) or a small town in North Carolina (sample 2). You did some research on property taxes in each state and chose two random samples shown in the table. The data represent the property taxes in dollars for a residence assessed at $250,000. Test whether $\mu_1 \neq \mu_2$ using level of significance $\alpha = 0.05$.

 propertytax

North Carolina		Ohio	
164	206	298	270
147	129	270	315
207	176	165	177
138	120	400	245
143	154	268	180
201	123	289	292
		285	291
		225	

35. Salaries for Grads. The National Association of Colleges (NAC) reported in 2003 that the mean starting salary for college graduates majoring in management information systems was $40,915 and for psychology majors was $27,454. Suppose the NAC data are based on surveys of size 144 for each major, with a standard deviation of $10,000 for the management information systems majors and $7000 for the psychology majors.

a. Explain why it is appropriate to apply t inference.

b. Construct and interpret a 95% confidence interval for $\mu_1 - \mu_2$.

c. Will a 99% confidence interval for $\mu_1 - \mu_2$ be wider or narrower? Explain your reasoning.

36. Park Usage. Suppose that planners for the town of The Woodlands, Texas, were interested in assessing usage of their parks. Random samples were taken of the number of daily visitors to Windvale Park and Cranebrook Park, with the statistics as reported here.

Windvale Park	$n_1 = 36$	$\bar{x}_1 = 110$	$s_1 = 60$
Cranebrook Park	$n_2 = 30$	$\bar{x}_2 = 150$	$s_2 = 75$

a. Construct and interpret a 95% confidence interval for $\mu_1 - \mu_2$.

b. Test at $\alpha = 0.05$ whether μ_1 is less than μ_2.

c. Explain whether the confidence interval in (a) could have been used to perform the hypothesis test in (b). Why or why not?

Coaching for the SAT. Use this information for Exercises 37–39. The College Board reports that a pretest and post-test study was done to investigate whether coaching had a significant effect on SAT scores. The improvement from pretest to post-test was 29 points for the coached sample of students, with a standard deviation of 59 points. For the noncoached students, the pretest to post-test improvement was 21 points with a standard deviation of 52 points.

37. Suppose we consider a sample of 100 students from each group. Perform a test at level of significance $\alpha = 0.05$ for whether the population mean coached SAT pretest–post-test improvement is greater than that for the noncoached students.

38. Refer to Exercise 37.
   a. Find a point estimate of the difference in population means.
   b. Find a 99% confidence interval for the difference in population means.
   c. Determine whether the population means differ, at level of significance $\alpha = 0.01$.

39. *What if* the sample sizes for each group were some number greater than $n = 100$.
   a. How would this affect the width of the confidence interval in Exercise 38(**b**)? Is this good? Explain.
   b. Would this change have any effect on our conclusion in the hypothesis test in Exercise 38(**c**)? Explain why or why not.

40. Nursing Support Services. A statistical study found that when nurses made home visits to pregnant teenagers to provide support services, discourage smoking, and otherwise provide care, the sample mean birth weight of the babies was higher for this treatment group (3285 grams) than for the control group (2922 grams) when the visits began before mid-gestation.[10] There were 21 patients in the treatment group and 11 in the control group. Suppose the birth weights for both groups follow a normal distribution. Assume that the population standard deviation in each sample is 500 grams.
   a. Construct and interpret a 95% Z confidence interval for $\mu_1 - \mu_2$.
   b. Test at level of significance $\alpha = 0.05$ whether the population birth weight differs between the two groups. Use the Z test.

c. Assess the strength of evidence against the null hypothesis.

41. Nursing Support Services. Refer to Exercise 40. *What if* the birth weights of the babies in each group are the same certain amount greater. Explain how this would affect the following.
   a. $\bar{x}_1 - \bar{x}_2$
   b. t_{data}
   c. *p*-value
   d. Conclusion

42. Phosphorus and Potassium. Use computer software to solve the following problems.

Nutrition
   a. Open the **Nutrition** data set. Explore the variable *phosphor,* which lists the amount of phosphorus (in milligrams) for each food item. Generate numerical summary statistics and graphs for the amount of phosphorus in the food. What is the sample mean amount of phosphorus? The sample standard deviation?
   b. Explore the variable *potass,* which lists the amount of potassium (in milligrams) for each food item. Generate numerical summary statistics and graphs for the amount of potassium in the food. What is the sample mean amount of potassium? The sample standard deviation?
   c. Is the independent sampling method the most appropriate way to test this hypothesis? Why or why not?
   d. Create a new variable in Excel or Minitab, **phos_pot,** which equals the amount of phosphorus minus the amount of potassium in each food item. Use a paired sample hypothesis test to test at level of significance $\alpha = 0.05$ whether the population mean difference differs from 0.

10.3   INFERENCE FOR TWO INDEPENDENT PROPORTIONS

OBJECTIVES   By the end of this section, I will be able to . . .

1   Perform and interpret Z tests for $p_1 - p_2$.

2   Compute and interpret Z intervals for $p_1 - p_2$.

3   Use Z intervals for $p_1 - p_2$ to perform two-tailed Z tests.

1   INDEPENDENT SAMPLE *Z* TESTS FOR $P_1 - P_2$

So far in this chapter, we have learned how to perform inference about population *means.* In this section, we learn how to perform hypothesis tests and construct confidence intervals about the difference between two population *proportions.* Recall that the sample proportion of success $\hat{p} = x/n$ is the ratio of the number of successes x to the number of trials n in a binomial experiment.

Here we consider two independent samples, each of which yields a sample proportion: $\hat{p}_1 = x_1/n_1$ and $\hat{p}_2 = x_2/n_2$. For example, a recent survey found the sample proportion of teenage boys (sample 1) and girls (sample 2) who post their last names in their online profiles to be

$$\hat{p}_1 = \frac{x_1}{n_1} = \frac{200}{500} = 0.400$$

and

$$\hat{p}_2 = \frac{x_1}{n_2} = \frac{96}{500} = 0.192$$

(See Example 10.14 for further details about these data.) Here we are interested in performing inference for the difference in population proportions $p_1 - p_2$, such as the difference in the proportions of *all* teenage boys and girls who post their last names in their online profiles. We use the difference in sample proportions $\hat{p}_1 - \hat{p}_2$ as our point estimate of the difference in population proportions $p_1 - p_2$, which is unknown. And just as in earlier sections where we investigated the sampling distribution of $\bar{x}_1 - \bar{x}_2$ to perform inference on $\mu_1 - \mu_2$, here we use the sampling distribution of $\hat{p}_1 - \hat{p}_2$ to help us perform inference about $p_1 - p_2$.

Developing Your Statistical Sense

Independent Samples Only

The inferential methods of this section are reserved for *independent* samples only. An example of a problem that would not use the methods of this section is the following. In the latest poll, suppose 45% supported the Democrat and 40% supported the Republican. Because each respondent had to choose between the Democratic candidate and the Republican candidate, their respective poll numbers are *not independent*.

The distribution of all possible values of $\hat{p}_1 - \hat{p}_2$ is called the **sampling distribution of $\hat{p}_1 - \hat{p}_2$**, with mean $p_1 - p_2$ and standard error $\sigma_{\hat{p}_1 - \hat{p}_2} = \sqrt{\dfrac{p_1(1-p_1)}{n_1} + \dfrac{p_2(1-p_2)}{n_2}}$.

Let x_1 and x_2 denote the number of successes, and let $n_1 - x_1$ and $n_2 - x_2$ denote the number of failures in sample 1 and sample 2, respectively. The sampling distribution of $\hat{p}_1 - \hat{p}_2$ is approximately normal when the number of successes and the number of failures in each sample are each at least 5, that is, when $x_1 \geq 5$, $(n_1 - x_1) \geq 5$, $x_2 \geq 5$, and $(n_2 - x_2) \geq 5$. Let $q_1 = 1 - p_1$, $q_2 = 1 - p_2$, $\hat{q}_1 = 1 - \hat{p}_1$ and $\hat{q}_2 = 1 - \hat{p}_2$.

Sampling Distribution of $\hat{p}_1 - \hat{p}_2$

When two random samples are drawn independently from two populations, then the quantity

$$Z = \frac{(\hat{p}_1 - \hat{p}_2) - (p_1 - p_2)}{\sqrt{\dfrac{p_1 q_1}{n_1} + \dfrac{p_2 q_2}{n_2}}}$$

has an approximately standard normal distribution when the following conditions are satisfied:

$$x_1 \geq 5, \quad (n_1 - x_1) \geq 5, \quad x_2 \geq 5, \quad (n_2 - x_2) \geq 5$$

and where $\hat{p}_1$ and n_1 represent the sample proportion and sample size of the sample taken from population 1 with population proportion p_1; $\hat{p}_2$ and n_2 represent the sample proportion and sample size of the sample taken from population 2 with population proportion p_2; and $q_1 = 1 - p_1$ and $q_2 = 1 - p_2$.

The three possible forms for the Z test for $p_1 - p_2$ are as follows.

$H_0 : p_1 = p_2$	$H_a : p_1 > p_2$	Right-tailed test
$H_0 : p_1 = p_2$	$H_a : p_1 < p_2$	Left-tailed test
$H_0 : p_1 = p_2$	$H_a : p_1 \neq p_2$	Two-tailed test

The null hypothesis asserts that $H_0: p_1 = p_2$. We denote this *common population proportion* as p. Since the null hypothesis is assumed true, the test statistic takes the following form:

$$Z_{\text{data}} = \frac{(\hat{p}_1 - \hat{p}_2) - (p_1 - p_2)}{\sqrt{\dfrac{p_1(1 - p_1)}{n_1} + \dfrac{p_2(1 - p_2)}{n_2}}} = \frac{(\hat{p}_1 - \hat{p}_2) - 0}{\sqrt{\dfrac{p_1(1 - p_1)}{n_1} + \dfrac{p_2(1 - p_2)}{n_2}}}$$

$$= \frac{(\hat{p}_1 - \hat{p}_2)}{\sqrt{\dfrac{p(1 - p)}{n_1} + \dfrac{p(1 - p)}{n_2}}} = \frac{(\hat{p}_1 - \hat{p}_2)}{\sqrt{p(1 - p)\left(\dfrac{1}{n_1} + \dfrac{1}{n_2}\right)}}$$

Since the common population proportion p is unknown, we estimate it using the following **pooled estimate of p:**

$$\hat{p}_{\text{pooled}} = \frac{x_1 + x_2}{n_1 + n_2}$$

Substituting this into the formula for the test statistic gives

$$Z_{\text{data}} = \frac{(\hat{p}_1 - \hat{p}_2)}{\sqrt{\hat{p}_{\text{pooled}} \cdot (1 - \hat{p}_{\text{pooled}})\left(\dfrac{1}{n_1} + \dfrac{1}{n_2}\right)}}$$

Z_{data} measures the distance between the sample proportions. Extreme values of Z_{data} indicate evidence against the null hypothesis.

Hypothesis Test for the Difference in Two Population Proportions: Critical-Value Method

Suppose we have two independent random samples taken from two populations with population proportions p_1 and p_2, and the required conditions are met: $x_1 \geq 5$, $(n_1 - x_1) \geq 5$, $x_2 \geq 5$, and $(n_2 - x_2) \geq 5$.

Step 1 **State the hypotheses.**
Use one of the forms from Table 10.14. State the meaning of p_1 and p_2.

Step 2 **Find Z_{crit} and state the rejection rule.**
Use Table 10.14.

Step 3 **Calculate Z_{data}**

$$Z_{\text{data}} = \frac{\hat{p}_1 - \hat{p}_2}{\sqrt{\hat{p}_{\text{pooled}} \cdot (1 - \hat{p}_{\text{pooled}})\left(\dfrac{1}{n_1} + \dfrac{1}{n_2}\right)}}$$

where

$$\hat{p}_{\text{pooled}} = \frac{x_1 + x_2}{n_1 + n_2}$$

Z_{data} follows an approximately standard normal distribution if the required conditions are satisfied.

Step 4 **State the conclusion and the interpretation.**
Compare Z_{data} with Z_{crit}.

Table 10.14 Critical regions and rejection rules for Z test for $p_1 - p_2$

	Form of Hypothesis Test		
	Right-tailed	**Left-tailed**	**Two-tailed**
Level of significance α	$H_0 : p_1 = p_2$ $H_a : p_1 > p_2$	$H_0 : p_1 = p_2$ $H_a : p_1 < p_2$	$H_0 : p_1 = p_2$ $H_a : p_1 \neq p_2$
0.10 0.05 0.01	$Z_{crit} = 1.28$ $Z_{crit} = 1.645$ $Z_{crit} = 2.33$	$Z_{crit} = -1.28$ $Z_{crit} = -1.645$ $Z_{crit} = -2.33$	$Z_{crit} = 1.645$ $Z_{crit} = 1.96$ $Z_{crit} = 2.58$
Critical region	α 0 Z_{crit} Noncritical region Critical region	α $-Z_{crit}$ 0 Critical region Noncritical region	$\alpha/2$ $\alpha/2$ $-Z_{crit}$ 0 Z_{crit} Critical region Noncritical region Critical region
	Reject H_0 if $Z_{data} \geq Z_{crit}$	Reject H_0 if $Z_{data} \leq Z_{crit}$	Reject H_0 if $Z_{data} \leq -Z_{crit}$ or $Z_{data} \geq Z_{crit}$

EXAMPLE 10.14

Z TEST FOR $p_1 - p_2$ USING THE CRITICAL-VALUE METHOD

Punchstock/Banana Stock

The Pew Internet and American Life Project (**www.pewinternet.org**) tracks the behavior of Americans on the Internet. In 2007, they published a report that described some of the behaviors of American teenagers in online social networks, such as Facebook. Teenagers who had online profiles were asked: "We'd like to know if your last name is posted to your profile or not." The results are shown in Table 10.15. Assume the samples are independent.

Table 10.15 Proportions of teenage boys and girls who post their last names in online profiles

	Boys	**Girls**
Number responding "yes"	$x_1 = 200$	$x_2 = 96$
Sample size	$n_1 = 500$	$n_2 = 500$
Sample proportion	$\hat{p}_1 = x_1/n_1$ $= 200/500$ $= 0.400$	$\hat{p}_2 = x_2/n_2$ $= 96/500$ $= 0.192$

a. Find the point estimate of the difference in the population proportions of boys and girls, $\hat{p}_1 - \hat{p}_2$.
b. Compute the pooled estimate of the common proportion, $\hat{p}_{pooled}$.
c. Calculate the value of the test statistic Z_{data}.
d. Test whether the population proportion of teenage boys who post their last name in their online profiles is greater than the population proportion of teenage girls who do so. Use the critical-value method at level of significance $\alpha = 0.01$.

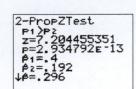

FIGURE 10.18 TI-83/84 results.

Solution

a. The point estimate is $\hat{p}_1 - \hat{p}_2 = 0.400 - 0.192 = 0.208$

b. $\hat{p}_{pooled} = \dfrac{x_1 + x_2}{n_1 + n_2} = \dfrac{200 + 96}{500 + 500} = 0.296$

c. $Z_{data} = \dfrac{\hat{p}_1 - \hat{p}_2}{\sqrt{\hat{p}_{pooled} \cdot (1 - \hat{p}_{pooled})\left(\dfrac{1}{n_1} + \dfrac{1}{n_2}\right)}} = \dfrac{0.400 - 0.192}{\sqrt{(0.296)(0.704)\left(\dfrac{1}{500} + \dfrac{1}{500}\right)}} \approx 7.204$

d. We check the conditions for performing the Z test for $p_1 - p_2$. We have: $x_1 = 200 \geq 5$, $x_2 = 96 \geq 5$, $n_1 - x_1 = 500 - 200 = 300 \geq 5$, and $n_2 - x_2 = 500 - 96 = 404 \geq 5$. We may thus proceed with the hypothesis test.

STEP 1 **State the hypotheses.**
The key words "greater than," together with the fact that sample 1 represents the boys, indicate that we have a right-tailed test:

$$H_0 : p_1 = p_2 \quad \text{versus} \quad H_a : p_1 > p_2$$

where p_1 and p_2 represent the population proportion of teenage boys and girls who post their last name in their online profiles, respectively.

STEP 2 **Find Z_{crit} and state the rejection rule.**
For a right-tailed test with level of significance $\alpha = 0.01$, Table 10.14 gives us $Z_{crit} = 2.33$ and our rejection rule: Reject H_0 if $Z_{data} \geq 2.33$.

STEP 3 **Calculate Z_{data}.**
From **(c)** we have $Z_{data} \approx 7.2$ (also see Figure 10.18).

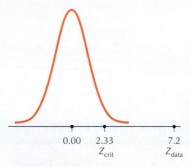

FIGURE 10.19 $Z_{data} = 7.2$ is extreme.

Now You Can Do Exercises 5–8.

STEP 4 **State the conclusion and the interpretation.**
$Z_{data} \approx 7.2 \geq 2.33$, therefore reject H_0 (see Figure 10.19). There is evidence at level of significance $\alpha = 0.01$ that the population proportion of teenage boys who post their last name in their online profiles is greater than the population proportion of teenage girls who do so.

We may also use the p-value method to perform the Z test for $p_1 - p_2$.

Hypothesis Test for the Difference in Two Population Proportions: p-Value Method

Suppose we have two independent random samples taken from two populations with population proportions p_1 and p_2, and the required conditions are met: $x_1 \geq 5$, $(n_1 - x_1) \geq 5$, $x_2 \geq 5$, and $(n_2 - x_2) \geq 5$.

Step 1 **State the hypotheses and the rejection rule.**
Use one of the forms from Table 10.15. State the meaning of p_1 and p_2. The rejection rule is *Reject H_0 if the p-value $\leq \alpha$.*

Step 2 **Calculate Z_{data}.**

$$Z_{data} = \dfrac{\hat{p}_1 - \hat{p}_2}{\sqrt{\hat{p}_{pooled} \cdot (1 - \hat{p}_{pooled})\left(\dfrac{1}{n_1} + \dfrac{1}{n_2}\right)}}$$

where $\hat{p}_{pooled} = \dfrac{x_1 + x_2}{n_1 + n_2}$. If the required conditions are satisfied, Z_{data} follows an approximately standard normal distribution.

Step 3 **Find the p-value.**
Either use technology or calculate the p-value using one of the forms in Table 10.16.

Step 4 **State the conclusion and the interpretation.**
Compare the p-value with α.

Table 10.16 p-Values for Z test for $p_1 - p_2$

Right-tailed test	Left-tailed test	Two-tailed test
$H_0 : p_1 = p_2$ $H_a : p_1 > p_2$	$H_0 : p_1 = p_2$ $H_a : p_1 < p_2$	$H_0 : p_1 = p_2$ $H_a : p_1 \neq p_2$
p-value $= P(Z > Z_{\text{data}})$ Area to right of Z_{data}	p-value $= P(Z < Z_{\text{data}})$ Area to left of Z_{data}	p-value $= P(Z > |Z_{\text{data}}|) + P(Z < -|Z_{\text{data}}|)$ $= 2 \cdot P(Z > |Z_{\text{data}}|)$ Sum of the two-tailed areas

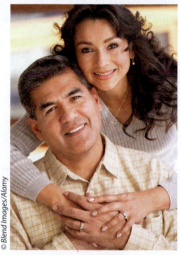

EXAMPLE 10.15

Z TEST FOR $p_1 - p_2$ USING THE p-VALUE METHOD

The General Social Survey tracks trends in American society through annual surveys. Married respondents were asked to characterize their feelings about being married. The results are shown here in a crosstabulation with gender. Test the hypothesis that the proportion of females who report being very happily married is smaller than the proportion of males who report being very happily married. Use the p-value method with level of significance $\alpha = 0.05$.

marriage

	Very happy	Pretty happy/ Not too happy	Total
Female	257	166	423
Male	242	124	366
Total	499	290	789

Solution

From the crosstabulation, we assemble the statistics in Table 10.17 for the independent random samples of men and women.

Table 10.17 Sample statistics of very happily married respondents

	Sample size	Number very happy	Sample proportion very happy
Females (sample 1)	$n_1 = 423$	$x_1 = 257$	$\hat{p}_1 = \dfrac{x_1}{n_1} = \dfrac{257}{423} \approx 0.6076$
Males (sample 2)	$n_2 = 366$	$x_2 = 242$	$\hat{p}_2 = \dfrac{x_2}{n_2} = \dfrac{242}{366} \approx 0.6612$

We first check whether the conditions for the Z test are valid: $x_1 = 257 \geq 5$, $(n_1 - x_1) = (423 - 257) = 166 \geq 5$, $x_2 = 242 \geq 5$, and $(n_2 - x_2) = (366 - 242) = 124 \geq 5$. We can therefore proceed.

© Blend Images/Alamy

STEP 1 **State the hypotheses and the rejection rule.**

Since we are interested in whether the proportion of females who report being very happily married *is smaller than* that of males and because the females represent sample 1, the hypotheses are

$$H_0 : p_1 = p_2 \qquad H_a : p_1 < p_2$$

where p_1 and p_2 represent the population proportions of all females and males, respectively, who report being very happily married. We will reject H_0 if the p-value $\leq \alpha = 0.05$.

STEP 2 **Find Z_{data}.**

First, use the data from Table 10.17 to find the values of $\hat{p}_{pooled}$.

$$\hat{p}_{pooled} = \frac{x_1 + x_2}{n_1 + n_2} = \frac{257 + 242}{423 + 366} \approx 0.63245$$

Then

$$Z_{data} = \frac{(0.6076 - 0.6612)}{\sqrt{0.63245 \cdot (1 - 0.63245)\left(\frac{1}{423} + \frac{1}{366}\right)}} \approx -1.56$$

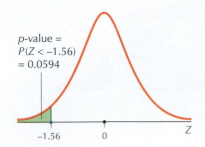

p-value =
$P(Z < -1.56)$
= 0.0594

FIGURE 10.20 *p*-Value for left-tailed *Z* test.

STEP 3 **Find the *p*-value.**

Since it is a left-tailed test, the p-value is given by Table 10.16 as $P(Z < Z_{data}) = P(Z < -1.56)$, as shown in Figure 10.20. This amounts to a Case 1 problem from Table 6.6 on page 289:

$$P(Z < -1.56) = 0.0594$$

Note: When the **p**-value is close to α, many data analysts prefer to simply assess the strength of evidence against the null hypothesis using criteria like those given in Table 9.6 (page 428).

Now You Can Do Exercises 9–12.

STEP 4 **State the conclusion and the interpretation.**

Since the p-value $= 0.0594$ is not less than or equal to $\alpha = 0.05$, we do not reject H_0. There is insufficient evidence that the proportion of females who report being very happily married is smaller than the proportion of males who do so.

---

2 INDEPENDENT SAMPLE *Z* INTERVAL FOR $p_1 - p_2$

We have learned how to perform *Z* tests for $p_1 - p_2$. Next we learn how to use sample statistics to estimate $p_1 - p_2$ using a confidence interval.

> **Confidence Interval for $p_1 - p_2$**
>
> For two independent random samples taken from two populations with population proportions p_1 and p_2, a **100(1 − α)% confidence interval for $p_1 - p_2$** is given by
>
> $$\hat{p}_1 - \hat{p}_2 \pm Z_{a/2} \sqrt{\frac{\hat{p}_1 \cdot \hat{q}_1}{n_1} + \frac{\hat{p}_2 \cdot \hat{q}_2}{n_2}}$$
>
> where $\hat{p}_1$ and n_1 represent the sample proportion and sample size of the sample taken from population 1 with population proportion p_1; $\hat{p}_2$ and n_2 represent the sample proportion and sample size of the sample taken from population 2 with population proportion p_2; the samples are drawn independently; and the following conditions are satisfied: $x_1 \geq 5$, $(n_1 - x_1) \geq 5$, $x_2 \geq 5$, and $(n_2 - x_2) \geq 5$.
>
> **Margin of Error E**
>
> The **margin of error** for a 100(1 − α)% confidence interval for $p_1 - p_2$ is given by
>
> $$E = Z_{a/2} \cdot \sqrt{\frac{\hat{p}_1 \cdot \hat{q}_1}{n_1} + \frac{\hat{p}_2 \cdot \hat{q}_2}{n_2}}$$

EXAMPLE 10.16

Z CONFIDENCE INTERVAL FOR $p_1 - p_2$

Use the sample statistics from Example 10.14 to do the following:
a. Calculate and interpret the margin of error E for confidence level 99%.
b. Construct and interpret a 99% confidence interval for $p_1 - p_2$.

Solution

The conditions for the confidence interval are the same as for the hypothesis test, and were checked in Example 10.14.
a. $\hat{q}_1 = 1 - \hat{p}_1 = 1 - 0.400 = 0.600 \qquad \hat{q}_2 = 1 - \hat{p}_2 = 1 - 0.192 = 0.808$
From Table 8.1 (page 358), the $Z_{\alpha/2}$ value for a 99% confidence level is 2.576.
Therefore, the margin of error is

$$E = Z_{\alpha/2} \cdot \sqrt{\frac{\hat{p}_1 \cdot \hat{q}_1}{n_1} + \frac{\hat{p}_2 \cdot \hat{q}_2}{n_2}} = (2.576)\sqrt{\frac{(0.400)(0.600)}{500} + \frac{(0.192)(0.808)}{500}} \approx 0.072$$

Since the margin of error is 0.072, we may estimate $p_1 - p_2$ to within 0.072 with 99% confidence.
b. The point estimate $\hat{p}_1 - \hat{p}_2 = 0.400 - 0.192 = 0.208$. The 99% confidence interval is therefore

$$\hat{p}_1 - \hat{p}_2 \pm E = 0.208 \pm 0.072 = (0.136, 0.280)$$

Now You Can Do Exercises 13–18.

We are 99% confident that the difference in population proportions of teenage boys and girls whose last name is posted to their profile lies between 0.136 and 0.280.

3 USE Z CONFIDENCE INTERVALS TO PERFORM Z TESTS FOR $p_1 - p_2$

Given a $100(1 - \alpha)\%$ Z confidence interval for $\hat{p}_1 - \hat{p}_2$, we may perform two-tailed Z tests for various hypothesized values of $p_1 - p_2$. If a proposed value lies outside the $100(1 - \alpha)\%$ Z confidence interval for $p_1 - p_2$, then the null hypothesis specifying this value would be rejected. Otherwise do not reject the null hypothesis.

EXAMPLE 10.17

USING A Z INTERVAL FOR $p_1 - p_2$ TO PERFORM Z TESTS ABOUT $p_1 - p_2$

Since this example asks whether $p_1 - p_2$ differs from (or is not equal to) a certain value, we can use the Z confidence interval to test the hypotheses. Example 10.16 provided a 99% Z confidence interval for $p_1 - p_2$, the difference in population proportions of teenage boys and girls whose last name is posted to their profile, as (0.136, 0.280). Test using level of significance $\alpha = 0.01$ whether the $p_1 - p_2$ differs from these values: (a) 0.1, (b) 0.2, (c) 0.3.

Solution

a. $H_0: p_1 - p_2 = 0.1$ versus $H_a: p_1 - p_2 \neq 0.1$.
The hypothesized value 0.1 lies outside the interval (0.136, 0.280), so we reject H_0.
b. $H_0: p_1 - p_2 = 0.2$ versus $H_a: p_1 - p_2 \neq 0.2$.
The hypothesized value 0.2 lies inside the interval, so we do not reject H_0.
c. $H_0: p_1 - p_2 = 0.3$ versus $H_a: p_1 - p_2 \neq 0.3$.
The hypothesized value 0.3 lies outside the interval, so we reject H_0.

Now You Can Do Exercises 19–22.

STEP-BY-STEP TECHNOLOGY GUIDE: *Z* Test and *Z* Interval $p_1 - p_2$

(Example 10.14 is used to illustrate the procedure.)

TI-83/84

Z Test for $p_1 - p_2$
Step 1   Press **STAT** and highlight **TESTS**.
Step 2   Select 6 (for the **2-Prop ZTest**).
Step 3   For **x1**, enter the number of successes in the first
sample, **200**.
Step 4   For **n1**, enter the size of the first sample, **500**.
Step 5   For **x2**, enter the number of successes in the second
sample, **96**.
Step 6   For **n2**, enter the size of the second sample, **500**.
Step 7   For p_1, choose the form of the hypothesis test.
For Example 10.14, choose $> p_2$ and press **ENTER** (Figure 10.21).
Step 8   Highlight **Calculate** and press **ENTER**. The results are
shown in Figure 10.18 in Example 10.14.

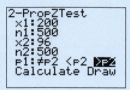

FIGURE 10.21

Z Interval for $p_1 - p_2$
Follow the same steps as for the two-sample *t* test in Section 10.2,
except "Select **B: 2-PropZInt**." Also, to select confidence level
(**C-Level**), enter **0.95** for 95%, for example.

EXCEL

Z Test and Z Interval for $p_1 - p_2$ Using the WHFStat Add-ins
Step 1   Load the **WHFStat Add-ins**.
Step 2   Select **Add-Ins > Macros > Testing a Proportion > Two
Samples**.

Step 3   For **Proportion 1**, enter n_1 for **Sample Size** and x_1 for
Number of Successes.
Step 4   For **Proportion 2**, enter n_2 for **Sample Size** and x_2
for **Number of Successes**. Select the **Confidence Level** and click **OK**.

MINITAB

Z Test and Z Interval for $p_1 - p_2$
Step 1   Click **Stat > Basic Statistics > 2 Proportions**.
Step 2   Select **Summarized Data**.
Step 3   For the **First** row, enter n_1 for **Trials** and x_1 for **Events**.

Step 4   For the **Second** row, enter n_2 for **Trials** and x_2 for
Events.
Step 5   Click **Options** and select the form of the alternative
hypothesis and a confidence level. Then click **OK** twice.

CRUNCHIT!

Z test and Z interval for $p_1 - p_2$
We will use the data from Example 10.14.
Step 1   Click **File . . .** then highlight **Load from
LaroseFundamentals2e . . . Chapter 10 . . .** and click on
Example 10.14.
Step 2   Click **Statistics** and select **Proportion . . . 2-sample**.
Select the **Summarized** tab. For **Sample 1** enter $n_1 = 500$ and
$x_1 = 200$, and for **Sample 2** enter $n_2 = 500$ and $x_2 = 96$.

For the hypothesis test:
Step 3   Select the **Hypothesis Test** tab, choose the correct form
of the **Alternative** hypothesis, and click **Calculate**.

For the confidence interval:
Step 3   Select the **Confidence Interval** tab, enter the
Confidence Interval Level, and click **Calculate**.

SECTION 10.3   Summary

1. The section discusses inferential methods for $p_1 - p_2$, the
difference between the proportions of two independent
populations. Two-sample *Z* tests for $p_1 - p_2$ are discussed.
These hypothesis tests may be carried out using either the
p-value method or the critical-value method.

2. $100(1 - \alpha)\%$ *Z* confidence intervals for $p_1 - p_2$ are
developed and illustrated.

3. We may use *Z* confidence intervals for $p_1 - p_2$ to conduct
two-tailed *Z* tests.

SECTION 10.3 Exercises

Clarifying the Concepts

1. $\hat{p}_{pooled}$ must always lie between which two quantities?

2. Does it make sense to use $\hat{p}_{pooled}$ when calculating confidence intervals for $p_1 - p_2$? Why or why not?

3. What does Z_{data} measure? What do extreme values of Z_{data} indicate?

4. What might we suggest if the p-value is very close to the level of significance α?

Practicing the Techniques

The summary statistics in Exercises 5–7 and 9–11 were taken from random samples that were drawn independently. Let n_1 and n_2 denote the size of samples 1 and 2, respectively. Let x_1 and x_2 denote the number of successes in samples 1 and 2, respectively.

For Exercises 5–7, perform the indicated hypothesis test using the critical-value method. Answer **(a)**–**(d)** for each exercise.
  a. State the hypotheses and find the critical value Z_{crit} and the rejection rule.
  b. Calculate $\hat{p}_{pooled}$.
  c. Calculate Z_{data}.
  d. Compare Z_{data} with Z_{crit}. State and interpret your conclusion.

5. Test at level of significance $\alpha = 0.10$ whether $p_1 \neq p_2$.

Sample 1	$n_1 = 100$	$x_1 = 80$
Sample 2	$n_2 = 40$	$x_2 = 30$

6. Test at level of significance $\alpha = 0.05$ whether $p_1 < p_2$.

Sample 1	$n_1 = 10$	$x_1 = 4$
Sample 2	$n_2 = 12$	$x_2 = 5$

7. Test at level of significance $\alpha = 0.01$ whether $p_1 > p_2$.

Sample 1	$n_1 = 200$	$x_1 = 60$
Sample 2	$n_2 = 250$	$x_2 = 40$

8. Refer to the data from Exercise 7. Test at level of significance $\alpha = 0.01$ whether $p_1 \neq p_2$.

For Exercises 9–11, perform the indicated hypothesis test using the p-value method. Answer **(a)**–**(e)** for each exercise.
  a. State the hypotheses and the rejection rule.
  b. Calculate $\hat{p}_{pooled}$.
  c. Calculate Z_{data}.
  d. Calculate the p-value.
  e. Compare the p-value with α. State and interpret your conclusion.

9. Test at level of significance $\alpha = 0.05$ whether $p_1 > p_2$.

Sample 1	$n_1 = 400$	$x_1 = 250$
Sample 2	$n_2 = 400$	$x_2 = 200$

10. Test at level of significance $\alpha = 0.05$ whether $p_1 < p_2$.

Sample 1	$n_1 = 1000$	$x_1 = 490$
Sample 2	$n_2 = 1000$	$x_2 = 620$

11. Test at level of significance $\alpha = 0.10$ whether $p_1 \neq p_2$.

Sample 1	$n_1 = 527$	$x_1 = 412$
Sample 2	$n_2 = 613$	$x_2 = 498$

12. Refer to the data from Exercise 11. Test at level of significance $\alpha = 0.10$ whether $p_1 < p_2$.

For Exercises 13–18, refer to the indicated data to answer **(a)**–**(d)**.
  a. We are interested in constructing a 95% confidence interval for $p_1 - p_2$. Is it appropriate to do so? Why or why not? If not appropriate, then do not perform **(b)**–**(e)**.
  b. Provide the point estimate of the difference in population proportions $p_1 - p_2$.
  c. Calculate the margin of error for a confidence level of 95%. What does this number mean?
  d. Construct and interpret a 95% confidence interval for $p_1 - p_2$.

13. Data from Exercise 5

14. Data from Exercise 6

15. Data from Exercise 7

16. Data from Exercise 9

17. Data from Exercise 10

18. Data from Exercise 11

For Exercises 19–22 a $100(1 - \alpha)\%$ Z confidence interval for $p_1 - p_2$ is given. Use the confidence interval to test using level of significance α whether $p_1 - p_2$ differs from each of the indicated hypothesized values.

19. A 95% Z confidence interval for $p_1 - p_2$ is $(0.5, 0.6)$. Hypothesized values are
  a. 0 **b.** 0.1 **c.** 0.57

20. A 99% Z confidence interval for $p_1 - p_2$ is $(0.01, 0.99)$. Hypothesized values are
  a. 0.2 **b.** 0 **c.** 0.999

21. A 90% Z confidence interval for $p_1 - p_2$ is $(0.1, 0.11)$. Hypothesized values are
  a. 0.151 **b.** 0.115 **c.** 0.105

22. A 95% Z confidence interval for $p_1 - p_2$ is $(0.43, 0.57)$. Hypothesized values are
  a. 0.41 **b.** 0.51 **c.** 0.61

Applying the Concepts

23. Online Photos. A Pew Internet and American Life Project (**www.pewinternet.org**) 2007 report stated that 74% of teenage boys posted their photo on their online profile, while 83% of teenage girls did so.[11] Assume that the sample sizes were each 500.

a. Is it appropriate to perform the Z test for the difference in population proportions? Why or why not?

b. Clearly state the meaning of p_1 and p_2.

c. Test whether the proportion of teenage boys posting their photo in their online profile differs from the proportion of teenage girls who do so, using level of significance $\alpha = 0.05$.

24. Medicare Recipients. The Centers for Medicare and Medicaid Services reported that 3305 of the 50,350 Medicare recipients living in Alaska were age 85 or over, and 73,289 of the 754,642 Medicare recipients living in Arizona were age 85 or over.

a. Find a point estimate of the difference in population proportions.

b. Clearly state the difference in meaning between p_1 and $\hat{p}_1$.

c. Test whether the population proportions differ, using level of significance $\alpha = 0.05$.

25. Women's Ownership of Businesses. The U.S. Census Bureau tracks trends in women's ownership of businesses. A random sample of 100 Ohio businesses showed 34 that were woman-owned. A sample of 200 New Jersey businesses showed 64 that were woman-owned. Test whether the population proportions of female-owned businesses in Ohio is greater than that of New Jersey, using level of significance $\alpha = 0.10$.

26. Fetal Cells and Breast Cancer. A number of fetal stem cells may cross the placenta from the fetus to the mother during pregnancy and remain in the mother's tissue for decades. A recent study shows that the presence of fetal cells in the mother may offer some protection against the onset of breast cancer.[12] Of the 54 women in the study with breast cancer, 14 had fetal cells. Of the 45 women without breast cancer, 25 had fetal cells. Test whether the population proportions of women with fetal cells is lower among women with breast cancer compared to women without breast cancer, using level of significance $\alpha = 0.01$.

27. Online Photos. Refer to Exercise 23 to answer the following questions.

a. Construct and interpret a 95% confidence interval for the difference in population proportions.

b. Use the confidence interval from (b) to test, using level of significance $\alpha = 0.05$, whether the population proportions differ.

c. Does your conclusion from (c) agree with your conclusion from Exercise 23(c)?

28. Medicare Recipients. Refer to Exercise 24 to answer the following questions.

a. Construct and interpret a 95% confidence interval for the difference in population proportions.

b. Use the confidence interval from (b) to test using level of significance $\alpha = 0.05$ whether the population proportions differ.

c. Does your conclusion from (c) agree with your conclusion from Exercise 24(c)?

29. Women's Ownership of Businesses. Refer to Exercise 25 to answer the following questions.

a. Construct and interpret a 90% confidence interval for the difference in population proportions.

b. Use the confidence interval from (a) to test using level of significance $\alpha = 0.10$ whether the population proportions differ.

c. Explain whether or not we could use the confidence interval from part (b) to perform the hypothesis test in Exercise 25(c). Why or why not?

30. Fetal Cells and Breast Cancer. Refer to Exercise 26 to answer the following questions.

a. Construct and interpret a 99% confidence interval for the difference in population proportions.

b. Use the confidence interval from (a) to test using level of significance $\alpha = 0.01$ whether the population proportions differ.

c. Explain whether or not we could use the confidence interval from (b) to perform the hypothesis test in Exercise 26(c). Why or why not?

31. Evidence for Alternative Medical Therapies? A company called QT, Inc., sells "ionized" bracelets, called Q-Ray bracelets, that it claims help to ease pain through balancing the body's flow of "electromagnetic energy." The Mayo Clinic decided to conduct a statistical experiment to determine whether the claims for the Q-Ray bracelets were justified.[13] At the end of four weeks, of the 305 subjects who wore the "ionized" bracelet, 236 (77.4%) reported improvement in their maximum pain index (where the pain was the worst). Of the 305 subjects who wore the placebo bracelet (a bracelet identical in every respect to the "ionized" bracelet except that there was no active ingredient— presumably, here, "ionization"), 234 (76.7%) reported improvement in their maximum pain index. Using level of significance $\alpha = 0.05$, test whether the population proportions reporting improvement differ between wearers of the ionized bracelet and wearers of the placebo bracelet.

Bringing It All Together

Males Listening to the Radio. Use the following information for Exercises 32–40. The Arbitron Corporation tracks trends in radio listening. In their publication *Radio Today*, Arbitron reported that 92% of 18- to 24-year-old males listen to the radio each week, while 87% of males 65 years and older listen to the radio each week. Suppose each sample size was 1000.

32. Is it appropriate to perform Z inference for the difference in population proportions? Why or why not?

33. Clearly describe what p_1 means and what p_2 means.

34. Explain what the difference is between p_1 and $\hat{p}_1$.

35. Calculate the margin of error for a 95% confidence interval for $p_1 - p_2$. Explain what this number means.

36. Construct and interpret a 95% confidence interval for $p_1 - p_2$.

37. Use the confidence interval from Exercise 36 to test, using level of significance $\alpha = 0.05$, whether $p_1 - p_2$ differs from the following.

    a. 0   **b.** 0.01   **c.** 0.05

38. Explain whether we could use the confidence interval from Exercise 36 to test whether the proportion of 18- of 24-year-old males who listen to the radio each week is greater than the proportion of males 65 years and older who do so. Why or why not?

39. Test using level of significance $\alpha = 0.05$ whether the proportion of 18- of 24-year-old males who listen to the radio each week is greater than the proportion of males 65 years and older who do so.

? 40. *What if*, instead of 1000, each sample size was 100. How would this change affect each of the following measures?

    a. Margin of error in Exercise 35.

    b. p-value in Exercise 39.

    c. Conclusion of the hypothesis test in Exercise 39.

CHAPTER 10   Formulas and Vocabulary

Section 10.1

- **$100(1 - \alpha)\%$ CONFIDENCE INTERVAL FOR μ_d** (p. 491).

$$\bar{x}_d \pm t_{\alpha/2}(s_d/\sqrt{n})$$

- **DEPENDENT SAMPLES** (p. 484)
- **INDEPENDENT SAMPLES** (p. 484)
- **MATCHED-PAIR SAMPLES** (p. 484)
- **TEST STATISTIC FOR THE PAIRED SAMPLE t TEST** (p. 486).

$$t_{\text{data}} = \frac{\bar{x}_d}{s_d/\sqrt{n}}$$

Section 10.2

- **$100(1 - \alpha)\%$ CONFIDENCE INTERVAL FOR $\mu_1 - \mu_2$** (p. 503).

$$(\bar{x}_1 - \bar{x}_2) \pm t_{\alpha/2}\sqrt{\frac{s_1^2}{n_1} + \frac{s_2^2}{n_2}}$$

- **MARGIN OF ERROR E** (p. 503). For a $100(1 - \alpha)\%$ confidence interval for $\mu_1 - \mu_2$,

$$E = t_{\alpha/2} \cdot \sqrt{\frac{s_1^2}{n_1} + \frac{s_2^2}{n_2}}$$

- **SAMPLING DISTRIBUTION OF $\bar{x}_1 - \bar{x}_2$** (pp. 498–499)
- The **pooled variance method** for t inference may be applied when the data analyst has reason to believe that the variances of the two populations are equal.
- When the **population standard deviation σ_1 and σ_2** are known, the data analyst may prefer to use Z inference for $\mu_1 - \mu_2$.
- Pooled estimate for the common variance σ^2 (p. 506):

$$s_{\text{pooled}}^2 = \frac{(n_1 - 1)s_1^2 + (n_2 - 1)s_2^2}{n_1 + n_2 - 2}$$

- Test statistic t_{data} for $\mu_1 - \mu_2$ using pooled variance (p. 506):

$$t_{\text{data}} = \frac{(\bar{x}_1 - \bar{x}_2)}{\sqrt{s_{\text{pooled}}^2\left(\frac{1}{n_1} + \frac{1}{n_2}\right)}}$$

- Pooled variance t confidence interval for μ (p. 507):

$$\bar{x}_1 - \bar{x}_2 \pm t_{\alpha/2}\sqrt{s_{\text{pooled}}^2\left(\frac{1}{n_1} + \frac{1}{n_2}\right)}$$

- Z test statistic for $\mu_1 - \mu_2$ when σ_1 and σ_2 are known (p. 508):

$$Z_{\text{data}} = \frac{\bar{x}_1 - \bar{x}_2}{\sqrt{\frac{\sigma_1^2}{n_1} + \frac{\sigma_2^2}{n_2}}}$$

- Z confidence interval for μ_1 and μ_2 when σ_1 and σ_2 are known (p. 509):

$$\bar{x}_1 - \bar{x}_2 \pm Z_{\alpha/2}\sqrt{\frac{\sigma_1^2}{n_1} + \frac{\sigma_2^2}{n_2}}$$

Section 10.3

- **$100(1 - \alpha)\%$ CONFIDENCE INTERVAL FOR $p_1 - p_2$** (p. 520).

$$\hat{p}_1 - \hat{p}_2 \pm Z_{\alpha/2}\sqrt{\frac{\hat{p}_1 \cdot \hat{q}_1}{n_1} + \frac{\hat{p}_2 \cdot \hat{q}_2}{n_2}}$$

- **MARGIN OF ERROR E** (p. 520). For a $100(1 - \alpha)\%$ confidence interval for $p_1 - p_2$,

$$E = Z_{\alpha/2} \cdot \sqrt{\frac{\hat{p}_1 \cdot \hat{q}_1}{n_1} + \frac{\hat{p}_2 \cdot \hat{q}_2}{n_2}}$$

- **POOLED ESTIMATE OF p** (p. 516).

$$\hat{p}_{\text{pooled}} = \frac{x_1 + x_2}{n_1 + n_2}$$

- **SAMPLING DISTRIBUTION OF $\hat{p}_1 - \hat{p}_2$** (p. 515)
- **TEST STATISTIC FOR THE INDEPENDENT SAMPLES Z TEST FOR $p_1 - p_2$** (p. 516).

$$Z_{\text{data}} = \frac{(\hat{p}_1 - \hat{p}_2)}{\sqrt{\hat{p}_{\text{pooled}} \cdot (1 - \hat{p}_{\text{pooled}})\left(\frac{1}{n_1} + \frac{1}{n_2}\right)}}$$

Section 10.1

1. Assume that a sample of differences for the matched pairs in the table follows a normal distribution, and carry out Steps (**a**) and (**b**).

Subject	1	2	3	4	5	6	7	8
Sample 1	100.7	110.2	105.3	107.1	95.6	109.9	112.3	94.7
Sample 2	104.4	112.5	105.9	111.4	99.8	109.9	115.7	97.7

    a. Calculate $\bar{x}_d$ and s_d.
    b. Construct a 95% confidence interval for μ_d.

2. For the data in Exercise 1, test whether $\mu_d < 0$, using the critical-value method and level of significance $\alpha = 0.05$.

3. For the data in Exercise 1, test whether $\mu_d < 0$, using the p-value method and level of significance $\alpha = 0.05$.

Section 10.2

Refer to the following summary statistics for two independent samples for Exercises 4–7.

Sample 1	$n_1 = 36$	$\bar{x}_1 = 14.4$	$s_1 = 0.01$
Sample 2	$n_2 = 81$	$\bar{x}_2 = 14.3$	$s_2 = 0.02$

4. We are interested in constructing a 95% confidence interval for $\mu_1 - \mu_2$. Explain why it is appropriate to do so.

5. Provide the point estimate of the difference in population means $\mu_1 - \mu_2$.

6. Calculate the margin of error for a confidence level of 95%.

7. Construct and interpret a 95% confidence interval for $\mu_1 - \mu_2$.

8. A random sample of 49 young persons with college degrees had a mean salary of $30,000 with a standard deviation of $5000. An independent random sample of 36 young persons without college degrees had a mean salary of $25,000 and a standard deviation of $4000.

    a. Test at level of significance $\alpha = 0.10$ whether the population mean salary of college graduates μ_1 is greater than the population mean salary of those without a degree μ_2.
    b. Construct and interpret a 90% confidence interval for $\mu_1 - \mu_2$.

Section 10.3

9. The Web site **www.internettrafficreport.com** reports on the current state of Internet data flow around the world. On August 1, 2004, the packet loss from 32 Asian Web sites was 16%, while the packet loss from 125 North American Web sites was 4%.

    a. Perform a hypothesis test of whether the population proportion from Asian Web sites is greater, at level of significance $\alpha = 0.05$, using the p-value method.
    b. Find a 90% confidence interval for the difference in population proportions.

10. The Centers for Disease Control, in their Pregnancy Risk Assessment Monitoring System, reported that 641 of 823 new mothers living in Florida and 658 of 824 new mothers living in North Carolina took their babies in for a checkup within one week of delivery.

    a. Find a 99% confidence interval for the difference in population proportions.
    b. Perform a hypothesis test of whether the population proportions differ, at level of significance $\alpha = 0.01$, using the p-value method.

True or False

1. True or false: In a dependent sampling method the subjects in the first sample determine the subjects for selection in the second sample.

2. True or false: The pooled estimate of p, $\hat{p}_{pooled} = (x_1 + x_2)/(n_1 + n_2)$, always lies between $\hat{p}_1$ and $\hat{p}_2$.

3. True or false: The test statistic Z_{data} measures the size of the typical error in using $\hat{p}_1 - \hat{p}_2$ to estimate $p_1 - p_2$.

Fill in the Blank

4. The conditions on paired sample data for performing a hypothesis test or constructing a confidence interval on paired sample data are that the population is _____ or the sample size is _____.

5. The notation E represents the _____ _____ _____ (three words).

6. _____ [notation] represents the sample mean of the set of n paired differences.

Short Answer

7. What is the notation used to indicate the difference in population means for two independent samples?

8. What statistic is used to estimate the common unknown population proportion?

9. If a $100(1 - \alpha)\%$ confidence interval for $\mu_1 - \mu_2$ contains 0, then with $100(1 - \alpha)\%$ confidence what can you conclude about the difference in the population means?

Calculations and Interpretations

10. Trying to quit smoking? Butt-Enders, a cigarette dependence reduction program, claims to lower the average number of cigarettes smoked for its participants. A sample of 10 participants consumed the following numbers of cigarettes on a randomly chosen day before and after attending Butt-Enders. Assume that the differences are normally distributed.

Participant	1	2	3	4	5
Before	40	20	60	30	50
After	20	0	40	30	20
Participant	6	7	8	9	10
Before	60	20	40	30	20
After	60	20	20	0	20

a. Find a 90% confidence interval for the population mean difference in number of cigarettes smoked.

b. Use your confidence interval to test at level of significance $\alpha = 0.10$ whether the population mean difference in number of cigarettes smoked differs from 0.

11. A family is trying to decide where to move. The choice has come down to Suburb A and Suburb B. A random sample of 40 households in Suburb A had a mean income of $50,000 and a standard deviation of $15,000. A random sample of 36 households in Suburb B had a mean income of $65,000 and a standard deviation of $20,000.

a. Test at level of significance $\alpha = 0.05$ whether the population mean income in Suburb A is less than the population mean income in Suburb B.

b. Construct and interpret a 95% confidence interval for $\mu_1 - \mu_2$.

Use this information for Exercises 12 and 13. A soft drink company recently performed a major overhaul of one of its bottling machines. Management is eager to determine whether the overhaul has resulted in an increase in productivity for the machine. One hundred "minute segments" are sampled at random from the updated machine (Sample 1) and a machine which was not updated (Sample 2), and the number of bottles processed is noted.

The mean and standard deviation of the number of bottles processed by each machine is given in the table.

Updated machine	$n_1 = 100$	$\bar{x}_1 = 200$	$s_1 = 30$
Non-updated machine	$n_2 = 100$	$\bar{x}_2 = 190$	$s_2 = 25$

12. Construct and interpret a 95% confidence interval for $\mu_1 - \mu_2$.

13. Refer to the previous exercise.

a. Test at level of significance $\alpha = 0.05$ whether μ_1 is greater than μ_2.

b. Explain whether the confidence interval in Exercise 12 could have been used to perform the hypothesis test in (a). Why or why not?

14. The U.S. Census Bureau reported that, for people 18–24 years old, the mean annual income for people who never married was $13,539 and for married people was $19,321. Suppose that this information came from a survey of 100 people from each group and that the sample standard deviations were $5000 for the people who never married and $8000 for the married people.

a. Test at level of significance $\alpha = 0.10$ whether the population mean income for never married people differs from that of married people.

b. If we construct a 90% confidence interval for $\mu_1 - \mu_2$, will the interval include 0? Explain why or why not.

c. Confirm your statements from (b).

15. The *2005 National Survey on Drug Use and Health* reported that, in 2004, 38.5% of 18–20 year olds reported having used an illicit drug within the past year, and 37.9% reported use in 2005. Assume $n_1 = n_2 = 1000$. Perform a hypothesis test of whether the population proportion of 18- to 20-year-olds who used an illicit drug decreased from 2004 to 2005, using level of significance $\alpha = 0.05$.

11 Further Inference Methods

Susan Wides/Getty Images

CASE STUDY

Online Dating

The Pew Internet and American Life Project reports that about 16 million people, representing 11% of the American Internet-using public, have visited a dating Web site, and 37% of Internet users who are currently seeking partners have gone to a dating Web site.[1] In this chapter, we apply the concepts and methodologies of categorical data analysis to investigate online dating. In Section 11.2, we examine whether women and men report different types of relationships, and whether women and men differ in how they self-report their physical appearance. ■

The Big Picture

Where we are coming from, and where we are headed . . .

- In Chapters 8–10, we learned how to perform inference for continuous variables. In Sections 11.1 and 11.2, we will learn methods for performing hypothesis tests for *multinomial* data, which are not continuous but *categorical*. These methods rely on the χ^2 distribution, which we learned in Chapters 8 and 9.

- Section 11.1 will cover the χ^2 goodness of fit test, while Section 11.2 introduces us to the χ^2 tests for independence and homogeneity of variance.

- Section 11.3 introduces us to *analysis of variance*, in which we compare the population means of several different groups and determine whether significant differences exist between these means.

- Finally, in Section 11.4 we will use inference methods in regression to examine whether there is evidence for a relationship between two continuous variables.

11.1 χ^2 GOODNESS OF FIT TEST

OBJECTIVES By the end of this section, I will be able to . . .

1 Explain what a multinomial random variable is and how to calculate expected frequencies.

2 Describe how a χ^2 goodness of fit test works.

3 Perform and interpret the results from the χ^2 goodness of fit test using the critical-value method, the *p*-value method, and the estimated *p*-value method.

According to **NetApplications.com**, the market share for the leading Internet browsers in May 2011 was as follows: Microsoft Internet Explorer, 55%; Firefox, 25%; others, 20%. Change is rapid in the online environment. Have these market shares changed since May 2011? How would we go about performing a hypothesis test to determine whether market shares have changed significantly? In Section 11.1, we examine this question using a new type of hypothesis test called a χ^2 *goodness of fit test*. We begin by first considering a new type of random variable that is used to represent categorical data.

1 THE MULTINOMIAL RANDOM VARIABLE

Recall from Chapter 1 that *categorical* (qualitative) variables take values that can be classified into categories. In Chapter 6, we considered binomial random variables, for which there are only two possible outcomes. Now let's consider the following type of random variable, which can have more than two possible values.

> **Multinomial Random Variable**
>
> A random variable is *multinomial* if it satisfies each of the following conditions:
> - Each independent trial of the experiment has k possible outcomes, $k = 2, 3, 4, \ldots$
> - The ith outcome (category) occurs with probability p_i, where $i = 1, 2, \ldots, k$ (that is, p_i is the population proportion for category i)
> - $\sum_{i=1}^{k} p_i = 1$ (Law of Total Probability)
>
> Data from a **multinomial random variable** are said to follow a *multinomial distribution*.

Note: The binomial distribution may be considered a special case of the multinomial distribution, with $k = 2$.

For example, suppose 30% of the residents of a particular town are Democrats, 30% are Republicans, and 40% are Independents. If we select $n = 100$ residents at random, then the number of Democrats, Republicans, and Independents observed follows a multinomial distribution, with

$$p_{\text{Democrats}} = 0.30, \quad p_{\text{Republicans}} = 0.30, \quad p_{\text{Independents}} = 0.40,$$

and

$$\sum_{i=1}^{3} p_i = 0.3 + 0.3 + 0.4 = 1$$

Now You Can Do Exercises 5–8.

Next, recall from Chapter 6 that the formula for finding the expected value (mean) of a binomial random variable having n trials and probability of success p is

$$\text{expected value} = n \cdot p$$

> For a multinomial random variable, the **expected frequency** of the ith category is
>
> $$\text{expected frequency}_i = E_i = n \cdot p_i$$
>
> where n represents the number of trials, and p_i represents the population proportion for the ith category.

EXAMPLE 11.1

BROWSER MARKET SHARE

According to **NetApplications.com**, the market share for the leading Internet browsers in May 2011 was as shown in Table 11.1.
a. If a random sample of size 100 is taken from the population in Table 11.1, verify that the result follows a multinomial distribution.
b. Find the expected frequency for each category in a series of 100 trials.

Table 11.1 Distribution of browser market share

Browser	Relative frequency
Microsoft Internet Explorer	0.55
Firefox	0.25
Other	0.20

Solution

a. There are $k = 3$ possible outcomes: Microsoft Internet Explorer, Firefox, and Other. Assigning probabilities using the relative frequency method, we have the following hypothesized proportions for each browser:

$$p_{\text{MS IE}} = 0.55, \quad p_{\text{Firefox}} = 0.25, \quad p_{\text{Other}} = 0.20$$

And

$$\sum_{i=1}^{3} p_i = 0.55 + 0.25 + 0.20 = 1$$

Since we assume the 1% Guideline (page 225) applies to the random sample, we may state the individual trials are independent. Therefore we have a multinomial distribution.

b. We have $n = 100$ trials (sample size $= 100$), so the expected frequencies are as provided in Table 11.2.

Table 11.2 Expected frequencies for browser preference in sample of size 100

Category	Expected frequency$_i = E_i = n \cdot p_i$
Microsoft Internet Explorer	$E_{\text{MS IE}} = 100 \cdot 0.55 = 55$
Firefox	$E_{\text{Firefox}} = 100 \cdot 0.25 = 25$
Other	$E_{\text{Other}} = 100 \cdot 0.20 = 20$

As a check on the calculations, we should have $\sum E_i = n$. In this case,

**Now You Can Do
Exercises 9a–12a.**

$$\sum E_i = 55 + 25 + 20 = 100 = n$$

What Do These Expected Frequencies Mean?	If we repeatedly took samples of 100 Internet users and asked about browser preference, the mean number of persons who preferred Firefox would approach 25 as the number of trials increased, *if the proportions given in Table 11.1 are correct.* Similarly, since 25% of the entire population of Internet users prefer Firefox, we would *expect* about 25% of any given sample of 100 Internet users to prefer Firefox, since the sample is a subset of the population. This of course begs the question: are the proportions in Table 11.1 still true? That is the type of question we will learn how to address here in Section 11.1.

2 WHAT IS A χ^2 GOODNESS OF FIT TEST?

Do the 2011 market shares still hold true today? In other words, has the distribution of the multinomial random variable *browser* given in Table 11.1 changed since May 2011? To determine this, we introduce a new type of hypothesis test, called a χ^2 **goodness of fit test.**

> **χ^2 Goodness of Fit Test**
>
> A χ^2 **goodness of fit test** is a hypothesis test used to determine whether a random variable follows a particular distribution. In a goodness of fit test, the hypotheses are
>
> H_0: The random variable follows a particular distribution.
> H_a: The random variable does not follow the distribution specified in H_0.

For Example 11.1, the null hypothesis completely specifies each of the probabilities in the relative frequency distribution, as follows:

$$H_0 : p_{\text{MS IE}} = 0.55, p_{\text{Firefox}} = 0.25, p_{\text{Other}} = 0.20$$

The alternative hypothesis simply denies the claim made by the null hypothesis:

$$H_a : \text{The random variable does not follow the distribution specified in } H_0.$$

In other words, H_a claims that the browser market shares have changed since May 2011.

Developing Your Statistical Sense

Fitting the Model to the Data

Now, a goodness of fit test sounds like something you do in a clothing store dressing room. Actually, the analogy to clothes is rather appropriate. Suppose winter is coming and you are in the market for a new pair of gloves. You find one pair that is especially attractive, but the gloves don't fit your hands. What do you do? You reject the ill-fitting gloves and search for a new pair. In statistics, the gloves represent the models and your hands represent the actual "hard data" observed in the sample.

The null hypothesis H_0 represents what is called a *model*, a working theory of how the population proportions are distributed. Our working model of how the market shares are distributed is stated in the null hypothesis:

$$\text{Model 1. } H_0 : p_{\text{MS IE}} = 0.55, p_{\text{Firefox}} = 0.25, p_{\text{Other}} = 0.20$$

Of course, we could also try other models if we think the market has changed, such as the following:

$$\text{Model 2. } H_0 : p_{\text{MS IE}} = 0.60, p_{\text{Firefox}} = 0.25, p_{\text{Other}} = 0.15$$

$$\text{Model 3. } H_0 : p_{\text{MS IE}} = 0.50, p_{\text{Firefox}} = 0.30, p_{\text{Other}} = 0.20$$

In hypothesis testing, we "try on" only one model at a time.

In statistics, a goodness of fit test determines if the actual "hard data" observed in the sample are consistent with the proportions stated in the null hypothesis. Market researchers would collect data on the actual preferences of a sample of 100 real Internet users in order to determine whether or not the market shares have changed. The sample is summarized in a set of *observed frequencies* of Internet users who prefer the various browsers. The χ^2 goodness of fit test then *compares these observed frequencies with the expected frequencies* found in Example 11.1.

> **How a Goodness of Fit Test Works**
>
> The goodness of fit test is based on a comparison of the *observed frequencies* (sample data) with the *expected frequencies* when H_0 is true. That is, we compare what we actually see with what we would expect to see if H_0 were true. If the difference between the observed and expected frequencies is large, we reject H_0.

The difference between the observed and expected frequencies is measured by the test statistic, χ^2_{data}. As usual, it comes down to how large a difference is large.

> **Test Statistic for the χ^2 Goodness of Fit Test**
>
> For a multinomial random variable with k categories and n trials, let O_i represent the observed frequency for category i, and let E_i represent the expected frequency for category i. Then the **test statistic for a goodness of fit test**
>
> $$\chi^2_{\text{data}} = \sum \frac{(O_i - E_i)^2}{E_i}$$
>
> approximately follows a χ^2 distribution with $k - 1$ degrees of freedom, if the following conditions are satisfied:
>
> **a.** None of the expected frequencies is less than 1.
>
> **b.** At most 20% of the expected frequencies are less than 5.

If the conditions are not satisfied, then it may be possible to combine two or more categories so that the conditions may then be fulfilled.

3 PERFORMING THE χ^2 GOODNESS OF FIT TEST

The χ^2 goodness of fit test may be performed using (a) the critical-value method or (b) the *p*-value method.

> **χ^2 Goodness of Fit Test: Critical-Value Method**
>
> **Step 1 State the hypotheses and check the conditions.**
>
> - The null hypothesis states that the multinomial random variable follows a particular distribution.
> - The alternative hypothesis states that the random variable does not follow that distribution.
>
> The following conditions must be met:
>
> **a.** None of the expected frequencies is less than 1.
>
> **b.** At most 20% of the expected frequencies are less than 5.
>
> The expected frequency for the ith category is $E_i = n \cdot p_i$ where n represents the number of trials and p_i represents the population proportion for the ith category.
>
> **Step 2 Find the χ^2 critical value χ^2_{crit} and state the rejection rule.** Use Table E in the Appendix. Reject H_0 if $\chi^2_{\text{data}} \geq \chi^2_{\text{crit}}$. (It is always a right-tailed test.)
>
> **Step 3 Calculate χ^2_{data}.**
>
> $$\chi^2_{\text{data}} = \sum \frac{(O_i - E_i)^2}{E_i}$$
>
> where O_i = observed frequency, and E_i = expected frequency.
>
> **Step 4 State the conclusion and the interpretation.** Compare χ^2_{data} with χ^2_{crit}.

Students may wish to review the characteristics of the χ^2 distribution (page 393) and the procedure for finding χ^2 critical values for a right-tailed test (page 464).

EXAMPLE 11.2

CRITICAL-VALUE METHOD FOR THE χ^2 GOODNESS OF FIT TEST

Test whether the Internet browser market shares have changed since May 2011, using the observed frequencies of browser preference from a survey of 100 Internet users in Table 11.3, and level of significance $\alpha = 0.05$.

Table 11.3 Observed frequencies of browser preference in sample of 100 Internet users

Browser	Observed frequency
Microsoft Internet Explorer	55
Firefox	35
Other	10

Solution

STEP 1 **State the hypotheses and check the conditions.**
The hypotheses are:

$$H_0 : p_{\text{MS IE}} = 0.55, \, p_{\text{Firefox}} = 0.25, \, p_{\text{Other}} = 0.20$$
$$H_a : \text{The random variable does not follow the distribution specified in } H_0.$$

Checking the conditions, the expected frequencies from Table 11.2 are

$$E_{\text{MS IE}} = 55 \quad E_{\text{Firefox}} = 25 \quad E_{\text{Other}} = 20$$

Since none of these expected frequencies is less than 1, and none of the expected frequencies is less than 5, the conditions for performing the goodness of fit test are satisfied.

All hypothesis tests in this chapter are right-tailed tests, so that we need to find χ^2_{crit} for the area to the right of the critical value only.

STEP 2 **Find the χ^2 critical value χ^2_{crit} and state the rejection rule.**
We have degrees of freedom $k - 1 = 3 - 1 = 2$ and $\alpha = 0.05$. Turning to the χ^2 table (Table E in the Appendix) in the column labeled $\chi^2_{0.05}$ and the row containing df $= 2$, we find $\chi^2_{\text{crit}} = \chi^2_{0.05} = 5.991$, as shown in Figure 11.1. The rejection rule is "Reject H_0 if $\chi^2_{\text{data}} \geq 5.991$."

Chi-Square (χ^2) Distribution

Area to the Right of Critical Value

Degrees of freedom	0.995	0.99	0.975	0.95	0.90	0.10	0.05	0.025
1	—	—	0.001	0.004	0.016	2.706	3.841	5.024
2	0.010	0.020	0.051	0.103	0.211	4.605	5.991	7.378
3	0.072	0.115	0.216	0.352	0.584	6.251	7.815	9.348

FIGURE 11.1 Finding the χ^2 critical value for df $= k - 1 = 2$ and level of significance $\alpha = 0.05$.

STEP 3 **Find the test statistic χ^2_{data}.**
The observed frequencies O_i are found in Table 11.3 and the expected frequencies are given in Table 11.2. Then

$$\chi^2_{\text{data}} = \sum \frac{(O_i - E_i)^2}{E_i} = \frac{(55 - 55)^2}{55} + \frac{(35 - 25)^2}{25} + \frac{(10 - 20)^2}{20} = 0 + 4 + 5 = 9$$

Table 11.4 gives the quantities needed to calculate χ^2_{data}.

Table 11.4 Calculating χ^2_{data}

Category	p_i	O_i	E_i	$O_i - E_i$	$(O_i - E_i)^2$	$\dfrac{(O_i - E_i)^2}{E_i}$
MS IE	0.55	55	55	0	0	$\dfrac{(55 - 55)^2}{55} = 0$
Firefox	0.25	35	25	10	100	$\dfrac{(35 - 25)^2}{25} = 4$
Other	0.20	10	20	-10	100	$\dfrac{(10 - 20)^2}{20} = 5$

STEP 4 **State the conclusion and the interpretation.**

Compare χ^2_{data} with χ^2_{crit}. $\chi^2_{data} = 9$ is greater than $\chi^2_{crit} = 5.991$, as shown in Figure 11.2. Therefore, we reject H_0.

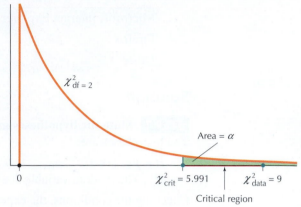

FIGURE 11.2 Reject H_0 when $\chi^2_{data} \geq \chi^2_{crit}$.

Now You Can Do
Exercises 19–22.

There is evidence that the random variable *browser* does not follow the distribution specified in H_0. In other words, there is evidence that the market shares for Internet browsers have changed.

Developing Your Statistical Sense

Be Careful How You Interpret the Conclusion

Note carefully what this conclusion says and what it doesn't say. The χ^2 goodness of fit test shows that there is evidence that the random variable does not follow the distribution specified in H_0. In particular, the conclusion does *not* state, for example, that Firefox's proportion is significantly greater. Informally, we can compare the observed frequency of 35 with the expected frequency of 25 for the Firefox browser and note that there appears to be evidence of an increase in market share for Firefox. But this is only informal and is not part of the hypothesis test. It is a common error in statistical analysis to form conclusions beyond what the hypothesis test is actually testing.

Next we turn to the *p*-value method. Since the χ^2 goodness of fit test is a right-tailed test, the *p*-value for the χ^2 statistic is defined as the area under the χ^2 curve to the right of the test statistic χ^2_{data}, as shown in Figure 11.3. That is,

$$p\text{-value} = P(\chi^2 > \chi^2_{data})$$

We can use technology to find the exact *p*-value for a particular value of χ^2_{data}. Or, alternatively, the *p*-value may be estimated using the χ^2 table.

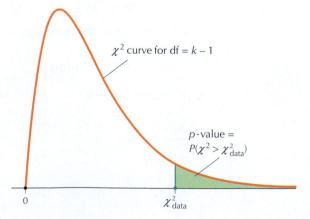

FIGURE 11.3
p-Value $= P(\chi^2 > \chi^2_{data})$.

χ^2 Goodness of Fit Test: *p*-Value Method

Step 1 **State the hypotheses and the rejection rule. Check the conditions.**
- The null hypothesis states that the multinomial random variable follows a particular distribution.
- The alternative hypothesis states that the random variable does not follow that distribution.
- Reject H_0 if the *p*-value $\leq \alpha$.

The following conditions must be met:
a. None of the expected frequencies is less than 1.
b. At most 20% of the expected frequencies are less than 5.

The expected frequency for the *i*th category is $E_i = n \cdot p_i$ where *n* represents the number of trials and p_i represents the population proportion for the *i*th category.

Step 2 **Calculate χ^2_{data}.**

$$\chi^2_{\text{data}} = \sum \frac{(O_i - E_i)^2}{E_i}$$

where O_i = observed frequency, and E_i = expected frequency.

Step 3 **Find the *p*-value.**

$$p\text{-value} = P(\chi^2 > \chi^2_{\text{data}}) \text{ (see Figure 11.3)}$$

Step 4 **State the conclusion and the interpretation.** Compare the *p*-value with α.

EXAMPLE 11.3

p-VALUE METHOD FOR THE χ^2 GOODNESS OF FIT TEST USING TECHNOLOGY

Table 11.5 2006 broadband adoption survey

Cable modem	DSL	Wireless/ Other
41%	50%	9%

Table 11.6 2009 broadband adoption survey

Cable modem	DSL	Wireless/ Other
410	330	260

The Pew Internet and American Life Project released the report *Home Broadband Adoption 2009*, which updated figures on the market share of cable modem, DSL, and wireless broadband from a 2006 survey (Table 11.5). The 2009 survey (Table 11.6) was based on a random sample of 1000 home broadband users. Test whether the population proportions have changed since 2006, using the *p*-value method, and level of significance $\alpha = 0.05$.

Solution

STEP 1 **State the hypotheses and the rejection rule. Check the conditions.**

$H_0 : p_{\text{Cable}} = 0.41, p_{\text{DSL}} = 0.50, p_{\text{Wireless/Other}} = 0.09$
$H_a :$ The random variable does not follow the distribution specified in H_0.

Reject H_0 if the *p*-value ≤ 0.05.

First we need to find the expected frequencies. We have $n = 1000$, so the expected frequencies are as shown here.

Expected frequencies for broadband access preference in sample of size $n = 1000$

Category	Expected frequency$_i = E_i = n \cdot p_i$
Cable	$E_{\text{Cable}} = 1000 \cdot 0.41 = 410$
DSL	$E_{\text{DSL}} = 1000 \cdot 0.50 = 500$
Wireless/Other	$E_{\text{Wireless/Other}} = 1000 \cdot 0.09 = 90$

Before we do the formal hypothesis test, let's try to figure out what the conclusion might be. Figure 11.4 is a clustered bar graph (see Section 2.1) of the observed and expected frequencies for each of the three categories. If H_0 were true, then, for each category, we would expect the green bars (observed frequencies) and yellow bars (expected frequencies) to have somewhat similar heights.

Note that the observed frequency for DSL is much lower than the expected frequency, while the observed frequency for wireless/other is much higher than the expected frequency. These both indicate evidence against the null hypothesis. Thus, we might expect to reject H_0.

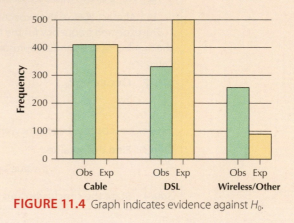

FIGURE 11.4 Graph indicates evidence against H_0.

Next check the requirements for this test. Since (a) none of the expected frequencies is less than 1 and (b) no more than 20% of the expected frequencies are less than 5, we may proceed. We use the instructions provided in the Step-by-Step Technology Guide at the end of this section.

STEP 2 Find the test statistic χ^2_{data}.
The TI-83/84 results in Figure 11.5 tell us that $\chi^2_{data} = 378.9111111 \approx 378.91$.

STEP 3 Find the *p*-value.
Figure 11.5 also tells us that

$$p\text{-value} = P(\chi^2 > 378.9111111) \approx 5.25409183E\text{-}83 \approx 0$$

Figure 11.6 illustrates why the *p*-value is so small. There is essentially no area to the right of $\chi^2_{data} = 378.91$ in the $\chi^2_{df=2}$ distribution.

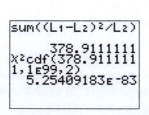

FIGURE 11.5 χ^2 test on TI-83/84.

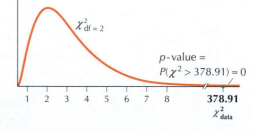

FIGURE 11.6 $\chi^2_{data} = 378.91$ is extreme.

STEP 4 State the conclusion and the interpretation.
Since the *p*-value is less than $\alpha = 0.05$, we reject H_0, which we expected. There is evidence at a level of significance $\alpha = 0.05$ that the proportions of broadband type in 2009 have changed since 2006.

Now You Can Do
Exercises 23–26.

EXAMPLE 11.4 **ESTIMATED *p*-VALUE METHOD FOR THE χ^2 GOODNESS OF FIT TEST**

Estimate the *p*-value from Example 11.3.

Solution

First find the row in the χ^2 table (Table E in the Appendix) for degrees of freedom $k - 1 = 3 - 1 = 2$. Then find where the value of χ^2_{data} would lie in relationship to the other χ^2 values in that row. Here, $\chi^2_{data} \approx 378.91$ is much greater than the largest value in that row, so the *p*-value must be much smaller than the area 0.005 associated with $\chi^2 = 10.597$ (Figure 11.7).

Degrees of freedom	Area to the Right of Critical Value			
	0.025	**0.01**	**0.005**	
1	5.024	6.635	7.879	
2	7.378	9.210	10.597	**378.91**

FIGURE 11.7 Estimating the *p*-value for the χ^2 goodness of fit test.

STEP-BY-STEP TECHNOLOGY GUIDE: The χ^2 Goodness of Fit Test

We illustrate the use of technology, once the observed and expected frequencies are known, for Example 11.3 (page 537).

TI-84

Step 1 Enter observed frequencies in list **L1** and expected frequencies in list **L2**.
Step 2 Press **STAT**, highlight **TESTS**, select **D: χ^2 GOF-Test**, and press **ENTER** (Figure 11.8).

Step 3 Highlight **df**, and enter degrees of freedom **2** (Figure 11.9).
Step 4 Highlight **Calculate** and press **ENTER**. The results are shown in Figure 11.10, including χ^2_{data} and the *p*-value.

FIGURE 11.8

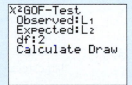

FIGURE 11.9

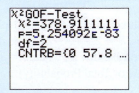

FIGURE 11.10

TI-83/84

To find χ^2_{data}:
Step 1 Enter observed frequencies in list **L1** and expected frequencies in list **L2**. Press **2nd QUIT**.
Step 2 Press **2nd LIST**, highlight **MATH**, select **5: sum(**, and press **ENTER** (Figure 11.11).

FIGURE 11.11

Step 3 Type the following: (L1–L2)²/L2 (see Figure 11.5 in Example 11.3) and press **ENTER**.
Step 4 The TI-83/84 then displays 378.9111111 as χ^2_{data} (see Figure 11.5, in Example 11.3).

To find the *p*-value:
Step 1 Select **2nd DISTR**, then χ^2 **cdf(**, and press **ENTER**.
Step 2 To get the *p*-value, that is, the area to the right of 378.9111111, enter **305.7526652, comma, 1ε99, comma, 2)**, as shown in Figure 11.5, in Example 11.3.

EXCEL

To find χ^2_{data}:
Step 1 Enter the observed and expected frequencies in rows 1 and 2 (Figure 11.12).
Step 2 In cell B3, enter: =(B1-B2)^2/B2 (Figure 11.12).

		B3			f_x =(B1-B2)^2/B2
	A	B	C	D	E
1	Observed	410	330	260	
2	Expected	410	500	90	
3		0			

FIGURE 11.12

Step 3 Copy the contents of cell B3 to cells C3 and D3.
Step 4 Select an empty cell, enter =SUM(B3:D3), and press **ENTER**. Excel then displays the value $\chi^2_{data} = 378.9111111$ (Figure 11.13).

		E3			f_x =SUM(B3:D3)
	A	B	C	D	E
1	Observed	410	330	260	
2	Expected	410	500	90	
3		0	57.8	321.111111	378.911111

FIGURE 11.13

To find the p-value:
Step 1 Select a cell and enter =**CHITEST(B1:D1,B2:D2)** and press **ENTER**.
Step 2 Excel then provides the p-value (Figure 11.14).

		A5			f_x =CHITEST(B1:D1,B2:D2)
	A	B	C	D	E
1	Observed	410	330	260	
2	Expected	410	500	90	
3		0	57.8	321.111111	378.911111
4					
5	5.25409E-83				

FIGURE 11.14

MINITAB

To find χ^2_{data}:
Step 1 Enter the observed frequencies (O) into **C1** and the hypothesized proportions into **C2**.
Step 2 Click **Stat > Tables > Chi-Square Goodness of Fit Test** (one variable).

Step 3 For observed counts, enter **C1**.
Step 4 Click **Specific Proportions** and enter **C2** in box.
Step 5 Click **OK**.

SECTION 11.1 Summary

1. A distribution is multinomial if (a) each independent trial has k possible outcomes, $k = 2,3,4,\cdots$; (b) the ith outcome (category) occurs with probability p_i, where $i = 1,2,\cdots,k$; and (c) $\sum_{i=1}^{k} p_i = 1$ (Law of Total Probability).

2. A goodness of fit test is a hypothesis test used to ascertain whether a random variable follows a particular distribution. In a goodness of fit test, the hypotheses are

H_0 : The random variable follows a particular distribution.
H_a : The random variable does not follow the distribution specified in H_0.

Compare the observed frequencies (actual data from the field) with the expected frequencies when H_0 is true. If the difference between the observed and expected frequencies is large, reject H_0.

3. The χ^2 goodness of fit test is performed using (a) the critical-value method or (b) the p-value method.

SECTION 11.1 Exercises

Clarifying the Concepts

1. What are the conditions required for a random variable to be multinomial?

2. Explain in your own words what is meant by a goodness of fit test.

3. Explain the meaning of the term *expected frequency*. (*Hint:* Use the idea of the long-run mean in your answer.)

4. State the hypotheses for a χ^2 goodness of fit test.

Practicing the Techniques

For Exercises 5–8, determine whether the distribution is multinomial.

5. A random sample of 12 residents is drawn from the town discussed on page 531 and their political party is ovserved.

6. We select 5 students from a group of 25 statistics students at random and without replacement, and we observe the student's class: freshman, sophomore, junior, or senior.

7. We choose 10 stocks at random and with replacement, and we observe the exchange that the stock is traded on: either the New York Stock Exchange, NASDAQ, London Stock Exchange, other Shenzhen Stock Exchange.

8. We pick 10 stocks at random and with replacement, and we observe the amount that the stock price increased or decreased since the last trading day.

For Exercises 9–12, the alternative hypothesis takes the form

H_a : The random variable does not follow the distribution specified in H_0.

  a. Find the expected frequencies.
  b. Determine whether the conditions for performing the χ^2 goodness of fit test are met.

9. $H_0 : p_1 = 0.50, p_2 = 0.25, p_3 = 0.25; n = 100$

10. $H_0 : p_1 = 0.2, p_2 = 0.3, p_3 = 0.4, p_4 = 0.1; n = 20$

11. $H_0 : p_1 = 0.9, p_2 = 0.05, p_3 = 0.04, p_4 = 0.01; n = 50$

12. $H_0 : p_1 = 0.4, p_2 = 0.35, p_3 = 0.10, p_4 = 0.10, p_5 = 0.05;$
$n = 200$

For Exercises 13–18, calculate the value of χ^2_{data}.

13.

O_i	E_i
10	12
12	12
14	12

14.

O_i	E_i
15	10
20	25
25	25

15.

O_i	E_i
20	25
30	25
40	30
40	50

16.

O_i	E_i
8	6
10	8
7	9
5	7

17.

O_i	E_i
1	6
10	6
8	6
0	6
11	6

18.

O_i	E_i
90	100
100	110
100	90
100	80
110	120

For Exercises 19–22, do the following.
  a. Calculate the expected frequencies and verify that the conditions for performing the χ^2 goodness of fit test are met.
  b. Find χ^2_{crit} for the χ^2 distribution with the given degrees of freedom. State the rejection rule.
  c. Calculate χ^2_{data}.
  d. Compare χ^2_{data} with χ^2_{crit}. State the conclusion and the interpretation.

19. $H_0 : p_1 = 0.4, p_2 = 0.3, p_3 = 0.3; O_1 = 50, O_2 = 25,$
$O_3 = 25$; level of significance $\alpha = 0.05$

20. $H_0 : p_1 = 1/3, p_2 = 1/3, p_3 = 1/3; O_1 = 40, O_2 = 30,$
$O_3 = 20$; level of significance $\alpha = 0.01$

21. $H_0 : p_1 = 0.4, p_2 = 0.35, p_3 = 0.10, p_4 = 0.10, p_5 = 0.05;$

$O_1 = 90, O_2 = 75, O_3 = 15, O_4 = 15, O_5 = 5$; level of significance $\alpha = 0.10$

22. $H_0 : p_1 = 0.3, p_2 = 0.2, p_3 = 0.2, p_4 = 0.2, p_5 = 0.1;$
$O_1 = 63, O_2 = 42, O_3 = 40, O_4 = 38, O_5 = 17$; level of significance $\alpha = 0.05$

For Exercises 23–26, do the following.
  a. State the rejection rule for the p-value method, calculate the expected frequencies, and verify that the conditions for performing the χ^2 goodness of fit test are met.
  b. Calculate χ^2_{data}.
  c. Find the p-value.
  d. Compare the p-value with level of significance α. State the conclusion and the interpretation.

23. $H_0 : p_1 = 0.50, p_2 = 0.50; O_1 = 40, O_2 = 60$; level of significance $\alpha = 0.05$

24. $H_0 : p_1 = 0.50, p_2 = 0.25, p_3 = 0.25; O_1 = 52, O_2 = 23,$
$O_3 = 25$; level of significance $\alpha = 0.10$

25. $H_0 : p_1 = 0.5, p_2 = 0.25, p_3 = 0.15, p_4 = 0.1;$
$O_1 = 90, O_2 = 55, O_3 = 40, O_4 = 15$; level of significance $\alpha = 0.10$

26. $H_0 : p_1 = 0.4, p_2 = 0.2, p_3 = 0.2, p_4 = 0.1, p_5 = 0.1;$
$O_1 = 90, O_2 = 45, O_3 = 40, O_4 = 15, O_5 = 10$; level of significance $\alpha = 0.05$

Applying the Concepts

27. Adult Education. The National Center for Education Statistics reported on the percentages of adults who enrolled in personal-interest courses, by the highest education level completed.[2] Of these, 8% had less than a high school diploma, 23% had a high school diploma, 32% had some college, 24% had a bachelor's degree, and 13% had a graduate or professional degree. A survey taken of 200 randomly selected adults who enrolled in personal-interest courses showed the following numbers for the highest education level completed. Test whether the distribution of education levels has changed, using level of significance $\alpha = 0.05$.

Less than high school	High school diploma	Some college	Bachelor's degree	Graduate or professional degree
12	40	62	54	32

28. Mall Restaurants. Based on monthly sales data, the International Council of Shopping Centers reported that the proportions of meals eaten at food establishments in shopping malls were as follows: fast food, 30%; food court, 46%; and restaurants, 24%. A survey of 100 randomly selected meals eaten at malls showed that 32 were eaten at fast-food places, 49 were eaten at food courts, and the rest were eaten at restaurants. Test whether the population proportions have changed, using level of significance $\alpha = 0.10$.

29. Spinal Cord Injuries. A study found that, of the minority patients who suffered spinal cord injury, 30% had a private health insurance provider, 55.6% used Medicare or Medicaid, and 14.4% had other arrangements.[3] Suppose that a sample of 1000 randomly selected minority patients with spinal cord injuries found that 350 had a private health insurance provider, 500 used Medicare or Medicaid, and 150 had other arrangements. Test whether the proportions have changed, using level of significance $\alpha = 0.05$.

30. The College Experience. A 2007 *New York Times* poll of Americans with at least a four-year college degree asked them how they would rate their overall experience as an undergraduate student. The results were 54% excellent, 39% good, 6% only fair, and 1% poor. A survey held this year of 500 randomly selected Americans with at least a four-year college degree found 275 rated their overall experience as an undergraduate student as excellent, 200 as good, 20 as only fair, and 5 as poor. Test whether the proportions have changed since 2007, using level of significance $\alpha = 0.05$.

31. University Dining. The university dining service believes there is no difference in student preference among the following four entrees: pizza, cheeseburgers, quiche, and sushi. A sample of 500 students showed that 250 preferred pizza, 215 preferred cheeseburgers, 30 preferred quiche, and 5 preferred sushi. Test at level of significance $\alpha = 0.01$ whether or not there is a difference in student preference among the four entrees.
(*Hint:* For the χ^2 test of no difference among the proportions, the null hypothesis states that all proportions are equal.)

32. Weekly Religious Services. A 2007 *New York Times* poll found that 31% of Americans attend religious services every week, 12% almost every week, 14% once or twice a month, 24% a few times a year, and 19% never. A survey taken this year of 100 randomly selected Americans showed 32 who attend religious services every week, 10 almost every week, 15 once or twice a month, 25 a few times a year, and 18 never. Test whether the population proportions have changed since 2007, using level of significance $\alpha = 0.10$.

33. Community College Advising. In 2007, the Community College Survey of Student Engagement found that 50% of students had met with an adviser by the end of their first four weeks at college, while 41% did not do so and 9% did not recall. A survey this year of 1000 randomly selected community college students had the following results.

Met with adviser by the end of first four weeks at college	Yes	No	Do not recall
Frequency	550	370	80

Test whether the population proportions have changed since 2007, using level of significance $\alpha = 0.05$.

34. Believing in Angels. Do you believe in angels? A Gallup Poll found that 78% of respondents believed in angels, 12% were not sure or had no opinion, and 10% didn't believe in angels. Suppose that a new survey of 1000 randomly selected people had the following results.

Believe in angels?	Yes	No	Not sure or no opinion
Frequency	820	110	70

Test whether the population proportions have changed, using level of significance $\alpha = 0.05$.

35. Believing in Angels. Refer to the previous exercise. *What if* the number of people responding "No" was less then 110. How would that affect the following, and why? Would the following increase, decrease, stay the same, or is there insufficient information to determine?
   a. χ^2_{data}
   b. *p*-Value
   c. Conclusion

11.2

χ^2 TESTS FOR INDEPENDENCE AND FOR HOMOGENEITY OF PROPORTIONS

OBJECTIVES   By the end of this section, I will be able to . . .

1   Explain what a χ^2 test for the independence of two variables is.

2   Perform and interpret a χ^2 test for the independence of two variables using the critical-value method and the p-value method.

3   Perform and interpret a test for the homogeneity of proportions.

1   INTRODUCTION TO THE χ^2 TEST FOR INDEPENDENCE

In Section 11.1, we learned that the χ^2 distribution could help us determine a model's goodness of fit to the data. Here, in Section 11.2, we will learn two more hypothesis tests that use the χ^2 distribution. Recall from Section 2.1 that a *contingency table,* also known as a *crosstabulation* or a *two-way table,* is a tabular summary of the relationship between two categorical variables. The categories of one variable label the rows, and the categories of the other variable label the columns. Each cell in the table contains the number of observations that fit the categories of that row and column. Table 11.7 is a contingency table based on the study *How Young People View Their Lives, Futures, and Politics: A Portrait of "Generation Next."*[4] The researchers asked 1500 randomly selected respondents, "How are things in your life?" Subjects were categorized by age and response. The researchers identified those aged 18–25 in 2007 as representing "Generation Next."

> The term *contingency table* derives from the fact that the table covers all possible combinations of the values for the two variables, that is, all possible contingencies.

Table 11.7 Contingency table showing relative frequencies of variable categories

Response	Age Group			Relative frequency
	Gen Nexter (18–25)	26+	Total	
Very happy	180	330	**510**	$\frac{510}{1500} = 0.34$
Pretty happy	378	435	**813**	$\frac{813}{1500} = 0.542$
Not too happy	42	135	**177**	$\frac{177}{1500} = 0.118$
Total	**600**	**900**	**1500**	
Relative frequency	$\frac{600}{1500} = 0.4$	$\frac{900}{1500} = 0.6$		

We can use contingency tables like Table 11.7 to determine whether two random variables are independent. Recall that two random variables are *independent* if the value of one variable does not affect the probabilities of the values of the other variable. For example, is a "Gen Nexter" (someone aged 18–25 in 2007) less likely to report that he or she is "very happy" and more likely to report that he or she is "pretty happy" than someone older? If so, then the response depends on age, so the variables *age group* and *response* are dependent.

> By "dependent" we simply mean that the variables are not independent.

To determine whether two categorical variables are independent, using the data in a contingency table, we use a χ^2 test for independence. Just like our χ^2 goodness of fit test from Section 11.1, the χ^2 **test for independence** is based on a comparison of the observed frequencies with the frequencies that are expected if the null hypothesis is assumed true.

χ^2 Test for Independence

To determine whether two categorical variables are independent, using the data from a contingency table, we use a χ^2 **test for independence.** The hypotheses take the form

H_0 : Variable A and Variable B are independent.
H_a : Variable A and Variable B are dependent.

We compare the observed frequencies with the frequencies that we expect if we assume that H_0 is correct. Large differences lead to the rejection of the null hypothesis.

Here, we are testing whether the variables *age group* and *response* are independent. Thus, the hypotheses are

H_0 : *Age group* and *response* are independent.
H_a : *Age group* and *response* are dependent.

H_0 states that a response to the survey question does not depend on the age group. H_a says that a response does depend on the age group. To calculate the expected frequencies, we begin by recalling the Multiplication Rule for Two Independent Events from Chapter 5 (page 222):

If A and B are any two independent events, $P(A \text{ and } B) = P(A)\,P(B)$.

To illustrate, let our events be defined as $A = 18$–25 age group, and $B =$ reported "very happy." Then, on the assumption that these events are independent, we have

$$P(\text{Gen Nexter and very happy}) = P(A \text{ and } B) = P(A)P(B) = \frac{600}{1500} \cdot \frac{510}{1500}$$
$$= 0.4 \cdot 0.34 = 0.136$$

Thus, the probability that a randomly chosen young person is both a Gen Nexter and is very happy is 0.136. Then, to find the expected frequency of this cell (Gen Nexters who are very happy), we multiply this probability 0.136 by the total sample size $n = 1500$, using the result from Section 11.1 that the expected frequency is

$$E = \text{expected frequency} = n \cdot p = 1500 \cdot 0.136 = 204$$

In other words, if the random variables *age group* and *response* are independent, then the expected frequency of Gen Nexters who report being very happy is

$$\text{expected frequency}_{\text{Gen Nexter and very happy}} = 1500 \cdot \frac{600}{1500} \cdot \frac{510}{1500} = 204$$

But note that two of the 1500s cancel, providing us with the shortcut

$$\text{expected frequency}_{\text{Gen Nexter and very happy}} = \frac{(600)(510)}{1500} = 204$$

Generalizing, this provides us with the following shortcut method for finding expected frequencies.

> **Expected Frequencies for a χ^2 Test for Independence**
>
> The expected frequencies for the cells of a contingency table in a χ^2 test for independence are given by
>
> $$\text{expected frequency} = \frac{(\text{row total})(\text{column total})}{\text{grand total}}$$

EXAMPLE 11.5

CALCULATING EXPECTED FREQUENCIES USING THE SHORTCUT METHOD

Calculate the expected frequencies from Table 11.7 using the shortcut method.

Solution

Table 11.8 contains the expected frequencies calculated using the shortcut method.

Table 11.8 Expected frequencies using the shortcut method

Response	Age Group		Total
	Gen Nexter (18–25)	**26+**	**Total**
Very happy	$\dfrac{(510)(600)}{1500} = 204$	$\dfrac{(510)(900)}{1500} = 306$	**510**
Pretty happy	$\dfrac{(813)(600)}{1500} = 325.2$	$\dfrac{(813)(900)}{1500} = 487.8$	**813**
Not too happy	$\dfrac{(177)(600)}{1500} = 70.8$	$\dfrac{(177)(900)}{1500} = 106.2$	**177**
Total	**600**	**900**	**1500**

Now You Can Do Exercises 5–10.

The χ^2 test for independence measures the difference between the observed frequencies and the expected frequencies using the following test statistic.

> **Test Statistic for the χ^2 Test for Independence**
>
> Let O_i represent the observed frequency in the ith cell, and E_i represent the expected frequency in the ith cell. Then the **test statistic for the independence of two categorical variables**
>
> $$\chi^2_{\text{data}} = \sum \frac{(O_i - E_i)^2}{E_i}$$
>
> approximately follows a χ^2 (chi-square) distribution with $(r - 1)(c - 1)$ degrees of freedom, where r is the number of categories in the row variable and c is the number of categories in the column variable, if the following conditions are satisfied:
>
> **a.** None of the expected frequencies is less than 1.
>
> **b.** At most 20% of the expected frequencies are less than 5.

2 PERFORMING THE χ^2 TEST FOR INDEPENDENCE

The χ^2 test for independence may be performed using either the critical-value method or the p-value method. We provide examples of each.

Caution: Do not include the row or column totals when counting the number of categories.

χ^2 Test for Independence: Critical-Value Method

Step 1 State the hypotheses and check the conditions.
   H_0 : Variable *A* and Variable *B* are independent.
   H_a : Variable *A* and Variable *B* are dependent.

The following conditions must be met:

a. None of the expected frequencies is less than 1.

b. At most 20% of the expected frequencies are less than 5.
The expected frequency for a given cell is

$$\text{expected frequency} = \frac{(\text{row total}) \cdot (\text{column total})}{\text{grand total}}$$

Step 2 Find the critical value χ^2_{crit} and state the rejection rule. Reject H_0 if $\chi^2_{\text{data}} \geq \chi^2_{\text{crit}}$. Use $(r - 1)(c - 1)$ degrees of freedom, where r is the number of categories in the row variable and c is the number of categories in the column variable.

Step 3 Calculate χ^2_{data}.

$$\chi^2_{\text{data}} = \sum \frac{(O_i - E_i)_2}{E_i}$$

where O_i = observed frequency and E_i = expected frequency for each cell.

Step 4 State the conclusion and the interpretation. Compare χ^2_{data} with χ^2_{crit}.

EXAMPLE 11.6 PERFORMING THE χ^2 TEST FOR INDEPENDENCE USING THE CRITICAL-VALUE METHOD

Using Table 11.7, test whether *age group* is independent of *response,* using level of significance $\alpha = 0.05$.

Solution

STEP 1 **State the hypotheses and check the conditions.**

   H_0 : *Age group* and *response* are independent.
   H_a : *Age group* and *response* are dependent.

We note from Table 11.8 that none of the expected frequencies are less than either 1 or 5. Therefore, the conditions are met, and we may proceed with the hypothesis test.

STEP 2 **Find the critical value χ^2_{crit} and state the rejection rule.**
The row variable, *response,* has three categories, so $r = 3$. The column variable, *age group,* has two categories, so $c = 2$. Thus,

$$\text{degrees of freedom} = (r - 1)(c - 1) = (3 - 1)(2 - 1) = 2$$

See Figure 11.1 (page 535) to review how to find χ^2_{crit}.

With level of significance $\alpha = 0.05$, this gives us $\chi^2_{\text{crit}} = 5.991$ from the χ^2 table. The rejection rule is therefore

$$\text{Reject } H_0 \text{ if } \chi^2_{\text{data}} \geq 5.991$$

STEP 3 **Calculate χ^2_{data}.**
The observed frequencies are found in Table 11.7 and the expected frequencies are found in Table 11.8. Then

$$\chi^2_{\text{data}} = \sum \frac{(O_i - E_i)^2}{E_i} = \frac{(180 - 204)^2}{204} + \frac{(330 - 306)^2}{306} + \frac{(378 - 325.2)^2}{325.2}$$

$$+ \frac{(435 - 487.8)^2}{487.8} + \frac{(42 - 70.8)^2}{70.8} + \frac{(135 - 106.2)^2}{106.2}$$

$$\approx 38.5192$$

STEP 4 **State the conclusion and the interpretation.**
Our χ^2_{data} of 38.5192 is greater than our χ^2_{crit} of 5.991 (see Figure 11.15), and so we reject H_0. The interpretation is: "There is evidence at level of significance $\alpha = 0.05$ that *age group and response* are dependent."

FIGURE 11.15
$\chi^2_{\text{data}} = 38.5192$ lies in the critical region.

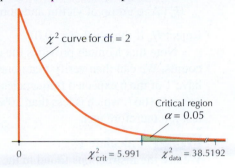

Now You Can Do
Exercises 11–14.

χ^2 Test for Independence: *p*-Value Method

Step 1 **State the hypotheses and the rejection rule. Check the conditions.**
H_0 : Variable A and Variable B are independent.
H_a : Variable A and Variable B are dependent.

Reject H_0 if the *p*-value $\leq \alpha$.
The following conditions must be met:

a. None of the expected frequencies is less than 1.

b. At most 20% of the expected frequencies are less than 5.

The expected frequency for a given cell is

$$\text{expected frequency} = \frac{(\text{row total})(\text{column total})}{\text{grand total}}$$

Step 2 **Calculate χ^2_{data}.**

$$\chi^2_{\text{data}} = \sum \frac{(O_i - E_i)^2}{E_i}$$

where O_i = observed frequency and E_i = expected frequency for each cell.

Step 3 **Find the *p*-value.**

$$p\text{-value} = P(\chi^2 > \chi^2_{\text{data}})$$

Step 4 **State the conclusion and the interpretation. Compare the *p*-value with α.**

EXAMPLE 11.7

χ^2 TEST FOR INDEPENDENCE USING THE *p*-VALUE METHOD AND TECHNOLOGY

homicideage

Table 11.9 contains the numbers of work-related homicides that took place in the United States in 2002, according to the Bureau of Labor Statistics, categorized by the age group of the victim and the type of homicide. Test whether homicide type and age group of victim are independent, using the TI-83/84, Minitab, the *p*-value method, and level of significance $\alpha = 0.01$.

Table 11.9 Contingency table of age group of victim versus type of homicide

Type of homicide	Age Group of Victim			Total
	Under 25	25 to 44	Over 44	
Shooting	31	258	180	469
Stabbing	5	21	37	63
Total	36	279	217	532

Solution

STEP 1 State the hypotheses and the rejection rule. Check the conditions.

H_0 : Age group of victim and homicide type are independent.
H_a : Age group of victim and homicide type are dependent.

Reject H_0 if the p-value ≤ 0.01.

Note that Minitab provides the expected counts (frequencies) below the observed counts. We can then verify that none of the expected frequencies is less than 1. We do have 1 of the 6 expected frequencies (4.26) with a value less than 5. But this represents $1/6 \approx 0.1667$, which is less than 20%, as required. The conditions for the χ^2 hypothesis test are therefore met.

STEP 2 Calculate χ^2_{data}.

We use the instructions found in the Step-by-Step Technology Guide at the end of this section. The TI-83/84 results in Figure 11.16 tell us that $\chi^2_{\text{data}} = 10.76001797$. The Minitab results in Figure 11.17 round this to "Chi-Sq" $= \chi^2_{\text{data}} = 10.760$.

```
Expected counts are printed below observed counts
Chi-Square contributions are printed below expected counts
                   Age 25
        Age < 25   - 44   Age > 44   Total
   1         31     258        180     469
          31.74  245.96     191.30
          0.017   0.589      0.668

   2          5      21         37      63
           4.26   33.04      25.70
          0.127   4.387      4.971

Total        36     279        217     532

Chi-Sq = 10.760, DF = 2, P-Value = 0.005
1 cells with expected counts less than 5.
```

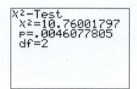

```
X²-Test
X²=10.76001797
P=.0046077805
df=2
```

FIGURE 11.16 TI-83/84 χ^2 results.

FIGURE 11.17 Minitab χ^2 results.

STEP 3 Find the p-value.

From the TI-83/84 results in Figure 11.16, we have

$$p\text{-value} = P(\chi^2 > \chi^2_{\text{data}}) = 0.0046077805$$

The Minitab results in Figure 11.17 round this to p-value $= 0.005$.

STEP 4 State the conclusion and the interpretation.

Since p-value $\approx 0.0046 < 0.01$, we reject H_0. There is evidence that the age group and homicide type are dependent.

Now You Can Do
Exercises 15–18.

3 TEST FOR THE HOMOGENEITY OF PROPORTIONS

Recall the two-sample Z test for $p_1 - p_2$ from Section 10.3, where we compared the proportions of two independent populations. When we extend that hypothesis test to k independent populations, we use a test statistic that follows a χ^2 distribution. Just as the null hypothesis for the two-sample test assumed no difference between the population proportions

- the null hypothesis for the k-sample test also assumes that all k proportions are equal, and

- the alternative hypothesis states that not all the population proportions are equal.

When performing the **test for the homogeneity of proportions,** we use the same steps as for the χ^2 test for independence.

Developing Your Statistical Sense

Difference Between χ^2 Test for Homogeneity and χ^2 Test for Independence

The difference between the test for homogeneity of proportions and the test for independence has to do with how the data are collected. If a single sample is taken and two variables are measured, then the test for independence is appropriate. If several (k) samples are taken and the sample proportion is measured for each sample, then the test for homogeneity of proportions is appropriate.

EXAMPLE 11.8

AIRLINE ON-TIME PERFORMANCE

Scott Olson/AFP/Getty Images

DATA FILE flyontime

The Bureau of Transportation Statistics (**www.bts.gov**) reports on the proportion of airline passenger flights that are on time, for each major airline. The January–April 2007 statistics for the three busiest carriers are shown in Table 11.10. Test whether the population proportions of on-time flights are the same for the three airlines, using the p-value method, Minitab, and level of significance $\alpha = 0.05$.

Table 11.10 Observed on-time statistics for three major airlines, January–April 2007

	Southwest	American	Skywest	Total
Number of on-time flights	146,607	68,939	60,298	275,844
Number of flights not on time	36,697	35,688	32,497	104,882
Total flights	183,304	104,627	92,795	380,726

What Results Might We Expect?

The observed sample proportions of on-time flights are as follows:

$$p_{\text{Southwest}} = \frac{146,607}{183,304} \approx 0.80 \quad p_{\text{American}} = \frac{68,939}{104,627} \approx 0.66 \quad p_{\text{Skywest}} = \frac{60,298}{92,795} \approx 0.65$$

The 80% on-time proportion of Southwest Airlines does seem to be somewhat higher than the on-time proportions of the other airlines. Thus, we would not be surprised if the hypothesis test found evidence that not all the population proportions were equal.

Solution

The Minitab results are shown here. We use the same steps as for the χ^2 test for independence.

```
Expected counts are printed below observed counts
Chi-Square contributions are printed below expected counts

         Southwest   American    Skywest    Total
   1       146607       68939      60298    275844
           132807.61   75804.46   67231.93
           1433.828    621.792    715.127

   2        36697       35688      32497    104882
           50496.39    28822.54   25563.07
           3771.027    1635.338   1880.815

Total      183304      104627     92795    380726

Chi-Sq = 10057.927, DF = 2, P-Value = 0.000
```

STEP 1   **State the hypotheses and the rejection rule. Check the conditions.**

$H_0 : p_{\text{Southwest}} = p_{\text{American}} = p_{\text{Skywest}}$
H_a : Not all the proportions in H_0 are equal.

Reject H_0 if the p-value ≤ 0.05.

None of the expected frequencies are less than either 1 or 5. Therefore, the conditions are met, and we may proceed with the hypothesis test.

STEP 2   **Find the test statistic χ^2_{data}.**
χ^2_{data} is shown as "Chi-Sq" = 10,057.927. There are $r = 2$ rows and $c = 3$ columns, so the degrees of freedom are $(r-1)(c-1) = (2-1)(3-1) = 2$.

STEP 3   **Find the p-value.**
Minitab provides the p-value, which is essentially 0.000.

STEP 4   **State the conclusion and the interpretation.**
The p-value of 0.000 is less than $\alpha = 0.05$. We therefore reject H_0, as expected. There is evidence at level of significance $\alpha = 0.05$ that not all population proportions of on-time flights are equal.

Now You Can Do Exercises 19–22.

CASE STUDY   **Online Dating**

Susan Wides/Getty Images

We look at two tests for independence in this Case Study. The first examines whether the type of relationship reported by respondents depends on the gender of the respondent. The second investigates whether the self-reported physical appearance of online daters depends on the person's gender.

Does the Reported Type of Relationship Depend on Gender?

The Pew Internet and American Life Project examined whether single men and women differed with respect to their current relationships. The observed frequencies are given in Table 11.11.

onlinedata

Table 11.11   Observed frequencies, online dating study

| | Gender | |
Type of relationship	Single men	Single women
In committed relationship	115	138
Not in committed relationship and not looking for partner	162	391
Not in committed relationship but looking for partner	89	54
Don't know/refused	19	18

We are interested in whether the type of relationship reported depends on the gender of the respondent. In other words, we will test whether the type of relationship is independent of gender. We will use the p-value method, with level of significance $\alpha = 0.05$, and we will follow the TI-83/84 instructions in the Step-by-Step Technology Guide on page 553 for the calculations.

What Results Might We Expect?

Table 11.11 and Figure 11.18 indicate that the proportion of men who are "looking" is greater than the proportion of women who are "looking." Similarly, the proportion of women who are "not looking" is greater than for men. This is evidence that the type of relationship depends on gender and that we might expect to reject the null hypothesis of independence.

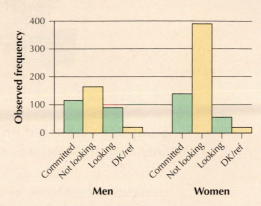

FIGURE 11.18 Graphical evidence indicates type of relationship depends on gender.

STEP 1 **State the hypotheses and the rejection rule. Check the conditions.**

H_0 : *Type of relationship* and *gender* are independent.
H_a : *Type of relationship* and *gender* are dependent.

Reject H_0 if the *p*-value ≤ 0.05.

Figure 11.19 shows the expected frequencies, none of which are less than 5. Thus, the conditions are met.

STEP 2 **Find χ^2_{data}.**
The TI-83/84 results in Figure 11.20 tell us

$$\chi^2_{data} = 61.12955651$$

STEP 3 **Find the *p*-value.**
Figure 11.20 also gives us the *p*-value:

$$p\text{-value} = 3.372011\text{E}{-}13 \approx 0.0000000000003372011$$

FIGURE 11.19 Expected frequencies.

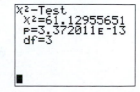

FIGURE 11.20 χ^2 results on TI-83/84.

STEP 4 **State the conclusion and the interpretation.**
Since the *p*-value $\le a = 0.05$, we reject H_0, as we expected. There is evidence that the type of relationship reported in the study depends on the gender of the respondent for level of significance $\alpha = 0.05$.

(continues)

Does Self-Reported Physical Appearance of Online Daters Depend on Gender?

A master's thesis from the Massachusetts Institute of Technology examined the characteristics and behavior of online daters.[5] Table 11.12 contains the self-reported physical appearance and gender of 52,817 users of an online dating service.

onlineappear

Table 11.12 Gender and self-reported physical appearance

	Physical Appearance				
	Very attractive	**Attractive**	**Average**	**Prefer not to answer**	**Total**
Female	3113	16,181	6093	3478	28,865
Male	1415	12,454	7274	2809	23,952
Total	4528	28,635	13,367	6287	52,817

Note from Table 11.12 that females seem to have higher proportions of those self-reporting as either attractive or very attractive, while males seem to have a higher proportion of those self-reporting as average. This is evidence that self-reported physical appearance does depend on gender and that we might expect to reject the null hypothesis of independence. We will test using the p-value method, with level of significance $\alpha = 0.01$, and Minitab. The hypotheses are

H_0 : *Self-reported physical appearance* and *gender* are independent.
H_a : *Self-reported physical appearance* and *gender* are dependent.

We reject H_0 if the p-value $\leq$ level of significance $\alpha = 0.01$.
The Minitab results in Figure 11.21 tell us

$$\chi^2_{data} = \text{"Chi-Sq"} = 847.702$$
$$p\text{-value} \approx 0$$

Figure 11.21 gives us the expected frequencies (highlighted in color), none of which are less than 5, allowing us to perform the hypothesis test. Since the p-value $\leq \alpha = 0.01$, we reject H_0, as we expected. There is evidence at level of significance $\alpha = 0.01$ that the self-reported physical appearance depends on the gender of the online dater.

```
Expected counts are printed below observed counts
Chi-Square contributions are printed below expected counts

              VA        Att       Ave       PNTA   Total
    F        3113      16181      6093       3478   28865
             2474.60   15649.30  7305.19    3435.91
             164.698   18.065    201.147    0.516

    M        1415      12454      7274       2809   23952
             2053.40   12985.70  6061.81    2851.09
             198.480   21.770    242.406    0.621
Total        4528      28635     13367      6287    52817

Chi-Sq = 847.702, DF = 3, P-Value = 0.000
```

FIGURE 11.21 Minitab results showing expected frequencies, χ^2_{data} and the *p*-value. ■

STEP-BY-STEP TECHNOLOGY GUIDE: Test for Independence or Test for the Homogeneity of Proportions

We demonstrate using Example 11.7 (page 547).

TI-83/84

Entering Matrix Data

Step 1   Press **2nd**, then **MATRIX**.

Step 2   Highlight **EDIT**, and press **ENTER**.

Step 3   Set the dimensions of **MATRIX[A]** (number of rows × number of columns). Table 11.9 has 2 rows and 3 columns, so enter **2**, press **ENTER**, enter **3**, and press **ENTER**.

Step 4   Press the down-arrow key. Enter the first number in the first cell, **31**, and press **ENTER**.

Step 5   Continue entering the data row by row until the matrix is complete (Figure 11.22).

```
MATRIX[A]  2 ×3
[ 31    258    180   ]
[ 5     21     37    ]
```

FIGURE 11.22

χ^2 Test for Independence or Test for Homogeneity of Proportions

Step 1   Enter the data into **Matrix[A]**.

Step 2   Press **STAT**, highlight **TESTS**, select **C: χ^2 Test**, and press **ENTER**.

Step 3   The expected frequencies are automatically generated and put into **MATRIX[B]**. Highlight **Calculate**, and press **ENTER**. The results are shown in Figure 11.16 in Example 11.7.

Step 4   To view the expected frequencies, press **2nd MATRIX**, highlight **EDIT**, choose **2** for **MATRIX[B]**, and press **ENTER**.

EXCEL

χ^2 Test for Independence or Test for Homogeneity of Proportions Using the WHFStat Macros

Step 1   Enter the data from Table 11.9, *including row and column totals,* in cells A1 to D3.

Step 2   Load the **WHFStat Macros**.

Step 3   Select **Add-Ins > Macros > Tables > Two Way Tables/Chi Squared Test**.

Step 4   Select cells A1 to D3 as the **Dataset Range**.

Step 5   Select **Chi-squared Test**, and click **OK**.

MINITAB

χ^2 Test for Independence or Test for Homogeneity of Proportions

Step 1   Enter the observed frequencies from Table 11.9 into the Minitab worksheet, as shown here.

Step 2   Click **Stat > Tables > Chi-Square Test**.

Step 3   Choose each of columns C1, C2, and C3 as the **Columns containing the table**. Then click **OK**. The results are shown in Figure 11.17 in Example 11.7.

C1	C2	C3
Age < 25	Age 25 - 44	Age > 44
31	258	180
5	21	37

CRUNCHIT!

Test for Independence
We will use the data from Example 11.7.

Step 1 Click **File . . .** then highlight **Load from LaroseFundamantals2e . . . Chapter 11 . . .** and click on **Example 11.7.**

Step 2 Click **Statistics** and select **Contingency tables . . . with counts.** For **Row Variable** select **Response.** For **Column Variable** select **Age Group.** For **Counts** select **Count.** Then click **Calculate.**

SECTION 11.2 Summary

1. To determine whether two categorical variables are independent, using the data from a contingency table, we use a χ^2 test for independence. The hypotheses take the form

H_0 : Variable A and Variable B are independent.

H_a : Variable A and Variable B are dependent.

2. The χ^2 test for independence is performed using the critical-value method, the exact p-value method, or the estimated p-value method. The observed frequencies are compared with the expected frequencies on the assumption that H_0 is correct. Large differences lead to the rejection of the null hypothesis.

3. The k-sample test, called the test for the homogeneity of proportions, determines whether all k population proportions are equal. The result uses a test statistic that follows a χ^2 distribution. The null hypothesis for the k-sample test assumes that all k population proportions are equal. The alternative hypothesis states that not all the population proportions are equal. When performing the test for the homogeneity of proportions, the same steps are used as for the χ^2 test for independence.

SECTION 11.2 Exercises

Clarifying the Concepts

1. Explain what a contingency table is.

2. Explain in your own words what is meant by a test for independence.

3. What is the difference between the χ^2 test for homogeneity of proportions and the two-sample Z test for the difference in proportions from Chapter 10?

4. Explain how the expected frequencies are calculated without using the shortcut method.

Practicing the Techniques

For Exercises 5–10, the observed frequencies are provided in a contingency table of two categorical variables. Find the expected frequencies, on the assumption that the variables are independent.

5.

	A1	A2
B1	10	20
B2	12	18

6.

	C1	C2
D1	50	100
D2	60	90

7.

	E1	E2	E3
F1	30	20	10
F2	35	24	8

8.

	G1	G2
H1	10	8
H2	8	10
H3	9	9

9.

	I1	I2	I3
J1	100	90	105
J2	50	60	55
J3	25	15	20

10.

	K1	K2	K3	K4
L1	40	70	90	100
L2	20	40	60	70
L3	30	65	65	70

For Exercises 11–14, test whether or not the variables are independent.
  a. State the hypotheses.
  b. Verify that the conditions for performing the χ^2 test for independence are met.
  c. Find χ^2_{crit} and state the rejection rule.
  d. Calculate χ^2_{data}.
  e. Compare χ^2_{data} with χ^2_{crit}. State the conclusion and the interpretation.

11. Exercise 5, level of significance $\alpha = 0.05$

12. Exercise 7, level of significance $\alpha = 0.10$

13. Exercise 9, level of significance $\alpha = 0.01$

14. Exercise 9, level of significance $\alpha = 0.10$

For Exercises 15–18, test whether or not the variables are independent.

    a. State the hypotheses and the rejection rule for the p-value method, and verify that the conditions for performing the χ^2 test for independence are met.

    b. Find χ^2_{data}.

    c. Calculate the p-value.

    d. Compare the p-value with α. State the conclusion and the interpretation.

15. Exercise 6, level of significance $\alpha = 0.05$

16. Exercise 8, level of significance $\alpha = 0.10$

17. Exercise 10, level of significance $\alpha = 0.01$

18. Exercise 10, level of significance $\alpha = 0.10$

For Exercises 19–22, test whether or not the proportions of successes are the same for all populations.

    a. State the hypotheses.

    b. Calculate the expected frequencies and verify that the conditions for performing the χ^2 test for homogeneity of proportions are met.

    c. Find χ^2_{crit} and state the rejection rule. Use level of significance $\alpha = 0.05$.

    d. Find χ^2_{data}.

    e. Compare χ^2_{data} with χ^2_{crit}. State the conclusion and the interpretation.

19.

	Sample 1	Sample 2	Sample 3
Successes	10	20	30
Failures	20	45	62

20.

	Sample 1	Sample 2	Sample 3
Successes	50	50	100
Failures	200	210	425

21.

	Sample 1	Sample 2	Sample 3	Sample 4
Successes	10	15	20	25
Failures	15	24	32	40

22.

	Sample 1	Sample 2	Sample 3	Sample 4
Successes	100	150	200	250
Failures	150	240	320	400

For Exercises 23–26, test whether or not the proportions of successes are the same for all populations.

    a. State the rejection rule for the p-value method using level of significance $\alpha = 0.05$, calculate the expected frequencies, and verify that the conditions for performing the χ^2 test for homogeneity of proportions are met.

    b. Find χ^2_{data}.

    c. Calculate the p-value.

    d. Compare the p-value with α. State the conclusion and the interpretation.

23.

	Sample 1	Sample 2	Sample 3
Successes	30	60	90
Failures	10	25	50

24.

	Sample 1	Sample 2	Sample 3
Successes	100	120	140
Failures	20	25	30

25.

	Sample 1	Sample 2	Sample 3	Sample 4
Successes	10	12	24	32
Failures	6	10	15	30

26.

	Sample 1	Sample 2	Sample 3	Sample 4
Successes	100	200	300	400
Failures	30	70	150	300

Applying the Concepts

27. Conditioning Mice. A psychologist is conducting research using white mice, brown mice, a classical conditioning stimulus, and an operant conditioning stimulus. The psychologist is interested in whether type of stimulus is independent of the type of mouse. One hundred mice were tested. The following table shows the number of each type of mice that completed their assigned task satisfactorily, given the type of stimulus. Test at level of significance $\alpha = 0.10$ whether type of stimulus and type of mouse are independent. 🔴 micecond

	Type of Stimulus		
Type of mouse	**Classical**	**Operant**	**Total**
White	20	40	60
Brown	10	30	40
Total	30	70	100

28. Cable TV Content Restrictions. A *Chicago Tribune* Poll asked, "Should government restrict violence and sexual content that appears on cable TV, or should government not impose restrictions?" The responses were categorized by political affiliation. Test whether the population proportion favoring restriction is the same for all three groups, using level of significance $\alpha = 0.05$. 🔴 tvcontent

	Restrict	**Not restrict/don't know**
Republicans	59	41
Independents	52	48
Democrats	53	47

29. Immigrant Origins and Preferences. Does the state where immigrants wish to settle depend on where the immigrant is coming from? The U.S. Department of Homeland Security tracks the continent of origin and the desired state of settlement for immigrants. Some of the data are shown here, in thousands. Test using the critical-value method whether continent of origin and state of settlement are independent, using level of significance $\alpha = 0.01$. immigrant

	California	**Florida**	**New York**
Europe	24.0	9.8	23.2
Asia	112.6	9.0	31.3
South America	8.0	16.1	17.7

30. Email, Phone, or in Person? What is the most effective way to handle a task at work: by email, by phone, or in person? Well, you probably say, it depends on the task. The Pew Internet and American Life Project Email at Work Survey surveyed 1000 randomly selected work email users, who chose the following methods as the best for handling certain work tasks. Test whether the proportions who favor email differ between the two tasks, using level of significance $\alpha = 0.05$ and the estimated p-value method. worktask

Task	**By email**	**By phone or in person**
Edit or review documents	670	330
Arrange meetings or appointments	630	370

31. Using Graphical Evidence. Sick of spam (unsolicited broadcast email)? Do you get more spam at your work, school, or home email address? The Pew Internet and American Life Project Email at Work Survey examined the proportion of spam in email users' work and home email accounts. Using only the information in the clustered bar graph below, would you conclude that the proportion of those who report "a lot of spam" is the same for work email and personal email? Why?

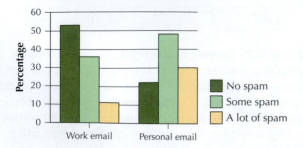

32. Spam, Spam, Spam. Continue your work from the previous exercise. The following contingency table shows the actual percentages in the graph above based on samples of size 100 for each of work email and personal email. Test whether the proportions who report "a lot of spam" are the same for work email and personal email, using level of significance a = 0.01. Does your conclusion agree with your conjecture in the previous exercise?

	None	**Some**	**A lot**
Work email	53%	36%	11%
Personal email	22%	48%	30%

33. Gender Differences in Computer/Video/Online Gaming. The Pew Internet and American Life Project collected data on the College Students Gaming Survey. Among the questions they asked 1720 randomly selected college students was "Which one of the following do you play the most: video games, computer games, or online games?" The results are summarized by gender in the following contingency table. games

	Video games	**Computer games**	**Internet games**
Male	616	221	139
Female	198	372	174

a. Before you carry out the hypothesis test, *what result might you expect?* Look over the data set carefully to see whether you can detect significant differences between the levels of the variables. Then see whether your hypothesis test bears out your intuition.

b. Test whether *gender* and *game type* are independent, using level of significance $\alpha = 0.01$.

34. Online Dating. A Pew Internet and American Life Project study reported that the proportion of urban residents who use online dating is 13%, while the proportion for suburban residents is 10% and the proportion for rural residents is 9%.[6] Test using level of significance $\alpha = 0.05$ whether there are differences among the population proportions of residents from the three categories who use online dating. Assume that each sample size was 1000. (*Hint:* The null hypothesis assumes that all proportions are equal.)

Use Technology for each of Exercises 35–38. goals

Bringing it All together

Goals of Middle School Students. Open the **Goals** data set. The subjects are students in grades 4, 5, and 6, from three school districts in Michigan. The students were asked which of the following was most important to them: good grades, athletic ability, or popularity. Information about the students' age, gender, race, and grade was also gathered, as well as whether their school was in an urban, suburban, or rural setting.[7]

35. How many observations are in the data set? How many variables?

36. Comparing gender and goals.
a. Looking at the data, do you think that boys and girls at this age differ in what is most important to them: grades, popularity, or sports? In other words, do you think that the variables *gender* and *goals* are dependent or independent?
b. Perform the χ^2 test for independence, using level of significance $\alpha = 0.05$.

37. Comparing goals and school setting.
a. Looking at the data, do you think that the setting of the school (urban, suburban, or rural) affects the goals of the students? Or do you think that it has no effect? In other words, do you think that the variables *urb_rur* and *goals* are independent or dependent?
b. Perform the χ^2 test for independence, using level of significance $\alpha = 0.10$.

38. Comparing grades and goals.
a. One thing we know for sure is that, as students get older, they get more serious and grades get more important to them (don't they?). So we would expect that the variables *grade* and *goals* would be dependent, wouldn't we? Is this borne out by looking at the data?
b. Perform the χ^2 test for independence, using level of significance $\alpha = 0.01$.

11.3 ANALYSIS OF VARIANCE

OBJECTIVES By the end of this section, I will be able to . . .

1 Describe the characteristics of the *F* distribution.

2 Explain how ANOVA works.

3 Perform analysis of variance.

In Sections 11.1 and 11.2, we used the χ^2 distribution to analyze categorical data. Here, in Section 11.3, we need to learn about a new distribution, the *F* distribution, which will help us with the analytic methods we will learn in Section 11.3.

1 *F* DISTRIBUTION

The **F distribution** was named in honor of the "grandfather of statistics," Sir Ronald A. Fisher. Like the χ^2 distribution, the *F* distribution is right-skewed, never takes negative values, and has an infinite number of different *F* curves (Figure 11.23). The shape of the curve depends on two different degrees of freedom.

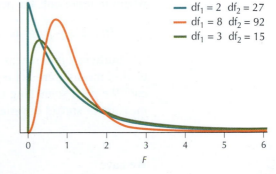

FIGURE 11.23
Shape of the F distribution for various degrees of freedom.

Note that the *F* distribution resembles the χ^2 distribution. This is not surprising since the values of the *F* distribution represent ratios of two χ^2 distributions. Moreover, the *F* distribution has two different degrees of freedom, which we shall call df_1 and df_2, derived from the degrees of freedom of the two χ^2 distributions represented in the ratio. Often, df_1 is called the *numerator degrees of freedom*, and df_2 is called the *denominator degrees of freedom*.

> **Properties of the F Curve**
>
> 1. The total area under the F curve equals 1.
> 2. The value of the F random variable is never negative, so the F curve sarts at 0. However, it extends indefinitely to the right. The curve approaches but never quite meets the horizontal axis.
> 3. Because of the characteristics described in (2), the F curve is right-skewed.
> 4. There is a different F curve for each different pair of degrees of freedom, df_1 and df_2.

Since the F distribution is continuous, we can find probabilities associated with values of F, and vice versa, just as we did with the normal, t, and χ^2 distributions.

Just as for any continuous distribution, probability is represented by the area below the F curve above an interval.

2 HOW ANALYSIS OF VARIANCE (ANOVA) WORKS

Analysis of variance (ANOVA) is a hypothesis test for determining whether three or more means of different populations are equal. ANOVA works by comparing the variability *between* the samples to the variability *within* the samples.

Suppose we are interested in determining whether there are significant differences in grade point averages (GPAs) among residents of three dormitories, A, B, and C. Table 11.13 displays three random samples of GPAs of ten residents from each dormitory.

Copyright Mark Richards/PhotoEdit

Table 11.13 Sample GPAs from Dorms A, B, and C

A	0.60	3.82	4.00	2.22	1.46	2.91	2.20	1.60	0.89	2.30
B	2.12	2.00	1.03	3.47	3.70	1.72	3.15	3.93	1.26	2.62
C	3.65	1.57	3.36	1.17	2.55	3.12	3.60	4.00	2.85	2.13

The sample mean GPA for Dormitory A is

$$\bar{x}_A = \frac{0.60 + 3.82 + 4.00 + 2.22 + 1.46 + 2.91 + 2.20 + 1.60 + 0.89 + 2.30}{10} = 2.2$$

Similarly, we can find the sample mean GPAs for the other dormitories: $\bar{x}_B = 2.5$ and $\bar{x}_C = 2.8$. We note that the sample means are not equal. The question is, Are the population means equal? Let μ_A, μ_B, and μ_C represent the population mean GPAs for Dormitories A, B, and C, respectively. We are interested in the following hypotheses, where μ_i represents the population mean GPA for dormitory i:

$$H_0: \ \mu_A = \mu_B = \mu_C \quad \text{versus} \quad H_a: \text{ not all the population means are equal}$$

Sufficient differences in the sample means would represent evidence that the population means were not equal. The question is, What represents "sufficiently" different? We need something to compare against, such as the *spread* of each sample. One measure of spread or variability is the *range*:

$$\text{range} = \text{max} - \text{min}$$

We have

$$\text{range (Dorm A)} = 4.00 - 0.60 = 3.40$$
$$\text{range (Dorm B)} = 3.93 - 1.03 = 2.90$$
$$\text{range (Dorm C)} = 4.00 - 1.17 = 2.83$$

These ranges are rather large spreads, and there is a considerable amount of overlap among the different dormitory GPAs, as shown in Figure 11.24.

Figure 11.24 shows the difference among the means for the three dorm GPAs compared with the spread of each dorm's GPAs, as measured by the range. The red triangles represent the sample means, $\bar{x}_A = 2.2$, $\bar{x}_B = 2.5$, and $\bar{x}_C = 2.8$. The spread of the sample means (shown by the red arrows) is much less than the spreads of the individual dorm GPAs (shown by the green arrows). Thus, the sample means $\bar{x}_A = 2.2$, $\bar{x}_B = 2.5$, and $\bar{x}_C = 2.8$ are not sufficiently different when compared against the spread of the GPAs. This graph would therefore not provide evidence to reject the null hypothesis that the population mean GPAs are all equal.

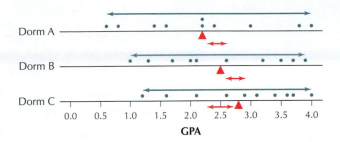

FIGURE 11.24

Comparison dotplot of GPAs for Dorms A, B, and C.

Now we make a similar comparison for the GPAs for Dormitories D, E, and F in Table 11.14.

Table 11.14 Sample GPAs from Dorms D, E, and F

D	2.16	2.23	2.09	2.17	2.25	2.19	2.24	2.28	2.25	2.14
E	2.45	2.34	2.58	2.49	2.60	2.42	2.55	2.62	2.45	2.50
F	2.80	2.75	2.93	2.68	2.88	2.75	2.87	2.81	2.73	2.80

The sample mean GPAs for Dormitories D, E, and F are the *same* as those for Dormitories A, B, and C, respectively: $\bar{x}_D = 2.2$, $\bar{x}_E = 2.5$, and $\bar{x}_F = 2.8$. Again we are interested in whether the population means are equal.

$$H_0: \ \mu_D = \mu_E = \mu_F \quad \text{versus} \quad H_a: \text{ not all the population means are equal}$$

Consider the comparison dotplot in Figure 11.25. There now seems to be better evidence for concluding that the three population means are not all equal. There is no overlap among the three samples because the spread *within* each dormitory is much smaller than for Dormitories A, B, and C.

$$\text{range (Dorm D)} = 2.28 - 2.09 = 0.19$$
$$\text{range (Dorm E)} = 2.62 - 2.34 = 0.28$$
$$\text{range (Dorm F)} = 2.93 - 2.68 = 0.25$$

Figure 11.25 on the next page shows the difference among the means for the three dorm GPAs compared with the range of each dorm's GPAs. The red triangles represent the sample means, $\bar{x}_D = 2.2$, $\bar{x}_E = 2.5$, and $\bar{x}_F = 2.8$. The spread of the sample means (red arrows) is much greater than the spreads of the individual dorm GPAs (green arrows). Thus, the sample means $\bar{x}_D = 2.2$, $\bar{x}_E = 2.5$, and $\bar{x}_F = 2.8$ are sufficiently different when compared against the range of the GPAs. This graph would, therefore, provide some evidence to reject the null hypothesis that the population mean GPAs are all equal.

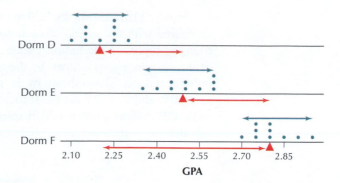

FIGURE 11.25
Comparison dotplot
of GPAs for Dorms D, E, and F.

Note that we arrived at *opposite conclusions* for the two sets of dormitories, *even though the sample means of the first group are identical to the sample means of the second group*. Here is the key difference:

- The within-sample spreads of Dormitories A, B, and C are *large*. Compared to these large spreads, the difference in sample means did not seem large;

- The within-sample spreads of Dormitories D, E, and F are *small*. Compared to these small spreads, the difference in sample means did seem large.

These are the types of comparisons that the ANOVA method makes.

Instead of using the range as the measure of spread, analysis of variance uses the standard deviation of the individual samples. Recall that samples with larger spread have larger standard deviations, just as they have larger ranges.

Developing Your Statistical Sense

How Does Analysis of Variance Work?

The key to how analysis of variance works is the following comparison. Compare

a. the variability in the sample means—that is, how large the differences are between the sample means (indicated by the lengths of the red arrows in Figures 11.24 and 11.25)—with

b. the variability within each sample—that is, the within-sample spreads (indicated by the lengths of the green arrows in Figures 11.24 and 11.25).

When (a) is much larger than (b), this is evidence that the population means are not all equal and that we should reject the null hypothesis. Thus, our analysis depends on measuring variability. And hence the term *analysis of variance*.

Just as for hypothesis-testing procedures from previous chapters, analysis of variance can be performed only if certain requirements are met.

> **Requirements for Performing Analysis of Variance**
>
> 1. Each of the k populations is normally distributed.
> 2. The variances (σ^2) of the populations are all equal.
> 3. The samples are independently drawn.

Our hypotheses for testing for the equality of the population mean GPA for Dormitories A, B, and C are

$$H_0 : \mu_A = \mu_B = \mu_C \quad \text{versus} \quad H_a : \text{not all the population means are equal}$$

Note: In analysis of variance, the null hypothesis *always* states that all the population means are equal and the alternative hypothesis *always* states that not all the population means are equal. Note that H_a is *not* stating that the population means are all different. For H_a to be true, it is sufficient for a single population mean to be different, even though all the other population means may be equal.

Let us stop for a moment to consider what these requirements and the hypotheses mean.

- If H_0 is true, then all three dormitories would have the same population mean GPA: $\mu_A = \mu_B = \mu_C = \mu$, where we denote the hypothesized common mean as μ.
- Requirement 1 states that each population is normally distributed.
- Requirement 2 states that all the population variances are equal. Let's call this common variance σ^2.

Putting all this together, H_0 assumes that the observations from each population come from the same normal distribution, with mean μ and variance σ^2.

Suppose we then take samples of size n from each group. Fact 3 in Chapter 7 states that the sampling distribution of $\bar{x}$ for a sample of size n taken from a normal population with mean μ and standard deviation σ (that is, variance σ^2) is also normal, with mean μ and standard deviation $\sigma/\sqrt{n}$ (that is, variance σ^2/n), as shown in Figure 11.26. Since each dormitory's GPA is assumed (under H_0) to come from the same sampling distribution, we would expect the sample means to be fairly close together.

On the other hand, if H_0 is not true, then not all the population means are equal (Figure 11.27). In this case, there is no sampling distribution common to all sample means, so we would not expect the sample means to be close together. Note in Figure 11.27 that each distribution nevertheless has the same shape (normal) and spread (i.e., variance) because of the requirements.

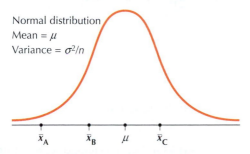

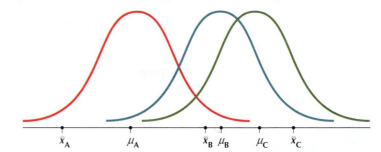

FIGURE 11.26 Common sampling distribution when H_0 is true.

FIGURE 11.27 No common sampling distribution when H_0 is not true.

Note: Normal probability plots were introduced in Chapter 7.

Procedure for Verifying the Requirements for Analysis of Variance

Step 1 **Normality.** Check that the data from each group are normally distributed, using normality probability plots.

Step 2 **Equal Variances.** Compute the sample standard deviation for each group to verify that the largest standard deviation is not larger than twice the smallest standard deviation.

Step 3 **Independence.** Verify that the samples drawn from each group are independently drawn.

EXAMPLE 11.9

VERIFY THE REQUIREMENTS FOR PERFORMING AN ANALYSIS OF VARIANCE

dormitory

Verify the requirements for performing an analysis of variance using the hypotheses

$$H_0 : \mu_A = \mu_B = \mu_C \quad \text{versus} \quad H_a : \text{not all the population means are equal}$$

where μ_i represents the population mean GPA for Dormitory i, using data from Table 11.13.

Solution

STEP 1 **Normality.**

To verify that each of the $k = 3$ populations is normally distributed, we examine normal probability plots of each sample, shown in Figure 11.28. Each plot indicates acceptable normality.

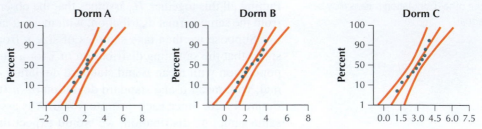

FIGURE 11.28 Normal probability plots verify normality requirement.

STEP 2 **Equal Variances.**

To find the standard deviation for Dorm A, we first find

$$
\begin{aligned}
\sum(x - \bar{x})^2 &= (0.60 - 2.2)^2 + (3.82 - 2.2)^2 + (4.00 - 2.2)^2 + (2.22 - 2.2)^2 \\
&\quad + (1.46 - 2.2)^2 + (2.91 - 2.2)^2 + (2.20 - 2.2)^2 + (1.60 - 2.2)^2 \\
&\quad + (0.89 - 2.2)^2 + (2.30 - 2.2)^2 \\
&= 11.5626
\end{aligned}
$$

Then

$$
s_A = \sqrt{\frac{\sum(x - \bar{x})^2}{n - 1}} = \sqrt{\frac{11.5626}{10 - 1}} \approx 1.133460777
$$

Note: We retain many decimal places when calculating s_A, s_B, and s_C because these values are used to calculate other quantities later on.

We similarly find $s_B \approx 1.030857248$ and $s_C \approx 0.9370284$. The largest, $s_A \approx 1.133460777$, is not larger than twice the smallest, $s_C \approx 0.9370284$. Thus, the equal variance requirement is satisfied.

STEP 3 **Independence.**

**Now You Can Do
Exercises 23a–25a.**

Since the students are randomly sampled from each dormitory, with the selection of students in one dormitory not affecting the selection of students sampled from the other dormitories, the independence assumption is also validated.

Assuming that H_0 is true, we estimate the common population mean μ using the **overall sample mean, $\bar{\bar{x}}$:**

Note: This form for $\bar{\bar{x}}$ is a weighted mean with the weights being the sample sizes.

$$
\bar{\bar{x}} = \frac{(n_1 \bar{x}_1 + n_2 \bar{x}_2 + \cdots + n_k \bar{x}_k)}{n_t}
$$

where there are k samples and n_t is the "total sample size" (sum of the k sample sizes). The overall sample mean $\bar{\bar{x}}$ is simply the mean of all the observations from all the samples. For the special case when all the sample sizes are equal, the overall sample mean $\bar{\bar{x}}$ is simply the mean of the k sample means,

$$
\bar{\bar{x}} = \frac{(\bar{x}_1 + \bar{x}_2 + \cdots + \bar{x}_k)}{k}
$$

EXAMPLE 11.10

CALCULATING THE OVERALL SAMPLE MEAN $\bar{\bar{x}}$

For the sample GPA data given in Table 11.13 for Dorms A, B, and C, calculate the overall sample mean, $\bar{\bar{x}}$.

Solution

We have $k = 3$ dormitories, with sample mean GPAs $\bar{x}_A = 2.2$, $\bar{x}_B = 2.5$, $\bar{x}_C = 2.8$. Also, $n_A = n_B = n_C = 10$, and $n_t = 10 + 10 + 10 = 30$. Thus,

$$\bar{\bar{x}} = \frac{(10(2.2) + 10(2.5) + 10(2.8))}{30} = 2.5$$

Since all the sample sizes are equal, we can also calculate $\bar{\bar{x}}$ as follows:

$$\bar{\bar{x}} = \frac{(2.2 + 2.5 + 2.8)}{3} = 2.5$$

**Now You Can Do
Exercises 7b–10b.**

What Does This Number Mean?	$\bar{\bar{x}} = 2.5$ is the mean GPA for all 30 students from all three samples. We can use $\bar{\bar{x}}$ as our estimate of the common population mean μ assumed in H_0.

Recall that analysis of variance works by comparing the variability in the sample means to the variability within each sample. We use the following statistics to measure these variabilities.

> The **mean square treatment (MSTR)** measures the variability in the sample means. MSTR is the sample variance of the sample means, weighted by sample size.
>
> $$MSTR = \frac{\sum n_i(\bar{x}_i - \bar{\bar{x}})^2}{k - 1}$$
>
> where n_i and $\bar{x}_i$ are the sample size and mean of the ith sample, $\bar{\bar{x}}$ is the overall sample mean, and there are k populations.
>
> The **mean square error (MSE)** measures the variability within the samples. MSE is the mean of the sample variances, weighted by sample size.
>
> $$MSE = \frac{\sum (n_i - 1)s_i^2}{n_t - k}$$
>
> where n_i and s_i^2 are the sample size and variance of the ith sample, n_t is the total sample size, and there are k populations.

The greater the distance between the sample means, the larger the MSTR.

The larger the standard deviation of the k samples, the larger the MSE.

We compare MSTR to MSE by taking the ratio of these two quantities.

> The **test statistic for analysis of variance** is
>
> $$F_{data} = \frac{MSTR}{MSE}$$
>
> F_{data} measures the variability among the sample means, compared to the variability within the samples. F_{data} follows an F distribution with $df_1 = k - 1$ and $df_2 = n_t - k$, when the following requirements are met: (1) each of the k populations is normally distributed, (2) the variances of the populations are all equal, and (3) the samples are independently drawn.

The term *mean square* represents a weighted mean of quantities that are squared. Each mean square itself consists of two parts: the *sum of squares* in the numerator and the *degrees of freedom* in the denominator. The numerator for MSTR is called the **sum of** squares treatment **(SSTR)**, and the numerator for MSE is called the **sum of squares error (SSE)**.

$$\text{MSTR} = \frac{\text{sum of squares treatment}}{\text{df}_1} = \frac{\text{SSTR}}{\text{df}_1} = \frac{\sum n_i\,(\bar{x}_i - \bar{\bar{x}})^2}{k - 1}$$

$$\text{MSE} = \frac{\text{sum of squares error}}{\text{df}_2} = \frac{\text{SSE}}{\text{df}_2} = \frac{\sum (n_i - 1)s_i^2}{n_t - k}$$

The **total sum of squares (SST)** is found by adding SSTR and SSE:

$$\text{SST} = \text{SSTR} + \text{SSE}$$

The ANOVA table shown in Table 11.15 is a convenient way to display the various statistics calculated during an analysis of variance. Note that the quantities in the mean square column equal the ratio of the two columns to its left.

Table 11.15 ANOVA table

Source of variation	Sum of squares	Degrees of freedom	Mean square	F-test statistic	p-value
Treatment	SSTR	$\text{df}_1 = k - 1$	$\text{MSTR} = \dfrac{\text{SSTR}}{k - 1}$	$F_{\text{data}} = \dfrac{\text{MSTR}}{\text{MSE}}$	$p(F > F_{\text{data}})$
Error	SSE	$\text{df}_2 = n_t - k$	$\text{MSE} = \dfrac{\text{SSE}}{n_t - k}$		
Total	SST				

EXAMPLE 11.11

CONSTRUCTING THE ANOVA TABLE

Use the summary statistics in Table 11.16 for the sample GPAs for Dorms A, B, and C to construct the ANOVA table.

Table 11.16 Summary statistics for sample GPAs for Dorms A, B, and C

	Dorm A	Dorm B	Dorm C
Mean	$\bar{x}_A = 2.2$	$\bar{x}_B = 2.5$	$\bar{x}_C = 2.8$
Standard deviation	$s_A \approx 1.133460777$	$s_B \approx 1.030857248$	$s_C \approx 0.9370284$
Sample size	$n_1 = 10$	$n_2 = 10$	$n_3 = 10$

Solution

We have $k = 3$ dormitories, and total sample size $n_t = 10 + 10 + 10 = 30$. Thus,

- $\text{SSTR} = \sum n_i(\bar{x}_i - \bar{\bar{x}})^2 = 10(2.2 - 2.5)^2 + 10(2.5 - 2.5)^2 + 10(2.8 - 2.5)^2$

$$= 10[(-0.3)^2 + (0)^2 + (0.3)^2] = 1.8$$

- $\text{SSE} \approx (10 - 1)(1.133460777)^2 + (10 - 1)(1.030857248)^2 + (10 - 1)(0.9370284)^2$

$$\approx 29.0288$$

- $\text{SST} = \text{SSTR} + \text{SSE} = 1.8 + 29.0288 = 30.8288$

- $\text{MSTR} = \dfrac{\text{SSTR}}{k-1} = \dfrac{1.8}{3-1} = 0.9$

- $\text{MSE} = \dfrac{\text{SSE}}{n_t - k} = \dfrac{29.0288}{30-3} = 1.0751407407$

- $F_{\text{data}} = \dfrac{\text{MSTR}}{\text{MSE}} = \dfrac{0.9}{1.0751407407} = 0.8370997079 \approx 0.84$

**Now You Can Do
Exercises 11d–14d, 19,
and 20.**

We summarize these calculations in the following ANOVA table with the results rounded for clarity.

Source of variation	Sum of squares	Degrees of freedom	Mean square	F-test statistic
Treatment	$\text{SSTR} = 1.8$	$\text{df}_1 = 3 - 1 = 2$	$\text{MSTR} = \dfrac{1.8}{2} = 0.9$	$F_{\text{data}} = \dfrac{0.9}{1.075} \approx 0.84$
Error	$\text{SSE} = 29.0288$	$\text{df}_2 = 30 - 3 = 27$	$\text{MSE} = \dfrac{29.0288}{27} \approx 1.075$	
Total	$\text{SST} = 30.8288$			

3 PERFORMING ONE-WAY ANOVA

Now that we know how it works, we next learn how to perform ANOVA.

Remember: H_a is not stating that the population means are all different.

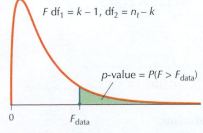

$F \ \text{df}_1 = k - 1,\ \text{df}_2 = n_t - k$

$p\text{-value} = P(F > F_{\text{data}})$

$0 \qquad F_{\text{data}}$

FIGURE 11.29 *p*-Value for the one-way ANOVA *F* test.

> **One-Way Analysis of Variance**
>
> We have taken random samples from each of *k* populations and want to test whether the population means of the *k* populations are all equal.
> Required conditions:
>
> 1. Each of the *k* populations is normally distributed.
> 2. The variances (σ^2) of the populations are all equal.
> 3. The samples are independently drawn.
>
> ***Step 1 State the hypotheses, and state the rejection rule.***
> $H_0: \mu_1 = \mu_2 = \cdots = \mu_k$ versus $H_a:$ not all the population means are equal
> where the μ's represent the population mean from each population. The rejection rule is *Reject H_0 if the p-value $\leq \alpha$.*
>
> ***Step 2 Calculate F_{data}.***
>
> $$F_{\text{data}} = \frac{\text{MSTR}}{\text{MSE}}$$
>
> where
> $$\text{MSTR} = \frac{\sum n_i (\bar{x}_i - \bar{\bar{x}})^2}{k-1} \quad \text{and} \quad \text{MSE} = \frac{\sum (n_i - 1)s_2^i}{n_t - k}$$
>
> F_{data} follows an *F* distribution with $\text{df}_1 = k - 1$ and $\text{df}_2 = n_t - k$ if the required conditions are satisfied, where n_t represents the total sample size.
>
> ***Step 3 Find the p-value.*** Use technology to find the $p\text{-value} = P(F > F_{\text{data}})$, as shown in Figure 11.29.
>
> ***Step 4 State the conclusion and the interpretation.*** Compare the *p*-value with α.

EXAMPLE 11.12 **PERFORMING ONE-WAY ANOVA**

Test using level of significance $\alpha = 0.05$ whether the population mean GPAs from Example 11.9 differ among the students in Dormitories A, B, and C.

What Result Might We Expect?

Recall that the comparison dotplot in Figure 11.24 (page 559) showed a large amount of overlap in the GPAs among the three dormitories. The large ranges illustrate the large within-dormitory spread of the GPAs for these dorms. When compared against this large within-sample variability, the variability in sample means may not seem large. Therefore, we might expect that the null hypothesis of no difference will *not be rejected*.

Solution

We already verified the requirements for performing the analysis of variance in Example 11.9.

STEP 1 **State the hypotheses, and state the rejection rule.**
Define the μ_i.

$$H_0 : \ \mu_A = \mu_B = \mu_C \quad \text{versus} \quad H_a : \text{ not all the population means are equal}$$

where μ_i represents the population mean GPA of students from dormitory i. The rejection rule is *Reject H_0 if the p-value $\leq \alpha$.*

STEP 2 **Calculate F_{data}.**
From Example 11.11, we have MSTR = 0.9, MSE = 1.0751407407, and

$$F_{data} = \frac{MSTR}{MSE} = \frac{0.9}{1.0751407407} = 0.8370997079$$

> **CAUTION** When calculating the p-value for analysis of variance, always retain as many decimal places in the value of F_{data} as you can. This will make the p-value as accurate as possible. Rounding F_{data} too much will make the p-value less accurate.

F_{data} follows an F distribution with $df_1 = k - 1 = 3 - 1 = 2$ and $df_2 = n_t - k = 30 - 3 = 27$.

STEP 3 **Find the p-value.**
We use the instructions provided in the Step-by-Step Technology Guide at the end of this section (page 570). From Figures 11.30 and 11.31, we have

$$p\text{-value} = P(F > F_{data}) = P(F > 0.8370997079) = 0.4438929572 \approx 0.4439$$

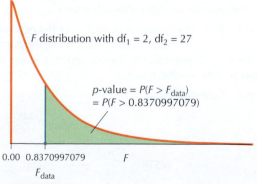

FIGURE 11.30 p-Value = $P(F > 0.8370997079)$.

FIGURE 11.31 TI-83/84 p-value.

STEP 4 **State the conclusion and the interpretation.**
Compare the p-value with α. Since the p-value of 0.4439 is not $\leq \alpha = 0.05$, we do not reject H_0. As expected, there is not enough evidence to conclude at level of significance $\alpha = 0.05$ that not all population mean GPAs are equal.

Now You Can Do
Exercises 15–18.

Alamy

| | | **EXAMPLE 11.13** | **PERFORMING ONE-WAY ANOVA USING TECHNOLOGY** |

Researchers from the Institute for Behavioral Genetics at the University of Colorado investigated the effect that the enzyme protein kinase C (PKC) has on anxiety in mice. The genotype for a particular gene in a mouse (or a human) consists of two *alleles* (copies) of each chromosome, one each from the father and mother. The investigators in the study separated the mice into three groups. In Group 0, neither of the mice's alleles for PKC produced the enzyme. In Group 1, one of the two alleles for PKC produced the enzyme and the other did not. In Group 2, both PKC alleles produced the enzyme. To measure the anxiety in the mice, scientists measured the time (in seconds) the mice spent in the "open-ended" sections of an elevated maze. It was surmised that mice spending more time in open-ended sections exhibit decreased anxiety. The data are provided in Table 11.17. Use technology to test at $\alpha = 0.01$ whether the population mean time spent in the open-ended sections of the maze was the same for all three groups.

 micemaze

Table 11.17 Time spent in open-ended section of maze

Group 0		Group 1		Group 2	
15.8	14.4	5.2	7.6	10.6	9.2
16.5	25.7	8.7	10.4	6.4	14.5
37.7	26.9	0.0	7.7	2.7	11.1
28.7	21.7	22.2	13.4	11.8	3.5
5.8	15.2	5.5	2.2	0.4	8.0
13.7	26.5	8.4	9.5	13.9	20.7
19.2	20.5	17.2	0.0	0.0	0.0
2.5		11.9		16.5	

What Result Might We Expect?

Figure 11.32 shows a plot of the time in open-ended sections for the mice in the three groups. Note that the Group 1 and Group 2 mice spent on average about the same amount of time in the open-ended sections but that Group 0 spent on average somewhat more time in the open-ended sections. This would tend to suggest that the null hypothesis that all three population means are equal should be rejected. Remember that to reject H_0, it is sufficient for just one of the population means to be different.

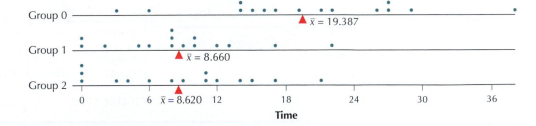

FIGURE 11.32 Evidence that the population mean of Group 0 is larger than the others.

Solution

We use the instructions provided in the Step-by-Step Technology Guide at the end of this section (page 570). We first verify whether the requirements are met.

- The normal probability plots in Figure 11.33 indicate acceptable normality.

- The group standard deviations are $s_0 \approx 9.0$, $s_1 \approx 6.0$, and $s_2 \approx 6.4$. Thus, the largest standard deviation is not greater than twice the smaller, which verifies the equal variances requirement.

- The selection of a mouse to a particular group did not affect the selection of mice to the other groups, so that the samples are independent.

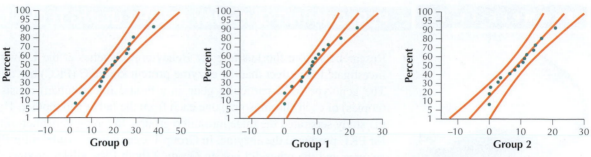

FIGURE 11.33 Normal probability plots.

Thus, we proceed with the one-way ANOVA.

$$H_0: \mu_{\text{Group 0}} = \mu_{\text{Group 1}} = \mu_{\text{Group 2}}$$
$$H_a: \text{Not all population means are equal}$$

where the μ's represent the population mean time spent in the open-ended sections of the maze for each group.

Figure 11.34 contains the results from the TI-83/84, showing where each statistic corresponds to the ANOVA table structure in Table 11.15. We have $F_{\text{data}} = 10.906$, with a p-value of "1.5320224E-4" = 0.00015320224. Since this p-value is less than $\alpha = 0.01$, we reject H_0. There is evidence at level of significance $\alpha = 0.01$ that the population mean times in the open-ended sections of the maze are not equal for all three groups.

Source of variation	Sum of squares	Degrees of freedom	Mean square	F-test statistic
Treatment	SSTR ≈ 1154.92	$df_1 = 2$	MSTR ≈ 577.46	$F_{\text{data}} \approx 10.906$
Error	SSE ≈ 2223.84	$df_2 = 42$	MSE ≈ 52.95	
Total	SST ≈ 3378.75			

```
One-way ANOVA
 F=10.90607167
 p=1.5320224E-4
 Factor
  df=2
  SS=1154.92044
↓ MS=577.460222
```
```
One-way ANOVA
↑ MS=577.460222
 Error
  df=42
  SS=2223.83733
  MS=52.9485079
 Sxp=7.27657254
```

FIGURE 11.34 Correspondence between TI-83/84 ANOVA output and the ANOVA table.

Figure 11.35 contains the Excel ANOVA results, and Figure 11.36 contains the Minitab ANOVA results. Values differ slightly due to rounding.

ANOVA

Source of Variation	SS	df	MS	F	P-value	F crit
Between Groups	1154.92	2	577.4602	10.90607	0.000153	3.219938
Within Groups	2223.84	42	52.94851			
Total	3378.76	44				

FIGURE 11.35 Excel ANOVA results.

Source	DF	SS	MS	F	P
Group	2	1154.9	577.5	10.91	0.000
Error	42	2223.8	52.9		
Total	44	3378.8			

FIGURE 11.36 Minitab ANOVA results.

One-way ANOVA may also be conducted using the critical-value method. The conditions are the same as for the p-value method.

EXAMPLE 11.14

PERFORMING ONE-WAY ANOVA USING THE CRITICAL-VALUE METHOD

micemaze

Use the data from Example 11.13 to test using the critical-value method and level of significance $\alpha = 0.01$ whether the population mean time spent in the open-ended sections of the maze was the same for all three groups.

Solution

The conditions for performing ANOVA were verified in Example 11.13.

STEP 1 **State the hypotheses.**

$H_0: \mu_{\text{Group } 0} = \mu_{\text{Group } 1} = \mu_{\text{Group } 2}$
$H_a:$ Not all population means are equal

where the μ's represent the population mean time spent in the open-ended sections of the maze for each group.

STEP 2 **Find the critical value F_{crit} and state the rejection rule.**

The one-way ANOVA test is a right-tailed test, so the F-critical value F_{crit} is the value of the F distribution for $df_1 = k - 1$ and $df_2 = n_t - k$ that has area α to the right of it (see Figure 11.37). Here, $df_1 = 3 - 1 = 2$ and $df_2 = 45 - 3 = 42$. To find F_{crit}, we may use the F tables or technology. To find our F_{crit} using Excel, enter **=FINV(0.01,2,42)** in cell A1, as shown in Figure 11.37. Thus, $F_{\text{crit}} = 5.149$. Since ANOVA is a right-tailed test, we will reject H_0 if $F_{\text{data}} \geq 5.149$.

FIGURE 11.37
Using Excel to find the F critical value.

	A1			f_x	=FINV(0.01,2,42)	
	A	B	C	D	E	F
1	5.149					

STEP 3 **Calculate F_{data}.**

From Example 11.13 we have $F_{\text{data}} = 10.906$.

STEP 4 **State the conclusion and interpretation.**

Since $F_{\text{data}} = 10.906 \geq F_{\text{crit}} = 5.149$ (Figure 11.38), we reject H_0. There is evidence that not all population mean times spent in the open-ended sections of the maze are equal.

FIGURE 11.38
$F_{\text{crit}} = 5.149$ has area of $\alpha = 0.01$ to the right of it.

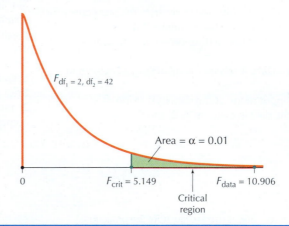

$F_{df_1 = 2, \, df_2 = 42}$

Area = α = 0.01

0      $F_{\text{crit}} = 5.149$ $F_{\text{data}} = 10.906$

Critical region

Developing Your Statistical Sense

Do Not Draw the Wrong Conclusion

Note that we did not conclude that all three population means are different. As long as one mean is sufficiently different from the other two, we would reject H_0. Our conclusion was simply that the population means were not all equal.

Also, we cannot yet formally conclude that Group 0 has a larger population mean time than the other groups, even though Figure 11.32 seems to indicate so. All we can formally conclude at this point is that not all the population means are equal. To learn *multiple comparisons,* which is the type of analysis needed to test whether the mean of Group 0 is larger than the others, please turn to *Discovering Statistics,* second edition.

The *One-Way ANOVA* applet allows you to experiment with various values for the sample means and the sample variability in order to see how changes in these values affect F_{data} and the p-value.

STEP-BY-STEP TECHNOLOGY GUIDE: Analysis of Variance

TI-83/84

Performing ANOVA
(Example 11.12, pages 565–566, is used to illustrate the procedure.)
Step 1 Enter the Dormitory A data in **L1**, the Dormitory B data in **L2**, and the Dormitory C data in **L3**.
Step 2 Press **STAT**, highlight **TESTS**, select **"ANOVA("**, and press **ENTER**.
Step 3 On the home screen, enter **"L1, L2, L3)"** and press **ENTER** (Figure 11.39).

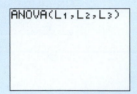

FIGURE 11.39

Finding the p-Value for a Given F_{data}.
(Example 11.12, pages 565–566, is used to illustrate the procedure.)
p-value $= P(F > F_{data}) = P(F > 0.8370997079)$, where $df_1 = 2$ and $df_2 = 27$.

Step 1 Press **2nd > DISTR**.
Step 2 Select **Fcdf(** and press **ENTER**.
Step 3 Enter **"0.8370997079, 1e99, 2, 27"** and press **ENTER**. The results are shown in Figure 11.31 (pages 565–566).

EXCEL

Performing ANOVA
(Example 11.12, pages 565–566, is used to illustrate the procedure.)
Step 1 Enter the Dormitory A data in column A, the Dormitory B data in column B, and the Dormitory C data in column C.
Step 2 Click **Data > Data Analysis > Anova: Single Factor**, and click **OK**.
Step 3 Select the input range of the data by clicking and dragging over the data in columns A, B, and C. Then click **OK**.

Finding the p-Value for a Given F_{data}.
(Example 11.12, pages 565–566, is used to illustrate the procedure.)
p-value $= P(F > F_{data}) = P(F > 0.8370997079)$, where $df_1 = 2$ and $df_2 = 27$.
Step 1 Select cell **A1**. Click the **Insert Function** icon f_x.
Step 2 For **Search for a Function**, type **FDIST** and click **OK**.
Step 3 For X, enter **0.8370997079**, for **Deg_freedom 1**, enter **2**, and for **Deg_freedom 2**, enter **27**. Then click **OK**. The cell now contains the p-value: 0.4438929572.

MINITAB

Performing ANOVA
(Example 11.12, pages 565–566, is used to illustrate the procedure.)
   Minitab accepts data in two different forms for performing ANOVA, *stacked* or *unstacked*. *Unstacked* refers to the data of each group being in a separate column. *Stacked* merges each group's data together in a single column, with the group numbers in a different column.

ANOVA (Stacked)
Step 1 Enter the **GPA** data for all three groups in C1 and the values for the categorical variable **Dorm** in C2.
Step 2 Click on **Stat > ANOVA > One-Way**.

Step 3 Choose the quantitative variable **GPA** as your **response** and the categorical variable **Dorm** as your **factor**. Then click **OK**.

ANOVA (Unstacked)
Step 1 Enter the Dormitory A data in **C1**, the Dormitory B data in **C2**, and the Dormitory C data in **C3**.
Step 2 Click Stat > ANOVA > One-Way (Unstacked).

Step 3 For **Responses (in separate columns)**, select columns C1–C3 and click **OK**.

Finding the p-Value for a Given F_{data}
(Example 11.12, pages 565–566, is used to illustrate the procedure.)

p-value $= P(F . F_{data}) = P(F > 0.8370997079)$, where $df_1 = 2$ and $df_2 = 27$.

Step 1 Click **Calc > Probability Distributions > F**.
Step 2 Select **Cumulative Probability**, enter **2** for **Numerator degrees of freedom** and **27** for **Denominator degrees of freedom**.
Step 3 Select **Input Constant**, enter **0.8370997079**, and click OK.

Step 4 Minitab then displays the cumulative probability $P(F < 0.8370997079) = 0.5561070428$. This cumulative probability represents the area to the left of 0.8370997079 (the unshaded area in Figure 11.30, page 566). Since the entire area under the curve equals 1, to get the p-value we need to subtract $P(F < 0.8370997079) = 0.5561070428$ from 1:

$$p\text{-value} = P(F > 0.8370997079) = 1 - P(F < 0.8370997079)$$
$$= 1 - 0.5561070428 = 0.4438929572$$

CRUNCHIT!

One-Way ANOVA
We will use the data from Example 11.13.

Step 1 Click **File ...** then highlight **Load from Larose Fundamentals 2e ... Chapter 12 ...** and click on **Example 11.13**.

Step 2 Click **Statistics** and select **ANOVA ... One-way**. Choose the **Columns** tab. Select each of the available columns and click **Calculate**.

| SECTION 11.3 | **Summary** |

1. Analysis of variance (ANOVA) is an inferential method for testing whether the means of different populations are equal. The null hypothesis always states that all the population means are equal, and the alternative hypothesis always states that not all the population means are equal. ANOVA works by comparing (a) the variability in the sample means and (b) the variability within each sample. If (a) is large *compared with* (b), this is evidence that the true means are not all equal and we should reject the null hypothesis.

2. ANOVA is usually performed using the p-value method and technology.

| SECTION 11.3 | **Exercises** |

Clarifying the Concepts

1. Does the overall sample mean always equal the mean of the sample means? Explain.

2. What does MSTR measure? What does MSE measure?

3. In your own words, explain how ANOVA works.

4. What are the required conditions for performing an analysis of variance?

5. A comparison dotplot of the SAT scores of three sororities shows no overlap at all between the groups. Does this represent evidence for or against the null hypothesis that all population means are equal?

6. True or false: If we reject the null hypothesis in an analysis of variance, then there is evidence that all the population mean sizes are different. If the statement is false, explain why it is false.

Practicing the Techniques

For Exercises 7–10, calculate the following measures.
  a. df_1 and df_2
  b. $\bar{\bar{x}}$
  c. SSTR
  d. SSE
  e. SST

7.

Sample A	Sample B	Sample C
$\bar{x}_A = 10$	$\bar{x}_B = 12$	$\bar{x}_C = 8$
$s_A = 1$	$s_B = 1$	$s_C = 1$
$n_A = 5$	$n_B = 5$	$n_C = 5$

8.

Sample A	Sample B	Sample C	Sample D
$\bar{x}_A = 10$	$\bar{x}_B = 12$	$\bar{x}_C = 8$	$\bar{x}_D = 14$
$s_A = 1$	$s_B = 1$	$s_C = 1$	$s_D = 1$
$n_A = 5$	$n_B = 5$	$n_C = 5$	$n_D = 5$

9.

Sample A	Sample B	Sample C	Sample D
$\bar{x}_A = 50$	$\bar{x}_B = 75$	$\bar{x}_C = 100$	$\bar{x}_D = 125$
$s_A = 5$	$s_B = 4$	$s_C = 6$	$s_D = 5$
$n_A = 100$	$n_B = 150$	$n_C = 200$	$n_D = 250$

10.

Sample A	Sample B	Sample C	Sample D
$\bar{x}_A = 0$	$\bar{x}_B = 10$	$\bar{x}_C = 20$	$\bar{x}_D = 10$
$s_A = 1.5$	$s_B = 2.25$	$s_C = 1.75$	$s_D = 2.0$
$n_A = 50$	$n_B = 100$	$n_C = 50$	$n_D = 100$

In Exercises 11–14, refer to the exercises cited and calculate the following measures.
  a. MSTR
  b. MSE
  c. F_{data}
  d. Construct the ANOVA table.

11. Exercise 7

12. Exercise 8

13. Exercise 9

14. Exercise 10

For Exercises 15–22, assume that the ANOVA assumptions are verified.

  For Exercises 15–18, test whether the population means differ, using $\alpha = 0.05$.
  a. State the hypotheses and the rejection rule.
  b. Calculate F_{data}. (*Hint:* You already calculated F_{data} in Exercises 11–14).
  c. Find the p-value.
  d. Compare the p-value with $\alpha = 0.05$. State the conclusion and the interpretation.

15. Data from Exercises 7 and 11

16. Data from Exercises 8 and 12

17. Data from Exercises 9 and 13

18. Data from Exercises 10 and 14

19. Part of an ANOVA table for an analysis of variance involving seven groups for a study follows. Each sample contained ten data values.

Source of variation	Sum of squares	Degrees of freedom	Mean square	F	p-value
Treatment	120	___	___	___	___
Error	315	___	___		
Total	___				

  a. Find all seven missing values in the table and fill in the blanks.
  b. Perform the appropriate hypothesis test using $\alpha = 0.05$.

20. Part of an ANOVA table for an analysis of variance involving three groups follows. Each sample contained six data values.

Source of variation	Sum of squares	Degrees of freedom	Mean square	F	p-value
Treatment	___	___	___	___	___
Error	90	___	___		
Total	150				

  a. Find all seven missing values in the table.
  b. Perform the appropriate hypothesis test using $\alpha = 0.01$.

21. Part of an ANOVA table follows.

Source of variation	Sum of squares	Degrees of freedom	Mean square	F	p-value
Treatment	___	4	10	1.0	___
Error	___	___	___		
Total	440				

  a. Find all five missing values in the table and fill in the blanks.
  b. Perform the appropriate hypothesis test using $\alpha = 0.10$.

22. Part of an ANOVA table follows.

Source of variation	Sum of squares	Degrees of freedom	Mean square	F	p-value
Treatment	___	2	___	2.0	___
Error	480	___	24		
Total	___				

  a. Find all five missing values in the table and fill in the blanks.
  b. Perform the appropriate hypothesis test using $\alpha = 0.05$.

Applying the Concepts

For Exercises 23–26, assume that the data are independently drawn random samples from normal populations.

    a. Verify the equal-variance assumption.
    b. Calculate the following measures.
        i. df_1 and df_2 **ii.** $\bar{\bar{x}}$ **iii.** SSTR   **iv.** SSE
        v. SST   **vi.** MSTR   **vii.** MSE   **viii.** F_{data}
    c. Construct the ANOVA table.
    d. Perform the appropriate one-way ANOVA using level of significance $\alpha = 0.05$.

23. Online, Hybrid, and Traditional Classrooms. A researcher randomly selected six students from each of three different treatment groups. The first group of students took elementary statistics online. The second group of students took the same course in the traditional in-class way. The third group of students took a hybrid course, which met once each week and also had an online component. The table shows the grade results. Researchers are interested in whether significant differences exist among the mean grades for the three groups. statclass

Online grades	Traditional grades	Hybrid grades
70	75	95
75	75	60
60	95	90
90	60	75
85	60	85
50	80	75
$\bar{x} = 71.6667$	$\bar{x} = 74.1667$	$\bar{x} = 80$
$s = 15.0555$	$s = 13.1972$	$s = 12.6491$

24. Store Sales. The district sales manager would like to determine whether there are significant differences in the mean sales among the four franchise stores in her district. Sales (in thousands of dollars) were tracked over 5 days at each of the four stores. The resulting data are summarized in the following table. storesales

Store A sales	Store B sales	Store C sales	Store D sales
10	20	3	30
15	20	7	25
10	25	5	30
20	15	10	35
20	20	4	30
$\bar{x} = 15$	$\bar{x} = 20$	$\bar{x} = 5.8$	$\bar{x} = 30$
$s = 5$	$s = 3.5355$	$s = 2.7749$	$s = 3.5355$

25. Education and Religious Background. The General Social Survey collected data on the number of years of education and the religious preference of the respondent. The summary statistics are shown here.

	n	Mean	Std. deviation
Protestant	1660	13.10	2.87
Catholic	683	13.51	2.74
Jewish	68	15.37	2.80
None	339	13.52	3.22
Other	141	14.46	3.18

26. The Full Moon and Emergency Room Visits. Is there a difference in emergency room visits before, during, and after a full moon? A study looked at the admission rate (number of patients per day) to the emergency room of a Virginia mental health clinic over a series of 12 full moons.[7] The data are provided in the table. Is there evidence of a difference in emergency room visits before, during, and after the full moon? fullmoon

Before		During		After	
6.4	11.5	5	13	5.8	13.5
7.1	13.8	13	16	9.2	13.1
6.5	15.4	14	25	7.9	15.8
8.6	15.7	12	14	7.7	13.3
8.1	11.7	6	14	11.0	12.8
10.4	15.8	9	20	12.9	14.5

27. ANOVA Can Be Applied to Two Populations. Researchers are interested in whether the mean heart rates of women and men differ. The following table provides summary statistics of random samples of pulse rates drawn from groups of women and men.

    a. Test using $\alpha = 0.05$ whether the population mean pulse rates differ.
    b. Which method of inference from an earlier chapter could we also use to solve this problem?

	Females	Males
n	65	65
$\bar{x}$	98.384	98.104
s	0.743	0.699

28. Store Sales. Refer to Exercise 24. *What if* the data are all wrong and all the stores actually have a sample mean sales of $30,000. Try to answer the following questions without touching your calculator.

a. Find the value of SSTR, MSTR, and F_{data}.

b. What would be the *p*-value of the ANOVA hypothesis test?

c. What would be the conclusion?

Bringing It All Together

Gas Mileage for European, Japanese, and American Cars.
Use this information for Exercises 29–31. The following figure shows a comparison boxplot of the vehicle mileage (in mpg) for random samples of automobiles manufactured in Europe, Japan, and the United States. The summary statistics are provided. We are interested in testing using $\alpha = 0.01$ whether population mean gas mileage differs among automobiles from the three regions. Assume that the assumptions are satisfied.

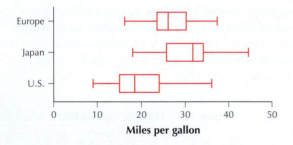

Miles per gallon

MPG	Sample 1: Europe	Sample 2: Japan	Sample 3: USA
Sample mean	$\bar{x}_1 = 27.603$	$\bar{x}_2 = 30.451$	$\bar{x}_3 = 20.033$
Sample standard deviation	$s_1 = 6.58$	$s_2 = 6.09$	$s_3 = 6.440$
Sample size	$n_1 = 68$	$n_2 = 79$	$n_3 = 245$

29. What Result Might We Expect?

a. Based on the graphical evidence in the comparison boxplot, what might be the conclusion? Explain your reasoning.

b. Perform the ANOVA, using whichever method you prefer.

c. Is your intuition from (a) supported?

30. Confidence Intervals as Further Clues in ANOVA.
Refer to Exercise 29. Suppose we construct a confidence interval for each of the population means. If at least one confidence interval does not overlap the others, then it is evidence against the null hypothesis.

a. Use a *t* interval from Section 8.2 to construct a 99% confidence interval for the population mean gas mileage of

i. European cars
ii. Japanese cars
iii. American cars

b. Is there one confidence interval from (a) that does not overlap the other two? If so, what does this mean in terms of the null hypothesis that all the population means are equal?

 31. Refer to the table of descriptive summaries of vehicle mileage. *What if* we discovered that we made a mistake in the data collection and that every Japanese vehicle tested actually had 1 mpg higher gas mileage than previously recorded. Explain how and why this change would affect the following measures—increase, decrease, or no change.

a. *n*   **f.** MSTR
b. *k*   **g.** MSE
c. SSTR   **h.** F_{data}
d. SSE   **i.** *p*-value
e. SST   **j.** Conclusion

32. Head Injuries and Vehicle Size.
This exercise uses the **Crash** data set, which contains information about the severity of injuries sustained by crash dummies when the National Transportation Safety Board crashed automobiles into a wall at 35 miles per hour. The variable *head_inj* contains a measure of the severity of the head injury sustained by crash dummies. The variable *size2* categorizes the type of vehicle, such as light, medium, heavy, pickup truck, MPV (SUV), and so on. The values of the variable *size2* are as follows: 1 5 compact car, 2 5 light car, 3 5 medium car, 4 5 heavy car, 5 5 minicompact car, 6 5 van, 7 5 pickup truck, and 8 5 MPV (SUV). Would you expect the population mean severity of head injuries suffered by the dummies to be the same across all the size categories? Use technology to perform the analysis of variance, using α 5 0.05. Comment on the results. crash

Use the *One-Way ANOVA* applet for Exercises 33 and 34.

33. Move the group means so that they are about the same by clicking and dragging the black dots so that they are about even horizontally.

a. What happens to the value of F (F_{data})?

b. Explain why this happens, using the concept of between-sample variability and the statistics SSTR, MSTR, and F_{data}.

34. Click Reset. Increase the **Pooled Standard Error**.

a. What happens to the value of F (F_{data})?

b. Explain why this change happens, using the concept of within-sample variability and the statistics SSE, MSE, and F_{data}.

11.4   INFERENCE IN REGRESSION

By the end of this section, I will be able to . . .

1   Explain the regression model and the regression model assumptions.

2   Perform the hypothesis test for the slope β_1 of the population regression equation.

3   Construct confidence intervals for the slope β_1.

4   Use confidence intervals to perform the hypothesis test for the slope β_1.

1   THE REGRESSION MODEL AND THE REGRESSION ASSUMPTIONS

Before we learn about the regression model and assumptions, let us review the correlation and regression topics that we learned in Chapter 4. Recall that the regression line approximates the relationship between two continuous variables and is described by the regression equation $\hat{y} = b_1 x + b_0$, where b_1 is the *slope* of the regression line, b_0 is the *y intercept*, x represents the *predictor variable*, y represents the *response variable*, and $\hat{y}$ represents the *estimated or predicted y-value*.

EXAMPLE 11.15

REVIEW OF REGRESSION TOPICS

textms

You may wish to refer to Section 4.1 for (a) and (b), and Section 4.2 for (c) and (d).

The Nielsen company has reported that the number of text messages that a person sends tends to decrease with age. Table 11.18 contains a random sample of 10 people, along with their age and the number of text messages they sent on the previous day.
a.   Construct and interpret a scatterplot of the response variable y versus the predictor variable x.
b.   Calculate and interpret the correlation coefficient r.
c.   Compute the regression equation $\hat{y} = b_1 x + b_0$. Interpret the meaning of the y intercept b_0 and the slope b_1 of the regression equation.
d.   Predict the number of text messages sent by a 20-year-old person, and calculate the prediction error (residual).

Table 11.18 Age and number of text messages

x = Age	y = Text messages	x = Age	y = Text messages
18	35	28	16
20	29	30	19
22	27	32	12
24	28	34	8
26	19	36	8

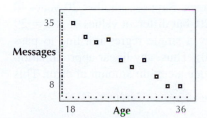

FIGURE 11.40 TI-83/84 scatterplot of messages versus age.

Solution

a.   Since the number of messages depends on age, and not vice versa, the predictor variable x is age and the response variable y is messages. Also, note that in **(d)** we are trying to predict the number of text messages, which tells us that messages is the response variable y since we never try to predict the known value of x. The TI-83/84 scatterplot is shown in Figure 11.40. As age increases, the number of messages tends to decrease.

FIGURE 11.41 TI-83/84 correlation and regression results.

b. Figure 11.41 shows the correlation coefficient $r \approx -0.9701$, calculated by the TI-83/84. Age and messages are negatively correlated. An increase in age is associated with a decrease in the number of messages.

c. Figure 11.41 shows that $a = b_1 = -1.5$ and $b = b_0 = 60.6$, and thus the regression equation is

$$\hat{y} = b_1 x + b_0 = (-1.5)\,(\text{age}) + 60.6$$

We can interpret b_0 and b_1 as follows:
- The y intercept $b_0 = 60.6$ is the estimated number of text messages sent by someone aged $x = 0$, which does not make sense because this value $x = 0$ lies far below the minimum value of x and therefore represents extrapolation.
- The slope $b_1 = -1.5$ means there is an estimated *decrease* of 1.5 in the number of text messages for each additional year of age.

d. For a 20-year-old person, the estimated number of daily text messages is

$$\hat{y} = b_1 x + b_0 = (-1.5)(20) + 60.6 = 30.6$$

The actual number of text messages sent by our 20-year-old in Table 11.18 is $y = 29$. Our prediction from **(c)** is $\hat{y} = 30.6$. Thus, our prediction error (or residual) is: $(y - \hat{y}) = (29 - 30.6) = -1.6$. Our 20-year-old sent slightly fewer text messages than expected.

Example 11.15 and our work in Chapter 4 on regression represented descriptive statistics. Next we turn to learning about *inference in regression*.

Note that the regression equation $\hat{y} = b_1 x + b_0 = (-1.5)(\text{age}) + 60.6$ depends on the sample. It is likely that a second sample will differ from the first, giving us a different regression line and different values for b_0 and b_1. In fact, for every different sample, b_0 and b_1 take different values since b_0 and b_1 are sample statistics. However, every sample comes from a population. Since we do not have data on the entire population, we are not able to calculate the population regression equation. The y intercept β_0 and slope β_1 of the population regression equation are unknown population parameters, just as μ and p are parameters in other contexts. Since the values of β_0 and β_1 are unknown, we need to perform inference to learn about them.

The **regression model** may be used to approximate the relationship between the predictor variable x and the response variable y for the *entire population* of (x, y) pairs.

Note that there is no "hat" on the y in the population regression equation because the equation represents a model of the relationship between the actual values of x and y, not an estimate of y.

> **Regression Model**
>
> The **population regression equation** is defined as
>
> $$y = \beta_1 x + \beta_0 + \varepsilon$$
>
> where β_0 is the y intercept of the population regression line, β_1 is the slope of the population regression line, and ε is the error term.

The 20-year-old in Table 11.18 sent 29 text messages. Suppose another 20-year-old sent 30 messages, so that both texters had age $x = 20$, but different values of y: $y = 29$ and $y = 30$. Then it would be impossible to draw a single regression line to pass through both $(x = 20, y = 29)$ and $(x = 20, y = 30)$. Thus, any linear approximation of the true relationship between x and y will introduce a certain amount of error. This is why the error term ε is needed.

Regression Model Assumptions

The regression model operates under a set of four assumptions that must be valid in order to perform the inference in this section.

> **Regression Model Assumptions**
>
> 1. **Zero-mean assumption.** The error term ε is a random variable, with a mean of 0. That is, the expected value of the random variable ε is 0: $E(\varepsilon) = 0$.
>
> 2. **Constant variance assumption.** The variance of ε, which is denoted as σ^2, is the same regardless of the value of x.
>
> 3. **Independence assumption.** The values of ε are independent of each other.
>
> 4. **Normality assumption.** The error term ε is a normal random variable.

To summarize, for each value of x, the values of y come from a normally distributed population with a mean on the population regression line $E(y) = \beta_1 x + \beta_0$ and constant standard deviation σ^2. Figure 11.42 illustrates how y is distributed for each value of x. Note that each normal curve has the same shape, indicating constant variance for each x.

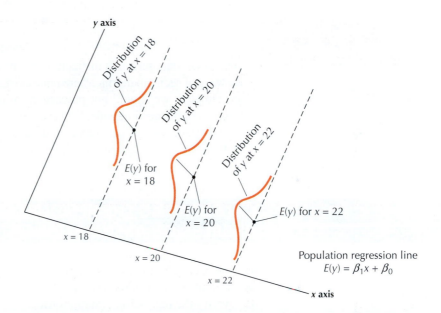

FIGURE 11.42
Illustrating the regression assumptions.

Verifying the Regression Assumptions

To check the regression model assumptions, we construct two graphs:

1. Scatterplot of the residuals (prediction errors $y - \hat{y}$) against the fitted values (**fitted values** refers to the predicted values, $\hat{y}$)

2. Normal probability plot of the residuals

Figure 11.43 shows four types of patterns that might be observed in the residuals versus fitted values plots.

- Plot (a) is a "healthy" plot, displaying no noticeable patterns.

- In plot (b) we see a curve, which indicates a violation of the independence assumption. Independence implies that knowing the value of a particular y does not help to predict the value of a different y. However, a curve suggests that knowing the value of a previous y helps in knowing the value of the next y.

- Plot (c) shows a "funnel" pattern, which contradicts the constant variance assumption. The residuals on the left are close together vertically (small variability), while the residuals on the right are far apart vertically (large variability).

- In plot (d) we see an increasing pattern, which violates the zero-mean assumption. The residuals on the left are all below the midline, so $E(y) < \beta_1 x + \beta_0$, while the residuals on the right are all above the midline, so $E(y) > \beta_1 x + \beta_0$.

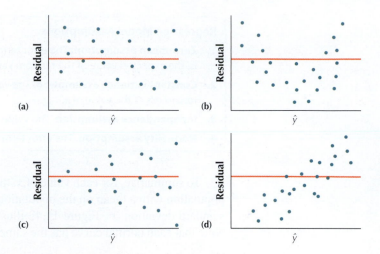

FIGURE 11.43
Patterns in the residuals versus predicted plots.

Developing Your Statistical Sense

Verifying the Regression Assumptions

With small data sets, it is difficult to ascertain whether or not patterns really exist. Be wary of seeing patterns where none exist. If one or more regression assumptions are violated, we should not proceed with inferential methods such as hypothesis tests or confidence intervals. However, even if one or more regression assumptions are violated, we can still report and interpret the descriptive regression statistics that we learned in Sections 4.2 and 4.3.

EXAMPLE 11.16

CALCULATING THE RESIDUALS AND VERIFYING THE REGRESSION ASSUMPTIONS

For the data in Example 11.15, do the following:
a. Calculate the residuals $y - \hat{y}$.
b. Verify the regression assumptions.

Solution

a. Table 11.19 contains the x and y data from Table 11.18, the fitted (predicted) values $\hat{y}$, and the residuals $y - \hat{y}$.

Table 11.19 Calculating the residuals

x = Age	y = Text messages	Fitted (predicted) values $\hat{y} = (-1.5)(\text{age}) + 60.6$	Residuals $y - \hat{y}$
18	35	33.6	1.4
20	29	30.6	−1.6
22	27	27.6	−0.6
24	28	24.6	3.4
26	19	21.6	−2.6
28	16	18.6	−2.6
30	19	15.6	3.4
32	12	12.6	−0.6
34	8	9.6	−1.6
36	8	6.6	1.4

b. The scatterplot in Figure 11.44 of the residuals versus fitted values shows no strong evidence of the unhealthy patterns shown in Figure 11.42. Thus, the independence assumption, the constant variance assumption, and the zero-mean assumption are verified. Also, the normal probability plot of the residuals in Figure 11.45 indicates no evidence of departures from normality in the residuals. Therefore we conclude that the regression assumptions are verified.

Now You Can Do
Exercises 7–14.

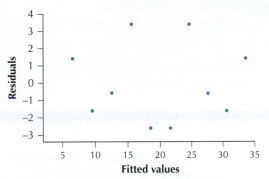

FIGURE 11.44 Scatterplot of residuals versus fitted values.

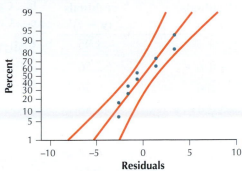

FIGURE 11.45 Normal probability plot of the residuals.

Once the regression assumptions have been verified, we may (a) perform hypothesis tests, and (b) construct confidence intervals for the population slope β_1.

2 HYPOTHESIS TESTS FOR SLOPE β_1

Suppose for a moment that, for the population regression equation $y = \beta_1 x + \beta_0 + \varepsilon$, the slope β_1 equals zero. Then the population regression equation would be

$$y = (0)x + \beta_0 + \varepsilon = \beta_0 + \varepsilon$$

That is,

- If β_1 *equals zero, then there is no relationship between x and y* because changing x in the equation $y = \beta_0 + \varepsilon$ does not affect y.

- If β_1 equals any other value, then there does exist a linear relationship between x and y.

This idea forms the basis for our inference in this section. To test whether there is a relationship between x and y, we begin with the hypothesis test to determine whether or not β_1 equals 0. The hypotheses are

$H_0: \beta_1 = 0$ There is no linear relationship between x and y.
$H_a: \beta_1 \neq 0$ There is a linear relationship between x and y.

Assuming $H_0: \beta_1 = 0$ is true, the test statistic t_{data} for this hypothesis test takes the following form.

Test Statistic t_{data}

$$t_{data} = \frac{b_1 - \beta_1}{s / \sqrt{\sum(x - \bar{x})^2}} = \frac{b_1 - 0}{s / \sqrt{\sum(x - \bar{x})^2}} = \frac{b_1}{s / \sqrt{\sum(x - \bar{x})^2}}$$

where b_1 represents the slope of the regression line, $s = \sqrt{\dfrac{SSE}{n - 2}}$ represents the standard error of the estimate (from Section 4.3), and $\sqrt{\sum(x - \bar{x})^2}$ represents the numerator of the sample variance of the x data (see page 103).

---

t_{data} consists of three quantities: b_1, s, and $\sqrt{\sum(x - \bar{x})^2}$. The next example shows how to calculate t_{data} by finding these three quantities.

EXAMPLE 11.17

CALCULATING t_{data}

Table 11.20 Calculating SSE

Residuals $y - \hat{y}$	Squared residuals $(y - \hat{y})^2$
1.4	1.96
−1.6	2.56
−0.6	0.36
3.4	11.56
−2.6	6.76
−2.6	6.76
3.4	11.56
−0.6	0.36
−1.6	2.56
1.4	1.96
	Sum = 46.4

All calculations up to the final result are expressed to nine decimal places.

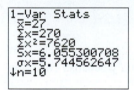

FIGURE 11.46 Summary statistics for the *x* (age) data.

Now You Can Do Parts (b)–(d) of Exercises 15–18 and parts (a)–(c) of Exercises 19–22.

Use the following steps to calculate the test statistic $t_{\text{data}} = \dfrac{b_1}{s/\sqrt{\Sigma(x - \bar{x})^2}}$ for the data in Table 11.20:

a. Find b_1, the slope of the regression line.
b. Calculate s, the standard error of the estimate.
c. Compute $\sqrt{\Sigma(x - \bar{x})^2}$, the numerator of the sample variance of the x data.

Solution

a. From Example 11.15, the slope of the regression line is $b_1 = -1.5$.
b. Recall from Section 4.3 (page 180) that

$$s = \sqrt{\frac{\text{SSE}}{n - 2}} = \sqrt{\frac{\Sigma(y - \hat{y})^2}{n - 2}} = \sqrt{\frac{\Sigma(\text{residual})^2}{n - 2}}$$

is the *standard error of the estimate.* Squaring each residual from Table 11.19 gives us the squared residuals in Table 11.20, and the sum of squared residuals, or sum of squares error, equal to

$$\text{SSE} = \Sigma(y - \hat{y})^2 = 46.4$$

Then the standard error of the estimate is $s = \sqrt{\dfrac{\text{SSE}}{n - 2}} = \sqrt{\dfrac{46.4}{8}} \approx 2.408318916$.

c. To compute $\Sigma(x - \bar{x})^2$, we note from page 103 that the sample variance of x is

$$s_x^2 = \frac{\Sigma(x - \bar{x})^2}{n - 1}$$

Multiplying each side of the equation by $n - 1$, we obtain an equation for the quantity $\Sigma(x - \bar{x})^2$:

$$\Sigma(x - \bar{x})^2 = (n - 1) \cdot s_x^2$$

The TI-83/84 output from Figure 11.46 shows that $s_x = 6.055300708$, and, since $n = 10$,

$$\Sigma(x - \bar{x})^2 = (n - 1) \cdot s_x^2 = (9)(6.055300708)^2 = 330$$

Therefore,

$$t_{\text{data}} = \frac{b_1}{s/\sqrt{\Sigma(x - \bar{x})^2}} = \frac{-1.5}{2.408318916/\sqrt{330}} \approx -11.3$$

Now that we have t_{data}, we can perform the hypothesis test for the slope β_1, as the next example shows using the critical-value method.

EXAMPLE 11.18

HYPOTHESIS TEST FOR SLOPE β_1 USING THE CRITICAL-VALUE METHOD

Test whether a linear relationship exists between age and text messages, using the data from Table 11.18 at level of significance $\alpha = 0.01$.

Solution

The regression assumptions were shown to be valid in Example 11.16. We may thus proceed with the hypothesis test.

STEP 1   **State the hypotheses.**

$H_0: \beta_1 = 0$ There is no linear relationship between age and text messages.
$H_a: \beta_1 \ne 0$ There is a linear relationship between age and text messages.

STEP 2   **Find the t critical value t_{crit} and the rejection rule.**
To find t_{crit}, use the t distribution table (Table D in the Appendix) for a two-tailed test and degrees of freedom df $= n - 2$. The rejection rule is

$$\text{Reject } H_0 \text{ if } t_{\text{data}} \ge t_{\text{crit}} \text{ or } t_{\text{data}} \le -t_{\text{crit}}.$$

Here, $n = 10$, so df $= 8$. For level of significance $\alpha = 0.01$, the t table gives us $t_{\text{crit}} = 3.355$. We will reject H_0 if $t_{\text{data}} \ge 3.355$ or $t_{\text{data}} \le -3.355$.

STEP 3   **Calculate t_{data}.**
From Example 11.17, we have

$$t_{\text{data}} = \frac{b_1}{s / \sqrt{\sum (x - \bar{x})^2}} \approx -11.3$$

**Now You Can Do
Exercises 15–18.**

STEP 4   **State the conclusion and the interpretation.**
Since $t_{\text{data}} \approx -11.3 \le -3.355$, we reject H_0. There is evidence at level of significance $\alpha = 0.01$ that $\beta_1 \ne 0$ and that there is a linear relationship between age and text messages.

The next example illustrates the steps for performing the hypothesis test for the slope β_1 using the p-value method.

EXAMPLE 11.19

HYPOTHESIS TEST FOR THE SLOPE β_1 USING THE p-VALUE METHOD AND TECHNOLOGY

shortmemory

In Section 4.3 we considered a study on short-term memory. Ten subjects were given a set of nonsense words to memorize within a certain amount of time and were later scored on the number of words they could remember. The results are repeated here in Table 11.21. Use the p-value method and technology to test using level of significance $\alpha = 0.01$ whether a linear relationship exists between time and score.

Solution

We begin by verifying the regression assumptions. The scatterplot of the residuals versus the fitted values in Figure 11.47 shows no strong evidence that the independence assumption, the constant variance assumption, or the zero-mean assumption is violated. Also, the normal probability plot of the residuals in Figure 11.48 offers evidence of the normality of the results. Therefore we conclude that the regression assumptions are verified, and proceed with the hypothesis test.

Table 11.21

Time (x)	Score (y)
1	9
1	10
2	11
3	12
3	13
4	14
5	19
6	17
7	21
8	24

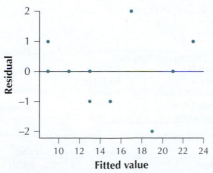

FIGURE 11.47 Residuals versus fitted values plot.

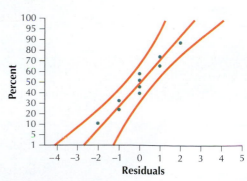

FIGURE 11.48 Normal probability plot of the residuals.

STEP 1 **State the hypotheses and the rejection rule.**

$H_0 : \beta_1 = 0$ There is no linear relationship between time and score.
$H_a : \beta_1 \neq 0$ There is a linear relationship between time and score.

Reject H_0 if the p-value ≤ 0.01.

STEP 2 **Calculate t_{data}.**

$$t_{data} = \frac{b_1}{s / \sqrt{\sum (x - \bar{x})^2}}$$

From page 178 in Section 4.3 we have $b_1 = 2$. From Example 4.14 on page 180 we have

$$s = \sqrt{\frac{12}{8}} \approx 1.224744871$$

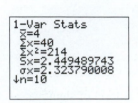

1-Var Stats
$\bar{x}=4$
$\Sigma x=40$
$\Sigma x^2=214$
$Sx=2.449489743$
$\sigma x=2.323790008$
$\downarrow n=10$

TI-83/84 summary statistics for
x (time) data.

From the TI-83/84 summary statistics, we have the standard deviation of the x (time) data to be $s_x = 2.449489743$. Thus, from Example 11.17:

$$\sum (x - \bar{x})^2 = (n - 1) \cdot s_x^2 \approx (9)2.449489743^2 = 54$$

Therefore,

$$t_{data} = \frac{b_1}{s / \sqrt{\sum (x - \bar{x})^2}} \approx \frac{2}{1.224744871 / \sqrt{54}} = 12$$

STEP 3 **Find the p-value.**

For instructions, see the Step-by-Step Technology Guide on page 581. The regression results (including the p-value) for the TI-83/84, Excel, and Minitab are shown in Figures 11.49, 11.50, and 11.51. (Differing results are due to rounding.)

Regression equation $\hat{y} = b_1 x + b_0$
(TI-83/84 expresses as $y = a + bx$)

$t_{data} = 12$

p-value of 2.1438667E-6 = 0.0000021439

Degrees of freedom, $n - 2 = 8$

$a = b_0 = 7$

$b = b_1 = 2$

Standard error of the estimate $s \approx 1.2247$

Coefficient of determination $r^2 \approx 0.9474$

Correlation coefficient $r \approx 0.9733$

LinRegTTest
y=a+bx
β≠0 and ρ≠0
t=12
P=2.1438667E-6
df=8
↓a=7

LinRegTTest
y=a+bx
β≠0 and ρ≠0
↑b=2
s=1.224744871
r²=.9473684211
r=.9733285268

FIGURE 11.49 TI-83/84 regression results.

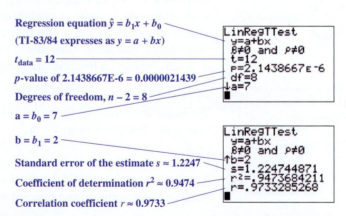

Regression Statistics	
Multiple R	0.9733
R Square	0.9474
Adjusted R Square	0.9408
Standard Error	1.2247
Observations	10

Correlation coefficient $r = 0.9733$
Coefficient of determination $r^2 = 0.9474$
Standard error of the estimate $s = 1.2247$
Sample size $n = 10$

	Coefficients	Standard Error	t Stat	P-value
Intercept	7	0.7710	9.0791	0.00001738
X Variable - Time	2	0.1667	12.0000	0.00000214

$b_0 = 7$
$b_1 = 2$
$t_{data} = 12$
p-value = 0.00000214

FIGURE 11.50 Excel regression results.

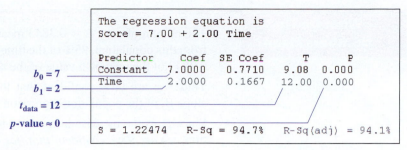

The regression equation is
Score = 7.00 + 2.00 Time

Predictor	Coef	SE Coef	T	P
Constant	7.0000	0.7710	9.08	0.000
Time	2.0000	0.1667	12.00	0.000

S = 1.22474   R-Sq = 94.7%   R-Sq(adj) = 94.1%

$b_0 = 7$
$b_1 = 2$
$t_{data} = 12$
$p\text{-value} \approx 0$

FIGURE 11.51 Minitab regression results.

Now You Can Do
Exercises 19–22,
parts (a)–(c).

STEP 4
Since the p-value of about 0.000 is $\leq \alpha = 0.01$, we reject H_0. There is evidence at level of significance $\alpha = 0.01$ for a linear relationship between time and score.

3   CONFIDENCE INTERVAL FOR SLOPE β_1

Recall that in Chapter 8 we constructed a confidence interval estimate for a population parameter, consisting of an interval of numbers that contain the parameter with a certain confidence level. Similarly, we can construct a confidence interval for the slope of the population regression equation β_1.

> **Confidence Interval for β_1**
>
> When the regression assumptions are met, a $100(1 - \alpha)\%$ confidence interval for β_1 is given by
>
> $$b_1 \pm t_{\alpha/2} \cdot \frac{s}{\sqrt{\sum (x - \bar{x})^2}}$$
>
> where b_1 is the point estimate of the slope β_1 of the population regression equation, s is the standard error of the estimate, and $t_{\alpha/2}$ has $n - 2$ degrees of freedom.
>
> **Margin of Error E**
>
> The margin of error for a $100(1 - \alpha)\%$ confidence interval for β_1 is given by
>
> $$E = t_{\alpha/2} \cdot \frac{s}{\sqrt{\sum (x - \bar{x})^2}}$$

Thus, the confidence interval for β_1 takes the form $b_1 \pm E$.

EXAMPLE 11.20

CONFIDENCE INTERVAL FOR THE SLOPE β_1

Construct a 95% confidence interval for the slope β_1 of the population regression equation for the memory-test data in Example 11.19.

Solution

The regression assumptions were verified in Example 11.19, where we found:
- $b_1 = 2$,
- $s = 1.224744871$, and
- $\sum (x - \bar{x})^2 = 54$.

From the t table (Appendix Table D), we find that, for 95% confidence, $t_{\alpha/2}$ for $n - 2 = 10 - 2 = 8$ degrees of freedom is $t_{\alpha/2} = 2.306$. So, our margin of error E is

$$E = t_{\alpha/2} \cdot \frac{s}{\sqrt{\sum (x - x)^2}} = (2.306) \left(\frac{1.224744874}{\sqrt{54}} \right) \approx 0.3843$$

Now You Can Do
Exercises 23–30.

The 95% confidence interval for β_1 is then given by

$$b_1 \pm E = 2 \pm 0.3843 = (1.6157, 2.3843)$$

What Do These Numbers Mean?

- The margin of error $E = 0.3843$ means that, when we repeatedly take samples from this population, 95% of the time the sample estimate b_1 will be within $E = 0.3843$ of the unknown value of the slope β_1 of the population regression line.

- Thus, we are 95% confident that the interval (1.6157, 2.3843) captures the slope β_1 of the population regression line.

- Since β_1 is the increase in memory-test score per added minute of memorization, *we are 95% confident that, for each additional minute of memorization, the increase in memory-test score will lie between 1.6157 and 2.3843 points.*

4 USING CONFIDENCE INTERVALS TO PERFORM THE *t* TEST FOR THE SLOPE β_1

As in earlier sections, we may use a $100(1 - \alpha)\%$ *t* confidence interval for the slope β_1 to perform the *t* test for β_1, which is a two-tailed test.

> **Equivalence of a Two-Tailed *t* Test About β_1 and a *t* Confidence Interval for β_1**
>
> - If a $100(1 - \alpha)\%$ t confidence interval for β_1 does not contain zero, then we would reject $H_0 : \beta_1 = 0$ for level of significance α, and conclude that a linear relationship exists between *x* and *y*.
> - If a $100(1 - \alpha)\%$ t confidence interval for β_1 does contain zero, then we would not reject $H_0 : \beta_1 = 0$ for level of significance α.

EXAMPLE 11.21 **USING CONFIDENCE INTERVALS TO PERFORM THE *t* TEST FOR THE SLOPE β_1**

 textms

a. Construct and interpret a 99% confidence interval for the slope β_1 for the text messaging data in Table 11.18.

b. Use the confidence interval in **(a)** to test whether a linear relationship exists between age and text messages, using level of significance $\alpha = 0.01$.

Solution

a. The regression assumptions were verified in Example 11.16. Also,
- In Example 11.15, we found $b_1 = -1.5$.
- In Example 11.17, we calculated $s = 2.408318916$, and $\sum(x - \bar{x})^2 = 330$.

From the *t* table, we find that, for 99% confidence, $t_{\alpha/2}$ for $n - 2 = 10 - 2 = 8$ degrees of freedom is $t_{\alpha/2} = 3.355$. So, our margin of error E is

$$E = t_{\alpha/2} \cdot \frac{s}{\sqrt{\sum(x - \bar{x})^2}} = (3.355)\left(\frac{2.408318916}{\sqrt{330}}\right) \approx 0.4448$$

The 99% confidence interval for β_1 is then given by

$$b_1 \pm E = -1.5 \pm 0.4448 = (-1.9448, -1.0552)$$

We are 99% confident that the interval $(-1.9448, -1.0552)$ captures the slope β_1 of the population regression line. That is, we are 99% confident that, for each additional year of age, the *decrease* in the number of text messages lies between 1.9448 and 1.0552.

b. The hypotheses are

$$H_0 : \beta_1 = 0 \quad \text{There is no linear relationship between age and text messages.}$$
$$H_a : \beta_1 \neq 0 \quad \text{There is a linear relationship between age and text messages.}$$

Since the confidence interval from **(a)** does not contain zero, we may conclude that a linear relationship exists between age and text messages, at level of significance $\alpha = 0.01$.

STEP-BY-STEP TECHNOLOGY GUIDE: Regression Analysis

Data from Example 11.19 (page 581) are used to illustrate the steps.

TI-83/84

Step 1 Enter the **X (Time)** data in **L1** and the **Y (Score)** data in **L2**.
Step 2 Press **STAT**, highlight **CALC**, and press **4** to choose **LinReg(ax+b)**. On the home screen, the following command appears: **LinReg(ax+b)**.
Step 3 Press **ENTER**. The output shows **y = ax+b, a=7, b=2**. The TI-83/84 denotes the slope β_1 as **a** and the y intercept b_0 as b. Thus the TI-83/84 is telling you that the estimated regression equation is $\hat{y} = 2x+7$.
Step 4 Now Press **STAT** again and press the right arrow key until **TESTS** is highlighted.

Step 5 Press the **down arrow** key until **E** is highlighted (for **LinRegTTest**).
Step 6 Press **ENTER**. The LinRegTTest menu appears.
Step 7 For **Xlist**, enter **L1** (or whichever list you entered the X data in).
Step 8 For **Ylist**, enter **L2** (or whichever list you entered the Y data in).
Step 9 For **Freq**, enter **1**, and for b & r highlight "$\neq 0$".
Step 10 Move the cursor over **Calculate**, make sure all your entries are correct, and press **ENTER**. The results are as shown in Figure 11.49 (page 582).

EXCEL

Step 1 Enter the "Time" variable in column **A** and the "Score" variable in column **B**.
Step 2 Click on **Data** > **Data Analysis** > **Regression** and click **OK**.
Step 3 For **Input Y Range**, select cells **B1 − B10**. For **Input X Range**, select cells **A1 − A10**.

Step 4 If you would like to verify the regression assumptions, then select **Residual Plots** and **Normal Probability Plots**.
Step 5 Click **OK**. The results are as shown in Figure 11.50 (page 582).

MINITAB

Step 1 Enter the "Time" variable in **C1** and the "Score" variable in **C2**.
Step 2 Click on **Stat** > **Regression** > **Regression**.
Step 3 Select "Score" as your **Response Variable** and "Time" as your **Predictor Variable**.

Step 4 If you would like to verify the regression assumptions, click the button labeled **Graphs** and select **Four in One**.
Step 5 Click **OK** twice. The results are as shown in Figure 11.51 (page 583).

SECTION 11.4   Summary

1. This section examines inferential methods for regression analysis. The regression model, or the (population) regression equation, is $y = \beta_1 x + \beta_0 + \varepsilon$, where β_0 is the y intercept of the population regression line, β_1 is the slope of the population regression line, and ε is the error term.

2. A hypothesis test may be performed to determine whether a linear relationship exists between x and y.

3. We can construct confidence intervals for the true value of the population regression slope β_1 since it is unknown.

SECTION 11.4   Exercises

Clarifying the concepts

1. What is the difference between the regression equation (calculated using the sample) and the population regression equation?

2. What are the four regression model assumptions?

3. How do we go about verifying the regression model assumptions?

4. What is the difference between b_0 and b_1 on the one hand and β_0 and β_1 on the other hand?

5. What does it mean for the relationship between x and y when β_1 equals 0?

6. What is the difference between s and s_x?

Practicing the Techniques

For Exercises 7–14, you are given the regression equation.
    a. Calculate the predicted values.
    b. Compute the residuals.
    c. Construct a scatterplot of the residuals versus the predicted values.
    d. Use technology to construct a normal probability plot of the residuals.
    e. Verify that the regression assumptions are valid.

7.

x	y
1	15
2	20
3	20
4	25
5	25

$\hat{y} = 2.5x + 13.5$

8.

x	y
0	10
5	20
10	45
15	50
20	75

$\hat{y} = 3.2x + 8$

9.

x	y
−5	0
−4	8
−3	8
−2	16
−1	16

$\hat{y} = 4x + 21.6$

10.

x	y
−3	−5
−1	−15
1	−20
3	−25
5	−30

$\hat{y} = -3x - 16$

11.

x	y
10	100
20	95
30	85
40	85
50	80

$\hat{y} = -0.5x + 104$

12.

x	y
0	11
20	11
40	16
60	21
80	26

$\hat{y} = 0.2x + 9$

13.

x	y
1	1
2	1
3	2
4	3
5	3

$\hat{y} = 0.6x + 0.2$

14.

x	y
1	6
2	5
2	4
2	3
3	2

$\hat{y} = -2x + 8$

For Exercises 15–18, follow these steps. Assume that the regression model assumptions are valid.
    a. Find t_{crit} for a two-tailed test with $\alpha = 0.05$ and $df = n - 2$.
    b. Calculate s.
    c. Compute $\Sigma (x - \bar{x})^2$.
    d. Calculate t_{data}.
    e. Perform the hypothesis test for the linear relationship between x and y, using the critical-value method and $\alpha = 0.05$.

15. Data in Exercise 7, where $b_1 = 2.5$

16. Data in Exercise 8, where $b_1 = 3.2$

17. Data in Exercise 9, where $b_1 = 4.0$

18. Data in Exercise 10, where $b_1 = -3$

For Exercises 19–22, follow these steps. Assume that the regression model assumptions are valid.
    a. Calculate s.
    b. Compute $\Sigma (x - \bar{x})^2$.
    c. Calculate t_{data}.
    d. Find $p\text{-value} = 2 \cdot P(t > |t_{\text{data}}|)$.
    e. Perform the hypothesis test for the linear relationship between x and y using the p-value method and $\alpha = 0.05$.

19. Data in Exercise 11, where $b_1 = -0.5$

20. Data in Exercise 12, where $b_1 = 0.2$

21. Data in Exercise 13, where $b_1 = 0.6$

22. Data in Exercises 14, where $b_1 = -2$

For Exercises 23–30, follow these steps. Assume that the regression model assumptions are valid.
    a. Find $t_{\alpha/2}$ for a 95% confidence interval for β_1.
    b. Find the margin of error E.
    c. Construct a 95% confidence interval for β_1.
    d. Use the confidence interval from **(c)** to perform the t test for β_1 at level of significance $\alpha = 0.05$.

23. Data in Exercise 7

24. Data in Exercise 8

25. Data in Exercise 9

26. Data in Exercise 10

27. Data in Exercise 11

28. Data in Exercise 12

29. Data in Exercise 13

30. Data in Exercise 14

Applying the Concepts

For Exercises 31–36, follow steps **(a)** and **(b)**.
    a. Verify the regression model assumptions. (*Hint:* You can use either Excel or Minitab; see the Step-by-Step Technology Guide on page pages 585–586.)

b. Perform the hypothesis test for the linear relationship between x and y, using level of significance $\alpha = 0.05$.

31. Volume and Weight. The following table contains the volume (x, in cubic meters) and weight (y, in kilograms) of five randomly chosen packages shipped to a local college. volweight

Volume (x)	Weight (y)
4	10
8	16
12	25
16	30
20	35

32. Family Size and Pets. Shown in the accompanying table are the number of family members (x) in a random sample taken from a suburban neighborhood, along with the number of pets (y) belonging to each family. familypet

Family size (x)	Pets (y)
2	1
3	2
4	2
5	3
6	3

33. World Temperatures. Listed in the following table are the low (x) and high (y) temperatures for a particular day, measured in degrees Fahrenheit, for a random sample of cities worldwide. worldtemp

City	Low (x)	High (y)
Kolkata	57	77
London	36	45
Montreal	7	21
Rome	39	55
San Juan	70	83
Shanghai	34	45

34. NCAA Power Ratings. The accompanying table shows the team's winning percentage (x) and power rating (y) for the 2011 NCAA Basketball Tournament, according to www.teamrankings.com. ncaa

School	Win%(x)	Rating (y)
Ohio State	91.9	121.0
Kansas	92.1	119.5
San Diego State	91.4	118.1
Duke	86.5	117.8
Connecticut	77.5	117.5
Pittsburgh	82.4	116.9
Kentucky	76.3	116.6
Notre Dame	79.4	116.3

35. Stock Prices. Would you expect there to be a relationship between the price (x) of a stock and its change in price (y) on a particular day? The table provides stock price and stock price change for June 1, 2011, for a random sample of 8 stocks. stocks

Stock	Price (x)	Change (y)
Bank of America	11.38	−0.36
Sirius XM Radio	2.3	−0.05
Microsoft	24.49	−0.52
General Electric	19.35	−0.28
Intel	22.34	−0.17
Pfizer	21.17	−0.28
Dell	15.75	−0.34
Lucent	5.69	0.02

For Exercises 36–40, do the following for the indicated data.
   a. Calculate the margin of error E for a 95% confidence interval for b_1.
   b. Construct a 95% confidence interval for β_1.
   c. Interpret the confidence interval.

36. Data from Exercise 31

37. Data from Exercise 32

38. Data from Exercise 33

39. Data from Exercise 34

40. Data from Exercise 35

41. Batting Average and Runs Scored. The table shows the top ten hitters in Major League Baseball for 2007. We are interested in estimating the number of runs scored (y) using the player's batting average (x). mlbhitters

Player	Team	Batting average (x)	Runs scored (y)
M. Ordonez	Detroit Tigers	.363	117
I. Suzuki	Seattle Mariners	.351	111
P. Polanco	Detroit Tigers	.341	105
M. Holliday	Colorado Rockies	.340	120
J. Posada	New York Yankees	.338	91
C. Jones	Atlanta Braves	.337	108
D. Ortiz	Boston Red Sox	.332	116
H. Ramirez	Florida Marlins	.332	125
E. Renteria	Atlanta Braves	.332	87
C. Utley	Philadelphia Phillies	.332	104

a. Construct a residuals versus predicted values plot. What type of pattern do you see?
b. Which regression assumption is violated?
c. Should we construct a confidence interval or perform a hypothesis test for the slope of the regression line?
d. Is it still appropriate to report the descriptive statistics we learned in Sections 4.2 and 4.3? Why?

42. Challenge Exercise. Suppose a regression analysis of y on x was found to be significant (that is, the null hypothesis was rejected) and the slope $b_1 > 0$. Consider the observation (max x, y), which represents the (x, y) data value for the maximum value of x in the data set. Suppose the residual for (max x, y) is negative. *What if* we increase max x by an arbitrary amount c so that the new data value is (max $x + c$, y). (All other data values in the data set are unchanged.) How will this increase affect the following measures? Will they increase, decrease, or remain unchanged, or is there insufficient information to determine the effect?

a. n e. MSE
b. SSE   f. MSR
c. SST   g. F
d. SSR

43. Challenge Exercise. Refer to Exercise 42. How and why will the change affect the following measures?

a. t_{data} d. p-value
b. r^2 e. Conclusion
c. s

Bringing It All Together

SAT Reading and Math Scores. Use this information for Exercises 44–48. The table shows the SAT scores for five students. We are interested in whether a linear relationship exists between the SAT Reading score (x) and the SAT Math score (y). statesat

Student	SAT Reading	SAT Math
Michael	497	510
Ashley	515	515
Tyler	518	523
Emily	501	514
Taylor	522	521

44. What Result Might We Expect? Consider the accompanying scatterplot of Math score versus Reading score. Is there evidence for or against the null hypothesis that no linear relationship exists? Explain. studentsat

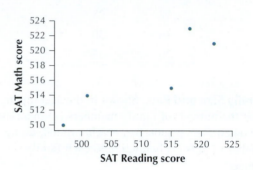

45. Consider the following graphics. Is there strong evidence that the regression assumptions are violated?

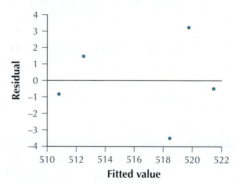

Plot of residuals versus fitted values.

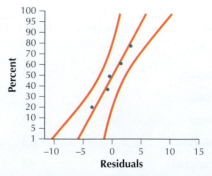

Normality plot of residuals.

46. Test whether a linear relationship exists between the SAT reading score and the SAT Math score using level of significance $\alpha = 0.10$.

47. Construct and interpret a 90% confidence interval for a slope β_1. statesat

48. Do your inferences in Exercises 45 and 46 agree with each other? Explain.

For Exercises 49–51 use technology to solve the following problems.
   a. Verify the regression model assumptions.
   b. Construct and interpret a 95% confidence interval for β_1.
   c. Based on the confidence interval constructed in **(b)**, would you expect the hypothesis test to reject the null hypothesis that $\beta_1 = 0$?

   d. Test at $\alpha = 0.05$ whether a linear relationship exists between x and y.

49. Open the **Darts** data set, which we used for the Chapter 3 Case Study. Use the Dow Jones Industrial Average (x) to estimate the pros' performance (y). Darts

50. Open the **Nutrition** data set. Estimate the number of calories per gram (y) using the amount of fat per gram (x). Nutrition

51. Open the **PulseandTemp** data set. Estimate body temperature (y) using heart rate (x). PulseandTemp

CHAPTER 11   Formulas and Vocabulary

Section 11.1
- **CONDITIONS FOR PERFORMING A GOODNESS OF FIT TEST** (p. 534)
- χ^2 **GOODNESS OF FIT TEST** (p. 532)
- **MULTINOMIAL RANDOM VARIABLE** (p. 531)
- **TEST STATISTIC FOR THE GOODNESS OF FIT TEST** (p. 534).

$$\chi^2_{data} = \sum \frac{(O_i - E_i)^2}{E_i}$$

Section 11.2
- χ^2 **TEST FOR INDEPENDENCE** (p. 544)
- **CONDITIONS FOR PERFORMING BOTH THE TEST FOR INDEPENDENCE AND THE TEST FOR THE HOMOGENEITY OF PROPORTIONS** (p. 545)
- **TEST FOR THE HOMOGENEITY OF PROPORTIONS** (p. 548)
- **TEST STATISTIC FOR BOTH THE TEST FOR INDEPENDENCE AND THE TEST FOR THE HOMOGENEITY OF PROPORTIONS** (p. 545).

$$\chi^2_{data} = \sum \frac{(O_i - E_i)^2}{E_i}$$

Section 11.3
- **ANALYSIS OF VARIANCE (ANOVA)** (p. 558)
- F_{crit} (p. 569)
- F critical values for a given area α to the left (p. 569).
- F_{data} (p. 563).

$$F_{data} = \frac{MSTR}{MSE}$$

- **HYPOTHESES FOR ANALYSIS OF VARIANCE** (p. 565).

$$H_0: \mu_1 = \mu_2 = \cdots = \mu_k$$

VERSUS

$$H_a: \text{not all the population means are equal}$$

- **MEAN SQUARE ERROR (MSE)** (p. 563).

$$MSE = \frac{\sum (n_i - 1)s_i^2}{n_t - k}$$

- **MEAN SQUARE TREATMENT (MSTR)** (p. 563).

$$MSTR = \frac{\sum n_i(\bar{x}_i - \bar{\bar{x}})^2}{k - 1}$$

- **OVERALL SAMPLE MEAN, $\bar{\bar{x}}$** (p. 562).

$$\bar{\bar{x}} = \frac{(n_1\bar{x}_1 + n_2\bar{x}_2 + \cdots + n_k\bar{x}_k)}{n_t}$$

- p-**VALUE** (p. 565)
- **SUM OF SQUARES ERROR (SSE)** (p. 564).

$$SSE = \sum (n_i - 1)s_i^2$$

- **SUM OF SQUARES TREATMENT (SSTR)** (pp. 563–564).

$$SSTR = \sum n_i(\bar{x}_i - \bar{\bar{x}})^2$$

- **TOTAL SUM OF SQUARES (SST)** (p. 564).

$$SST = SSTR + SSE$$

Section 11.4
- **CONFIDENCE INTERVAL FOR SLOPE β_1** (p. 583).

$$b_1 \pm t \cdot \frac{s}{\sqrt{\sum (x - \bar{x})^2}}$$

- **FITTED VALUES** (p. 577)
- **MARGIN OF ERROR E** (p. 583)
- **POPULATION REGRESSION EQUATION** (p. 576)
- **REGRESSION MODEL** (p. 576)
- **REGRESSION MODEL ASSUMPTIONS** (p. 577)
- **TEST STATISTIC t_{data}** (p. 578).

$$t_{data} = \frac{b_1}{s/\sqrt{\sum (x - \bar{x})^2}}$$

Section 11.1

For Exercises 1–3, perform the χ^2 goodness of fit test.

1. ALCOHOL ABUSE AND DEPENDENCE IN COLLEGE. A report found that 25% of college students had abused alcohol in the last 12 months, while a further 6% (not counted in the 25%) were alcohol-dependent.[9] Suppose that a new survey of 1000 randomly selected college students finds 275 who had abused alcohol in the last 12 months and a further 50 (not counted in the 275) who are alcohol-dependent. Test whether the population proportions have changed, using level of significance $\alpha = 0.10$.

2. TRULY RANDOM LOTTERY DRAWING? Have you ever wondered whether lottery drawings are truly random? For example, the accompanying histogram shows the frequencies of the third digit in the Maryland lottery's Pick 3 game (218 drawings from September 1989 to April 1990). In a Pick 3 game, you choose a three-digit number between 000 and 999, and if your number comes up, you win the cash prize. Notice that 1 appears as the third digit least of all the digits, and quite a bit less often than some of the other digits. Does the relative scarcity of 1s indicate that the system is flawed?

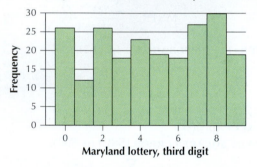

Frequency histogram of third digits in Maryland lottery's Pick 3 game.

The relative frequency distribution of the third digit is shown in the following table. We would, of course, expect each digit to show up 10% of the time. Test whether the population proportions of digits are all 0.10, using level of significance $\alpha = 0.05$. marylandlott

Digit	Count	Percent
0	26	11.93
1	12	5.50
2	26	11.93
3	18	8.26
4	23	10.55
5	19	8.72
6	18	8.26
7	27	12.39
8	30	13.76
9	19	8.72
N =	218	

3. ALTERNATIVE MEDICINE USE. A study examined the prevalence of alternative medicine usage by age group among persons with diabetes.[10] In the study, 5.7% of the subjects were aged 18–34 years, 20.7% were aged 35–49 years, 38.8% were aged 50–64 years, and 34.8% were age 65 or older. Suppose that a study conducted this year found that, of the 1000 randomly selected respondents with diabetes, 70 were 18–34 years old, 220 were 35–49 years old, 440 were 50–64 years old, and 270 were over age 65. Test using level of significance $\alpha = 0.05$ whether the proportions have changed.

4. SEPTEMBER 11 AND PEARL HARBOR. The terrorist attacks on New York City and Washington, D.C., on September 11, 2001, were often compared to the Japanese attack on Pearl Harbor on December 7, 1941. In an NBC News Terrorism Poll, the following question was asked: Would you say that Tuesday's attacks are more serious than, equal to, or not as serious as the Japanese attack on Pearl Harbor? This poll was conducted on September 12, 2001, and the results are given in the accompanying table. Were there systematic differences in the way men and women responded to this question? In other words, are the variables *poll response* and *gender* independent? Perform the χ^2 test for independence between *poll response* and *gender,* using level of significance $\alpha = 0.01$. terroristpoll

	Gender		
	Male	**Female**	**Total**
More serious	200	212	412
Equal	70	84	154
Not as serious	23	6	29
Not sure	11	12	23
Total	304	314	618

5. HAPPINESS IN MARRIAGE. The General Social Survey tracks trends in American society. The accompanying crosstabulation shows the responses to a question that asked people to characterize their feelings about being married. Test whether happiness in marriage is independent of gender, using level of significance $\alpha = 0.05$. happymarriage

Respondents' gender	Happiness in Marriage			
	Very happy	**Pretty happy**	**Not too happy**	**Total**
Male	242	115	9	366
Female	257	149	17	423
Total	499	264	26	789

6. PREGNANCY AND HIV TESTING. A study examined the proportions of pregnant women in the United States who have had an HIV test in the past 12 months.[11] The proportions for the Northeast, Midwest, South, and West were 56.8%, 49.3%, 58.5%, and 50.2%. Test whether the population proportions of pregnant women who have had an HIV test in the past 12 months are the same across all four regions, using level of significance $\alpha = 0.01$. Assume that each sample size equals 1000.

7. THE DIGITAL DIVIDE: ACCOUNTING FOR INCOME. It is well known that a greater proportion of whites than blacks use the Internet. This is one aspect of what is known as the "digital divide." However, what if we control for income? That is, suppose that we consider only whites, blacks, and Hispanics of a certain annual income range, say, more than $50,000. The Pew Internet and American Life Project conducted a survey in which the following proportions of respondents with incomes above $50,000 were found to be using the Internet. Test whether the digital divide exists after accounting for income. That is, test whether or not there is a significant difference in Internet use levels among the races. Use level of significance $\alpha = 0.05$. Assume each sample size equals 400.

Whites	Blacks	Hispanics
82%	65%	82%

SECTION 11.3

8. For the following data, assume that the ANOVA assumptions are met, and calculate the measures in (a)–(h).

Sample A	Sample B	Sample C	Sample D
$\bar{x}_A = 0$	$\bar{x}_B = 10$	$\bar{x}_C = 20$	$\bar{x}_D = 10$
$s_A = 1.5$	$s_B = 2.25$	$s_C = 1.75$	$s_D = 2.0$
$n_A = 50$	$n_B = 100$	$n_C = 50$	$n_D = 100$

    a. df_1 and df_2
    b. $\bar{\bar{x}}$
    c. SSTR
    d. SSE
    e. SST
    f. MSTR
    g. MSE
    h. F_{data}

9. Construct the ANOVA table for the statistics in Exercise 1.

For Exercises 10–11, assume that the ANOVA assumptions are met and perform the appropriate analysis of variance using $\alpha = 0.05$.

10. DIFFERENCES IN MEDICAL TREATMENTS. A psychologist is interested in investigating whether differences in mean client improvement exist for three medical treatments. Seven clients undergoing each medical treatment were asked to rate their level of satisfaction on a

scale of 0 to 100. The data are provided in the following table. ⬤ medicaltreatmt

Medical treatment 1	Medical treatment 2	Medical treatment 3
75	75	100
100	100	100
0	25	50
50	75	90
50	50	75
40	75	75
25	60	90

11. CUSTOMER SATISFACTION. The district sales manager of a local chain store would like to determine whether there are significant differences in the mean customer satisfaction among the four franchise stores in her district. Customer satisfaction data were gathered over seven days at each of the four stores. The resulting data are summarized in the accompanying table. ⬤ customersatisfy

Customer satisfaction in four stores

Store A	Store B	Store C	Store D
50	60	25	75
40	45	30	60
60	70	50	80
60	70	30	90
50	60	40	70
45	65	25	85
55	70	45	95
$\bar{x}_A = 51.43$	$\bar{x}_B = 62.86$	$\bar{x}_C = 35.00$	$\bar{x}_D = 79.29$
$s_A = 7.48$	$s_B = 9.06$	$s_C = 10.00$	$s_D = 12.05$

Section 11.4

For Exercises 12–14, test whether there is a linear relationship between x and y, using level of significance $\alpha = 0.05$.

12. EDUCATION AND EARNINGS. The U.S. Census Bureau reports the mean annual earnings of American citizens according to the number of years of education. We are interested in the relationship between earnings (y, in thousands of dollars) and years of education (x). ⬤ eduearn

Education (x)	Annual earnings (y)
8	18.6
10	18.9
12	27.3
13	29.7
14	34.2
16	51.2
18	60.4

13. High School GPA and College GPA. The college admissions office would like to determine if there is a relationship between the high school grade point average and the first-year college grade point average of first-year college students, using the data in the following table. 🐾 gpa

GPA Student	High school GPA (x)	First-year college (y)
1	2.4	2.6
2	2.5	1.9
3	2.9	2.7
4	2.7	2.5
5	3.0	2.4
6	3.5	2.9
7	3.0	2.7
8	3.6	3.1
9	3.4	3.0
10	3.9	3.3

14. Used Cars: Price versus Age. Do you think you can predict the price of a used car based on how old it is? The table shows the price (in thousands of dollars) and the

age (in years) of 10 previously owned vehicles of the same make and model. 🐾 ageprice

Car	Age (x)	Price (y)
1	1	18.0
2	2	16.0
3	3	15.5
4	4	13.5
5	4	14.5
6	5	10.5
7	5	12.0
8	6	9.5
9	7	8.5
10	8	7.0

For Exercises 15–17, construct and interpret a 95% confidence interval for β_1.

15. Data in Exercise 12

16. Data in Exercise 13

17. Data in Exercise 14

CHAPTER 11 | **Quiz**

True or False

1. True or false: The F curve is symmetric.
2. True or false: In a χ^2 test for independence, the degrees of freedom equals $k - 1$.
3. True or false: If we reject the null hypothesis in an ANOVA, we conclude that there is evidence that all the population means are different.

Fill in the Blank

4. In the test for the homogeneity of proportions, the null hypothesis states that all k population proportions are_____.
5. In ANOVA the _____ _____ _____ [three words] measures the variability in the sample means.
6. In ANOVA the _____ _____ _____ [three words] measures the variability within the samples.

Short Answer

7. In ANOVA what do we use for an estimate of the overall population mean?
8. In the test for the homogeneity of proportions, which hypothesis states that not all population proportions are equal?
9. How does one calculate the degrees of freedom for the χ^2 test for independence?

Calculations and Interpretations

10. Illicit Drug Use Among Young People. Monitoring the Future (**www.monitoringthefuture.org**), at the University of Michigan, is an "an ongoing study of the behaviors, attitudes, and values of American secondary school students, college students, and young adults." They reported the lifetime prevalence of the use of any illicit drug among 8th-graders, 10th-graders, and 12th-graders, as shown in the table. Test using level of significance $\alpha = 0.01$ for differences among the proportions of children in those grades who have ever used an illicit drug.

	8th-graders	10th-graders	12th-graders
Have used an illicit drug	3,655	6,527	7,461
Have never used an illicit drug	13,345	9,873	7,139

11. Beef Cattle and Farm Size. The National Agricultural Statistics Service publishes data on farm products in the United States.[11] The accompanying table shows the number of beef cattle on smaller-scale operations (farms having fewer than 50 head) for three states. Test whether the proportions of cattle on smaller farms are the same across all three states, using level of significance $\alpha = 0.05$.

	Texas	Oklahoma	Pennsylvania
Beef cattle on smaller scale operations	103,000	3,600	11,400
Beef cattle on operations that are not smaller scale	28,000	44,400	600

For Exercises 12–14, perform the appropriate analysis of variance using $\alpha = 0.05$.

12. GAS MILEAGE AND NUMBER OF CYLINDERS. When it comes to getting good gas mileage, does the number of cylinders in your engine make a difference? The following table provides the summary statistics regarding miles per gallon for 4-cylinder, 6-cylinder, and 8-cylinder cars.

	4 cylinders	6 cylinders	8 cylinders
n	199	83	103
$\bar{x}$	29.3	20.0	15.0
s	5.7	3.8	2.9

13. HOURS WORKED AND MARITAL STATUS. The General Social Survey tracks demographic trends. Here we are interested in whether the mean number of hours worked differs by marital status. The summary statistics are shown here.

	N	Mean	Std. Deviation
MARRIED	964	42.76	14.08
WIDOWED	72	40.13	14.28
DIVORCED	342	43.69	13.93
SEPARATED	79	41.66	15.71
NEVER MARRIED	478	41.03	14.03

14. CALORIES IN BREAKFAST CEREALS. A dietary researcher is interested in whether differences exist in the mean number of calories in breakfast cereals made by different manufacturers. The summary statistics for the samples from three manufacturers appear in the following table.

	Kellogg's	Quaker	Ralston Purina
n	23	8	8
$\bar{x}$	109	95	115
s	22	29	23

For Exercises 15 and 16, construct and interpret a 95% confidence interval for the slope β_1 of the regression line.

15. MEN'S HEIGHTS AND WEIGHTS. The university medical unit is collecting data on the heights and weights of the male students on campus. A random sample of six male students showed the following heights (in inches) and weights (in pounds).

Student	Height (x)	Weight (y)
2	68	145
3	69	160
5	70	165
6	71	180
8	72	180
10	75	210

16. RATIO ACCOUNTING GRADES. An accounting professor is trying to predict the performance of her students in the second semester of the introductory accounting course by their performance in the first semester. The first-semester grade and second-semester grade were recorded for a random sample of eight students taking the two-semester course at a local college. The results are shown in the table.

Student	First-semester grade (x)	Second-semester grade (y)
2	80	90
3	50	75
5	90	80
6	75	80
7	50	60
8	95	90
11	60	55
12	75	70

17. For the data in Exercise 15, perform the hypothesis test for the linear relationship between x and y using $\alpha = 0.05$

18. For the data in Exercise 16, perform the hypothesis test for the linear relationship between x and y using $\alpha = 0.05$

ANSWERS TO ODD-NUMBERED EXERCISES AND CHAPTER QUIZZES

Chapter 1

Section 1.1
1. (a) "No car." (b) "I did not have a car or a way to leave."
3. Answers will vary.
5. Note the large differences in the comparative heights of the rectangles that measure responses of sadness, anger, and disbelief.
7. (a) About 36,000,000 (b) About 7600
9. About 5400

Section 1.2
1. Answers will vary.
3. Elements
5. Categorical variable
7. A population is the collection of all elements (persons, items, or data) of interest in a particular study. A sample is a subset of the population from which the information is collected.
9. The value of a parameter is constant but usually unknown. The value of a statistic may vary from sample to sample but is usually known.
11. Students Michael, Ashley, Christopher, and Jessica
13. Freshman, sophomore, junior, and senior
15. Hospitals City, Memorial, Children's, Eldercare, and County
17. General and specialized
19. Height, siblings, and Math SAT
21. Number of floors, HMO ranking, number of patients per nurse, year opened
23. Siblings, Math SAT
25. Number of floors, HMO ranking, year opened
27. Gender
29. Math SAT
31. Type
33. Year opened
35. (a) Quantitative (b) Interval
37. (a) Quantitative (b) Ratio
39. (a) Quantitative (b) Ratio
41. (a) Qualitative (b) Ordinal
43. (a) Qualitative (b) Nominal
45. (a) Qualitative (b) Nominal
47. (a) Quantitative (b) Ratio
49. Population: all home sales in Tarrant County, Texas; sample: 100 home sales selected
51. Population: all students at Portland Community College; sample: 50 selected Portland Community College students.
53. Descriptive statistics; the variable describes a sample.
55. Statistical inference; the sample was used to draw a conclusion about the entire population.
57. (a) Elements: Endangered species Pygmy rabbit, Florida panther, Red wolf, and West-Indian manatee; Variables: Year listed as endangered, Estimated number remaining, and Range. (b) Qualitative variables: Range; Quantitative variables: Year listed as endangered and estimated number remaining. (c) Year listed as endangered—interval; estimated number remaining—ratio, range—nominal. (d) Year listed as endangered—discrete, Estimated number remaining—discrete. (e) 1973, 50, Florida.

59. (a) Elements: States Texas, Missouri, Minnesota, Ohio, and South Dakota; Variables: Proportion of GE corn and most prevalent type. (b) Qualitative variables: Most prevalent type; Quantitative variables: Proportion of GE corn (c) Proportion of GE corn—ratio; most prevalent type—nominal (d) Proportion of GE corn—continuous (e) 79%, Herbicide-tolerant
61. (a) Elements: Commodities—oil, gold, and coffee; variables—*price per share* and *percent change*. (b) Qualitative variables: None; Quantitative variables: *price per share* and *percent change* (c) *price per share* and *percent change* represent ratio data. (d) *price per share* and *percent change* are continuous. (e) $1699.40, + 0.04%.
63. They compared the average lifetime of a sample of their own light bulb to the reported average lifetimes of other current models of light bulbs.
65. (a) Campuses Arizona State, Ohio State, Central Florida, University of Minnesota, and University of Texas (b) Location, enrollment, and rank (c) Location (d) Enrollment and rank (e) Location—nominal; enrollment—ratio; rank—ordinal

Section 1.3
1. Convenience sampling usually only includes a select group of people. For example, surveying people at a mall on a workday during working hours would probably include few if any people who work full time.
3. Answers will vary; could have chosen a random sample of houses and apartments and surveying the people door to door, for instance.
5. A sample for which every element has an equal chance of being included.
7. Cluster sampling
9. Convenience sampling
11. Target population: All college students; Potential population: All students working out at the gymnasium on the Monday night Brandon was there.
13. Target population: All small businesses; Potential population: Small businesses near the state university.
15. Vague terminology
17. Neither simple nor clear
19. (a) Observational (b) response variable: how often they attend religious services; predictor variable: whether or not the family is large (at least four children)
21. (a) Experimental (b) response variable: performance of the electronics equipment; predictor variable: whether or not a piece of equipment has a new computer processor
23. Answers will vary.
25. Answers will vary.
27. Level of insect damage to crops
29. The new pesticide
31. LDL cholesterol level in the bloodstream
33. New medication
35. Randomization
37. Answers will vary. For instance, the poll by Ann Landers was extremely biased. Only people who read the Ann Landers column and felt strongly about the poll responded to this poll. The *Newsday* poll was done professionally, and therefore the sample used was more likely to be representative of the population.

39. Desired response type is open to interpretation: preference or yes/no.
41. Predictor variable: patient diet, Mediterranean or Western; response variable: risk for a second heart attack.
43. **(a)** The 305 subjects that wore the placebo bracelet
(b) The subjects were randomly assigned to wear either the placebo bracelet or the ionized bracelet. **(c)** There are 305 subjects in both the treatment and the control groups.
45. This study is an experimental study because the subjects were randomly assigned to either a treatment or a control.
47. Answers will vary.

Chapter 1 Review

1. **(a)** Cars Subaru Forester, Honda CR-V, Nissan Rogue, and Mitsubishi Outlander **(b)** Cylinders, passengers, base price, and customer satisfaction **(c)** Customer satisfaction
(d) Cylinders, passengers, and base price
(e) Cylinders—ratio; passengers—ratio; base price—ratio; customer satisfaction—ordinal
3. 4, 5, $20,295, above average
5. **(a)** All registered voters in the United States **(b)** People on the lists of people who owned cars and had telephones **(c)** All people on the lists of people who owned cars and had telephones **(d)** Not similar; answers will vary.
7. **(a)** Replication **(b)** Surveying only four dentists is not likely to get a sample representative of the population of all dentists.
9. No; there may be other factors that determine a child's cognitive skills.

Chapter 1 Quiz

1. False
2. False
3. collecting
4. observation
5. sample
6. Observational study
7. Experimental study
8. Predictor variable: drug given to an elderly patient with Alzheimer's, new or placebo; response variable: whether or not the patient's Alzheimer's symptoms are reduced.
9. **(a)** All statistics students **(b)** The students in the statistics class who were selected for the sample **(c)** Left-handed or not; qualitative **(d)** No; not likely to be very far away from the population proportion since enrollment in a specific statistics class is not dependent on being left-handed or not.
10. Different people have different interpretations of the words *often, occasionally, sometimes,* and *seldom.*

Chapter 2

Section 2.1

1. We use graphical and tabular form to summarize data in order to organize it in a format where we can better assess the information. If we just report the raw data, it may be extremely difficult to extract the information contained in the data.
3. True.
5. The sample size, n.
7. The row totals, the column totals
9. When the sample sizes are substantially different

11.

Variable: political party affiliation	Frequency
Democrat	7
Independent	6
Republican	7
Total	20

13.

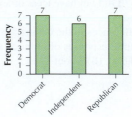

Political party affiliation

15.

Variable: blood type	Frequency
A	11
AB	1
B	3
O	10
Total	25

17.

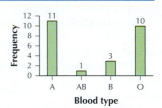

Blood type

19.

Variable: major	Frequency
Business	4
Math	4
Psychology	4
Total	12

21.

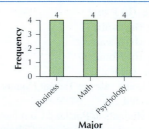

Major

23.

Variable: gender	Frequency
Female	7
Male	5
Total	12

25.

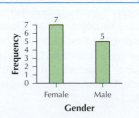

Gender

27.

	Female	Male	Total
Business	2	2	4
Math	3	1	4
Psychology	2	2	4
Total	7	5	12

29.

Variable: class	**Frequency**
Freshman	3
Sophomore	4
Junior	3
Senior	4
Total	14

31.

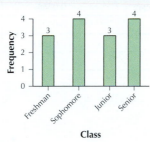

33.

Variable: handedness	**Frequency**
Left	4
Right	10
Total	14

35.

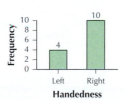

37.

	Freshman	Sophomore	Junior	Senior	Total
Left	1	1	1	1	4
Right	2	3	2	3	10
Total	3	4	3	4	14

39. No. There are actually two categorical variables—*level of education* and *whether or not the person owns a cell phone*. The percents are percents of each category of level of education who own cell phones and not the percent of the whole group who own cell phones.

41. (a) Several times a day; 43.4% **(b)** Every few weeks; 5.1%

43. (a) Fractures; 26% **(b)** Traumatic brain injury; 9% **(c)** Yes. It would have to be one of the injuries included in the category "Other injuries."

45. (a)–(b)

Continent	**Frequency**	**Relative frequency**
Africa	1	0.10
Asia	5	0.50
Europe	1	0.10
North America	2	0.20
South America	1	0.10

(c)

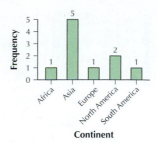

(d)

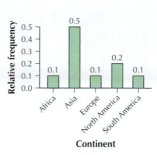

(e)

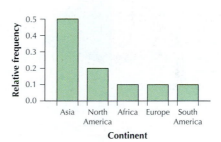

(f)

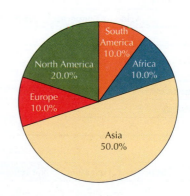

47. (a)–(b)

Main use	**Frequency**	**Relative frequency**
Industry	2	0.20
Irrigation	6	0.60
Not reported	2	0.20

(c)

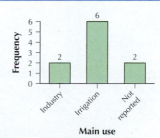

(d)

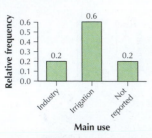

(e)

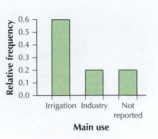

(f)

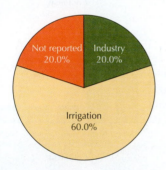

55. (a) Relative frequency distribution of *vehicle type*

Variable: vehicle type	Relative frequency
SUVs	0.3130
Compact cars	0.1083
Midsize cars	0.1015
Subcompact cars	0.0931
Standard pickup trucks	0.0897
Large cars	0.0643
Station wagons	0.0525
Small pickup trucks	0.0499
Two seaters	0.0431
Minicompact cars	0.0364
Vans	0.0321
Minivans	0.0161
Total	1.00

(b)

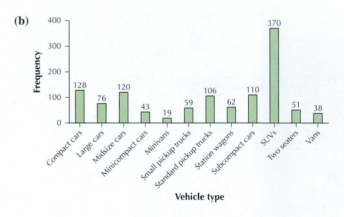

49.

	Arid	Temperate	Tropical	Total
Africa	0	0	1	1
Asia	4	1	0	5
Europe	0	1	0	1
North America	0	2	0	2
South America	1	0	0	1
Total	5	4	1	10

(c)

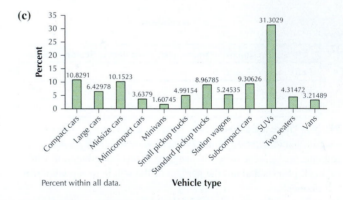

51.

	Industry	Irrigation	Not reported	Total
Arid	0	5	0	5
Temperate	2	0	2	4
Tropical	0	1	0	1
Total	2	6	2	10

53.

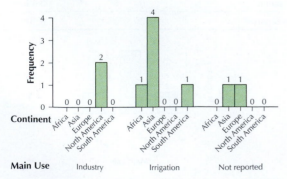

(d)

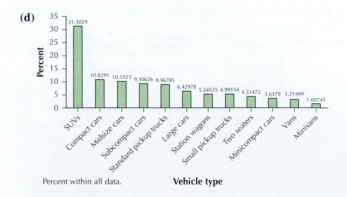

(e)

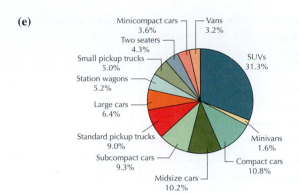

63.

Response to "How much do you enjoy shopping?"	Frequency	Relative frequency
A lot	1338	1338/4514 ≈ 0.2964
Some	1255	1255/4514 ≈ 0.2780
Only a little	1159	1159/4514 ≈ 0.2568
Not at all	717	717/4514 ≈ 0.1588
Don't know/refused	45	45/4514 ≈ 0.0100
Total	4514	1.0000

65. See answer 63.

57. (a)

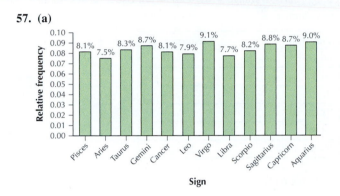

67.

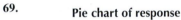
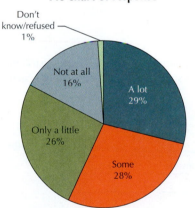

(b)

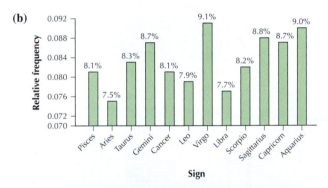

69.

The graph in **(b)** uses an adjusted scale, which is misleading. Use this graph to magnify the small variability in percentages.

59. Missing values are in red

"How much do you enjoy shopping?"	Gender		Total
	Male	**Female**	
A lot	388	950	1338
Some	528	673	1255
Only a little	662	497	1159
Not at all	497	220	717
Don't know/refused	20	25	45
Total	2149	2365	4514

61. (a) Women **(b)** Women **(c)** Men **(d)** Men

71.

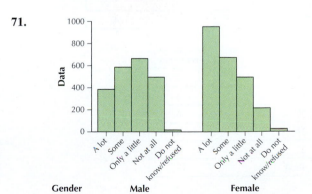

73. (a) Girls: 0.525; boys: 0.475

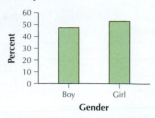

(b) Grades: 51.67%; popular: 29.50%; sports: 18.83%

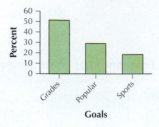

75. (a) and (b)

Class	Frequency	Relative frequency
Freshman	5	0.25
Sophomore	5	0.25
Junior	5	0.25
Senior	5	0.25

77. Answers will vary.

Section 2.2

1. Both: frequency distribution, relative frequency distribution; quantitative data only: histograms, frequency polygons, stem-and-leaf displays, dotplot.

3. Between 5 and 20

5. Answers will vary.

7. Answers will vary.

9.

Number of game consoles	Frequency
0	9
1	10
2	5
Total	24

11.

Age	Frequency
18	2
19	4
20	6
21	4
22	2
Total	18

13.

Age	Frequency
18–19	6
20–21	10
22–23	2
Total	18

15. Using 6 classes: Range $= 87 - 61 = 26$. Use 6 classes, so the class width $= \dfrac{26}{6} = 4.33$. Use class width $= 5$.

Pulse rate	Frequency
60–64	2
65–69	4
70–74	3
75–79	5
80–84	5
85–89	1
Total	20

Using 5 classes: Range $= 87 - 61 = 26$. Use 5 classes, so the class width $= \dfrac{26}{5} = 5.2$. Use class width $= 6$.

Pulse rate	Frequency
60–65	3
66–71	3
72–77	5
78–83	7
84–89	2
Total	20

17. Using 6 classes:

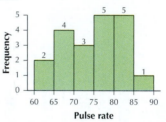

Using 5 classes:

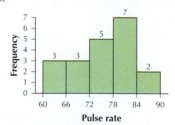

19. Using 6 classes:

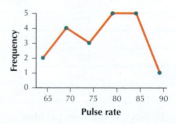

Using 5 classes:

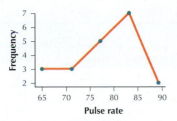

Stem-and-leaf display.

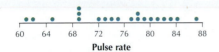

6	125999
7	23457889
8	012347

21.

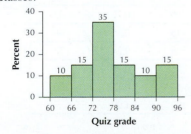

23. Using 6 classes:

Quiz grades	Relative frequency
60–65	0.10
66–71	0.15
72–77	0.35
78–83	0.15
84–89	0.10
90–95	0.15
Total	1.00

Using 5 classes:

Quiz grades	Relative frequency
62–68	0.10
69–75	0.40
76–82	0.25
83–89	0.10
90–96	0.15
Total	1.00

25. Using 6 classes:

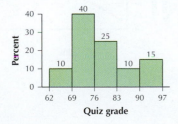

Using 5 classes:

27.

6	2599
7	022455679
8	0257
9	245

29. (a) 4 **(b)** 1 and 6 **(c)** 15 times **(d)** 15% of the times
31. (a) 46 **(b)** 33 (not including a frequency of 0) **(c)** highest: 49; lowest: 33. **(d)** left-skewed
33. (a) Divide the frequency values by the total frequency—classes not affected **(b)** change the scale along the relative frequency (vertical) axis by multiplying the relative frequency values by the total frequency—shape of distribution not affected **(c)** 19
35. (a) 0 **(b)** 0 **(c)** $25 to $27.5 has the largest relative frequency, 4/19 = 0.2105. **(d)** 3 **(e)** 0
37. Data set: 23 24 25 26 27 28 28 29 30 31 31 32 32 32
39. Histogram with five classes

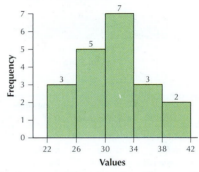

41. (a) 15 **(b)** 37.5 **(c)** 52.5 **(d)** 67.5 to 82.5 **(e)** 22.5 to 37.5
43. (a) 2000 **(b)** 1000 **(c)** 1000 to 3000 **(d)** 17,000 to 19,000

45.

Classes	Frequency	Relative frequency
550–599	1	1/12 ≈ 0.0833
600–649	1	1/12 ≈ 0.0833
650–699	1	1/12 ≈ 0.0833
700–749	0	0/12 = 0.0000
750–799	3	3/12 = 0.2500
800–849	3	3/12 = 0.2500
850–899	1	1/12 ≈ 0.0833
900–949	2	2/12 ≈ 0.1667

47.

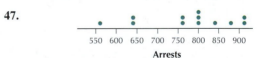

49. (a) Range = 98 − 57 = 41. Use 6 classes, so the class width = $\frac{41}{6}$ = 6.833. Use class width = 7.

Exam score	Frequency
57–63	3
64–70	2
71–77	7
78–84	4
85–91	2
93–98	2
Total	20

(b)

Exam score	Relative frequency
57–63	0.15
64–70	0.10
71–77	0.35
78–84	0.20
85–91	0.10
92–98	0.10
Total	1.00

(c)

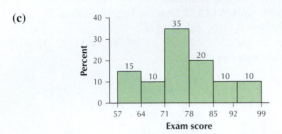

51.

	Dotplot	Histogram	Stem-and-leaf	Frequency polygon
(a) Symmetry and skewness	Appropriate to use	Appropriate to use	Appropriate to use for small ranges of data	Appropriate to use
(b) Construct using pencil and paper	Easily done for small ranges of data	Easily done for small ranges of data	Easily done for small ranges of data	Easily done for small ranges of data
(c) Retain complete knowledge of the data	Appropriate	Appropriate only if the data are ungrouped	Appropriate	Appropriate only if the data are ungrouped
(d) Presentation in front of non-statisticians	Appropriate	Appropriate	Appropriate	Appropriate

53. 961; 22
55. Yes; fats and oils.
57. One whole cheesecake (2053 grams of cholesterol)
59. (a) 2 **(b)** 4.00, 4.30
61. Answers will vary.

Section 2.3

1. A frequency distribution gives the frequency counts for each class (grouped or ungrouped). A cumulative frequency distribution gives the number of values which are less than or equal to the upper limit of a given class for grouped data or it gives the number of values which are less than or equal to a given number for ungrouped data.
3. Ogive.

5. Time series data.

7.

Age	Frequency	Relative frequency	Cumulative frequency
17.0–18.9	4	0.2	4
19.0–20.9	10	0.5	14
21.0–22.9	6	0.3	20
	$n = 20$	1.0	

9.

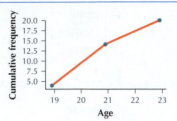

11.

Height (inches)	Frequency	Relative frequency	Cumulative frequency
60.0–63.9	3	0.12	3
64.0–67.9	10	0.40	13
68.0–71.9	10	0.40	23
72.0–75.9	2	0.08	25
	$n = 25$	1.00	

13.

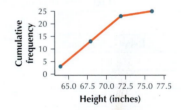

15.

Value of single die roll	Frequency	Relative frequency	Cumulative frequency
1	13	0.13	13
2	20	0.20	33
3	15	0.15	48
4	24	0.24	72
5	15	0.15	87
6	13	0.13	100
Total	100	1.00	

17.

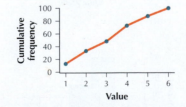

19.

Stock prices (dollars)	Frequency	Relative frequency	Cumulative frequency
5.00–7.49	1	0.0526	1
7.50–9.99	1	0.0526	2
10.00–12.49	2	0.1053	4
12.50–14.99	1	0.0526	5
15.00–17.49	2	0.1053	7
17.50–19.99	0	0	7
20.00–22.49	3	0.1579	10
22.50–24.99	3	0.1579	13
25.00–27.49	4	0.2105	17
27.50–29.49	2	0.1053	19
Total	$n = 19$	1.0000	

21.

23. Using 6 classes:

Pulse rate	Frequency	Relative frequency	Cumulative frequency
60–64	2	0.10	2
65–69	4	0.20	6
70–74	3	0.15	9
75–79	5	0.25	14
80–84	5	0.25	19
85–89	1	0.05	20
	20	1.00	

Using 5 classes:

Pulse rate	Frequency	Relative frequency	Cumulative frequency
60–65	3	0.15	3
66–71	3	0.15	6
72–77	5	0.25	11
78–83	7	0.35	18
84–89	2	0.10	20
	20	1.00	

25. Using 6 classes:

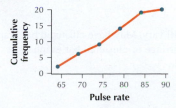

Using 5 classes:

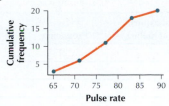

27. Using 6 classes:

Quiz score	Frequency	Relative frequency	Cumulative frequency
60–65	2	0.10	2
66–71	3	0.15	5
72–77	7	0.35	12
78–83	3	0.15	15
84–89	2	0.10	17
90–95	3	0.15	20
	20	1.00	

Using 5 classes:

Quiz score	Frequency	Relative frequency	Cumulative frequency
62–68	2	0.10	2
69–75	8	0.40	10
76–82	5	0.25	15
83–89	2	0.10	17
90–96	3	0.15	20
	20	1.00	

29. Using 6 classes:

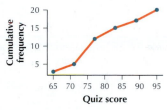

Using 5 classes:

31.

33. (a) 0.8 (b) 2.39 (c) 1.99

35.

Agricultural exports (in billions of dollars)	Frequency	Cumulative frequency
$0–$1.9	3	3
$2.0–$3.9	9	12
$4.0–$5.9	6	18
$6.0–$7.9	1	19
$8.0–$9.9	0	19
$10.0–$11.9	0	19
$12.0–$13.9	1	20
Total	20	

(a) 12
(b) 18
(c) 2

37.

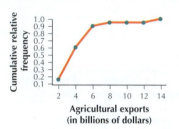

39. (a)

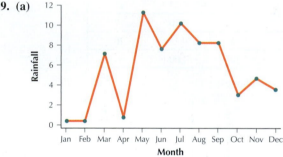

(b) Summer

Section 2.4

1. Answers will vary.
3. Figure 2.33
5. Table 2.23 gives the actual number of cars stolen.
7. (a) Biased distortion or embellishment; omitting the zero on the relevant scales; inaccuracy in relative lengths of bars in a bar chart. (b) A Pareto chart or pie chart can be used.
9. (a) The number of people living with AIDS is increasing.
(b) Using two dimensions (area) to emphasize a one-dimensional difference.
(c)

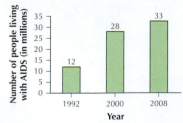

11. (a)

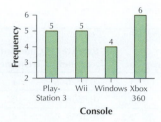

(b) Manipulating the scale, omitting the 0 on the vertical scale

(c)

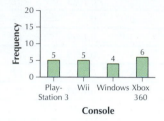

(d) Manipulating the scale
13. Answers will vary.

Chapter 2 Review

1. No, because the variable is categorical.

3.

Part of speech	Frequency
Adjective	1
Adverb	2
Article	3
Conjunction	3
Preposition	9
Pronoun	7
Verb	6
Total	31

5.

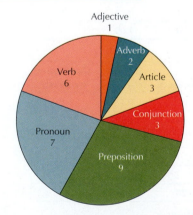

7. 0.6612
9. 0.0246
11. Answers will vary. May have clustered bar graph by happiness of marriage or clustered bar graph by sex.
13. 62%; middle

15.

Average size of household	Frequency
2.25–2.34	1
2.35–2.44	0
2.45–2.54	12
2.55–2.64	23
2.65–2.74	10
2.75–2.84	3
2.85–2.94	0
2.95–3.04	1
3.05–3.14	0
3.15–3.24	1
Total	51

17.

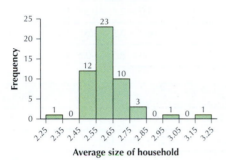

19.

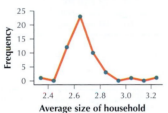

21. (a)–(b) Cannot be done because the variable is qualitative.

23.

Chapter 2 Quiz

1. False
2. True
3. sample size.
4. frequency distribution
5. Symmetric
6. Right skewed

7–10.

Vowels	Frequency	Cumulative frequency	Relative frequency	Cumulative relative frequency
a	73	73	0.1931	0.1931
e	130	203	0.3439	0.5370
i	74	277	0.1958	0.7328
o	74	351	0.1958	0.9286
u	27	378	0.0714	1.0000

11.

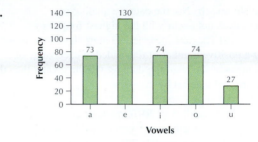

12.

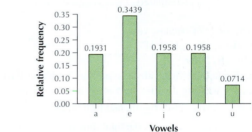

13.

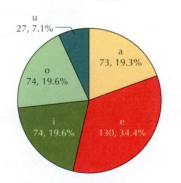

14. and 15. Can't construct because variable is qualitative.

Chapter 3

Section 3.1

1. A value that locates the center of the data set
3. Because the mean depends in part on the sum of all data values, an outlier will skew the mean (pull it in one direction or another). Since the median simply depends on position in an ordered list, it is not sensitive to outliers.
5. Sample size (n)
7. x_i
9. Sample mean ($\bar{x}$)
11. Median
13. (a) 5 **(b)** 18
15. (a) 7 **(b)** 81.429
17. (a) 7 **(b)** 75
19. (a) 5 **(b)** 1576.8
21. 18
23. 80

25. 20

27. 75

29. $6.12

31. $25

33. 25

35. 8

37. Median = 3

39. (a) 6 cylinders (b) 5 cylinders (c) 4 cylinders

41. (a) 24 mpg (b) 21.5 mpg (c) 18 mpg

43. (a) 624.667 (b) 614.5

45. (a) 604.167 (b) 602.5

47. (a) English; no. (b) No, the data are qualitative.
(c) Economics does not occur with the highest frequency.

49. (a) $25.17 (b) The new mean is $10 more than the original mean. (c) If a positive number of d is added to each value of a data set, the mean of the resulting data set will be greater than the mean of the original data set by d.

51. (a) $15.95 (b) Stieg Larsson

53. Mean = 2009.5, median = 2010, mode = 2011

55. Mean = 5.5 years, median = 5 years, mode = 4 years

57. (a) Female (b) Approximately 73 (c) Approximately 74
(d) Female; yes

59. (a) 78 (b) 79 (c) Females; yes

61. (a) 74.9 (b) 75.25 (c) 75.417

63. (a) $14.98 (b) 2009

65. Since she walked the first mile at a speed of 5 mph for the first mile, her time for walking the first mile was $\frac{1}{5}$ hours. Similarly, her time to walk the second, third, fourth, and fifth miles was $\frac{1}{4}, \frac{1}{3}, \frac{1}{2}$, and $\frac{1}{1}$ hours, respectively. Thus the total time it took Emily to walk 5 miles is $\frac{1}{5} + \frac{1}{4} + \frac{1}{3} + \frac{1}{2} + \frac{1}{1} = \Sigma\frac{1}{x}$ hours. Therefore her average speed is $\frac{\text{distance}}{\text{time}} = \frac{n}{\Sigma\frac{1}{x}}$, which is the harmonic mean. The arithmetic mean is just the average of the 5 rates.

67. Answers will vary.

69. Answers will vary.

71. (a) Answers will vary. (b) The mean increases. (c) The median remains the same.

Section 3.2

1. Deviation for a data value gives the distance the value is from the mean.

3. Benefit—simple to calculate, Drawbacks—quite sensitive to extreme values, does not use all of the data values.

5. Benefit—uses all of the numbers in a data set. Drawback—can be time-consuming to calculate.

7. False

9. When all of the data values are the same

11. 25

13. 0

15. 3.0

17. 0

19. 10

21. 10

23. (a) 10

(b)

x	$x - \mu$
5	$5 - 10 = -5$
25	$25 - 10 = 15$
0	$0 - 10 = -10$
10	$10 - 10 = 0$

25. (a) 10

(b)

x	$x - \mu$
10	$10 - 10 = 0$
10	$10 - 10 = 0$
10	$10 - 10 = 0$
10	$10 - 10 = 0$
10	$10 - 10 = 0$

27. (a) 2.5

(b)

x	$x - \mu$
1.0	$1.0 - 2.5 = -1.5$
3.0	$3.0 - 2.5 = 0.5$
4.0	$4.0 - 2.5 = 1.5$
2.0	$2.0 - 2.5 = -0.5$

29. (a)

x	$x - \mu$	$(x - \mu)^2$
5	$5 - 10 = -5$	25
25	$25 - 10 = 15$	225
0	$0 - 10 = -10$	100
10	$10 - 10 = 0$	0
		$\Sigma(x - \mu)^2 = 350$

(b) 87.5

31. (a)

x	$x - \mu$	$(x - \mu)^2$
10	$10 - 10 = 0$	0
10	$10 - 10 = 0$	0
10	$10 - 10 = 0$	0
10	$10 - 10 = 0$	0
10	$10 - 10 = 0$	0
		$\Sigma(x - \mu)^2 = 0$

(b) 0

33. (a)

x	$x - \mu$	$(x - \mu)^2$
1.0	$1.0 - 2.5 = -1.5$	2.25
3.0	$3.0 - 2.5 = 0.5$	0.25
4.0	$4.0 - 2.5 = 1.5$	2.25
2.0	$2.0 - 2.5 = -0.5$	0.25
		$\Sigma(x - \mu)^2 = 5.0$

(b) 1.25

35. 9.4

37. 0

39. 1.12

41. (a) 0 (b) 0 (c) The data values typically differ from the mean $\bar{x} = 3.14159$ by 0 units.

43. (a) 19.2 (b) 4.4 (c) The data values typically differ from the mean $\bar{x} = 14.5$ by about 4.4 units.

45. (a) 19.2 (b) 4.4 (c) The data values typically differ from the mean $\bar{x} = -14.5$ by about 4.4 units.

47. About 68%

49. About 99.7%

51. About 95%

53. About 2.5%

55. At least 75%.

57. At least 93.75%

59. At least 75%

61. At least 84%

63. (i)—(d); (ii)—(b); (iii)—(c); (iv)—(a)

65. (a) 8 cylinders (b) 9.6 cylinders^2 (c) 3.098 cylinders

67. (a) 30 mpg (b) 116 mpg^2 (c) 10.770 mpg

69. (a) range = largest data value − smallest data value = 674 − 585 = 89 (b) 1104.3 (c) 33.2

71. (a) range = largest data value − smallest data value = 620 − 595 = 25. (b) 88.6 (c) 9.4

73. Zooplankton: 6.86, phytoplankton: 9.96 (a) phytoplankton (b) phytoplankton

75. Range for Colony A = 73; range for Colony B = 91 (a) Colony B (b) Colony B

77. (a) 95.04 wins squared (b) 9.7 wins

79. (a) The sample consisting of the New York Yankees and the Baltimore Orioles will yield the largest sample standard deviation. (b) The sample consisting of the Tampa Bay Rays and the Boston Red Sox will yield the smallest sample standard deviation.

81. (a) Can not be found since $k = 1$ (b) At least 55.6% (c) At most 44%

83. (a) About 68% (b) Between 68% and 95% (c) Between 2.5% and 16%

85. SAT Mathematics test: Range = 89, Variance = 1104.267; SAT Reading test: Range = 23, Variance = 89.9; SAT Writing test: Range = 25, Variance = 88.567; Yes.

87. (a) It would not affect any of the measures of spread. (b) SAT Mathematics test: Range = 89, Variance = 1104.27, Standard deviation = 33.231; SAT Writing test: Range = 25, Variance = 88.567, Standard deviation = 9.411

89. (a) Range = 15; standard deviation = 5.48. (b) Adding a positive constant to each value in a data set will not change the value of the original range or standard deviation.

91. (a) Cylinders: CV = 51.64%; Engine size: CV = 52.89%; City mpg: CV = 44.88% (b) Engine size, City mpg

93. (a) Cylinders: MAD = 2; Engine size: MAD = 1.189; City mpg: MAD = 8.333 (b) City mpg, Engine size

95. (a) Skewness = 0 (b) Skewness = 3 (c) Skewness = −3 (d) Skewness = −1.5 (e) Skewness = 0 (f) Skewness = 0.6

97. (a) Pros: Skewness = 0.182; Darts: Skewness = 0.197; DJIA: Skewness = −0.077 (b) Pros and Darts are slightly right-skewed, DJIA are slightly left-skewed.

99. (a) SAT Mathematics test: $s = 33.231$; SAT Reading test: $s = 9.482$; SAT Writing test: $s = 9.411$

(b)

	Range	Sample variance	Sample standard deviation	Coefficient of variation	Mean absolute deviation
SAT Mathematics	89	1104.267	33.231	5.32%	26.556
SAT Reading	23	89.9	9.482	1.57%	7.667
SAT Writing	25	88.567	9.411	1.56%	7.167

(c) Yes (d) The variability of the scores on the SAT Mathematics test is greater than the variability of the scores on the other two tests.

101. Answers will vary.

103. Answers will vary.

Section 3.3

1. These formulas will provide only estimates because we will not know the exact data values.

3. The weighted mean of this data set is $\bar{x} = \dfrac{\Sigma(w \cdot x)}{\Sigma w} = \dfrac{(1 \cdot 2) + (1 \cdot 7) + (1 \cdot 4)}{1 + 1 + 1} = \dfrac{2 + 7 + 4}{3} = \dfrac{13}{3} = 4.333333333 \approx 4.3$. This is also the sample mean of the sample consisting of the data values 2, 7, 4.

5. 69

7. 3.2

9.

Class limits	Midpoints
0–1.99	1
2–3.99	3
4–5.99	5
6–7.99	7
8–9.99	9

11. 14.2857

13. Estimated standard deviation = 6.226998; estimated variance = 38.7755.

15. (a)

Age	Frequency	Midpoints
0–4.99	63,422	2.5
5–17.99	240,629	11.5
18–64.99	540,949	41.5

(b) Estimated mean = 30.0298 years (c) Estimated standard deviation = 15.455909 years; estimated variance = 238.8851 years squared

17. Estimated mean = 135.5224; estimated standard deviation = 95.6874

19. $58.72

Section 3.4

1. Positive z-score: the data value is above the mean. Negative z-score: the data value is below the mean. z-score of zero: the data value is equal to the mean.

3. Answers will vary.

5. It is possible for the 1st percentile to equal the 99th percentile if all of the data values are the same.

7. False

9. Right-skewed with a few values much larger than the rest; median line of box plot closer to the line for Q3 than the line for Q1.

11. Not possible. Q1, the 25th percentile, will always be less than or equal to Q3, the 75th percentile. Thus the IQR = Q3 − Q1 is always greater than or equal to zero.

13. 3.5

15. (a) −1.5 (b) David's blood sugar level lies 1.5 standard deviations below the mean blood sugar level of 100 mg/dl.

17. 80 mg/dl

19. Juan: $z = 1$; Luis: $z = 1$; They both did the same.

21. Outlier

23. Not an outlier

25. $14.50

27. $5

29. $20

31. 100%
33. 8%
35. 17%
37. $11
39. $7.00
41. **(a)** 109.167 calories **(b)** 120.812 calories **(c)** 97.522 calories **(d)** 114.989 calories
43. **(a)** 105 **(b)** 110 calories **(c)** 115 calories **(d)** 130
45. **(a)** 105 calories **(b)** 110 calories **(c)** 115 calories **(d)** 10 calories
47. **(a)** 90 calories **(b)** Bran Chex and Bran Flakes
49. **(a)** 5.073 million **(b)** 15.151 million **(c)** −5.005 million **(d)** 8.433 million
51. **(a)** Valerian with 2.1 million **(b)** Ginseng with 8.8 million **(c)** 2 million **(d)** 14.7 million
53. **(a)** Bee pollen with 2.8 million **(b)** Fish oil with 4.2 million **(c)** Garlic with 7.1 million **(d)** 4.3 million
55. **(a)** 7.7 million **(b)** Ginkgo biloba
57. **(a)** −1.304 **(b)** 1.477 **(c)** −0.261
59. No outliers
61. **(a)** 5.75 **(b)** 6.1 **(c)** 5.3
63. **(a)** 5.3 **(b)** 5.75 **(c)** 6.1 **(d)** 0.8

Section 3.5

1. False
3. **(a)** The median will be about the same distance from Q1 and Q3, and the upper and lower whiskers will be about the same length. **(b)** The median is closer to Q1 than to Q3, and the upper whisker is much longer than the lower whisker. **(c)** The median is closer to Q3 than to Q1, and the lower whisker is much longer than the upper whisker.
5. Any data value located 1.5 (IQR) or more below Q1 or 1.5 (IQR) or more above Q3 is considered an outlier.
7. Q1 = 65 inches, Q2 = median = 68 inches, Q3 = 70 inches
9. Minimum = 64 inches, Q1 = 65 inches, Q2 = median = 68 inches, Q3 = 70 inches, maximum = 78 inches
11. Outlier
13. Q1 = 15 minutes, Q2 = median = 15 minutes, Q3 = 22.5 minutes
15. Minimum = 10 minutes, Q1 = 15 minutes, Q2 = median = 15 minutes, Q3 = 22.5 minutes, maximum = 50 minutes
17. Outlier
19. Q1 = 68, Q2 = median = 76, Q3 = 85.5
21. Min = 51, Q1 = 68, median = 76, Q3 = 85.5, max = 98
23. Not an outlier
25. **(a)** Right-skewed **(b)** Minimum = 0, Q1 = 1, Q2 = median = 3, Q3 = 7.5, maximum = 12
27. **(a)** Right-skewed **(b)** Minimum = 5, Q1 = 10, Q2 = median = 15, Q3 = 25, maximum = 45
29. x
31. Min = 8.33, Q1 = 13.69, median = 23.375, Q3 = 37.79, max = 55.46
33. Q1 − 1.5 * IQR = −22.46 and Q3 + 1.5 * IQR = 73.94. There are no values outside this interval, so there are no outliers.
35. Min = −4.09, Q1 = −0.14, median = −0.015, Q3 = 0.08, max = 0.1.
37. Q1 − 1.5 * IQR = −0.47 and Q3 + 1.5 * IQR = 0.41. The change −4.09 is an outlier.
39. Min = 2,000,000; Q1 = 2,800,000; median = 4,200,000; Q3 = 7,100,000; max = 14,700,000
41. Q1 − 1.5 * IQR = −3.65 and Q3 + 1.5 * IQR = 13.55. Usage of 14,700,000 is the only outlier.

43. Mean: 5,073,000; standard deviation: 3,359,300
45. Zooplankton: Minimum = −6.6, Q1 = −3, Q2 = median = −1.54, Q3 = −0.64, maximum = 0.26; phytoplankton: minimum = 0.65, Q1 = 1.57, Q2 = median = 2.55, Q3 = 3.04, maximum = 10.61
47. Zooplankton: left-skewed; phytoplankton: right-skewed
49. Zooplankton: mean = −2.048; phytoplankton: mean = 3.123. They concur.
51. Zooplankton: IQR = 2.36; phytoplankton: IQR = 1.47. Zooplankton more variable
53. Zooplankton: −6.60 is moderately unusual; phytoplankton: 10.61 is moderately unusual.
55. Mean = 1.784 mg, standard deviation = 3.138 mg, min = 0.000 mg, Q1 = 0.300 mg, median = 0.800 mg, Q3 = 1.700 mg, max = 37.600 mg. Range = 37.600 mg − 0.000 mg = 37.600 mg. IQR = 1.700 mg − 0.300 mg = 1.400 mg
57. The boxplot is very right-skewed.

Chapter 3 Review

1. 3.1227
3. 2.55
5. The mode, since the value with the largest frequency is unaffected by the deletion of values 90 or less.
7. Mean = 396.8; range = 803
9. 276.2
11. 3.3133
13. 16.5
15. 59.5
17. 1.44
19. Since the largest and the smallest ragweed pollen indices have z-scores that are between −2 and 2, there are no outliers and no moderately unusual values.
21. 90%
23. Q1 = 25, Q2 = 34.5, Q3 = 48
25. No outliers
27. No outliers, yes
29. At least 75%
31. 75 mph

Chapter 3 Quiz

1. False
2. False
3. False
4. outlier
5. center
6. mean
7. robust measures
8. mode
9. Zero
10. Class midpoint
11. **(a)** Mean = 87,453 **(b)** Median = 98,008
12. **(a)** Range = 86,910 **(b)** Standard deviation = 33,857
13. Estimated mean = 61.6527; estimated standard deviation = 18.4518.
14. **(a)** 1.5 **(b)** −1 **(c)** 1 **(d)** −1.5 **(e)** 0
15. **(a)** 60 **(b)** Between 34% and 81.5% **(c)** No, furthermore we must assume that one of the values of k is less than 1. **(d)** Between 2.5% and 16%.
16. **(a)** 501.5 **(b)** 512 **(c)** 518
17. IQR = 16.5
18. Min = 499, Q1 = 501.5, median = 512, Q3 = 518, max = 523.

19. Q1 − 1.5 * IQR = 476.75 and Q3 + 1.5 * IQR = 542.75. All the SAT scores lie between 476.75 and 542.75, so there are no outliers.

20.

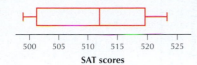

Chapter 4

Section 4.1

1. Scatterplot

3. Between −1 and 1, inclusive

5. Often, the value of the x variable can be used to predict or estimate the value of the y variable.

7. They decrease.

9.

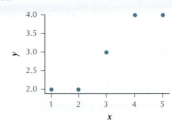

11.

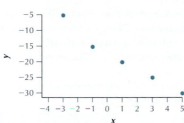

13. (a) Strong negative linear relationship (b) They decrease.

15. (a) Moderate positive linear relationship (b) They increase.

17. (a) Perfect negative linear relationship (b) They decrease.

19. 0.9487

21. −0.9686

23. The variables x and y are strongly positively correlated. As x increases, y increases.

25. The variables x and y are strongly negatively correlated. As x increases, y decreases.

27. x and y are positively correlated.

29. x and y are negatively correlated.

31. i

33. iii

35. (a) (1,1), (2,3), (3,3), (4,4), (5,6), (6,6), (7,7), (8,7), (9,9), (10,11) (b) Minitab: Pearson correlation of x and y = 0.978. TI-83/84: $r = 0.9781316853$

37. (a) (1,7), (2,8), (3,7), (4,6), (5,6), (6,5), (7,6), (8,5), (9,7), (10,6) (b) Minitab: Pearson correlation of x and y = −0.522. TI-83/84: $r = -0.5222329679$

39. x and y are positively correlated.

41. Weakly negatively correlated

43. (a)

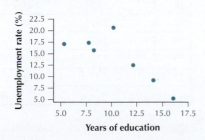

(b) Negative

45. Negatively correlated

47. Not correlated

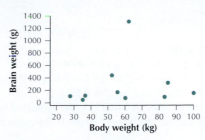

49. Brain weight and body weight are not correlated. As body weight increases, brain weight tends to remain the same. Yes

51. (a) $\bar{x} = 9.16666667$; $\bar{y} = 5.5$; $s_x = 9.432214303$; $s_y = 4.679743583$; $r = -0.7453498716$. Minitab: Correlations: Hip-Hop CDs owned, Country CDs owned; Pearson correlation of Hip-Hop CDs owned and Country CDs owned = −0.745. TI-83/84: $r = -0.7453498716$. (b) Yes (c) The variables number of Hip-Hop CDs owned and number of Country CDs owned are negatively correlated. As the number of Country CDs owned increases, the number of Hip-Hop CDs owned decreases.

53. (a) The dots form the same pattern. The only difference is that the dots are shifted 5 units up.

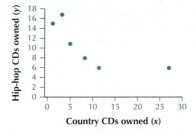

(b) $r = -0.7453$ (c) They are the same. (d) The correlation coefficient remains unchanged when a constant is added to each y data value.

55. Positively correlated

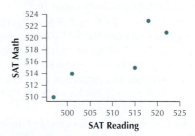

57. SAT Reading scores and SAT Math scores are positively correlated. As the SAT Reading score increases, the SAT Math score increases. Yes.

59. Answers will vary.

61. (a) Positively correlated (b) Negatively correlated (c) Not correlated

Section 4.2

1. To approximate the relationship between two numerical variables using the regression line and the regression equation

3. We can find the predicted value of y by plugging a given value of x into the regression equation and simplifying.

5. Extrapolation is the process of making predictions based on x-values that are beyond the range of the x-values in our data set.

7. Negative

9. Positive

11. Positive

13. (a) $b_1 = 3.4$ (b) $b_0 = -1.5$ (c) $\hat{y} = 3.4x - 1.5$

15. (a) $b_1 = 4$ (b) $b_0 = 21.6$ (c) $\hat{y} = 4x + 21.6$

17. (a) $b_1 = 0.01$ (b) $b_0 = 2.47$ (c) $\hat{y} = 0.01x + 2.47$

19. (a) $b_1 = 0.5$ (b) $b_0 = 5$ (c) $\hat{y} = 0.5x + 5$

21. (a) For each increase of 1 unit in x, the estimated value of y increases by 3.4 units. (b) When x equals 0, the estimated value of y is -1.5.

23. (a) For each increase of 1 unit in x, the estimated value of y increases by 4 units. (b) When x equals 0, the estimated value of y is 21.6.

25. (a) For each increase of 1 unit in x, the estimated value of y increases by 0.01 unit. (b) When x equals 0, the estimated value of y is 2.47.

27. (a) For each increase of 1 unit in x, the estimated value of y increases by 0.5 unit. (b) When x equals 0, the estimated value of y is 5.

29. 8.7

31. 13.6

33. 2.57

35. 5

37. (a) 0.3 (b) The data point lies above the regression line, so the actual value of y is larger than predicted given $x = 3$.

39. (a) 2.4 (b) The data point lies above the regression line, so the actual value of y is larger than predicted given $x = 2$.

41. Can't do since prediction represents extrapolation.

43. Does not represent extrapolation

45. Does not represent extrapolation

47. Does not represent extrapolation

49. Does not represent extrapolation

51. (a) $b_1 = -1.24$, $b_0 = 26.19$ (b) The estimated unemployment rate is -1.24 times the number of years of education plus 26.19. (c) For each increase of 1 year of education, the estimated unemployment rate decreases by 1.24%. (d) When the number of years of education equals 0, the estimated unemployment rate is 26.19%.

53. (a) 0.43; 298.86 (b) $\hat{y} = 0.43x + 298.86$. The estimated SAT Math score is equal to 0.43 times the SAT Reading score plus 298.86. (c) The slope $b_1 = 0.43$ means that the estimated SAT Math score increases by 0.43 point for every increase of 1 point in the SAT Reading score. (d) The y intercept $b_0 = 298.86$ means that the estimated SAT Math score is 298.86 when the SAT Reading score is 0.

55. (a) 13.79 (b) 7.59 (c) 20 years is outside of the range of the data set. (d) 6.81, above the regression line. The observed unemployment rate of 20.6 is greater than the predicted unemployment rate of 13.79 for 10 years of education.

57. (a) 514.29 (b) The SAT Reading score can't be 0, so this situation will never happen. (c) No, a mean SAT Reading score of 400 is out of the range of the data set. (d) -0.29. The observed mean SAT Math score of 514 for New Jersey is less than the predicted mean SAT Math score of 514.29.

59. (a) Decrease (b) No change (c) Increases if slope is positive, decreases if slope is negative (d)–(e) No change

61. (a) Then State A has 0.282% more households headed by women than State B. (b) Then State C has 1.41% fewer households headed by women than State D.

63. (a) It decreases.

(b)

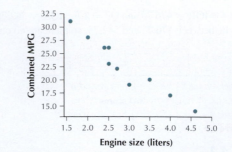

(c) Engine size and combined mpg are negatively correlated. Yes.

65. (a) $r = -0.9585$; yes (b) Engine size is negatively correlated with combined mpg; yes (c) As the engine size of a car increases the combined mpg decreases.

67. (a) 21.94 mpg (b) -2.94, below. The observed combined mpg of 19 for the Chevrolet Equinox is less than the predicted combined mpg of 21.94.

69. Answers will vary.

71. Answers will vary.

Section 4.3

1. The standard error of the estimate s is a measure of the size of the typical difference between the predicted value of y and the observed value of y.

3. SSE measures the prediction errors. SSE is the sum of the squared prediction errors. Since we want our prediction errors to be small, we want SSE to be as small as possible.

5. Measure of the variability in y. The variance s^2 of the y's.

7. No

9. 64% of the variability in the variable y is accounted for by the linear relationship between x and y.

11. (a) and (b)

x	y	$\hat{y} = 3.4x - 1.5$	$(y - \hat{y})$	$(y - \hat{y})^2$
1	2	1.9	0.1	0.01
2	5	5.3	-0.3	0.09
3	9	8.7	0.3	0.09
4	12	12.1	-0.1	0.01
				SSE = 0.2

13. (a)

x	y	Predicted value $\hat{y} = 21.6 + 4x$	Residual $(y - \hat{y})$	(Residual)^2 $(y - \hat{y})^2$
-5	0	1.6	-1.6	2.56
-4	8	5.6	2.4	5.76
-3	8	9.6	-1.6	2.56
-2	16	13.6	2.4	5.76
-1	16	17.6	-1.6	2.56

(b) SSE = 19.2

15. (a) and (b)

x	y	$\hat{y} = 0.01x + 2.47$	$(y - \hat{y})$	$(y - \hat{y})^2$
5	2	2.52	-0.52	0.2704
10	3	2.57	0.43	0.1849
15	3	2.62	0.38	0.1444
20	3	2.67	0.33	0.1089
25	2	2.72	-0.72	0.5184
30	3	2.77	0.23	0.0529
				SSE = 1.2799

17. $s = 0.3162$

19. The typical error in prediction is 2.5298.
21. 0.5657; TI-83/84: 0.5648
23. (a) $s^2 = 19.33333333$ (b) SST = 58 (c) SSR = 57.8
25. (a) $s^2 = 44.8$ (b) SST = 179.2 (c) SSR = 160
27. (a) $s^2 = 0.2666666667$ (b) SST = 1.3333 (c) SSR = 0.0534
29. (a) $r^2 = 0.9966$ (b) 99.66% of the variability in the variable y is accounted for by the linear relationship between x and y. (c) $r = 0.9983$
31. (a) $r^2 = 0.8929$ (b) 89.29% of the variability in the variable y is accounted for by the linear relationship between x and y. (c) $r = 0.9449$
33. (a) $r^2 = 0.0401$, from the TI-83/84: $r^2 = 0.0429$ (b) 4.01% (4.29%) of the variability in the variable y is accounted for by the linear relationship between x and y. (c) $r = 0.2001$; TI-83/84: $r = 0.2070$
35. (a) and (b)

x = Years of education	y = Unemployment rate	$\hat{y} = -1.24x + 26.19$	$(y - \hat{y})$	$(y - \hat{y})^2$
5	16.8	19.99	−3.19	10.1761
7.5	17.1	16.89	0.21	0.0441
8	15.3	16.27	−0.97	0.9409
10	20.6	13.79	6.81	46.3761
12	11.7	11.31	0.39	0.1521
14	8.1	8.83	−0.73	0.5329
16	3.8	6.35	−2.55	6.5025
				SSE = 64.7247

37. (a) and (b)

x = Mean SAT Reading score	y = Mean SAT Math score	$\hat{y} = 0.43x + 298.86$	$(y - \hat{y})$	$(y - \hat{y})^2$
497	510	512.57	−2.57	6.6049
515	515	520.31	−5.31	28.1961
518	523	521.6	1.4	1.96
501	514	514.29	−0.29	0.0841
522	521	523.32	−2.32	5.3824
				SSE = 42.2275

39. (a) $s^2 = 33.96952381$; SST = 203.8171429 (b) SSR = 139.0924429 (c) $r^2 = 0.6837$, TI-83/84: $r^2 = 0.6824$, 68.24% (68.37%) of the variability in the variable y = the unemployment rate is accounted for by the linear relationship between x = years of education and y = the unemployment rate. (d) $r = -0.8269$. TI-83/84: $r = -0.8269$. TI-83/84: $r = 0.8261$.
41. (a) $s^2 = 28.3$; SST = 113.2 (b) SSR = 70.9725 (c) $r^2 = 0.6270$; TI-83/84: $r^2 = 0.7730$, 62.70% (77.30%) of the variability in the variable y = mean SAT Math score is accounted for by the linear relationship between x = mean SAT Reading score and y = mean SAT Math score. (d) $r = 0.7918$; TI-83/84: $r = 0.8792$.
43. (a)

Low (x)	High (y)	$\hat{y} = 1.05x + 11.9$	$(y - \hat{y})$	$(y - \hat{y})^2$
57	77	71.75	5.25	27.5625
36	45	49.7	−4.7	22.09
7	21	19.25	1.75	3.0625
39	55	52.85	2.15	4.6225
70	83	85.4	−2.4	5.76
34	45	47.6	−2.6	6.76
				SSE = 69.8575

(b) $s = 4.1790$; TI-83/84: $s = 4.1767$. If we know the low temperature (x) for a particular day in 2006, then our estimate of the high temperature (y) for that day will typically differ from the actual high temperature by 4.1790 (4.1767) degrees Fahrenheit. (c) $s^2 = 524.2666667$; SST = 2621.333333 (d) SSR = 2551.475833 (e) $r^2 = 0.9734$, 97.34% of the variability in the variable y = high temperature is accounted for by the linear relationship between x = low temperature and y = high temperature. (f) $r = 0.9866$
45. (a) $x = 10$ years of education; $y = 20.6$ = unemployment rate. It doesn't follow the trend of the higher the number of years of education, the lower the unemployment rate. (b) Since $r^2 = 0.6824$, 68.24% of the variability in the variable y = unemployment rate is accounted for by the linear relationship between x = years of education and y = unemployment rate. Hence the statement is not true. (c) Since the absolute values of the residuals for 5, 10, and 16 years of education are more than 1%, this claim is not always true. (d) Since $b_1 = -1.24$, we can say that each additional year of education drops the predicted unemployment rate by 1.24%.
47. (a)

x = Engine size (liters)	y = Combined (city/highway) gas mileage (MPG)	$\hat{y} = -5.49x + 38.41$	$(y - \hat{y})$	$(y - \hat{y})^2$
1.6	31	29.626	1.374	1.887876
2.0	28	27.43	0.57	0.3249
2.5	26	24.685	1.315	1.727925
2.5	23	24.685	−1.685	2.839225
2.4	26	25.234	0.766	0.586756
2.7	22	23.587	−1.587	2.518569
3.0	19	21.94	−2.94	8.6436
3.5	20	19.195	0.805	0.648025
4.0	17	16.45	0.55	0.3025
4.6	14	13.156	0.844	0.712336
				SSE = 20.193012

(b) SSE is the sum of the squared residuals. Since we know that $\hat{y} = -5.49x + 38.41$ is the regression line, according to the least-squares criterion, no other possible straight line would result in a smaller SSE. (c) Chevrolet Equinox. It has much less combined mpg than expected. (d) Since the residual for the Suburu Forester is negative, the actual combined mpg is lower than expected.
49. (a) $s^2 = 27.6$, SST = 248.4 (b) SSR = 228.206988, SSR measures the amount of improvement in the accuracy of our estimates using the regression equation compared with relying only on the y-values and ignoring the x information. (c) $r^2 = 0.9187$, 91.87% of the variability in the variable y = combined mpg is accounted for by the linear relationship between x = engine size and y = combined mpg.
51. Since $(\bar{x}, \bar{y})$ is on the regression line, the slope and the y intercept would remain the same.
53. 38.1744186 mpg

55. (a)

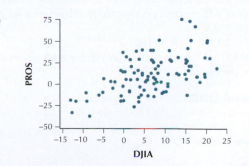

(b) $\hat{y} = 1.49x + 0.83$. The estimated increase (in percent) in the Pros stock portfolio equals 1.49 times the increase in the DJIA plus 0.83. **(c)** $r^2 = 0.289$, so 28.9% of the variability in the Pros price increase is accounted for by the linear relationship between the Pros price increase and the DJIA. **(d)** $s = 18.8545$. The typical difference between the predicted Pros price increase and the actual Pros price increase is 18.8545%. **(e)** $r = \sqrt{r^2} = \sqrt{0.289} = 0.5376$

57. (a)

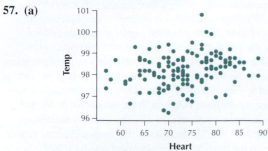

(b) $\hat{y} = 0.0263x + 96.3$. The estimated body temperature equals 0.0263 times the heart rate, plus 96.3. **(c)** $r^2 = 0.064$, so 6.4% of the variability in body temperature is accounted for by the linear relationship between body temperature and heart rate. **(d)** $s = 0.7120$. The typical difference between the predicted body temperature and the actual body temperature is 0.7120. **(e)** $r = \sqrt{r^2} = \sqrt{0.064} = 0.2530$

59. Answers will vary.
61. Answers will vary.
63. Answers will vary.
65. Answers will vary.

Chapter 4 Review

1.

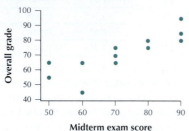

3. $r = 0.838$.
5. Midterm exam scores and overall grades are positively correlated. Low (high) midterm exam scores are associated with low (high) overall grades.
7. The predicted overall grade ($\hat{y}$) in elementary statistics is 0.77 times the midterm exam score (x) plus 15.99.
9. The predicted overall grade in elementary statistics for a student with a midterm grade of 0 is 15.99.
11. Two students have a midterm exam score of 50. The prediction error for the overall grade of 65 is 10.51 and the prediction error for the overall grade of 55 is 0.51. In both instances the predicted value is less than the actual value. Since no student had a midterm exam score of 100, we can't calculate the prediction error for $x = 100$.
13. SSE = 598.9
15. SST = 2006.3. SSR = 1407.3.
17. $r = 0.8373$. Midterm exam scores and overall grades are positively correlated.

Chapter 4 Quiz

1. False
2. False
3. estimate

4. unit
5. extrapolation
6. negative
7.

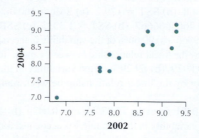

8. Positive
9. $\hat{y} = 0.75x + 2.04$
10. SSR = 3.7726, SSE = 0.5899, SST = 4.3625.
11. $s = 0.2429$. The typical difference between the predicted 2004 percentage and the actual 2004 percentage is 0.2429 percentage points.
12. $r^2 = 0.865$, meaning that 86.5% of the variability in 2004 percentage is accounted for by the 2002 percentage.
13. $r = 0.9300$. The 2002 percentage and the 2004 percentage are positively correlated.
14. (a) $-0.1980, 0.3020$. **(b)** $0.1456, -0.0544$. **(c)** 0.0511.

Chapter 5

Section 5.1

1. Answers will vary; chance, likelihood.
3. Answers will vary.
5. The experiment has equally likely outcomes.
7. We consider all available information, tempered by our experience and intuition, and then assign a probability value that expresses our estimate of the likelihood that the outcome will occur.
9. First find out how many students are at your college and find out how many of them like hip-hop music. Then calculate the relative frequency of students who like hip-hop music. Use the relative frequency method.
11. No, probability for females is greater than 1.
13. No, sum of probabilities is greater than 1.
15. It is a probability model.
17. 1/13
19. 1/52
21. 1/6
23. 1/2
25. 1/3
27. outcome; event; event; event; event; event
29.

Outcomes

Even number → Even number → (Even, Even)

Even number → Odd number → (Even, Odd)

Odd number → Even number → (Odd, Even)

Odd number → Odd number → (Odd, Odd)

31. $1/4 = 0.25$

33. 1/4 = 0.25

35. Let L = tossing a number less than 4 and G = tossing a number greater than or equal to 4.

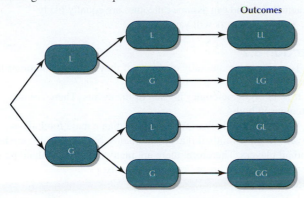

37. 1/4 = 0.25

39.

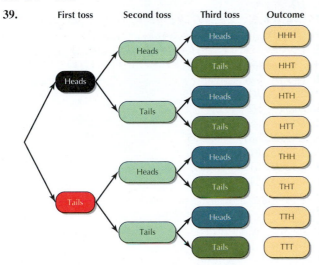

41. We can follow the branches to get all possible outcomes.

43. 1/8 = 0.125

45. 3/8 = 0.375

47.

Number of heads	Probability
0	1/8 = 0.125
1	3/8 = 0.375
2	3/8 5 0.375
3	1/8 = 0.125

49. 4/36 = 1/9

51. 1/36

53. 0

55. Sum of 7

57. 40/100 = 2/5 = 0.4

59. 20/100 = 1/5 = 0.2

61. The relative frequency method

63. 100/200 = 1/2 = 0.5

65. 40/200 = 2/10 = 0.2

67.

Favorite color	Probability
Red	30/100 = 0.3
Blue	25/100 = 0.25
Green	20/100 = 0.2
Black	10/100 = 0.1
Violet	10/100 = 0.1
Yellow	5/100 = 0.05

69. (a)

First person	Second person	Outcome
Cheeseburger	Cheeseburger	Cheeseburger, cheeseburger
	Hot dog	Cheeseburger, hot dog
	Veggie burger	Cheeseburger, veggie burger
Hot dog	Cheeseburger	Hot dog, cheeseburger
	Hot dog	Hot dog, hot dog
	Veggie burger	Hot dog, veggie burger
Veggie burger	Cheeseburger	Veggie burger, cheeseburger
	Hot dog	Veggie burger, hot dog
	Veggie burger	Veggie burger, veggie burger

(b) {Cheeseburger and cheeseburger, Cheeseburger and hot dog, Cheeseburger and veggie burger, Hot dog and cheeseburger, Hot dog and hot dog, Hot dog and veggie burger, Veggie burger and cheeseburger, Veggie burger and hot dog, Veggie burger and veggie burger}

71. (a) 0.33 **(b)** 0.67 **(c)** Relative frequency method

73. (a)

	Frequency	Relative frequency
Girls	18	18/44 = 0.4091
Boys	26	26/44 = 0.5909
Total	44	44/44 = 1.0000

(b)

Outcome	Probability
Girl	18/44 = 0.4091
Boy	26/44 = 0.5909

(c) Both P (Girl) = 18/44 = 0.4091 and P (Boy) = 26/44 = 0.5909 are between 0 and 1. P (Girl) + P (Boy) = 18/44 + 26/44 = 0.4091 + 0.5909 = 44/44 = 1.0000.

75. (a) 5/18 **(b)** 13/18 **(c)** $1.39

77.

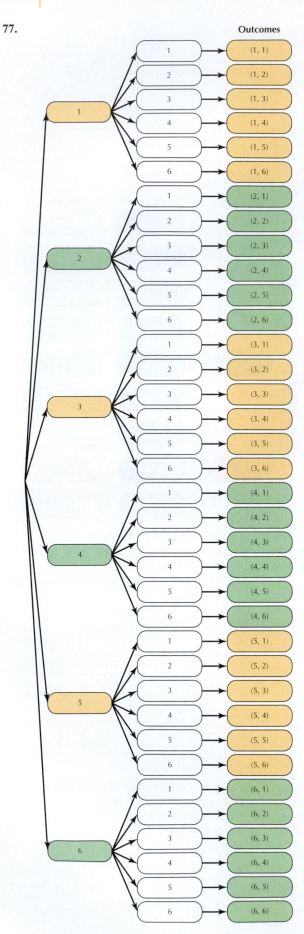

Outcomes

(1, 1) (1, 2) (1, 3) (1, 4) (1, 5) (1, 6)
(2, 1) (2, 2) (2, 3) (2, 4) (2, 5) (2, 6)
(3, 1) (3, 2) (3, 3) (3, 4) (3, 5) (3, 6)
(4, 1) (4, 2) (4, 3) (4, 4) (4, 5) (4, 6)
(5, 1) (5, 2) (5, 3) (5, 4) (5, 5) (5, 6)
(6, 1) (6, 2) (6, 3) (6, 4) (6, 5) (6, 6)

79. Events can consist of more than one outcome, but outcomes can't consist of more than one event.

81. 1/9. Classical probability method; have the sample space but no actual data and can assume outcomes are equally likely

83. (a)–(d) Answers will vary.

Section 5.2

1. Two events are mutually exclusive if they have no outcomes in common.

3. It is all of the outcomes in each of the events. There are no outcomes in both.

5. You are more likely to select a male than a male football player. All male football players are males, but most males are not football players. Therefore, there are many more males than male football players at any college or university.

7. 5/6

9. 1/2

11. 1/2

13. $\{K\spadesuit, K\clubsuit, K\heartsuit, K\diamondsuit, A\heartsuit, 2\heartsuit, 3\heartsuit, 4\heartsuit, 5\heartsuit, 6\heartsuit, 7\heartsuit, 8\heartsuit, 9\heartsuit,$
$10\heartsuit, J\heartsuit, Q\heartsuit, A\diamondsuit, 2\diamondsuit, 3\diamondsuit, 4\diamondsuit, 5\diamondsuit, 6\diamondsuit, 7\diamondsuit, 8\diamondsuit, 9\diamondsuit, 10\diamondsuit, J\diamondsuit, Q\diamondsuit\}$

15. $\{A\heartsuit, 2\heartsuit, 3\heartsuit, 4\heartsuit, 5\heartsuit, 6\heartsuit, 7\heartsuit, 8\heartsuit, 9\heartsuit, 10\heartsuit, J\heartsuit, Q\heartsuit, K\heartsuit,$
$A\diamondsuit, 2\diamondsuit, 3\diamondsuit, 4\diamondsuit, 5\diamondsuit, 6\diamondsuit, 7\diamondsuit, 8\diamondsuit, 9\diamondsuit, 10\diamondsuit, J\diamondsuit, Q\diamondsuit, K\diamondsuit\}$

17. $\{K\heartsuit\}$

19. 28/52 = 7/13

21. 26/52 = 1/2

23. 1/52

25. 23,952/52,817 ≈ 0.4535

27. 7,274/52,817 ≈ 0.1377

29. 4,528/52,817 ≈ 0.0857

31. 1,415/52,817 ≈ 0.0268

33. 1

35. 5/6

37. 4/6 = 2/3

39. 0

41. 1/6

43. 2/6 = 1/3

45. 5/18

47. 1/18

49. 1/3

51. 2/52 = 1/26

53. 6/13

55. 0

57. 3/13

59. 10/13

61. 11/26

63. 1/8

65. 3/8

67. {BBB, BBG, BGB, GBB, GGB, GBG, BGG, GGG}

69. 3/8

71. 3/4

73. 2/9

75. (a) 3/5 **(b)** 3/4 **(c)** 2/5

77. (a) 1/2 **(b)** 7/10 **(c)** 9/10

79. (a) 1/2 **(b)** 7/13 **(c)** 11/26 **(d)** 0 **(e)** 3/4

81. (a) 1966/3691 ≈ 0.5326 **(b)** 2104/3691 ≈ 0.5700
(c) 1220/3691 ≈ 0.3305 **(d)** 2850/3691 ≈ 0.7721

83. (a) 0 **(b)** 0.5022

85. (a) 1/3 **(b)** 8/12 = 2/3; 1 − 1/3 = 2/3

87. (a) 1/4 **(b)** 1/3

Section 5.3

1. (a) Yes. **(b)** The probability of winning the football game depends on whether or not the star quarterback can play in the game.

3. For $P(A \mid B)$, we assume that the event B has occurred, and now need to find the probability of event A, given event B. On the other hand, for $P(A \cap B)$, we do not assume that event B has occurred, and instead need to determine the probability that both events occurred.

5. Answers will vary.

7. (a) Independent; sampling with replacement **(b)** Dependent; sampling without replacement

9. $50/200 = 1/4 = 0.25$

11. $100/200 = 1/2 = 0.5$

13. $40/200 = 1/5 = 0.2$

15. $60/200 = 3/10 = 0.3$

17. $40/100 = 2/5 = 0.4$

19. $60/100 = 3/5 = 0.6$

21. $40/50 = 4/5 = 0.8$

23. $60/150 = 2/5 = 0.4$

25. $178/288 \approx 0.6181$

27. $161/178 \approx 0.9045$

29. Dependent

31. Independent

33. Dependent

35. 0.27

37. 0.125

39. 0.2

41. 0.05

43. 0.04

45. $1/4 = 0.25$

47. $25/102 \approx 0.2451$

49. 2 is 0.1% of 2000, so we can use the 1% Guideline to approximate the probabilities.

51. 0.015625

53. Dependent

57. They are independent only if $P(X) = 0$ or $P(Y) = 0$. Otherwise they are dependent.

59. If the intersection of W and Z is empty, then $P(W \cap Z) = 0$. They are independent only if $P(W) = 0$ or $P(Z) = 0$. Otherwise they are dependent.

61. $(1/3)^3$

63. $(1/3)^5$

65. $1 - (1/3)^3$

67. $1 - (1/3)^5$

69. $(1/2)^5$

71. 0.24

73. 0.4

75. 0.1

77. 0.2

79. 0.1667

81. 0.2

83. 0.5

85. 0.8

87. 0.7

89. Yes. $P(C \text{ and } D) = 0.21 = (0.7)(0.3) = P(C) P(D)$, $P(C \mid D) = 0.7 = P(C)$, and $P(D \mid C) = 0.3 = P(D)$.

91. $P(E \text{ and } F)$

93. 1/6

95. 1/2

97. (a) 0.0016 **(b)** 0.0000001024 **(c)** 0.15065344

99. (a) Without replacement; the only way to make sure that we sample two different computers is to sample without replacement. **(b)** 1/10 **(c)** 1/11 **(d)** 109/110 **(e)** 1/110 **(f)** Either reject the batch if at least one computer is defective or increase the sample size.

101. (a) 1/2 **(b)** 9/19 **(c)** 4/9

103. (a) 0.3430 **(b)** 0.3236

105. No, P (more serious than Pearl Harbor | female) $= 0.6752 \neq 0.6667 = P$ (more serious than Pearl Harbor) and P (more serious than Pearl Harbor | male) $= 0.6579 \neq 0.6667 = P$ (more serious than Pearl Harbor).

107. (a) 1/3 **(b)** 1/10 **(c)** 1/6 **(d)** 1/15

109. No; $P(C) = 1/2 \neq 5/9 = P(C \mid F)$, $P(C) = 1/2 \neq 5/12 = P(C \mid M)$, $P(F \text{ and } C) = 1/3 \neq 3/10 = P(F) P(C)$, and $P(M \text{ and } C) = 1/6 \neq 1/5 = P(M) P(C)$.

Section 5.4

1. Tree diagram

3. In a permutation, order is important. In a combination, order is not important.

5. Answers will vary.

7.

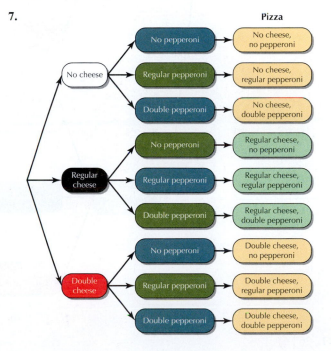

9.

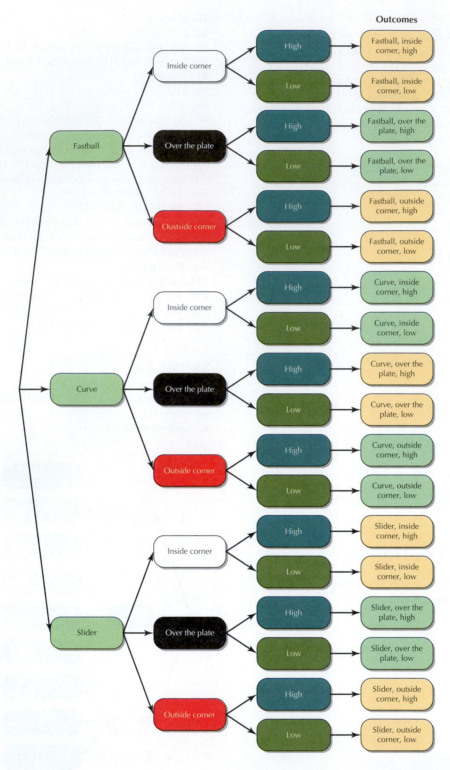

11. 26^4
13. 20
15. 24
17. 720
19. 1
21. 1
23. 12
25. 210
27. 6720
29. 100

31. 93,326,215,443,944,152,681,699,238,856,266,700,490,715,
968,264,381,621,468,592,963,895,217,599,993,229,915,608,941,
463,976,156,518,286,253,697,920,827,223,758,251,185,210,916,
864,000,000,000,000,000,000,000,000

33. 35
35. 165
37. 11
39. 1
41. $5!/(2!1!1!1!) = 60$
43. $_7C_3 = 7!/3!\cdot4! = 7!/4!\cdot3! = {}_7C_4$

45. {Amy, Bob, Chris}, {Amy, Chris, Bob}, {Bob, Amy, Chris}, {Bob, Chris, Amy}, {Chris, Amy, Bob}, {Chris, Bob, Amy}, {Amy, Bob, Danielle}, {Amy, Danielle, Bob}, {Bob, Amy, Danielle}, {Bob, Danielle, Amy}, {Danielle, Amy, Bob}, {Danielle, Bob, Amy}, {Amy, Chris, Danielle}, {Amy, Danielle, Chris}, {Chris, Amy, Danielle}, {Chris, Danielle, Amy}, {Danielle, Amy, Chris}, {Danielle, Chris, Amy}, {Bob, Chris, Danielle}, {Bob, Danielle, Chris}, {Chris, Bob, Danielle}, {Chris, Danielle, Bob}, {Danielle, Bob, Chris}, {Danielle, Chris, Bob}. $_4P_3 = 24$

47. {Amy, Bob, Chris}, {Amy, Chris, Bob}, {Chris, Amy, Bob}, {Chris, Bob, Amy}, {Bob, Amy, Chris}, and {Bob, Chris, Amy} are all different permutations but the same combination.

49. $r!$

51. (a)

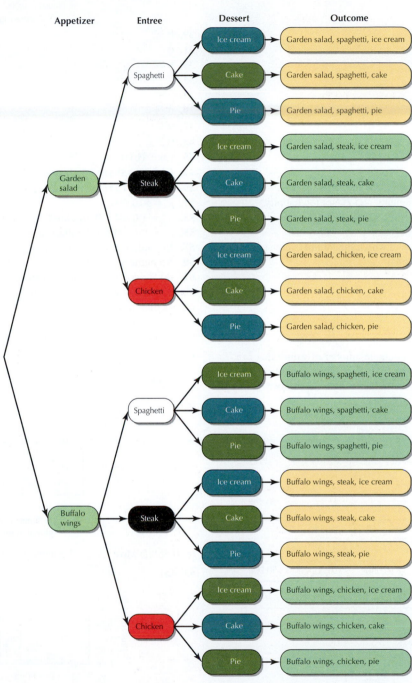

(b) 18

53. 3,628,800

55. 720

57. 20

59. 300

61. 20

63. 184,756

65. 6720

Chapter 5 Review

1. 3/8

3. 0

5. 1/2

7. (a) 0.213 **(b)** 0.656 **(c)** 0

9. 0

11. (a) 1/6 **(b)** 1/6

13. Men's TV channel, since $P(\text{Dog} \mid \text{Male}) = 5/12 > 5/18 = P(\text{Dog} \mid \text{Female})$

15. 60

Chapter 5 Quiz

1. False
2. True
3. 0, 1
4. or, and
5. 0.5
6. 1
7. With replacement
8. Intersection of A and B.
9. (a) 1/9 (b) 8/9 (c) 5/18 (d) 1/18 (e) 1/3
10. 0.2
11. 0.2125
12. (a) 1/4 (b) 3/13 (c) 1/13 (d) 1/2 (e) 1/52 (f) 1/26
13. (a) 0.5361 (b) 0.4639 (c) 0.0330
14. (a) 0.0215 (b) 0.0114
15. No, $P(\text{Not too happily married}) = 0.0330 \neq 0.0402 = P(\text{Not too happily married} \mid \text{Female})$ and $P(\text{Not too happily married}) = 0.0330 \neq 0.0246 = P(\text{Not too happily married} \mid \text{Male})$
16. 4
17. (a) Permutation; the order in which the numbers are selected is important. (b) 6840 (c) 1/6840

Chapter 6

Section 6.1

1. Answers will vary.
3. Discrete: takes finite or a countable number of values that can be graphed as separate points on the number line; continuous takes infinitely many values that form an interval on the number line.
5. $\Sigma P(X) = 1$ and $0 \leq P(X) \leq 1$.
7. Discrete
9. Continuous
11. Discrete
13. {0, 1, 2, 3, 4, 5, 6, 7, 8, 9, 10, 11, 12, 13, 14, 15}
15. {0, 1, 2, 3, 4}

17.

X = Number of CDs	0	1	2	3	4
P(X)	0.06	0.24	0.38	0.22	0.10

19.

X = Money gained	−$10,000	$10,000	$50,000
P(X)	1/3	1/2	1/6

21. No, the probabilities don't add up to 1.
23. No, $P(X = 1)$ is negative.

25.

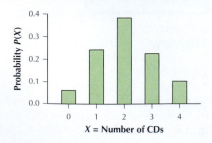

27.

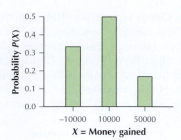

29. 0.32
31. 0
33. 0.40
35. 0
37. 2/3
39. 0
41. 0.5
43. 0.2
45. (a) 2 (b) 0
47. (a) $10,000 (b) $50,000
49. $\mu = 2.06$ CDs
51. $\mu = \$10,000$
53. $\sigma^2 = 1.0964$ CDs squared, $\sigma = 1.0471$ CDs
55. $\sigma^2 = 400,000,000$ dollars squared, $\sigma = \$20,000$
57. No outliers, no moderately unusual values
59. No outliers, $50,000 is moderately unusual.
61. (a) We don't know the number of games that will be played in the finals before the finals begin. This introduces an element of chance into the experiment, thereby making the number of games a random variable. (b) There are only a finite number of possibilities for the number of games that can be played in the finals.

(c)

X = Games	4	5	6	7
P(X)	0.25	0.2	0.25	0.3

(d)

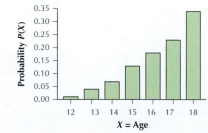

(e) 0.45 (f) Most likely: 7 games; Least likely: 5 games
63. (a)

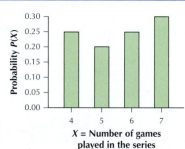

(b) 0.43 (c) 0.25 (d) The answer to (b) includes the probability that $X = 16$ and the answer to (c) does not. (e) Most likely: 18 years; Least likely: 12 years
65. (a) $\mu = 2.46$ courses. If we were to consider an infinite number of faculty at all degree-granting institutions of higher learning in the United States in the fall 2010 semester, the mean number of courses taught would be 2.46 courses.

(b) $\sigma^2 = 1.3684$ courses squared, $\sigma = 1.1698$ courses
(c) $Z = 2.1713$, moderately unusual
67. (a)

X = Sum of dice	2	3	4	5	6	7	8	9	10	11	12
P(X)	1/36	1/18	1/12	1/9	5/36	1/6	5/36	1/9	1/12	1/18	1/36

(b) The mean is about 7.

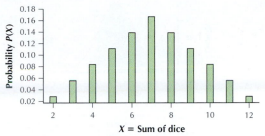

(c) $\mu = 7$. If we were to consider tossing two dice an infinite number of times, the mean sum of the dice would be 7.
(d) 7. The estimate is equal to the actual value.
(e)

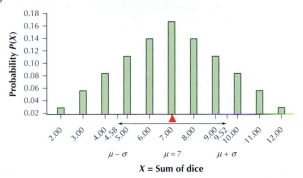

69. (a) No. The mean is 2 but the most likely value is 0.

X	0	2	8
P(X)	0.6	0.2	0.2

(b) Symmetric, one mode

Section 6.2

1. (i) Each trial of the experiment has only two possible mutually exclusive outcomes (or is defined in such a way that the number of outcomes is reduced to two). One outcome is denoted a success and the other a failure. **(ii)** There is a fixed number of trials, known in advance of the experiment. **(iii)** The experimental outcomes are independent of each other. **(iv)** The probability of observing a success remains the same from trial to trial.
3. If you perform an experiment n times, you can't have more than n successes. For example, if you flip a coin 10 times you can't get 11 heads.
5. Not binomial; the events "Person A comes to party" and "Person B comes to party" may not be independent.
7. Binomial, $X =$ number of correct answers, $n = 8$, $p = 1/4 = 0.25$, $1 - p = 3/4 = 0.75$
9. Not binomial; not a fixed number of trials
11. Not binomial, trials are not independent, sample is more than 1% of the population.
13. Binomial; $n = 2$, $X =$ number of games won, $p = 0.25$, $1 - p = 0.75$

15. 0.3955
17. 0.1172
19. 0.2301
21. 0.6328
23. 0.1611
25. 0.8891
27. 0.9744
29. $1/8 = 0.125$
31. $3/8 = 0.375$
33. $7/8 = 0.875$
35. 0.0154
37. 0.4823
39. 0.5177
41. 0.0778
43. 0.6826
45. 0 (TI-83/84: 0.00001139)
47. 0.0059
49. (a) $\mu = 1.5$ heads. If we repeat the experiment of tossing a fair coin 3 times an infinite number of times, record the number of heads in each performance of the experiment, and take the mean of all of the performances of this experiment, the mean number of heads will equal $\mu = 1.5$. **(b)** $\sigma^2 = 0.75$ head squared **(c)** $\sigma = 0.8660$ head
51. (a) $\mu = 2$ vehicles. If we repeat this experiment an infinite number of times, record the number of vehicles obeying the speed limit in each sample, and take the mean of all of these samples, the mean number of vehicles obeying the speed limit will equal $\mu = 2$. **(b)** $\sigma^2 = 1.2$ vehicles squared **(c)** $\sigma = 1.0954$ vehicles

53. (a)

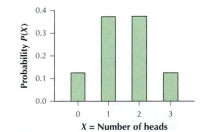

(b) 1 and 2 heads
55. (a)

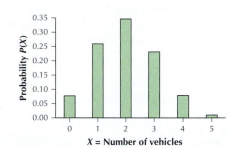

(b) 2 vehicles
57. (a) It fulfills the requirements: (i) There are only two possible outcomes for each trial: correct answer or incorrect answer. (ii) We know in advance that the quiz will have 5 questions. (iii) Since you are randomly guessing the answer to each question, the trials are independent. (iv) Since each question has 4 responses, the probability of guessing correctly remains the same from question to question. **(b)** $n = 5$, $p = 1/4 = 0.25$ **(c)** 0.1035 **(d)** 0.8965
59. (a) $p = 0.378$ is not in the table **(b)** 0.0313 **(c)** 0.0395
61. (a) $\mu = 1.25$ correct answers. If we repeat this experiment an infinite number of times, record the number of correct answers for each quiz taken, and take the mean of all of the quizzes, the mean

number of correct answers will equal $\mu = 1.25$. $\sigma^2 = 0.9375$ correct answer squared, $\sigma = 0.9682$ correct answer **(b)** Five correct answers is considered an outlier; 4 correct answers is considered moderately unusual. **(c)** Mode is 1 correct answer.

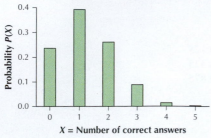

63. (a) $\mu = 5.67$ vowels. If we repeat this experiment an infinite number of times, record the number of vowels for each sample, and take the mean of all of the samples, the mean number of vowels will equal $\mu = 5.67$. $\sigma^2 = 3.5267$ vowels squared, $\sigma = 1.8780$ vowels **(b)** $Z = -1.4217$, not unusual
65. (a) 10 students **(b)** 10 students **(c)** 10 students
67. (a) $1287/2{,}598{,}960 \approx 0.0005$ **(b)** $27{,}885/2{,}598{,}960 \approx 0.0107$ **(c)** $29{,}172/2{,}598{,}960 \approx 0.0112$ **(d)** $211{,}926/2{,}598{,}960 \approx 0.0815$ **(e)** $2{,}357{,}862/2{,}598{,}960 \approx 0.9072$

Section 6.3

1. The probability that X equals some particular value is zero.
3. Area under the normal distribution curve above an interval.
5. False
7. 0.5
9. 0.65
11. 0.01
13. 0.5
15. 0.1
17. About 0.68
19. About 0.025
21. About 0.997
23. About 0.0235
25. 0
27. 0.5
29. Greater than 0.5. Since $X = 4285$ is greater than the mean of 3285 and the area to the left of $\mu = 3285$ is 0.5, the area to the left of $X = 4285$ is greater than the area to the left of $X = 3285$.
31. About 0.997
33. A has mean 10; B has mean 25. The peak of a normal curve is at the mean; from the graphs we see that the mean of A is less than the mean of B.
35. $\mu = 0, \sigma = 1$
37. $\mu = 10, \sigma = 2$
39. (a) 0.25 **(b)** 0.25 **(c)** 0

41. (a)

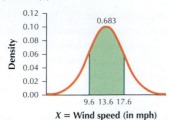

(b) About 0.68
43. About 0.16
45. (a) About 0.025 **(b)** About 0.68

47. (a) 9.5 minutes **(b)** 9 minutes **(c)** 9.75 minutes
(d) 0.5 minute **(e)** 1 minute **(f)** 0.25 minute

Section 6.4

1. $\mu = 0$
3. True
5. 0.9750
7. 0.4821
9. 0.1359
11. (a)

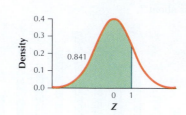

(b) 0.8413
13. (a)

(b) 0.9987
15. (a)

(b) 0.0047
17. (a)

(b) 0.4207
19. (a)

(b) 0.1020
21. (a)

(b) 0.9987

23. (a)

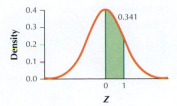

(b) 0.3413

25. (a)

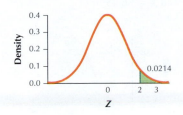

(b) 0.0214

27. (a)

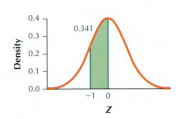

(b) 0.3413

29. (a)

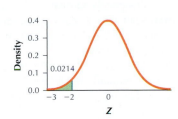

(b) 0.0214

31. (a)

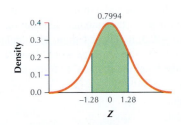

(b) 0.7994

33. (a)

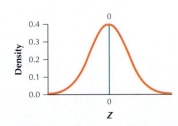

(b) 0

35. (a)

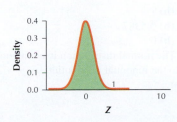

(b) 1

37. (a)

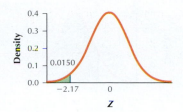

(b) 0.0150

39. (a)

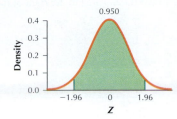

(b) 0.9500

41. (a)

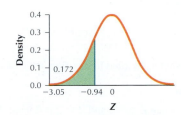

(b) 0.1725

43. (a)

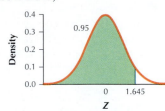

(b) 0.5000

45. Less than 0; $Z = -0.43$; -0.43 is less than 0
47. Less than 0; $Z = -0.45$; -0.45 is less than 0
49. 1.65 (TI-83/84: 1.645)

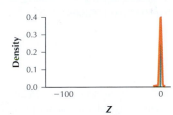

51. 2.05

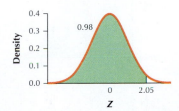

53. Less than 0; $Z = -0.87$; -0.87 is less than 0.
55. Less than 0; $Z = -2.03$; -2.03 is less than 0

57. -1.28

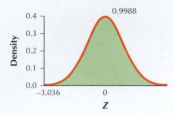

59. -3.036 (Using the table, both -3.03 and -3.04 have area to left of them equal to 0.0012 and area to the right of them as 0.9988.)

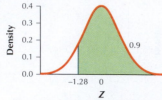

61. -1.28 and 1.28
63. -2.33 and 2.33
65. $Z = 0$
67. $Z = 2.58$
69. Therefore Nicholas scored higher than 84.13% of the test takers.
71. $Z = -0.13$
73. The area between $Z = -2$ and $Z = 2$ is 0.9544. By the Empirical Rule, the area between $Z = -2$ and $Z = 2$ is about 0.95.
75. (a) 0.0668 (b) 0.9332 (c) 0.8664
77. $Z = -2.58$ and $Z = 2.58$.
79. -0.67; 0; 0.67

Section 6.5

1. To standardize things means to make them all the same, uniform, or equivalent. To standardize a normal random variable X, we transform X into the standard normal random variable Z using the formula $Z = \dfrac{X - \mu}{\sigma}$. We do this so that we can use the standard normal table to find the probabilities.
3. 0.5
5. 0.8413
7. 0.0062
9. 0.9332
11. 0.8400
13. 0.0049
15. $X = 86.45$
17. 89.6

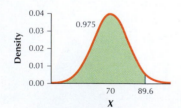

19. 46.7

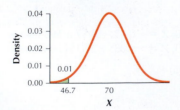

21. 44.2

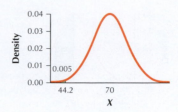

23. $X = 53.55$ and $X = 86.45$
25. 46.7 and 93.3

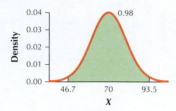

27. (a) 0.5 (b) 0.1587 (c) 0.1359
29. (a) 0.1423 (b) 0.1423 (c) 26.67% (d) $X = 27.6$ mph
(e) Z-score is -2.27; moderately unusual
31. (a) 18.29 ounces (b) 11.71 ounces (c) 11.71 ounces and 18.29 ounces
33. (a) 21.28 mph (TI-83/84: 21.29 mph) (b) 5.92 mph
(TI-83/84: 5.91 mph) (c) 5.92 (5.91) mph and 21.28 (21.29) mph
(d) Z-score is -2.27; moderately unusual
35. (a) 0.0139 (b) 0.5249 (c) 4.87 days; the mean equals the median. (d) The Z-score for $X = 8$ days is 3.23. Since $3.23 \geq 3$, a hospital stay of 8 days is unusual.
37. (a) 0.0013 (b) 0.3989 (TI-83/84: 0.4004) (c) 0.702 million shares (TI-83/84: 0.704 million) (d) Outlier, $Z = 41.5$
39. (a) 0.0062 (b) 0.0228 (c) 0.7506 (d) 0.2963

Section 6.6

1. For certain values of n and p, it may be inconvenient to calculate probabilities for the binomial distribution. For example, if $n = 100$ and $p = 0.5$, it may be tedious to calculate $P(X > 57)$, which, in the absence of technology, would involve 44 applications of the binomial probability formula.
3. Appropriate
5. Not appropriate
7. Not appropriate
9. 0.1272
11. 0.4364
13. 0.4364
15. 0.3616
17. 0.0992
19. 0.6772
21. 0.0853
23. 0.0992
25. (a) 0.0558 (b) 0.5279
27. (a) 0.0037 (b) 1
29. (a) 0.4721 (b) 0.5387
31. (a) 0.5517 (b) 0
33. (a) No (b) The normal distribution is not a good approximation to the binomial distribution ($n \cdot p = 2 < 10$), so not appropriate.

Chapter 6 Review

1. (a) 0.15 **(b)** 0.60

3. (a) 0.1887 **(b)** 0.8113 **(c)** 0.7880

5. 0

7. Less than 0.5. Since the area to the right of the mean $\mu = 106$ mm is 0.5 and $X = 110$ mm is greater than the mean $\mu = 106$ mm, the area to the right of $X = 110$ mm is less than the area to the right of the mean $\mu = 106$.

9. (a)

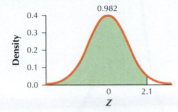

(b) 0.9821

11. (a)

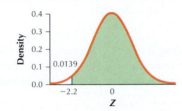

(b) 0.9861

13. (a)

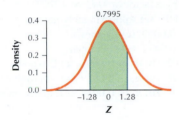

(b) 0.7995

15. (a) 0.7967 **(b)** 1.88% **(c)** 0.2967 **(d)** $X = 72.46$ mph
(e) The Z-score for $X = 55$ mph is -3.33. Since $-3.33 \le -3$, a driver from South Dakota who never drives faster than 55 mph on the Interstate is unusual.

17. (a) 0.2420 **(b)** 0.0705 **(c)** 266.55 **(d)** Moderately unusual, $Z = 2.5$

Chapter 6 Quiz

1. True

2. False

3. False

4. 0.5

5. 0

6. 0

7. discrete

8. binomial

9. $\mu = 0, \sigma = 1$

10. (a)

$X =$ Amount won	0	5
$P(X)$	13/18	5/18

(b) \$1.39 **(c)** \$1.39

11. (a) 0.0962 **(b)** 19 CEOs **(c)** $\mu = 19$ CEOs, $Var(X) = 15.39$, $SD(X) = 3.9230$. The expected number of CEOs who drive luxury cars in a random sample of 100 CEOs is 19. **(d)** Z-score is 5.35; unusual

12. (a) 0.1003 **(b)** 33.22% **(c)** \$4329.50 **(d)** Z-score is -2.05; moderately unusual

Chapter 7

Section 7.1

1. Statistical inference refers to learning about population characteristics by studying the same characteristics in a sample.

3. For a given sample size n, it is normal with mean μ and standard deviation $\sigma_{\bar{x}} = \sigma/\sqrt{n}$.

5. $\mu_{\bar{x}} = 100, \sigma_{\bar{x}} = 4$

7. $\mu_{\bar{x}} = 0, \sigma_{\bar{x}} = 3.3333$

9. $\mu_{\bar{x}} = -10, \sigma_{\bar{x}} = 0.5$

11. 0.1587

13. 0.6826

15. 0.9772

17. 112.34

19. 114.7

21. 87.66 and 112.34

23. Normal (50, 2)

25. 0.1587

27. 53.29

29. 46.71 and 53.29

31. Normal with mean $\mu = 10$ and standard deviation $\sigma_{\bar{x}} = 1$

33. 0.1587

35. 11.96

37. 8.04 and 11.96

39. (a) $\mu_{\bar{x}} = 1.7$ seconds, $\sigma_{\bar{x}} = 0.1$ second **(b)** 0.1587
(c) 0.1587

41. (a) 0.1056 **(b)** 0.1359 **(c)** 0.4013

43. (a) 1.86 seconds **(b)** 1.54 seconds **(c)** 1.54 seconds and 1.86 seconds

45. (a) \$48.4 million and \$151.6 million; TI-83/84: \$48.48 million and \$151.52 million

(b)

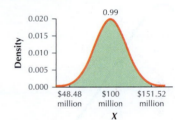

47. (a) $_5C_2 = 10$ **(b)** $\mu = 6.9$ **(c)** $\sigma = 0.7483$

49. (a) $\mu_{\bar{x}} = 6.9$ **(b)** Yes. Fact 1: $\mu_{\bar{x}} = \mu$

51. 0.4483

53. (a) 0.3015 **(b)** Sample means are less variable than individual observations, so 500 is more standard deviations below $\mu_{\bar{x}}$ than below μ.

55. (a–b) Increase **(c)** Decrease

Section 7.2

1. To determine whether or not the data are normally distributed. If the points either cluster around a straight line or nearly all fall within the curved bounds, then it is likely that the data set is normal. If there are systematic deviations off the straight line, then that is evidence against the claim that the data set is normal.

3. $n = 30$

5. 4 times as large

7. Not acceptable

9. Acceptable

11. (a) 516 **(b)** 38.6667, Normal

13. (a) 80 **(b)** 1; Approximately normal
15. (a) 50 miles per gallon **(b)** 1.5 miles per gallon; Unknown
17. 0.2676 (TI-83/84: 0.2674)
19. 0.9772
21. Not possible. The variable is not normally distributed and the sample size is less than 30.
23. 8.08
25. $3.00
27. $58,355
29. (a) 0.0918 (TI-83/84: 0.0912) **(b)** 0.8164 (TI-83/84: 0.8176)
31. (a) 0.0026 **(b)** 0.4974 **(c)** Since $n \geq 30$, the distribution of the sample mean is approximately normal by the Central Limit Theorem.
33. (a) 214.34 **(b)** 189.66
35. (a) 111.1 computers **(b)** 136.9 computers

(c)

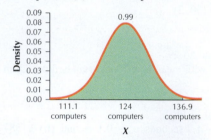

37. Against normality. There are several points outside of the curved lines and most of the points are close to the upper curved line.
39. 0.9544 (TI-83/84: 0.9545)
41. (a) $n = 2$

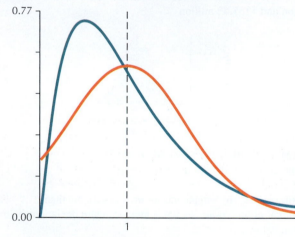

(b) $n = 5$

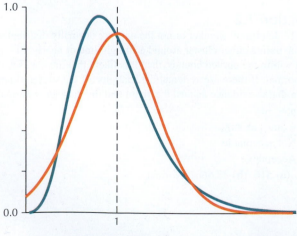

(c) $n = 30$

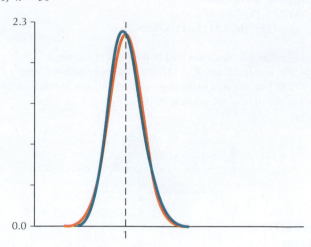

Section 7.3

1. If we take a sample of size n, the sample proportion $\hat{p}$ is $\hat{p} = x/n$, where x represents the number of individuals in the sample that have the particular characteristic. Examples will vary.
3. $\sigma_{\hat{p}} = \sqrt{p \cdot (1 - p)/n}$
5. It decreases by a factor of $1/\sqrt{2} \approx 0.7071$.
7. (a) 0.5 **(b)** 0.05 **(c)** Approximately normal
9. (a) 0.01 **(b)** 0.0099 **(c)** Unknown
11. (a) 0.9 **(b)** 0.0474 **(c)** Unknown
13. (a) $\mu_{\hat{p}} = 0.02$ **(b)** $\sigma_{\hat{p}} \approx 0.0099$ **(c)** Unknown
15. (a) $\mu_{\hat{p}} = 0.98$ **(b)** $\sigma_{\hat{p}} \approx 0.0089$ **(c)** Approximately normal
17. (a) $\mu_{\hat{p}} = 0.99$ **(b)** $\sigma_{\hat{p}} \approx 0.0044$ **(c)** Approximately normal
19. 10
21. 50
23. 500
25. 0.1587
27. Not possible; sampling distribution of $\hat{p}$ is unknown.
29. Not possible; sampling distribution of $\hat{p}$ is unknown.
31. Not possible since $np = (200)(0.02) = 4 < 5$.
33. 0.564
35. 0.962
37. 0.052
39. (a) $\mu_{\hat{p}} = 0.25$, $\sigma_{\hat{p}} \approx 0.0722$ **(b)** Approximately normal (0.25, 0.0722)
(c) 0.4443 (TI-83/84: 0.4449)
41. (a) $\mu_{\hat{p}} = 0.75$, $\sigma_{\hat{p}} \approx 0.0968$ **(b)** 0.7324 (TI-83/84: 0.7323) **(c)** 0.0959 (TI-83/84: 0.0954)
43. (a) 0.1312, 0.3688

(b)

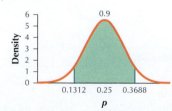

(c) For $\hat{p} = 2/36 \approx 0.0556$, $Z = -2.69$. Thus $\hat{p} = 2/36$ is considered moderately unusual. **(d)** Sample proportions between 0 and 0.0334 inclusive and between 0.4666 and 1 inclusive would be considered outliers.
45. (a) 0.5003, 0.9997 (TI-83/84: 0.5007, 0.9993)

(b)

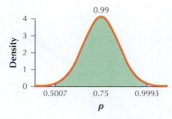

(c) For $\hat{p} = 14/20 = 0.7$, $Z = -0.5165$. Thus $\hat{p} = 0.7$ is neither moderately unusual nor an outlier.
47. (a) Remain unchanged. Since $\mu_{\hat{p}} = p$, $\mu_{\hat{p}}$ does not depend on the sample size. Therefore an increase in the sample size would not affect $\mu_{\hat{p}}$. **(b)** Decrease. Since $\sigma_{\hat{p}} = \sqrt{p \cdot (1 - p)/n}$, an increase in the denominator would result in a decrease in the fraction, which would result in a decrease in $\sigma_{\hat{p}}$. **(c)** Decrease. Since $0.04 - 0.13 = -0.09$ is negative, and $\sigma_{\hat{p}}$ is positive and decreases, $Z = 0.04 - 0.13/\sigma_{\hat{p}}$ will decrease. Therefore the area to the left of $Z = 0.04 - 0.13/\sigma_{\hat{p}}$ will decrease. Since $P(\hat{p} < 0.04)$ is equal to this area, $P(\hat{p} < 0.04)$ will decrease. **(d)** Increase. Since $0.10 - 0.13 = -0.03$ is negative, $0.15 - 0.13 = 0.02$ is positive, and $\sigma_{\hat{p}}$ is positive and decreases, $Z = 0.10 - 0.13/\sigma_{\hat{p}}$ will decrease and $Z = 0.15 - 0.13/\sigma_{\hat{p}}$ will increase. Thus the area between these two values will increase. Since $P(0.10 < \hat{p} < 0.15)$ is equal to this area, $P(0.10 < \hat{p} < 0.15)$ will increase.
(e) Decrease. Since $0.49 - 0.13 = 0.36$ is positive, $0.45 - 0.13 = 0.32$ is positive, and $\sigma_{\hat{p}}$ is positive and decreases, both $Z = 0.49 - 0.13/\sigma_{\hat{p}}$ and $Z = 0.45 - 0.13/\sigma_{\hat{p}}$ will increase. Both of these values will be farther out on the right tail of the standard normal distribution, where the curve is closer to the Z axis. Therefore, the area between these two values of Z will decrease. Since $P(0.45 < \hat{p} < 0.49)$ is equal to this area, it will decrease. **(f)** Increase. Since $\sigma_{\hat{p}}$ decreases, $1.96 \, \sigma_{\hat{p}}$ decreases. Since the 2.5th percentile is $0.13 - 1.96 \, \sigma_{\hat{p}}$, the 2.5th percentile increases. **(g)** Decrease. Since $\sigma_{\hat{p}}$ decreases, $1.96 \, \sigma_{\hat{p}}$ decreases. Since the 97.5th percentile is $0.13 + 1.96 \, \sigma_{\hat{p}}$, the 97.5th percentile decreases.
49. (a) 0.5 **(b)** 0 **(c)** 0 **(d)** 0.5
51. The results of Exercises 49 and 50 do not support this claim. The 97.5th percentile for the males is less than the 2.5th percentile for the females. Also $P(p < 0.41)$ and $P(p > 0.65)$ are both very different for males and females.

Chapter 7 Review

1. $\mu_{\bar{x}} = 10$, $\sigma_{\bar{x}} = 1$
3. $\mu_{\bar{x}} = 10$, $\sigma_{\bar{x}} = 5/7 \approx 0.7143$
5. $\mu_{\bar{x}} = 50$ and $\sigma_{\bar{x}} = 10$.
7. 0.1056
9. 0.7888
11. 0.8185
13. 100
15. We have $np = (40)(0.1) = 4 < 5$, so the sampling distribution of $\hat{p}$ is unknown. Thus $P(\hat{p} < 0.12)$ can't be found.
17. 0.0153
19. (a) 0.2578 (TI-83/84: 0.2590) **(b)** 0.2422 (TI-83/84: 0.2410) **(c)** 0.7422 (TI-83/84: 0.7410) **(d)** 0.0437, 0.1963

Chapter 7 Quiz

1. True
2. False
3. Sampling error
4. Approximately normal

5. No
6. $np \geq 5$ and $n(1 - p) \geq 5$
7. (a) 0.1587 **(b)** 0.9500 **(c)** 0.1056
8. (a) 45.15 grams **(b)** 34.85 grams **(c)** 34.85 grams and 45.15 grams
9. (a) 0.0228 **(b)** 0.0228 **(c)** 0.9544
10. (a) 68.77 inches **(b)** 67.23 inches **(c)** 67.23 and 68.77 inches
11. (a) 0.5 **(b)** 0.2422 (TI-83/84: 0.2406) **(c)** 0.0174, 0.1146

Chapter 8

Section 8.1

1. A range of values is more likely to contain μ than a point estimate is to be exactly equal to μ. We have no measure of confidence that our point estimate is close to μ. A confidence level for a confidence interval means that if we take sample after sample for a very long time, then in the long run, the percent of intervals that will contain the population mean μ will equal the confidence level.
3. We are 95% confident that the population mean football score lies between 15 and 25.
5. $\bar{x} \pm E$ is shorthand for writing the two values $\bar{x} - E$ and $\bar{x} + E$. $\pm$ is shorthand notation for writing two numbers.
7. (a) $Z_{\alpha/2}$ increases. **(b)** Since the confidence level is $(1 - \alpha) \times 100\%$, as the confidence level increases, $1 - \alpha$ increases. Thus α and $\alpha/2$ will decrease. Since $\alpha/2$ is the area underneath the standard normal curve to the right of $Z_{\alpha/2}$, a decrease in $\alpha/2$ will result in an increase in $Z_{\alpha/2}$.
9. Increases, Decreases
11. $\bar{x} = 2$
13. $\bar{x} = 14$
15. No
17. Yes
19. We can use the Z interval.
21. $Z_{\alpha/2} = 2.576$
23. $Z_{\alpha/2} = 1.96$
25. $Z_{\alpha/2} = 1.645$
27. (a) 0.5 **(b)** $Z_{\alpha/2} = 1.96$ **(c)** (34.02, 35.98). We are 95% confident that the true mean μ lies between 34.02 and 35.98.
29. (a) 2 **(b)** $Z_{\alpha/2} = 1.96$ **(c)** (11.08, 18.92). We are 95% confident that the true mean μ lies between 11.08 and 18.92.
31. (a) 1 **(b)** $Z_{\alpha/2} = 1.96$ **(c)** (18.04, 21.96). We are 95% confident that the true mean μ lies between 18.04 and 21.96.
33. (a) 0.98 **(b)** We can estimate μ to within 0.98 with 95% confidence.
35. (a) 3.92 **(b)** We can estimate μ to within 3.92 with 95% confidence.
37. (a) 1.96 **(b)** We can estimate μ to within 1.96 with 95% confidence.
39. (a) (9.342, 10.658). We are 90% confident that the true mean μ lies between 9.342 and 10.658. **(b)** (9.216, 10.784). We are 95% confident that the true mean μ lies between 9.216 and 10.784. **(c)** (8.9696, 11.0304). We are 99% confident that the true mean μ lies between 8.9696 and 11.0304. **(d)** The confidence interval for a given sample size becomes wider as the confidence level increases.
41. 1
43. 5
45. 5
47. 11

49. (a) 69 gallons **(b)** 3.65 gallons **(c)** $Z_{\alpha/2} = 1.96$ **(d)** (61.84, 76.16). We are 95% confident that μ lies between 61.84 gallons and 76.16 gallons.

51. (a) 107 seconds **(b)** 19.5 seconds **(c)** $Z_{\alpha/2} = 1.96$ **(d)** (68.78, 145.22). We are 95% confident that the true mean length of time that boys remain engaged with a science exhibit at a museum μ lies between 68.78 seconds and 145.22 seconds.

53. (a) 7.16 gallons. We can estimate μ to within 7.16 gallons with 95% confidence. **(b)** 3 **(c)** 62

55. (a) $E = 38.22$ seconds. We can estimate μ, the mean length of time that boys remain engaged with a science exhibit at a museum, to within 38.22 seconds with 95% confidence. **(b)** 59 **(c)** 9604 days, approximately 26.31 years

57. (a) The normal probability plot indicates an acceptable level of normality.

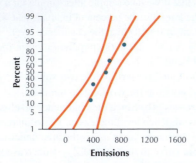

(b) (415.067, 709.333); TI-83/84: (415.08, 709.32). We are 90% confident that the population mean carbon emissions lies between 415.067 (415.08) million tons and 709.333 (709.32) million tons. **(c)** $E = 147.133$ million tons. We can estimate the population mean emissions level of all nations to within 147.133 million tons with 90% confidence. **(d)** 44 nations

59. (a) The normal probability plot indicates an acceptable level of normality.

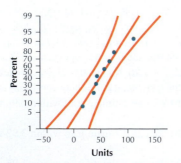

(b) (27.510, 82.240); TI-83/84: (27.554, 82.196). We are 99% confident that the population mean number of Wii games that are sold in the United States each week lies between 27.510 (27.554) thousand games and 82.240 (82.196) thousand games. **(c)** $E = 27.365$ thousand games. We can estimate the population mean number of Wii games sold to within 27.365 thousand games with 99% confidence. **(d)** 239 games

61. (a) 7 bits **(b)** 2 bits **(c)** $Z_{\alpha/2} = 1.96$ **(d)** $n = 97$ **(e)** $n = 385$

63. (a) 6199 small firms **(b)** See the histogram in **(e)**.

(c)

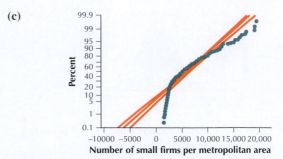

Since the majority of the points lie outside of the curved lines, the normality assumption is not valid.

(d) (3188.95, 9209.05). We are 95% confident that the average number of small firms per metropolitan area lies between 3188.95 and 9209.05.

(e)

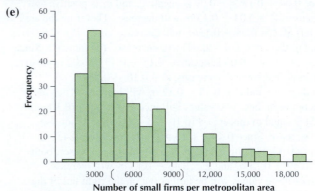

65. Answers will vary.

Section 8.2

1. In most real-world problems, the population standard deviation σ is unknown, so we can't use the Z interval.

3. The t curve approaches closer and closer to the Z curve.

5. (a) $t_{\alpha/2} = 1.833$ **(b)** $t_{\alpha/2} = 2.262$ **(c)** $t_{\alpha/2} = 3.250$

7. (a) The value of $t_{\alpha/2}$ increases as the confidence level increases. **(b)** The larger the value of $1 - \alpha$, the larger the value of $t_{\alpha/2}$ will have to be in order to have an area of $1 - \alpha$ between $-t_{\alpha/2}$ and $t_{\alpha/2}$. $t_{\alpha/2} = 1.833$ for a 90% confidence interval with 9 degrees of freedom; $t_{\alpha/2} = 2.262$ for a 95% confidence interval with 9 degrees of freedom; $t_{\alpha/2} = 3.250$ for a 99% confidence interval with 9 degrees of freedom.

9. (a) $\bar{x} = 2$, $s = 1$ **(b)** $t_{\alpha/2} = 2.776$ **(c)** (0.759, 3.241); TI-83/84: (0.758, 3.242). We are 95% confident that the population mean lies between 0.759 (0.758) and 3.241 (3.242).

11. (a) $\bar{x} = 14$, $s = 3$ **(b)** $t_{\alpha/2} = 2.776$ **(c)** (10.276, 17.724); TI-83/84: (10.275, 17.725). We are 95% confident that the population mean lies between 10.276 (10.275) and 17.724 (17.725).

13. (a) $t_{\alpha/2} = 2.064$ **(b)** (7.936, 12.064)

(c)

15. (a) $t_{\alpha/2} = 3.182$ **(b)** (40.454, 59.546)

(c)

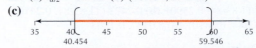

17. (a) $t_{\alpha/2} = 1.860$ **(b)** $(-23.720, -16.280)$

(c)

19. (a) $t_{\alpha/2} = 1.987$ **(b)** $(98.013, 101.987)$
(c)

21. (a) $t_{\alpha/2} = 2.660$ **(b)** $(32.340, 37.660)$
(c)

23. (a) $t_{\alpha/2} = 1.664$ **(b)** $(-21.1093, -18.8907)$
(c)

25. (a) $t_{\alpha/2} = 2.009$ **(b)** $t_{\alpha/2} = 2.0054$
27. (a) $t_{\alpha/2} = 1.684$ **(b)** $t_{\alpha/2} = 1.68$
29. (a) $(95.937, 104.063)$. We are 95% confident that the population mean lies between 95.937 and 104.063. **(b)** $(95.944, 104.056)$. We are 95% confident that the population mean lies between 95.944 and 104.056.
31. (a) $(9.503, 10.497)$. We are 90% confident that the population mean lies between 9.503 and 10.497. **(b)** $(9.505, 10.495)$. We are 90% confident that the population mean lies between 9.505 and 10.495.
33. Since the distribution of the population is unknown, Case 1 does not apply. Since the sample size of $n = 25$ is small ($n < 30$), Case 2 does not apply. Thus we cannot construct the indicated confidence interval.
35. Case 1 $(9.3387, 10.6613)$
37. Since the distribution of the population is unknown, Case 1 does not apply. Since the sample size of $n = 16$ is small ($n < 30$), Case 2 does not apply. Thus we cannot construct the indicated confidence interval.
39. Case 2 $(47.97, 52.03)$
41. $E = 1.241$. We can estimate the population mean to within 1.241 with 95% confidence.
43. $E = 3.724$. We can estimate the population mean to within 3.724 with 95% confidence.
45. $E = 2.064$. We can estimate the population mean to within 2.064 with 95% confidence.
47. $E = 9.546$. We can estimate the population mean to within 9.546 with 95% confidence.
49. (a) $t_{\alpha/2} = 1.987$ **(b)** $(3.790, 6.810)$; TI-83/84: $(3.792, 6.808)$. We are 95% confident that the population mean length of stay in hospital for sickle-cell anemia patients lies between 3.790 (3.792) days and 6.810 (6.808) days.
51. (a) $t_{\alpha/2} = 1.994$ **(b)** $(113.09, 126.91)$. We are 95% confident that the true mean revenue collected from all parking meters μ lies between \$113.09 and \$126.91.
53. (a) $E = 1.510$ days. We can estimate the population mean length of stay in hospital for sickle-cell anemia to within 1.510 days with 95% confidence. **(b)** It will decrease.
55. (a) $E = \$6.91$. We can estimate μ, the true mean revenue collected from all parking meters, to within \$6.91 with 95% confidence. **(b)** Increasing the sample size and decreasing the confidence level. Increasing the sample size is more desirable. A lower confidence level means we are less confident that the population mean lies in our interval.
57. (a) $(376.455, 747.945)$; TI-83/84: $(376.47, 747.93)$. We are 90% confident that the population mean carbon emissions lies between 376.455 (376.47) million tons and 747.945 (747.93) million tons. **(b)** $E = 185.745$ million tons. We can estimate the population mean carbon emissions to within 185.745 million tons with 90% confidence. **(c)** Increasing the sample size and

decreasing the confidence level. Increasing the sample size is more desirable. A lower confidence level means we are less confident that the population mean lies in our interval.
59. (a) $(19.380, 90.370)$; TI-83/84: $(19.375, 90.375)$. We are 99% confident that the population mean number of units sold per Wii in the United States lies between 19.380 (19.375) thousand games and 90.370 (90.375) thousand games. **(b)** $E = 35.495$ thousand Wii games. We can estimate the population mean number of units sold per Wii game in the United States to within 35.495 thousand Wii games with 99% confidence. **(c)** Increase the sample size.
61. (a) See the graph. All the data points lie between the curved lines. In fact all the points lie close to the center line. Thus the distribution appears to be normal. **(b)** $t_{\alpha/2} = 2.015$ **(c)** $E = 3.276$ miles per gallon. We can estimate μ, the true mean city gas mileage for hybrid cars, within 3.276 miles per gallon with 90% confidence. **(d)** $(27.057, 33.609)$. We are 90% confident that the true mean city gas mileage for hybrid cars μ lies between 27.057 miles per gallon and 33.609 miles per gallon.
63. (a)

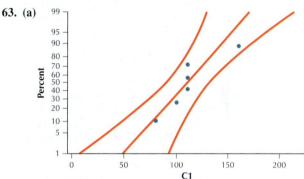

(b) Yes, the points do not appear to lie in a straight line. **(c)** Since the data do not appear to be normal, Case 1 does not apply. Since the sample size of $n = 6$ is small ($n < 30$), Case 2 does not apply. Thus a t interval cannot be used.
65. (a) An increase in the sample size will result in a decrease in $t_{\alpha/2}$. **(b)** Since the margin of error is $E = t_{\alpha/2}(s/\sqrt{n})$ and the sample size n occurs in the denominator, an increase in the sample size will result in a decrease in $t_{\alpha/2}$ and a decrease in the margin of error. **(c)** Since the width of the confidence interval is $2E$, an increase in the sample size will result in a decrease in E, which will result in a decrease in the width of the confidence interval.
67. 2392.25
69. 1.895
71. $(2208.2785, 2576.2215)$. We are 90% confident that μ lies between 2208.2785 and 2576.2215 cigarettes per capita.

Section 8.3

1. No, unless there is some reason to suspect that the value of p has changed.
3. $\hat{p} = 2/5 = 0.4$
5. $\hat{p} = 14/25 = 0.56$
7. (a) $Z_{\alpha/2} = 1.96$ **(b)** $n\hat{p} = (100)(0.2) = 20 \geq 5$ and $n(1 - \hat{p}) = (100)(1 - 0.2) = 80 \geq 5$. Thus the conditions for constructing a confidence interval for p are met. **(c)** $(0.1216, 0.2784)$. We are 95% confident that the population proportion lies between 0.1216 and 0.2784.

(d)

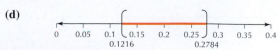

9. (a) $Z_{\alpha/2} = 1.96$ **(b)** $n\hat{p} = (100)(0.05) = 5 \geq 5$ and $n(1 - \hat{p}) = (100)(1 - 0.05) = 95 \geq 5$. Thus the conditions for constructing a confidence interval for p are met. **(c)** $(0.0073, 0.0927)$. We are 95% confident that the population proportion lies between 0.0073 and 0.0927

(d)

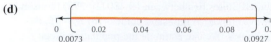

11. (a) $Z_{\alpha/2} = 1.645$ **(b)** $n\hat{p} = (25)(0.2) = 5 \geq 5$ and $n(1 - \hat{p}) = (25)(1 - 0.2) = 20 \geq 5$. Thus the conditions for constructing a confidence interval for p are met. **(c)** $(0.0684, 0.3316)$. We are 90% confident that the population proportion lies between 0.0684 and 0.3316.

(d)

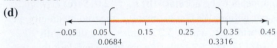

13. (a) $Z_{\alpha/2} = 2.576$ **(b)** $n\hat{p} = (25)(0.2) = 5 \geq 5$ and $n(1 - \hat{p}) = (25)(1 - 0.2) = 20 \geq 5$. Thus the conditions for constructing a confidence interval for p are met. **(c)** $(-0.0061, 0.4061)$. We are 99% confident that the population proportion lies between -0.0061 and 0.4061.

(d)

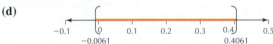

15. (a) $Z_{\alpha/2} = 1.96$ **(b)** $\hat{p} = X/n = 12/25 = 0.48$, $n\hat{p} = (25)(0.48) = 12 \geq 5$ and $n(1 - \hat{p}) = (25)(1 - 0.48) = 13 \geq 5$. Thus the conditions for constructing a confidence interval for p are met. **(c)** $(0.2842, 0.6758)$. We are 95% confident that the population proportion lies between 0.2842 and 0.6758.

(d)

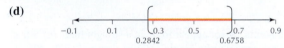

17. (a) $Z_{\alpha/2} = 2.576$ **(b)** $\hat{p} = X/n = 50/100 = 0.5$, $n\hat{p} = (100)(0.5) = 50 \geq 5$ and $n(1 - \hat{p}) = (100)(1 - 0.50) = 50 \geq 5$. Thus the conditions for constructing a confidence interval for p are met. **(c)** $(0.3712, 0.6288)$. We are 99% confident that the population proportion lies between 0.3712 and 0.6288.

(d)

19. (a) $Z_{\alpha/2} = 1.96$ **(b)** $\hat{p} = X/n = 26/64 = 0.40625$, $np = (64)(0.40625) = 26 \geq 5$ and $n(1 - p) = (64)(1 - 0.40625) = 38 \geq 5$. Thus the conditions for constructing a confidence interval for p are met. **(c)** $(0.2859, 0.5266)$. We are 95% confident that the population proportion lies between 0.2859 and 0.5266.

(d)

21. (a) $Z_{\alpha/2} = 1.645$ **(b)** $\hat{p} = X/n = 18/49 \approx 0.3673$, $n\hat{p} = (49)(0.3673) = 17.9977 \geq 5$ and $n(1 - \hat{p}) = (49)(1 - 0.3673) = 31.0023 \geq 5$. Thus the conditions for constructing a confidence interval for p are met. **(c)** $(0.2540, 0.4806)$; TI-83/84: $(0.2541, 0.4806)$. We are 90% confident that the population proportion lies between 0.2540 (0.2541) and 0.4806.

(d)

23. $E = 0.0784$
25. $E = 0.0427$
27. $E = 0.1316$
29. $E = 0.2061$
31. (a) 0.3099 **(b)** 0.098 **(c)** 0.0310 **(d)** 0.0098
33. (a) Since the margin of error is $E = Z_{\alpha/2} \cdot \sqrt{\hat{p}(1 - \hat{p})/n}$, an increase in the sample size while $\hat{p}$ remains constant results in a decrease in the margin of error. **(b)** Since the width of the confidence interval is $2E$, an increase in the sample size while $\hat{p}$ remains constant results in a decrease in the width of the confidence interval.
35. 897
37. 385
39. 5
41. 752
43. 1844
45. 17,074
47. Increases
49. (a) $Z_{\alpha/2} = 2.576$ **(b)** $\hat{p} = X/n = 183/830 \approx 0.2205$, $np = (830)(0.2205) = 183.015 \geq 5$ and $n(1 - p) = (830)(1 - 0.2205) = 646.985 \geq 5$. Thus the conditions for constructing a confidence interval for p are met. **(c)** $(0.1834, 0.2576)$. We are 99% confident that the population proportion of millennials who are married lies between 0.1834 and 0.2576.

(d)

51. (a) 1.645 **(b)** We have $n\hat{p} = 25(0.83) = 20.75 \geq 5$, but $n(1 - \hat{p}) = 25(1 - 0.83) = 4.25 < 5$, so we cannot use the Z interval for p. **(c)** We have $n\hat{p} = 25(0.83) = 20.75 \geq 5$, but $n(1 - \hat{p}) = 25(1 - 0.83) = 4.25 < 5$, so we cannot use the Z interval for p.
53. (a) $E = 0.0371$ **(b)** We can estimate the proportion of American millennials who are married to within 0.0371 with 99% confidence.
55. (a)–(b) It is not appropriate to calculate the margin of error.
57. (a) $E = 0.0295$. We can estimate the population proportion of all Hawaiians who are thriving to within 0.0295 with 95% confidence. **(b)** $(0.6255, 0.6845)$. We are 95% that the true population proportion of Hawaiians who are thriving lies between 0.6255 and 0.6845.
59. $(0.5565, 0.7435)$
61. (a) Decrease **(b)** Unchanged **(c)** Decrease
63. We have $n\hat{p} = 40(0.975) = 39 \geq 5$ but $n(1 - \hat{p}) = 40(1 - 0.975) = 1 < 5$. Thus we cannot use the Z interval for p.
65. (a) Decrease in $Z_{\alpha/2}$ from 1.96 to 1.645. **(b)** Decrease in the margin of error from 0.0748 to 0.0628. **(c)** Decrease in the width of the confidence interval from 0.1496 to 0.1256.

Section 8.4

1. The population must be normal.
3. To use this method, the distribution has to be symmetric and the χ^2 curve is not symmetric.
5. False. The χ^2 curve is *not* symmetric. It is right-skewed.
7. True
9. $\chi^2_{1-\alpha/2} = \chi^2_{0.95} = 13.848$ and $\chi^2_{\alpha/2} = \chi^2_{0.05} = 36.415$.
11. $\chi^2_{1-\alpha/2} = \chi^2_{0.995} = 9.886$ and $\chi^2_{\alpha/2} = \chi^2_{0.005} = 45.559$.
13. $\chi^2_{1-\alpha/2} = \chi^2_{0.975} = 5.629$ and $\chi^2_{\alpha/2} = \chi^2_{0.025} = 26.119$.
15. For a given sample size, $\chi^2_{1-\alpha/2}$ decreases and $\chi^2_{\alpha/2}$ increases as the confidence level increases.
17. Lower bound $= 6.59$, upper bound $= 17.33$
19. Lower bound $= 5.27$, upper bound $= 24.28$

21. Lower bound $= 2.47$, upper bound $= 4.40$

23. As the confidence level increases but the sample size stays the same, the lower bound for the confidence interval for σ^2 decreases and the upper bound for the confidence interval for σ^2 increases.

25. Lower bound $= 4.73$, upper bound $= 33.33$

27. Lower bound $= 5.78$, upper bound $= 21.33$

29. Lower bound $= 2.32$, upper bound $= 4.99$

31. As the sample size increases but the confidence level stays the same, the lower bound of a confidence interval for σ^2 increases and the upper bound of a confidence interval for σ^2 decreases.

33. (c) Lower bound $= \sqrt{\dfrac{(n-1)s^2}{\chi_{\alpha/2}^2}} = \sqrt{\dfrac{(10-1)8.942222222}{19.023}} = $

$2.056858804 \approx 2.057$

Upper bound $= \sqrt{\dfrac{(n-1)s^2}{\chi_{1-\alpha/2}^2}} = \sqrt{\dfrac{(10-1)\,8.942222222}{2.700}} = $

$\sqrt{29.80740741} = 5.45961605 \approx 5.460$

We are 95% confident that the population standard deviation of the amount of biomass consumed by power plants lies between 2.057 and 5.460 trillion BTU.

35. (a) BTU squared (b) BTU (c) BTU

37. (0.218, 0.855). We are 95% confident that the population standard deviation σ lies between 0.218 million dollars and 0.855 million dollars.

Chapter 8 Review

1. (a) $\sigma/\sqrt{n} = 2$ (b) $Z_{\alpha/2} = 1.96$ (c) $E = 3.92$. We can estimate μ to within 3.92 with 95% confidence. (d) (46.08, 53.92). We are 95% confident that the true mean μ lies between 46.08 and 53.92.

3. (a) 7 points (b) 0.2981 point (c) 1.645 (d) 0.4904 point. We can estimate μ to within 0.4904 point with 90% confidence. (e) (6.5096, 7.4904). We are 90% confident that the true mean increase in IQ points for all children after listening to a Mozart piano sonata for about 10 minutes μ lies between 6.5106 points and 7.4904 points.

5. 35

7. 68

9. (20.289, 23.711)

11. (a) (2162.65, 2621.85) (b) (2052.56, 2731.94) (c) The interval in (a) is more precise than the interval in (b) but the interval in (b) has higher confidence of containing μ.

13. (a) 1.96 (b) We have $n\hat{p} = 500(0.99) = 495 \geq 5$ and $n(1-\hat{p}) = 500(1-0.99) = 5 \geq 5$. Thus we can use the Z interval for p. (c) 0.0087. We can estimate p to within $E = 0.0087$ with 95% confidence. (d) (0.9813, 0.9987). Thus we are 95% confident that the true proportion lies between 0.9813 and 0.9987.

15. 664

17. 5

19. 752

21. Lower bound $= 70.278$, upper bound $= 155.798$

23. Lower bound $= 8.383$, upper bound $= 12.482$

25. Lower bound $= 30.537$, upper bound $= 104.367$. We are 95% confident that σ, the population standard deviation of total union membership per state, lies between 30.537 and 104.367 thousand.

Chapter 8 Quiz

1. False

2. True

3. 4

4. less

5. α is a probability.

6. Either the population is normal or the sample size is large ($n \geq 30$).

7. (a) $E = \$705$. We can estimate μ, the mean cost of a college education, to within $705 with 90% confidence. (b) (29,795, 31,205). We are 90% confident that the true mean cost of a college education lies between $29,795 and $31,205.

8. (a) $E = 49.35$ pounds. We can estimate μ, the mean femur load number in a frontal crash for the passenger in a 2005 Ford Equinox SUV, to within 49.35 pounds with 90% confidence. (b) (953.65, 1052.35). We are 90% confident that the true mean femur load number in a frontal crash for the passenger in a 2005 Ford Equinox SUV lies between 953.65 pounds and 1052.35 pounds.

9. (a) $E = 0.0142$. We can estimate p, the true proportion of all Americans who attended a religious service in response to the attacks on the World Trade Center and the Pentagon, to within 0.0142 with 95% confidence. (b) (0.2513, 0.2797). We are 95% confident that the true proportion of all Americans who attended a religious service in response to the attacks on the World Trade Center and the Pentagon lies between 0.2513 and 0.2797.

10. (a) $E = 0.0386$. We can estimate p, the true proportion of all Québecois who favor independence for the Province of Quebec, to within 0.0386 with 99% confidence. (b) (0.3014, 0.3786)

11. (a) lower bound $= 1.340$, upper bound $= 6.428$. We are 95% confident that the population standard deviation σ lies between 1.340 and 6.248 hours. (b) lower bound $= 2.680$, upper bound $= 12.856$. We are 95% confident that the population standard deviation σ lies between 2.680 and 12.856 hours.

12. 752

Chapter 9

Section 9.1

1. The null hypothesis is assumed to be true unless the sample evidence indicates that the alternative hypothesis is true instead. It represents what has been tentatively assumed about the value of the parameter. It is the status quo hypothesis. The alternative hypothesis represents an alternative claim about the value of the parameter. The researcher concludes that the alternative hypothesis is true only if the evidence provided by the sample data indicates that it is true.

3.

Form	Null hypothesis		Alternative hypothesis
1	$H_0 : \mu = \mu_0$	vs.	$H_a : \mu > \mu_0$
2	$H_0 : \mu = \mu_0$	vs.	$H_a : \mu < \mu_0$
3	$H_0 : \mu = \mu_0$	vs.	$H_a : \mu \neq \mu_0$

5. A Type I error occurs when one rejects H_0 when H_0 is true. A Type II error occurs when one does not reject H_0 when H_0 is false.

7. No. It depends on how many standard deviations the sample mean of 90 is below the population mean of 100 and the level of significance of the test.

9. $H_0 : \mu = 10$ vs. $H_a : \mu > 10$

11. $H_0 : \mu = 0$ vs. $H_a : \mu \neq 0$

13. $H_0 : \mu = 36$ vs. $H_a : \mu \neq 36$

15. (a) $H_0 : \mu = 43.9$ vs. $H_a : \mu > 43.9$ (b) A Type II error was made.

17. (a) $H_0 : \mu = 700$ vs. $H_a : \mu < 700$ (b) No error was made.

19. (a) $H_0 : \mu = 1{,}602{,}000{,}000$ vs. $H_a : \mu > 1{,}602{,}000{,}000$ (b) Conclude that the mean is greater than 1.602 billion when the population mean is actually greater than 1.602 billion, and conclude that the mean is equal to 1.602 billion when the population is actually equal to 1.602 billion.

(c) Concluding that the mean is greater than 1.602 billion when the population mean is actually equal to 1.602 billion
(d) Concluding that the mean is equal to 1.602 billion when the population mean is actually greater than 1.602 billion
21. (a) $H_0 : \mu = 3.24$ vs. $H_a : \mu > 3.24$ (b) Conclude that the mean is greater than \$3.24 when it actually is greater than \$3.24, and conclude that the mean is equal to \$3.24 when it actually is equal to \$3.24. (c) Concluding that the mean is greater than \$3.24 when it actually is equal to \$3.24 (d) Concluding that the mean is equal to \$3.24 when it actually is greater than \$3.24
23. (a) $H_0 : \mu = 673$ vs. $H_a : \mu < 673$ (b) Conclude that the mean is less than 673 when it actually is less than 673, and conclude that the mean is equal to 673 when it actually is equal to 673. (c) Concluding that the mean is less than 673 when it actually is equal to 673 (d) Concluding that the mean is equal to 673 when it actually is less than 673.

Section 9.2

1. When the observed value of $\bar{x}$ is unusual or extreme in the sampling distribution of $\bar{x}$ that assumes H_0 is true, we should reject H_0. Otherwise, we should not reject H_0.
3. A statistic generated from a data set for the purpose of testing a statistical hypothesis
5. The value of Z that separates the critical region from the noncritical region
7. The critical region for a right-tailed test lies in the right (upper) tail.
9. $Z_{data} = 1$
11. $Z_{data} = 2.5$
13. $Z_{data} = -1.5$
15. $Z_{data} = -5$
17. (a) $Z_{crit} = 1.28$

(b)

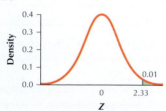

(c) Reject H_0 if $Z_{data} \geq 1.28$.
19. (a) $Z_{crit} = 2.33$

(b)

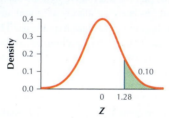

(c) Reject H_0 if $Z_{data} \geq 2.33$.
21. (a) $Z_{crit} = -1.645$

(b)

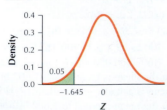

(c) Reject H_0 if $Z_{data} \leq -1.645$.

23. (a) $Z_{crit} = 1.96$

(b)

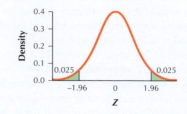

(c) Reject H_0 if $Z_{data} \leq -1.96$ or $Z_{data} \geq 1.96$
25. (a) It increases. (b) It becomes smaller.
27. (a) $H_0 : \mu = 10$ vs. $H_a : \mu > 10$ (b) $Z_{crit} = 1.28$. Reject H_0 if $Z_{data} \geq 1.28$. (c) $Z_{data} = 1$ (d) Since $Z_{data} = 1$ is not ≥ 1.28, the conclusion is do not reject H_0. There is insufficient evidence at the 0.10 level of significance that the population mean is greater than 10.
29. (a) $H_0 : \mu = 10$ vs. $H_a : \mu > 10$ (b) $Z_{crit} = 2.33$. Reject H_0 if $Z_{data} \geq 2.33$. (c) $Z_{data} = 2.5$ (d) Since $Z_{data} = 2.5$ is ≥ 2.33, the conclusion is reject H_0. There is evidence at the 0.01 level of significance that the population mean is greater than 10.
31. (a) $H_0 : \mu = 7$ vs. $H_a : \mu < 7$ (b) $Z_{crit} = -1.645$. Reject H_0 if $Z_{data} \leq -1.645$. (c) $Z_{data} = -1.5$ (d) Since $Z_{data} = -1.5$ is not ≤ -1.645, the conclusion is do not reject H_0. There is insufficient evidence at the 0.05 level of significance that the population mean is less than 7.
33. (a) $H_0 : \mu = 100$ vs. $H_a : \mu \neq 100$ (b) $Z_{crit} = 1.96$. Reject H_0 if $Z_{data} \leq -1.96$ or if $Z_{data} \geq 1.96$. (c) $Z_{data} = -5$ (d) Since $Z_{data} = -5$ is ≤ -1.96, the conclusion is reject H_0. There is evidence at the 0.05 level of significance that the population mean differs from 100.
35. (a) $H_0 : \mu = 80$ vs. $H_a : \mu > 80$ (b) $Z_{crit} = 1.645$. Reject H_0 if $Z_{data} \geq 1.645$. (c) $Z_{data} = 1$.

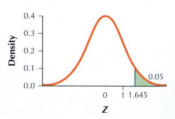

(d) Since $Z_{data} = 1$ is not ≥ 1.645, the conclusion is do not reject H_0. There is insufficient evidence at the 0.05 level of significance that the population mean number of connections to community pages, groups, and events is greater than 80.
37. (a) $H_0 : \mu = 10$ vs. $H_a : \mu \neq 10$ (b) $Z_{crit} = 2.58$. Reject H_0 if $Z_{data} \leq -2.58$ or if $Z_{data} \geq 2.58$. (c) $Z_{data} = 1$.

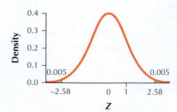

(d) Since $Z_{data} = 1$ is not ≤ -2.58 and not ≥ 2.58, the conclusion is do not reject H_0. There is insufficient evidence at the 0.01 level of significance that the population mean number of text messages per day differs from 10.

39. (a) $H_0 : \mu = 3.70$ vs. $H_a : \mu > 3.70$ **(b)** $Z_{\text{crit}} = 1.645$. Reject H_0 if $Z_{\text{data}} \geq 1.645$. **(c)** $Z_{\text{data}} = 2$.

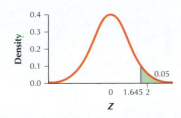

(d) Since $Z_{\text{data}} = 2$ is ≥ 1.645, the conclusion is reject H_0. There is evidence at the 0.05 level of significance that the population mean price of regular gasoline is greater than $3.70 per gallon. Therefore we can conclude at the 0.05 level of significance that the population mean price for a gallon of regular gasoline has risen since June 2011.

41. (a) $H_0 : \mu = 175$ vs. $H_a : \mu \neq 175$ **(b)** $Z_{\text{crit}} = 1.645$. Reject H_0 if $Z_{\text{data}} \leq -1.645$ or if $Z_{\text{data}} \geq 1.645$. **(c)** $Z_{\text{data}} = 8$.

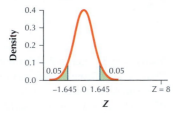

(d) Since $Z_{\text{data}} = 8$ is ≥ 1.645, the conclusion is reject H_0. There is evidence at the 0.10 level of significance that the population mean height of Americans has changed from 175 centimeters.

43. The histogram indicates that the data are extremely right-skewed and therefore not normally distributed. Thus Case 1 does not apply. Since the sample size of $n = 16$ is small ($n < 30$), Case 2 does not apply. Thus it is not appropriate to apply the Z test.

45. (a) Decrease **(b)** Unchanged **(c)** Depends on new value of $\bar{x}$.

47. (a) $H_0 : \mu \leq 60$ vs. $H_a : \mu > 60$ **(b)** $t_{\text{crit}} = 1.771$. Reject H_0 if $t_{\text{data}} > 1.771$ **(c)** $t_{\text{data}} = 2.50$ **(d)** Since $t_{\text{data}} > 1.771$, we reject H_0. There is evidence that the population mean response time is greater than 60 milliseconds.

Section 9.3

1. False

3. It gives us extra information about whether H_0 was barely rejected or not rejected or whether it was a no-brainer decision to reject or not reject H_0.

5. False

7. 0.1587

9. 0.0013

11. 0.0062

13. 0.0038 (TI-83/84: 0.0037)

15. It decreases.

17. (a) $H_0 : \mu = 98.6$ vs. $H_a : \mu > 98.6$. Reject H_0 if the p-value $\leq \alpha = 0.05$. **(b)** $Z_{\text{data}} = 0.5$. **(c)** 0.3085 **(d)** Since the p-value = 0.3085 is not ≤ 0.05, we therefore do not reject H_0. There is insufficient evidence at level of significance $\alpha = 0.05$ that the population mean is greater than 98.6.

19. (a) $H_0 : \mu = -0.1$ vs. $H_a : \mu > -0.1$. Reject H_0 if the p-value $\leq \alpha = 0.05$. **(b)** $Z_{\text{data}} = 2$. **(c)** 0.0228 **(d)** Since the p-value = 0.0228 is ≤ 0.05, we therefore reject H_0. There is evidence at level of significance $\alpha = 0.05$ that the population mean is greater than -0.1.

21. (a) $H_0 : \mu = -50$ vs. $H_a : \mu \neq -50$. Reject H_0 if p-value $\leq \alpha = 0.05$ **(b)** $Z_{\text{data}} = 2.6667$. **(c)** 0.0077 **(d)** Since the p-value = 0.0077 is ≤ 0.05, we therefore reject H_0. There is evidence at level of significance $\alpha = 0.05$ that the population mean differs from 50.

23. No evidence

25. Solid evidence

27. Very strong evidence

29.

	Value of μ_0	Form of hypothesis test, with $\alpha = 0.05$	Where μ_0 lies in relation to 95% confidence interval (2.7, 6.9)	Conclusion of hypothesis test
(a)	-3	$H_0 : \mu = -3$ vs. $H_a : \mu \neq -3$	Outside	Reject H_0
(b)	-2	$H_0 : \mu = -2$ vs. $H_a : \mu \neq -2$	Inside	Do not reject H_0
(c)	0	$H_0 : \mu = 0$ vs. $H_a : \mu \neq 0$	Inside	Do not reject H_0
(d)	5	$H_0 : \mu = 5$ vs. $H_a : \mu \neq 5$	Inside	Do not reject H_0
(e)	7	$H_0 : \mu = 7$ vs. $H_a : \mu \neq 7$	Outside	Reject H_0

31.

	Value of μ_0	Form of hypothesis test, with $\alpha = 0.10$	Where μ_0 lies in relation to 90% confidence interval $(-10, -5)$	Conclusion of hypothesis test
(a)	-3	$H_0 : \mu = -3$ vs. $H_a : \mu \neq -3$	Outside	Reject H_0
(b)	-8	$H_0 : \mu = -8$ vs. $H_a : \mu \neq -8$	Inside	Do not reject H_0
(c)	-11	$H_0 : \mu = -11$ vs. $H_a : \mu \neq -11$	Outside	Reject H_0
(d)	0	$H_0 : \mu = 0$ vs. $H_a : \mu \neq 0$	Outside	Reject H_0
(e)	7	$H_0 : \mu = 7$ vs. $H_a : \mu \neq 7$	Outside	Reject H_0

33.

	Value of μ_0	Form of hypothesis test, with $\alpha = 0.05$	Where μ_0 lies in relation to 95% confidence interval (0, 1)	Conclusion of hypothesis test
(a)	1.5	$H_0 : \mu = 1.5$ vs. $H_a : \mu \neq 1.5$	Outside	Reject H_0
(b)	-1	$H_0 : \mu = -1$ vs. $H_a : \mu \neq -1$	Outside	Reject H_0
(c)	0.5	$H_0 : \mu = 0.5$ vs. $H_a : \mu \neq 0.5$	Inside	Do not reject H_0
(d)	0.9	$H_0 : \mu = 0.9$ vs. $H_a : \mu \neq 0.9$	Inside	Do not reject H_0
(e)	1.2	$H_0 : \mu = 1.2$ vs. $H_a : \mu \neq 1.2$	Outside	Reject H_0

35. (a) $H_0 : \mu = 43.9$ vs. $H_a : \mu > 43.9$. Reject H_0 if the p-value ≤ 0.10. **(b)** 19.61 **(c)** ≈ 0 **(d)** Since the p-value $\leq \alpha$, reject. H_0. There is evidence that the population mean referral rate is greater than 43.9 per 1000 children.

37. (a) $H_0 : \mu = 700$ vs. $H_a : \mu < 700$. Reject H_0 if the p-value ≤ 0.10. **(b)** -20 **(c)** ≈ 0 **(d)** Since the p-value $\leq \alpha$, reject H_0. There is evidence that the population mean number of meals prepared and eaten at home is less than 700.

39. (a) $H_0 : \mu = 2$ vs. $H_a : \mu \neq 2$. Reject H_0 if p-value $\leq \alpha = 0.05$. **(b)** $Z_{\text{data}} = 1.2$ **(c)** 0.2302 (TI-83/84: 0.2301) **(d)** Since the p-value $= 0.2302$ is not ≤ 0.05, we therefore do not reject H_0. There is insufficient evidence at level of significance $\alpha = 0.05$ that the population mean daily number of shares traded differs from 2 billion shares.

41. (a) $H_0 : \mu = 3$ vs. $H_a : \mu < 3$. Reject H_0 if the p-value ≤ 0.01. **(b)** -13.5 **(c)** p-value ≈ 0 **(d)** Since the p-value $\leq \alpha$, reject H_0. There is evidence that the population mean time hybrid cars take to recoup their initial cost is less than 3 years.

43. (a) (23,160, 24,840) **(b)** (i) Since $\mu_0 = 24,000$ lies in the confidence interval, we do not reject H_0. (ii) Since $\mu_0 = 23,000$ does not lie in the confidence interval, we reject H_0. (iii) Since $\mu_0 = 23,200$ lies in the confidence interval, we do not reject H_0. (iv) Since $\mu_0 = 25,000$ does not lie in the confidence interval, we reject H_0.

45. (a) Remains the same **(b)** Remains the same **(c)** Remains the same **(d)** Increases **(e)** Remains the same

47. $H_0 : \mu = 3.14$ vs. $H_a : \mu < 3.14$. $Z_{\text{data}} = -1.35$. 0.0885. Since the p-value $= 0.0885$ is not ≤ 0.05, we therefore do not reject H_0. There is insufficient evidence at level of significance $\alpha = 0.05$ that the population mean family size is less than 3.14 persons.

49. (a) Decrease from -1.5 to -2.25 **(b)** Decrease from 0.0668 to 0.0122 **(c)** Since the p-value is less than α, we reject H_0. There is evidence that the true mean family size in America is less than 3.15 persons.

51. (a) There is insufficient evidence that the true mean heart rates for all women is less than 78 beats per minute and there is insufficient evidence that the true mean heart rate for all woman is different than 78 beats per minute. **(b)** The p-value for **(c)** is twice the p-value in **(b)**. If α is between these two p-values, then the conclusion for the one-tailed test will be "Reject H_0" and the conclusion for the two-tailed test will be "Do not reject H_0." **(c)** There is no evidence against the null hypothesis in **(b)** and **(c)**.

53. (a) Decrease **(b)** Decrease **(c)** Decrease **(d)** Depends on new value of σ.

55. There is solid evidence against the null hypothesis.

Section 9.4

1. The population standard deviation σ is known.

3. (a) $H_0 : \mu = 22$ vs. $H_a : \mu < 22$ **(b)** $t_{\text{crit}} = -1.697$. Reject H_0 if $t_{\text{data}} \leq -1.697$.

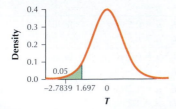

(c) $t_{\text{data}} = -2.7839$ **(d)** Since $t_{\text{data}} = -2.7839$ is ≤ -1.697, the conclusion is reject H_0. There is evidence at the 0.05 level of significance that the population mean is less than 22.

5. (a) $H_0 : \mu = 11$ vs. $H_a : \mu \leq 11$ **(b)** $t_{\text{crit}} = 2.602$. Reject H_0 if $t_{\text{data}} \geq 2.602$.

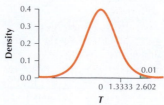

(c) $t_{\text{data}} = 1.3333$ **(d)** Since $t_{\text{data}} = 1.3333$ is not ≥ 2.602, the conclusion is do not reject H_0. There is insufficient evidence at the 0.01 level of significance that the population mean is greater than 11.

7. (a) $H_0 : \mu = 100$ vs. $H_a : \mu > 100$ **(b)** $t_{\text{crit}} = 2.492$. Reject H_0 if $t_{\text{data}} \geq 2.492$.

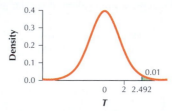

(c) $t_{\text{data}} = 2$ **(d)** Since $t_{\text{data}} = 2$ is not ≥ 2.492, the conclusion is do not reject H_0. There is insufficient evidence at the 0.01 level of significance that the population mean is greater than 100.

9. (a) $H_0 : \mu = 102$ vs. $H_a : \mu \neq 102$ **(b)** $t_{\text{crit}} = 1.990$. Reject H_0 if $t_{\text{data}} \leq -1.990$ or if $t_{\text{data}} \geq 1.990$.

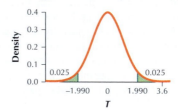

(c) $t_{\text{data}} = 3.6$ **(d)** Since $t_{\text{data}} = 3.6$ is ≥ 1.990, the conclusion is reject H_0. There is evidence at the 0.05 level of significance that the population mean differs from 102.

11. (a) $H_0 : \mu = 1000$ vs. $H_a : \mu \neq 1000$ **(b)** $t_{\text{crit}} = 1.711$. Reject H_0 if $t_{\text{data}} \leq -1.711$ or if $t_{\text{data}} \geq 1.711$.

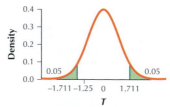

(c) $t_{\text{data}} = -1.25$ **(d)** Since $t_{\text{data}} = -1.25$ is not ≤ -1.711 and not ≥ 1.711, the conclusion is do not reject H_0. There is insufficient evidence at the 0.10 level of significance that the population mean differs from 1000.

13. (a) $H_0 : \mu = 9$ vs. $H_a : \mu \neq 9$ **(b)** $t_{\text{crit}} = 1.690$. Reject H_0 if $t_{\text{data}} \leq -1.690$ or if $t_{\text{data}} \geq 1.690$.

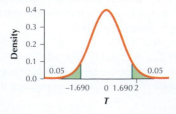

(c) $t_{data} = 2$ **(d)** Since $t_{data} = 2$ is ≥ 1.690, the conclusion is reject H_0. There is evidence at the 0.10 level of significance that the population mean differs from 9.

15. (a) $H_0 : \mu = 10$ vs. $H_a : \mu < 10$. Reject H_0 if the p-value $\leq \alpha = 0.01$. **(b)** $t_{data} = -5.4$ **(c)** 0 **(d)** Since the p-value $= 0$ is $\leq \alpha = 0.01$, the conclusion is reject H_0. There is evidence at the 0.01 level of significance that the population mean is less than 10.

17. (a) $H_0 : \mu = 100$ vs. $H_a : \mu > 100$. Reject H_0 if the p-value $\leq \alpha = 0.10$. **(b)** $t_{data} = 2$ **(c)** 0.0285 **(d)** Since the p-value $= 0.0285$ is $\leq \alpha = 0.10$, the conclusion is reject H_0. There is evidence at the 0.10 level of significance that the population mean is greater than 100.

19. (a) $H_0 : \mu = 200$ vs. $H_a : \mu > 200$. Reject H_0 if the p-value $\leq \alpha = 0.05$. **(b)** $t_{data} = 120$ **(c)** 0 **(d)** Since the p-value $= 0$ is $\leq \alpha = 0.05$, the conclusion is reject H_0. There is evidence at the 0.05 level of significance that the population mean is greater than 200.

21. (a) $H_0 : \mu = 25$ vs. $H_a : \mu \neq 25$. Reject H_0 if the p-value $\leq \alpha = 0.01$. **(b)** $t_{data} = 0$ **(c)** 1 **(d)** Since the p-value $= 1$ is not $\leq \alpha = 0.01$, the conclusion is do not reject H_0. There is insufficient evidence at the 0.01 level of significance that the population mean differs from 25.

23. (a) $H_0 : \mu = 3.14$ vs. $H_a : \mu \neq 3.14$. Reject H_0 if the p-value $\leq \alpha = 0.10$. **(b)** $t_{data} = 0.18$ **(c)** 0.8616 **(d)** Since the p-value $= 0.8616$ is not $\leq \alpha = 0.10$, the conclusion is do not reject H_0. There is insufficient evidence at the 0.10 level of significance that the population mean differs from 3.14.

25. (a) $H_0 : \mu = 0$ vs. $H_a : \mu \neq 0$. Reject H_0 if the p-value $\leq \alpha = 0.05$. **(b)** $t_{data} = 6$ **(c)** 0.0003 **(d)** Since the p-value $= 0.0003$ is $\leq \alpha = 0.05$, the conclusion is reject H_0. There is evidence at the 0.05 level of significance that the population mean differs from 0.

27. p-value < 0.005

29. p-value < 0.01

31.

	Value of μ_0	Form of hypothesis test, with $\alpha = 0.05$	Where μ_0 lies in relation to 95% confidence interval (1, 4)	Conclusion of hypothesis test
(a)	0	$H_0 : \mu = 0$ vs. $H_a : \mu \neq 0$	Outside	Reject H_0
(b)	2	$H_0 : \mu = 2$ vs. $H_a : \mu \neq 2$	Inside	Do not reject H_0
(c)	5	$H_0 : \mu = 5$ vs. $H_a : \mu \neq 5$	Outside	Reject H_0

33.

	Value of μ_0	Form of hypothesis test, with $\alpha = 0.10$	Where μ_0 lies in relation to 90% confidence interval $(-20, -10)$	Conclusion of hypothesis test
(a)	-21	$H_0 : \mu = -21$ vs. $H_a : \mu \neq -21$	Outside	Reject H_0
(b)	-5	$H_0 : \mu = -5$ vs. $H_a : \mu \neq -5$	Outside	Reject H_0
(c)	-12	$H_0 : \mu = -12$ vs. $H_a : \mu \neq -12$	Inside	Do not reject H_0

35.

	Value of μ_0	Form of hypothesis test, with $\alpha = 0.05$	Where μ_0 lies in relation to 95% confidence interval $(-1, 1)$	Conclusion of hypothesis test
(a)	1.5	$H_0 : \mu = 1.5$ vs. $H_a : \mu \neq 1.5$	Outside	Reject H_0
(b)	-1.5	$H_0 : \mu = -1.5$ vs. $H_a : \mu \neq -1.5$	Outside	Reject H_0
(c)	0	$H_0 : \mu = 0$ vs. $H_a : \mu \neq 0$	Inside	Do not reject H_0

37. Critical-value method: $H_0 : \mu = 15,200$ vs. $H_a : \mu > 15,200$. $t_{crit} = 1.660$. Reject H_0 if $t_{data} \geq 1.660$. $t_{data} = 3.2$. Since $t_{data} = 3.2$ is ≥ 1.660, the conclusion is reject H_0. There is evidence at the 0.05 level of significance that the population mean cost of a stay in the hospital for women aged 18–44 is greater than \$15,200. Therefore we can conclude at level of significance 0.05 that the population mean cost of a stay in the hospital for American women aged 18–24 has increased since 2010. **p-value method:** $H_0 : \mu = 15,200$ vs. $H_a : \mu > 15,200$. Reject H_0 if the p-value $\leq \alpha = 0.05$. $t_{data} = 3.2$. p-value $= 0.0007$. Since the p-value $= 0.0007$ is $\leq \alpha = 0.05$, the conclusion is reject H_0. There is evidence at the 0.05 level of significance that the population mean cost of a stay in the hospital for women aged 18–44 is greater than \$15,200. Therefore we can conclude at level of significance 0.05 that the population mean cost of a stay in the hospital for American women aged 18–24 has increased since 2010.

39. Critical-value method: $H_0 : \mu = 130$ vs. $H_a : \mu < 130$. $t_{crit} = -1.662$. Reject H_0 if $t_{data} \leq -1.662$. $t_{data} = -4$. Since $t_{data} = -4$ is ≤ -1.662, the conclusion is reject H_0. There is evidence at the 0.05 level of significance that the population mean number of Facebook friends is less than 130. **p-value method:** $H_0 : \mu = 130$ vs. $H_a : \mu < 130$. Reject H_0 if p-value $\leq \alpha = 0.05$. $t_{data} = -4$. p-value $= 0$. Since the p-value $= 0$ is $\leq \alpha = 0.05$, the conclusion is reject H_0. There is evidence at the 0.05 level of significance that the population mean number of Facebook friends is less than 130.

41. No. The distribution of the variable is not normal and the sample size is less than 30.

43. (a) Case 1 applies, so we can apply the t test. **(b)** $H_0 : \mu = 30$ vs. $H_a : \mu > 30$. Reject H_0 if p-value ≤ 0.01. $t_{data} = 3.54$. p-value $= 0.0031570524$. Since p-value ≤ 0.01, we reject H_0. There is evidence that the population mean gas mileage is greater than 30 mpg.

45. (a) Decrease **(b)** Unchanged **(c)** Increase **(d)** We don't know what the conclusion will be. **(e)** Will result in a conclusion of "Do not reject H_0." **(f)** We don't know what the strength of the evidence against the null hypothesis will be.

47. (a) There is evidence that the population mean tuition and fees at community colleges this year is greater than \$2272. **(b)** We would not reject H_0. This is a Type II error. Answers will vary. **(c)** Answers will vary.

49.

```
Descriptive Statistics: TOT_POP

Variable    N   N*    Mean  SE Mean   StDev  Minimum    Q1
Median     Q3  Maximum
TOT_POP    790   0   18305     9284  260938     1000  1901
4013     9059  7322564
```

Section 9.5

1. $\hat{p}$ is the sample proportion and p is population proportion.

3. Answers will vary.

5. Between 0 and 1 inclusive: $0 \leq p_0 \leq 1$

7. 2.8868

9. 7.2169

11. -4.47

13. 0

15. **(a)** We have $np_0 = 225(0.5) = 112.5 \geq 5$ and $n(1 - p_0) = 225(1 - 0.5) = 112.5 \geq 5$, so we can use the Z test for proportions. **(b)** $H_0 : p = 0.5$ vs. $H_a : p < 0.5$ **(c)** $Z_{crit} = -1.645$. Reject H_0 if $Z_{data} \leq -1.645$. **(d)** -1.67 **(e)** Since $Z_{data} \leq -1.645$, we reject H_0. There is evidence that the population proportion is less than 0.5.

17. **(a)** We have $np_0 = 400(0.6) = 240 \geq 5$ and $n(1 - p_0) = 400(1 - 0.6) = 160 \geq 5$, so we can use the Z test for proportions. **(b)** $H_0 : p = 0.6$ vs. $H_a : p > 0.6$ **(c)** $Z_{crit} = 1.645$. Reject H_0 if $Z_{data} \geq 1.645$. **(d)** 2.04 **(e)** Since $Z_{data} \geq 1.645$, we reject H_0. There is evidence that the population proportion is greater than 0.6.

19. **(a)** We have $np_0 = 100(0.4) = 40 \geq 5$ and $n(1 - p_0) = 100(1 - 0.4) = 60 \geq 5$, so we can use the Z test for proportions. **(b)** $H_0 : p = 0.4$ vs. $H_a : p > 0.4$. Reject H_0 if the p-value ≤ 0.05. **(c)** 0.82 **(d)** p-value $= 0.2061$ **(e)** Since the p-value is not ≤ 0.05, we do not reject H_0. There is insufficient evidence that the population proportion is greater than 0.4.

21. **(a)** We have $np_0 = 900(0.5) = 450 \geq 5$ and $n(1 - p_0) = 900(1 - 0.5) = 450 \geq 5$, so we may use the Z test for proportions. **(b)** $H_0 : p = 0.5$ vs. $H_a : p \neq 0.5$. Reject H_0 if the p-value ≤ 0.05. **(c)** 1.67 **(d)** p-value $= 0.095$ **(e)** Since the p-value is not ≤ 0.05, we do not reject H_0. There is insufficient evidence that the population proportion is not equal to 0.5.

23.

	Value of p_0	Form of hypothesis test, with $\alpha = 0.05$	Where p_0 lies in relation to 95% confidence interval $(0.1, 0.9)$	Conclusion of hypothesis test
(a)	0	$H_0 : p = 0$ vs. $H_a : p \neq 0$	Outside	Reject H_0
(b)	1	$H_0 : p = 1$ vs. $H_a : p \neq 1$	Outside	Reject H_0
(c)	0.5	$H_0 : p = 0.5$ vs. $H_a : p \neq 0.5$	Inside	Do not reject H_0

25.

	Value of p_0	Form of hypothesis test, with $\alpha = 0.10$	Where p_0 lies in relation to 90% confidence interval $(0.1, 0.2)$	Conclusion of hypothesis test
(a)	0.09	$H_0 : p = 0.09$ vs. $H_a : p \neq 0.09$	Outside	Reject H_0
(b)	0.9	$H_0 : p = 0.9$ vs. $H_a : p \neq 0.9$	Outside	Reject H_0
(c)	0.19	$H_0 : p = 0.19$ vs. $H_a : p \neq 0.19$	Inside	Do not reject H_0

27. $np_0 = 500(0.172) = 86 \geq 5$ and $n(1 - p_0) = 500(1 - 0.172) = 414 \geq 5$, so we may use the Z test for proportions. $H_0 : p = 0.172$ vs. $H_a : p \neq 0.172$. Reject H_0 if p-value ≤ 0.10. $Z_{data} = -0.12$. p-value $= 0.9044$. Since p-value is not ≤ 0.10, we do not reject H_0. There is insufficient evidence that the population proportion of Americans who identified themselves as Baptists is not equal to 0.172.

29. $np_0 = 400(0.071) = 28.4 \geq 5$ and $n(1 - p_0) = 400(1 - 0.071) = 371.6 \geq 5$, so we may use the Z test for proportions. $H_0 : p = 0.071$ vs. $H_a : p \neq 0.071$. Reject H_0 if p-value ≤ 0.01. $Z_{data} = 1.28$. p-value $= 0.2006$. Since p-value is not ≤ 0.01, we do not reject H_0. There is insufficient evidence that the population proportion of Americans aged 20–24 is not equal to 0.071.

31. **(a)** We have $np_0 = 400(0.123) = 49.2 \geq 5$ and $n(1 - p_0) = 400(1 - 0.123) = 350.8 \geq 5$, so we can use the Z test for proportions. **(b)** $H_0 : p = 0.123$ vs. $H_a : p > 0.123$. Reject H_0 if p-value ≤ 0.05. $Z_{data} = -0.43$. p-value $= 0.3336$. Since p-value is not ≤ 0.05, we do not reject H_0. There is insufficient evidence that the population proportion of California residents of Asian ethnicity is greater than 0.123.

33. **(a)** Yes. We have $np_0 = 100(0.153) = 15.3 \geq 5$ and $n(1 - p_0) = 100(1 - 0.153) = 84.7 \geq 5$. **(b)** $H_0 : p = 0.153$ vs. $H_a : p \neq 0.153$. Reject H_0 if the P-value ≤ 0.01. $Z_{data} = 2.14$. p-value $= 0.0324$. Since the p-value is not ≤ 0.01, we do not reject H_0. There is insufficient evidence that the population proportion of Hispanic families that had a household income of at least \$75,000 is not equal to 0.153.

35. **(a)** There is no evidence against the null hypothesis. **(b)** Do not reject H_0 because the two methods for performing the hypothesis test are equivalent. **(c)** Since the conclusion is do not reject H_0, the 95% confidence interval will contain 0.456.

37. **(a)** Type II; answers will vary. **(b)** Since we did not reject H_0, our hypothesis test does not support this headline.

39. **(a)** Unchanged **(b)–(c)** Decrease **(d)** Unchanged **(e)** Depends on new value of $\hat{p}$.

41. **(a)** Decrease **(b)** Increase by a factor of $\sqrt{2}$ **(c)** Decrease **(d)** Unchanged **(e)** The conclusion will now be to reject H_0.

Section 9.6

1. Answers will vary.

3. No, σ will never be less than 0.

5. Answers will vary.

7. $H_0 : \sigma = 10$ vs. $H_a : \sigma > 10$

9. $H_0 : \sigma = 3$ vs. $H_a : \sigma \neq 3$

11. $\chi^2_{data} = 60$

13. $\chi^2_{data} = 10.417$

15. $\chi^2_{data} = 6.125$

17. $\chi^2 = \chi^2_{0.05} = 31.410$

19. $\chi^2_{\alpha/2} = \chi^2_{0.025} = 27.488$ and $\chi^2_{1 - \alpha/2} = \chi^2_{0.975} = 6.262$

21. $\chi^2_{1 - \alpha} = \chi^2_{0.90} = 2.833$

23. **(a)** Reject H_0 if $\chi^2_{data} \geq 31.410$ **(b)** Since $\chi^2_{data} \geq 31.410$, we reject H_0. There is evidence that the population standard deviation is greater than 1.

25. **(a)** Reject H_0 if $\chi^2_{data} \leq 6.262$ or $\chi^2_{data} \geq 27.488$. **(b)** Since χ^2_{data} is not ≤ 6.262 and χ^2_{data} is not ≥ 27.488, we do not reject H_0. There is insufficient evidence that the population standard deviation is different from 3.

27. **(a)** Reject H_0 if $\chi^2_{data} \leq 2.833$. **(b)** Since χ^2_{data} is not ≤ 2.833, we do not reject H_0. There is insufficient evidence that the population standard deviation is less than 20.

29. **(a)**

(b) p-value $= 7.121750863 \times 10^{-6}$ **(c)** Since the p-value ≤ 0.05, we reject H_0. There is evidence that the population standard deviation is greater than 1.

31. (a)

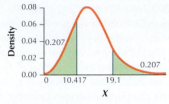

(b) p-value $= 0.4145552434$ **(c)** Since the p-value is not ≤ 0.05, we do not reject H_0. There is insufficient evidence that the population standard deviation is different from 3.

33. (a)

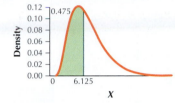

(b) p-value $= 0.4747679539$ **(c)** Since the p-value is not ≤ 0.05, we do not reject H_0. There is insufficient evidence that the population standard deviation is less than 20.

35.

	Value of σ_0	Form of hypothesis test, with $\alpha = 0.05$	Where σ_0 lies in relation to 95% confidence interval (1, 4)	Conclusion of hypothesis test
(a)	0	$H_0 : \sigma = 0$ vs. $H_a : \sigma \neq 0$	Outside	Reject H_0
(b)	2	$H_0 : \sigma = 2$ vs. $H_a : \sigma \neq 2$	Inside	Do not reject H_0
(c)	5	$H_0 : \sigma = 5$ vs. $H_a : \sigma \neq 5$	Outside	Reject H_0

37.

	Value of σ_0	Form of hypothesis test, with $\alpha = 0.10$	Where σ_0 lies in relation to 90% confidence interval (100, 200)	Conclusion of hypothesis test
(a)	150	$H_0 : \sigma = 150$ vs. $H_a : \sigma \neq 150$	Inside	Do not reject H_0
(b)	250	$H_0 : \sigma = 250$ vs. $H_a : \sigma \neq 250$	Outside	Reject H_0
(c)	0	$H_0 : \sigma = 0$ vs. $H_a : \sigma \neq 0$	Outside	Reject H_0

39. (a) 982.75 **(b)** $H_0 : \sigma = 36.5$ vs. $H_a : \sigma > 36.5$. Reject H_0 if the p-value ≤ 0.01. p-value ≈ 0. Since the p-value ≤ 0.01, we reject H_0. There is evidence that the population standard deviation of DDT level in the breast milk of Hispanic women in the Yakima valley is greater than 36.5 parts per billion.
41. (a) 0.00001497113333 **(b)** $H_0 : \sigma = 30,000$ vs. $H_a : \sigma \neq 30,000$. Reject H_0 if the p-value ≤ 0.05. p-value ≈ 0. Since the p-value ≤ 0.05, we reject H_0. There is evidence that the population standard deviation of union membership differs from 30,000. **(c)** No, since the conclusion for the hypothesis test is reject H_0.
43. p-Value method: $H_0 : \sigma = 50$ vs. $H_a : \sigma > 50$. Reject H_0 if p-value ≤ 0.05. $\chi^2_{\text{data}} = 104$. p-value $= 0.3721497012$.

Since p-value is not ≤ 0.05, we do not reject H_0. There is insufficient evidence that the population standard deviation of test scores for boys is greater than 50 points. **Critical-value method:** $H_0 : \sigma = 50$ vs. $H_a : \sigma > 50$. $\chi^2_\alpha = \chi^2_{0.05} = 124.342$. Reject H_0 if $\chi^2_{\text{data}} \geq 124.342$. $\chi^2_{\text{data}} = 104$. Since χ^2_{data} is not ≥ 124.342, we do not reject H_0. There is insufficient evidence that the population standard deviation of test scores for boys is greater than 50 points.

Section 9.7
1. A Type II error is not rejecting H_0 when H_0 is false.
3. The probability of rejecting H_0 when H_0 is false
5. (a) 51.024

(b)

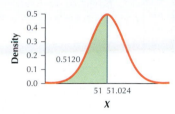

(c) 0.5120 **(d)** 0.4880
7. (a) 51.024

(b)

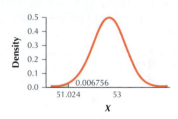

(c) 0.0068 **(d)** 0.9932
9. (a) 51.024

(b)

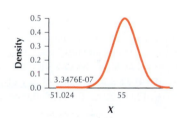

(c) TI-83/84: 0.0000003353 **(d)** 0.9999996647
11. (a) 96.71

(b)

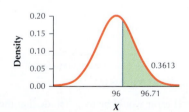

(c) 0.3613 **(d)** 0.6387
13. (a) 96.71

(b)

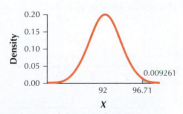

(c) 0.0093 **(d)** 0.9907

15. (a) 96.71
(b)

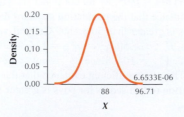

(c) 0.000006658 **(d)** 0.999993342
17.

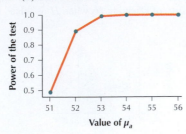

19. (a) A Type II error would be to conclude that the population mean daily number of shares traded is 1.6 billion when it actually is more than 1.6 billion. **(b)** TI-83/84: 0.8520 **(c)** TI-83/84: 0.6718 **(d)** TI-83/84: 0.4384 **(e)** TI-83/84: 0.2251 **(f)** 0.1480, 0.3282, 0.5616, 0.7749

(g)

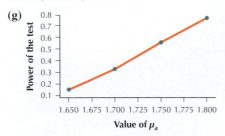

21. (a) A Type II error would be concluding that the population mean salary for entry-level accountants is $41,560 when it is actually different from $41,560. **(b)** TI-83/84: 0.8577 **(c)** TI-83/84: 0.1788 **(d)** TI-83/84: 0.00175 **(e)** TI-83/84: 0.0000004334 **(f)** 0.1423, 0.8212, 0.99825, 0.9999995666

(g)

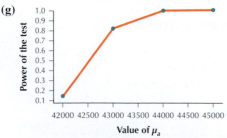

23. (a) A Type II error would be to conclude that the population mean time that it takes owners of hybrid cars to recoup their initial cost is 3 years when it actually is less than 3 years.
(b) TI-83/84: 0.0000001173 **(c)** TI-83/84: 0 **(d)** TI-83/84: 0
(e) TI-83/84: 0 **(f)** 0.9999998827, 1, 1, 1

(g)

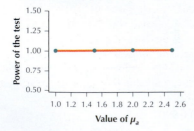

Chapter 9 Review

1. $H_0 : \mu = 12$ vs. $H_a : \mu < 12$
3. $H_0 : \mu = 0$ vs. $H_a : \mu < 0$
5. (a) $H_0 : \mu = 202.7$ vs. $H_a : \mu < 202.7$ **(b)** We conclude that (1) the population mean number of speeding-related fatalities is less than 202.7 when it actually is and (2) the mean number of speeding-related fatalities is greater than or equal to 202.7 when it actually is. **(c)** The population mean number of speeding-related fatalities is less than 202.7 when it actually is greater than or equal to 202.7. **(d)** The population mean number of speeding-related fatalities is greater than or equal to 202.7 when it actually is less than 202.7.
7. -1
9. -10
11. (a) 1.28 **(b)** Reject H_0 if $Z_{\text{data}} \geq 1.28$.
(c)

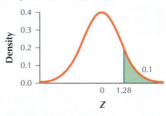

(d) Since $Z_{\text{data}} \geq 1.28$, we reject H_0. There is evidence that the population mean is greater than μ_0.
13. (a) $H_0 : \mu = 673$ vs. $H_a : \mu < 673$ **(b)** -1.645; reject H_0 if $Z_{\text{data}} \leq -1.645$. **(c)** $Z_{\text{data}} = -5.52$

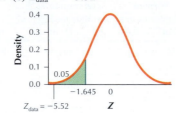

(d) Since $Z_{\text{data}} \leq -1.645$, we reject H_0. There is evidence that the population mean credit score in Florida is less than 673.
15. (a) $H_0 : \mu = 500$ vs. $H_a : \mu \neq 500$. Reject H_0 if the p-value $\leq$ 0.05. **(b)** 4 **(c)** $6.337206918 \times 10^{-5}$

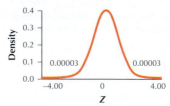

(d) Since the p-value ≤ 0.05, reject H_0. There is evidence that the population mean is different than 500.
17. $H_0 : \mu = 6{,}096$ vs. $H_a : \mu > 6{,}096$. Reject H_0 if the p-value $\leq$ 0.01. $Z_{\text{data}} = 11.9$; p-value $= 6.09738351 \times 10^{-33}$. Since the p-value ≤ 0.01, reject H_0. There is evidence that the population mean per capita annual expenditures on health care is greater than $6096.
19. $t_{\text{crit}} = 1.415$

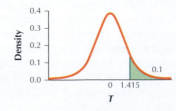

21. $t_{crit} = 2.998$

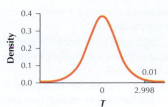

23. $H_0 : \mu = 9$ vs. $H_a : \mu \neq 9$. $t_{crit} = 1.753$. Reject H_0 if $t_{data} \leq -1.753$ or $t_{data} \geq 1.753$. $t_{data} = 1.33$. Since t_{data} is not ≤ -1.753 and t_{data} is not ≥ 1.753, we do not reject H_0. There is insufficient evidence that the population mean is different from 9.

25. (a) We have $np_0 = 1000(0.8) = 800 \geq 5$ and $n(1 - p_0) = 1000(1 - 0.8) = 200 \geq 5$. **(b)** $H_0 : p = 0.8$ vs. $H_a : p > 0.8$ **(c)** $Z_{crit} = 1.28$. Reject H_0 if $Z_{data} \geq 1.28$. **(d)** $Z_{data} = 2.37$ **(e)** Since $Z_{data} \geq 1.28$, we reject H_0. There is evidence that the population proportion is greater than 0.8.

27. (a) We have $np_0 = 100(0.4) = 40 \geq 5$ and $n(1 - p_0) = 100(1 - 0.4) = 60 \geq 5$. **(b)** $H_0 : p = 0.4$ vs. $H_a : p \neq 0.4$ **(c)** $Z_{crit} = 2.58$. Reject H_0 if $Z_{data} \leq -2.58$ or $Z_{data} \geq 2.58$. **(d)** $Z_{data} = 3.06$ **(e)** Since $Z_{data} \geq 2.58$, we reject H_0. There is evidence that the population proportion is not equal to 0.4.

29. (a) We have $np_0 = 100(0.25) = 25 \geq 5$ and $n(1 - p_0) = 100(1 - 0.25) = 75 \geq 5$. **(b)** $H_0 : p = 0.25$ vs. $H_a : p < 0.25$. Reject H_0 if the p-value ≤ 0.05. **(c)** 0 **(d)** 0.5 **(e)** Since the p-value is not ≤ 0.05, we do not reject H_0. There is insufficient evidence that the population proportion is less than 0.25.

31. (a) $H_0 : \sigma = 6$ vs. $H_a : \sigma > 6$ **(b)** $\chi_\alpha^2 = \chi_{0.05}^2 = 30.144$. Reject H_0 if $\chi_{data}^2 \geq 30.144$ **(c)** $\chi_{data}^2 = 42.75$

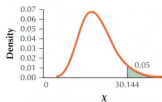

(d) Since $\chi_{data}^2 \geq 30.144$, we reject H_0. There is evidence that the population standard deviation is greater than 6.

33. (a) $H_0 : \sigma = 35$ vs. $H_a : \sigma < 35$. Reject H_0 if the p-value ≤ 0.05. **(b)** 6.857 **(c)** p-value $= 0.5560805474$

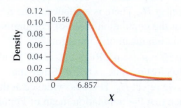

(d) Since the p-value is not ≤ 0.05, we do not reject H_0. There is insufficient evidence that the population standard deviation is less than 35.

35. $H_0 : \sigma = 50$ vs. $H_a : \sigma \neq 50$. Reject H_0 if the p-value ≤ 0.01. $\chi_{data}^2 = 0.690$. p-value $= 0.094887$. Since the p-value is not ≤ 0.01, we do not reject H_0. There is insufficient evidence that the population standard deviation differs from 50.

37. (a) $\bar{x}_{critical, lower} = 95.1625$, $\bar{x}_{critical, upper} = 104.8375$

(b)

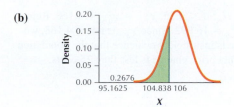

(c) 0.2676 **(d)** 0.7324

39. (a) $\bar{x}_{critical, lower} = 95.1625$, $\bar{x}_{critical, upper} = 104.8375$

(b)

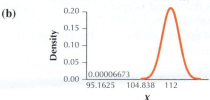

(c) 0.00006675 **(d)** 0.99993325

41.

Chapter 9 Quiz

1. False
2. True
3. True
4. I
5. small
6. α
7. $np_0 \geq 5$ and $n(1 - p_0) \geq 5$
8. A small p-value indicates that there is strong evidence against the null hypothesis. A large p-value indicates that there is no evidence against the null hypothesis.
9. No
10. (a) $H_0 : \mu = 1.14$ vs. $H_a : \mu < 1.14$. -1.645. Reject H_0 if $Z_{data} \leq -1.645$. $Z_{data} = -1.68$. Since $Z_{data} \leq -1.645$, we reject H_0. There is evidence that the population mean fee charged by banks when you withdraw funds from an ATM machine not owned by your bank is less than \$1.14. **(b)** Type I error, Type II error
11. (a) No, since the population standard deviation is not known. $H_0 : \mu = 32$ vs. $H_a : \mu \neq 32$. 1.690. Reject H_0 if $t_{data} \leq -1.690$ or $t_{data} \geq 1.690$. $\chi_{data}^2 = 1.80$. Since $t_{data} \geq 1.690$, we reject H_0. There is evidence that the population mean years of potential life lost in alcohol-related fatal automobile accidents is different from 32 years. **(b)** p-value $= 0.0805$, so there is moderate evidence against the null hypothesis.
12. $H_0 : p = 0.127$ vs. $H_a : p > 0.127$. Reject H_0 if the p-value ≤ 0.05. $Z_{data} = 0.93$. p-value $= 0.1762$. Since $0.15 \leq p$-value, there is no evidence against the null hypothesis that the population proportion of preterm births is less than or equal to 0.127. Since the p-value ≤ 0.05, we do not reject H_0. There is insufficient evidence that the population proportion of preterm births is greater than 0.127.

13. $H_0 : \sigma = 0.25$ vs. $H_a : \sigma < 0.25$. $\chi^2_{1-\alpha} = \chi^2_{0.90} = 4.168$. Reject H_0 if $\chi^2_{data} \leq 4.168$. $\chi^2_{data} = 16.992$. Since χ^2_{data} is not ≤ 4.168, we do not reject H_0. There is insufficient evidence that the population standard deviation of net price change is less than 25 cents.

Chapter 10

Section 10.1

1. When the subjects selected for the first sample do not determine the subjects in the second sample

3. Matched pairs or paired samples

5. Since both samples of games were based on the same players, this is an example of dependent sampling.

7. Since the same students are taking both tests, this is an example of dependent sampling.

9. $\bar{x}_d = 1.1$, $s_d = 0.9618$

11. $\bar{x}_d = -7.1429$, $s_d = 2.6726$

13. $\bar{x}_d = -0.00625$, $s_d = 0.2095$

15. $H_0 : \mu_d = 0$ vs. $H_a : \mu_d > 0$. $t_{crit} = 2.132$. Reject H_0 if $t_{data} \geq 2.132$. $t_{data} = 2.557$. Since $t_{data} = 2.557$ is ≥ 2.132, we reject H_0. There is evidence at the $\alpha = 0.05$ level of significance that the population mean difference is greater than 0.

17. $H_0 : \mu_d = 0$ vs. $H_a : \mu_d < 0$. $t_{crit} = -1.440$. Reject H_0 if $t_{data} \leq -1.440$. $t_{data} = -7.071$. Since $t_{data} = -7.071$ is ≤ -1.440, we reject H_0. There is evidence at the $\alpha = 0.10$ level of significance that the population mean difference is less than 0.

19. $H_0 : \mu_d = 0$ vs. $H_a : \mu_d \neq 0$. Reject H_0 if the p-value ≤ 0.05. $t_{data} = -0.084$. p-value $= 0.9351$. Since the p-value $= 0.9351$ is not ≤ 0.05, we do not reject H_0. There is insufficient evidence at the $\alpha = 0.05$ level of significance that the population mean difference is not equal to 0.

21. $(-0.0940, 2.294)$. We are 95% confident that the population mean difference lies between -0.0040 and 2.294.

23. $(-9.106, -5.180)$. We are 90% confident that the population mean difference lies between -9.106 and -5.180.

25. $(-0.181, 0.169)$. We are 95% confident that the population mean difference lies between -0.181 and 0.169.

27. (a) $H_0 : \mu_d = 0$ vs. $H_a : \mu_d \neq 0$. $\mu_0 = 0$ lies inside of the interval $(-5, 5)$, so we do not reject H_0 at the $\alpha = 0.05$ level of significance. **(b)** $H_0 : \mu_d = -6$ vs. $H_a : \mu_d \neq -6$. $\mu_0 = -6$ lies outside of the interval $(-5, 5)$, so we reject H_0 at the $\alpha = 0.05$ level of significance. **(c)** $H_0 : \mu_d = 4$ vs. $H_a : \mu_d \neq 4$. $\mu_0 = 4$ lies inside of the interval $(-5, 5)$, so we do not reject H_0 at the $\alpha = 0.05$ level of significance.

29. (a) $H_0 : \mu_d = -10$ vs. $H_a : \mu_d \neq -10$. $\mu_0 = -10$ lies outside of the interval $(10, 20)$, so we reject H_0 at the $\alpha = 0.10$ level of significance. **(b)** $H_0 : \mu_d = 25$ vs. $H_a : \mu_d \neq 25$. $\mu_0 = 25$ lies outside of the interval $(10, 20)$, so we reject H_0 at the $\alpha = 0.10$ level of significance. **(c)** $H_0 : \mu_d = 0$ vs. $H_a : \mu_d \neq 0$. $\mu_0 = 0$ lies outside of the interval $(10, 20)$, so we reject H_0 at the $\alpha = 0.10$ level of significance.

31. (a) $\bar{x}_d = 753.75$, $s_d = 986.1658$ **(b)** There is insufficient evidence that 2007 models are on average more expensive.

33. (a) $\bar{x}_d = 15.14$, $s_d = 3.7787$ **(b) Critical-value method:** $H_0 : \mu_d = 0$ vs. $H_a : \mu_d \neq 0$. $t_{crit} = 3.250$. Reject H_0 if $t_{data} \leq -3.250$ or if $t_{data} \geq 3.250$. $t_{data} = 12.670$. Since $t_{data} = 12.670$ is ≥ 3.250, we reject H_0. There is evidence at the $\alpha = 0.01$ level of significance that the population mean difference between high and low temperatures is different from 0. **p-value method:** $H_0 : \mu_d = 0$ vs. $H_a : \mu_d \neq 0$. Reject H_0 if the p-value ≤ 0.01. $t_{data} = 12.670$. p-value ≈ 0. Since the p-value ≈ 0 is ≤ 0.01, we

reject H_0. There is evidence at the $\alpha = 0.01$ level of significance that the population mean difference between high and low temperatures is different from 0.

35. $(-815.2398, 2322.7398)$. We are 95% confident that the population mean difference in car prices lies between $-\$815.2398$ and $\$2322.7398$.

37. (a) $(12.437, 17.843)$. We are 95% confident that the population mean difference between high and low temperatures lies between 12.437 and 17.843 degrees. **(b)** $\mu_0 = 0$ lies outside of the interval $(12.437, 17.8430)$, so we reject H_0. This is the same conclusion we reached in the hypothesis test in Exercise 33.

39. (a) $(-15.4467, 0.6467)$ **(b) Critical value method:** $H_0 : \mu_d = 0$ versus $H_0 : \mu_d \neq 0$. df $= n - 1 = 10 - 1 = 9$, $\alpha = 0.10$, $t_{crit} = 1.833$. Reject H_0 if $t_{data} \leq -1.833$ or if $t_{data} \geq 1.833$. $t_{data} = -1.69$. Do not reject H_0. There is insufficient evidence that the 2007 math test scores for eighth graders differ from the 1995 math test scores for eighth graders. **p-value method:** $H_0 : \mu_d = 0$, $H_a : \mu_d \neq 0$. Reject H_0 if p-value ≤ 0.10. $t_{data} = -1.69$. p-value $= 0.1261351394$. Do not reject H_0. There is insufficient evidence that the 2007 math test scores for eighth graders differ from the 1995 math test scores for eighth graders.

41. Because we are taking home sales of the same counties in 2006 and 2007

43. There is insufficient evidence that the population mean number of home sales differed from 2006 to 2007.

45. The conclusion would remain the same.

Section 10.2

1. The two populations are normally distributed. The sample sizes are large (at least 30).

3. (a) $H_0 : \mu_1 = \mu_2$ vs. $H_a : \mu_1 \neq \mu_2$ **(b)** $t_{crit} = 1.690$. Reject H_0 if $t_{data} \leq -1.690$ or $t_{data} \geq 1.690$. **(c)** $t_{data} = 4.243$. **(d)** Since $t_{data} \geq 1.690$, we reject H_0. There is evidence that the population mean for Population 1 is different from the population mean for Population 2.

5. (a) $H_0 : \mu_1 = \mu_2$ vs. $H_a : \mu_1 < \mu_2$ **(b)** $t_{crit} = -2.423$. Reject H_0 if $t_{data} \leq -2.423$. **(c)** $t_{data} = -5.077$ **(d)** Since $t_{data} = -5.077$ is ≤ -2.423, we reject H_0. There is evidence at the $\alpha = 0.01$ level of significance that the population mean of Population 1 is less than the population mean of Population 2.

7. (a) $H_0 : \mu_1 = \mu_2$ vs. $H_a : \mu_1 \neq \mu_2$. Reject H_0 if the p-value ≤ 0.10. **(b)** $t_{data} = -2.492$. **(c)** p-value $= 0.0162$. **(d)** Since the p-value is ≤ 0.10, we reject H_0. There is evidence at the $\alpha = 0.10$ level of significance that the population mean of Population 1 is different from the population mean of Population 2.

9. (a) $H_0 : \mu_1 = \mu_2$ vs. $H_a : \mu_1 < \mu_2$. Reject H_0 if the p-value ≤ 0.05. **(b)** $t_{data} = -13.868$ **(c)** p-value ≈ 0. **(d)** Since the p-value ≈ 0 is ≤ 0.05, we reject H_0. There is evidence at the $\alpha = 0.05$ level of significance that the population mean of Population 1 is less than the population mean of Population 2.

11. (a) $\bar{x}_1 - \bar{x}_2 = 2$ **(b)** $E = 0.797$. We can estimate the difference in the population means of Population 1 and Population 2 to within 0.797 with 90% confidence. **(c)** $(1.203, 2.797)$. We are 90% confident that the difference in the population means of Population 1 and Population 2 lies between 1.203 and 2.797.

13. (a) $\bar{x}_1 - \bar{x}_2 = -10$ **(b)** $E = 5.326$. We can estimate the difference in the population means of Population 1 and Population 2 to within 5.326 with 99% confidence. **(c)** $(-15.326, -4.674)$. We are 99% confident that the difference in the population means of Population 1 and Population 2 lies between -15.326 and -4.674.

15. (a) $\bar{x}_1 - \bar{x}_2 = -1$ (b) $E = 0.811$. We can estimate the difference in the population means of Population 1 and Population 2 to within 0.811 with 95% confidence. (c) $(-1.811, -0.189)$. We are 95% confident that the difference in the population means of Population 1 and Population 2 lies between -1.811 and -0.189.

17. (a) $H_0: \mu_1 - \mu_2 = 0$ vs. $H_a: \mu_1 - \mu_2 \neq 0$. $\mu_0 = 0$ lies outside of the interval $(10, 15)$, so we reject H_0 at the $\alpha = 0.05$ level of significance. (b) $H_0: \mu_1 - \mu_2 = 12$ vs. $H_a: \mu_1 - \mu_2 \neq 12$. $\mu_0 = 12$ lies inside of the interval $(10, 15)$, so we do not reject H_0 at the $\alpha = 0.05$ level of significance. (c) $H_0: \mu_1 - \mu_2 \neq 16$ vs. $H_a: \mu_1 - \mu_2 \neq 16$. $\mu_0 = 16$ lies outside of the interval $(10, 15)$, so we reject H_0 at the $\alpha = 0.05$ level of significance.

19. (a) $H_0: \mu_1 - \mu_2 = -10.1$ vs. $H_a: \mu_1 - \mu_2 \neq -10.1$. $\mu_0 = -10.1$ lies outside of the interval $(-10, 10)$, so we reject H_0 at the $\alpha = 0.10$ level of significance. (b) $H_0: \mu_1 - \mu_2 = -9.9$ vs. $H_a: \mu_1 - \mu_2 \neq -9.9$. $\mu_0 = -9.9$ lies inside of the interval $(-10, 10)$, so we do not reject H_0 at the $\alpha = 0.10$ level of significance. (c) $H_0: \mu_1 - \mu_2 = 0$ vs. $H_a: \mu_1 - \mu_2 \neq 0$. $\mu_0 = 0$ lies inside of the interval $(-10, 10)$, so we do not reject H_0 at the $\alpha = 0.10$ level of significance.

21. $H_0: \mu_1 = \mu_2$ vs. $H_a: \mu_1 > \mu_2$. $t_{\text{crit}} = 1.294$. Reject H_0 if $t_{\text{data}} \geq 1.294$. $s_{\text{pooled}}^2 = 110.5$. $t_{\text{data}} \approx 0.807$. Since $t_{\text{data}} \approx 0.807$ is not ≥ 1.294, we do not reject H_0. There is insufficient evidence at the $\alpha = 0.10$ level of significance that the population mean of Population 1 is greater than the population mean of Population 2.

23. $(-2.940, 6.940)$. We are 95% confident that the difference in the population means of Population 1 and Population 2 lies between -2.940 and 6.940.

25. $H_0: \mu_1 = \mu_2$ vs. $H_a: \mu_1 > \mu_2$. $Z_{\text{crit}} = 1.645$. Reject H_0 if $Z_{\text{data}} \geq 1.645$. $Z_{\text{data}} \approx 2.757$. Since $Z \approx 2.757$ is ≥ 1.645, we reject H_0. There is evidence at the $\alpha = 0.05$ level of significance that the population mean of Population 1 is greater than the population mean of Population 2.

27. $(0.289, 1.711)$. We are 95% confident that the difference in the population means of Population 1 and Population 2 lies between 0.289 and 1.711.

29. Since both sample sizes are less than 30 and the distribution of both populations is unknown, it is not appropriate to use Welch's t test.

31. (a) -1192 (b) $2,426.795$ (c) $(-3,618.795, 1,234.795)$. We are 95% confident that the interval captures the difference of the population mean incomes for Sacramento County and Los Angeles County, California. (d) $H_0: \mu_1 = \mu_2$ vs. $H_a: \mu_1 < \mu_2$. $t_{\text{crit}} = -1.690$. Reject H_0 if $t_{\text{data}} \leq -1.690$. $t_{\text{data}} = -0.997$. Since $t_{\text{data}} = -0.997$ is not ≤ -1.690, we do not reject H_0. There is insufficient evidence at the $\alpha = 0.05$ level of significance that the population mean income in Sacramento County, California, in 2004 was less than the population mean income in Los Angeles County, California, in 2004. (e) The confidence interval in (c) could not have been used to perform the hypothesis test in (d) because the hypothesis test in (d) is a one-tailed test and confidence intervals can only be used to perform two-tailed tests.

33. (a) $(-1.047, 4.247)$. We are 95% confident that the interval captures the difference in the population mean number of children per teacher in the towns of Cupertino, California, and Santa Rosa, California. (b) $H_0: \mu_1 - \mu_2 = 0$ vs. $H_a: \mu_1 - \mu_2 \neq 0$. $\mu_0 = 0$ lies inside of the interval $(-1.047, 4.247)$, so we do not reject H_0. There is insufficient evidence at the $\alpha = 0.01$ level of significance that the population mean number of children per teacher in the town of Cupertino, California, differs from the population

mean number of children per teacher in the town of Santa Rosa, California.

35. (a) Since both sample sizes are large ($n_1 \geq 30$ and $n_2 \geq 30$), Case 2 applies. (b) $(11,442.85, 15,479.15)$. We are 95% confident that the interval captures the difference between the population mean starting salary for college graduates majoring in information systems and the population mean starting salary for college graduates majoring in psychology. (c) Wider; the higher the confidence level, the wider the confidence interval

37. $H_0: \mu_1 = \mu_2$ vs. $H_a: \mu_1 > \mu_2$. Reject H_0 if p-value ≤ 0.05. $t_{\text{data}} = 1.017$. p-value $= 0.1558$. Since p-value is not ≤ 0.05, we do not reject H_0. There is insufficient evidence that the population coached SAT score improvement is greater than the population noncoached SAT score improvement. **Critical-value method:** $H_0: \mu_1 = \mu_2$ vs. $H_a: \mu_1 > \mu_2$. $t_{\text{crit}} = 1.662$. Reject H_0 if $t_{\text{data}} \geq 1.662$. $t_{\text{data}} = 1.017$. Since $t_{\text{data}} = 1.017$ is not ≥ 1.662, we do not reject H_0. There is insufficient evidence at the $\alpha = 0.05$ level of significance that the population mean coached SAT improvement is greater than the population mean noncoached improvement.

39. (a) Since the width of the confidence interval is

$$2 \cdot t_{\alpha/2} \cdot \sqrt{\frac{s_1^2}{n_1} + \frac{s_2^2}{n_2}},$$ an increase in the sample sizes will result

in a decrease in the width of the confidence interval. This is good because smaller confidence intervals give a more precise estimate. (b) It depends on how large the new sample sizes are.

41. (a)–(d) Unchanged

Section 10.3

1. $\hat{p}_1$ and $\hat{p}_2$

3. Z_{data} measures the standardized distance between sample proportions. Extreme values of Z_{data} indicate evidence against the null hypothesis.

5. (a) $H_0: p_1 = p_2$ vs. $H_a: p_1 \neq p_2$; $Z_{\text{crit}} = 1.645$. Reject H_0 if $Z_{\text{data}} \leq -1.645$ or $Z_{\text{data}} \geq 1.645$. (b) 0.7857 (c) 0.65 (d) Since $Z_{\text{data}} \geq -1.645$ and $Z_{\text{data}} \leq 1.645$, we do not reject H_0. There is insufficient evidence that the population proportion from Population 1 is different from the population proportion from Population 2.

7. (a) $H_0: p_1 = p_2$ vs. $H_a: p_1 > p_2$. $Z_{\text{crit}} = 2.33$. Reject H_0 if $Z_{\text{data}} \geq 2.33$. (b) $\hat{p}_{\text{pooled}} = 100/450 \approx 0.2222$. (c) $Z_{\text{data}} = 3.550$. (d) Since $Z_{\text{data}} = 3.550$ is ≥ 2.33, we reject H_0. There is evidence at the $\alpha = 0.01$ level of significance that the population proportion of Population 1 is greater than the population proportion of Population 2.

9. (a) $H_0: p_1 = p_2$ vs. $H_a: p_1 > p_2$. Reject H_0 if p-value ≤ 0.05. (b) $\hat{p}_{\text{pooled}} = 450/800 = 0.5625$ (c) $Z_{\text{data}} = 3.563$ (d) p-value $= 0.0002$ (e) Since p-value $= 0.0002$ is ≤ 0.05, we reject H_0. There is evidence at the $\alpha = 0.05$ level of significance that the population proportion of Population 1 is greater than the population proportion of Population 2.

11. (a) $H_0: p_1 = p_2$ vs. $H_a: p_1 \neq p_2$. Reject H_0 if p-value ≤ 0.10. (b) $\hat{p}_{\text{pooled}} = 910/1140 \approx 0.7982$ (c) $Z_{\text{data}} \approx -1.284$ (d) p-value ≈ 0.1991 (e) Since p-value ≈ 0.1991 is not ≤ 0.10, we do not reject H_0. There is insufficient evidence at the $\alpha = 0.10$ level of significance that the population proportion of Population 1 is different from the population proportion of Population 2.

13. (a) $x_1 = 80 \geq 5$, $n_1 - x_1 = 20 \geq 5$, $x_2 = 30 \geq 5$, and $n_2 - x_2 = 10 \geq 5$, so it is appropriate. (b) 0.05 (c) 0.1554. The point estimate $\hat{p}_1 - \hat{p}_2$ will lie within $E = 0.1554$ of the difference in population proportions $p_1 - p_2$ 95% of the time. **(d)** $(-0.1054,$ $0.2054)$. We are 95% confident that the difference in population proportions lies between -0.1054 and 0.2054.

15. (a) $x_1 = 60 \geq 5$, $n_1 - x_1 = 140 \geq 5$, $x_2 = 40 \geq 5$, and $n_2 - x_2 = 210 \geq 5$, so it is appropriate. **(b)** 0.14 **(c)** 0.078. The point estimate $\hat{p}_1 - \hat{p}_2$ will lie within $E = 0.078$ of the difference in population proportions $p_1 - p_2$ 95% of the time. **(d)** $(0.062,$ $0.218)$. We are 95% confident that the difference in population proportions lies between 0.062 and 0.218.

17. (a) $x_1 = 490 \geq 5$, $n_1 - x_1 = 510 \geq 5$, $x_2 = 620 \geq 5$, and $n_2 - x_2 = 380 \geq 5$, so it is appropriate. **(b)** -0.13 **(c)** 0.0431. The point estimate $\hat{p}_1 - \hat{p}_2$ will lie within $E = 0.0431$ of the difference in population proportions $p_1 - p_2$ 95% of the time. **(d)** $(-0.1731, -0.0869)$. We are 95% confident that the difference in population proportions lies between -0.1731 and -0.0869.

19. (a) $H_0 : p_1 - p_2 = 0$ vs. $H_a : p_1 - p_2 \neq 0$. The hypothesized value of 0 lies outside the interval $(0.5, 0.6)$, so we reject H_0 at the $\alpha = 0.05$ level of significance. **(b)** $H_0 : p_1 - p_2 = 0.1$ vs. $H_a : p_1 - p_2 \neq 0.1$. The hypothesized value of 0.1 lies outside the interval $(0.5, 0.6)$, so we reject H_0 at the $\alpha = 0.05$ level of significance. **(c)** $H_0 : p_1 - p_2 = 0.57$ vs. $H_a : p_1 - p_2 \neq 0.57$. The hypothesized value of 0.57 lies inside the interval $(0.5, 0.6)$, so we do not reject H_0 at the $\alpha = 0.05$ level of significance.

21. (a) $H_0 : p_1 - p_2 = 0.151$ vs $H_a : p_1 - p_2 \neq 0.151$. The hypothesized value of 0.151 lies outside of the interval $(0.1, 0.11)$, so we reject H_0 at the $\alpha = 0.10$ level of significance. **(b)** $H_0 : p_1 - p_2 = 0.115$ vs. $H_a : p_1 - p_2 \neq 0.115$. The hypothesized value of 0.115 lies outside of the interval $(0.1, 0.11)$, so we reject H_0 at the $\alpha = 0.10$ level of significance. **(c)** $H_0 : p_1 - p_2 = 0.105$ vs. $H_a : p_1 - p_2 \neq 0.105$. The hypothesized value of 0.105 lies inside of the interval $(0.1, 0.11)$, so we do not reject H_0 at the $\alpha = 0.10$ level of significance.

23. (a) $x_1 = 0.74 (500) = 370 \geq 5$, $(n_1 - x_1) = 130 \geq 5$, $x_2 = 0.83 (500) = 415 \geq 5$, and $(n_2 - x_2) = 85 \geq 5$. Therefore it is appropriate to perform the Z test for the difference in population proportions. **(b)** p_1 is the population proportion of teenage boys who posted their photo on their online profile and p_2 is the population proportion of teenage girls who posted their photo on their online profile. **(c) Critical-value method:** $H_0 : p_1 = p_2$ vs. $H_a : p_1 \neq p_2$. $Z_{crit} = 1.96$. Reject H_0 if $Z_{data} \leq -1.96$ or if $Z_{data} \geq 1.96$. $\hat{p}_{pooled} = 785/1000 = 0.785$. $Z_{data} = -3.464$. Since $Z_{data} = -3.464$ is ≤ -1.96, we reject H_0. There is evidence at the $\alpha = 0.05$ level of significance that the population proportion of teenage boys who posted their photo on their online profile differs from the population proportion of teenage girls who posted their photo on their online profile. **p-value method:** $H_0 : p_1 = p_2$ vs. $H_a : p_1 \neq p_2$. Reject H_0 if the p-value ≤ 0.05. $\hat{p}_{pooled} = 785/1000 = 0.785$. $Z_{data} = -3.464$. p-value $= 0.0005$. Since the p-value $= 0.0005$ is ≤ 0.05, we reject H_0. There is evidence at the $\alpha = 0.05$ level of significance that the population proportion of teenage boys who posted their photo on their online profile differs from the population proportion of teenage girls who posted their photo on their online profile.

25. Critical-value method: $H_0 : p_1 = p_2$ vs. $H_a : p_1 > p_2$. $Z_{crit} = 1.28$. Reject H_0 if $Z_{data} \geq 1.28$. $\hat{p}_{pooled} = 98/300 \approx 0.3267$. $Z_{data} = 0.348$. Since $Z_{data} = 0.348$ is not ≥ 1.28, we do not reject H_0. There is insufficient evidence at the $\alpha = 0.10$ level of significance that the population proportion of Ohio businesses that are owned by women is greater than the population proportion of New Jersey

businesses that are owned by women. **p-value method:** $H_0 : p_1 = p_2$ vs. $H_a : p_1 > p_2$. Reject H_0 if the p-value ≤ 0.10. $\hat{p}_{pooled} = 98/300 \approx 0.3267$. $Z_{data} = 0.348$. p-value $= 0.3639$. Since the p-value $= 0.3639$ is not ≤ 0.10, we do not reject H_0. There is insufficient evidence at the $\alpha = 0.10$ level of significance that the population proportion of Ohio businesses that are owned by women is greater than the population proportion of New Jersey businesses that are owned by women.

27. (a) $(-0.1409, -0.0391)$. TI-83/84: $(-0.1406, -0.0394)$. We are 95% confident that the difference of the population proportion of teenage boys who post their photo on their online profile and the population proportion of teenage girls who post their photo on their online profile lies between $-0.1409(-0.1406)$ and -0.0391 (-0.0394). **(b)** $H_0 : p_1 = p_2$ vs. $H_a : p_1 \neq p_2$. The hypothesized value of 0 lies outside of the interval in **(a)**, so we reject H_0. There is evidence that the population proportion of teenage boys who post their photo on their online profile differs from the population proportion of teenage girls who post their photo on their online profile. **(c)** Yes, it agrees.

29. (a) $(-0.0745, 0.1145)$. TI-83/84: $(-0.0749, 0.1150)$. We are 90% confident that the difference of the population proportion of Ohio businesses that are owned by women and the population proportion of New Jersey businesses that are owned by women lies between $-0.0745(-0.0749)$ and $0.1145 (0,1150)$. **(b)** $H_0 : p_1 = p_2$ vs. $H_a : p_1 \neq p_2$. Our hypothesized value of 0 lies inside the interval in **(a)**, so we do not reject H_0. There is insufficient evidence that the population proportion of Ohio businesses that are owned by women differs from the population proportion of New Jersey businesses that are owned by women. **(c)** No, it is a one-sided test and confidence intervals can only be used to perform two-sided tests.

31. $H_0 : p_1 = p_2$ vs. $H_a : p_1 \neq p_2$. Reject H_0 if p-value ≤ 0.05. $\hat{p}_{pooled} = 0.7705$. $Z_{data} = 0.21$. p-value $= 0.8336$. Since p-value is not ≥ 0.05, we do not reject H_0. There is insufficient evidence that the proportion of the people who wore the ionized bracelets who reported improvement in their maximum pain index is different from the proportion of the people who wore the placebo bracelets who reported improvement in their maximum pain index.

33. $p_1 =$ the population proportion of 18- to 24-year-old males who listen to the radio each week and $p_2 =$ the population proportion of males age 65 or older who listen to the radio each week.

35. 0.0269. The point estimate of the difference in the population proportion of 18- to 24-year-old males who listen to the radio each week and the population proportion of males 65 years and older who listen to the radio each week will lie within $E = 0.0269$ of the difference in population proportions $p_1 - p_2$ 95% of the time.

37. (a) $H_0 : p_1 - p_2 = 0$ vs. $H_a : p_1 - p_2 \neq 0$. The hypothesized value of 0 does not lie in the interval from Exercise 37, so we reject H_0. There is evidence that the difference in the population proportion of 18- to-24-year-old males who listen to the radio each week and the population proportion of males 65 years and older who listen to the radio each week differs from 0. **(b)** $H_0 : p_1 - p_2 = 0.01$ vs. $H_a : p_1 - p_2 \neq 0.01$. The hypothesized value of 0.01 does not lie in the interval from Exercise 37, so we reject H_0. There is evidence that the difference in the population proportion of 18- to 24-year-old males who listen to the radio each week and the population proportion of males 65 years and older who listen to the radio each week differs from 0.01. **(c)** $H_0 : p_1 - p_2 = 0.05$ vs. $H_a : p_1 - p_2 \neq 0.05$. The hypothesized value of 0.05 lies in the interval from Exercise 37, so we do not reject H_0. There is insufficient evidence that the difference in the population proportion of 18- to 24-year-old males who listen to the radio each week and the population proportion of males 65 years and older who listen to the radio each week differs from 0.05.

39. Critical-value method: $H_0 : p_1 = p_2$ vs. $H_a : p_1 > p_2$. $Z_{\text{crit}} = 1.645$. Reject H_0 if $Z_{\text{data}} \geq 1.645$. $\hat{p}_{\text{pooled}} = 1790/2000 = 0.895$. $Z_{\text{data}} = 3.647$. Since $Z_{\text{data}} = 3.647$ is ≥ 1.645, we reject H_0. There is evidence at the $\alpha = 0.05$ level of significance that the population proportion of 18- to 24-year-old males who listen to the radio each week is greater than the population proportion of males 65 years and older who listen to the radio each week. **p-value method:** $H_0 : p_1 = p_2$ vs. $H_a : p_1 > p_2$. Reject H_0 if the p-value ≤ 0.05. $\hat{p}_{\text{pooled}} = 1790/2000 = 0.895$. $Z_{\text{data}} = 3.647$. p-value $= 0.00013$. Since the p-value $= 0.00013$ is ≤ 0.05, we reject H_0. There is evidence at the $\alpha = 0.05$ level of significance that the population proportion of 18- to 24-year-old males who listen to the radio each week is greater than the population proportion of males 65 years and older who listen to the radio each week.

Chapter 10 Review

1. (a) $\bar{x}_d = -2.6875$, $s_d = 1.6146$ (b) $(-4.0376, -1.3374)$

3. $H_0 : \mu_d = 0$ vs. $H_a : \mu_d < 0$. Reject H_0 if p-value ≤ 0.05. $t_{\text{data}} = -4.708$. p-value $= 0.0010939869$. Since the p-value ≤ 0.05, we reject H_0. There is evidence that the population mean of the differences is less than 0.

5. 0.1

7. (0.094, 0.106). We are 95% confident that the interval captures the difference in population means.

9. (a) $H_0 : p_1 = p_2$ vs. $H_a : p_1 \neq p_2$. Reject H_0 if the p-value ≤ 0.05. $\hat{p}_{\text{pooled}} = 10/157 \approx 0.0636942675$. $Z_{\text{data}} = 2.40$. p-value $= 0.0163$. Since the p-value $= 0.0163$ is ≤ 0.05, we reject H_0. There is evidence at the $\alpha = 0.05$ level of significance that the population proportion of packet loss from Asian Web sites differs from the population proportion of packet loss from North American Web sites. (b) (0.0096, 0.2304)

Chapter 10 Quiz

1. True

2. True

3. False

4. normal; large (greater than or equal to 30)

5. margin of error

6. $\bar{x}_d$

7. $\mu_1 - \mu_2$

8. $\hat{p}_{\text{pooled}}$

9. No difference

10. (a) (6.6680, 21.3320) (b) Since 0 does not lie in the confidence interval, we reject H_0. There is evidence that the population mean difference in the number of cigarettes smoked before and after attending Butt-Enders is different from 0.

11. (a) **Critical-value method:** $H_0 : \mu_1 = \mu_2$ vs. $H_a : \mu_1 < \mu_2$. $t_{\text{crit}} = -1.690$. Reject H_0 if $t_{\text{data}} \leq -1.690$. $t_{\text{data}} = -3.667$. Since $t_{\text{data}} = -3.667$ is ≤ -1.690, we reject H_0. There is evidence at the $\alpha = 0.05$ level of significance that the population mean income in Suburb A is less than the population mean income in Suburb B. **p-value method:** $H_0 : \mu_1 = \mu_2$ vs. $H_a : \mu_1 < \mu_2$. Reject H_0 if p-value ≤ 0.05. $t_{\text{data}} = -3.667$. p-value $= 0.0004$. Since p-value $= 0.0004$ is ≤ 0.05, we reject H_0. There is evidence at the $\alpha = 0.05$ level of significance that the population mean income in Suburb A is less

than the population mean income in Suburb B. (b) $(-23,304.69, -6,695.31)$. We are 95% confident that the interval captures the difference of the population mean income of Suburb A and the population mean income of Suburb B.

12. (2.2406, 17.7594). We are 95% confident that the interval captures the difference of the population mean number of bottles processed by the updated machine and the population mean number of bottles processed by the non-updated machine.

13. (a) Since $t_{\text{data}} \geq 1.662$, we reject H_0. There is evidence that the population mean number of bottles processed by the updated machine is greater than the population mean number of bottles processed by the non-updated machine. (b) Since confidence intervals can be used only to perform two-tailed tests and the hypothesis test in (a) is a one-tailed test, the confidence interval in Exercise 12 cannot be used to perform the hypothesis test in (a).

14. (a) $H_0 : \mu_1 = \mu_2$ vs. $H_a : \mu_1 \neq \mu_2$. $t_{\text{crit}} = 1.662$. Reject H_0 if $t_{\text{data}} \leq -1.662$ or $t_{\text{data}} \geq 1.662$. $t_{\text{data}} = -6.129$. Since $t_{\text{data}} \leq -1.662$, we reject H_0. There is evidence that the population mean income of people 18 to 24 years old who never married is different from the population mean income of people 18 to 24 years old who are married. (b) No, the conclusion of the two-tailed hypothesis test for $\alpha = 0.10$ is "Reject H_0." (c) $(-\$7349.928, -\$4214.072)$. The confidence interval does not include 0.

15. $H_0 : p_1 = p_2$; $H_a : p_1 < p_2$. Reject H_0 if the p-value ≤ 0.05. $Z_{\text{data}} = 0.28$; p-value $= 0.3897$. Since the p-value is not ≥ 0.05, we do not reject H_0. There is insufficient evidence that the population proportion of 18- to 20-year-olds who used an illicit drug decreased from 2004 to 2005.

Chapter 11

Section 11.1

1. (1) Each independent trial of the experiment has k possible outcomes, $k = 2,3, \ldots$ (2) The ith outcome (category) occurs with probability p_i, where $i = 1, 2, \ldots, k$ (3) $\sum^k p_i = 1$.

3. It is the long-run mean of that random variable after an arbitrarily large number of trials.

5. Multinomial

7. Multinomial

9. (a) $E_1 = 50$, $E_2 = 25$, $E_3 = 25$ (b) Conditions are met.

11. (a) $E_1 = n \cdot p_1 = (100)(0.9) = 90$, $E_2 = n \cdot p_2 = (100)(0.05) = 5$, $E_3 = n \cdot p_3 = (100)(0.04) = 4$, $E_4 = n \cdot p_4 = (100)(0.01) = 1$ (b) the conditions are not met.

13. 0.667

15. 7.333

17. 17.667

19. (a) $E_1 = 40$, $E_2 = 30$, $E_3 = 30$; conditions are met. (b) $\chi^2_{\text{crit}} = \chi^2_{0.05} = 5.991$. Reject H_0 if $\chi^2_{\text{data}} \geq 5.991$. (c) 4.167 (d) Since χ^2_{data} is not ≥ 5.991, we do not reject H_0. There is insufficient evidence that the random variable does not follow the distribution specified in H_0.

21. (a) $E_1 = 80$, $E_2 = 70$, $E_3 = 20$, $E_4 = 20$, $E_5 = 10$; conditions are met. (b) $\chi^2_{\text{crit}} = \chi^2_{0.10} = 7.779$. Reject H_0 if $\chi^2_{\text{data}} \geq 7.779$. (c) 6.607 (d) Since χ^2_{data} is not ≥ 7.779, we do not reject H_0. There is insufficient evidence that the random variable does not follow the distribution specified in H_0.

23. (a) Reject H_0 if the p-value ≤ 0.05. $E_1 = 50$, $E_2 = 50$; conditions are met. (b) 4 (c) p-value $= 0.0455$. (d) Since the p-value ≤ 0.05, we reject H_0. There is evidence that the random variable does not follow the distribution specified in H_0.

25. (a) Reject H_0 if the p-value ≤ 0.10. $E_1 = 100$, $E_2 = 50$, $E_3 = 30$, $E_4 = 20$; conditions are met. (b) 6.083 (c) p-value $= 0.1076$. (d) Since the p-value is not ≤ 0.10, we do not reject H_0. There is insufficient evidence that the random variable does not follow the distribution specified in H_0.

27. Since χ^2_{data} is not ≥ 9.488, we do not reject H_0. There is insufficient evidence that the distribution of education levels has changed since 2005.

29. $H_0: p_{\text{phip}} = 0.30$, $p_{\text{mm}} = 0.556$, $p_{\text{other}} = 0.144$. H_a: The random variable does not follow the distribution specified in H_0. $E_{\text{phip}} = 300$, $E_{\text{mm}} = 556$, $E_{\text{other}} = 144$. Since none of the expected frequencies is less than 1 and none of the expected frequencies is less than 5, the conditions for performing the χ^2 goodness of fit test are met. $\chi^2_{\text{crit}} = \chi^2_{0.05} = 5.991$. Reject H_0 if $\chi^2_{\text{data}} \geq 5.991$. $\chi^2_{\text{data}} = 14.224$. Since $\chi^2_{\text{data}} \geq 5.991$, we reject H_0. There is evidence that the population proportions of minority patients who suffered spinal cord injuries, who had a private health insurance provider, Medicare, Medicaid, or other arrangements, have changed.

31. $H_0: p_{\text{pizza}} = 0.25$, $p_{\text{cheeseburger}} = 0.25$, $p_{\text{quiche}} = 0.25$, $p_{\text{sushi}} = 0.25$. H_a: The random variable does not follow the distribution specified in H_0. $E_{\text{pizza}} = 125$, $E_{\text{cheeseburger}} = 125$, $E_{\text{quiche}} = 125$, $E_{\text{sushi}} = 125$. Since none of the expected frequencies is less than 1 and none of the expected frequencies is less than 5, the conditions for performing the χ^2 goodness of fit test are met. $\chi^2_{\text{crit}} = \chi^2_{0.01} = 11.345$. Reject H_0 if $\chi^2_{\text{data}} \geq 11.345$. $\chi^2_{\text{data}} = 377.2$. Since $\chi^2_{\text{data}} \geq 11.345$, we reject H_0. There is evidence that there is a difference in student preference among the four entries.

33. $H_0: p_{\text{sawadv}} = 0.50$, $p_{\text{notseeadv}} = 0.41$, $p_{\text{notrecall}} = 0.09$. H_a: The random variable does not follow the distribution specified in H_0. $E_{\text{sawadv}} = 500$, $E_{\text{notseeadv}} = 410$, $E_{\text{notrecall}} = 90$. Since none of the expected frequencies is less than 1 and none of the expected frequencies is less than 5, the conditions for performing the χ^2 goodness of fit test are met. Reject H_0 if p-value ≤ 0.05. $\chi^2_{\text{data}} = 10.014$. p-value $= 0.0069$. Since p-value ≤ 0.05, we reject H_0. There is evidence that the population proportions have changed since 2007.

35. (a)–(c) Insufficient information

Section 11.2

1. Tabular summary of the relationship between two categorical variables

3. The two-sample Z test for the difference in proportions from Chapter 10 is for comparing proportions of two independent populations, and the χ^2 test for homogeneity of proportions is for comparing proportions of k independent populations.

5.

	A1	A2	Total
B1	11	19	30
B2	11	19	30
Total	22	38	60

7.

	E1	E2	E3	Total
F1	30.71	20.79	8.50	60
F2	34.29	23.21	9.50	67
Total	65	44	18	127

9.

	I1	I2	I3	Total
J1	99.2788	93.6058	102.1154	295
J2	55.5288	52.3558	57.1154	165
J3	20.1923	19.0385	20.7692	60
Total	174.9999	165.0001	180	520

11. (a) H_0: Variable A and Variable B are independent. H_a: Variable A and Variable B are not independent.

(b)

	A1	A2	Total
B1	11	19	30
B2	11	19	30
Total	22	38	60

Since none of the expected frequencies is less than 1 and none of the expected frequencies is less than 5, the conditions for performing the χ^2 test for independence are met. (c) 3.841. Reject H_0 if $\chi^2_{\text{data}} \geq 3.841$. (d) 0.2871 (e) Since χ^2_{data} is not ≥ 3.841, we do not reject H_0. There is insufficient evidence that variable A and variable B are not independent.

13. (a) H_0: Variable I and Variable J are independent. H_a: Variable I and Variable J are not independent.

(b)

	I1	I2	I3	Total
J1	99.2788	93.6058	102.1154	295
J2	55.5288	52.3558	57.1154	165
J3	20.1923	19.0385	20.7692	60
Total	174.9999	165.0001	180	520

Since none of the expected frequencies is less than 1 and none of the expected frequencies is less than 5, the conditions for performing the χ^2 test for independence are met. (c) 13.277. Reject H_0 if $\chi^2_{\text{data}} \geq 13.277$. (d) 4.000 (e) Since χ^2_{data} is not ≥ 13.277, we do not reject H_0. There is insufficient evidence that variable I and variable J are not independent.

15. (a) H_0: Variable C and Variable D are independent. H_a: Variable C and Variable D are not independent. Reject H_0 if the p-value ≤ 0.05.

	C1	C2	Total
D1	55	95	150
D2	55	95	150
Total	110	190	300

Since none of the expected frequencies is less than 1 and none of the expected frequencies is less than 5, the conditions for performing the χ^2 test for independence are met. (b) 1.4354 (c) p-value $= 0.2309$ (d) Since the p-value is not ≤ 0.05, we do not reject H_0. There is insufficient evidence that variable C and variable D are not independent.

17. (a) H_0: Variable K and Variable L are independent. H_a: Variable K and Variable L are not independent. Reject H_0 if p-value ≤ 0.01.

	K1	K2	K3	K4	Total
L1	37.5	72.92	89.58	100	300
L2	23.75	46.18	56.74	63.33	190
L3	28.75	55.90	68.68	76.67	230
Total	90	175	215	240	720

Since none of the expected frequencies is less than 1 and none of the expected frequencies is less than 5, the conditions for performing the χ^2 test for independence are met. **(b)** $\chi^2_{data} =$ 4.906 **(c)** p-value = 0.5560 **(d)** Since p-value is not ≤ 0.01, we do not reject H_0. There is insufficient evidence that variable K and variable L are not independent.

19. (a) $H_0 : p_1 = p_2 = p_3$. H_a : Not all the proportions in H_0 are equal.

(b)

	Sample 1	Sample 2	Sample 3	Total
Successes	9.63	20.86	29.52	60.01
Failures	20.37	44.14	62.48	126.99
Total	30	65	92	187

Since none of the expected frequencies is less than 1 and none of the expected frequencies is less than 5, the conditions for performing the χ^2 test for homogeneity of proportions are met. **(c)** 5.991. Reject H_0 if $\chi^2_{data} \geq 5.991$. **(d)** 0.0846 **(e)** Since χ^2_{data} is not ≥ 5.991, we do not reject H_0. There is insufficient evidence that not all the proportions in H_0 are equal.

21. (a) $H_0 : p_1 = p_2 = p_3 = p_4$. H_a : Not all the proportions in H_0 are equal.

(b)

	Sample 1	Sample 2	Sample 3	Sample 4	Total
Successes	9.67	15.08	20.11	25.14	70
Failures	15.33	23.92	31.89	39.86	111
Total	25	39	52	65	181

Since none of the expected frequencies is less than 1 and none of the expected frequencies is less than 5, the conditions for performing the χ^2 test for homogeneity of proportions are met. **(c)** 7.815. Reject H_0 if $\chi^2_{data} \geq 7.815$. **(d)** 0.0213 **(e)** Since χ^2_{data} is not ≥ 7.815, we do not reject H_0. There is insufficient evidence that not all the proportions in H_0 are equal.

23. (a) $H_0 : p_1 = p_2 = p_3$. H_a : Not all the proportions in H_0 are equal. Reject H_0 if the p-value ≤ 0.05.

	Sample 1	Sample 2	Sample 3	Total
Successes	27.17	57.74	95.09	180
Failures	12.83	27.26	44.91	85
Total	40	85	140	265

Since none of the expected frequencies is less than 1 and none of the expected frequencies is less than 5, the conditions for performing the χ^2 test for homogeneity of proportions are met. **(b)** 2.0442 **(c)** p-value = 0.3598. **(d)** Since the p-value is not ≤ 0.05, we do not reject H_0. There is insufficient evidence that not all the proportions in H_0 are equal.

25. (a) $H_0 : p_1 = p_2 = p_3 = p_4$. H_a : Not all the proportions in H_0 are equal. Reject H_0 if the p-value ≤ 0.05.

	Sample 1	Sample 2	Sample 3	Sample 4	Total
Successes	8.98	12.35	21.88	34.79	78
Failures	7.02	9.65	17.12	27.21	61
Total	16	22	39	62	139

Since none of the expected frequencies is less than 1 and none of the expected frequencies is less than 5, the conditions for performing the χ^2 test for homogeneity of proportions are met. **(b)** 1.264 **(c)** p-value = 0.7377. **(d)** Since the p-value is not ≤ 0.05, we do not reject H_0. There is insufficient evidence that not all the proportions in H_0 are equal.

27. H_0 : *Type of stimulus* and *type of mouse* are independent. H_a : *Type of stimulus* and *type of mouse* are not independent. Reject H_0 if p-value ≤ 0.05. Since none of the expected frequencies is less than 1 and none of the expected frequencies is less than 5, the conditions for performing the χ^2 test for independence are met. $\chi^2_{data} = 0.7937$. p-value = 0.3730. Since p-value is not ≤ 0.05, we do not reject H_0. There is insufficient evidence that *type of stimulus* and *type of mouse* are not independent.

29. H_0 : *Continent of origin* and *state of settlement* are independent. H_a : *Continent of origin* and *state of settlement* are not independent. Since none of the expected frequencies is less than 1 and none of the expected frequencies is less than 5, the conditions for performing the χ^2 test for independence are met. $\chi^2_{crit} = \chi^2_{0.05} = 13.277$. Reject H_0 if $\chi^2_{data} \geq 13.277$. $\chi^2_{data} = 54.874$. Since $\chi^2_{data} \geq 13.277$, we reject H_0. There is evidence that *continent of origin* and *state of settlement* are not independent.

31. No. For the work email group the dark green bar is longer, and for the personal email group the light green bar is longer. This means that no spam is more common for the work email group and that some spam is more common for the personal email group.

33. (a) The type of game with the highest frequency for males is video games and the type of game with the highest frequency for females is computer games. We see some evidence that the most frequently played type of of game depends in part on gender and that the two variables may not be independent. We thus might expect to reject H_0. **(b)** Since the p-value ≤ 0.01, we reject H_0. There is evidence that *gender* and *type of game* are not independent.

35. 478 observations, 11 variables.

37. (a) Dependent **(b)** Since the p-value ≈ 0.001, p-value ≤ 0.10. Thus we reject H_0. There is evidence that *urb_rural* and *goals* are not independent.

Section 11.3

1. No. If the sample sizes are not all the same, then we need to calculate the overall sample mean by calculating the weighted mean of the sample means where the weights are the sample sizes.

3. Answers will vary.

5. Against.

7. (a) $df_1 = 2$, $df_2 = 12$ **(b)** 10 **(c)** 40 **(d)** 12 **(e)** 52

9. (a) $df_1 = 3$, $df_2 = 696$ **(b)** 96.42857143 **(c)** 491,071.4286 **(d)** 18,248 **(e)** 509,319.4286

11. (a) 20 **(b)** 1 **(c)** 20

(d)

Source of variation	Sum of squares	Degrees of freedom	Mean square	F
Treatments	40	2	20	20
Error	12	12	1	
Total	52			

13. (a) 163,690.4762 **(b)** 26.2183908 **(c)** 6243.343

(d)

Source of variation	Sum of squares	Degrees of freedom	Mean square	F
Treatments	491,071.4286	3	163,690.4762	6243.3435
Error	18,248	696	26.2184	
Total	509,319.4286			

15. (a) $H_0 : \mu_1 = \mu_2 = \mu_3$. H_a : Not all the population means are equal. Reject H_0 if the p-value ≤ 0.05. **(b)** 20 **(c)** p-value $= 0.00015$. **(d)** Since the p-value ≤ 0.05, we reject H_0. There is evidence that not all the population means are equal.

17. (a) $H_0 : \mu_A = \mu_B = \mu_C = \mu_D$. H_a : Not all the population means are equal. Reject H_0 if the p-value ≤ 0.05. **(b)** $F_{\text{data}} = 6243.3435$ **(c)** p-value $= 0$ **(d)** Since p-value ≤ 0.05, we reject H_0. There is evidence that not all of the population means are equal.

19. (a) Missing values are in red.

Source of variation	Sum of squares	Degrees of freedom	Mean square	F-test statistic
Treatment	SSTR = 120	$df_1 = 6$	MSTR = 20	$F_{\text{data}} = 44$
Error	SSE = 315	$df_2 = 693$	MSE = 0.4545454545	
Total	SST = 435			

(b) $H_0 : \mu_1 = \mu_2 = \mu_3 = \mu_4 = \mu_5 = \mu_6 = \mu_7$. H_a : Not all the population means are equal. Reject H_0 if the p-value ≤ 0.05. $F_{\text{data}} = 44$; p-value ≈ 0. Since the p-value ≤ 0.05, we reject H_0. There is evidence that not all the population means are equal.

21. (a) Missing values are in red.

Source of variation	Sum of squares	Degrees of freedom	Mean square	F-test statistic
Treatment	SSTR = 40	$df_1 = 4$	MSTR = 10	$F_{\text{data}} = 1.0$
Error	SSE = 400	$df_2 = 40$	MSE = 10	
Total	SST = 440			

(b) $H_0 : \mu_1 = \mu_2 = \mu_3 = \mu_4 = \mu_5$. H_a : Not all the population means are equal. $F_{\text{crit}} = 2.06$. Reject H_0 if $F_{\text{data}} \geq 2.06$. $F_{\text{data}} = 1.0$. Since F_{data} is not ≥ 2.06, we do not reject H_0. There is insufficient evidence that not all the population means are equal.

23. (a) The largest sample standard deviation $S_{\text{online}} = 15.0555$ is not more than twice the smallest sample standard deviation $2 \cdot S_{\text{hybrid}} = 2(12.6491) = 25.2982$. **(b)** (i) $df_1 = 2$, $df_2 = 15$; (ii) $\bar{\bar{x}} = 75.27777778$; (iii) SSTR $= 219.4444444$; (iv) SSE $= 2804.16667$; (v) SST $= 3023.611111$; (vi) MSTR $= 109.7222222$; (vii) MSE $= 186.9444444$; (viii) $F_{\text{data}} = 0.5869$

(c)

Source of variation	Sum of squares	Degrees of freedom	Mean square	F
Treatment	219.4444444	2	109.7222222	0.5869
Error	2804.16667	15	186.9444444	
Total	3023.611111			

(d) $H_0 : \mu_{\text{online}} = \mu_{\text{traditional}} = \mu_{\text{hybrid}}$ vs. H_a : Not all the population means are equal. Reject H_0 if $F_{\text{data}} \geq 6.36$. $F_{\text{data}} = 0.5869$. Since $F_{\text{data}} = 0.5869$ is not ≥ 6.36, we do not reject H_0. There is insufficient evidence that not all the population mean grades of the three classes are equal.

25. (a) The largest sample standard deviation $s_{\text{none}} = 3.22$ is not more than twice the smallest sample standard deviation $2 \cdot s_{\text{Catholic}} = 2(2.74) = 5.48$. **(b)** (i) $df_1 = 4$, $df_2 = 2886$ (ii) $\bar{\bar{x}} = 13.36583535$ (iii) SSTR $= 581.5002576$ (iv) SSE $= 24{,}230.7355$ (v) SST $= 24{,}812.23576$ (vi) MSTR $= 145.3750644$ (vii) MSE $= 8.395958247$ (viii) $F_{\text{data}} = 17.31488654$

(c)

Source of variation	Sum of squares	Degrees of freedom	Mean square	F
Treatment	581.5002576	4	145.3750644	17.31488654
Error	24,230.7355	2886	8.395958247	
Total	24,812.23576			

(d) $H_0 : \mu_{\text{Protestant}} = \mu_{\text{Catholic}} = \mu_{\text{Jewish}} = \mu_{\text{none}} = \mu_{\text{other}}$ versus H_a : Not all the population means are equal. Reject H_0 if $F_{\text{data}} \geq 2.38$. $F_{\text{data}} = 17.31488654$. Since $F_{\text{data}} = 17.31488654$ is ≥ 2.38, we reject H_0. There is evidence that not all the population mean number of years are equal.

27. (a) $H_0 : \mu_{\text{Females}} = \mu_{\text{Males}}$. H_a : The two population means are not equal. $\mu_{\text{Females}} = $ the population mean heart rate for females; $\mu_{\text{Males}} = $ the population mean heart rate for males. Reject H_0 if the p-value ≤ 0.05. 4.896939413, p-value $= 0.0287$. Since the p-value is ≤ 0.05, we reject H_0. There is evidence that the population mean heart rates are not equal. **(b)** Inference for Two Independent Means, Section 10.2.

29. (a) Since the boxplot for the gas mileage of automobiles manufactured in the United States does not overlap the other boxplots, the conclusion might be to reject H_0. **(b)** $H_0 : \mu_{\text{Europe}} = \mu_{\text{Japan}} = \mu_{\text{USA}}$. H_a : Not all of the population means are equal. Reject H_0 if the p-value ≤ 0.01. $F_{\text{data}} = 96.6250761$. p-value $= 8.53843292 \times 10^{-35}$. Since the p-value is ≤ 0.01, we reject H_0. There is evidence that not all the population mean gas mileages are equal. **(c)** Yes

31. (a)–(b) No change **(c)** Increase **(d)** No change **(e)–(f)** Increase **(g)** No change **(h)** Increase **(i)** Decrease **(j)** No change

33. (a, b) Decreases

Section 11.4

1. The regression equation is calculated from a sample and is valid only for values of x in the range of the sample data. The population regression equation may be used to approximate the relationship between the predictor variable x and the response variable y for the entire population of (x, y) pairs.

3. We construct a scatterplot of the residuals against the fitted values and a normal probability plot of the residuals. We must make sure that the scatterplot contains no strong evidence of any unhealthy patterns and that the normal probability plot indicates no evidence of departures from normality in residuals.

5. There is no relationship between x and y.

7. (a) and **(b)**

x	y	Predicted value $\hat{y} = 13.5 + 2.5x$	Residual $(y - \hat{y})$
1	15	16	-1
2	20	18.5	1.5
3	20	21	-1
4	25	23.5	1.5
5	25	26	-1

(c) and **(d)** See Student Solutions Manual. **(e)** The scatterplot of the residuals contains an unhealthy pattern, so the regression assumptions are not verified.

9. (a) and (b)

x	y	Predicted value $\hat{y} = 21.6 + 4x$	Residual $(y - \hat{y})$
−5	0	1.6	−1.6
−4	8	5.6	2.4
−3	8	9.6	−1.6
−2	16	13.6	2.4
−1	16	17.6	−1.6

(c) and (d) See Student Solutions Manual. **(e)** The scatterplot of the residuals contains an unhealthy pattern, so the regression assumptions are not verified.

11. (a) and (b)

x	y	Predicted value $\hat{y} = 104 - 0.5x$	Residual $(y - \hat{y})$
10	100	99	1
20	95	94	1
30	85	89	−4
40	85	84	1
50	80	79	1

(c) and (d) See Student Solutions Manual. **(e)** The scatterplot of the residuals contains an unhealthy pattern, so the regression assumptions are not verified.

13. (a) and (b)

x	y	$\hat{y} = 0.6x + 0.2$	$y - \hat{y}$
1	1	0.8	0.2
2	1	1.4	−0.4
3	2	2	0
4	3	2.6	0.4
5	3	3.2	−0.2

(c)

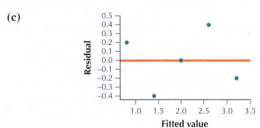

(d)

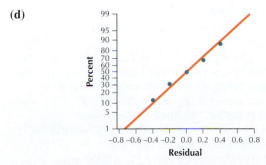

(e) The scatterplot in **(c)** of the residuals versus fitted values shows no strong evidence of unhealthy patterns. Thus, the independence assumption, the constant variance assumption, and the zero-mean assumption are verified. Also, the normal

probability plot of the residuals in **(d)** indicates no evidence of departure from normality of the residuals. Therefore we conclude that the regression assumptions are verified.

15. (a) $t_{\text{crit}} = 3.182$ **(b)** $s = 1.58113883$ **(c)** $\Sigma(x - \bar{x})^2 = 10$ **(d)** $t_{\text{data}} = 5$ **(e)** $H_0 : \beta_1 = 0$: There is no linear relationship between x and y. $H_a : \beta_1 \neq 0$: There is a linear relationship between x and y. Reject H_0 if $t_{\text{data}} \geq 3.182$ or $t_{\text{data}} \leq -3.182$. Since $t_{\text{data}} = 5 \geq 3.182$, we reject H_0. There is evidence at level of significance $\alpha = 0.05$ that $\beta_1 \neq 0$ and that there is a linear relationship between x and y.

17. (a) $t_{\text{crit}} = 3.182$ **(b)** $s = 2.529822128$. **(c)** $\Sigma(x - \bar{x})^2 = 10$ **(d)** $t_{\text{data}} = 5$ **(e)** $H_0 : \beta_1 = 0$: There is no linear relationship between x and y. $H_a : \beta_1 \neq 0$: There is a linear relationship between x and y. Reject H_0 if $t_{\text{data}} \geq 3.182$ or $t_{\text{data}} \leq -3.182$. Since $t_{\text{data}} = 5 \geq 3.182$, we reject H_0. There is evidence at level of significance $\alpha = 0.05$ that $\beta_1 \neq 0$ and that there is a linear relationship between x and y.

19. (a) $s = 2.581988897$ **(b)** $\Sigma(x - \bar{x})^2 = 1000$ **(c)** $t_{\text{data}} = -6.1237$ **(d)** p-value $= 0.0088$ **(e)** $H_0 : \beta_1 = 0$: There is no linear relationship between x and y. $H_a : \beta_1 \neq 0$: There is a linear relationship between x and y. Reject H_0 if p-value ≤ 0.05. Since p-value $= 0.0088 \leq 0.05$, we reject H_0. There is evidence at level of significance $\alpha = 0.05$ that $\beta_1 \neq 0$ and that there is a linear relationship between x and y.

21. (a) $s = 0.3651483717$ **(b)** $\Sigma(x - \bar{x})^2 = 10$ **(c)** $t_{\text{data}} = 5.1962$ **(d)** p-value $= 0.0138$ **(e)** $H_0 : \beta_1 = 0$: There is no linear relationship between x and y. $H_a : \beta_1 \neq 0$: There is a linear relationship between x and y. Reject H_0 if p-value ≤ 0.05. Since p-value $= 0.0138 \leq 0.05$, we reject H_0. There is evidence at level of significance $\alpha = 0.05$ that $\beta_1 \neq 0$ and that there is a linear relationship between x and y.

23. (a) $t_{\alpha/2} = 3.182$ **(b)** $E = 1.591$ **(c)** $(0.909, 4.091)$ **(d)** $H_0 : \beta_1 = 0$: There is no linear relationship between x and y. $H_a : \beta_1 \neq 0$: There is a linear relationship between x and y. Since the confidence interval from **(c)** does not contain zero, we may conclude that $\beta_1 \neq 0$ and that a linear relationship exists between x and y, at level of significance $\alpha = 0.05$.

25. (a) $t_{\alpha/2} = 3.182$ **(b)** $E = 2.5456$ **(c)** $(1.4544, 6.5456)$ **(d)** $H_0 : \beta_1 = 0$: There is no linear relationship between x and y. $H_a : \beta_1 \neq 0$: There is a linear relationship between x and y. Since the confidence interval from **(c)** does not contain zero, we may conclude that $\beta_1 \neq 0$ and that a linear relationship exists between x and y, at level of significance $\alpha = 0.05$.

27. (a) $t_{\alpha/2} = 3.182$ **(b)** $E = 0.2598$ **(c)** $(-0.7598, -0.2402)$ **(d)** $H_0 : \beta_1 = 0$: There is no linear relationship between x and y. $H_a : \beta_1 \neq 0$: There is a linear relationship between x and y. Since the confidence interval from **(c)** does not contain zero, we may conclude that $\beta_1 \neq 0$ and that a linear relationship exists between x and y, at level of significance $\alpha = 0.05$.

29. (a) $t_{\alpha/2} = 3.182$ **(b)** $E = 0.3674$ **(c)** $(0.2326, 0.9674)$. TI-83/84: $(0.2325, 0.9675)$ **(d)** $H_0 : \beta_1 = 0$: There is no linear relationship between x and y. $H_a : \beta_1 \neq 0$: There is a linear relationship between x and y. Since the confidence interval from **(c)** does not contain zero, we may conclude that $\beta_1 \neq 0$ and that a linear relationship exists between x and y, at level of significance $\alpha = 0.05$.

31. (a) See Student Solutions Manual. The scatterplot of the residuals contains an unhealthy pattern, so the regression assumptions are not verified. **(b)** $H_0 : \beta_1 = 0$: There is no

relationship between volume (x) and weight (y). $H_a : \beta_1 \neq 0$: There is a linear relationship between volume (x) and weight (y). Reject H_0 if the p-value ≤ 0.05. Since the p-value ≤ 0.05, we reject H_0. There is evidence for a linear relationship between volume (x) and weight (y).
33. (a) See the Instructor's Guide with Solutions. The scatterplot of the residuals contains no strong evidence of unhealthy patterns and the normal probability plot indicates no evidence of departures from normality in the residuals. Therefore we conclude that the regression assumptions are verified. **(b)** Since $t_{\text{data}} \geq 2.776$, we reject H_0. There is evidence for a linear relationship between *Low* (x) and *High* (y).

35. (a)

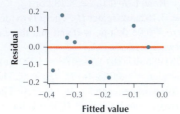

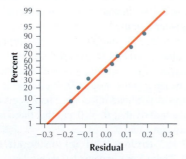

The scatterplot above of the residuals versus fitted values shows no strong evidence of unhealthy patterns. Thus, the independence assumption, the constant variance assumption, and the zero-mean assumption are verified. Also, the normal probability plot of the residuals above indicates no evidence of departure from normality of the residuals. Therefore we conclude that the regression assumptions are verified.
(b) $H_0 : \beta_1 = 0$: There is no linear relationship between price (x) and change (y). $H_a : \beta_1 \neq 0$: There is a linear relationship between price (x) and change (y). Reject H_0 if p-value ≤ 0.05. $t_{\text{data}} = -2.4412$. Since p-value $= 0.0504$, which is not ≤ 0.05, we do not reject H_0. There is insufficient evidence at level of significance $\alpha = 0.05$ that $\beta_1 \neq 0$ and that there is a linear relationship between price (x) and change (y).
37. (a) $E = 0.3182$ **(b)** $(0.1818, 0.8182)$ **(c)** We are 95% confident that the interval $(0.1818, 0.8182)$ captures the population slope β_1 of the relationship between *Family Size* and *Pets*.
39. (a) $E = 0.1393$ **(b)** $(0.0536, 0.3322)$. TI-83/84: $(0.0536, 0.3323)$ **(c)** We are 95% confident that the interval $(0.0536, 0.3322)$ $((0.0536, 0.3323))$ captures the slope β_1 of the regression line. That is, we are 95% confident that, for each additional percent of games won, the increase in the rating of the team lies between 0.0536 and 0.3322 (0.3323).
41. (a) See Student Solutions Manual. The residuals vs. predicted values plot shows a funnel pattern. **(b)** The funnel pattern in the residuals vs. predicted values plot violates the constant variance assumption. **(c)** No, because one of the regression assumptions is violated. **(d)** Yes. It is appropriate to perform the descriptive

statistics we learned in Section 4.2 and 4.3. These are just calculations based on the data. It is inferential statistics that it is not appropriate to perform. We cannot make any inferences using the regression equation if the regression equation is not valid.
43. (a–b) Decrease **(c–d)** Increase **(e)** Depends on the new p-value.
45. No, the regression assumptions are not violated.
47. $(0.1125, 0.7403)$ TI-83/84: $(0.1125, 0.7404)$. We are 90% confident that the interval $(0.1125, 0.7403)$ $((0.1125, 0.7404))$ captures the slope β_1 of the regression line. That is, we are 90% confident that, for each additional point on the SAT Reading score, the increase in the SAT Math score lies between 0.1125 and 0.7403 (0.7404).
49. (a) See the Instructor's Guide with Solutions. The scatterplot of the residuals contains no strong evidence of unhealthy patterns and the normal probability plot indicates no evidence of departures from normality in the residuals. Therefore we conclude that the regression assumptions are verified. **(b)** $(1.0203, 1.9577)$. We are 95% confident that the interval $(1.0203, 1.9577)$ captures the population slope β_1 of the relationship between *Dow Jones Industrial Average* (x) and *pros' performance* (y). **(c)** Since 0 does not lie in the confidence interval, we would expect to reject the null hypothesis that $\beta_1 = 0$. **(d)** $H_0 : \beta_1 = 0$. There is no relationship between *Dow Jones Industrial Average* (x) and *pros' performance* (y). $H_a : \beta_1 \neq 0$. There is a linear relationship between *Dow Jones Industrial Average* (x) and *pros' performance* (y). Reject H_0 if p-value ≤ 0.05. $t_{\text{data}} = 6.31$. p-value ≈ 0. Since the p-value ≤ 0.05, we reject H_0. There is evidence for a linear relationship between *Dow Jones Industrial Average* (x) and *pros' performance* (y).
51. (a) See the Instructor's Guide with Solutions. The scatterplot of the residuals contains no strong evidence of unhealthy patterns and the normal probability plot indicates no evidence of departures from normality in the residuals. Therefore we conclude that the regression assumptions are verified. **(b)** $(0.0087, 0.0439)$. We are 95% confident that the interval $(0.0087, 0.0439)$ captures the population slope β_1 of the relationship between *heart rate* and *body temperature*. **(c)** Since 0 does not lie in the confidence interval, we would expect to reject the null hypothesis that $\beta_1 = 0$. **(d)** $H_0 : \beta_1 = 0$. There is no relationship between *heart rate* (x) and *body temperature* (y). $H_a : \beta_1 \neq 0$. There is a linear relationship between *heart rate* (x) and *body temperatue* (y). Reject H_0 if p-value ≤ 0.05. $t_{\text{data}} = 2.97$. p-value $= 0.004$. Since the p-value ≤ 0.05, we reject H_0. There is evidence for a linear relationship between *heart rate* (x) and *body temperature* (y).

Chapter 11 Review

1. $H_0 : p_{\text{abusedalcohol}} = 0.25, p_{\text{alcoholdependent}} = 0.06, p_{\text{other}} = 0.69$. H_a: The random variable does not follow the distribution specified in H_0. $E_{\text{abusedalcohol}} = 250, E_{\text{alcoholdependent}} = 60, E_{\text{other}} = 690$. Since none of the expected frequencies is less than 1 and none of the expected frequencies is less than 5, the conditions for performing the χ^2 goodness of fit test are met. Reject H_0 if p-value ≤ 0.10. $\chi^2_{\text{data}} = 4.493$. p-value $= 0.1057687682$. Since p-value is not ≤ 0.10, we do not reject H_0. There is insufficient evidence that the population proportions have changed since 2002.
3. $H_0 : p_{18-34} = 0.057, p_{35-49} = 0.207, p_{50-64} = 0.388, p_{\text{over65}} = 0.348$. H_a: The random variable does not follow the distribution specified in H_0. $E_{18-34} = 57, E_{35-49} = 207, E_{50-64} = 388, E_{\text{over65}} = 348$. Since none of the expected frequencies is less than 1 and none of the expected frequencies is less than 5, the conditions

for performing the χ^2 goodness of fit test are met. $\chi^2_{crit} = \chi^2_{0.05} =$
7.815. Reject H_0 if $\chi^2_{data} \geq 7.815$. $\chi^2_{data} = 28.233$. Since $\chi^2_{data} \geq$
7.815, we reject H_0. There is evidence that the proportions have
changed since 2006.

5. H_0: *Happiness in marriage* and *gender* are independent.
H_a: *Happiness in marriage* and *gender* are not independent. Since
none of the expected frequencies is less than 1 and none of the
expected frequencies is less than 5, the conditions for performing
the χ^2 test for independence are met. $\chi^2_{crit} = \chi^2_{0.05} = 5.991$. Reject
H_0 if $\chi^2_{data} \geq 5.991$. $\chi^2_{data} = 3.190$. Since χ^2_{data} is not ≥ 5.991, we
do not reject H_0. There is insufficient evidence that *happiness in
marriage* and *gender* are not independent.

7. H_0: $p_{Whites} = p_{Blacks} = p_{Hispanics}$. H_a: Not all the proportions in H_0
are equal. Reject H_0 if p-value ≤ 0.05. Since none of the expected
frequencies is less than 1 and none of the expected frequencies
is less than 5, the conditions for performing the χ^2 test for
homogeneity of proportions are met. $\chi^2_{data} = 42.658$. p-value $\approx$
0. Since p-value ≤ 0.05, we reject H_0. There is evidence that
Internet use levels is not the same for all races.

9.

Source of variation	Sum of squares	Degrees of freedom	Mean square	F-test statistic
Treatment	SSTR = 10,000	$df_1 = 3$	MSTR = 3333.3333	$F_{data} =$ 852.4117985
Error	SSE = 1157.5	$df_2 = 296$	MSE = 3.910472973	
Total	SST = 11,157.5			

11. H_0: $\mu_A = \mu_B = \mu_C = \mu_D$. H_a: Not all the population means
are equal. $\mu_A =$ the population mean customer satisfaction at
Store A, $\mu_B =$ the population mean customer satisfaction at Store
B, $\mu_C =$ the population mean customer satisfaction at Store C,
and $\mu_D =$ the population mean customer satisfaction at Store D.
Reject H_0 if the p-value < 0.05. $F_{data} = 25.47$. p-value ≈ 0. Since
the p-value ≤ 0.05, we reject H_0. There is evidence that not all the
population means are equal.

Source	df	SS	MS	F	P
Factor	3	7321.4	2440.5	25.47	0.000
Error	24	2300.0	95.8		
Total	27	9621.4			

13. H_0: $\beta_1 = 0$: There is no linear relationship between *High school
GPA* (x) and *First-year college GPA* (y). H_a: $\beta_1 \neq 0$ There is a linear
relationship between *High school GPA* (x) and *First-year college GPA*
(y). Reject H_0 if $t_{data} \geq 2.306$ or $t_{data} \leq -2.306$. Since $t_{data} = 4.5727$
≥ 2.306, we reject H_0. There is evidence at level of significance $\alpha =$
0.05 that $\beta_1 \neq 0$ and that there is a linear relationship between *High
school GPA* (x) and *First-year college GPA* (y).

Chapter 11 Quiz

1. True

2. False

3. False

4. 1, 5

5. equal

6. expected frequency

7. $\bar{\bar{x}}$

8. H_a, the alternative hypothesis

9. Degrees of freedom $= (r - 1)(c - 1)$, where $r =$ the number
of categories in the row variable and $c =$ the number of categories
in the column variable.

10. $E_1 = 48$, $E_2 = 40$, $E_3 = 32$, $E_4 = 24$, $E_5 = 9.6$, $E_6 = 6.4$.
Conditions are met. $\chi^2_{crit} = 11.071$. Reject H_0 if $\chi^2_{data} \geq 11.071$.
$\chi^2_{data} = 2.917$. Since χ^2_{data} is not ≥ 11.071, we do not reject H_0.
There is sufficient evidence that the random variable does not
follow the distribution specified in H_0.

11. $E_1 = 20$, $E_2 = 20$, $E_3 = 20$, $E_4 = 20$, $E_5 = 20$. Conditions are
met. $\chi^2_{crit} = 13.277$. Reject H_a if $\chi^2_{data} \geq 13.277$. $\chi^2_{data} = 0.5$. Since
χ^2_{data} is not ≥ 13.277, we do not reject H_0. There is insufficient
evidence that the random variable does not follow the distribution
specified in H_0.

12. $E_1 = 60$, $E_2 = 50$, $E_3 = 40$, $E_4 = 30$, $E_5 = 12$, $E_6 = 8$.
Conditions are met. $\chi^2_{crit} = 11.071$. Reject H_a if $\chi^2_{data} \geq 11.071$.
$\chi^2_{data} = 5.5$. Since χ^2_{data} is not ≥ 11.071, we do not reject H_0. There
is insufficient evidence that the random variable does not follow
the distribution specified in H_0.

13. (a) The higher the grade level, the higher the proportion
of students who have used an illicit drug. **(b)** H_0: $p_{8th-graders} =$
$p_{10th-graders} = p_{12th-graders}$. H_a: Not all the proportions in H_0 are
equal. Reject H_0 if p-value ≤ 0.01. Since none of the expected
frequencies is less than 1 and none of the expected frequencies
is less than 5, the conditions for performing the χ^2 test for
homogeneity of proportions are met. $\chi^2_{data} = 3060.14226$.
p-value ≈ 0. Since p-value ≤ 0.01, we reject H_0. There is
evidence that the proportions of children in those grades that
have ever used an illicit drug are not all the same.

14. H_0: *Gender* and *sport preference* are independent. H_a:
Gender and *sport preference* are not independent. Reject H_0 if
p-value ≤ 0.05. Since none of the expected frequencies is less
than 1 and none of the expected frequencies is less than 5, the
conditions for performing the χ^2 test for independence are met.
$\chi^2_{data} = 19.857$. p-value $= 0.00004876$. Since p-value ≤ 0.05, we
reject H_0. There is evidence that *gender* and *sport preference* are
not independent.

15.

x	y	Predicted value $\hat{y} = 8.8649x - 454.5946$	Residual $(y - \hat{y})$	(Residual)² $(y - \hat{y})^2$
68	145	148.2186	-3.2186	10.35938596
69	160	157.0835	2.9165	8.50597225
70	165	165.9484	-0.9484	0.89946256
71	180	174.8133	5.1867	26.90185689
72	180	183.6782	-3.6782	13.52915524
75	210	210.2729	-0.2729	0.07447441

(6.9243, 10.8055). We are 95% confident that the interval (6.9243,
10.8055) captures the population slope β_1 of the relationship
between *weight* and *height*.

16.

x	y	xy	x^2	y^2
80	90	7200	6400	8100
50	75	3750	2500	5625
90	80	7200	8100	6400
75	80	6000	5625	6400
50	60	3000	2500	3600
95	90	8550	9025	8100
60	55	3300	3600	3025
75	70	5250	5625	4900
$\sum x = 575$	$\sum y = 600$	$\sum xy = 44{,}250$	$\sum x^2 = 43{,}375$	$\sum y^2 = 46{,}150$

x	y	Predicted value $\hat{y} = 0.5496x + 35.4962$	Residual $(y - \hat{y})$	(Residual)^2 $(y - \hat{y})^2$
80	90	79.4642	10.5358	111.00308164
50	75	62.9762	12.0238	144.57176644
90	80	84.9602	−4.9602	24.60358404
75	80	76.7162	3.2838	10.78334244
50	60	62.9762	−2.9762	8.85776644
95	90	87.7082	2.2918	5.25234724
60	55	68.4722	−13.4722	181.50017284
75	70	76.7162	−6.7162	45.10734244

(0.2074, 0.8918). We are 95% confident that the interval (0.2074, 0.8918) captures the population slope β_1 of the relationship between *first-semester grade* and *second-semester grade*.

17. H_0: $\beta_1 = 0$. There is no linear relationship between height (x) and weight (y). H_1: $\beta_1 \neq 0$. There is a linear relationship between *height* (x) and *weight* (y). Reject H_0 if p-value ≤ 0.05. $t_{data} \approx 12.68$. df $= n - 2 = 6 - 2 = 4$. p-value $= 0$. Since p-value is ≤ 0.05, we reject H_0. There is evidence for a linear relationship between *height* (x) and *weight* (y).

18. H_0: $\beta_1 = 0$. There is no linear relationship between *first-semester grade* (x) and *second-semester grade* (y). H_1: $\beta_1 \neq 0$. There is a linear relationship between *first-semester grade (x)* and *second-semester grade (y)*. Reject H_0 if p-value ≤ 0.05. $t_{data} \approx 3.70$. df $= n - 2 = 8 - 2 = 6$. p-value $= 0.0100883789$. Since p-value is ≤ 0.05, we reject H_0. There is evidence for a linear relationship between *first-semester grade (x)* and *second-semester grade (y)*.

TABLES APPENDIX

Table A Random numbers

10480	15011	01536	02011	81647	91646	67179	14194	62590	36207	20969	99570	91291	90700
22368	46573	25595	85393	30995	89198	27982	53402	93965	34095	52666	19174	39615	99505
24130	48360	22527	97265	76393	64809	15179	24830	49340	32081	30680	19655	63348	58629
42167	93093	06243	61680	07856	16376	39440	53537	71341	57004	00849	74917	97758	16379
37570	39975	81837	16656	06121	91782	60468	81305	49684	60672	14110	06927	01263	54613
77921	06907	11008	42751	27756	53498	18602	70659	90655	15053	21916	81825	44394	42880
99562	72905	56420	69994	98872	31016	71194	18738	44013	48840	63213	21069	10634	12952
96301	91977	05463	07972	18876	20922	94595	56869	69014	60045	18425	84903	42508	32307
89579	14342	63661	10281	17453	18103	57740	84378	25331	12566	58678	44947	05584	56941
85475	36857	43342	53988	53060	59533	38867	62300	08158	17983	16439	11458	18593	64952
28918	69578	88231	33276	70997	79936	56865	05859	90106	31595	01547	85590	91610	78188
63553	40961	48235	03427	49626	69445	18663	72695	52180	20847	12234	90511	33703	90322
09429	93969	52636	92737	88974	33488	36320	17617	30015	08272	84115	27156	30613	74952
10365	61129	87529	85689	48237	52267	67689	93394	01511	26358	85104	20285	29975	89868
07119	97336	71048	08178	77233	13916	47564	81056	97735	85977	29372	74461	28551	90707
51085	12765	51821	51259	77452	16308	60756	92144	49442	53900	70960	63990	75601	40719
02368	21382	52404	60268	89368	19885	55322	44819	01188	65255	64835	44919	05944	55157
01011	54092	33362	94904	31273	04146	18594	29852	71585	85030	51132	01915	92747	64951
52162	53916	46369	58586	23216	14513	83149	98736	23495	64350	94738	17752	35156	35749
07056	97628	33787	09998	42698	06691	76988	13602	51851	46104	88916	19509	25625	58104
48663	91245	85828	14346	09172	30168	90229	04734	59193	22178	30421	61666	99904	32812
54164	58492	22421	74103	47070	25306	76468	26384	58151	06646	21524	15227	96909	44592
32639	32363	05597	24200	13363	38005	94342	28728	35806	06912	17012	64161	18296	22851
29334	27001	87637	87308	58731	00256	45834	15398	46557	41135	10367	07684	36188	18510
02488	33062	28834	07351	19731	92420	60952	61280	50001	67658	32586	86679	50720	94953
81525	72295	04839	96423	24878	82651	66566	14778	76797	14780	13300	87074	79666	95725
29676	20591	68086	26432	46901	20849	89768	81536	86645	12659	92259	57102	80428	25280
00742	57392	39064	66432	84673	40027	32832	61362	98947	96067	64760	64584	96096	98253
05366	04213	25669	26422	44407	44048	37937	63904	45766	66134	75470	66520	34693	90449
91921	26418	64117	94305	26766	25940	39972	22209	71500	64568	91402	42416	07844	69618
00582	04711	87917	77341	42206	35126	74087	99547	81817	42607	43808	76655	62028	76630
00725	69884	62797	56170	86324	88072	76222	36086	84637	93161	76038	65855	77919	88006
69011	65797	95876	55293	18988	27354	26575	08625	40801	59920	29841	80150	12777	48501
25976	57948	29888	88604	67917	48708	18912	82271	65424	69774	33611	54262	85963	03547
09763	83473	73577	12908	30883	18317	28290	35797	05998	41688	34952	37888	38917	88050
91567	42595	27958	30134	04024	86385	29880	99730	55536	84855	29080	09250	79656	73211
17955	56349	90999	49127	20044	59931	06115	20542	18059	02008	73708	83517	36103	42791
46503	18584	18845	49618	02304	51038	20655	58727	28168	15475	56942	53389	20562	87338
92157	89634	94824	78171	84610	82834	09922	25417	44137	48413	25555	21246	35509	20468
14577	62765	35605	81263	39667	47358	56873	56307	61607	49518	89656	20103	77490	18062
98427	07523	33362	64270	01638	92477	66969	98420	04880	45585	46565	04102	46880	45709
34914	63976	88720	82765	34476	17032	87589	40836	32427	70002	70663	88863	77775	69348
70060	28277	39475	46473	23219	53416	94970	25832	69975	94884	19661	72828	00102	66794
53976	54914	06990	67245	68350	82948	11398	42878	80287	88267	47363	46634	06541	97809
76072	29515	40980	07391	58745	25774	22987	80059	39911	96189	41151	14222	60697	59583
90725	52210	83974	29992	65831	38857	50490	83765	55657	14361	31720	57375	56228	41546
64364	67412	33339	31926	14883	24413	59744	92351	97473	89286	35931	04110	23726	51900
08962	00358	31662	25388	61642	34072	81249	35648	56891	69352	48373	45578	78547	81788
95012	68379	93526	70765	10593	04542	76463	54328	02349	17247	28865	14777	62730	92277
15664	10493	20492	38391	91132	21999	59516	81652	27195	48223	46751	22923	32261	85653

Reprinted with permission from W. H. Beyer, *Handbook of Tables for Probability and Statistics*, 2nd ed. Copyright CRC Press, Boca Raton, Fla., 1986.

Table B Binomial distribution

						p				
n	X	0.10	0.15	0.20	0.25	0.30	0.35	0.40	0.45	0.50
2	0	0.8100	0.7225	0.6400	0.5625	0.4900	0.4225	0.3600	0.3025	0.2500
	1	0.1800	0.2550	0.3200	0.3750	0.4200	0.4550	0.4800	0.4950	0.5000
	2	0.0100	0.0225	0.0400	0.0625	0.0900	0.1225	0.1600	0.2025	0.2500
3	0	0.7290	0.6141	0.5120	0.4219	0.3430	0.2746	0.2160	0.1664	0.1250
	1	0.2430	0.3251	0.3840	0.4219	0.4410	0.4436	0.4320	0.4084	0.3750
	2	0.0270	0.0574	0.0960	0.1406	0.1890	0.2389	0.2880	0.3341	0.3750
	3	0.0010	0.0034	0.0080	0.0156	0.0270	0.0429	0.0640	0.0911	0.1250
4	0	0.6561	0.5220	0.4096	0.3164	0.2401	0.1785	0.1296	0.0915	0.0625
	1	0.2916	0.3685	0.4096	0.4219	0.4116	0.3845	0.3456	0.2995	0.2500
	2	0.0486	0.0975	0.1536	0.2109	0.2646	0.3105	0.3456	0.3675	0.3750
	3	0.0036	0.0115	0.0256	0.0469	0.0756	0.1115	0.1536	0.2005	0.2500
	4	0.0001	0.0005	0.0016	0.0039	0.0081	0.0150	0.0256	0.0410	0.0625
5	0	0.5905	0.4437	0.3277	0.2373	0.1681	0.1160	0.0778	0.0503	0.0312
	1	0.3280	0.3915	0.4096	0.3955	0.3602	0.3124	0.2592	0.2059	0.1562
	2	0.0729	0.1382	0.2048	0.2637	0.3087	0.3364	0.3456	0.3369	0.3125
	3	0.0081	0.0244	0.0512	0.0879	0.1323	0.1811	0.2304	0.2757	0.3125
	4	0.0004	0.0022	0.0064	0.0146	0.0284	0.0488	0.0768	0.1128	0.1562
	5		0.0001	0.0003	0.0010	0.0024	0.0053	0.0102	0.0185	0.0312
6	0	0.5314	0.3771	0.2621	0.1780	0.1176	0.0754	0.0467	0.0277	0.0156
	1	0.3543	0.3993	0.3932	0.3560	0.3025	0.2437	0.1866	0.1359	0.0938
	2	0.0984	0.1762	0.2458	0.2966	0.3241	0.3280	0.3110	0.2780	0.2344
	3	0.0146	0.0415	0.0819	0.1318	0.1852	0.2355	0.2765	0.3032	0.3125
	4	0.0012	0.0055	0.0154	0.0330	0.0595	0.0951	0.1382	0.1861	0.2344
	5	0.0001	0.0004	0.0015	0.0044	0.0102	0.0205	0.0369	0.0609	0.0938
	6			0.0001	0.0002	0.0007	0.0018	0.0041	0.0083	0.0156
7	0	0.4783	0.3206	0.2097	0.1335	0.0824	0.0490	0.0280	0.0152	0.0078
	1	0.3720	0.3960	0.3670	0.3115	0.2471	0.1848	0.1306	0.0872	0.0547
	2	0.1240	0.2097	0.2753	0.3115	0.3177	0.2985	0.2613	0.2140	0.1641
	3	0.0230	0.0617	0.1147	0.1730	0.2269	0.2679	0.2903	0.2918	0.2734
	4	0.0026	0.0109	0.0287	0.0577	0.0972	0.1442	0.1935	0.2388	0.2734
	5	0.0002	0.0012	0.0043	0.0115	0.0250	0.0466	0.0774	0.1172	0.1641
	6		0.0001	0.0004	0.0013	0.0036	0.0084	0.0172	0.0320	0.0547
	7				0.0001	0.0002	0.0006	0.0016	0.0037	0.0078
8	0	0.4305	0.2725	0.1678	0.1001	0.0576	0.0319	0.0168	0.0084	0.0039
	1	0.3826	0.3847	0.3355	0.2670	0.1977	0.1373	0.0896	0.0548	0.0312
	2	0.1488	0.2376	0.2936	0.3115	0.2965	0.2587	0.2090	0.1569	0.1094
	3	0.0331	0.0839	0.1468	0.2076	0.2541	0.2786	0.2787	0.2568	0.2188
	4	0.0046	0.0185	0.0459	0.0865	0.1361	0.1875	0.2322	0.2627	0.2734
	5	0.0004	0.0026	0.0092	0.0231	0.0467	0.0808	0.1239	0.1719	0.2188
	6		0.0002	0.0011	0.0038	0.0100	0.0217	0.0413	0.0703	0.1094
	7			0.0001	0.0004	0.0012	0.0033	0.0079	0.0164	0.0313
	8					0.0001	0.0002	0.0007	0.0017	0.0039

Note: Blank entries indicate a binomial probability of less than 0.00005.

(Continued)

Table B Binomial distribution (*continued*)

n	X	0.10	0.15	0.20	0.25	0.30	0.35	0.40	0.45	0.50
						p				
9	0	0.3874	0.2316	0.1342	0.0751	0.0404	0.0207	0.0101	0.0046	0.0020
	1	0.3874	0.3679	0.3020	0.2253	0.1556	0.1004	0.0605	0.0339	0.0176
	2	0.1722	0.2597	0.3020	0.3003	0.2668	0.2162	0.1612	0.1110	0.0703
	3	0.0446	0.1069	0.1762	0.2336	0.2668	0.2716	0.2508	0.2119	0.1641
	4	0.0074	0.0283	0.0661	0.1168	0.1715	0.2194	0.2508	0.2600	0.2461
	5	0.0008	0.0050	0.0165	0.0389	0.0735	0.1181	0.1672	0.2128	0.2461
	6	0.0001	0.0006	0.0028	0.0087	0.0210	0.0424	0.0743	0.1160	0.1641
	7			0.0003	0.0012	0.0039	0.0098	0.0212	0.0407	0.0703
	8				0.0001	0.0004	0.0013	0.0035	0.0083	0.0176
	9						0.0001	0.0003	0.0008	0.0020
10	0	0.3487	0.1969	0.1074	0.0563	0.0282	0.0135	0.0060	0.0025	0.0010
	1	0.3874	0.3474	0.2684	0.1877	0.1211	0.0725	0.0403	0.0207	0.0098
	2	0.1937	0.2759	0.3020	0.2816	0.2335	0.1757	0.1209	0.0763	0.0439
	3	0.0574	0.1298	0.2013	0.2503	0.2668	0.2522	0.2150	0.1665	0.1172
	4	0.0112	0.0401	0.0881	0.1460	0.2001	0.2377	0.2508	0.2384	0.2051
	5	0.0015	0.0085	0.0264	0.0584	0.1029	0.1536	0.2007	0.2340	0.2461
	6	0.0001	0.0012	0.0055	0.0162	0.0368	0.0689	0.1115	0.1596	0.2051
	7		0.0001	0.0008	0.0031	0.0090	0.0212	0.0425	0.0746	0.1172
	8			0.0001	0.0004	0.0014	0.0043	0.0106	0.0229	0.0439
	9					0.0001	0.0005	0.0016	0.0042	0.0098
	10							0.0001	0.0003	0.0010
12	0	0.2824	0.1422	0.0687	0.0317	0.0138	0.0057	0.0022	0.0008	0.0002
	1	0.3766	0.3012	0.2062	0.1267	0.0712	0.0368	0.0174	0.0075	0.0029
	2	0.2301	0.2924	0.2835	0.2323	0.1678	0.1088	0.0639	0.0339	0.0161
	3	0.0853	0.1720	0.2362	0.2581	0.2397	0.1954	0.1419	0.0923	0.0537
	4	0.0213	0.0683	0.1329	0.1936	0.2311	0.2367	0.2128	0.1700	0.1208
	5	0.0038	0.0193	0.0532	0.1032	0.1585	0.2039	0.2270	0.2225	0.1934
	6	0.0005	0.0040	0.0155	0.0401	0.0792	0.1281	0.1766	0.2124	0.2256
	7		0.0006	0.0033	0.0115	0.0291	0.0591	0.1009	0.1489	0.1934
	8		0.0001	0.0005	0.0024	0.0078	0.0199	0.0420	0.0762	0.1208
	9			0.0001	0.0004	0.0015	0.0048	0.0125	0.0277	0.0537
	10					0.0002	0.0008	0.0025	0.0068	0.0161
	11						0.0001	0.0003	0.0010	0.0029
	12								0.0001	0.0002
15	0	0.2059	0.0874	0.0352	0.0134	0.0047	0.0016	0.0005	0.0001	
	1	0.3432	0.2312	0.1319	0.0668	0.0305	0.0126	0.0047	0.0016	0.0005
	2	0.2669	0.2856	0.2309	0.1559	0.0916	0.0476	0.0219	0.0090	0.0032
	3	0.1285	0.2184	0.2501	0.2252	0.1700	0.1110	0.0634	0.0318	0.0139
	4	0.0428	0.1156	0.1876	0.2252	0.2186	0.1792	0.1268	0.0780	0.0417
	5	0.0105	0.0449	0.1032	0.1651	0.2061	0.2123	0.1859	0.1404	0.0916
	6	0.0019	0.0132	0.0430	0.0917	0.1472	0.1906	0.2066	0.1914	0.1527
	7	0.0003	0.0030	0.0138	0.0393	0.0811	0.1319	0.1771	0.2013	0.1964
	8		0.0005	0.0035	0.0131	0.0348	0.0710	0.1181	0.1647	0.1964
	9		0.0001	0.0007	0.0034	0.0116	0.0298	0.0612	0.1048	0.1527
	10			0.0001	0.0007	0.0030	0.0096	0.0245	0.0515	0.0916
	11				0.0001	0.0006	0.0024	0.0074	0.0191	0.0417
	12					0.0001	0.0004	0.0016	0.0052	0.0139
	13						0.0001	0.0003	0.0010	0.0032
	14								0.0001	0.0005
	15									

Note: Blank entries indicate a binomial probability of less than 0.00005.

Table B Binomial distribution (*continued*)

						p					
n	*X*	0.10	0.15	0.20	0.25	0.30	0.35	0.40	0.45	0.50	
18	0	0.1501	0.0536	0.0180	0.0056	0.0016	0.0004	0.0001			
	1	0.3002	0.1704	0.0811	0.0338	0.0126	0.0042	0.0012	0.0003	0.0001	
	2	0.2835	0.2556	0.1723	0.0958	0.0458	0.0190	0.0069	0.0022	0.0006	
	3	0.1680	0.2406	0.2297	0.1704	0.1046	0.0547	0.0246	0.0095	0.0031	
	4	0.0700	0.1592	0.2153	0.2130	0.1681	0.1104	0.0614	0.0291	0.0117	
	5	0.0218	0.0787	0.1507	0.1988	0.2017	0.1664	0.1146	0.0666	0.0327	
	6	0.0052	0.0301	0.0816	0.1436	0.1873	0.1941	0.1655	0.1181	0.0708	
	7	0.0010	0.0091	0.0350	0.0820	0.1376	0.1792	0.1892	0.1657	0.1214	
	8	0.0002	0.0022	0.0120	0.0376	0.0811	0.1327	0.1734	0.1864	0.1669	
	9		0.0004	0.0033	0.0139	0.0386	0.0794	0.1284	0.1694	0.1855	
	10		0.0001	0.0008	0.0042	0.0149	0.0385	0.0771	0.1248	0.1669	
	11			0.0001	0.0010	0.0046	0.0151	0.0374	0.0742	0.1214	
	12				0.0002	0.0012	0.0047	0.0145	0.0354	0.0708	
	13					0.0002	0.0012	0.0045	0.0134	0.0327	
	14						0.0002	0.0011	0.0039	0.0117	
	15							0.0002	0.0009	0.0031	
	16								0.0001	0.0006	
	17									0.0001	
	18										
20	0	0.1216	0.0388	0.0115	0.0032	0.0008	0.0002				
	1	0.2702	0.1368	0.0576	0.0211	0.0068	0.0020	0.0005	0.0001		
	2	0.2852	0.2293	0.1369	0.0669	0.0278	0.0100	0.0031	0.0008	0.0002	
	3	0.1901	0.2428	0.2054	0.1339	0.0716	0.0323	0.0123	0.0040	0.0011	
	4	0.0898	0.1821	0.2182	0.1897	0.1304	0.0738	0.0350	0.0139	0.0046	
	5	0.0319	0.1028	0.1746	0.2023	0.1789	0.1272	0.0746	0.0365	0.0148	
	6	0.0089	0.0454	0.1091	0.1686	0.1916	0.1712	0.1244	0.0746	0.0370	
	7	0.0020	0.0160	0.0545	0.1124	0.1643	0.1844	0.1659	0.1221	0.0739	
	8	0.0004	0.0046	0.0222	0.0609	0.1144	0.1614	0.1797	0.1623	0.1201	
	9	0.0001	0.0011	0.0074	0.0271	0.0654	0.1158	0.1597	0.1771	0.1602	
	10		0.0002	0.0020	0.0099	0.0308	0.0686	0.1171	0.1593	0.1762	
	11			0.0005	0.0030	0.0120	0.0336	0.0710	0.1185	0.1602	
	12			0.0001	0.0008	0.0039	0.0136	0.0355	0.0727	0.1201	
	13				0.0002	0.0010	0.0045	0.0146	0.0366	0.0739	
	14					0.0002	0.0012	0.0049	0.0150	0.0370	
	15						0.0003	0.0013	0.0049	0.0148	
	16							0.0003	0.0013	0.0046	
	17								0.0002	0.0011	
	18									0.0002	
	19										
	20										

Note: Blank entries indicate a binomial probability of less than 0.00005.

(Continued)

Table B Binomial distribution (*continued*)

n	X	0.55	0.60	0.65	0.70	0.75	0.80	0.85	0.90	0.95
2	0	0.2025	0.1600	0.1225	0.0900	0.0625	0.0400	0.0225	0.0100	0.0025
	1	0.4950	0.4800	0.4550	0.4200	0.3750	0.3200	0.2550	0.1800	0.0950
	2	0.3025	0.3600	0.4225	0.4900	0.5625	0.6400	0.7225	0.8100	0.9025
3	0	0.0911	0.0640	0.0429	0.0270	0.0156	0.0080	0.0034	0.0010	0.0001
	1	0.3341	0.2880	0.2389	0.1890	0.1406	0.0960	0.0574	0.0270	0.0071
	2	0.4084	0.4320	0.4436	0.4410	0.4219	0.3840	0.3251	0.2430	0.1354
	3	0.1664	0.2160	0.2746	0.3430	0.4219	0.5120	0.6141	0.7290	0.8574
4	0	0.0410	0.0256	0.0150	0.0081	0.0039	0.0016	0.0005	0.0001	
	1	0.2005	0.1536	0.1115	0.0756	0.0469	0.0256	0.0115	0.0036	0.0005
	2	0.3675	0.3456	0.3105	0.2646	0.2109	0.1536	0.0975	0.0486	0.0135
	3	0.2995	0.3456	0.3845	0.4116	0.4219	0.4096	0.3685	0.2916	0.1715
	4	0.0915	0.1296	0.1785	0.2401	0.3164	0.4096	0.5220	0.6561	0.8145
5	0	0.0185	0.0102	0.0053	0.0024	0.0010	0.0003	0.0001		
	1	0.1128	0.0768	0.0488	0.0284	0.0146	0.0064	0.0022	0.0005	
	2	0.2757	0.2304	0.1811	0.1323	0.0879	0.0512	0.0244	0.0081	0.0011
	3	0.3369	0.3456	0.3364	0.3087	0.2637	0.2048	0.1382	0.0729	0.0214
	4	0.2059	0.2592	0.3124	0.3601	0.3955	0.4096	0.3915	0.3281	0.2036
	5	0.0503	0.0778	0.1160	0.1681	0.2373	0.3277	0.4437	0.5905	0.7738
6	0	0.0083	0.0041	0.0018	0.0007	0.0002	0.0001			
	1	0.0609	0.0369	0.0205	0.0102	0.0044	0.0015	0.0004	0.0001	
	2	0.1861	0.1382	0.0951	0.0595	0.0330	0.0154	0.0055	0.0012	0.0001
	3	0.3032	0.2765	0.2355	0.1852	0.1318	0.0819	0.0415	0.0146	0.0021
	4	0.2780	0.3110	0.3280	0.3241	0.2966	0.2458	0.1762	0.0984	0.0305
	5	0.1359	0.1866	0.2437	0.3025	0.3560	0.3932	0.3993	0.3543	0.2321
	6	0.0277	0.0467	0.0754	0.1176	0.1780	0.2621	0.3771	0.5314	0.7351
7	0	0.0037	0.0016	0.0006	0.0002	0.0001				
	1	0.0320	0.0172	0.0084	0.0036	0.0013	0.0004	0.0001		
	2	0.1172	0.0774	0.0466	0.0250	0.0115	0.0043	0.0012	0.0002	
	3	0.2388	0.1935	0.1442	0.0972	0.0577	0.0287	0.0109	0.0026	0.0002
	4	0.2918	0.2903	0.2679	0.2269	0.1730	0.1147	0.0617	0.0230	0.0036
	5	0.2140	0.2613	0.2985	0.3177	0.3115	0.2753	0.2097	0.1240	0.0406
	6	0.0872	0.1306	0.1848	0.2471	0.3115	0.3670	0.3960	0.3720	0.2573
	7	0.0152	0.0280	0.0490	0.0824	0.1335	0.2097	0.3206	0.4783	0.6983
8	0	0.0017	0.0007	0.0002	0.0001					
	1	0.0164	0.0079	0.0033	0.0012	0.0004	0.0001			
	2	0.0703	0.0413	0.0217	0.0100	0.0038	0.0011	0.0002		
	3	0.1719	0.1239	0.0808	0.0467	0.0231	0.0092	0.0026	0.0004	
	4	0.2627	0.2322	0.1875	0.1361	0.0865	0.0459	0.0185	0.0046	0.0004
	5	0.2568	0.2787	0.2786	0.2541	0.2076	0.1468	0.0839	0.0331	0.0054
	6	0.1569	0.2090	0.2587	0.2965	0.3115	0.2936	0.2376	0.1488	0.0515
	7	0.0548	0.0896	0.1373	0.1977	0.2670	0.3355	0.3847	0.3826	0.2793
	8	0.0084	0.0168	0.0319	0.0576	0.1001	0.1678	0.2725	0.4305	0.6634

Note: Blank entries indicate a binomial probability of less than 0.00005.

Table B Binomial distribution (*continued*)

n	X	0.55	0.60	0.65	0.70	0.75	0.80	0.85	0.90	0.95
9	0	0.0008	0.0003	0.0001						
	1	0.0083	0.0035	0.0013	0.0004	0.0001				
	2	0.0407	0.0212	0.0098	0.0039	0.0012	0.0003			
	3	0.1160	0.0743	0.0424	0.0210	0.0087	0.0028	0.0006	0.0001	
	4	0.2128	0.1672	0.1181	0.0735	0.0389	0.0165	0.0050	0.0008	
	5	0.2600	0.2508	0.2194	0.1715	0.1168	0.0661	0.0283	0.0074	0.0006
	6	0.2119	0.2508	0.2716	0.2668	0.2336	0.1762	0.1069	0.0446	0.0077
	7	0.1110	0.1612	0.2162	0.2668	0.3003	0.3020	0.2597	0.1722	0.0629
	8	0.0339	0.0605	0.1004	0.1556	0.2253	0.3020	0.3679	0.3874	0.2985
	9	0.0046	0.0101	0.0207	0.0404	0.0751	0.1342	0.2316	0.3874	0.6302
10	0	0.0003	0.0001							
	1	0.0042	0.0016	0.0005	0.0001					
	2	0.0229	0.0106	0.0043	0.0014	0.0004	0.0001			
	3	0.0746	0.0425	0.0212	0.0090	0.0031	0.0008	0.0001		
	4	0.1596	0.1115	0.0689	0.0368	0.0162	0.0055	0.0012	0.0001	
	5	0.2340	0.2007	0.1536	0.1029	0.0584	0.0264	0.0085	0.0015	0.0001
	6	0.2384	0.2508	0.2377	0.2001	0.1460	0.0881	0.0401	0.0112	0.0010
	7	0.1665	0.2150	0.2522	0.2668	0.2503	0.2013	0.1298	0.0574	0.0105
	8	0.0763	0.1209	0.1757	0.2335	0.2816	0.3020	0.2759	0.1937	0.0746
	9	0.0207	0.0403	0.0725	0.1211	0.1877	0.2684	0.3474	0.3874	0.3151
	10	0.0025	0.0060	0.0135	0.0282	0.0563	0.1074	0.1969	0.3487	0.5987
12	0	0.0001								
	1	0.0010	0.0003	0.0001						
	2	0.0068	0.0025	0.0008	0.0002					
	3	0.0277	0.0125	0.0048	0.0015	0.0004	0.0001			
	4	0.0762	0.0420	0.0199	0.0078	0.0024	0.0005	0.0001		
	5	0.1489	0.1009	0.0591	0.0291	0.0115	0.0033	0.0006		
	6	0.2124	0.1766	0.1281	0.0792	0.0401	0.0155	0.0040	0.0005	
	7	0.2225	0.2270	0.2039	0.1585	0.1032	0.0532	0.0193	0.0038	0.0002
	8	0.1700	0.2128	0.2367	0.2311	0.1936	0.1329	0.0683	0.0213	0.0021
	9	0.0923	0.1419	0.1954	0.2397	0.2581	0.2362	0.1720	0.0852	0.0173
	10	0.0339	0.0639	0.1088	0.1678	0.2323	0.2835	0.2924	0.2301	0.0988
	11	0.0075	0.0174	0.0368	0.0712	0.1267	0.2062	0.3012	0.3766	0.3413
	12	0.0008	0.0022	0.0057	0.0138	0.0317	0.0687	0.1422	0.2824	0.5404
15	0									
	1	0.0001								
	2	0.0010	0.0003	0.0001						
	3	0.0052	0.0016	0.0004	0.0001					
	4	0.0191	0.0074	0.0024	0.0006	0.0001				
	5	0.0515	0.0245	0.0096	0.0030	0.0007	0.0001			
	6	0.1048	0.0612	0.0298	0.0116	0.0034	0.0007	0.0001		
	7	0.1647	0.1181	0.0710	0.0348	0.0131	0.0035	0.0005		
	8	0.2013	0.1771	0.1319	0.0811	0.0393	0.0138	0.0030	0.0003	
	9	0.1914	0.2066	0.1906	0.1472	0.0917	0.0430	0.0132	0.0019	
	10	0.1404	0.1859	0.2123	0.2061	0.1651	0.1032	0.0449	0.0105	0.0006
	11	0.0780	0.1268	0.1792	0.2186	0.2252	0.1876	0.1156	0.0428	0.0049

Note: Blank entries indicate a binomial probability of less than 0.00005.

(Continued)

Table B Binomial distribution (*continued*)

n	X	0.55	0.60	0.65	0.70	0.75	0.80	0.85	0.90	0.95
						p				
	12	0.0318	0.0634	0.1110	0.1700	0.2252	0.2501	0.2184	0.1285	0.0307
	13	0.0090	0.0219	0.0476	0.0916	0.1559	0.2309	0.2856	0.2669	0.1348
	14	0.0016	0.0047	0.0126	0.0305	0.0668	0.1319	0.2312	0.3432	0.3658
	15	0.0001	0.0005	0.0016	0.0047	0.0134	0.0352	0.0874	0.2059	0.4633
18	0									
	1									
	2	0.0001								
	3	0.0009	0.0002							
	4	0.0039	0.0011	0.0002						
	5	0.0134	0.0045	0.0012	0.0002					
	6	0.0354	0.0145	0.0047	0.0012	0.0002				
	7	0.0742	0.0374	0.0151	0.0046	0.0010	0.0001			
	8	0.1248	0.0771	0.0385	0.0149	0.0042	0.0008	0.0001		
	9	0.1694	0.1284	0.0794	0.0386	0.0139	0.0033	0.0004		
	10	0.1864	0.1734	0.1327	0.0811	0.0376	0.0120	0.0022	0.0002	
	11	0.1657	0.1892	0.1792	0.1376	0.0820	0.0350	0.0091	0.0010	
	12	0.1181	0.1655	0.1941	0.1873	0.1436	0.0816	0.0301	0.0052	0.0002
	13	0.0666	0.1146	0.1664	0.2017	0.1988	0.1507	0.0787	0.0218	0.0014
	14	0.0291	0.0614	0.1104	0.1681	0.2130	0.2153	0.1592	0.0700	0.0093
	15	0.0095	0.0246	0.0547	0.1046	0.1704	0.2297	0.2406	0.1680	0.0473
	16	0.0022	0.0069	0.0190	0.0458	0.0958	0.1723	0.2556	0.2835	0.1683
	17	0.0003	0.0012	0.0042	0.0126	0.0338	0.0811	0.1704	0.3002	0.3763
	18		0.0001	0.0004	0.0016	0.0056	0.0180	0.0536	0.1501	0.3972
20	0									
	1									
	2									
	3	0.0002								
	4	0.0013	0.0003							
	5	0.0049	0.0013	0.0003						
	6	0.0150	0.0049	0.0012	0.0002					
	7	0.0366	0.0146	0.0045	0.0010	0.0002				
	8	0.0727	0.0355	0.0136	0.0039	0.0008	0.0001			
	9	0.1185	0.0710	0.0336	0.0120	0.0030	0.0005			
	10	0.1593	0.1171	0.0686	0.0308	0.0099	0.0020	0.0002		
	11	0.1771	0.1597	0.1158	0.0654	0.0271	0.0074	0.0011	0.0001	
	12	0.1623	0.1797	0.1614	0.1144	0.0609	0.0222	0.0046	0.0004	
	13	0.1221	0.1659	0.1844	0.1643	0.1124	0.0545	0.0160	0.0020	
	14	0.0746	0.1244	0.1712	0.1916	0.1686	0.1091	0.0454	0.0089	0.0003
	15	0.0365	0.0746	0.1272	0.1789	0.2023	0.1746	0.1028	0.0319	0.0022
	16	0.0139	0.0350	0.0738	0.1304	0.1897	0.2182	0.1821	0.0898	0.0133
	17	0.0040	0.0123	0.0323	0.0716	0.1339	0.2054	0.2428	0.1901	0.0596
	18	0.0008	0.0031	0.0100	0.0278	0.0669	0.1369	0.2293	0.2852	0.1887
	19	0.0001	0.0005	0.0020	0.0068	0.0211	0.0576	0.1368	0.2702	0.3774
	20			0.0002	0.0008	0.0032	0.0115	0.0388	0.1216	0.3585

Note: Blank entries indicate a binomial probability of less than 0.00005.

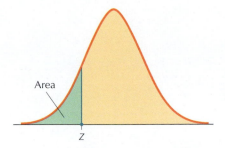

Area

Z

Table C Standard normal distribution

Z	0.00	0.01	0.02	0.03	0.04	0.05	0.06	0.07	0.08	0.09
−3.4	0.0003	0.0003	0.0003	0.0003	0.0003	0.0003	0.0003	0.0003	0.0003	0.0002
−3.3	0.0005	0.0005	0.0005	0.0004	0.0004	0.0004	0.0004	0.0004	0.0004	0.0003
−3.2	0.0007	0.0007	0.0006	0.0006	0.0006	0.0006	0.0006	0.0005	0.0005	0.0005
−3.1	0.0010	0.0009	0.0009	0.0009	0.0008	0.0008	0.0008	0.0008	0.0007	0.0007
−3.0	0.0013	0.0013	0.0013	0.0012	0.0012	0.0011	0.0011	0.0011	0.0010	0.0010
−2.9	0.0019	0.0018	0.0018	0.0017	0.0016	0.0016	0.0015	0.0015	0.0014	0.0014
−2.8	0.0026	0.0025	0.0024	0.0023	0.0023	0.0022	0.0021	0.0021	0.0020	0.0019
−2.7	0.0035	0.0034	0.0033	0.0032	0.0031	0.0030	0.0029	0.0028	0.0027	0.0026
−2.6	0.0047	0.0045	0.0044	0.0043	0.0041	0.0040	0.0039	0.0038	0.0037	0.0036
−2.5	0.0062	0.0060	0.0059	0.0057	0.0055	0.0054	0.0052	0.0051	0.0049	0.0048
−2.4	0.0082	0.0080	0.0078	0.0075	0.0073	0.0071	0.0069	0.0068	0.0066	0.0064
−2.3	0.0107	0.0104	0.0102	0.0099	0.0096	0.0094	0.0091	0.0089	0.0087	0.0084
−2.2	0.0139	0.0136	0.0132	0.0129	0.0125	0.0122	0.0119	0.0116	0.0113	0.0110
−2.1	0.0179	0.0174	0.0170	0.0166	0.0162	0.0158	0.0154	0.0150	0.0146	0.0143
−2.0	0.0228	0.0222	0.0217	0.0212	0.0207	0.0202	0.0197	0.0192	0.0188	0.0183
−1.9	0.0287	0.0281	0.0274	0.0268	0.0262	0.0256	0.0250	0.0244	0.0239	0.0233
−1.8	0.0359	0.0351	0.0344	0.0336	0.0329	0.0322	0.0314	0.0307	0.0301	0.0294
−1.7	0.0446	0.0436	0.0427	0.0418	0.0409	0.0401	0.0392	0.0384	0.0375	0.0367
−1.6	0.0548	0.0537	0.0526	0.0516	0.0505	0.0495	0.0485	0.0475	0.0465	0.0455
−1.5	0.0668	0.0655	0.0643	0.0630	0.0618	0.0606	0.0594	0.0582	0.0571	0.0559
−1.4	0.0808	0.0793	0.0778	0.0764	0.0749	0.0735	0.0721	0.0708	0.0694	0.0681
−1.3	0.0968	0.0951	0.0934	0.0918	0.0901	0.0885	0.0869	0.0853	0.0838	0.0823
−1.2	0.1151	0.1131	0.1112	0.1093	0.1075	0.1056	0.1038	0.1020	0.1003	0.0985
−1.1	0.1357	0.1335	0.1314	0.1292	0.1271	0.1251	0.1230	0.1210	0.1190	0.1170
−1.0	0.1587	0.1562	0.1539	0.1515	0.1492	0.1469	0.1446	0.1423	0.1401	0.1379
−0.9	0.1841	0.1814	0.1788	0.1762	0.1736	0.1711	0.1685	0.1660	0.1635	0.1611
−0.8	0.2119	0.2090	0.2061	0.2033	0.2005	0.1977	0.1949	0.1922	0.1894	0.1867
−0.7	0.2420	0.2389	0.2358	0.2327	0.2296	0.2266	0.2236	0.2206	0.2177	0.2148
−0.6	0.2743	0.2709	0.2676	0.2643	0.2611	0.2578	0.2546	0.2514	0.2483	0.2451
−0.5	0.3085	0.3050	0.3015	0.2981	0.2946	0.2912	0.2877	0.2843	0.2810	0.2776
−0.4	0.3446	0.3409	0.3372	0.3336	0.3300	0.3264	0.3228	0.3192	0.3156	0.3121
−0.3	0.3821	0.3783	0.3745	0.3707	0.3669	0.3632	0.3594	0.3557	0.3520	0.3483
−0.2	0.4207	0.4168	0.4129	0.4090	0.4052	0.4013	0.3974	0.3936	0.3897	0.3859
−0.1	0.4602	0.4562	0.4522	0.4483	0.4443	0.4404	0.4364	0.4325	0.4286	0.4247
−0.0	0.5000	0.4960	0.4920	0.4880	0.4840	0.4801	0.4761	0.4721	0.4681	0.4641

(Continued)

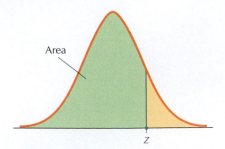

Area

Z

Table C Standard normal distribution (*continued*)

Z	0.00	0.01	0.02	0.03	0.04	0.05	0.06	0.07	0.08	0.09
0.0	0.5000	0.5040	0.5080	0.5120	0.5160	0.5199	0.5239	0.5279	0.5319	0.5359
0.1	0.5398	0.5438	0.5478	0.5517	0.5557	0.5596	0.5636	0.5675	0.5714	0.5753
0.2	0.5793	0.5832	0.5871	0.5910	0.5948	0.5987	0.6026	0.6064	0.6103	0.6141
0.3	0.6179	0.6217	0.6255	0.6293	0.6331	0.6368	0.6406	0.6443	0.6480	0.6517
0.4	0.6554	0.6591	0.6628	0.6664	0.6700	0.6736	0.6772	0.6808	0.6844	0.6879
0.5	0.6915	0.6950	0.6985	0.7019	0.7054	0.7088	0.7123	0.7157	0.7190	0.7224
0.6	0.7257	0.7291	0.7324	0.7357	0.7389	0.7422	0.7454	0.7486	0.7517	0.7549
0.7	0.7580	0.7611	0.7642	0.7673	0.7704	0.7734	0.7764	0.7794	0.7823	0.7852
0.8	0.7881	0.7910	0.7939	0.7967	0.7995	0.8023	0.8051	0.8078	0.8106	0.8133
0.9	0.8159	0.8186	0.8212	0.8238	0.8264	0.8289	0.8315	0.8340	0.8365	0.8389
1.0	0.8413	0.8438	0.8461	0.8485	0.8508	0.8531	0.8554	0.8577	0.8599	0.8621
1.1	0.8643	0.8665	0.8686	0.8708	0.8729	0.8749	0.8770	0.8790	0.8810	0.8830
1.2	0.8849	0.8869	0.8888	0.8907	0.8925	0.8944	0.8962	0.8980	0.8997	0.9015
1.3	0.9032	0.9049	0.9066	0.9082	0.9099	0.9115	0.9131	0.9147	0.9162	0.9177
1.4	0.9192	0.9207	0.9222	0.9236	0.9251	0.9265	0.9279	0.9292	0.9306	0.9319
1.5	0.9332	0.9345	0.9357	0.9370	0.9382	0.9394	0.9406	0.9418	0.9429	0.9441
1.6	0.9452	0.9463	0.9474	0.9484	0.9495	0.9505	0.9515	0.9525	0.9535	0.9545
1.7	0.9554	0.9564	0.9573	0.9582	0.9591	0.9599	0.9608	0.9616	0.9625	0.9633
1.8	0.9641	0.9649	0.9656	0.9664	0.9671	0.9678	0.9686	0.9693	0.9699	0.9706
1.9	0.9713	0.9719	0.9726	0.9732	0.9738	0.9744	0.9750	0.9756	0.9761	0.9767
2.0	0.9772	0.9778	0.9783	0.9788	0.9793	0.9798	0.9803	0.9808	0.9812	0.9817
2.1	0.9821	0.9826	0.9830	0.9834	0.9838	0.9842	0.9846	0.9850	0.9854	0.9857
2.2	0.9861	0.9864	0.9868	0.9871	0.9875	0.9878	0.9881	0.9884	0.9887	0.9890
2.3	0.9893	0.9896	0.9898	0.9901	0.9904	0.9906	0.9909	0.9911	0.9913	0.9916
2.4	0.9918	0.9920	0.9922	0.9925	0.9927	0.9929	0.9931	0.9932	0.9934	0.9936
2.5	0.9938	0.9940	0.9941	0.9943	0.9945	0.9946	0.9948	0.9949	0.9951	0.9952
2.6	0.9953	0.9955	0.9956	0.9957	0.9959	0.9960	0.9961	0.9962	0.9963	0.9964
2.7	0.9965	0.9966	0.9967	0.9968	0.9969	0.9970	0.9971	0.9972	0.9973	0.9974
2.8	0.9974	0.9975	0.9976	0.9977	0.9977	0.9978	0.9979	0.9979	0.9980	0.9981
2.9	0.9981	0.9982	0.9982	0.9983	0.9984	0.9984	0.9985	0.9985	0.9986	0.9986
3.0	0.9987	0.9987	0.9987	0.9988	0.9988	0.9989	0.9989	0.9989	0.9990	0.9990
3.1	0.9990	0.9991	0.9991	0.9991	0.9992	0.9992	0.9992	0.9992	0.9993	0.9993
3.2	0.9993	0.9993	0.9994	0.9994	0.9994	0.9994	0.9994	0.9995	0.9995	0.9995
3.3	0.9995	0.9995	0.9995	0.9996	0.9996	0.9996	0.9996	0.9996	0.9996	0.9997
3.4	0.9997	0.9997	0.9997	0.9997	0.9997	0.9997	0.9997	0.9997	0.9997	0.9998

Table D *t*-Distribution

		Confidence level				
		80%	**90%**	**95%**	**98%**	**99%**
		Area in one tail				
		0.10	**0.05**	**0.025**	**0.01**	**0.005**
		Area in two tails				
		0.20	**0.10**	**0.05**	**0.02**	**0.01**
df	**1**	3.078	6.314	12.706	31.821	63.657
	2	1.886	2.920	4.303	6.965	9.925
	3	1.638	2.353	3.182	4.541	5.841
	4	1.533	2.132	2.776	3.747	4.604
	5	1.476	2.015	2.571	3.365	4.032
	6	1.440	1.943	2.447	3.143	3.707
	7	1.415	1.895	2.365	2.998	3.499
	8	1.397	1.860	2.306	2.896	3.355
	9	1.383	1.833	2.262	2.821	3.250
	10	1.372	1.812	2.228	2.764	3.169
	11	1.363	1.796	2.201	2.718	3.106
	12	1.356	1.782	2.179	2.681	3.055
	13	1.350	1.771	2.160	2.650	3.012
	14	1.345	1.761	2.145	2.624	2.977
	15	1.341	1.753	2.131	2.602	2.947
	16	1.337	1.746	2.120	2.583	2.921
	17	1.333	1.740	2.110	2.567	2.898
	18	1.330	1.734	2.101	2.552	2.878
	19	1.328	1.729	2.093	2.539	2.861
	20	1.325	1.725	2.086	2.528	2.845
	21	1.323	1.721	2.080	2.518	2.831
	22	1.321	1.717	2.074	2.508	2.819
	23	1.319	1.714	2.069	2.500	2.807
	24	1.318	1.711	2.064	2.492	2.797
	25	1.316	1.708	2.060	2.485	2.787
	26	1.315	1.706	2.056	2.479	2.779
	27	1.314	1.703	2.052	2.473	2.771
	28	1.313	1.701	2.048	2.467	2.763
	29	1.311	1.699	2.045	2.462	2.756
	30	1.310	1.697	2.042	2.457	2.750
	31	1.309	1.696	2.040	2.453	2.744
	32	1.309	1.694	2.037	2.449	2.738
	33	1.308	1.692	2.035	2.445	2.733
	34	1.307	1.691	2.032	2.441	2.728
	35	1.306	1.690	2.030	2.438	2.724
	36	1.306	1.688	2.028	2.435	2.719
	37	1.305	1.687	2.026	2.431	2.715
	38	1.304	1.686	2.024	2.429	2.712
	39	1.304	1.685	2.023	2.426	2.708
	40	1.303	1.684	2.021	2.423	2.704
	50	1.299	1.676	2.009	2.403	2.678
	60	1.296	1.671	2.000	2.390	2.660
	70	1.294	1.667	1.994	2.381	2.648
	80	1.292	1.664	1.990	2.374	2.639
	90	1.291	1.662	1.987	2.368	2.632
	100	1.290	1.660	1.984	2.364	2.626
	1000	1.282	1.646	1.962	2.330	2.581
	z	1.282	1.645	1.960	2.326	2.576

Table E Chi-square (χ^2) distribution

Degrees of freedom	Area to the right of critical value									
	0.995	0.99	0.975	0.95	0.90	0.10	0.05	0.025	0.01	0.005
1	—	—	0.001	0.004	0.016	2.706	3.841	5.024	6.635	7.879
2	0.010	0.020	0.051	0.103	0.211	4.605	5.991	7.378	9.210	10.597
3	0.072	0.115	0.216	0.352	0.584	6.251	7.815	9.348	11.345	12.838
4	0.207	0.297	0.484	0.711	1.064	7.779	9.488	11.143	13.277	14.860
5	0.412	0.554	0.831	1.145	1.610	9.236	11.071	12.833	15.086	16.750
6	0.676	0.872	1.237	1.635	2.204	10.645	12.592	14.449	16.812	18.548
7	0.989	1.239	1.690	2.167	2.833	12.017	14.067	16.013	18.475	20.278
8	1.344	1.646	2.180	2.733	3.490	13.362	15.507	17.535	20.090	21.955
9	1.735	2.088	2.700	3.325	4.168	14.684	16.919	19.023	21.666	23.589
10	2.156	2.558	3.247	3.940	4.865	15.987	18.307	20.483	23.209	25.188
11	2.603	3.053	3.816	4.575	5.578	17.275	19.675	21.920	24.725	26.757
12	3.074	3.571	4.404	5.226	6.304	18.549	21.026	23.337	26.217	28.299
13	3.565	4.107	5.009	5.892	7.042	19.812	22.362	24.736	27.688	29.819
14	4.075	4.660	5.629	6.571	7.790	21.064	23.685	26.119	29.141	31.319
15	4.601	5.229	6.262	7.261	8.547	22.307	24.996	27.488	30.578	32.801
16	5.142	5.812	6.908	7.962	9.312	23.542	26.296	28.845	32.000	34.267
17	5.697	6.408	7.564	8.672	10.085	24.769	27.587	30.191	33.409	35.718
18	6.265	7.015	8.231	9.390	10.865	25.989	28.869	31.526	34.805	37.156
19	6.844	7.633	8.907	10.117	11.651	27.204	30.144	32.852	36.191	38.582
20	7.434	8.260	9.591	10.851	12.443	28.412	31.410	34.170	37.566	39.997
21	8.034	8.897	10.283	11.591	13.240	29.615	32.671	35.479	38.932	41.401
22	8.643	9.542	10.982	12.338	14.042	30.813	33.924	36.781	40.289	42.796
23	9.260	10.196	11.689	13.091	14.848	32.007	35.172	38.076	41.638	44.181
24	9.886	10.856	12.401	13.848	15.659	33.196	36.415	39.364	42.980	45.559
25	10.520	11.524	13.120	14.611	16.473	34.382	37.652	40.646	44.314	46.928
26	11.160	12.198	13.844	15.379	17.292	35.563	38.885	41.923	45.642	48.290
27	11.808	12.879	14.573	16.151	18.114	36.741	40.113	43.194	46.963	49.645
28	12.461	13.565	15.308	16.928	18.939	37.916	41.337	44.461	48.278	50.993
29	13.121	14.257	16.047	17.708	19.768	39.087	42.557	45.722	49.588	52.336
30	13.787	14.954	16.791	18.493	20.599	40.256	43.773	46.979	50.892	53.672
40	20.707	22.164	24.433	26.509	29.051	51.805	55.758	59.342	63.691	66.766
50	27.991	29.707	32.357	34.764	37.689	63.167	67.505	71.420	76.154	79.490
60	35.534	37.485	40.482	43.188	46.459	74.397	79.082	83.298	88.379	91.952
70	43.275	45.442	48.758	51.739	55.329	85.527	90.531	95.023	100.425	104.215
80	51.172	53.540	57.153	60.391	64.278	96.578	101.879	106.629	112.329	116.321
90	59.196	61.754	65.647	69.126	73.291	107.565	113.145	118.136	124.116	128.299
100	67.328	70.065	74.222	77.929	82.358	118.498	124.342	129.561	135.807	140.169

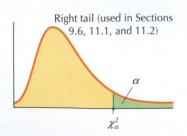

Right tail (used in Sections 9.6, 11.1, and 11.2)

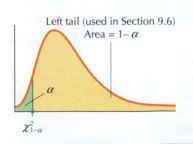

Left tail (used in Section 9.6)
Area = $1-\alpha$

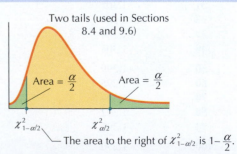

Two tails (used in Sections 8.4 and 9.6)

Area = $\frac{\alpha}{2}$

Area = $\frac{\alpha}{2}$

$\chi^2_{1-\alpha/2}$

$\chi^2_{\alpha/2}$

The area to the right of $\chi^2_{1-\alpha/2}$ is $1-\frac{\alpha}{2}$.

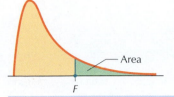
Area
F

Table F *F*-Distribution critical values

	Area in right tail	1	2	3	4	5	6	7	8
					df₁				
1	0.100	39.86	49.59	53.59	55.83	57.24	58.20	58.91	59.44
	0.050	161.45	199.50	215.71	224.58	230.16	233.99	236.77	238.88
	0.025	647.79	799.50	864.16	899.58	921.85	937.11	948.22	956.66
	0.010	4052.20	4999.50	5403.40	5624.60	5763.60	5859.00	5928.40	5981.10
	0.001	405284.00	500000.00	540379.00	562500.00	576405.00	585937.00	592873.00	598144.00
2	0.100	8.53	9.00	9.16	9.24	9.29	9.33	9.35	9.37
	0.050	18.51	19.00	19.16	19.25	19.30	19.33	19.35	19.37
	0.025	38.51	39.00	39.17	39.25	39.30	39.33	39.36	39.37
	0.010	98.50	99.00	99.17	99.25	99.30	99.33	99.36	99.37
	0.001	998.50	999.00	999.17	999.25	999.30	999.33	999.36	999.37
3	0.100	5.54	5.46	5.39	5.34	5.31	5.28	5.27	5.25
	0.050	10.13	9.55	9.28	9.12	9.01	8.94	8.89	8.85
	0.025	17.44	16.04	15.44	15.10	14.88	14.73	14.62	14.54
	0.010	34.12	30.82	29.46	28.71	28.24	27.91	27.67	27.49
	0.001	167.03	148.50	141.11	137.10	134.58	132.85	131.58	130.62
4	0.100	4.54	4.32	4.19	4.11	4.05	4.01	3.98	3.95
	0.050	7.71	6.94	6.59	6.39	6.26	6.16	6.09	6.04
	0.025	12.22	10.65	9.98	9.60	9.36	9.20	9.07	8.98
	0.010	21.20	18.00	16.69	15.98	15.52	15.21	14.98	14.80
	0.001	74.14	61.25	56.18	53.44	51.71	50.53	49.66	49.00
5	0.100	4.06	3.78	3.62	3.52	3.45	3.40	3.37	3.34
	0.050	6.61	5.79	5.41	5.19	5.05	4.95	4.88	4.82
	0.025	10.01	8.43	7.76	7.39	7.15	6.98	6.85	6.76
	0.010	16.26	13.27	12.06	11.39	10.97	10.67	10.46	10.29
	0.001	47.18	37.12	33.20	31.09	29.75	28.83	28.16	27.65
6	0.100	3.78	3.46	3.29	3.18	3.11	3.05	3.01	2.98
	0.050	5.99	5.14	4.76	4.53	4.39	4.28	4.21	4.15
	0.025	8.81	7.26	6.60	6.23	5.99	5.82	5.70	5.60
	0.010	13.75	10.92	9.78	9.15	8.75	8.47	8.26	8.10
	0.001	35.51	27.00	23.70	21.92	20.80	20.03	19.46	19.03
7	0.100	3.59	3.26	3.07	2.96	2.88	2.83	2.78	2.75
	0.050	5.59	4.74	4.35	4.12	3.97	3.87	3.79	3.73
	0.025	8.07	6.54	5.89	5.52	5.29	5.12	4.99	4.90
	0.010	12.25	9.55	8.45	7.85	7.46	7.19	6.99	6.84
	0.001	29.25	21.69	18.77	17.20	16.21	15.52	15.02	14.63
8	0.100	3.46	3.11	2.92	2.81	2.73	2.67	2.62	2.59
	0.050	5.32	4.46	4.07	3.84	3.69	3.58	3.50	3.44
	0.025	7.57	6.06	5.42	5.05	4.82	4.65	4.53	4.43
	0.010	11.26	8.65	7.59	7.01	6.63	6.37	6.18	6.03
	0.001	25.41	18.49	15.83	14.39	13.48	12.86	12.40	12.05

df₂ (left margin label)

(Continued)

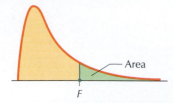

Area

F

Table F *F*-Distribution critical values (*continued*)

						df$_1$			
	Area in right tail	9	10	15	20	30	60	120	1000
1	0.100	59.86	60.19	61.22	61.74	62.26	62.79	63.06	63.30
	0.050	240.54	241.88	245.95	248.01	250.10	252.20	253.25	254.19
	0.025	963.28	968.63	984.87	993.10	1001.4	1009.8	1014.0	1017.7
	0.010	6022.5	6055.8	6157.3	6208.7	6260.6	6313.0	6339.4	6362.7
	0.001	602284.0	605621.0	615764.0	620908.0	626099.0	631337.0	633972.0	636301.0
2	0.100	9.38	9.39	9.42	9.44	9.16	9.47	9.48	9.49
	0.050	19.38	19.40	19.43	19.45	19.46	19.48	19.49	19.49
	0.025	39.39	39.40	39.43	39.45	39.46	39.48	39.49	39.50
	0.010	99.39	99.40	99.43	99.45	99.47	99.48	99.49	99.50
	0.001	999.39	999.40	999.43	999.45	999.47	999.48	999.49	999.50
3	0.100	5.24	5.23	5.20	5.18	5.17	5.15	5.14	5.13
	0.050	8.81	8.79	8.70	8.66	8.62	8.57	8.55	8.53
	0.025	14.47	14.42	14.25	14.17	14.08	13.99	13.95	13.91
	0.010	27.35	27.23	26.87	26.69	26.50	26.32	26.22	26.14
	0.001	129.86	129.25	127.37	126.42	125.45	124.47	123.97	123.53
4	0.100	3.94	3.92	3.87	3.84	3.82	3.79	3.78	3.76
	0.050	6.00	5.96	5.86	5.80	5.75	5.69	5.66	5.63
	0.025	8.90	8.84	8.66	8.56	8.46	8.36	8.31	8.26
	0.010	14.66	14.55	14.20	14.02	13.84	13.65	13.56	13.47
	0.001	48.47	48.05	46.76	46.10	45.43	44.75	44.40	44.09
5	0.100	3.32	3.30	3.24	3.21	3.17	3.14	3.12	3.11
	0.050	4.77	4.74	4.62	4.56	4.50	4.43	4.40	4.37
	0.025	6.68	6.62	6.43	6.33	6.23	6.12	6.07	6.02
	0.010	10.16	10.05	9.72	9.55	9.38	9.20	9.11	9.03
	0.001	27.24	26.92	25.91	25.39	24.87	24.33	24.06	23.82
6	0.100	2.96	2.94	2.87	2.84	2.80	2.76	2.74	2.72
	0.050	4.10	4.06	3.94	3.87	3.81	3.74	3.70	3.67
	0.025	5.52	5.46	5.27	5.17	5.07	4.96	4.90	4.86
	0.010	7.98	7.87	7.56	7.40	7.23	7.06	6.97	6.89
	0.001	18.69	18.41	17.56	17.12	16.67	16.21	15.98	15.77
7	0.100	2.72	2.70	2.63	2.59	2.56	2.51	2.49	2.47
	0.050	3.68	3.64	3.51	3.44	3.38	3.30	3.27	3.23
	0.025	4.82	4.76	4.57	4.47	4.36	4.25	4.20	4.15
	0.010	6.72	6.62	6.31	6.16	5.99	5.82	5.74	5.66
	0.001	14.33	14.08	13.32	12.93	12.53	12.12	11.91	11.72
8	0.100	2.56	2.54	2.46	2.42	2.38	2.34	2.32	2.30
	0.050	3.39	3.35	3.22	3.15	3.08	3.01	2.97	2.93
	0.025	4.36	4.30	4.10	4.00	3.89	3.78	3.73	3.68
	0.010	5.91	5.81	5.52	5.36	5.20	5.03	4.95	4.87
	0.001	11.77	11.54	10.84	10.48	10.11	9.73	9.53	9.36

df$_2$

Table F *F*-Distribution critical values (continued)

<table>
<tr><th colspan="12" style="text-align:center">df₁</th></tr>
<tr><th></th><th>Area in
right tail</th><th>1</th><th>2</th><th>3</th><th>4</th><th>5</th><th>6</th><th>7</th><th>8</th><th>9</th><th>10</th></tr>
<tr><td rowspan="5">9</td><td>0.100</td><td>3.36</td><td>3.01</td><td>2.81</td><td>2.69</td><td>2.61</td><td>2.55</td><td>2.51</td><td>2.47</td><td>2.44</td><td>2.42</td></tr>
<tr><td>0.050</td><td>5.12</td><td>4.26</td><td>3.86</td><td>3.63</td><td>3.48</td><td>3.37</td><td>3.29</td><td>3.23</td><td>3.18</td><td>3.14</td></tr>
<tr><td>0.025</td><td>7.21</td><td>5.71</td><td>5.08</td><td>4.72</td><td>4.48</td><td>4.32</td><td>4.20</td><td>4.10</td><td>4.03</td><td>3.96</td></tr>
<tr><td>0.010</td><td>10.56</td><td>8.02</td><td>6.99</td><td>6.42</td><td>6.06</td><td>5.80</td><td>5.61</td><td>5.47</td><td>5.35</td><td>5.26</td></tr>
<tr><td>0.001</td><td>22.86</td><td>16.39</td><td>13.90</td><td>12.56</td><td>11.71</td><td>11.13</td><td>10.70</td><td>10.37</td><td>10.11</td><td>9.89</td></tr>
<tr><td rowspan="5">10</td><td>0.100</td><td>3.29</td><td>2.92</td><td>2.73</td><td>2.61</td><td>2.52</td><td>2.46</td><td>2.41</td><td>2.38</td><td>2.35</td><td>2.32</td></tr>
<tr><td>0.050</td><td>4.96</td><td>4.10</td><td>3.71</td><td>3.48</td><td>3.33</td><td>3.22</td><td>3.14</td><td>3.07</td><td>3.02</td><td>2.98</td></tr>
<tr><td>0.025</td><td>6.94</td><td>5.46</td><td>4.83</td><td>4.47</td><td>4.24</td><td>4.07</td><td>3.95</td><td>3.85</td><td>3.78</td><td>3.72</td></tr>
<tr><td>0.010</td><td>10.04</td><td>7.56</td><td>6.55</td><td>5.99</td><td>5.64</td><td>5.39</td><td>5.20</td><td>5.06</td><td>4.94</td><td>4.85</td></tr>
<tr><td>0.001</td><td>21.04</td><td>14.91</td><td>12.55</td><td>11.28</td><td>10.48</td><td>9.93</td><td>9.52</td><td>9.20</td><td>8.96</td><td>8.75</td></tr>
<tr><td rowspan="5">12</td><td>0.100</td><td>3.18</td><td>2.81</td><td>2.61</td><td>2.48</td><td>2.39</td><td>2.33</td><td>2.28</td><td>2.24</td><td>2.21</td><td>2.19</td></tr>
<tr><td>0.050</td><td>4.75</td><td>3.89</td><td>3.49</td><td>3.26</td><td>3.11</td><td>3.00</td><td>2.91</td><td>2.85</td><td>2.80</td><td>2.75</td></tr>
<tr><td>0.025</td><td>6.55</td><td>5.10</td><td>4.47</td><td>4.12</td><td>3.89</td><td>3.73</td><td>3.61</td><td>3.51</td><td>3.44</td><td>3.37</td></tr>
<tr><td>0.010</td><td>9.33</td><td>6.93</td><td>5.95</td><td>5.41</td><td>5.06</td><td>4.82</td><td>4.64</td><td>4.50</td><td>4.39</td><td>4.30</td></tr>
<tr><td>0.001</td><td>18.64</td><td>12.97</td><td>10.80</td><td>9.63</td><td>8.89</td><td>8.38</td><td>8.00</td><td>7.71</td><td>7.48</td><td>7.29</td></tr>
<tr><td rowspan="5">15</td><td>0.100</td><td>3.07</td><td>2.70</td><td>2.49</td><td>2.36</td><td>2.27</td><td>2.21</td><td>2.16</td><td>2.12</td><td>2.09</td><td>2.06</td></tr>
<tr><td>0.050</td><td>4.54</td><td>3.68</td><td>3.29</td><td>3.06</td><td>2.90</td><td>2.79</td><td>2.71</td><td>2.64</td><td>2.59</td><td>2.54</td></tr>
<tr><td>0.025</td><td>6.20</td><td>4.77</td><td>4.15</td><td>3.80</td><td>3.58</td><td>3.41</td><td>3.29</td><td>3.20</td><td>3.12</td><td>3.06</td></tr>
<tr><td>0.010</td><td>8.68</td><td>6.36</td><td>5.42</td><td>4.89</td><td>4.56</td><td>4.32</td><td>4.14</td><td>4.00</td><td>3.89</td><td>3.80</td></tr>
<tr><td>0.001</td><td>16.59</td><td>11.34</td><td>9.34</td><td>8.25</td><td>7.57</td><td>7.09</td><td>6.74</td><td>6.47</td><td>6.26</td><td>6.08</td></tr>
<tr><td rowspan="5">20</td><td>0.100</td><td>2.97</td><td>2.59</td><td>2.38</td><td>2.25</td><td>2.16</td><td>2.09</td><td>2.04</td><td>2.00</td><td>1.96</td><td>1.94</td></tr>
<tr><td>0.050</td><td>4.35</td><td>3.49</td><td>3.10</td><td>2.87</td><td>2.71</td><td>2.60</td><td>2.51</td><td>2.45</td><td>2.39</td><td>2.35</td></tr>
<tr><td>0.025</td><td>5.87</td><td>4.46</td><td>3.86</td><td>3.51</td><td>3.29</td><td>3.13</td><td>3.01</td><td>2.91</td><td>2.84</td><td>2.77</td></tr>
<tr><td>0.010</td><td>8.10</td><td>5.85</td><td>4.94</td><td>4.43</td><td>4.10</td><td>3.87</td><td>3.70</td><td>3.56</td><td>3.46</td><td>3.37</td></tr>
<tr><td>0.001</td><td>14.82</td><td>9.95</td><td>8.10</td><td>7.10</td><td>6.46</td><td>6.02</td><td>5.69</td><td>5.44</td><td>5.24</td><td>5.08</td></tr>
<tr><td rowspan="5">25</td><td>0.100</td><td>2.92</td><td>2.53</td><td>2.32</td><td>2.18</td><td>2.09</td><td>2.02</td><td>1.97</td><td>1.93</td><td>1.89</td><td>1.87</td></tr>
<tr><td>0.050</td><td>4.24</td><td>3.39</td><td>2.99</td><td>2.76</td><td>2.60</td><td>2.49</td><td>2.40</td><td>2.34</td><td>2.28</td><td>2.24</td></tr>
<tr><td>0.025</td><td>5.69</td><td>4.29</td><td>3.69</td><td>3.35</td><td>3.13</td><td>2.97</td><td>2.85</td><td>2.75</td><td>2.68</td><td>2.61</td></tr>
<tr><td>0.010</td><td>7.77</td><td>5.57</td><td>4.68</td><td>4.18</td><td>3.85</td><td>3.63</td><td>3.46</td><td>3.32</td><td>3.22</td><td>3.13</td></tr>
<tr><td>0.001</td><td>13.88</td><td>9.22</td><td>7.45</td><td>6.49</td><td>5.89</td><td>5.46</td><td>5.15</td><td>4.91</td><td>4.71</td><td>4.56</td></tr>
<tr><td rowspan="5">50</td><td>0.100</td><td>2.81</td><td>2.41</td><td>2.20</td><td>2.06</td><td>1.97</td><td>1.90</td><td>1.84</td><td>1.80</td><td>1.76</td><td>1.73</td></tr>
<tr><td>0.050</td><td>4.03</td><td>3.18</td><td>2.79</td><td>2.56</td><td>2.40</td><td>2.29</td><td>2.20</td><td>2.13</td><td>2.07</td><td>2.03</td></tr>
<tr><td>0.025</td><td>5.34</td><td>3.97</td><td>3.39</td><td>3.05</td><td>2.83</td><td>2.67</td><td>2.55</td><td>2.46</td><td>2.38</td><td>2.32</td></tr>
<tr><td>0.010</td><td>7.17</td><td>5.06</td><td>4.20</td><td>3.72</td><td>3.41</td><td>3.19</td><td>3.02</td><td>2.89</td><td>2.78</td><td>2.70</td></tr>
<tr><td>0.001</td><td>12.22</td><td>7.96</td><td>6.34</td><td>5.46</td><td>4.90</td><td>4.51</td><td>4.22</td><td>4.00</td><td>3.82</td><td>3.67</td></tr>
<tr><td rowspan="5">100</td><td>0.100</td><td>2.76</td><td>2.36</td><td>2.14</td><td>2.00</td><td>1.91</td><td>1.83</td><td>1.78</td><td>1.73</td><td>1.69</td><td>1.66</td></tr>
<tr><td>0.050</td><td>3.94</td><td>3.09</td><td>2.70</td><td>2.46</td><td>2.31</td><td>2.19</td><td>2.10</td><td>2.03</td><td>1.97</td><td>1.93</td></tr>
<tr><td>0.025</td><td>5.18</td><td>3.83</td><td>3.25</td><td>2.92</td><td>2.70</td><td>2.54</td><td>2.42</td><td>2.32</td><td>2.24</td><td>2.18</td></tr>
<tr><td>0.010</td><td>6.90</td><td>4.82</td><td>3.98</td><td>3.51</td><td>3.21</td><td>2.99</td><td>2.82</td><td>2.69</td><td>2.59</td><td>2.50</td></tr>
<tr><td>0.001</td><td>11.50</td><td>7.41</td><td>5.86</td><td>5.02</td><td>4.48</td><td>4.11</td><td>3.83</td><td>3.61</td><td>3.44</td><td>3.30</td></tr>
<tr><td rowspan="5">200</td><td>0.100</td><td>2.73</td><td>2.33</td><td>2.11</td><td>1.97</td><td>1.88</td><td>1.80</td><td>1.75</td><td>1.70</td><td>1.66</td><td>1.63</td></tr>
<tr><td>0.050</td><td>3.89</td><td>3.04</td><td>2.65</td><td>2.42</td><td>2.26</td><td>2.14</td><td>2.06</td><td>1.98</td><td>1.93</td><td>1.88</td></tr>
<tr><td>0.025</td><td>5.10</td><td>3.76</td><td>3.18</td><td>2.85</td><td>2.63</td><td>2.47</td><td>2.35</td><td>2.26</td><td>2.18</td><td>2.11</td></tr>
<tr><td>0.010</td><td>6.76</td><td>4.71</td><td>3.88</td><td>3.41</td><td>3.11</td><td>2.89</td><td>2.73</td><td>2.60</td><td>2.50</td><td>2.41</td></tr>
<tr><td>0.001</td><td>11.15</td><td>7.15</td><td>5.63</td><td>4.81</td><td>4.29</td><td>3.92</td><td>3.65</td><td>3.43</td><td>3.26</td><td>3.12</td></tr>
<tr><td rowspan="5">1000</td><td>0.100</td><td>2.71</td><td>2.31</td><td>2.09</td><td>1.95</td><td>1.85</td><td>1.78</td><td>1.72</td><td>1.68</td><td>1.64</td><td>1.61</td></tr>
<tr><td>0.050</td><td>3.85</td><td>3.00</td><td>2.61</td><td>2.38</td><td>2.22</td><td>2.11</td><td>2.02</td><td>1.95</td><td>1.89</td><td>1.84</td></tr>
<tr><td>0.025</td><td>5.04</td><td>3.70</td><td>3.13</td><td>2.80</td><td>2.58</td><td>2.42</td><td>2.30</td><td>2.20</td><td>2.13</td><td>2.06</td></tr>
<tr><td>0.010</td><td>6.66</td><td>4.63</td><td>3.80</td><td>3.34</td><td>3.04</td><td>2.82</td><td>2.66</td><td>2.53</td><td>2.43</td><td>2.34</td></tr>
<tr><td>0.001</td><td>10.89</td><td>6.96</td><td>5.46</td><td>4.65</td><td>4.14</td><td>3.78</td><td>3.51</td><td>3.30</td><td>3.13</td><td>2.99</td></tr>
</table>

df₂

(Continued)

Table F *F*-Distribution critical values (continued)

		df₁									
	Area in right tail	12	15	20	25	30	40	50	60	120	1000
9	0.100	2.38	2.34	2.30	2.27	2.25	2.23	2.22	2.21	2.18	2.16
	0.050	3.07	3.01	2.94	2.89	2.86	2.83	2.80	2.79	2.75	2.71
	0.025	3.87	3.77	3.67	3.60	3.56	3.51	3.47	3.45	3.39	3.34
	0.010	5.11	4.96	4.81	4.71	4.65	4.57	4.52	4.48	4.40	4.32
	0.001	9.57	9.24	8.90	8.69	8.55	8.37	8.26	8.19	8.00	7.84
10	0.100	2.28	2.24	2.20	2.17	2.16	2.13	2.12	2.11	2.08	2.06
	0.050	2.91	2.85	2.77	2.73	2.70	2.66	2.64	2.62	2.58	2.54
	0.025	3.62	3.52	3.42	3.35	3.31	3.26	3.22	3.20	3.14	3.09
	0.010	4.71	4.56	4.41	4.31	4.25	4.17	4.12	4.08	4.00	3.92
	0.001	8.45	8.13	7.80	7.60	7.47	7.30	7.19	7.12	6.94	6.78
12	0.100	2.15	2.10	2.06	2.03	2.01	1.99	1.97	1.96	1.93	1.91
	0.050	2.69	2.62	2.54	2.50	2.47	2.43	2.40	2.38	2.34	2.30
	0.025	3.28	3.18	3.07	3.01	2.96	2.91	2.87	2.85	2.79	2.73
	0.010	4.16	4.01	3.86	3.76	3.70	3.62	3.57	3.54	3.45	3.37
	0.001	7.00	6.71	6.40	6.22	6.09	5.93	5.83	5.76	5.59	5.44
15	0.100	2.02	1.97	1.92	1.89	1.87	1.85	1.83	1.82	1.79	1.76
	0.050	2.48	2.40	2.33	2.28	2.25	2.20	2.18	2.16	2.11	2.07
	0.025	2.96	2.86	2.76	2.69	2.64	2.59	2.55	2.52	2.46	2.40
	0.010	3.67	3.52	3.37	3.28	3.21	3.13	3.08	3.05	2.96	2.88
	0.001	5.81	5.54	5.25	5.07	4.95	4.80	4.70	4.64	4.47	4.33
20	0.100	1.89	1.84	1.79	1.76	1.74	1.71	1.69	1.68	1.64	1.61
	0.050	2.28	2.20	2.12	2.07	2.04	1.99	1.97	1.95	1.90	1.85
	0.025	2.68	2.57	2.46	2.40	2.35	2.29	2.25	2.22	2.16	2.09
	0.010	3.23	3.09	2.94	2.84	2.78	2.69	2.64	2.61	2.52	2.43
	0.001	4.82	4.56	4.29	4.12	4.00	3.86	3.77	3.70	3.54	3.40
25	0.100	1.82	1.77	1.72	1.68	1.66	1.63	1.61	1.59	1.56	1.52
	0.050	2.16	2.09	2.01	1.96	1.92	1.87	1.84	1.82	1.77	1.72
	0.025	2.51	2.41	2.30	2.23	2.18	2.12	2.08	2.05	1.98	1.91
	0.010	2.99	2.85	2.70	2.60	2.54	2.45	2.40	2.36	2.27	2.18
	0.001	4.31	4.06	3.79	3.63	3.52	3.37	3.28	3.22	3.06	2.91
50	0.100	1.68	1.63	1.57	1.53	1.50	1.46	1.44	1.42	1.38	1.33
	0.050	1.95	1.87	1.78	1.73	1.69	1.63	1.60	1.58	1.51	1.45
	0.025	2.22	2.11	1.99	1.92	1.87	1.80	1.75	1.72	1.64	1.56
	0.010	2.56	2.42	2.27	2.17	2.10	2.01	1.95	1.91	1.80	1.70
	0.001	3.44	3.20	2.95	2.79	2.68	2.53	2.44	2.38	2.21	2.05
100	0.100	1.61	1.56	1.49	1.45	1.42	1.38	1.35	1.34	1.28	1.22
	0.050	1.85	1.77	1.68	1.62	1.57	1.52	1.48	1.45	1.38	1.30
	0.025	2.08	1.97	1.85	1.77	1.71	1.64	1.59	1.56	1.46	1.36
	0.010	2.37	2.22	2.07	1.97	1.89	1.80	1.74	1.69	1.57	1.45
	0.001	3.07	2.84	2.59	2.43	2.32	2.17	2.08	2.01	1.83	1.64
200	0.100	1.58	1.52	1.46	1.41	1.38	1.34	1.31	1.29	1.23	1.16
	0.050	1.80	1.72	1.62	1.56	1.52	1.46	1.41	1.39	1.30	1.21
	0.025	2.01	1.90	1.78	1.70	1.64	1.56	1.51	1.47	1.37	1.25
	0.010	2.27	2.13	1.97	1.87	1.79	1.69	1.63	1.58	1.45	1.30
	0.001	2.90	2.67	2.42	2.26	2.15	2.00	1.90	1.83	1.64	1.43
1000	0.100	1.55	1.49	1.43	1.38	1.35	1.30	1.27	1.25	1.38	1.08
	0.050	1.76	1.68	1.58	1.52	1.47	1.41	1.36	1.31	1.24	1.11
	0.025	1.96	1.85	1.72	1.64	1.58	1.50	1.45	1.41	1.29	1.13
	0.010	2.20	2.06	1.90	1.79	1.72	1.61	1.54	1.50	1.35	1.16
	0.001	2.77	2.54	2.30	2.14	2.02	1.87	1.77	1.69	1.49	1.22

df₂

Table G Critical values for correlation coefficient

n	
3	0.997
4	0.950
5	0.878
6	0.811
7	0.754
8	0.707
9	0.666
10	0.632
11	0.602
12	0.576
13	0.553
14	0.532
15	0.514
16	0.497
17	0.482
18	0.468
19	0.456
20	0.444
21	0.433
22	0.423
23	0.413
24	0.404
25	0.396
26	0.388
27	0.381
28	0.374
29	0.367
30	0.361

NOTES AND DATA SOURCES

Chapter 1

1. T. J. Scanlon, R. N. Luben, F. L. Scanlon, and N. Singleton, "Is Friday the 13th bad for your health?" *British Medical Journal* 307 (December 1993).

2. U.S. Census Bureau, *The Population Profile of the United States: 2000*, **www.consensus.gov/population/www/pop-profile/profile2000**.

3. Pew Internet and American Life Project, "Cyberbullying and online teens," June 2007, **www.pewinternet.org**.

4. National Agricultural Statistics Service.

5. Iain McGregor and Wayne Hall, "MDMA (Ecstasy) neurotoxicity: assessing and communicating the risks," *Lancet* 355 (9217, May 20, 2000): 1818–21.

6. Michel de Lorgeril, Patricia Salen, Jean-Louis Martin, Isabelle Monjaud, Jacques Delaye, and Nicole Mamelle, "Mediterranean diet, traditional risk factors, and the rate of cardiovascular complications after myocardial infarction, final report of the Lyon Diet Heart Study," *Circulation: Journal of the American Heart Association* 99 (1999): 779–85. The American Heart Association (**www.americanheart.org**) identifies the following characteristics as common to most Mediterranean diets. There is a "high consumption of fruits, vegetables, bread and other cereals, potatoes, beans, nuts and seeds. Olive oil is an important monounsaturated fat source. Dairy products, fish and poultry are consumed in low to moderate amounts, and little red meat is eaten."

7. U.S. Department of Health and Human Services, *The Health Consequences of Involuntary Exposure to Tobacco Smoke: A Report of the Surgeon General—Executive Summary* (U.S. Department of Health and Human Services, Centers for Disease Control and Prevention, Coordinating Center for Health Promotion, National Center for Chronic Disease Prevention and Health Promotion, Office on Smoking and Health, 2006).

8. R. L. Bratton et al., "Effect of 'ionized' wrist bracelets on musculoskeletal pain: a randomized, double-blind, placebo-controlled trial," *Mayo Clinic Proceedings* 77 (2002):1164–68.

Chapter 2

1. Roper Center, University of Connecticut.

2. M. A. Chase and G. M. Dummer, "The role of sports as a social determinant for children," *Research Quarterly for Exercise and Sport* 63 (1992): 418–24.

3. U.S. Bureau of Labor Statistics.

Chapter 3

1. For more on clickstream analysis, see Zdravko Markov and Daniel Larose, *Data Mining the Web: Uncovering Patterns in Web Content, Structure, and Usage* (John Wiley and Sons, 2007).

2. U.S. Census Bureau.

3. Michael Brett and Charles Goldman, "A meta-analysis of the freshwater trophic cascade," *Proceedings of the National Academy of Sciences* 93 (July 1996).

4. Dr. Peter Nonacs, "Foraging habits of thatch ants," Department of Statistics, University of California at Los Angeles and the Sierra Nevada Aquatic Research Laboratory, **www.stat.ucla.edu/datasets/**.

5. Children's Bureau, Administration for Children and Families, U.S. Department of Health and Human Services.

6. B. S. Glenn et al., "Changes in systolic blood pressure associated with lead in blood and bone," *Epidemiology* 17 (September 2006).

7. National Center for Education Statistics, 2005.

8. National Center for Health Statistics, *Health*, 2006.

Chapter 4

1. *Global Digital Communication: Texting, Social Networking Popular Worldwide*, Pew Research Center Global Attitudes Project, December 2011, **http://www.pewglobal.org/files/2011/12/Pew -Global-Attitudes-Technology-Report-FINAL-December-20-2011.pdf**

2. See note 1.

3. *Crime in the United States, 2004*, **www.fbi.gov**.

Chapter 5

1. Amanda Lenhart et al., *Writing, Technology, and Teens*, Pew Internet and American Life Project, December 2007.

2. U.S. Census Bureau, 2004 American Community Survey.

3. Andrew Rocco Tresolini Fiore, "Romantic regressions: an analysis of behavior in online dating systems," master's thesis, Massachusetts Institute of Technology, 2004.

4. Kristen Purcell, Roger Enner, and Nicole Henderson, *The Rise of Apps Culture*, Pew Research Center's Internet and American Life Project. **www.pewinternet.org**.

5. Washington Initiative (**greaterwashington.org**).

6. *Profile of Hired Farmworkers, a 2008 Update/ERR-60*, Economic Research Service/USDA.

Chapter 6

1. *How Americans Use Text Messaging*, by Aaron Smith, Pew Internet and American Life Project, Pew Research Center, Washington, D.C., **http://pewinternet.org/Reports/2011/Cell-Phone -Texting-2011.aspx**

2. U.S. National Center for Education Statistics. The category "5 or more" has been changed to "5" for this exercise.

3. Gunter Hitsch, Ali Hortacsu, and Dan Ariely, "What makes you click: an empirical analysis of online dating"; available online at **www .aeaweb.org/annual_mtg_papers/2006/0106_0800_0502.pdf**.

4. **www.networkworld.com/news/2010/062310-20-percent-of -android-apps.html**.

5. Stephen J. Blumberg and Julian V. Luke, *Wireless Substitution: Early Release of Estimates from the National Health Interview Survey*, July–December 2009, National Center for Health Statistics, Centers for Disease Control and Prevention.

6. *Women in Management: Analysis of Female Managers' Representation, Characteristics, and Pay*, Government Accountability Office publication GAO-10-892R, September 20, 2010.

7. D. L. Olds, C. R. Henderson Jr., R. Tatelbaum, et al., "Improving the delivery of prenatal care and outcomes of pregnancy: a randomized trial of nurse home visitation," *Pediatrics* 77 (1986): 16–28.

8. Allen J. Wilcox, National Institutes of Health, "The analysis of birth weight and infant mortality," *International Journal of Epidemiology* (December 2001). **eb.niehs.nih.gov/bwt/subcfreq.htm**.

9. Lynn Unruh and Myron Fottler, "Patient turnover and nursing staff adequacy," *Health Services Research*, April 2006.

10. See note 1.

11. Harvard School of Public Health, survey of 5046 adults in hurricane high-risk areas, June–July 2007.

12. The Associated Press/Ipsos Poll actually contacted 1000 adults in June 2007.

13. Barbara Alving et al., "Trends in blood pressure among children and adolescents," *Journal of the American Medical Association* 291 (May 2004): 2107–13.

14. Phillida Bunkle and John Lepper, "Women's participation in gambling: whose reality? A public health issue," paper presented to the European Association for the Study of Gambling Conference, Barcelona, Spain, October 2002.

Chapter 7

1. United States Geological Survey, neic.usgs.gov/neis/qed/.

2. A small business is defined by the SBA as having fewer than 20 employees.

3. Sloan Burke, Michele Wallen, Karen Vail-Smith, and David Knox, "Using technology to control intimate partners: An exploratory study of college undergraduates," *Computers in Human Behavior* 27 (3, May 2011): 1162–67.

4. Murray Mittleman et al., "Determinants of myocardial onset study," *Circulation: Journal of the American Heart Association*, June 1999.

Chapter 8

1. Adapted from A. Johnson, "Results from analyzing metals in 1999 Spokane River fish and crayfish samples," Quantitative Environmental Learning Project, Washington State Department of Ecology report 00-03-017, www.seattlecentral.edu/qelp/sets/021/021.html.

2. Kevin Crowley et al., "Parents explain more often to boys than girls during shared scientific thinking," *Psychological Science* 12 (3, May 2001): 258–61.

3. U.S. Energy Information Administration, 2005.

4. Florida Department of Financial Services, 2011.

5. www.vgchartz.com, April 1, 2011.

6. National Weather Service.

7. George Miller, "The magical number seven, plus or minus two: some limits on our capacity for processing information," *Psychological Review* 63 (1956): 81–97.

8. Mary C. Meyer, "Wider shoes for wider feet?" *Journal of Statistics Education* 14 (1, 2006), www.amstat.org/publications/jse/v14n1/datasets.meyer.html.

9. Robert J. Pianta et al., "Teaching: opportunities to learn in America's elementary classroom," *Science* 315 (March 30, 2007): 1795–96.

10. See Note 3.

11. See Note 4.

12. See Note 5.

13. See Note 6.

14. Community College Survey of Student Engagement (CCSSE), 2007, www.ccsse.org. The survey reported that 178 of 307 (57.98045603%) students worked with classmates outside class to prepare a group assignment during the current academic year. The sample results in Example 8.16 (174 of 300, or 58%) were chosen for ease of calculation.

15. Christopher Reynolds, "Prey tell," *American Demographics* 25 (8, October 2003): 48.

16. www.gallup.com/poll/146885/Positivity-Optimism-Norm-Thriving-States.aspx.

17. http://hosted.ap.org/specials/interactives/wdc/debt_stress/index.html.

18. Mildred Cho and Lisa Bero, "The quality of drug studies published in symposium proceedings," *Annals of Internal Medicine*, 124 (5, March 1996): 485–89.

19. See Note 3.

20. See Note 4.

Chapter 9

1. Press release, August 23, 2007: "Consumers report eating at home more in the wake of high gas prices," NPD Group, Inc., 900 West Shore Road, Port Washington, NY 11050.

2. "When it comes to height, Americans no longer stand tallest," *Research News*, The Ohio State University, researchnews.osu.edu/.

3. *Digital Transactions News*, September 2007.

4. K. Marien, A. Conseur, and M. Sanderson, "The effect of fish consumption on DDT and DDE levels in breast milk among Hispanic immigrants," *Journal of Human Lactation* 14 (3, 1998): 237–42.

5. C. J. Earle, L. B. Brubaker, and G. Segura, International Tree Ring Data Base, NOAA/NGDC Paleoclimatology Program, Boulder, CO.

6. See Note 3.

7. Caroline Davis, Elizabeth Blackmore, Deborah Katzman, and John Fox, "Anorexia nervosa case study," paper presented at Statistical Society of Canada Annual Conference, Montreal, 2004. We have reversed the research question from that of the original case study.

8. health.usnews.com/sections/health/west-hospitals.

9. Courtesy American Heritage Center, University of Wyoming.

10. Data courtesy of OzDASL (Australian Data and Story Library) at statsci.org. The original source is Cara Dubois, ed., *Lowie's Selected Papers in Anthropology* (University of California Press, 1960).

11. http://moconews.net/article/419-average-number-of-apps-downloaded-to-iphone-40-android-25/.

12. *Vital Signs: Current Cigarette Smoking, Morbidity and Mortality Weekly Report*, September 10, 2010. http://www.cdc.gov/mmwr/preview/mmwrhtml/mm5935a3.htm.

13. Mary Madden and Amanda Lenhart "Online dating," Pew Internet and American Life Project, 2006.

14. Barry Kosmin and Egon Mayer, "Principal investigators," American Religious Identification Survey, Graduate Center, City University of New York.

15. Brady Hamilton, Joyce Martin, and Stephanie Ventura, "Births: preliminary data for 2005," *National Vital Statistics Reports* 55 (11), U.S. Department of Health and Human Services.

16. "Patterns and trends in nonmedical prescription pain reliever use: 2002 to 2005," in *NSDUH Report*, Substance Abuse and Mental Health Services Administration, April 6, 2007.

17. Jeff Humphries, "The multicultural economy: minority buying power in the new century," Selig Center for Economic Growth, Terry College of Business, University of Georgia, 2006.

18. "Trends in the prevalence of alcohol use among eighth graders: Monitoring the Future Study, 1991–2003," NIAAA, National Institutes of Health.

19. "Fact sheet: National Survey on Environmental Management of Asthma and Children's Exposure to Environmental Tobacco Smoke," U.S. Environmental Protection Agency, May 17, 2005.

20. Based on data from the U.S. Census Bureau.

21. Energy Information Administration, "Annual electric generator report," Form EIA-906.

22. See Note 5.

23. See Note 6.

24. U.S. Bureau of Labor Statistics.

25. Mary C. Meyer, "Wider shoes for wider feet?" *Journal of Statistics Education* 14 (1, 2006).

26. Steve Strand, Ian Deary, and Pauline Smith, "Sex differences in cognitive abilities test scores: a UK national picture," *British Journal of Educational Psychology* 76 (2006): 463–80.

27. Siobhan Banks and David Dinges, "Behavioral and physiological consequences of sleep restriction," *Journal of Clinical Sleep Medicine* 15 (2007): 519–28.

28. U.S. Census Bureau.

29. "A nation online: entering the broadband age," Economics and Statistics Administration, U.S. Department of Commerce.

30. U.S. Bureau of Justice Statistics.

31. Joyce A. Martin et al., "Births: final data for 2005," *National Vital Statistics Reports*, 56 (6, December 5, 2007).

Chapter 10

1. Kelley, H. H., "The warm-cold variable in first impression of persons," *Journal of Personality* 18 (1950): 431–39.

2. A. Towler and R. L. Dipboye, "The effect of instructor reputation and need for cognition on student behavior," poster presented at American Psychological Society conference, May 1998.

3. K. J. Thomas et al., "Randomized controlled trial of a short course of traditional acupuncture compared with usual care for persistent non-specific low back pain," *British Medical Journal* 23 (September 2006).

4. Karin Olson and John Hanson, "Using reiki to manage pain," *Cancer Prevention and Control* 1 (2, 1997): 108–13.

5. "Highway safety projects—before and after study update," *Measures, Markers, and Mileposts*, Washington State Department of Transportation, December 2005.

6. P. A. Mackowiak, S. S. Wasserman, and M. M. Levine, "A critical appraisal of 98.6 degrees F, the upper limit of the normal body temperature, and other legacies of Carl Reinhold August Wunderlich," *Journal of the American Medical Association* 268 (1992): 1578–80.

7. George W. Snedecor and William G. Cochran, *Statistical Methods*, 8th Ed. (Iowa State University Press, 1989).

8. See Barry K. Moser and Gary R. Stevens, "Homogeneity of variance in the two-sample means test," *American Statistician* 46 (1, February 1992): 19–21.

9. Steven Reinberg, "U.S. kids using media almost 8 hours a day: survey finds few parents set rules as use of 'smart' phones,

computers soars," *Bloomberg Business Week: Executive Health*, January 20, 2010. **www.businessweek.com/lifestyle/content /healthday/635134.html**.

10. D. L. Olds, C. R. Henderson Jr, R. Tatelbaum et al., "Improving the delivery of prenatal care and outcomes of pregnancy: a randomized trial of nurse home visitation," *Pediatrics* 77 (1986): 16–28.

11. Amanda Lenhart and Mary Madden, "Teens, privacy, and online social networks: how teens manage their online identities and personal information in the age of MySpace," Pew Internet and American Life Project, April 2007.

12. Vijayakrishna K. Gadi et al., "Case-control study of fetal microchimerism and breast cancer," *PLoS one* 3 (March 5, 2008). (plos one, doi; 10:1371/journal.pone.0001706).

13. R. L. Bratton et al., "Effect of 'ionized' wrist bracelets on musculoskeletal pain: a randomized, double-blind, placebo-controlled trial," *Mayo Clinic Proceedings* 77 (2002): 1164–68.

Chapter 11

1. Mary Madden and Amanda Lenhart, *Online Dating*, Pew Internet and American Life Project, 2005.

2. U.S. Department of Education, National Center for Education Statistics, Adult Education Survey of the 2005 National Household Education Surveys Program.

3. Derek M. Burnett et al., "Impact of minority status following traumatic spinal cord injury," *NeuroRehabilitation* 17 (2002): 187–94.

4. Pew Research Center for the People and the Press, *How Young People View Their Lives, Futures, and Politics: A Portrait of "Generation Next"* (Washington, D.C., 2007).

5. Andrew Rocco Tresolini Fiore, "Romantic regressions: an analysis of behavior in online dating systems," master's thesis, Program in Media Arts and Sciences, Massachusetts Institute of Technology, 2004.

6. See Note 1.

7. S. Blackman and D. Catalina, "The moon and the emergency room," *Perceptual and Motor Skills* 37 (1973): 624–26.

8. J. R. Knight, H. Wechsler, M. Kuo, M. Seibring, E. R. Weitzman, and M. Schuckit, "Alcohol abuse and dependence among U.S. college students," *Journal of Studies on Alcohol* 63, (3, 2002): 263–70.

9. Donald Garrow and Leonard Egede, "National patterns and correlates of complementary and alternative medicine use in adults with diabetes," *Journal of Alternative and Complementary Medicine* 12 (2006): 895–902.

10. J. E. Anderson and S. Sansom, "HIV testing in a national sample of pregnant US women: who is not getting tested?" *AIDS Care* 19 (March 2007): 375–80.

11. National Agricultural Statistics Service, *Agricultural Statistics*, **www.usda.gov/nass** 2006.

Page numbers in **boldface** indicate definitions; those followed by *f* indicate figures; those followed by *t* indicate tables.

Table E Chi-square (χ^2) distribution

Degrees of freedom	Area to the right of critical value									
	0.995	0.99	0.975	0.95	0.90	0.10	0.05	0.025	0.01	0.005
1	—	—	0.001	0.004	0.016	2.706	3.841	5.024	6.635	7.879
2	0.010	0.020	0.051	0.103	0.211	4.605	5.991	7.378	9.210	10.597
3	0.072	0.115	0.216	0.352	0.584	6.251	7.815	9.348	11.345	12.838
4	0.207	0.297	0.484	0.711	1.064	7.779	9.488	11.143	13.277	14.860
5	0.412	0.554	0.831	1.145	1.610	9.236	11.071	12.833	15.086	16.750
6	0.676	0.872	1.237	1.635	2.204	10.645	12.592	14.449	16.812	18.548
7	0.989	1.239	1.690	2.167	2.833	12.017	14.067	16.013	18.475	20.278
8	1.344	1.646	2.180	2.733	3.490	13.362	15.507	17.535	20.090	21.955
9	1.735	2.088	2.700	3.325	4.168	14.684	16.919	19.023	21.666	23.589
10	2.156	2.558	3.247	3.940	4.865	15.987	18.307	20.483	23.209	25.188
11	2.603	3.053	3.816	4.575	5.578	17.275	19.675	21.920	24.725	26.757
12	3.074	3.571	4.404	5.226	6.304	18.549	21.026	23.337	26.217	28.299
13	3.565	4.107	5.009	5.892	7.042	19.812	22.362	24.736	27.688	29.819
14	4.075	4.660	5.629	6.571	7.790	21.064	23.685	26.119	29.141	31.319
15	4.601	5.229	6.262	7.261	8.547	22.307	24.996	27.488	30.578	32.801
16	5.142	5.812	6.908	7.962	9.312	23.542	26.296	28.845	32.000	34.267
17	5.697	6.408	7.564	8.672	10.085	24.769	27.587	30.191	33.409	35.718
18	6.265	7.015	8.231	9.390	10.865	25.989	28.869	31.526	34.805	37.156
19	6.844	7.633	8.907	10.117	11.651	27.204	30.144	32.852	36.191	38.582
20	7.434	8.260	9.591	10.851	12.443	28.412	31.410	34.170	37.566	39.997
21	8.034	8.897	10.283	11.591	13.240	29.615	32.671	35.479	38.932	41.401
22	8.643	9.542	10.982	12.338	14.042	30.813	33.924	36.781	40.289	42.796
23	9.260	10.196	11.689	13.091	14.848	32.007	35.172	38.076	41.638	44.181
24	9.886	10.856	12.401	13.848	15.659	33.196	36.415	39.364	42.980	45.559
25	10.520	11.524	13.120	14.611	16.473	34.382	37.652	40.646	44.314	46.928
26	11.160	12.198	13.844	15.379	17.292	35.563	38.885	41.923	45.642	48.290
27	11.808	12.879	14.573	16.151	18.114	36.741	40.113	43.194	46.963	49.645
28	12.461	13.565	15.308	16.928	18.939	37.916	41.337	44.461	48.278	50.993
29	13.121	14.257	16.047	17.708	19.768	39.087	42.557	45.722	49.588	52.336
30	13.787	14.954	16.791	18.493	20.599	40.256	43.773	46.979	50.892	53.672
40	20.707	22.164	24.433	26.509	29.051	51.805	55.758	59.342	63.691	66.766
50	27.991	29.707	32.357	34.764	37.689	63.167	67.505	71.420	76.154	79.490
60	35.534	37.485	40.482	43.188	46.459	74.397	79.082	83.298	88.379	91.952
70	43.275	45.442	48.758	51.739	55.329	85.527	90.531	95.023	100.425	104.215
80	51.172	53.540	57.153	60.391	64.278	96.578	101.879	106.629	112.329	116.321
90	59.196	61.754	65.647	69.126	73.291	107.565	113.145	118.136	124.116	128.299
100	67.328	70.065	74.222	77.929	82.358	118.498	124.342	129.561	135.807	140.169

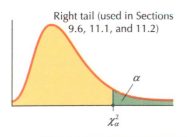

Right tail (used in Sections 9.6, 11.1, and 11.2)

α

χ^2_α

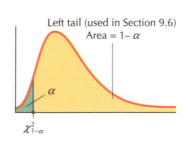

Left tail (used in Section 9.6)
Area = $1-\alpha$

α

$\chi^2_{1-\alpha}$

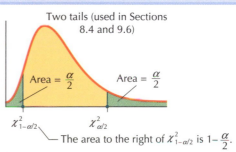

Two tails (used in Sections 8.4 and 9.6)

Area = $\frac{\alpha}{2}$ Area = $\frac{\alpha}{2}$

$\chi^2_{1-\alpha/2}$ $\chi^2_{\alpha/2}$

The area to the right of $\chi^2_{1-\alpha/2}$ is $1-\frac{\alpha}{2}$.

Table D t-Distribution

df		80%	90%	95%	98%	99%
				Confidence level		
				Area in one tail		
		0.10	0.05	0.025	0.01	0.005
				Area in two tails		
		0.20	0.10	0.05	0.02	0.01
df	1	3.078	6.314	12.706	31.821	63.657
	2	1.886	2.920	4.303	6.965	9.925
	3	1.638	2.353	3.182	4.541	5.841
	4	1.533	2.132	2.776	3.747	4.604
	5	1.476	2.015	2.571	3.365	4.032
	6	1.440	1.943	2.447	3.143	3.707
	7	1.415	1.895	2.365	2.998	3.499
	8	1.397	1.860	2.306	2.896	3.355
	9	1.383	1.833	2.262	2.821	3.250
	10	1.372	1.812	2.228	2.764	3.169
	11	1.363	1.796	2.201	2.718	3.106
	12	1.356	1.782	2.179	2.681	3.055
	13	1.350	1.771	2.160	2.650	3.012
	14	1.345	1.761	2.145	2.624	2.977
	15	1.341	1.753	2.131	2.602	2.947
	16	1.337	1.746	2.120	2.583	2.921
	17	1.333	1.740	2.110	2.567	2.898
	18	1.330	1.734	2.101	2.552	2.878
	19	1.328	1.729	2.093	2.539	2.861
	20	1.325	1.725	2.086	2.528	2.845
	21	1.323	1.721	2.080	2.518	2.831
	22	1.321	1.717	2.074	2.508	2.819
	23	1.319	1.714	2.069	2.500	2.807
	24	1.318	1.711	2.064	2.492	2.797
	25	1.316	1.708	2.060	2.485	2.787
	26	1.315	1.706	2.056	2.479	2.779
	27	1.314	1.703	2.052	2.473	2.771
	28	1.313	1.701	2.048	2.467	2.763
	29	1.311	1.699	2.045	2.462	2.756
	30	1.310	1.697	2.042	2.457	2.750
	31	1.309	1.696	2.040	2.453	2.744
	32	1.309	1.694	2.037	2.449	2.738
	33	1.308	1.692	2.035	2.445	2.733
	34	1.307	1.691	2.032	2.441	2.728
	35	1.306	1.690	2.030	2.438	2.724
	36	1.306	1.688	2.028	2.435	2.719
	37	1.305	1.687	2.026	2.431	2.715
	38	1.304	1.686	2.024	2.429	2.712
	39	1.304	1.685	2.023	2.426	2.708
	40	1.303	1.684	2.021	2.423	2.704
	50	1.299	1.676	2.009	2.403	2.678
	60	1.296	1.671	2.000	2.390	2.660
	70	1.294	1.667	1.994	2.381	2.648
	80	1.292	1.664	1.990	2.374	2.639
	90	1.291	1.662	1.987	2.368	2.632
	100	1.290	1.660	1.984	2.364	2.626
	1000	1.282	1.646	1.962	2.330	2.581
	z	1.282	1.645	1.960	2.326	2.576

Table E Chi-square (χ^2) distribution

Degrees of freedom	\multicolumn{10}{c}{Area to the right of critical value}
	0.995	0.99	0.975	0.95	0.90	0.10	0.05	0.025	0.01	0.005
1	—	—	0.001	0.004	0.016	2.706	3.841	5.024	6.635	7.879
2	0.010	0.020	0.051	0.103	0.211	4.605	5.991	7.378	9.210	10.597
3	0.072	0.115	0.216	0.352	0.584	6.251	7.815	9.348	11.345	12.838
4	0.207	0.297	0.484	0.711	1.064	7.779	9.488	11.143	13.277	14.860
5	0.412	0.554	0.831	1.145	1.610	9.236	11.071	12.833	15.086	16.750
6	0.676	0.872	1.237	1.635	2.204	10.645	12.592	14.449	16.812	18.548
7	0.989	1.239	1.690	2.167	2.833	12.017	14.067	16.013	18.475	20.278
8	1.344	1.646	2.180	2.733	3.490	13.362	15.507	17.535	20.090	21.955
9	1.735	2.088	2.700	3.325	4.168	14.684	16.919	19.023	21.666	23.589
10	2.156	2.558	3.247	3.940	4.865	15.987	18.307	20.483	23.209	25.188
11	2.603	3.053	3.816	4.575	5.578	17.275	19.675	21.920	24.725	26.757
12	3.074	3.571	4.404	5.226	6.304	18.549	21.026	23.337	26.217	28.299
13	3.565	4.107	5.009	5.892	7.042	19.812	22.362	24.736	27.688	29.819
14	4.075	4.660	5.629	6.571	7.790	21.064	23.685	26.119	29.141	31.319
15	4.601	5.229	6.262	7.261	8.547	22.307	24.996	27.488	30.578	32.801
16	5.142	5.812	6.908	7.962	9.312	23.542	26.296	28.845	32.000	34.267
17	5.697	6.408	7.564	8.672	10.085	24.769	27.587	30.191	33.409	35.718
18	6.265	7.015	8.231	9.390	10.865	25.989	28.869	31.526	34.805	37.156
19	6.844	7.633	8.907	10.117	11.651	27.204	30.144	32.852	36.191	38.582
20	7.434	8.260	9.591	10.851	12.443	28.412	31.410	34.170	37.566	39.997
21	8.034	8.897	10.283	11.591	13.240	29.615	32.671	35.479	38.932	41.401
22	8.643	9.542	10.982	12.338	14.042	30.813	33.924	36.781	40.289	42.796
23	9.260	10.196	11.689	13.091	14.848	32.007	35.172	38.076	41.638	44.181
24	9.886	10.856	12.401	13.848	15.659	33.196	36.415	39.364	42.980	45.559
25	10.520	11.524	13.120	14.611	16.473	34.382	37.652	40.646	44.314	46.928
26	11.160	12.198	13.844	15.379	17.292	35.563	38.885	41.923	45.642	48.290
27	11.808	12.879	14.573	16.151	18.114	36.741	40.113	43.194	46.963	49.645
28	12.461	13.565	15.308	16.928	18.939	37.916	41.337	44.461	48.278	50.993
29	13.121	14.257	16.047	17.708	19.768	39.087	42.557	45.722	49.588	52.336
30	13.787	14.954	16.791	18.493	20.599	40.256	43.773	46.979	50.892	53.672
40	20.707	22.164	24.433	26.509	29.051	51.805	55.758	59.342	63.691	66.766
50	27.991	29.707	32.357	34.764	37.689	63.167	67.505	71.420	76.154	79.490
60	35.534	37.485	40.482	43.188	46.459	74.397	79.082	83.298	88.379	91.952
70	43.275	45.442	48.758	51.739	55.329	85.527	90.531	95.023	100.425	104.215
80	51.172	53.540	57.153	60.391	64.278	96.578	101.879	106.629	112.329	116.321
90	59.196	61.754	65.647	69.126	73.291	107.565	113.145	118.136	124.116	128.299
100	67.328	70.065	74.222	77.929	82.358	118.498	124.342	129.561	135.807	140.169

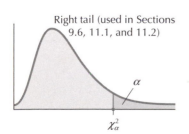

Right tail (used in Sections 9.6, 11.1, and 11.2)

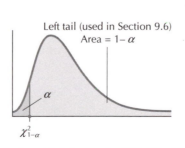

Left tail (used in Section 9.6)
Area = $1 - \alpha$

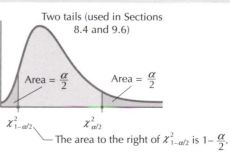

Two tails (used in Sections 8.4 and 9.6)

Area = $\dfrac{\alpha}{2}$ Area = $\dfrac{\alpha}{2}$

$\chi^2_{1-\alpha/2}$ $\chi^2_{\alpha/2}$

The area to the right of $\chi^2_{1-\alpha/2}$ is $1 - \dfrac{\alpha}{2}$.

IMPORTANT FORMULAS
for Larose, Discovering the Fundamentals of Statistics Second Edition
© 2013 by W.H. Freeman and Company

Chapter 3   Describing Data Numerically

- Sample mean (p. 83): $\bar{x} = \sum x/n$
- Population mean (p. 84): $\mu = \sum x/N$
- Range (p. 98): Largest data value – smallest data value
- Population variance (p. 101): $\sigma^2 = \dfrac{\sum(x-\mu)^2}{N}$ or
  $\sigma^2 = \dfrac{\sum x^2 - \left(\sum x\right)^2/N}{N}$
- Population standard deviation (p. 101):
  $\sigma = \sqrt{\sigma^2} = \sqrt{\dfrac{\sum(x-\mu)^2}{N}}$ or $\sigma = \sqrt{\dfrac{\sum x^2 - \left(\sum x\right)^2/N}{N}}$
- Sample variance (p. 103): $s^2 = \dfrac{\sum(x-\bar{x})^2}{n-1}$ or
  $s^2 = \dfrac{\sum x^2 - \left(\sum x\right)^2/n}{n-1}$
- Sample standard deviation (p. 103):
  $s = \sqrt{s^2} = \sqrt{\dfrac{\sum(x-\bar{x})^2}{n-1}}$ or $s = \sqrt{\dfrac{\left(\sum x^2 - \sum x\right)^2/n}{n-1}}$
- Weighted mean (p. 115):
  $\bar{x} = \dfrac{\sum(w \cdot x)}{\sum w}$
- Estimated mean for data grouped into a frequency distribution
  (p. 116): $\bar{x} = \dfrac{\sum(f \cdot x)}{\sum f}$
- Estimated variance for data grouped into a frequency distribution
  (p. 117): $s^2 = \dfrac{\sum(x-\bar{x})^2 \cdot f}{\sum f}$

- Estimated standard deviation for data grouped into a frequency
  distribution (p. 117): $s = \sqrt{s^2} = \sqrt{\dfrac{\sum(x-\bar{x})^2 \cdot f}{\sum f}}$
- Percentile (position of pth percentile) (p. 125): $i = (p/100)n$
- Z-score for sample data (p. 121): $\dfrac{x-\bar{x}}{s}$
- Z-score for population data (p. 121): $\dfrac{x-\mu}{\sigma}$
- Chebyshev's Rule (p. 107): *At least* $\left(1 - \dfrac{1}{k^2}\right)100\%$ of the values
  from any data set will fall within k standard deviations of the mean,
  where $k > 1$.
- Calculating a data value, given its z-score (p. 123):
  - For a sample: $x = z\text{-score} \cdot s + \bar{x}$
  - For a population: $x = z\text{-score} \cdot \sigma + \mu$
- The Empirical Rule: If the data distribution is bell-shaped (p. 105):
  - About 68% of the data values will fall within one standard
    deviation of the mean.
  - About 95% of the data values will fall within two standard
    deviations of the mean.
  - About 99.7% of the data values will fall within three standard
    deviations of the mean.
- Interquartile range (p. 131): IQR = Q3 − Q1
- Percentile Rank (p. 127):
  percentile rank of data value $x = \dfrac{\text{number of values in data set} \leq x}{\text{total number of values in data set}} \cdot 100$
- Five-number summary (p. 135): Minimum, Q1, Median, Q3,
  Maximum
- Lower fence (for box plot) (p. 136): Q1 − 1.5(IQR)
- Upper fence (for box plot) (p. 136): Q3 + 1.5(IQR)

Chapter 4   Describing the Relationship Between Two Variables

- Correlation coefficient r (p. 153 or 157):
  $r = \dfrac{\sum(x-\bar{x})(y-\bar{y})}{(n-1)s_x s_y}$ or $r = \dfrac{\sum xy - \left(\sum x \sum y\right)/n}{(n-1)\, s_x s_y}$

- Regression equation (regression line) (p. 164): $\hat{y} = b_0 + b_1 x$

- Slope of the regression line (p. 164 or p. 166): $b_1 = \dfrac{r \cdot s_y}{s_x}$ or
  $b_1 = \dfrac{\sum xy - \left(\sum x \sum y\right)/n}{\sum x^2 - \left(\sum x\right)^2/n}$

- y-Intercept (p. 167): $b_0 = \bar{y} - (b_1 \cdot \bar{x})$

- Prediction error or residual (p. 169): $(y - \hat{y})$
- SSE, sum of squares error (p. 179): $\sum(y - \hat{y})^2$
- Standard error of the estimate (p. 180):
  $s = \sqrt{\text{MSE}} = \sqrt{\dfrac{\text{SSE}}{n-2}}$
- SST, sum of squares total (p. 181 or p. 184): SST = $(n-1)s^2$ or
  SST = $\sum y^2 - \left(\sum y\right)^2/n$
- The coefficient of determination (p. 183): r^2 = SSR/SST
- SSR, sum of squares regression (p. 182): SSR = $\sum(\hat{y} - \bar{y})^2$
- Correlation coefficient r (p. 185): Can be expressed as
  $r = \pm\sqrt{r^2}$, taking the positive or negative sign of the slope b_1.

Chapter 5   Probability

- Classical method for assigning probabilities (p. 197):
  $P(E) = \dfrac{\text{number of outcomes in } E}{\text{number of outcomes in sample space}} = \dfrac{N(E)}{N(S)}$
- Relative frequency method of assigning probabilities (p. 202):
  $P(E) \approx \dfrac{\text{frequency of } E}{\text{number of trials of experiment}}$
- Probabilities for complements (p. 210): $P(A) + P(A^C) = 1$,
  $P(A) = 1 - P(A^C)$ and $P(A^C) = 1 - P(A)$
- Addition Rule (p. 211):
  $P(A \text{ or } B) = P(A \cup B) = P(A) + P(B) - P(A \cap B)$
- Addition Rule for mutually exclusive events (p. 213):
  $P(A \cup B) = P(A) + P(B)$

- Conditional probability (p. 218):
  $P(B|A) = \dfrac{P(A \cap B)}{P(A)} = \dfrac{N(A \cap B)}{N(A)}$
- Independent events (p. 219): Events A and B are *independent* if
  $P(A|B) = P(A)$ or if $P(B|A) = P(B)$.
- Multiplication Rule (p. 220): $P(A \cap B) = P(B)\, P(A|B)$, or
  $P(A \cap B) = P(A)\, P(B|A)$
- Alternate method for determining independence (p. 222):
  - If $P(A)\, P(B) = P(A \cap B)$, then events A and B are *independent*.
  - If $P(A)\, P(B) \neq P(A \cap B)$, then events A and B are *dependent*.
- Multiplication Rule for n independent events (p. 227):
  $P(A \cap B \cap C \cap \ldots) = P(A)\, P(B)\, P(C) \ldots$

Table D *t*-Distribution

		80%	90%	Confidence level 95%	98%	99%
				Area in one tail		
		0.10	0.05	0.025	0.01	0.005
				Area in two tails		
		0.20	0.10	0.05	0.02	0.01
df	1	3.078	6.314	12.706	31.821	63.657
	2	1.886	2.920	4.303	6.965	9.925
	3	1.638	2.353	3.182	4.541	5.841
	4	1.533	2.132	2.776	3.747	4.604
	5	1.476	2.015	2.571	3.365	4.032
	6	1.440	1.943	2.447	3.143	3.707
	7	1.415	1.895	2.365	2.998	3.499
	8	1.397	1.860	2.306	2.896	3.355
	9	1.383	1.833	2.262	2.821	3.250
	10	1.372	1.812	2.228	2.764	3.169
	11	1.363	1.796	2.201	2.718	3.106
	12	1.356	1.782	2.179	2.681	3.055
	13	1.350	1.771	2.160	2.650	3.012
	14	1.345	1.761	2.145	2.624	2.977
	15	1.341	1.753	2.131	2.602	2.947
	16	1.337	1.746	2.120	2.583	2.921
	17	1.333	1.740	2.110	2.567	2.898
	18	1.330	1.734	2.101	2.552	2.878
	19	1.328	1.729	2.093	2.539	2.861
	20	1.325	1.725	2.086	2.528	2.845
	21	1.323	1.721	2.080	2.518	2.831
	22	1.321	1.717	2.074	2.508	2.819
	23	1.319	1.714	2.069	2.500	2.807
	24	1.318	1.711	2.064	2.492	2.797
	25	1.316	1.708	2.060	2.485	2.787
	26	1.315	1.706	2.056	2.479	2.779
	27	1.314	1.703	2.052	2.473	2.771
	28	1.313	1.701	2.048	2.467	2.763
	29	1.311	1.699	2.045	2.462	2.756
	30	1.310	1.697	2.042	2.457	2.750
	31	1.309	1.696	2.040	2.453	2.744
	32	1.309	1.694	2.037	2.449	2.738
	33	1.308	1.692	2.035	2.445	2.733
	34	1.307	1.691	2.032	2.441	2.728
	35	1.306	1.690	2.030	2.438	2.724
	36	1.306	1.688	2.028	2.435	2.719
	37	1.305	1.687	2.026	2.431	2.715
	38	1.304	1.686	2.024	2.429	2.712
	39	1.304	1.685	2.023	2.426	2.708
	40	1.303	1.684	2.021	2.423	2.704
	50	1.299	1.676	2.009	2.403	2.678
	60	1.296	1.671	2.000	2.390	2.660
	70	1.294	1.667	1.994	2.381	2.648
	80	1.292	1.664	1.990	2.374	2.639
	90	1.291	1.662	1.987	2.368	2.632
	100	1.290	1.660	1.984	2.364	2.626
	1000	1.282	1.646	1.962	2.330	2.581
	z	1.282	1.645	1.960	2.326	2.576

IMPORTANT FORMULAS
for Larose, Discovering the Fundamentals of Statistics Second Edition
© 2013 by W.H. Freeman and Company

- Factorial symbol $n!$ (p. 237): $0! = 1$; $1! = 1$;
  $n! = n(n-1)(n-2)\cdots 3 \cdot 2 \cdot 1$
- Permutation of r items chosen from n distinct items (p. 257):

  $_nP_r = \dfrac{n!}{(n-r)!}$

- Combination of r items chosen from n distinct items (p. 239):

  $_nC_r = \dfrac{n!}{r!(n-r)!}$

- Permutations of nondistinct items (p. 241): $\dfrac{n!}{n_1! \cdot n_2! \cdot \ldots \cdot n_k!}$

Chapter 6 Random Variables and the Normal Distribution

- Mean μ of a discrete random variable X (p. 258):

  $\mu = \sum X \cdot P(X)$

- Variance of a discrete random variable X (p. 261):

  $\sigma^2 = \sum (X - \mu)^2 \cdot P(X)$ or

  $\sigma^2 = \sum (X^2 \cdot P(X)) - \mu^2$

- Standard deviation of a discrete random variable X (p. 261):

  $\sigma = \sqrt{\sum (X - \mu)^2 \cdot P(X)}$ or $\sigma = \sqrt{\sum (X^2 \cdot P(X)) - \mu^2}$

- The binomial probability distribution formula (p. 270):
  $P(X) = (_nC_X)\, p^n\, (1-p)^{n-X}$
- Mean of a binomial random variable (p. 273): $\mu = n \cdot p$
- Variance of a binomial random variable (p. 273):

  $\sigma^2 = n \cdot p \cdot (1-p)$

- Standard deviation of a binomial random variable (p. 273):

  $\sigma = \sqrt{n \cdot p \cdot (1-p)}$

- Standardizing a normal random variable (p. 299): $Z = \dfrac{X - \mu}{\sigma}$
- Calculating the X-value, given a Z-value (p. 288): $X = Z\sigma + \mu$

Chapter 7 Sampling Distributions

- Mean and standard deviation of the sampling distribution of the sample mean $\bar{x}$ (p. 324):

  $\mu_{\bar{x}} = \mu$, $\qquad \sigma_{\bar{x}} = \dfrac{\sigma}{\sqrt{n}}$

- Standardizing a normal sampling distribution for means (p. 326):

  $Z = \dfrac{\bar{x} - \mu_{\bar{x}}}{\sigma_{\bar{x}}} = \dfrac{\bar{x} - \mu}{\sigma/\sqrt{n}}$

- Central Limit Theorem for Means (p. 334): Given a population with mean μ and standard deviation σ, the sampling distribution of the sample mean $\bar{x}$ becomes approximately normal $(\mu, \sigma/\sqrt{n})$ as the sample size gets larger, regardless of the shape of the population.

- Mean and standard deviation of the sampling distribution of the sample proportion $\hat{p}$ (p. 343):

  $\mu_{\hat{p}} = p$, $\qquad \sigma_{\hat{p}} = \sqrt{\dfrac{p \cdot (1-p)}{n}}$

- Central Limit Theorem for Proportions (p. 345): The sampling distribution of the sample proportion $\hat{p}$ follows an approximately normal distribution with mean $\mu_{\hat{p}} = p$ and standard deviation

  $\sigma_{\hat{p}} = \sqrt{\dfrac{p \cdot (1-p)}{n}}$ when both the following conditions are satisfied: (1) $np \geq 5$ and (2) $n(1-p) \geq 5$.

- Standardizing a normal sampling distribution for proportions

  (p. 346): $Z = \dfrac{\hat{p} - \mu_{\hat{p}}}{\sigma_{\hat{p}}} = \dfrac{\hat{p} - p}{\sqrt{\dfrac{p(1-p)}{n}}}$

Chapter 8 Confidence Intervals

- $100(1 - \alpha)\%$ Z confidence interval for μ (p. 357):

  Lower Bound $= \bar{x} - Z_{\alpha/2}(\sigma/\sqrt{n})$, Upper Bound $= \bar{x} + Z_{\alpha/2}(\sigma/\sqrt{n})$

  provided either the original population is normal, and σ is known, or the sample size is large ($n \geq 30$), and σ is known.
- Sample size for estimating the population mean (p. 364):

  $n = \left(\dfrac{(Z_{\alpha/2})\sigma}{E}\right)^2$

  where $Z_{\alpha/2}$ is associated with the desired confidence level, and E is the desired margin of error. Round up to the next integer.
- $100(1 - \alpha)\%$ t confidence interval for μ (p. 373):

  Lower Bound $= \bar{x} - t_{\alpha/2}(s/\sqrt{n})$, Upper Bound $= \bar{x} + t_{\alpha/2}(s/\sqrt{n})$

  where $t_{\alpha/2}$ is based on $n - 1$ degrees of freedom and either the population is normal or the sample size is large ($n \geq 30$).
- $100(1 - \alpha)\%$ Z confidence interval for p (p. 383): Lower Bound $=$

  $\hat{p} - Z_{\alpha/2}\sqrt{\dfrac{\hat{p} \cdot \hat{q}}{n}}$, Upper Bound $= \hat{p} + Z_{\alpha/2}\sqrt{\dfrac{\hat{p} \cdot \hat{q}}{n}}$

  The Z interval for p may be used only if $both$ of the following conditions apply: $n\hat{p} \geq 5$ and $n(1 - \hat{p}) \geq 5$.
- Sample size for estimating a population proportion when $\hat{p}$ is known (p. 387):

  $n = \hat{p}(1 - \hat{p})\left(\dfrac{Z_{\alpha/2}}{E}\right)^2$

where $Z_{\alpha/2}$ is associated with the desired confidence level, and E is the desired margin of error. Round up to the next integer.
- Sample size for estimating a population proportion when $\hat{p}$ is not known (p. 387):

  $n = \left(\dfrac{(0.5)(Z_{\alpha/2})}{E}\right)^2$

  where $Z_{\alpha/2}$ is associated with the desired confidence level, and E is the desired margin of error. Round up to the next integer.
- $100(1 - \alpha)\%$ χ^2 confidence interval for the population variance σ^2

  (p. 395): Lower Bound $= \dfrac{(n-1)s^2}{\chi^2_{\alpha/2}}$, Upper Bound $= \dfrac{(n-1)s^2}{\chi^2_{1-\alpha/2}}$

  where $\chi^2_{1-\alpha/2}$ and $\chi^2_{\alpha/2}$ are the critical values for a χ^2 distribution with $n - 1$ degrees of freedom, and provided that the sample is taken from a normal population.
- $100(1 - \alpha)\%$ χ^2 confidence interval for the population standard deviation σ (p. 395):

  Lower Bound $= \sqrt{\dfrac{(n-1)s^2}{\chi^2_{\alpha/2}}}$, Upper Bound $= \sqrt{\dfrac{(n-1)s^2}{\chi^2_{1-\alpha/2}}}$

  where $\chi^2_{1-\alpha/2}$ and $\chi^2_{\alpha/2}$ are the critical values for a χ^2 distribution with $n - 1$ degrees of freedom, and provided that the sample is taken from a normal population.

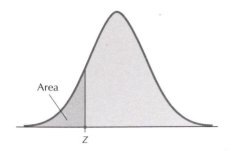

Table C Standard normal distribution

Z	0.00	0.01	0.02	0.03	0.04	0.05	0.06	0.07	0.08	0.09
−3.4	0.0003	0.0003	0.0003	0.0003	0.0003	0.0003	0.0003	0.0003	0.0003	0.0002
−3.3	0.0005	0.0005	0.0005	0.0004	0.0004	0.0004	0.0004	0.0004	0.0004	0.0003
−3.2	0.0007	0.0007	0.0006	0.0006	0.0006	0.0006	0.0006	0.0005	0.0005	0.0005
−3.1	0.0010	0.0009	0.0009	0.0009	0.0008	0.0008	0.0008	0.0008	0.0007	0.0007
−3.0	0.0013	0.0013	0.0013	0.0012	0.0012	0.0011	0.0011	0.0011	0.0010	0.0010
−2.9	0.0019	0.0018	0.0018	0.0017	0.0016	0.0016	0.0015	0.0015	0.0014	0.0014
−2.8	0.0026	0.0025	0.0024	0.0023	0.0023	0.0022	0.0021	0.0021	0.0020	0.0019
−2.7	0.0035	0.0034	0.0033	0.0032	0.0031	0.0030	0.0029	0.0028	0.0027	0.0026
−2.6	0.0047	0.0045	0.0044	0.0043	0.0041	0.0040	0.0039	0.0038	0.0037	0.0036
−2.5	0.0062	0.0060	0.0059	0.0057	0.0055	0.0054	0.0052	0.0051	0.0049	0.0048
−2.4	0.0082	0.0080	0.0078	0.0075	0.0073	0.0071	0.0069	0.0068	0.0066	0.0064
−2.3	0.0107	0.0104	0.0102	0.0099	0.0096	0.0094	0.0091	0.0089	0.0087	0.0084
−2.2	0.0139	0.0136	0.0132	0.0129	0.0125	0.0122	0.0119	0.0116	0.0113	0.0110
−2.1	0.0179	0.0174	0.0170	0.0166	0.0162	0.0158	0.0154	0.0150	0.0146	0.0143
−2.0	0.0228	0.0222	0.0217	0.0212	0.0207	0.0202	0.0197	0.0192	0.0188	0.0183
−1.9	0.0287	0.0281	0.0274	0.0268	0.0262	0.0256	0.0250	0.0244	0.0239	0.0233
−1.8	0.0359	0.0351	0.0344	0.0336	0.0329	0.0322	0.0314	0.0307	0.0301	0.0294
−1.7	0.0446	0.0436	0.0427	0.0418	0.0409	0.0401	0.0392	0.0384	0.0375	0.0367
−1.6	0.0548	0.0537	0.0526	0.0516	0.0505	0.0495	0.0485	0.0475	0.0465	0.0455
−1.5	0.0668	0.0655	0.0643	0.0630	0.0618	0.0606	0.0594	0.0582	0.0571	0.0559
−1.4	0.0808	0.0793	0.0778	0.0764	0.0749	0.0735	0.0721	0.0708	0.0694	0.0681
−1.3	0.0968	0.0951	0.0934	0.0918	0.0901	0.0885	0.0869	0.0853	0.0838	0.0823
−1.2	0.1151	0.1131	0.1112	0.1093	0.1075	0.1056	0.1038	0.1020	0.1003	0.0985
−1.1	0.1357	0.1335	0.1314	0.1292	0.1271	0.1251	0.1230	0.1210	0.1190	0.1170
−1.0	0.1587	0.1562	0.1539	0.1515	0.1492	0.1469	0.1446	0.1423	0.1401	0.1379
−0.9	0.1841	0.1814	0.1788	0.1762	0.1736	0.1711	0.1685	0.1660	0.1635	0.1611
−0.8	0.2119	0.2090	0.2061	0.2033	0.2005	0.1977	0.1949	0.1922	0.1894	0.1867
−0.7	0.2420	0.2389	0.2358	0.2327	0.2296	0.2266	0.2236	0.2206	0.2177	0.2148
−0.6	0.2743	0.2709	0.2676	0.2643	0.2611	0.2578	0.2546	0.2514	0.2483	0.2451
−0.5	0.3085	0.3050	0.3015	0.2981	0.2946	0.2912	0.2877	0.2843	0.2810	0.2776
−0.4	0.3446	0.3409	0.3372	0.3336	0.3300	0.3264	0.3228	0.3192	0.3156	0.3121
−0.3	0.3821	0.3783	0.3745	0.3707	0.3669	0.3632	0.3594	0.3557	0.3520	0.3483
−0.2	0.4207	0.4168	0.4129	0.4090	0.4052	0.4013	0.3974	0.3936	0.3897	0.3859
−0.1	0.4602	0.4562	0.4522	0.4483	0.4443	0.4404	0.4364	0.4325	0.4286	0.4247
−0.0	0.5000	0.4960	0.4920	0.4880	0.4840	0.4801	0.4761	0.4721	0.4681	0.4641

- Z test statistic for $\mu_1 - \mu_2$ when σ^1 and σ^2 are known (p. 508):

$$Z_{data} = \frac{\bar{x}_1 - \bar{x}_2}{\sqrt{\dfrac{\sigma_1^2}{n_1} + \dfrac{\sigma_2^2}{n_2}}}$$

- Z confidence interval for μ_1 and μ_2 when σ^1 and σ^2 are known (p. 509):

$$\bar{x}_1 - \bar{x}_2 \pm Z_{a/2} \sqrt{\frac{\sigma_1^2}{n_1} + \frac{\sigma_2^2}{n_2}}$$

- Test statistic for the independent samples Z test for $p_1 - p_2$ (p. 516):

$$Z_{data} = \frac{(\hat{p}_1 - \hat{p}_2)}{\sqrt{\hat{p}_{pooled} \cdot (1 - \hat{p}_{pooled})\left(\dfrac{1}{n_1} + \dfrac{1}{n_2}\right)}}$$

when the following conditions are satisfied: $x_1 \geq 5$, $(n_1 - x_1) \geq 5$, $x_2 \geq 5$, and $(n_2 - x_2) \geq 5$, and where $\hat{p}_{pooled} = \dfrac{x_1 + x_2}{n_1 + n_2}$.

Chapter 11   Further Inference Methods

- The expected frequency of the ith category when testing goodness of fit (p. 531): $E_i = n \cdot p_i$ where n is the number of trials, and p_i is the population proportion for the ith category.
- Test statistic for the goodness of fit test (p. 534):

$$\chi^2_{data} = \sum \frac{(O_i - E_i)^2}{E_i}$$ assuming the following conditions are true: (a) None of the expected frequencies is less than 1, and (b) at most 20% of the expected frequencies are less than 5. Use $k - 1$ degrees of freedom for the goodness of fit test, and $(r - 1)(c - 1)$ degrees of freedom for the test for independence or homogeneity of proportions.

- Expected frequencies for a χ^2 test for independence or for testing homogeneity of proportions (p. 547):

$$\text{Expected frequency} = \frac{(\text{row total})(\text{column total})}{\text{grand total}}$$

- Overall sample mean, $\bar{\bar{x}}$ (p. 562): The mean of all the observations from all the samples:

$$\bar{\bar{x}} = \frac{(n_1\bar{x}_1 + n_2\bar{x}_2 + \cdots + n_k\bar{x}_k)}{n_t}$$

- Test statistic for performing an analysis of variance (p. 563):

$$F_{data} = \frac{\text{MSTR}}{\text{MSE}}$$

- Mean square error (MSE) (p. 563):

$$\text{MSE} = \frac{\sum (n_i - 1)s_i^2}{n_t - k}$$

- Mean square treatment (MSTR) (p. 563):

$$\text{MSTR} = \frac{\sum n_i (\bar{x}_i - \bar{\bar{x}})^2}{k - 1}$$

ANOVA table

Source of variation	Sum of squares	Degrees of freedom	Mean square	F-test statistic
Treatment	SSTR	$df_1 = k - 1$	$MSTR = \dfrac{SSTR}{k - 1}$	$F_{data} = \dfrac{MSTR}{MSE}$
Error	SSE	$df_2 = n_t - k$	$MSE = \dfrac{SSE}{n_t - k}$	
Total	SST			

- The regression model, or the regression equation (p. 576): $y = \beta_0 + \beta_1 x + \varepsilon$, where: β_0 is the y intercept of the population regression line, β_1 is the slope of the population regression line, and ε is the error term.

- Confidence interval for the true slope β_1 of the regression line (p. 583): $b_1 \pm (t_{crit})(s_{b_1})$, where t_{crit} is based on $n - 2$ degrees of freedom.
- Test statistic (p. 579): $t_{data} = b_1/s_{b_1}$

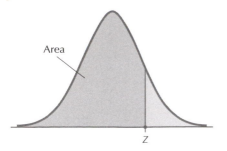

Area

Z

Table C Standard normal distribution (*continued*)

Z	0.00	0.01	0.02	0.03	0.04	0.05	0.06	0.07	0.08	0.09
0.0	0.5000	0.5040	0.5080	0.5120	0.5160	0.5199	0.5239	0.5279	0.5319	0.5359
0.1	0.5398	0.5438	0.5478	0.5517	0.5557	0.5596	0.5636	0.5675	0.5714	0.5753
0.2	0.5793	0.5832	0.5871	0.5910	0.5948	0.5987	0.6026	0.6064	0.6103	0.6141
0.3	0.6179	0.6217	0.6255	0.6293	0.6331	0.6368	0.6406	0.6443	0.6480	0.6517
0.4	0.6554	0.6591	0.6628	0.6664	0.6700	0.6736	0.6772	0.6808	0.6844	0.6879
0.5	0.6915	0.6950	0.6985	0.7019	0.7054	0.7088	0.7123	0.7157	0.7190	0.7224
0.6	0.7257	0.7291	0.7324	0.7357	0.7389	0.7422	0.7454	0.7486	0.7517	0.7549
0.7	0.7580	0.7611	0.7642	0.7673	0.7704	0.7734	0.7764	0.7794	0.7823	0.7852
0.8	0.7881	0.7910	0.7939	0.7967	0.7995	0.8023	0.8051	0.8078	0.8106	0.8133
0.9	0.8159	0.8186	0.8212	0.8238	0.8264	0.8289	0.8315	0.8340	0.8365	0.8389
1.0	0.8413	0.8438	0.8461	0.8485	0.8508	0.8531	0.8554	0.8577	0.8599	0.8621
1.1	0.8643	0.8665	0.8686	0.8708	0.8729	0.8749	0.8770	0.8790	0.8810	0.8830
1.2	0.8849	0.8869	0.8888	0.8907	0.8925	0.8944	0.8962	0.8980	0.8997	0.9015
1.3	0.9032	0.9049	0.9066	0.9082	0.9099	0.9115	0.9131	0.9147	0.9162	0.9177
1.4	0.9192	0.9207	0.9222	0.9236	0.9251	0.9265	0.9279	0.9292	0.9306	0.9319
1.5	0.9332	0.9345	0.9357	0.9370	0.9382	0.9394	0.9406	0.9418	0.9429	0.9441
1.6	0.9452	0.9463	0.9474	0.9484	0.9495	0.9505	0.9515	0.9525	0.9535	0.9545
1.7	0.9554	0.9564	0.9573	0.9582	0.9591	0.9599	0.9608	0.9616	0.9625	0.9633
1.8	0.9641	0.9649	0.9656	0.9664	0.9671	0.9678	0.9686	0.9693	0.9699	0.9706
1.9	0.9713	0.9719	0.9726	0.9732	0.9738	0.9744	0.9750	0.9756	0.9761	0.9767
2.0	0.9772	0.9778	0.9783	0.9788	0.9793	0.9798	0.9803	0.9808	0.9812	0.9817
2.1	0.9821	0.9826	0.9830	0.9834	0.9838	0.9842	0.9846	0.9850	0.9854	0.9857
2.2	0.9861	0.9864	0.9868	0.9871	0.9875	0.9878	0.9881	0.9884	0.9887	0.9890
2.3	0.9893	0.9896	0.9898	0.9901	0.9904	0.9906	0.9909	0.9911	0.9913	0.9916
2.4	0.9918	0.9920	0.9922	0.9925	0.9927	0.9929	0.9931	0.9932	0.9934	0.9936
2.5	0.9938	0.9940	0.9941	0.9943	0.9945	0.9946	0.9948	0.9949	0.9951	0.9952
2.6	0.9953	0.9955	0.9956	0.9957	0.9959	0.9960	0.9961	0.9962	0.9963	0.9964
2.7	0.9965	0.9966	0.9967	0.9968	0.9969	0.9970	0.9971	0.9972	0.9973	0.9974
2.8	0.9974	0.9975	0.9976	0.9977	0.9977	0.9978	0.9979	0.9979	0.9980	0.9981
2.9	0.9981	0.9982	0.9982	0.9983	0.9984	0.9984	0.9985	0.9985	0.9986	0.9986
3.0	0.9987	0.9987	0.9987	0.9988	0.9988	0.9989	0.9989	0.9989	0.9990	0.9990
3.1	0.9990	0.9991	0.9991	0.9991	0.9992	0.9992	0.9992	0.9992	0.9993	0.9993
3.2	0.9993	0.9993	0.9994	0.9994	0.9994	0.9994	0.9994	0.9995	0.9995	0.9995
3.3	0.9995	0.9995	0.9995	0.9996	0.9996	0.9996	0.9996	0.9996	0.9996	0.9997
3.4	0.9997	0.9997	0.9997	0.9997	0.9997	0.9997	0.9997	0.9997	0.9997	0.9998

Chapter 9   Hypothesis Testing

- The test statistic used for the Z test for the mean (p. 414):

$$Z_{data} = \frac{\bar{x} - \mu_0}{\sigma_{\bar{x}}} = \frac{\bar{x} - \mu_0}{\sigma/\sqrt{n}}$$

- The test statistic used for the t test for the mean (p. 437):

$$t_{data} = \frac{\bar{x} - \mu_0}{s_{\bar{x}}} = \frac{\bar{x} - \mu_0}{s/\sqrt{n}}$$

The three possible forms for the hypotheses for a test for μ

Form	Null and alternative hypotheses
Right-tailed test	$H_0: \mu = \mu_0$ versus $H_a: \mu > \mu_0$
Left-tailed test	$H_0: \mu = \mu_0$ versus $H_a: \mu < \mu_0$
Two-tailed test	$H_0: \mu = \mu_0$ versus $H_a: \mu \neq \mu_0$

Finding the p-value

Type of hypothesis test

Right-tailed test
$H_0: \mu = \mu_0$ versus $H_a: \mu > \mu_0$
p-value $= P(Z > Z_{data})$
Area to right of Z_{data}

Left-tailed test
$H_0: \mu = \mu_0$ versus $H_a: \mu < \mu_0$
p-value $= P(Z < Z_{data})$
Area to left of Z_{data}

Two-tailed test
$H_0: \mu = \mu_0$ versus $H_a: \mu \neq \mu_0$
p-value $= P(Z > |Z_{data}|) + P(Z < -|Z_{data}|)$
$\qquad = 2 \cdot P(Z > |Z_{data}|)$
Sum of the two tail areas.

- The test statistic used for the Z test for the proportion (p. 452):

$$Z_{data} = \frac{(\hat{p} - p_0)}{\sigma_{\hat{p}}} = \frac{(\hat{p} - p_0)}{\sqrt{\frac{p_0(1 - p_0)}{n}}}$$

- The test statistic used for the χ^2 test for σ (p. 464):

$$\chi^2_{data} = \frac{(n - 1)s^2}{\sigma_0^2}$$

- Rejection rule for performing a hypothesis test using the p-value method (p. 424): Reject H_0 when the p-value $\leq \alpha$. Otherwise, do not reject H_0.

Rejection rules for Z test for the mean

Form of test		Rejection rules: "Reject H_0 if…"
Right-tailed	$H_0: \mu = \mu_0$ vs. $H_a: \mu > \mu_0$	$Z_{data} \geq Z_{crit}$
Left-tailed	$H_0: \mu = \mu_0$ vs. $H_a: \mu < \mu_0$	$Z_{data} \leq Z_{crit}$
Two-tailed	$H_0: \mu = \mu_0$ vs. $H_a: \mu \neq \mu_0$	$Z_{data} \geq Z_{crit}$ or $Z_{data} \leq -Z_{crit}$

The three possible forms for the hypotheses for a test for p

Form	Null and alternative hypotheses
Right-tailed test, one-tailed test	$H_0: p = p_0$ versus $H_a: p > p_0$
Left-tailed test, one-tailed test	$H_0: p = p_0$ versus $H_a: p < p_0$
Two-tailed test	$H_0: p = p_0$ versus $H_a: p \neq p_0$

The three possible forms for the hypotheses for a test for σ

Form	Null and alternative hypotheses
Right-tailed test, one-tailed test	$H_0: \sigma = \sigma_0$ versus $H_a: \sigma > \sigma_0$
Left-tailed test, one-tailed test	$H_0: \sigma = \sigma_0$ versus $H_a: \sigma < \sigma_0$
Two-tailed test	$H_0: \sigma = \sigma_0$ versus $H_a: \sigma \neq \sigma_0$

Chapter 10   Two-Sample Inference

- $100(1 - \alpha)\%$ confidence interval for μ_d (matched-pair data) (p. 491): Lower Bound: $\bar{x}_d - (t_{\alpha/2})(s_d/\sqrt{n})$, Upper Bound: $\bar{x}_d + (t_{\alpha/2})(s_d/\sqrt{n})$, where $\bar{x}_d$ and s_d represent the sample mean and sample standard deviation of the differences, and $t_{\alpha/2}$ is found using $n - 1$ degrees of freedom.

- Test statistic for the paired sample t test (p. 486): $t_{data} = \dfrac{\bar{x}_d}{s_d/\sqrt{n}}$

- $100(1 - \alpha)\%$ confidence interval for $\mu_1 - \mu_2$ (p. 503):

Lower Bound: $(\bar{x}_1 - \bar{x}_2) - t_{\alpha/2}\sqrt{\dfrac{s_1^2}{n_1} + \dfrac{s_2^2}{n_2}}$

Upper Bound: $(\bar{x}_1 - \bar{x}_2) + t_{\alpha/2}\sqrt{\dfrac{s_1^2}{n_1} + \dfrac{s_2^2}{n_2}}$ where $t_{\alpha/2}$ is found using degrees of freedom the smaller of $n_1 - 1$ and $n_2 - 1$.

- $100(1 - \alpha)\%$ confidence interval for $p_1 - p_2$ (p. 520):

Lower Bound: $\hat{p}_1 - \hat{p}_2 \pm (Z_{\alpha/2})\sqrt{\dfrac{\hat{p}_1 \cdot \hat{q}_1}{n_1} + \dfrac{\hat{p}_2 \cdot \hat{q}_2}{n_2}}$

Upper Bound: $\hat{p}_1 - \hat{p}_2 \pm (Z_{\alpha/2})\sqrt{\dfrac{\hat{p}_1 \cdot \hat{q}_1}{n_1} + \dfrac{\hat{p}_2 \cdot \hat{q}_2}{n_2}}$

- Pooled estimate for the common variance σ^2 (p. 506):

$$s_{pooled}^2 = \frac{(n_1 - 1)s_1^2 + (n_2 - 1)s_2^2}{n_1 + n_2 - 2}$$

- Test statistic t_{data} for $\mu_1 - \mu_2$ using pooled variance:

$$t_{data} = \frac{(\bar{x}_1 - \bar{x}_2)}{\sqrt{s_{pooled}^2\left(\frac{1}{n_1} + \frac{1}{n_2}\right)}}$$

- Pooled variance t confidence interval for μ (p. 507):

$$\bar{x}_1 - \bar{x}_2 \pm t_{\alpha/2}\sqrt{s_{pooled}^2\left(\frac{1}{n_1} + \frac{1}{n_2}\right)}$$